Edwin R Maxson

A treatise on the practice of medicine

Edwin R Maxson

A treatise on the practice of medicine

ISBN/EAN: 9783741101144

Manufactured in Europe, USA, Canada, Australia, Japa

Cover: Foto ©berggeist007 / pixelio.de

Manufactured and distributed by brebook publishing software
(www.brebook.com)

Edwin R Maxson

A treatise on the practice of medicine

A

TREATISE

<small>ON THE</small>

PRACTICE OF MEDICINE.

<small>BY</small>

EDWIN R. MAXSON, M. D.,

FORMERLY LECTURER ON THE INSTITUTES AND PRACTICE OF MEDICINE IN THE
GENEVA MEDICAL COLLEGE.

<small>PHILADELPHIA.</small>
LINDSAY & BLAKISTON,
1861.

HENRY B. ASHMEAD, BOOK AND JOB PRINTER,
Nos. 1102 and 1104 Sansom Street

PREFACE.

THIS Work, which has been written during the past two years, has been under contemplation for several years; and consists in part of a course of lectures delivered in the Geneva Medical College. It has been arranged and prepared for publication, in part from an earnest solicitation of members of the medical class, for whose opinion I entertain the highest respect.

It will be seen that I have glanced at the anatomy and physiology as I have taken up the diseases peculiar to each part of the human system. This I have done in part to make the work more valuable to those practitioners of medicine who have not time to review anatomy and physiology; but more especially to keep the mind of the reader fixed on the diseased part, and its conditions; thus rendering the work not only more valuable, but I trust more interesting.

I have attempted to draw up the work without even the shadow of empiricism, by taking the human system in health as the standard; and then noting the deviations from that standard.

constituting the various morbid conditions or diseases. By taking this course I have been enabled to arrive at clear indications of treatment, from direct pathological conditions, for every prescription which I have made.

This course by no means precludes the benefit of experience in the use of remedies; as remedies which are indicated from pathological conditions are always those which experience finds the most successful. I have preferred, then, to arrive at them in this way, rather than empirically, as it tends to lead the mind of the student and practitioner of medicine to prescribe for *conditions*, without reference to names.

Among the remedies which I have suggested, it will be seen that I have included the most approved new ones; among which are the valuable preparations of Tilden & Co., and others, which are now coming into very general use, with heretofore no work on the Practice of Medicine even referring to them. This I trust may render the work especially valuable to students and practitioners generally.

While I have attempted to make the work as original as a practical treatise on this subject can well be made, I have examined carefully the standard works on the Practice of Medicine, and especially the Cyclopædia of Practical Medicine, revised by Professor Dunglison, as well as the more recent valuable works of Professors Watson, Dickson, Wood and Bennett. And while I have preferred in the main to give my own opinion, it should be remembered that I have availed myself of the opinions of these and other distinguished authors, with whose observations I have carefully compared my own.

And besides, I have exchanged opinions freely with eminent physicians, in order that any new light connected with the Science or Practice of Medicine might have due bearing in its arrangement.

And further, I am under special obligations to Professor John William Draper, M.D., LL.D., of the University of New York, Professor James Hadley, M.D., of Buffalo, Professor W. D. Wilson, D.D., of Geneva, and to my preceptor and former partner, Wm. V. V. Rosa, A.M., M.D., of Watertown, N. Y.; all of whom have kindly made important suggestions in relation to the plan and execution of this work, without being in the least responsible for any of its imperfections.

As the work consists in part of a course of lectures, not designed for publication, it is possible that I may in some instances have incorporated the ideas of others, even in their own language, without giving due credit. If such has been the case, it was unintentional on my part, and unavoidable under the circumstances.

I hope, on the other hand, that I may not be regarded as having deviated too widely from the beaten track of my illustrious predecessors, especially as the Science of Medicine is progressive; every year throwing new light upon our science, which must needs be brought in, or have its bearing upon a treatise of this character.

My professional career thus far has been one of activity, having been engaged in an extensive practice, which has absorbed nearly all my reasonable waking hours; during which I have gathered up the ideas here incorporated, in a form, I trust, however, to be understood. And while I may not have given my work the polish of style which it might have received had more leisure been afforded me, I confidently believe it will be none the less valuable as a practical treatise, for having been arranged and written in the midst of an exciting and extensive practice.

And now, with these few explanations, I throw the work upon its own merits, and unqualifiedly offer it to my brethren of the medical profession. And if it is found to be an interesting and valuable work for the medical student, and as a reference for practitioners of medicine generally, I shall feel amply repaid for the arduous labor which has brought it forth. And, finally, should it have the good fortune to reach a second edition, I shall gladly incorporate into it anything which may be indicated by the progressive advance of Medical Science.

E. R. MAXSON.

GENEVA, April, 1861.

CONTENTS.

CHAPTER I.
OF DISEASE.

PA

Section I.—Nature of Disease   17
II.—Causes of Disease   22
III.—Symptoms of Disease   34
IV.—Diagnosis of Disease   43
V.—Treatment of Disease   46

CHAPTER II.
OF IRRITATION, CONGESTION AND INFLAMMATION.

Section I.—Irritation   51
II.—Congestion   54
III.—Inflammation   57

CHAPTER III.
OF FEVER.

Section I.—The Pathology of Fever   67
II.—The Causes of Fever   78
III.—The Phenomena of Fever   86

CHAPTER IV.
GENERAL FEVERS.

Section I.—Intermittent Fever   94
II.—Bilious Remittent Fever   102
III.—Simple Continued Fever   111

PAGE

Section IV.—Enteric Continued Fever 120
V.—Typhus Continued Fever 129
VI.—Yellow Fever . 138
VII.—Plague . 148
VIII.—Diphtheria 154

CHAPTER V.

EXANTHEMATOUS FEVERS.

Section I.—Variola—(Small-pox) . 162
II.—Vaccina—(Cow-pox) . 171
III.—Varicella—(Chicken-pox) 174
IV.—Rubeola—(Measles) . 175
V.—Scarlatina—(Scarlet Fever) 180
VI.—Erysipelas . 186
VII.—Purpura . 197
VIII.—Glanders—(Equina) 200
IX.—Dengue—(Dandy Fever). 203

CHAPTER VI.

GENERAL INFLAMMATORY DISEASES.

Section I.—Acute Rheumatism 206
II.—Chronic Rheumatism 212
III.—Gout—(Arthritis) 216

CHAPTER VII.

DISEASES OF THE NERVOUS SYSTEM.

Section I.—Cephalagia—(Headache) 224
II.—Meningitis . 226
III.—Cerebritis 236
IV.—Tuberculous Meningitis 240
V.—Spinal Meningitis 244
VI.—Myelitis . 247
VII.—Cerebro-spinal Meningitis 250
VIII.—Apoplexy 254
IX.—Paralysis 261
X.—Epilepsy 268

PAGE

SECTION XI.—CATALEPSY. . . . 276

XII.—CHOREA—(*St. Vitus's Dance*) . 280

XIII.—INSANITY 283

XIV.—MANIA A POTU—(*Delirium Tremens*) . 298

XV.—ECLAMPSIA—(*Convulsions*) 304

XVI.—HYSTERIA . . . 308

XVII.—SPINAL IRRITATION . . 313

XVIII.—TETANUS—(*Locked-Jaw*) . . . 315

XIX.—HYDROPHOBIA—(*Canine Rabies*) . . 321

XX.—NEURALGIA 325

XXI.—AMAUROSIS . . 331

XXII.—SINGULTUS—(*Hiccough*) 332

CHAPTER VIII.

DISEASES OF THE DIGESTIVE SYSTEM.

SECTION I.—STOMATITIS—(*Sore Mouth*) 334

II.—GLOSSITIS . . . 338

III.—PHARYNGITIS—(*Sore Throat*) . . 339

IV.—TONSILLITIS—(*Quinsy*) . 342

V.—PAROTITIS—(*Mumps*) . 344

VI.—OESOPHAGITIS . . . 345

VII.—ACUTE GASTRITIS . 346

VIII.—CHRONIC GASTRITIS . . . 349

IX.—CANCER OF THE STOMACH 351

X.—PERITONEAL ENTERITIS . 354

XI.—MUCOUS ENTERITIS . 356

XII.—DYSENTERY—(*Bloody Flux*) . . 359

XIII.—MALIGNANT DYSENTERY . . . 363

XIV.—CANCER OF THE INTESTINES . . . 366

XV.—ACUTE PERITONITIS . . . 367

XVI.—CHRONIC PERITONITIS . . . 370

XVII.—ACUTE HEPATITIS . . . 372

XVIII.—CHRONIC HEPATITIS . . 377

XIX.—SPLENITIS 379

XX.—DYSPEPSIA—(*Indigestion*) . . . 381

XXI.—DIARRHŒA . . 385

XXII.—CHOLERA MORBUS . 388

XXIII.—MALIGNANT CHOLERA . 390

XXIV.—CHOLERA INFANTUM . . 394

PAGE

SECTION XXV.—FLATULENT COLIC ... 397

XXVI.—BILIOUS COLIC ... 399

XXVII.—LEAD COLIC 401

XXVIII.—INTUSSUSCEPTION 405

XXIX.—CONSTIPATION ... 407

XXX.—INTESTINAL WORMS ... 408

XXXI.—HEMORRHOIDS—(*Piles*) ... 412

XXXII.—JAUNDICE—(*Icterus*) ... 416

CHAPTER IX.

DISEASES OF THE RESPIRATORY SYSTEM.

SECTION I.—AUSCULTATION AND PERCUSSION ... 420

II.—PLEURITIS—(*Pleurisy*) 428

III.—PNEUMONIA—(*Pneumonitis*) ... 432

IV.—BILIOUS PNEUMONIA 438

V.—CATARRH 441

VI.—LARYNGITIS ... 444

VII.—TRACHEITIS—(*Rattles*) 447

VIII.—LARYNGO-TRACHEITIS—(*Croup*) ... 449

IX.—ACUTE BRONCHITIS 452

X.—CHRONIC BRONCHITIS 455

XI.—ASTHMA 458

XII.—HOOPING-COUGH—(*Pertussis*) 461

XIII.—TUBERCULAR PHTHISIS—(*Consumption*) 463

XIV.—APNŒA—(*Asphyxia*) 471

XV.—PNEUMOTHORAX 478

XVI.—EMPHYSEMA 479

CHAPTER X.

DISEASES OF THE CIRCULATORY SYSTEM.

SECTION I.—AUSCULTATION OF THE HEART ... 481

II.—PERICARDITIS 488

III.—ENDOCARDITIS ... 493

IV.—CARDITIS 496

V.—ORGANIC DISEASES OF THE HEART 498

VI.—SYMPATHETIC AFFECTIONS OF THE HEART 506

VII.—NEURALGIA OF THE HEART ... 511

		PAGE
SECTION VIII.—SYNCOPE		512
IX.—ARTERITIS		514
X.—PHLEBITIS		516
XI.—CRURAL PHLEBITIS		518
XII.—HEMORRHAGE		520
XIII.—EPISTAXIS		523
XIV.—HÆMATEMESIS		524
XV.—INTESTINAL HEMORRHAGE		526
XVI.—HÆMATURIA		528
XVII.—HÆMOPTYSIS		530
XVIII.—METRORRHAGIA—(*Uterine Hemorrhage*)		534
XIX.—SCORBUTUS—(*Scurvy*)		537
XX.—ANÆMIA—(*Chlorosis*)		540
XXI.—HYDROPS—(*Dropsy*)		543
XXII.—ANASARCA		547
XXIII.—ASCITES		548
XXIV.—HYDROTHORAX		552
XXV.—HYDROPERICARDIUM		555
XXVI.—HYDROCEPHALUS		557
XXVII.—SCROFULA		561
XXVIII.—BRONCHOCELE—(*Goitre*)		566

CHAPTER XI.

DISEASES OF THE EYE.

		PAGE
SECTION I.—CATARRHAL OPHTHALMIA		568
II.—PURULENT OPHTHALMIA		571
III.—SCROFULOUS OPHTHALMIA		574
IV.—RHEUMATIC OPHTHALMIA		576
V.—CORNEITIS		578
VI.—IRITIS		580
VII.—EXOPHTHALMIA		583

CHAPTER XII.

DISEASES OF THE EAR.

		PAGE
SECTION I.—GENERAL OTITIS		586
II.—EXTERNAL OTITIS		590
III.—INTERNAL OTITIS		593

	PAGE
SECTION IV.—OTORRHŒA	595
V.—OTALGIA—(*Ear-ache*)	597
VI.—NERVOUS DEAFNESS	598

CHAPTER XIII.

DISEASES OF THE SKIN.

SECTION I.—RASHES AND ERUPTIONS	601
II.—RED RASH	605
III.—ROSE RASH—(*Roseola*)	606
IV.—NETTLE RASH	608
V.—PAPULAR ERUPTIONS	609
VI.—VESICULAR ERUPTIONS	611
VII.—PUSTULAR ERUPTIONS	614
VIII.—SCALY ERUPTIONS	617
IX.—ANIMALCULAR ERUPTIONS	620
X.—CRYPTOGAMOUS ERUPTIONS	621

CHAPTER XIV.

DISEASES OF THE URINARY ORGANS.

SECTION I.—NEPHRITIS	626
II.—ALBUMINURIA	630
III.—NEPHRALGIA	634
IV.—ACUTE CYSTITIS	635
V.—CHRONIC CYSTITIS	637
VI.—DIABETES MELLITUS	638
VII.—DIABETES INSIPIDUS	641
VIII.—LITHIASIS—(*Gravel*)	643
IX.—SUPPRESSION OF URINE	650
X.—RETENTION OF URINE	652
XI.—DYSURIA—(*Strangury*)	653
XII.—INCONTINENCE OF URINE	654

CHAPTER XV.

DISEASES OF THE GENITAL ORGANS.

SECTION I.—SPERMATORRHŒA	656
II.—GONORRHŒA	663

                                                      PAGE
Section III.—Syphilis   .   666
        IV.—Metroperitonitis .   670
         V.—Chronic Metritis   .   .   675
        VI.—Disease of the Cervix Uteri   677
       VII.—Leucorrhœa   .   .   680
      VIII.—Amenorrhœa   684
        IX.—Dysmenorrhœa .   688
Conclusion   .   691
Index   .   -   .   693

A TREATISE ON THE PRACTICE OF MEDICINE.

CHAPTER I.

OF DISEASE.

SECTION I.—NATURE OF DISEASE.

BY *disease*, is here meant a deviation from health ; and hence by the *nature of disease*, I mean the nature of deviations from the standard of health ; in order to understand which, it is proper that we should take a glance here at the physical, intellectual and moral constitution of man.

Man as he came forth from the hand of the Creator was a perfectly constituted being ; the mind was the essential part, the body being a mere instrument of the mind. The mind was probably constituted with an intuitive consciousness of self and notion of God ; but to gain a knowledge of the material world and for other important purposes, the body was an indispensable appendage. As the mind then gained a knowledge of the material world through the bodily senses, and by the aid of knowledge thus obtained, improved its knowledge of self and notions of God, it was a physical, intellectual and moral necessity that the body should be a perfect instrument of the mind.

The necessity of perfect bodies, however, becomes more apparent when we remember that if our bodies are imperfect, we fail to acquire through the senses, a correct knowledge of the material world ; and as a consequence fail to acquire correct knowledge of ourselves, as well as notions of God. Thus failing to understand ourselves and the material world, as well as to form correct notions of God, we fail to appreciate our *relations*, and hence our *obligations* to ourselves, to God, and to all creatures and things ; thus rendering physical degeneracy, or disease of our bodies, not only a physical but an intellectual and moral calamity.

If then our *physical, intellectual* and *moral natures* require that we have perfect bodies, what principles can be more important than those which teach the laws of health and the best method of restoring to health our bodies when in a state of disease ? To suppose that man was created physically imperfect, or that he must necessarily have become so, would be indirectly to charge the Creator with folly and would be in the highest degree absurd ; and equally absurd would be the notion that anything except man's own imprudence in deviating from the laws of health, or of his being, could have led to the present state of physical degeneracy.

It is probable then, if man had always obeyed the laws of his being, that we might have escaped pain, disease and death; and like Enoch and Elijah, "having walked with God," we should have been removed to a more elevated state or genial clime; and thus escape the really unnatural passage, "through the valley of the shadow of death." This appears the more probable when we remember that human life has been gradually shortened from the first deviation from the laws of health to the present time. And whether the first deviation from the laws of health, which occurred in Eden, consisted merely in eating the unwholesome "forbidden fruit;" or whether it was from moral wrong in partaking of that which had been forbidden, or whether, as is probable, it was from both; it matters little, for we are so constituted that every deviation from the laws of our being, physical, intellectual or moral is a violation of the laws of health.

It is unreasonable to suppose that accidents or even the elements in nature which now appear to produce disease, would have harmed man if he had retained his original purity and always obeyed the laws of his being.

In fact if we reflect upon the origin of every disease to which we are now liable, many of which we inherit, we shall find that they have arisen one after another as our multiplied deviations from the laws of health have led to them; every new disease leading the way to or predisposing to that which has followed, as gluttony, drunkenness, licentiousness, tobacco, &c., have led to scrofula, and scrofula to consumption. And this is equally true of the contagious diseases; one imprudence after another having led to such a degree of physical derangement that a particular species of disease is produced, and the system while laboring under this particular derangement, generates a contagion which propagates the disease. Thus it was that licentiousness led to the venereal disease; the system after the disease was once produced generating the virus or contagion which has propagated the disease. Hence we see that instead of the elements being at war with us, it is man that is at war with himself and with the elements. And it is by that warfare and reckless disregard of the laws of our being that we inherit, and are now liable to the vast variety of diseases from which unfortunately we suffer.

This may be thought foreign to our subject, but I think it is not; especially as we should know whether the ills to which we are liable are a necessary result of the order of Providence; or whether they result from our own imprudence and were, therefore, entirely unnecessary. For if our sickness, pain and death, were a direct result of the warring of the elements against our life and health, there would be little left for us to do in the matter; except to fortify ourselves and contend as best we could with the order of Providence. But if, as we have seen, all our sickness, pain and physical degeneracy is either the direct or indirect result of imprudence, we have vast and fearful responsibilities in this matter. And the more weighty appear these responsibilities when we remember that not only our physical but our intellectual and moral natures are implicated in this degeneracy.

Now such being the case, in relation to our responsibility in this matter, it becomes us, as Christian Medical Philosophers, to see what may be done on our part, by way of correction, to lessen the ills to which mankind are now liable; not only by way of prevention; but also to the cure of disease. And if we have arrived at such a state of physical degeneracy, that we may never regain our native physical perfection, let us do what we can towards reaching that state; and may we not hope that the good time may yet arrive, when "the child shall die an hundred years old?" and when disease and pain will be comparatively unknown? Until this state of things can be reached, the human family will suffer from a variety of ills, which, for the sake of convenience, we will call disease.

Now, we can form no correct notion of disease, till we understand our physical organization in health; and also, the natural functions of all the organs and parts of our bodies in a state of perfect health. So that a correct knowledge of ANATOMY and PHYSIOLOGY, is the only possible way for us to acquire a knowledge of deviations from the standard of health, or what we call disease. But as Diogenes, more than two thousand years ago, in wandering with a torch, in open day-light, about the streets of Corinth, declared himself unable to find a *man;* so we at the present day, may search in vain for a perfect physical organization, the deviations from which to form a correct knowledge of disease. In fact, in order to get a perfect standard, we must go back in imagination to the infancy of our race; when the human body, fresh from the hand of the Creator, was perfect in all its parts.

Disease then, is nothing in itself; but consists in a deviation from a state of health. Consequently he who does not understand what a healthy condition should be, can by no possibility arrive at a rational conclusion in relation to disease. We must take then, the human system, in the most perfect state in which we find it at the present day, together with our imagination of what it should be, as our standard from which to calculate disease; and we should bear in mind also, that man is a triune being; with mind, body, and a moral nature, and that a healthy physical organization is indispensible to the perfect development of the intellectual and moral man.

Anatomy then, and Physiology, is the foundation, not only of MEDICAL, but to a certain extent of INTELLECTUAL and MORAL PHILOSOPHY; and though we may not be able to comprehend the minute relation between the mind and body, we must bear in mind the fact of this relation, at every step of our investigation, into the *nature, causes, symptoms, diagnosis and treatment of disease.*

If we examine our bodies, we shall find that they are just that kind of self-regulating instruments of the mind, which we should suppose an ALL-WISE BEING would have invented. Or at least they were so, in their original state of perfection; the fluids and solid tissues, being in due proportion to each other. The supply of nourishment furnished by the food, being exactly sufficient to supply the wastes of the system. While then, this balance was preserved, and the fluids and solid tissues retained their normal state, as to quality and quantity, the human body was in a state of health; and might have continued so, and probably retained the health and vigor of youth; never decaying, never wearing out, and thus man, as I have already suggested, might probably have remained *physically immortal.*

But as we have already seen, we must take the human system as it is; bruised and shattered by the numberless deviations from the laws of health, since the creation of man; every one of which has tended to mar this handy-work of God, till we now inherit scarcely constitution enough to enable us to live on; even with prudence on our part to seventy years, while those who are imprudent, or inherit a greater degree of physical degeneracy or imperfection, may scarcely continue even a miserable existence through the periods of youth, childhood, or even infancy.

Thus has the human system gone on degenerating, for near six thousand years, till few of us inherit constitutions that enable us to live past three score years and ten, while the average duration of human life is probably less than half that period.

Now this hereditary physical imperfection, or predisposition to disease or premature death or decay, differs in different individuals, some inheriting it as we have seen, in a greater and others in a less degree, and some tending to dissolution in one way, and others in another. And while, as we have already seen, we have to take mainly the system in health, as it is at the present time, as the standard from which to calculate disease; we cannot proceed a single step, in our inquiry into the *nature, causes, symptoms, diagnosis* and *treatment* of disease, without understanding, and taking into account the inherited imperfection of constitution, or predisposition to disease, in each particular case.

In relation to the *nature* of hereditary or inherited physical imperfection, or predisposition to disease; we have already seen, that it varies materially in different individuals, not only in degree, but in kind; some tending to an early death in one way, and others in another. But it is probable that in every case, the leading or prevailing physical imperfection of both parents are entailed upon the offspring, rendering the whole physical organization, including fluids and solids, more or less imperfect; but the leading imperfection in the constitution of the child, being similar to that of one or both parents; from which, it should be remembered, every particle of the material, fluid and solid, of which the new-born infant is composed, has been derived.

Hence it is, no doubt, that children are born *scrofulous, tuberculous, syphilitic,* &c.; literally fulfilling the declaration of the ALMIGHTY, that "the iniquities of the fathers," shall be "visited upon the children;" and that too by a necessary law of nature, which is really however little more than another term, to indicate a decree of the ALMIGHTY.

Having thus inquired into the nature of hereditary imperfection or disease, or predisposition to disease; we are now prepared, keeping the facts here arrived at in mind, to arrive at rational conclusions, in our examination into the *nature* of *acquired disease,* which now demands our consideration.

A person may be said to be in health, when the fluids and solid tissues of the body are in such a state, that all the functions of the body are carried on harmoniously; taking into account of course, hereditary

physical imperfection, or predisposition to disease. **But all variations**
from this standard, involving a derangement of structure or function,
and causing suffering, **or** endangering life, we may call disease.

For the sake of convenience, disease involving a change of the composition of the fluids, or **structure of the** solid tissues may be called
organic; while those derangements in the functions of the body, not involving any apparent change **in the composition** of the fluids, **or structure of** the solids, may be called functional. It is possible that all diseases may **involve a** change **in the** composition **of the** fluids, **or** structure
of the solids; but in all cases in which the organic change is not perceptible, the derangement may be regarded as *functional.*

Disease again, **may be** either *general* or *local;* **in the** one **case,** involving more or less, every organ and tissue of the body; while in the other,
it is confined more especially, to some particular organ or tissue.

The *fevers,* and general *inflammatory* affections, may **be called** *general,* while those diseases **of** whatever character, which involve only particular parts, organs, or tissues, **as the** brain, lungs, liver, skin, &c.,
properly fall under the head of *local disease;* though attended usually,
with more **or** less general disturbance.

Now to understand the nature of *general disease,* it is **necessary to**
bear in mind the composition of the blood, and other fluids of the body,
as well as the structure of the different tissues, **and the** relation existing
between the solids **and fluids.** When this is done, it will be impossible
for us to refer any *general disease,* exclusively to either the blood or
solid tissues. For though the cause, whatever it may chance to be, may
operate primarily upon the blood; its influence is immediately felt upon
the brain and nervous system. And as the character of the blood, as
well as condition of the solids, depend upon a due amount of nervous
influence; we soon have a general derangement of the fluids and solids,
of either an organic or functional character.

On the other hand, if the cause acts primarily upon the solids; as for
instance, a fright acting through the mind upon the brain and nervous
system, the blood becomes instantly involved, in consequence of the
derangement which occurs in the generation and distribution of the nervous influence, affecting the circulation and also the secretions generally.
Again, when we remember that the formation of the blood depends upon
the condition of the solid tissues of the digestive apparatus, as well as
of the nervous system, and that the integrity of all the solid tissues depend upon a constant and due supply from the blood; we see at once
the utter impossibility of referring any *general* or even local disease,
exclusively to either the fluids or solids.

In all *general* or *local diseases* then, whether *organic* or *functional,*
the derangement involves more or less the blood and solid tissues, which-
ever may have been the medium, through which the morbid agent or
influence has primarily operated.

Now, we may have *general* or *local disease* from every possible deviation from the laws of health, and rules of propriety, as well as from
hereditary and other influences, in a certain degree beyond our control,
at the present day; some of which act upon, and change more or less
the quality, quantity and character of the blood, directly; while others

act directly upon the **solid** tissues, as moral influences, impairing the
vitality of the cerebro-nervous system, hindering a due generation and
proper distribution of the vital force, **or nervous** influence; and thus
producing disease: **causes** then operating in either of these ways, may
lead to *irritation, congestion* and *inflammation,* to the *febrile,* and *general inflammatory affections;* and also, to functional or organic diseases
of the *nervous, digestive, respiratory* and **circulatory systems,** or their
structures, as well as to diseases of the *eye,* **ear, skin, urinary** and *genital
organs.*

Now, when we bear in mind, that the human system is composed of
the *fluids;* and the solid tissues, consisting of the *cellular, muscular,
nervous,* &c., so combined as to form the structures of the different organs of the body; and remember that all the organs depend upon a due
amount and proper distribution of the *nervous influence,* for the performance of their functions; we can scarcely overrate the influence which
the *brain and nervous system* has, in all diseases of an organic or functional character; and local as well as general. It must however be borne
in mind, at every step, that the brain and nervous system depend upon
a due amount of stimulus from the blood, to enable them to generate
and distribute sufficient *nervous influence* to enable the *organs* of the
body to carry on their functions.

We may conclude then, that *disease* consists in a derangement of the
system, either *general* or *local,* involving more or less the *fluids* and
solid tissues of the body, and especially the *blood* and *nervous system,*
and embracing every possible deviation, whether *organic* or *functional,*
from the standard of health.

Having now completed what I have to say here on the *nature* of dis-
ease; I shall leave this subject and in the succeeding sections of this
chapter shall consider the *causes, symptoms, diagnosis* and *treatment* of
disease. In the following chapter I shall consider *irritation, congestion*
and *inflammation;* in the third, the *pathology, causes* and *phenomena* of
fever; in the fourth, *general fevers;* in the fifth, *exanthematous fevers,*
and the sixth, the *general inflammatory diseases.*

Having thus considered the more general diseases to which the human
system is liable; I shall proceed to take up the diseases involving more
especially particular parts, in the following order:

In the seventh chapter, *diseases of the nervous system;* in the eighth,
diseases of the digestive system; in the ninth, *diseases of the respiratory
system;* in the tenth, *diseases of the circulatory system;* in the eleventh,
diseases of the eye; in the twelfth, *diseases of the ear;* in the thirteenth,
diseases of the skin; in the fourteenth, *diseases of the urinary organs;*
and in the fifteenth, *diseases of the genital organs;* ending with a short
conclusion, which will complete this TREATISE on the PRACTICE OF
MEDICINE.

SECTION II.—CAUSES OF DISEASE.

By *causes of disease;* technically *etiology;* I mean here any and every
agent or influence capable of deranging, either directly or indirectly, any
function or structure of the body. It includes of course, along with

hereditary imperfection, every imprudence to which the human family are addicted, which always predispose to, and often excite disease: and also all those agents in nature, which, in consequence mainly of this predisposition, are exciting causes of disease.

Now it is evident that every cause which is capable of predisposing the system to disease, is capable, under other circumstances, of acting as an exciting cause of disease ; and also that the exciting causes, under circumstances which favor it, predispose the system to disease. With this qualification, I prefer to consider the causes of disease as *predisposing* and *exciting ;* and will, therefore, proceed first to consider the more important abuses of the body, and imprudences which always predispose to, and often excite disease; under the head of *predisposing causes ;* and then will take up those agents or influences in nature, which more generally act as exciting causes, under the head of *exciting causes of disease.*

It is highly proper, however, in order to appreciate all the *direct* and *indirect* causes of disease ; that we bear in mind continually the anatomy, physiology and functions of the body in health : and also salutary influences ; such as *cleanliness*, proper *food* and *drink, clothing, temperance, virtue, contentment, &c.*, which tend to perpetuate health ; that we may the more readily appreciate the imprudences and influences, which have a pernicious effect, and produce disease.

PREDISPOSING CAUSES OF DISEASE.

By *predisposing* causes of disease, I mean here all those disturbing influences which operate upon the human system to weaken the powers of vital resistance, and thus render the system susceptible of being thrown into a state of disease, by various elements, and influences in nature, which except for this predisposition, might, in most cases at least have remained harmless.

It will not be expected, of course, that I shall enumerate all the predisposing causes of disease; but I will endeavor to consider those imprudences and abuses of the body, which are of practical importance, and which I am confident, from careful observation in numerous cases for nearly twenty years, not only predispose the system to acute disease ; but also lay the foundation of *physical, intellectual* and *moral* weakness; as well as for permature decay and death. And as most of these causes are avoidable ; and as many of them would be avoided by thousands in the community, if they were sufficiently enlightened, I cannot exaggerate the importance of this subject ; and would urge it upon my brethren of the Medical Profession, as Christian Philosophers, and as benefactors of mankind.

HEREDITARY PREDISPOSITION.—Hereditary predisposition or tendency to disease, consist in whatever defect or imperfection we may be born with ; the result of every departure from the laws of health of our ancestors, even back to the first pair in "Eden." And, also, we may include those accidental influences in nature, which have had a pernicious effect, in consequence mainly, of acquired predispositions.

I need not say that the vices of every preceding age, have had their influence in lessening the powers of vital resistance ; in shortening human life, and in rendering the human system, even in new-born infants, immeasurably below its original standard of perfection. Hence it is, no doubt, that human life has an average now of less than forty years ; while our early ancestors, who did not inherit the result of imprudences practised by more than one or two generations, lived in some cases, near one thousand years.

Now while ten thousand influences must have been operating for nearly six thousand years, to bring about this state of hereditary physical imperfection, or predisposition to disease ; I think it must be admitted that *filth*, improper *clothing*, improper and insufficient *food*, *intoxicating liquors*, *licentiousness* in its various forms, the violent angry *passions*, and latterly tobacco, have contributed largely, and probably mainly, to bring about this hereditary tendency to disease ; for which the individuals composing this generation are in no way responsible. It is probable also, that the marriage of near relatives, since hereditary imperfections have prevailed, has also had a very pernicious influence, in cases in which it has been practised.

Now such being the origin, and some of the causes of hereditary tendency to disease ; we should expect, as we really find ; that this predisposition would vary materially in different individuals, families, and races of men ; depending much upon the customs, habits, and imprudences of their progenitors. And thus it is, no doubt, that we find some individuals, families and races of men, predisposed to one form of disease, and others to another.

But this reflection upon the past is important, only as it points out indication to us for the present and future ; a matter for which we are all more or less responsible, and in which we should be deeply interested. This in fact appears the more important, when we remember that the same chain of pernicious influences, with a corresponding shortening of human life, that has attended them thus far, would in less than five hundred years more, exterminate the human family.*

Hereditary predisposition to disease is a matter, then, never to be overlooked ; and while we cannot hinder that which is past, let us in all our intercourse with our fellow-men strive to inculcate those principles of righteousness, which may show a diminution of this predisposition in succeeding generations.

FILTH.—Filth, or want of cleanliness, is another predisposing cause of disease. This, it is true, is a cause which does not, directly, very extensively prevail in civilized communities ; but indirectly it affects all classes : for we have only to remember that a very small den of filth may become a pest, by the impurity which it imparts to the air, which may operate perhaps insidiously upon the most cleanly, who would themselves gladly obey the laws of health and life.

Filth operates directly upon the systems of those addicted to the disgusting habit, by hindering a due exhalation from the skin ; in consequence of which the various functions of body are rendered exceedingly liable to become deranged, by very slight exciting causes. This may be noticed in all cases in which there is an endemic or epidemic influence

* Men lived about 1000 years, near 5000 years ago ; now the average of human life is less than 40 years ; hence human life would be reduced, at the same rate, below the age for generation, in less than 500 years more.

operating upon a community; the filthy, as a general rule, being by far the most frequent victims of the influence of the general exciting cause. Strict personal cleanliness, then, should always be enjoined; not only to lessen the predisposition to disease among those addicted to it, but also as a security for those who maintain strict cleanliness themselves, and would gladly obey the laws of health and life.

FOOD.—Food becomes a predisposing cause of disease, if it be *improper, too much, too little*, or if it be taken *irregularly;* all of which deviations from that which is proper require our careful consideration.

Improper food may be so from its indigestible properties; in which case it not only fails to afford the system the nourishment it requires, but it also acts as an irritant to the alimentary mucous membrane, causing irritation, inflammation, &c., and leading or predisposing to *dyspepsia*, chronic *diarrhœa*, and a long train of gastric, hepatic and other diseases too numerous to mention. Or food may be digestible, and yet be highly pernicious, from the fact of its being of an unwholesome character, and wanting the nutritive properties; as, for instance, putrid meats. In such cases, the system not only fails to receive its required nourishment—which of course lessens the powers of vital resistance—but the putrid or unwholesome material which it furnishes the system depraves or decomposes the blood; thus rendering the system tenfold more liable to attacks of acute disease from slight exciting causes.

Too much food, by hindering the digestive process, and also irritating the gastro-intestinal mucous membrane, predisposes to disease of the digestive system; or if digested, it may convey too much of the stimulating properties into the blood, and thus strongly predispose the system to various acute inflammatory affections, from comparatively slight exciting causes.

Insufficient amount of food, by cutting off the necessary supplies for the blood, soon renders the nervous system irritable, and lessening the powers of vital resistance, strongly predisposes the system to attacks, from slight accidental causes, of neuralgic, congestive, or passive inflammatory diseases.

Irregularity in taking food is, according to my observation, one of the most prevalent and fearful predisposing causes of disease. The stomach will receive and digest a reasonable amount of food, taken at regular intervals—three or four times in twenty-four hours—and in that case, if the system is in an ordinary state of health, the stomach will call for about the amount of food it can well digest, or that the system requires.

If, however, food be taken at irregular hours, or between meals, as is too generally the case, especially with children, the stomach, for want of rest, becomes unable to properly digest the food taken; the gastro-intestinal mucous membrane becomes irritable; acids are generated in the stomach to excess; the appetite becomes irregular; too much or too little food is taken; a due supply of nutriment for the blood is cut off; the nervous system becomes irritable; all the fluids and solid structures of the body become more or less deranged, and the patient becomes strongly predisposed to scrofula, consumption, dropsy, epilepsy, catalepsy, and, in fact, almost every variety of disease to which the human system is liable. If now to the sum of physical wretchedness which the habit of irregular eating produces, we add the inconvenience which attends it, and then take into account the fact that there is no excuse for it, the prevalence of the habit is one of the most wonderful and alarming of the age; and, without doubt, is one of the most frequent exciting, as well as predisposing, causes of disease.

CLOTHING.—Improper clothing, by affording the system insufficient protection from *winds, moisture* and *cold*, predisposes the system to various acute inflammatory affections. If the amount of clothing be too much, the system by degrees loses its capacity of adjusting its temperature to atmospheric changes; and hence, if from inattention to the matter, as the temperature of the air falls suddenly, the clothing be not correspondingly changed, there is a strong liability at least that some inflammatory or catarrhal affection may be the result. If, on the other hand, the clothing is defective, or insufficient to protect the system, the cutaneous exhalation is liable to be greatly diminished, and the perspirable matter being retained in the blood, predisposes the system to various functional and even organic affections.

LICENTIOUSNESS.—*Licentiousness,* whether it be from *masturbation,* as too often practised by the young and unmarried, or from *Onanism,* or from excesses in *sexual intercourse,* as too often practised by the married, and by the grossly licentious unmarried, is an alarming predisposing cause of disease.

Masturbation, as thousands of degraded objects of pity clearly demonstrate, impoverishes the blood, irritates the genital organs in both sexes, exhausts the nervous energy, debilitates the brain and whole nervous system, produces general physical, intellectual and moral weakness, or depravity, and strongly predisposes the unfortunate sufferer to *melancholia, hypochondria* and *insanity;* as well as to a vast train of functional and organic diseases.

*Onanism,** by interrupting the free discharge of the seminal fluid from the male organ, generally sooner or later produces an irritation of the seminal ducts, of the vesiculæ seminales, and also, in many cases, of the prostate gland; and being only a species of masturbation, and in every respect at variance with the order of nature, affects the genital organs of both sexes, and, in fact, the whole organization of the system, in a manner very similar to that already described as the result of masturbation, and thus of course becomes an alarming predisposing cause of disease.

Excessive sexual indulgence, though less injurious than *masturbation,* as practiced by the unmarried, or than that form of masturbation termed *Onanism,* practised by the married and by the grossly licentious, is in too many cases a dangerous draft upon the blood and nervous energy; and thus, no doubt, is very often a predisposing cause of disease. And irritating as it does the female genital organs, it soon affects the nervous system, and becomes scarcely a less evil to the female than to the male thus practicing.

Thus we see that licentiousness, in all its forms, is a fearful predisposing cause of disease.

* See the practice of Onan, in the Bible, Genesis **xxxviii.** chapter and 9th verse.

INTOXICATING LIQUORS.—*Intoxicating drinks,* whether taken in the form of fermented or distilled liquors, are a very frequent predisposing cause of disease. If taken in small quantity, and at regular intervals, as for instance after each meal, they are very liable, sooner or later, to irritate the mucous membrane of the stomach; and being long continued, the nervous system becomes debilitated to such an extent, that the system is scarcely able to carry on its functions without it. If now the liquor be continued, in order to keep the system up, it has to be increased in quantity. This gradually increases the gastric irritation, and nervous debility, till the system becomes so far reduced, that a suspension of the liquor from any cause often leaves the patient in a state of *delirium,* literally, " madness from the bowl." Thus we see, that the use of intoxicating liquors leads to delirium tremens, and it is notoriously true, that protracted drunkenness leads to a great variety of physical disease, as well as to intellectual weakness, and moral depravity.

TOBACCO.—*Tobacco,* containing as it does three deadly poisonous principles: *Nicotia, Nicotianin* and an *Empyreumatic oil,* is a fearful cause of disease, especially as it is so generally used by smoking, snuffing and chewing; the poisonous principles entering the circulation, and directly impairing vitality, by disqualifying the brain for generating sufficient vital force or nervous influence; and also disqualifying the nerves for a proper distribution of that which is generated. In this way, it produces nervous tremors, and strongly predisposes to paralysis of the lower limbs, as well as to a host of functional, organic and *mental,* diseases.

ANGER.—*Anger,* especially if it be of a violent character or grade; or if it be indulged for a protracted period, is sure to produce a functional, if not an organic disturbance in the system; and predisposes not only to a variety of physical diseases; but also fearfully endangers, the intellectual and moral powers.

Now having enumerated the prominent predisposing causes of disease, and having given a brief explanation of the manner in which these causes thus act to predispose to disease. I want it distinctly understood, that they may all, under other circumstances, or if long continued, become exciting causes of the diseases to which we have seen they predispose. Bearing these facts in mind, we are prepared to pass on to the consideration of those elements or influences in nature, which more generally act as exciting causes of disease.

EXCITING CAUSES OF DISEASE.

HEAT.—A moderate degree of heat is essential to vegetable as well as to animal life; and it only requires the exercise of a sound discretion, for every individual to guard himself against its pernicious influences. Without the exercise of this discretion however, heat may become a frightful exciting cause of disease. This we see illustrated in " *sun stroke,*" caused by an undue exposure to the piercing rays of the sun, in persons especially predisposed to the affection.

Again, we see it illustrated in its direct affects when applied to the human body in water, steam, or as is latterly more frequent, in the rapid combustion of various inflammable fluids, so generally, and as I believe improperly used for lights; a fearful case of which fell under my care, a few weeks since in this village. Thus we see that heat, an essential life giving principle or agent, becomes through carelessness and imprudence, and the failure to exercise a sound discretion, a fearful exciting cause of disease.

COLD.—*Cold* is a negative condition, and implies an absence of caloric or heat; hence we may see at a glance, that a due degree of cold is indispensable, for all living beings and things. And a judicious exposure to it, by human beings is always salutary, and always a source of comfort. It is only then from an undue exposure to cold, and that too more generally by persons weakened by previous improper exposures to heat, or other influences, that cold becomes an exciting cause of disease.

By the improper sudden exposure of the body to cold, after it has been overheated, an interruption of the perspiratory function is liable to follow, attended with some acute inflammatory affection, thus rendering cold, injudiciously applied, an exciting cause of disease. Again, cold may be improperly applied to the system for a long time, or the body may thus be exposed to its influence, till its sedative or depressing effects may so far interrupt the various functions of the body that *scrofula, dropsy, consumption*, &c., may be the result, as we see illustrated in many of the men who accompanied the late illustrious DR. KANE, in his northern expedition in search of Sir John Franklin.

One of these noble men, whose constitution suffered, but did not give out during that long voyage, fell under my care a few months since.

Finally, cold may, by its intensity, and its prolonged application, produce a depression which will interrupt the circulation, depress the muscular power, benumb the brain and whole nervous system, and finally, interrupting one function after another, it leads to an irresistible drowsiness, which is followed by death. Thus we see that *cold*, while it is essential to life and health, in a due degree, becomes, if the body be imprudently exposed to its influence, an exciting cause of disease; and may even produce fatal effects.

WATER.—Water again is an element indispensable to the very existence of all living beings; and for the comfort and well-being of man, it is indispensable as a drink, and convenient at least, for ablution; and yet strange as it may seem, its improper use, in various ways and forms, renders it a frequent exciting cause of disease. Too much water taken into the stomach when the body has been overheated, especially if it be cold, is liable to produce alarming effects. Or the habit of taking too much water at meals, or at other times, is liable to dilute the gastric juice to such a degree, as to render it incompetent to dissolve the food; and hence dyspepsia, with all its consequences is the result. I had an alarming illustration of this, in a little boy of this village, a few weeks since, which was cured only after I prohibited his drinking large draughts of water, frequently during the day and night.

Again, water may be drank too hot or too cold; and in either case, is liable to produce the most fearful results. Or by being applied too hot or too cold, or too frequently to the surface of the body, it is liable to produce too great stimulating or depressing effects, and thus to disturb the distribution of the nervous influence; or at least, to call the skin into undue activity; causing it to throw off offensive waste matters from the system, which should pass by the kidneys or through the alimentary canal, thus rendering the skin more filthy, than if only washed with water of a medium temperature; and at proper intervals. Thus we see that water, which is indispensable to the life and comfort of man, is liable when improperly used or applied, to become an exciting cause of disease.

ELECTRICITY.—*Electricity* is a principle which pervades the human body as well as almost every object or substance in nature; and might very likely have been a universal source of health and comfort, if our systems had retained their original state of perfection, and provided a due amount of caution and prudence were always exercised in avoiding it in too violent a shock.

It is a well known fact that damp air is always in a low electrical state; and that very dry air is highly charged with electricity. Now through carelessness or inattention if the air of an apartment becomes excessively dry, the electricity with which it becomes charged may produce too stimulating effects upon the nervous system; and thus it may become an exciting cause of various acute inflammatory diseases. On the other hand, if people expose themselves in low places or apartments where the air is damp, and of course in a low electrical state, the system imparts its electricity to the surrounding air to produce an equilibrium; and thus a portion of the necessary stimulus for the nervous system being removed, if it be continued for a long time is very liable to derange the various functions of the body, and thus it may lead to *dropsical*, *scrofulous* and *tuberculous* diseases.

In relation to the effects of violent shocks of electricity, as it reaches us from the clouds, a sound discretion should always be exercised as in every other means of self protection, and with the knowledge we have of the protection afforded by conductors placed where persons or buildings are exposed; its fearful effects from such a source and in such a form may generally be averted.

LIGHT.—*Light* is another agent in nature of inestimable value; originating by the will of the ALMIGHTY in "the morning of creation," it has opened to the mind, through the organ of sight, the handy-works of God by which we are everywhere surrounded. This agent probably favors a healthy action in the animal functions, as it is well known to in vegetable life. Light is the proper stimulant for the eye, and only becomes an irritant when admitted in undue brilliancy, or when acting upon the organ previously debilitated or irritated from some other cause.

AERIAL POISONS.—Various gasses, as carbonic acid, carbonic oxide, carburetted hydrogen, &c., are liable to be present in the air in certain localities, and a little caution may be necessary in avoiding them.

Carbonic Acid, it is well known, is one of the products of respiration and combustion, and is the gas from which vegetables and trees obtain their carbon or charcoal; they absorbing this gas through their leaves, retaining the carbon and throwing off the oxygen in a pure state to support animal life. Now the proper constituents of atmospheric air, is one part of oxygen to four parts of nitrogen, by volume. If then carbonic acid or the other noxious gasses be mingled with it to any considerable extent, it is liable to produce disease either by want of a due amount of oxygen in the air respired to support the vital flame, or the gas may produce a positive injurious effect and thus cause disease.

As carbonic acid is the most practically important of these gases, it is proper that we should remember that its action in a diluted state is probably by hindering the due escape of the same gas from the lungs, where, it will be remembered, it is being formed by a union of the oxygen of the air with the carbon of the blood. As a consequence of this, carbonic acid accumulates in the blood, and, passing to the brain, produces its narcotic effect, with stupor, and, if long continued, may cause a suspension of the voluntary and vital functions, from the cerebral oppression.

It is probable, however, that carbonic acid, in a pure or concentrated state, causes death by producing a spasm of the glottis, leading to complete asphyxia.

Now, as carbonic acid is heavier than common air, it may generally be avoided, at least in the concentrated form in which it is likely to produce disease.

In wells, cisterns, deep pits, &c., in which it is liable to accumulate in a dangerous degree of concentration, it is always proper to let down a lighted candle before the laborers are admitted, and if it be found that there is sufficient oxygen to support combustion, there will be enough to support the vital flame.

VEGETABLE POISONS.—*Vegetable poisons* exist in various degrees of concentration; all of which, if taken into the system in undue quantity, are liable to produce disease, and some of them even death. It is not necessary, however, for me to enumerate them here, as the good sense with which God has endowed His creatures, generally enables them to avoid their pernicious influences—if we except one, and that the most disgusting of them all, " *Tobacco.*"

MINERAL POISONS.—Various mineral poisons may excite disease, such as *lead, arsenic, copper*, &c.; but here again a sound discretion may generally avoid their noxious effects. I need not say that all the poisonous minerals, and especially *lead, arsenic* and *copper*, should be avoided as conductors for water, or as culinary or household dishes or utensils. And I believe further, that lead, even as a paint, should be avoided as far as possible. I am compelled to this conclusion in relation to the above-named and other poisonous minerals, from a careful observation of the fearful diseases of which, I am confident, they are so frequent exciting causes.

PARASITES.—Various *animal* and *vegetable* parasites are liable to cover or infest the human body. But a minute account of them here would be of no practical value. One important fact in relation to them, should, however, be borne in mind, and it is this: if these parasites are to be regarded as causes of disease, whether vegetable or animal, their presence in or upon the human body is always attributable to a predisposition produced by some previous imprudence; and generally, as I believe, to previous filthy habits. This position appears the more certain, when we remember that vegetables thrive but poorly, or will not even grow at all, where there is no soil to nourish them; and that animals will not thrive, or even live at all, where their right of possession is stoutly disputed, by a " stronger than they."

KOINO-MIASMATA OR MALARIA.—*Koino-miasmata, malaria,* or, in other words, the *paludal poison,* is no doubt a powerful exciting cause of disease. This agent, whether it consists of *sulphuretted hydrogen, animalcules, microscopic fungi,* or of some unknown principle or compound, is freely generated wherever water, with decomposing vegetable matter, is exposed to a temperature of from 60° to 80° Fahrenheit. The paludal poison is generated more copiously as the marshes dry up; the vegetable decomposition probably going on most rapidly at such times; the miasmata or malaria being also carried into the air with the ascending vapor, for which they appear to have a strong affinity. This fact we have seen illustrated in the beautiful and usually healthy village of Geneva, at the foot of Seneca Lake, in the State of New York. Five years ago the rains filled our lake so that it overflowed the lowlands at its foot, a little to the east of the village. This appeared to produce little disease with us the first season, but the two succeeding years, as the water settled in the lake, leaving the overflowed lands, consisting mainly of vegetable matter, to dry up, ague, bilious remittent fever, bilious diarrhœa, &c., were very prevalent in and about the locality. Now, as the outlet of the lake has been dredged, and a large ditch cut through the lowlands, and as the soil of these lands has become dry, the miasmatic diseases have mostly disappeared.

It is probable that miasmata are taken into the air, in connection with moisture, for which, as I have already intimated, they evidently have a strong affinity, and being considerably heavier than common air, they are carried along by winds, or gentle breezes, near the surface of the ground, being interrupted often by a forest, buildings, or even high fences, till gradually their interception by various objects, and their dilution with pure air, render their influence comparatively harmless. Thus it is, in my opinion, that koino-miasmata seldom penetrate far into a large city, or extend their influence far from their source in a woodland region.

The paludal poison appears to have the power to support vegetation, and hence it is in part, no doubt, that miasmatic diseases are less frequent in the early summer, when vegetables are in full growth, than in autumn, when vegetation no longer appropriates it.

Now, of whatever koino-miasmata, malaria, or the paludal poison may consist, it probably enters the blood through the lungs, stomach, and perhaps the cutaneous absorbents, and thus becomes a fearful exciting cause of disease. Of the manner in which it operates, the diseases which it produces, the distances which it may be carried in the air to produce them, the best means of avoiding its pernicious influence, &c., I shall take up more fully when we come to consider the *causes of fever,* in a succeeding chapter. It may be proper, however, as of general application, to state here, that localities in which marsh miasmata are general, should be improved by ditching, cultivation, &c.; and further, that persons necessarily building in such localities, should seek out the most elevated site possible; and that persons living in such localities should keep in-doors at evening and early morning; and, finally, that they should sleep as high from the ground as possible, thus to avoid as far as may be this unpleasant miasmatic agent.

Before leaving this subject however, it may be proper to raise the question; how far, if at all, koino-miasmata would have been injurious to the human system; provided the laws of health had always been observed, and our constitutions had never been depraved. There may be room here for an honest difference of opinion. But as the system gradually becomes accustomed to its influence, and, as it is said, even requires it, after long years of exposure to it, I incline to the opinion, that if the human system had not been contaminated and debilitated by other unnecessary influences, that it would either have remained harmless to us, or else we should have retained good sense enough to have kept beyond its reach.

IDIO-MIASMATA OR ANIMAL EFFLUVIA.—By *Idio-miasmata*, or *animal affluvia* or *poison*, I mean here the noxious effluvia which result from the decomposition of exhalations and excretions from the bodies of persons of filthy habits, when crowded together in filthy confined apartments, &c., but do not include under this head, those emanations which are the result of a secretory process; and which have the power of producing the same disease in others, by which they have been produced, namely the *contagions*.

Idio-miasmata or the *animal effluvia*, are generally exhalations from the feces, and also from decomposed urine, perspirable matter, saliva, &c., in filthy confined apartments, where these matters are allowed to accumulate. These effluvia may accumulate at any time, and in any place, where the matters from which they exhale, are allowed to accumulate in quantities sufficient. It is generally, however, in cold seasons of the year, when filthy apartments are kept most closed, that these effluvia accumulate in sufficient degree of contraction to produce disease. It is doubtless in the main owing to this, that the low typhous fevers, or diseases which they produce, more generally prevail in cold seasons of the year.

The exact nature or composition of idio-miasmata or the animal effluvia, is not definitely known; but it appears highly probable, that sulphuretted hydrogen, is at least a constituent part. It is quite certain however, that they enter the blood, through the lungs, stomach, and probably the skin; and that while they directly depress the nervous system, they also partially decompose or corrupt the blood, and thus become an exciting cause of disease.

In relation to these effluvia, it is hardly necessary to say, that the circumstances, which lead to their exhalation, should always be avoided, and this I believe may generally be done, by a rigid enforcement, of proper sanitary regulations. I believe in this respect, that an improvement might be made in our sanitary laws, by a careful reference to those sanitary regulations which God gave to Moses and Aaron, for the Israelites; and which were rigidly enforced, during their journeyings in the wilderness.*

CONTAGION.—By *contagion*, I mean here any product of a peculiar disease, whether in a solid, fluid or aeriform state, which is capable of producing the same disease in another person; and of propagating itself as well as the disease of which it is the cause and effect, "through

* See Leviticus XVth, and other Chapters of Leviticus in the Bible.

any number of unprotected individuals," the disease thus propagated being properly called contagious.

Contagion may be the product of secretion or exhalation; and it acts in a solid or liquid form, either by simple contact as in gonorrhœa and syphilis, or else by application to an abraded surface or wound as in vaccination and hydrophobia; while in the aeriform state, through the atmosphere, it probably acts mainly by inhalation through the lungs. It should further be remembered that contagious matter may be the result of local action in a diseased individual, as in syphilis, &c., or it may be exhaled from the whole surface of the body, as in typhus, scarlatina, measles, &c.

Contagion in the aeriform state may not be conveyed to any very considerable distance by the air in a degree of concentration sufficient to produce its disease, but it is quite certain that the effluvia, as well as the solid and liquid forms, may become attached to cloths, and especially to filthy articles, and thus be conveyed to great distances; thus retaining also the power of communicating its disease for a considerable time.

The period of incubation in contagious diseases varies from two or three days to several weeks; and the diseases are generally attended with febrile action, very generally run a tolerably definite course, and in many of them the system becomes insusceptible to a second attack. Thus we see that the *contagions* become a frequent exciting cause of disease, subjecting almost every individual born into the world, at the present day, to a train of diseases; through which if he pass unscathed he may do well. This naturally leads us to inquire into the origin of contagious diseases; and to see, as Christian philosophers, whether they might have been or may yet be avoided.

Now it is absurd to suppose that God made these contagions, as for instance that of syphilis, and then created a predisposition in the system to them; thus directly forcing them upon us without fault on our part. We must then take the other horn of the dilemma and conclude, as every circumstance appears to imply, that imprudence in deviating from the laws of health has created not only the predisposition, but has also caused each of the contagious diseases to originate one after another till we have arrived at the present catalogue. If then this be so, wider deviations from the laws of health, longer continued, will swell the catalogue. On the other hand, better sanitary regulations, better morals, and a more rigid observance of the laws of health, by the community in general, might greatly diminish the number of contagious diseases and perhaps in the end might exterminate them.

This theory in relation to the origin of contagious diseases, appears the more probable, when we remember that most, if not all, the contagious diseases, even now very frequently arise without an exposure of the person to the contagion generated in another patient, the result no doubt in such cases of various imprudences; the system once in the diseased state generating a contagion which propagates the disease.

ENDEMIC INFLUENCES.—By *endemic*, from εν "in," and δημος, "the people," is here meant that influence which gives character to the diseases of any locality, the term implying that it originates in the locality or among its inhabitants or people. It is evident that an

3

endemic influence consists of the sum total of all the pernicious influences which are operating upon the health of the community in any locality referred to. In order then to arrive at the endemic influence in any locality, we should consider the climate, elevation, winds, soils and productions, as well as the cleanliness, morals, intelligence, prudence, &c., of its inhabitants; and we may form an idea in our minds of a sort of compound endemic influence, which may not very generally be erroneous, and perhaps never so, if we could appreciate exactly all the influences which are operating.

This, in my opinion, constitutes all there is in an endemic influence; and the indications for its removal, or alleviation, consists in correcting, as far as may be corrected, all the natural influences by ditching, draining, cultivating, &c.; and also securing to the inhabitants, cleanliness, temperance, virtue, industry, cheerfulness; and in short, a strict observance of the laws of health in every respect.

EPIDEMIC INFLUENCE.—By *epidemic*, from ɛπι, "upon," and δημος, "the people," is meant as the term implies, an influence coming upon the people more extensive in its effects than an endemic, and liable to extend, as some have done, nearly or quite around the globe.

Various theories have been formed to account for epidemic influences, most of which have placed the cause in some mysterious principle beyond the power of man to reach or modify; as though the ALMIGHTY had placed us here, and let loose upon us occasional emissaries of destruction without our fault or even ability to palliate or resist.

Now this may be so; but I incline to the opinion, that more light on this subject and a more careful observation of the causes of disease generally, will yet bring home to us the unpleasant fact that the imprudences of mankind in deviating from the laws of health, produces the predisposition to epidemics, as well as to all other diseases; and that the elements in nature, which appear to act as causes of disease, are generally, if not universally so, in consequence of this predisposition, the result of man's imprudence. If this be so; we have, in order to ascertain the constituents of an epidemic influence, to take into account all the imprudences of the people, as far as the influence may extend, and then add to this the pernicious influences, which the various elements in nature, are unnecessarily allowed to exert upon them; and we shall have formed in our mind, if we have exactly appreciated all the influences, a correct notion of the *epidemic influence*.

If this be so, it may account for the fact that in nearly all epidemics the first cases are the most fatal. For it is well known that epidemics, as a rule, sweep off the more intemperate portions of the human family; and as those most predisposed by intemperance, would in that view of the case, be the ones first attacked, so would also such cases be more certainly fatal.

SECTION III.—SYMPTOMS OF DISEASE.

By *symptoms of disease*, I mean here the phenomena developed by disease in any organ or function of the human body; which may be sensible to the patient, or perceptible to a careful observer. I need not say, that it would be impossible to give in this place, more than a general view of the symptoms of disease. And as I shall consider *inflammation* in the following chapter, and the *phenomena of fever* in the third chapter, I shall pass over the symptoms of inflammation and fever generally, altogether in this place; and will confine myself here to a consideration of the general symptoms developed by the *nervous system*, the *digestive system*, the *circulatory system*, the *respiratory system*, by the *eye*, the *ear*, the *skin*, the *urinary organs*, and the *genital organs.*

Neither shall I attempt to give the particular symptoms peculiar to each disease, with which the different parts of these various systems are liable to be affected; but only such general symptoms as might not necessarily be called up, when we come to treat of these diseases. By thus taking in this place such a view of the general symptoms developed by these various systems, in a state of disease, I trust that the particular symptoms may be more readily appreciated, when we come to take up particular diseases in their proper order.

Before proceeding to inquire into the general symptoms developed in the human system in a state of disease, it is proper to bear in mind the anatomy and physiology of the human body; and also to remember the relations of the different parts, as well as the phenomena exhibited in health, by comparison with which, the phenomena or symptoms developed in disease, will be more readily appreciated. Almost any person of common observation will readily detect variations from a state of health, by the *countenance, positions, motions, mental conditions, restlessness, irritability,* &c. of persons with whom they associate; and I need not say that the general symptoms thus developed, in all variations from a healthy condition, should always be carefully noticed by the medical man.

SYMPTOMS DEVELOPED BY THE NERVOUS SYSTEM.—As the brain, and whole nervous system is the medium through which the mind communicates with the body, or the part upon which the mind directly acts, and as it is the medium through which we experience pain, the minute nerves pervading every tissue, transmitting impressions from every part to the mind, and carrying again the influence of the will to the voluntary muscles of every part; we can see that its symptoms from diseased action must be very numerous and peculiar. It should also be remembered that the brain is the generator of a vital force, or nervous influence, which being transmitted through the nerves, enables the different organs of the body to carry on their functions. It is also through the nervous system, that the mind becomes conscious of pain and tenderness.

The brain and whole nervous system, however, depend upon a due amount of stimulus from the blood to enable them to carry on their functions in a proper manner. It is evident then that diseases of the nervous system develop a train of symptoms which may exhibit themselves through every organ and tissue of the body. I will here, however, speak only of the general symptoms which attend an augmented or increased action, and a diminished action, and attempt to show the general phenomena of each.

Increased action of the nervous system, if it be moderate, is attended with increased activity of all the functions of the body; the increased amount of nervous influence exciting the digestive process, accelerating the circulation, increasing respiration, giving lustre to the eye, delicacy to the ear, a glow of health to the skin, activity to the urinary and genital organs, and increased activity to the mental faculties.

If, however, the increased action becomes very great, instead of having the pleasant train of symptoms I have enumerated, there is developed a frightful train.

The functions of the digestive system become interrupted or impaired, the respiration becomes rapid, the circulatory system becomes tumultuously excited, the eyes become red and glassy, the ear becomes painfully sensitive to sounds, the skin may become dry and hot, the urinary secretion is nearly suspended, the patient becomes wakeful, restless and delirious, the wildest insanity may occur, the cephelaga becomes intolerable, and if serious organic changes occur in the brain, the patient dies either in a state of intense agony or passes into a fatal coma and thus expires. Such, according to my observation, are the general symptoms developed in cases of increased morbid action in the nervous system; including of course organic changes in the brain itself.

Diminished action of the nervous system, on the other hand, is attended with diminished action in all the functions of the body. The digestive organs become inactive or inefficient, respiration becomes defective, the circulation becomes languid, the eye becomes dull and heavy, the countenance becomes drawn and sallow, the skin is pale and exsanguinated, the functions of the urinary and genital organs are morbidly deranged, the intellect becomes dull; and all the motions of the patient indicate a condition of languor, inactivity and gloom. If, however, the nervous depression is very great, all these general symptoms become very much aggravated; digestion is nearly suspended, respiration becomes hardly perceptible, the blood recedes from the extreme vessels, and internal congestions or passive inflammations occur; the countenance becomes pale and haggard, the muscles become thin and flabby, the mind becomes irritable, gloomy and hypochondriacal; all the energies of the system become exhausted, and the patient restless, and dragging himself down in bed, either in a state of muttering delirium or profound coma, expires.

With this general view of the symptoms developed by increased and diminished morbid action of the nervous system, we may be able to keep in our minds, as we take a glance at the general symptoms developed by the other systems of the body; the important agency of the vital force, or nervous influence, in almost every symptom that is developed in every organ and tissue of the body.

The *position* of the patient in cases of disease of the nervous system, attended with increased action, is apt to be very uncertain or changeable; now assuming this attitude, and presently that, the motions in many cases becoming furious; but in diseases of the nervous system attended with depression, or diminished nervous action, the patient is apt to lay upon the back, and in cases of great prostration to drag down towards the foot of the bed.

The *countenance* again in cases of increased action of the nervous system is apt to become flushed, and if it becomes very great the eyes may become wild and glassy. In cases of diminished action or depression, however, of the nervous system the countenance is generally pale, the eyes sunken and heavy, and there is a general expression of gloom; but the most prominent trait of the countenance marking nervous depression, is a leaden colored streak under the eyes; first named I believe, by M Jadelot, "the oculo-zygomatic trait."

This peculiar trait I have more especially noticed in patients whose nervous systems have become greatly depressed by licentious habits; whether from masturbation, onanism or excessive sexual indulgence. I speak of this peculiar trait, as an observance of it is often of great value in affording a clue to the causes which may have been operating to produce the nervous depression; especially as patients are often in such cases slow to confess or even to admit the truth.

With this glance at the general symptoms developed by the nervous system, let us pass on to the consideration of the general symptoms belonging to the digestive system; remembering that every disease in all the systems, organs, or structures of the human body, must necessarily be attended with morbidly increased or diminished action of the nervous system.

SYMPTOMS DEVELOPED BY THE DIGESTIVE SYSTEM.—The symptoms developed in diseases of the digestive system, are exhibited by the tongue, appetite, thirst, nausea, appearance of the alvine discharges, pain, position, countenance, &c.

In most cases of *excitation*, whether inflammatory or from irritation, attended with increased morbid action of the nervous system, there is liable to be a red, dry, or coated tongue, loss of appetite, thirst, nausea, vomiting, diarrhœa, &c.; the functions of all the organs of the body affected sympathetically, being morbidly excited, and more or less perverted. In most cases of depression, in diseases of the digestive system, there is liable to be a flabby, and perhaps a darkly-coated tongue, loss of appetite, with colic, and perhaps acid eructations, colliquative diarrhœa, internal congestions, wasting of flesh, and finally, a depressed action of all the organs of the body, and especially of those parts which sympathize more directly with the digestive system.

The *position* of the patient in diseases of the digestive system, attended with inflammatory action, is generally rather stationary; but little motion being apparently desirable, as it usually increases the pain and general uneasiness. In spasmodic affections, however, unattended with inflammation, the patient is apt to be tossing about on the bed, as the motion evidently affords partial relief. Perhaps the most marked position assumed, is that upon the back, with the limbs drawn up, in cases of general inflammation of the peritoneum and abdominal viscera.

The *countenance*, in diseases of the digestive system, is either flushed or pale; is liable to have a sallow, greasy, or yellowish cast, and to have a general expression of gloom or despondency. There are, also, a few particular traits which should always be noticed; such as the blueish color of the white of the eye, in diseases of the spleen; the yellowish, in diseases of the liver; the paleness around the mouth, in gastro-intestinal inflammation or irritation; and the peculiar gloomy expression usually exhibited, if the duodenum be the seat of the intestinal inflammation.

It is hardly necessary to state, in conclusion, that the sympathetic disturbances in the functions of all the organs of the body, from diseases of the digestive system, are very considerable, depending, however, upon the seat, nature and extent of the disease, and also upon the sympathetic relation of each organ with the diseased part, &c.

SYMPTOMS DEVELOPED BY THE RESPIRATORY SYSTEM.—The general symptoms developed by diseases of the respiratory system are dyspnœa, cough, expectoration, and the sounds elicited by auscultation and percussion, as well as the positions assumed, countenance, &c. In *spasmodic* affections of the respiratory system, attended usually with diminished nervous action or energy, the dyspnœa may be very considerable; the cough is apt to be hollow and dry; in many cases there is no expectoration; and the sympathetic disturbances of the system depend upon the seat and severity of the local spasmodic affection.

In *inflammatory* and organic diseases, generally, the dyspnœa is liable to be considerable; the cough, in the early stages dry, is apt to be attended, in the latter stages, with a mucous, bloody, or purulent expectoration; and the sympathetic affections or disturbances, in the functions of the various organs of the body, are always more or less considerable. In no class of diseases is the necessity of inquiring into the condition of the nervous action or energy more important than in diseases of the respiratory system, whatever may be the seat, or apparent character of the disease. This is rendered so, from the liability of passive inflammation, to which the lungs especially, by the character of their structure, are peculiarly predisposed.

The *positions* assumed in diseases of the respiratory system, depend upon the nature and seat of the disease. In *pneumonia*, if but one lung be involved in the inflammation, the patient lies on the affected side; if both, the patient lies on the back, with the shoulders elevated. In *pleurisy*, the patient lies on the well side, as it creates less pain; while in bronchitis the patient generally lays on the back, with the shoulders considerably elevated.

The *countenance*, in this class of diseases, varies with the seat and nature of the disease. In pneumonia, the face is livid, and has an anxious expression; in bronchitis, the face is pale, and the countenance exhibits a hurried aspect; while in tubercular phthisis, the countenance has the hectic flush, and wears a hopeful expression. With this general glance we will pass on to the consideration of the general symptoms developed in diseases of the circulatory system, leaving the consideration of the sounds elicited by auscultation, percussion, &c., till we come to take up the special diseases of the respiratory system, in a succeeding chapter of this treatise, in their proper place.

SYMPTOMS DEVELOPED BY THE CIRCULATORY SYSTEM.—The symptoms developed by diseases of the circulatory system, include the actions of the heart; the frequency, quickness, volume, tension, and strength of the pulse, the condition of the veins and capillaries, and the state of the blood.

It is proper to remember that the pulse of a new-born infant beats from 130 to 150 times per minute; that of a middle aged person, from 70 to 80, while that of the aged, may vary from 50 to 70. It should also, be borne in mind, that sex, position, and various accidental circumstances are liable to vary the frequency of the pulsations considerably from this standard, which circumstances should always be noticed and taken into the account.

In *functional* diseases of the circulatory system, attended with increased or exalted nervous action, the pulse is liable to be frequent, quick, tolerably strong and tense, while in functional diseases, attended with diminished nervous action, or depressed vital power, the pulse may more generally be frequent, and perhaps quick, but it is likely to be not very strong or tense. The pulse, in inflammatory diseases of the circulatory system, if the inflammation be active, may be frequent, quick, full, tense and strong; if however, the inflammation be passive, being attended with depression of the vital power, the pulse may be frequent or slow, but it is not generally very strong or tense. Finally, in organic diseases of the circulatory system, whether the disease be of the heart or arteries, the pulse is liable to become very irregular and in most cases intermittent.

The *capillary* and *venous* circulation correspond in most cases with the cardiac and arterial, while the state of the blood which should always be examined when drawn, by the microscope if necessary, will generally be found weak or strong, according to the exalted or diminished action of the nervous system. Why this is generally the case, we need not wonder, when we remember that the blood enables the brain and nervous system to generate and distribute the nervous influence or vital force.

The *positions* assumed in diseases of the circulatory system, are, some of them peculiar. In hydropericardium and hydrothorax, the patient is found with the head and shoulders considerably elevated; or in hydrothorax, if the fluid occupies but one side of the chest, the patient may be found lying on the affected side, in some cases.

The *countenance* in dropsical cases has a bloated or tumid appearance, and in cases of organic disease of the heart, attended with considerable interruption to the circulation through the lungs, the countenance wears an anxious expression, and sometimes there may be seen a pale or leaden colored line, extending from the angles of the lips to the margin of the chin. This trait of the countenance, so significant of organic disease of the heart, has been called "the labial trait."

The sympathetic derangement of the various organs of the body, with diseases of the circulatory system, are very considerable, and as the arteries, veins and capillaries penetrate every organ and tissue of the body; general diseases of the circulatory system, as well as of the nervous system, must necessarily involve more or less, every organ and tissue of the body.

SYMPTOMS DEVELOPED BY THE EYE.—The symptoms developed in diseases of the eye, are pain, redness, intolerance of light, or want of sensitiveness to light, &c., depending upon the nature of the disease. In *active* inflammation of the organ, attended with increased action or excitation of the nervous system, there is apt to be intolerance of light, pain, redness, and more or less cephalgia, &c.; but in *passive* inflammatory ophthalmia, attended with diminished nervous action, or vital depression, there may be intolerance of light, pain and redness, but in addition to these symptoms, the cornea is apt to present a cloudy appearance.

In *amaurosis*, a disease depending upon diminished vital action in the optic nerve, and attended in many cases with general nervous depression, the eye may appear to the observer, nearly natural, but the patient experiences partial or total loss of sight, and the disease may or may not be attended with pain. In organic diseases of the eye generally, the appearance of the eye usually indicates the nature of the organic changes.

Exophthalmia, or a protrusion of the eyes from their sockets, depending upon general debility, with relaxation of the muscles of the eye, and a congested condition of the tissues in the back part of the orbit, forming the cushion of the eye, is usually a disease of females, though very rare; and so far as my observation has extended, it usually occurs in connection with *bronchocele*, in anaemic females, suffering from amenorrhoea. The protrusion of the eye-balls, in some cases giving the appearance of enlargement, is liable to deceive the careless observer; but by careful attention to all the symptoms, together with the complications, the nature of the disease need not be mistaken.

The *position* assumed by the patient with ophthalmia, is liable to be on the back, with the head slightly elevated. The *countenance* in this class of diseases, is well understood by every observer; but the countenance and position in walking, in children, with scrofulous ophthalmia should be born in mind; especially the bowing or bending forward of the head, and squinting through the eyebrows. The sympathetic disturbance of the different organs, with diseases of the eye, are very considerable, the brain however, being the part more especially involved, in most cases.

SYMPTOMS DEVELOPED BY THE EAR.—The symptoms developed in diseases of the ear, are a sense of roaring, pain, intolerance of sound, and deafness, &c. In cases of inflammation of the structures of the middle and internal ear, attended with nervous excitation, or increased nervous action, there is pain, more or less severe, intolerance of sound, with great acuteness of hearing, in some cases, and generally severe sympathetic cephalalgia. In cases of otitis, attended with nervous depression, the pain may be in some cases dull, and not unfrequently there is dullness of hearing, or even deafness.

If wax has accumulated in the meatus auditorious to any considerable extent, it is apt to produce a sense of roaring, and it is hardly necessary to say, that a loss of action in the auditory nerve, produces nervous deafness. The *position* of the patient in otitis, is apt to be very changeable, and the *countenance* is apt to be indicative of pain and anxiety. It should also be remembered, that while the different organs of the body may sympathize more or less with diseases of the ear, the brain has the most intimate sympathetic relation.

SYMPTOMS DEVELOPED BY THE SKIN.—The general symptoms developed by the skin, are its temperature, moisture or dryness, color and a class of rashes and eruptions. In cases of great debility, the skin is apt to be pale, cool, and in some cases, it is covered with a clammy perspiration. In sthenic conditions, attended with increased action of the nervous and circulatory function, the skin is flushed and warm, and in most cases dry. A great variety of rashes and eruptions appear upon the skin; some of them indicative of the exenthematous fevers; while others are more local affections of the skin. Especially is this the case with the animalcular and cryptogamous eruptions; for a clear diagnosis of which the microscope becomes of great value.

The *position* of the patient in exanthematous affections is apt to be very changeable, on account of the intense itching which is liable to attend: and the *countenance*, so far as it is changed, is indicative of that very peculiar sensation; the face of course being marked by the rash or eruption in many cases. The different organs, structures, and tissues of the body are liable to suffer by sympathy, in many diseases of the skin; and while the mucous membranes generally suffer most, the alimentary mucous membrane is especially liable to suffer, so far as my observation has extended.

SYMPTOMS DEVELOPED BY THE URINARY ORGANS.—The general symptoms developed in diseases of the urinary organs, include those attending diseases of the kidneys, ureters, bladders, and urethræ; and also the quantity and quality of the urine, as well as the manner of its discharge.

In inflammatory diseases of the kidneys, there is apt to be a dull, heavy pain; if, however, the ureters, bladder, or urethræ be the seat of the inflammation, the pain may be acute, and of a biting or stinging character. In neuralgic affections of the urinary organs, the pain is sharp and acute, as in neuralgia generally.

The quantity of urine in inflammatory affections of the kidneys is usually small; and if the ureters, bladder, or urethra be involved in the inflammation, there is liable to be difficulty, with more or less pain, in voiding it.

In nervous diseases of the urinary organs, if attended with general diminution of vital force, the quantity of urine secreted may be morbidly increased, and there may be incontinence; or, if the disease be of the urinary passages, and spasmodic, there may be temporary retention of urine.

In *albuminuria*, or granular disease of the kidneys, and perhaps in other forms of organic disease of these glands, the urine is liable to be charged with albumen, which may be detected by exposure to heat. The urine is liable also to contain acid or alkaline substances, depending generally upon an excess of acids or alkalies in the fluids of the body; and these substances may cohere or unite in the urinary passages, and form stone in the cavity of the kidneys or bladder; or they may be suspended in the urine till it is voided, and then it may separate and form a sediment in the vessel, if allowed to stand.

Bile and various other ingredients, not found in healthy urine, may be separated from the blood by the kidneys; rendering an examination of the urine, in most diseases, as by heat, the microscope, etc., if necessary, a matter of great importance.

The *position* assumed by patients with nephritis is peculiar, the body being usually bent forward; and if but one kidney be involved, being turned a little to the affected side. The *countenance* is not very peculiar, though, in addition to the expression of pain, when it exists, the face is apt to have a dingy, bloated appearance, in some cases at least.

The sympathetic relations existing between the kidneys and the various organs of the body are very general; but the stomach appears to sympathize, perhaps more strongly, than any other part—considerable nausea and vomiting usually occurring in cases of nephritis. There is, however, a strong sympathy between the urinary and genital organs, as we should naturally suppose, from their immediate connection.

SYMPTOMS DEVELOPED BY THE GENITAL ORGANS.—The general symptoms developed in diseases of the genital organs, of course vary with the *sex.* In *males,* they differ with the nature and seat of the disease, and depend, also, in many cases, upon the general conditions of the system. In inflammatory diseases of these organs in males, attended usually with increased action of the nervous and circulatory functions, there is liable to be pain, more or less severe; and in many cases there is a morbid excitation of function, with or without seminal or other discharges. In cases of mere irritation, however, with debility of the organs, and perhaps attended with a general depression or diminished vital action, there may or may not be pain or any abnormal discharges; but in too many cases, if some form of licentiousness has been practiced, there is liable to be morbid seminal or other discharges.

The general symptoms developed in diseases of the genital organs in females are either inflammatory, organic, or nervous; and the symptoms, of course, depend upon the nature, seat, and character of the disease. In inflammatory and organic diseases generally, there is apt to be pain, tenderness, and in many cases a morbid discharge; while in nervous diseases, attended with local and general debility, if there be pain, it is of a sharp, darting character. Suppression or retention of the menses, as well as uterine hemorrhages, may be connected with a sthenic or asthenic condition of the system; the local and general disease being in the one case *active,* and in the other *passive;* depending generally upon the state of the blood, and also upon a consequent increased or diminished action of the nervous system.

The positions assumed and expressions of countenance in diseases of the genital organs are not very particular, if we except hysterical affections, in which disease the patient is either tossing about, or else laying apparently almost lifeless: the countenance in the one case exhibiting the wildest excitement, while in the other it is much as in profound sleep.

The sympathetic relations of the genital organs with the other organs of the body are very general, especially in the female; sympathetic disturbance of the nervous system being, as I believe, the cause of most of the phenomena of hysteria.

Having thus taken a glance at the general symptoms developed in diseases of the *nervous system,* the *digestive system,* the *respiratory system,* the *circulatory system,* of the *eye,* of the *ear,* of the *urinary* and *genital organs,* we are prepared to pass on to the consideration of the *diagnosis* and *treatment* of disease, in the two succeeding sections of this chapter. Before, however, we leave this subject, it may be well to remember that pain depending upon inflammation, in the structures generally, is apt to be continuous, and more or less increased by pressure; while spasmodic pain is often intermittent, and may be more or less relieved by pressure.

It is also proper to remember here, as of general application, that in inflammation of the serous membrane, as in the pleura, the pain is lancinating; in the fibrous structures, it is dull; in the nervous, darting and acute; while in the parenchymatous and cellular, generally, it is apt to be dull, throbbing, and heavy.

Thus have I completed what I had to say here on the general symptoms of disease; leaving the particular symptoms to be called up when we come to take up the particular diseases of the various organs and structures of the body, in their proper place.

SECTION IV.—DIAGNOSIS OF DISEASE.

By *diagnosis*, from δια, and γινωσκω, 'I know,' is here meant to know or discriminate; hence, by diagnosis of disease, is meant to know or discriminate disease; and as disease consists merely in a deviation from the standard of health, *diagnosis* is really ascertaining the deviations from a state of health. It is true that various names have been given to these deviations; and hence, we say, this and that disease. With this explanation, we are prepared to pass on to a consideration of the diagnosis of disease. It is proper, however, to state, before we proceed, that the term *diagnosis* is used technically; not only to know disease, but also to distinguish one disease from another.

Now from the very nature of the case, it is absolutely impossible to form a correct diagnosis of disease, without a knowledge of anatomy and physiology. For how can we arrive at a deviation from a state of health, without first knowing what a healthy condition should be? In fact, attempting to form a correct diagnosis or notion of disease without a previous knowledge of the human system, and of the nature and operations of its functions, would be just as absurd as it would be for a man, unacquainted with the machinery of a watch, to attempt to account for any failure in its keeping correct time.

With a correct understanding, however, of anatomy and physiology, and of the causes of disease, as well as of the means of preserving health, the man of good common sense may generally form a correct diagnosis; thus making *pathology*, or "*diseased physiology*," to the close observer, a plain matter of common sense; following as plainly as any effect follows a cause. With these preliminary considerations, we are prepared to pass on to the consideration of the best manner of forming a diagnosis; by the history of the case; an examination of the symptoms, &c.

Calls — It is hardly necessary to say, that calls to attend the sick should always be attended to without delay; even in cases not apparently alarming. For it shows a reasonable desire on the part of the physician, to gratify the feelings, at least, of the patient; and it leaves the doctor at leisure to attend important calls which require haste, should they be received. While, then, great haste and bluster should not be exhibited, due promptness is always commendable; and is often a matter of great account in the mind of the patient.

Question the messenger. — In calls to visit the sick, it is generally proper, especially in the country, to inquire of the messenger who is sick, and what appears to be the matter. This enables the physician to prepare himself with whatever he may be likely to need, by way of instruments, medicines, &c.; a consideration of importance to country practitioners at least. Besides, if the messenger is an intelligent person, the answers may enable the physician to calculate, with some degree of certainty, at what time he may be able to return. And further, indefinite as such answers must always be, they may afford a clue to a correct diagnosis in the case.

Inquire of the Friends.—When a very sick patient is visited for the first time, I think it is generally well to inquire of the friends or family in relation to the case, and, if possible, to get from some one of them a general history of the case. This enables the physician to come directly at the important particulars in a manner to give the patient the least possible worriment. This is especially important if the patient be a child, or a very sick adult. In adults, however, if the patient is not very sick, it may not be necessary, or even advisable.

Examination of the Patient.—In the examination of the patient, it is best to ascertain the age, occupation, and residence, if not already known. The *age* enables the physician to appreciate any influence that particular periods of life may be exerting—an item of importance in many cases, in female patients, at the age of puberty, or at the critical period of life. The *occupation*, too, may have had an important bearing in modifying or even producing the disease. The *residence* enables the physician to appreciate any endemic or local influence which may have produced or modified the disease. All these circumstances being taken into account, the examination may be proceeded with in a rational manner.

In all cases, it is best next to inquire how long the patient has been sick or poorly, and to get from the patient a history of the case. It may be best, however, to interpose questions, to ascertain facts and circumstances which may not appear clear, from the account of the patient as he proceeds with the history, as it keeps up a continuous chain of circumstances which otherwise might appear broken or indefinite.

Having thus got a history of the case, it is best to inquire into the causes which may have been operating, as it may, in part, account for the first deviation from health; and it may also account for many peculiarities in the history of the case, which might otherwise be comparatively unaccountable. In many cases the patient may have formed correct notions as to the cause of his sickness; if not, an inquiry into the habits of the patient may enable the physician to, at least.

In *chronic* or *acute* cases of disease, having proceeded thus far, it is best to mark the countenance, position, and motions of the patient; and having appreciated whatever they may indicate, a general examination of the organs and functions of the body should be made, beginning, in chronic cases, with the brain and nervous system, and passing on to the digestive, respiratory, and circulatory systems, and, if necessary, inquiring into the condition of the eye, ear, and skin, as well as of the urinary and genital organs.

In very *acute diseases*, the attention may be at once fixed upon the diseased organ or part, in which case it may be best to examine that part first, and then to notice the complications involving the other parts, organs, or functions of the body. This would relieve the patient from too much worriment, and might be amply sufficient, in many acute diseases, especially in cases in which the natural complications with the local disease might readily be inferred.

In examining cases it is best to ascertain if there be ·pain or tenderness, and also the character of the pain, if it be experienced. This is of great importance in cases of disease affecting mainly internal organs, and should, therefore, never be omitted. The absence of pain, or even tenderness, should not, however, be considered as an evidence of the absence of disease, or of inflammation in all cases.

The *pulse* should be examined to ascertain the condition of the circulatory system, and it is always proper to note the frequency of the pulsations, as well as the quickness, volume, tension, strength, regularity, &c., as indicative of the power of the heart and arteries, and, indirectly, of the degree of vital power. In making an examination of the pulse the arm should lie in a semi-flexed position, and two or three fingers being placed gently upon the radial artery, the pulsations should be noticed for a minute at least.

Having thus ascertained the age, occupation, and residence of the patient, when not previously known; and also learned the time of attack, as well as the history of the case; and noticed the countenance, position, and motions of the patient; and inquired in relation to the pain and tenderness; and, lastly, having examined the pulse, the attention should be turned to the part or parts in which the disease appears to be located—and I need not say that every symptom which can have a bearing upon a correct diagnosis, should be carefully inquired into and appreciated.

In completing the examination, a careful *inspection* of the diseased part, as well as *palpation* in various ways, may become necessary. In many cases, also, *pressure, succussion, percussion, auscultation, mensuration,* and *chemical,* as well as *microscopical* examinations may be required, depending, of course, upon the nature and seat of the disease.

Having thus ascertained the seat and nature of the disease, in the manner I have suggested, it becomes a matter of great importance to estimate correctly the degree of vital force, or the amount of nervous energy which remains. For by taking into account the degree of nervous influence generated by the brain, and distributed by the nerves, a correct conclusion may generally be formed as to whether the disease be of an active or passive character, or in other words, whether it be attended with augmented or diminished action of the nervous system.

Finally, when a correct diagnosis has thus been arrived at, whether the disease be of some part of the *nervous, digestive, respiratory,* or *circulatory* systems; or of the *eye, ear, skin;* or of some portion of the *urinary* or *genital* organs; we should not only carefully arrive at the exact condition of the diseased part, but we should estimate correctly, not only the sympathetic relations, but should also discriminate carefully between sympathetic disturbance and primary disease.

Thus have I completed what I had to say here on the diagnosis of disease; and I hope that the suggestions here made may be sufficient to put the student upon the right track; that when he comes to assume the fearful responsibility of examining and prescribing for the sick, he may feel no embarrassment in relation to the best method to be pursued. I will, however, in conclusion again suggest the importance, when necessary, of microscopic examinations, especially of the saliva, milk, blood, pus, sputum, vomited matters, fœces, &c. It is necessary, however, in order to appreciate an abnormal appearance of these, or any other secretion or excretion, &c., to understand their normal appearance in a state of health. Thus does the evidence afforded by the microscope become a plain matter of common sense, as well as the evidence afforded by every symptom which it becomes necessary to examine, in forming a diagnosis, for the purpose of arriving at the indications in the treatment of disease.

Having now taken a glance at the *nature, causes, symptoms,* and *diagnosis* of disease, I shall proceed, in the following section, to take into consideration the *treatment* of *disease,* which will complete this chapter on *disease.*

SECTION V.—TREATMENT OF DISEASE.

By *treatment of disease,* or *therapeutics,* I mean here, to attend upon or alleviate the sick; and also the modus operandi of remedies, as well as the rules for applying them, &c. I wish however, to offer only a few general suggestions here, and will therefore consider, first, the duties of the physician in attending upon the sick, and will then inquire into the best method of fulfilling indications, &c.; and, lastly, give some general rules which may serve as a guide in administering or applying remedies, &c.

To attend upon the sick is one of the most responsible positions in which a man can be placed. The physician, then, should be a man of strict morals, stern integrity, good education, and withal, a man of good common sense. With these essential qualifications he stands forth among his fellow men, a teacher of righteousness, second to no man, or class of men, not even the clergy excepted. The physician, then, thus qualified—and let none other assume to be such—offers his services directly, by inserting a notice in a paper, or indirectly, by placing a sign at his door, it matters not which. In either case he is responsible to the community for the faithful performance of his duties.

The duties of the physician thus qualified and situated, are substantially these:—To observe himself, and teach others, so far as he is able, the *Laws of Health,* so that people may never get sick, if he can help it. To use his best endeavors, when called on, in case of sickness, to have the invalid restored to health as soon as possible. And, finally, to give such advice to all his patients as shall enable them, as far as possible, when once restored to health, to keep well.

Now in relation to the best method of giving advice to the community, that they may avoid getting sick, as well as to the advice proper for each patient restored, that they may avoid further sickness; I will leave it to the good sense of my fellow students of the Medical Profession, and will only offer a few suggestions in relation to the treatment of the sick; or as I have termed it, the treatment of disease. As then I have already suggested, in a previous section, that a correct diagnosis should always be formed, so that the Physician may have a correct understanding of the disease of his patient, or of the deviations of his system from a state of health; so nothing should be prescribed, as a remedy, without first getting a clear *Indication.*

INDICATIONS.—By *indications*, I mean any manifestation offered by disease; of what is proper for its removal. Indications then may point to any remedial measure, such as removing the cause of the disease, correcting bad habits, exercise, fresh air, cheerfulness, and in fact, to any and everything, that could tend to restore health. When then indications are once discovered they should be fulfilled, and in just so far as they can be, without the administration of medicines. For medicine, it should be remembered, is at best little better than a *necessary evil*. It is then for the good sense, of every medical man, to judge how far he can fulfill the indications, furnished by the condition of the patient, without the administration of drugs. It is certain, however, that just so far as it can be done, by removing the causes, correcting bad habits, regulating the diet, by exercise, cleanliness, cheerfulness, and in short, by enforcing a strict observance of the laws of health, it should be done. At any rate, I would never prescribe medicine for any patient, unless that which was indicated, in these respects, were strictly complied with.

After doing, however, all that can be done of a hygienic character, if any indications remain to be fulfilled, medicines should be prescribed if they can be found exactly adapted to fulfill the indications. But before resorting to medicines, it is well to remember, that disease is always, in one sense, a condition of debility; and that if depletion is ever required in the treatment of disease, it is only so to prevent a worse evil, and that it is always done at the expense of vitality even when it be imperatively required. It is well then to avoid depletion as far as it is consistent with the welfare of the patient; but if a suffering organ require it, it must be resorted to, even though it be at the expense of the general strength.

When there are indications which demand medicine, if it is possible to fulfill them by remedies that will not produce or leave a bad impression upon the fluids or solids of the system, such remedies should be used. And if the indications can be fulfilled by a remedy or medicine that will improve permanently the blood and general system, such remedies should by all means be preferred and used instead. But to save life, when nothing short will do, it is right not only to deplete, but also to use remedies, the permanent effects of which may not be the most pleasant. This becomes a duty, as most people prefer to live, even though it be "at a poor dying rate." When, however, extreme measures become thus necessary, every possible precaution should be taken to administer them in such a way as to render them the safest possible; and yet to fulfill the indications for which they are administered.

Among the most prominent *Indications* that are liable to be presented, in diseased conditions, are, for *depletion, repletion, dilution, stimulation, sedation, revulsion, suppression, alteration, chemical action, mechanical influence, &c.*; concerning the proper fulfillment of which, it is well that we should consider, as we pass along. I will therefore take them up in the order in which I have named them.

Depletion.—When depletion becomes necessary, if it can be done by cutting off the supply of food, it should thus be accomplished; but if not, and the abstraction of blood becomes necessary, the local should be resorted to if it will answer, but if not, and there is no other alternative, general bleeding should be resorted to, to prevent a worse calamity.

Repletion.—If a tonic becomes indicated, in consequence of weakness, or poverty of the blood; if a regulated and nourishing diet will correct it, with proper exercise, air, &c., that is the best prescription; but if that is not sufficient, it is best to use in addition, such a tonic or tonics as will supply the blood and solids, with that in which they are deficient.

Dilution.—If there be an indication to dilute the blood, it should always be done by pure water, or something not irritating to the alimentary mucous membrane, and which, on entering the blood will dilute it, without rendering it an irritant to the various tissues of the body.

Stimulation.—If the indication is to stimulate, that may sometimes be accomplished by stimulating food, good air, cheerfulness, &c.; but if that is not sufficient, it is well to resort to such a remedy as shall not irritate, or in any way permanently impair the tone, vigor, or functions of the system.

Sedation.—If a sedative be indicated, let every source of excitement or irritation be carefully removed, and then, with soft words and gentle manners, allay excitability as far as is possible. If, however, these measures do not entirely fulfill the indications, use in addition, such a sedative as will quiet arterial or nervous excitability, without permanently impairing vitality or deranging the functions of the body.

Revulsion.—If counter-irritation be indicated, it should be done by the most convenient, cleanly and effectual means; always taking care to avoid unnecessary torture, as well as to prevent as far as may be, ugly scars, especially on the fair sex, or about the face in any case.

Suppression.—If the indication is to suppress some morbid condition, as that of excessive fear, rational assurances should be held out of safety; and that assurance once established, may entirely suppress or destroy the morbid state of fear. So, if an indication arises to produce a new morbid condition, in order to suppress some seated disease, care should always be taken that the morbid condition produced to cure the disease, be not worse than the original morbid condition.

Alteration.—If alteratives are indicated, it is well to inquire just what alteration is needed, in the fluids or solids, or both; and then to select the most safe, convenient, and reliable remedy, that will produce the alteration required, without itself in any way impairing vitality, or deranging any of the organs or functions of the body.

Chemical action.—If the indication is for some chemical agent, as to neutralize an acid in the stomach, or to render harmless some poison taken into the stomach, care should be taken to select such an agent as combining with the irritant will produce a compound of the most harmless character possible; or, which if possible, shall itself act remedially. Thus, if a child has acid in the stomach, in such excess as to require neutralizing, if the bowels are constipated, soda may be indicated, because uniting with the acid, a laxative salt is formed, which may regulate the bowels; but, if on the other hand, the child has a diarrhœa, chalk may be indicated, because, uniting with the acid it forms a compound, slightly astringent, which itself may aid in correcting the diarrhœa.

Mechanical influence.—When mechanical influences are indicated, such as position, compression, friction, &c.; care should be taken that it be accomplished, with the least possible inconvenience to the patient, and that in fulfilling the indication, no other morbid condition be produced, or if any, the least possible, under the circumstances.

Having now finished what I had to say, on the duties of the Physician, in attending upon the sick, and also, thrown out a few suggestions in relation to the best method of fulfilling indications, &c., it only remains for me to give some general rules in relation to medicines, and I am through with what I proposed to offer on therapeutics, or the treatment of disease. As then, we have already seen that diagnosis is essential, in order to arrive at the indications in every case; so a knowledge of MATERIA MEDICA, becomes an absolute necessity, in order to select remedies, to fulfill indications. I need not say then, that *materia medica* should be studied with exquisite care, so that each remedy and class of remedies, shall be as familiar as the hours of the day.

MATERIA MEDICA.—Various classifications of medicines have been attempted, in order to render the study easy and systematic. Among the classes into which medicines or remedies have been arranged, are narcotics, antispasmodics, tonics, astringents, emetics, cathartics, emenagogues, diuretics, diaphoretics, expectorants, sialagogues, errhines, epispastics, refrigerants, antacids, diluents, and emolients, &c. Now, with a correct knowledge of each medicine or agent, included in the above classes, or in each list, of any other classification; indications arising in any disease, will readily suggest the class from which the remedy is to be selected. Then, by running over in the mind, the medicines in that class, the particular medicine best adapted to fulfill the indication, may be selected and prescribed.

Care should always be taken that none but genuine drugs be selected or used, as others might not only fail to fulfill the indications, but would also be liable to do positive injury. In administering medicines, they should be given as little combined as possible, and they should also be given in such a form as to be as pleasant to the taste as may be. And, in relation to the doses or quantity to be given at one time, if the customary dose is to be deviated from, in any case, I think it is generally best to give less, and then to repeat the dose if necessary, especially if the patient be a female, or of a slender constitution.

In proportioning the dose to the *age*, the greatest possible care should be exercised, as carelessness in this respect might lead not only to a failure in the effect, but also, if too large a quantity be given, to the most serious consequences. A rule which I have always followed, and which, at least, has the advantage of convenience, is as follows: Divide the age of the patient by itself, increased by twelve, and the quotient gives the fraction of an adult dose that the child requires.

Thus, if it be required to find the dose for a child three years old, add twelve to three, the age of the child, and we have fifteen for a dividend; which, being divided by three, the age of the child, gives a quotient of five, showing that the dose for a child three years old is one-fifth of that for an adult. Now if the dose of the medicine to be given would be for an adult ten grains, the dose for the child three years old would be one-fifth of ten, which is two, thus showing that the dose of that particular medicine is two grains for a child three years old. The same rule will apply to any age, and no matter what the adult dose may chance to be. It is,

4

therefore, a good rule, and I would recommend its adoption and application in arriving at the dose for patients under twenty years of age, after which an adult dose may generally be required.

Now I will only add a few suggestions in relation to *prescriptions, quiet* of the patient, and *food*, and I am through with what I had to say on the treatment of disease in this place.

Prescriptions.—If prescriptions are written, they should be made plain, so that the druggist need not make a mistake; and prescriptions should never be put up by any except careful, sober, responsible men, that proper drugs may be used, and due care be exercised in putting them up. If medicines are prepared by the physician, they should be left under the care and directions of a responsible nurse, that no mistakes may occur in administering them; and the hours of administering them should be arranged as conveniently for the nurse as may be, and yet be the best possible for the patient for whom they are prescribed.

Quiet.—Patients suffering from severe disease should be kept quiet, and not be disturbed by company any further than is absolutely necessary for the care of the patient. And as nurses and friends of the patient are generally poor judges in relation to these matters, and as the responsibility of keeping out, in some cases, near relatives, is more than the nurse might, in all cases, be able to endure alone, I believe the physician should give, in all cases in which it is necessary, imperative orders in relation to this matter, that may not be departed from, and thus secure the patient from destruction by intruding friends.

Nourishment.—Now as disease is, in one sense, a state of debility, it is not best, as a general rule, to starve patients, even though they may be laboring under acute disease. For it should be remembered that if a person in health should go without food for several days, great prostration, with disturbance of all the functions of the body must be the result; and, if the fasting be continued for a sufficient time, death is the inevitable result. If, now, this be so, how shall a patient, reduced by disease, endure to be deprived of food or nourishment? The truth is, they will not endure it for any great length of time, and therefore, the experiment should not be tried. If then, as is often the case, there be no appetite for solid food, broths or milk, diluted with an equal part of toast water, may be given; and as soon as the appetite returns for food, toast, or some other suitable form of food, should be given, at meal hours, through the whole period of convalescence.

CHAPTER II.

OF IRRITATION, CONGESTION AND INFLAMMATION.

SECTION I.—IRRITATION.

BY *irritation* I mean here a morbid disturbance in the vital actions, in the whole or any part of the system, short of inflammation, involving primarily the nerves, and being attended with more or less pain. Irritation, then, may be associated with a sthenic condition of the system, attended with augmented or increased nervous action; or it may occur in an asthenic condition, attended with diminished nervous action or depression of vital power. Or irritation may be comparatively local, not associated with any very marked general increased or diminished action of the nervous function, as in the case of local injuries, &c. I will therefore proceed to consider irritation under the heads of *active or sthenic, passive or asthenic,* and *local,* as being the most convenient.

Active or Sthenic Irritation.—By active or sthenic irritation, then, I mean that which is attended with an active or sthenic condition of the system, or in which there is a general morbid augmented action of the nervous function. And in order to appreciate this condition, it is only necessary to remember that when the blood is in a healthy state, and its action upon the brain and whole nervous system is such that the brain generates a due amount of vital force or nervous influence, and this influence is properly distributed by the nerves to the various organs and structures of the body, there is no increase or diminution of nervous excitability; and hence there is no general irritability of the system, and there is not likely to be set up any local irritation.

Let, however, an irritant be introduced into the blood of a stimulating character, or let there arise in the mind any violent excitement, as of anger, and the brain and whole nervous system may be excited to irritation; and this irritation may be slight, or it may be very considerable, and attended with a general disturbance of the functions of the body. Now I cannot better illustrate this general irritation, than by taking a strong man, in a fit of violent anger, with an excited expression of countenance, flashing eyes, rapid motions of the body, &c.

The same condition of general irritation may be produced by any irritant introduced into the blood, and if there were no special point in the system more predisposed to take on disease than another, this condition of irritation might be succeeded by a general inflammatory fever; or if the cause were gradually removed, or if it should spend its influence, the general irritation might gradually subside, and the nervous action might be reduced to its normal state. More generally, however, in cases of general irritation of this character, there is some point in the system more predisposed than others, in which case there is liable to be developed local inflammation; or in case the cause spends its influence, without producing general inflammatory fever or local inflammation, there is liable to be left some special point of local irritation. Such local irritation may be the result of permanent impressions produced in the minute nerves of the extreme arteries, veins, and capillaries; and although it may be permanent, I believe it more generally subsides after a time. Such are my views in relation to general irritation, attended with increased or augmented nervous action, and such also, I believe, its liability to be succeeded by general inflammatory fever, local inflammation, or permanent local irritation. Let us, then, pass on to consider *passive* or *asthenic* irritation.

Passive or Asthenic Irritation.—We have seen that an elevation of nervous action above the standard of health, is liable to produce an active or sthenic irritation; we are now to see that a diminished nervous action below the standard of health, or nervous depression, will produce general irritability of a passive character. This variety of general irritation cannot be better illustrated than by a person in great fear, or laboring under the influence of protracted grief. The same condition may also be produced by various depressing agents introduced into the blood, as, for instance, the typhus, contagion, &c.

Now, whatever may be the depressing agent or influence, vitality is depressed, and the brain does not generate sufficient nervous influence, and that which is generated is but poorly distributed by the nerves; in consequence of which the circulation is carried on imperfectly, and the blood receding from the surface and extreme vessels accumulates in the internal organs and parts. The accumulation of blood in the brain, and along the pinal cord, may increase the general irritation already produced by the nervous depression or diminished nervous action. In such cases of general irritation the face becomes pale, the eyes heavy, the hand trembles, the mind is irritable, and if the cause continues to operate, some kind of general fever, or some local inflammation of a passive character is liable to succeed.

In such cases, if there be no special organ or part more predisposed to take on disease than another, the system is liable to rally, if the cause is removed, and thus the general irritation may subside. If, however, the cause continues to operate, and the agent be some one of the causes of fevers, they are very liable to be developed sooner or later. If, however, there be some special point in the system more predisposed than others, a passive inflammation may be set up, and that, too, with or even before the development of general fever. Finally, in cases of general asthenic irritation, in which there is produced no general fever or local passive inflammation, there is a liability to a permanent general or local irritation, caused in most cases by some change in the brain, spinal cord, or nervous ganglia. Such, then, according to my observation, is general asthenic or passive irritation, and such, also, the liability of its being succeeded by some form of fever, passive inflammation, or else of its being perpetuated as a general or local irritation, from some permanent change in the brain, spinal cord, or nervous ganglia.

Local Irritation.—We have seen that local irritation may remain in some organ or structure of the body; the result of general irritation, either active or passive, and may become permanent. Local irritation may also be produced in any organ or tissue of the body, and too, from a great variety of direct causes. Thus we see the eye irritated from wind, dust, &c.; the skin irritated by various acrid substances, the mouth and stomach irritated by hot drinks, &c.; the intestinal mucous membrane by crude articles of food, the bladder by cantharides, the brain by want of sleep, and in fact there is not an organ or tissue of the body but what may be thus accidentally irritated.

Now in all these cases of local irritation there is evidently a morbid excitability of the nerves produced, attended with a sense of heat or pain; and the irritation may pass on to a sufficient disturbance of the capillary vessels to develop inflammation or even symptomatic fever, or if the cause or causes be removed, the irritation may subside, and the part may assume a normal condition; or finally, the irritation may remain permanently and yet develop nothing further, being little more than an inconvenience, so far as the general health is concerned. Finally, when we take into account the cerebro-spinal and ganglionic systems of nerves, and remember how they connect the different organs of the body, we should expect, as we really find, that an irritation of one organ is liable to produce sympathetic irritation of another; and this fact should always be born in mind in tracing *sympathetic* irritations to the *primary*.

Nature of Irritation.—It is only necessary to say, under this head, that irritation consists in a morbid excitability of the whole nervous system or else of the nerves of a particular part; the nervous action in some cases being increased or augmented above the standard of health, while in others, the nervous action is diminished or depressed below the standard of health; the irritation probably in all cases being attended with a sense of heat, burning or pain.

Irritation of an active or sthenic character, if it pass on to inflammation, develops an inflammation of an active character. If it pass on to a general fever, it is liable to be of an inflammatory character. On the other hand, if passive or asthenic irritation pass on to a general fever, it is liable to be of an intermittent, nervous, or low typhus type; and if it pass on to local inflammation it is liable to be of a passive character.

Finally, it is well to bear in mind the fact that all fevers, as well as all inflammations are preceded by irritation; but it must also be remembered, that all irritations need not necessarily pass on to either fever or inflammation, especially if the cause or causes be removed; the irritation in such cases either terminating by resolution, or else becoming permanent, but being only *irritations*. It should also be remembered that irritation is liable to disturb the functions of a part or the whole of the body; depending upon the degree of disturbance in the generation and distribution of the **nervous** *influence*.

Causes.—General irritation of the system may be produced, as we have already seen, by any exciting or depressing agent, acting directly upon the nervous system, or indirectly, by changing the blood. Hence we have, as causes of irritation, heat, cold, moisture, electricity, koino-miasmata, idio-miasmata, contagions, epidemic **and** endemic influences, &c.. as general causes, or as causes of general irritation. We have also causes which act locally—as dust or wind to the eyes, injuries of the surface from any cause, hot or irritating drinks, coarse articles of food, injuries of the brain, or along the spine, &c. It should be remembered, however, that the general causes may leave local irritation, or that local causes may produce general irritation.

Treatment.—The treatment of irritation, whether general or local, consists in removing the cause as soon as possible, and then in correcting the morbid disturbance or derangement in the nervous action. If the irritation be general, and the cause has been one of excitement, increasing the nervous action, saline cathartics, cooling drinks, low diet and plenty of proper exercise may be sufficient to let down the nervous excitability to a normal state. If, however, the irritation be of a passive character, the nervous action being depressed below the standard of health, then gentle exercise, fresh air, good diet, and if necessary, camphor, quinine or iron may be required to bring the depressed nervous action up to the standard of health.

In cases of local irritation of an active character, after removing the cause, cold applications may be of service. If, however, the local irritation be of a passive character, warm applications may be required.

Finally, in cases of sympathetic irritation, depending as it generally does upon a primary difficulty in the brain, or along the spine, cups or blisters should be applied along the spine, at the seat of the primary difficulty, and this being removed, the sympathetic irritation will generally subside.

Now having completed here what I had to say on *irritation*, we have opened the way for the consideration of *congestion* and *inflammation*, which I shall take up in the two succeeding sections, which will complete this chapter. The subjects of irritation, congestion and inflammation having been thus considered in this chapter, we shall be prepared to pass on in the next to the consideration of the *pathology*, *causes* and *phenomena* of fevers.

I make this suggestion here in relation to the course of taking up these subjects, in order to keep in mind the importance of each, and to call up the important bearing which each one of them has upon the succeeding, along our path of *general pathology* and *therapeutics*.

SECTION II.—CONGESTION.

By congestion is here meant an undue accumulation of blood in the extreme arteries, veins or capillaries of any organ or structure of the body. The accumulation may take place in the extreme arteries and capillaries of a part, in consequence of increased morbid action of the heart and arteries, in sthenic conditions of the system, in which case the congestion may be called *active*, especially as it generally follows active or sthenic irritation of the part. Or congestion may occur from an accumulation of blood in the minute veins of a part as well as of the capillaries; the result either of general debility alone, or of debility, attended with some local or general obstruction to the passage of blood through the veins or right side of the heart, in which case the congestion may be called passive, especially as it very often follows asthenic or *passive* irritation of the organ or congested part. I will, therefore, proceed to consider congestion as *active* and *passive*.

Active Congestion.—By active congestion, then, I mean **that which** occurs in a sthenic condition of the system, and which generally follows active irritation ; the accumulation of blood being the result, in some cases, of the active irritation, the blood being hurried to the point, by the irritation exciting the arteries, and also, more or less, the heart itself. Or the congestion may occur in **consequence** of a general active irritation of the nervous system, giving **a morbid,** excited action to the **heart; the active** local congestion being **produced in** consequence of some **predisposition** of the particular part to **take an** active arterial congestion. ·

When active congestion **thus** occurs there is **a fullness of the** part, and **there is** apt to be more or less disturbance in the function of the organ **or part.** If the congestion be slight, there may be an increased or morbid **activity of** function, **but if the congestion** be considerable, **there is a liability of** a partial or entire suspension of function of the congested **part ; as we** see sometimes in active congestion **of** the brain, the **suspension** of vitality being the result.

Active congestion may terminate by resolution if the cause be removed, or is allowed to spend its force, or the congestion may pass on to active inflammation, or a rupture of the blood vessels, and laceration of the tissues may occur ; and we may have apolexy, if the congestion be of the lungs or brain. It should be born in mind then, that active inflammation is always preceded by active congestion of the part ; but that active congestion need not necessarily pass on to inflammation, as resolution may occur. **With this glance** at *active,* let us pass on to the consideration of *passive* congestion.

Passive Congestion.—By passive congestion, as I have already intimated, I mean that which occurs in asthenic conditions of the system ; **the accumulation taking place in the capillaries and** extreme veins, and depending upon debility with or without some special obstruction to a free passage of blood through the heart or venous trunks.

Passive congestion may follow passive irritation, and it may terminate **by resolution** or apolexy, or it may pass on to passive inflammation, as most or perhaps all cases of passive inflammation are preceded by passive congestion of the part. In cases of passive congestion there is more or less fullness of the part with perhaps a feeling of numbness, and if it be of the lungs, we have the whole train of symptoms growing out of disturbed respiration ; or if it be of the brain, of cephalic and general nervous depression.

It is **also proper to bear in** mind the fact, that we may have congestion depending upon local causes, and not necessarily connected with any very general disturbance of the system ; the general nervous and circulatory function being in a comparatively normal state. Thus we may have cephalic congestion from a continued or even temporary stooping attitude. We may also have congestion of the hemorrhoidal veins from constipation, or of all the portal veins, from some obstruction to a free passage of blood through the liver, &c.

Congestions then, whether active or passive, or more especially local and not attended with any very marked general exaltation or depression of nervous, vascular or vital power, is liable, if the cause be removed, to terminate by resolution; or it may pass on to pulmonary or cephalic apoplexy, or if the congestion be active, it may be followed by active, or if it be passive, by passive inflammation. With this general view, we are prepared to pass on to consider the *causes* of congestion.

Causes.—The causes of active congestion are, like those of active irritation, influences which excite morbid, nervous, and vascular action. These causes may act directly upon the nervous system, to excite the circulation, as violent anger; or the cause may act through the blood, to irritate, and accelerate nervous and vascular action. It is probable, then, that mental excitement, too stimulating articles of food and drinks, and, perhaps, a high electrical state of the atmosphere, are among the most frequent causes of active congestion.

Passive congestions, on the other hand, may be produced by a great variety of depressing agents, some of which may act directly upon the nervous system, as that of fear, melancholy, &c.; while others may act through the blood, as we see illustrated in koino-miasmata, idio-miasmata, the contagions, &c. It may also be produced by bad or insufficient food, as well as by any and every imprudence which either directly or indirectly depresses vitality. And besides, as we have already seen, various accidental local causes may operate, such as constipation of the bowels, positions of the body, ligatures about the limbs, &c.

Diagnosis.—Active congestion may generally be distinguished from inflammation by careful attention to the following differences. In active congestion there is slight fullness, a dull, heavy pain, or feeling of numbness, and but slight, if any elevation of temperature of the part, and little or no general heat or special dryness of the skin; while in active inflammation there is pain, tenderness, heat, thirst, swelling, redness, and often more or less general febrile excitement.

Passive congestion may be distinguished from passive inflammation by the absence of pain, tenderness, heat, thirst, redness, &c. There is a point, however, at which congestions become inflammations, if they pass on to that condition, in which it becomes necessary to notice carefully all the symptoms, in order to discriminate between them.

Treatment.—In cases of active congestion, after removing the cause, a saline cathartic, cooling drinks, a low diet, &c., may be sufficient to overcome the difficulty. If not, however, general or local bleeding, cold applications, &c., may be required. In cases of passive congestion, after removing the cause, camphor, carbonate of ammonia, wine whey, quinine, iron, &c., may be required, and the patient should be allowed a liberal diet, good air, moderate exercise, &c.

Various local applications may be required, depending upon the nature and seat of the congestion. If it be of the brain, the warm foot-bath, and sinapisms to the feet may be indicated. If the congestion be of the lungs or other internal organs, sinapisms may also be of great service. Finally, in cases of general congestion, of a passive character, in which it becomes necessary to produce reaction at once, in order to prevent a suspension of vitality, I know of nothing so effectual as the application of a strong infusion of capsicum, in vinegar, along the spine, a little warm.

It seems hardly necessary to say, that, in cases of congestion depending upon local causes, it becomes necessary to remove the cause, in order to overcome the difficulty. If the congestion be of the brain, and depends upon some occupation which requires stooping, the labor should be suspended for a time. If the congestion be of the lower limbs, and the cause be the wearing of elastics too tight, they should be loosened or removed entirely. If the congestion be of the hemorrhoidal veins, and constipation be the cause, the use of bread made of unbolted flour, and regularity at stool each day, at a particular hour, may overcome the constipation. If not, a teaspoonful of sulphur and cream of tartar, of each equal parts, may be given each morning, till the constipation and hemorrhoidal congestion are overcome.

Finally, in all cases in which, from any peculiar formation of the body, there is a predisposition to congestions, especially of the brain, as is the case with persons of large head and short neck, it is important to guard against the causes which would be likely to produce it in either of its forms.

SECTION III.—INFLAMMATION.

By inflammation I mean that peculiar morbid condition of the tissues of the body generally characterized by pain, increased heat, redness and swelling; all of which phenomena, however, are liable to great variation, depending upon its character, and also upon the nature of the structure in which the inflammation occurs.

Symptoms or Phenomena.—Redness is the most invariable symptom or phenomena of inflammation. It generally occurs in inflamed parts, and is owing, in part, to enlargement of the capillaries to such a degree that the blood may be seen, while in the ordinary condition of parts the blood is scarcely seen in the minute capillaries, or if it is it does not appear red, as it does when the capillaries are enlarged.

It is probable, also, that increased redness may depend, in part, upon an increase in the number of red corpuscles in the vessels of the inflamed part; and it may be augmented, in some cases, by extravasation, and perhaps by the passage from the arteries to the veins of arterial blood, unchanged.

The redness may vary from a bright red to a deep crimson, or nearly purple, and is generally most intense in some one point, and gradually diminishes as it recedes, till it terminates in a healthy color. This redness generally continues after death, and forms one of the evidences of internal inflammation in post mortem examinations. Along with the redness, there is generally felt by the patient more or less throbbing, which is probably owing to an irritation which has been communicated to the larger arteries in the immediate vicinity of the inflammation, and perhaps augmented by increased density of the inflamed structures.

Increased heat is another symptom of inflammation, which is sometimes absent. It is probable that the sensation of heat, in inflamed parts, depends on two causes—a slight accumulation of heat, and also an increased sensibility of the nerves, by which the sensation of heat in the part is greatly increased, and hence there is not that elevation of temperature which the sensation of the part would indicate.

Yet it is probable that in the first stages of inflammation, the increased amount of blood in the part serves to elevate the temperature, especially if the inflammation be on the surface of the body, for here the blood is constantly receiving oxygen, and throwing off carbonic acid, and most likely, in the combustion which occurs in the capillaries heat is evolved, as in the lungs, and hence in part its increase. It should be remembered, also, that the closure of the exhalent tubes, if the inflammation affects the surface of the body, may, by stopping evaporation of perspirable matter from the surface, cause to be retained in the part more or less heat, which would have passed off in a latent state if the evaporation of perspirable matter had continued.

According to the experiments of Prof. Gross,* the variation in the temperature of the blood is from 92° to 104°, the average being 96°; and he also states, that "the temperature of a scrofulous tumor has been found raised as much as $5\frac{1}{4}$° above the general heat of the body."

Pain, though a general, is not an invariable symptom of inflammation. As a general rule, the looser the structure inflamed the less violent is the sensation of pain. Inflammation of the lungs, or of the alimentary mucous membrane, may pass on to even a fatal termination unattended with pain.

The pain arises from an irritation of the nerves of the part; and that irritation may be in part from the cause, direct or remote, of the inflammation, increased often, more or less, by the swelling of the surrounding tissues, and also, probably, by a thickened and tense state of the neurilemma. The character of the pain is modified by the nature of the inflamed structure. In inflammation of the mucous membranes, if pain attends, it is of a burning or stinging character, and not very severe. It is important to bear this in mind, as inflammation of the alimentary mucous membrane may exist with very little pain, and but slight tenderness to pressure. This slight peculiar pain in the mucous membrane is doubtless owing to the looseness of the mucous structures, allowing the distention of the small capillaries, without very materially irritating the small nerves that surround them, as well as those which supply the capillaries themselves.

The pain in the pleura, and other serous membranes, is lancinating, and generally exceedingly acute. This is doubtless owing to the firmness of the serous membranes, so that any enlargement of the small capillaries produces a mechanical irritation of the nerves which intervene and surround them, or which enter into their structure.

Pain in the ligaments or fibrous structures is of a dull, aching or gnawing character, owing to the peculiarity of their structure. The capillaries are fewer, and probably of a smaller size, than in other tissues, in consequence of which their distention does not so materially interfere with one another. And further, the firmness of the fibrous structures prevents a rapid distention of the capillaries, so that the change which takes place is gradual; hence the gnawing character of the pain which is experienced in inflammation of the fibrous structures.

In inflammation of the cellular tissue the pain is generally not very severe at first, but may become so during its latter stages. And this is

* See Gross' Elements of Pathological Anatomy, pp. 42, 43.

as we should suppose, when we remember that the looseness of the cellular tissue allows the small vessels to be distended without materially
irritating the small intervening nerves at first. But as the distention
becomes considerable, and the effused fluid accumulates, the pain is increased, according to the degree of inflammation, and consequent irritation of the nerves of the inflamed part.

The pain experienced in inflammation of the skin, is of a peculiar,
smarting, itching or teasing character, just as we should suppose when
we take into account the peculiar structure of the dermoid tissue.

For the slightly distended capillaries, pressing upon each other, and
the intervening nerves, produce a smarting, stinging sensation, while the
pressure on the exhalent tubes, very likely in part, produces by their
obstruction, the peculiar itching, teasing sensation.

This is rendered quite certain, by the fact that such a sensation is
produced, to some extent, when the exhalents are obstructed by sudden
exposure to cold.

The violence of the pain in an inflamed part, controls, to a great extent,
the sympathetic febrile reaction which the local inflammation produces.

Thus, in acute bronchitis, or in inflammation of the alimentary mucous
membrane, the pain is generally not very severe, and there is not a very
vigorous reaction of the heart and arteries.

But in inflammation of the pleura, and other serous membranes, in
which there is intense pain, the arterial reaction is generally very great.

Swelling is another symptom of inflammation which varies according
to the tissue involved, and the degree and extent of the inflammation,
being generally slight in the mucous membranes, and often very considerable in the glandular structures. The swelling which occurs immediately on the occurrence of a local irritation or inflammation, is in consequence of dilatation of the capillaries; but if inflammation continues,
and effusion takes place, the effused serum will of course increase the
swelling.

Diagnosis.—Inflammation of external parts may generally be distinguished from irritation, or neuralgia, by the redness, heat and swelling,
and perhaps tenderness; some, or all of which symptoms, generally
attend, rendering the diagnosis clear and decided at once. But to distinguish inflammation of internal parts, it becomes necessary to take
into account the peculiar structure of the suspected tissue or organ, and
then, by noticing carefully all the symptoms, a correct diagnosis may
generally be formed.

By keeping in mind the character of pain peculiar to each tissue,
inflammation may generally be distinguished from mere irritation or neuralgia, by keeping in mind the following facts:—In inflammation, the
pain is generally more continuous, seated and throbbing than in neuralgia. In inflammation, there is often tenderness on pressure, which
symptom is not present in neuralgia, at least to so great an extent.

Position, too, is often an important diagnostic symptom in internal
inflammation, the patient generally assuming such a posture as will produce the least tension of the inflamed part, while in neuralgia the patient
is seldom still, appearing more or less relieved by tossing about, and
getting in almost every position. This we see illustrated in abdominal

inflammation, as the patient lies on his back, with his knees drawn up, in order to take off the tension of the abdominal muscles.

Inflammation usually produces, also, a general febrile excitement, proportioned, more or less, to the intensity of the pain; the pain being generally, as we have seen, as the density of the structure inflamed, while in neuralgia there is generally little, if any, sympathetic fever set up.

By taking all these peculiarities into account, and also the functional derangements which occur, as well as the nature of the exciting cause, and by using a good degree of common sense and discrimination, the diagnosis between inflammation and neuralgia may be rendered clear, and the presence or absence of internal inflammation, in any tissue or organ, put beyond a reasonable doubt.

Causes.—The causes of inflammation are so numerous that an attempt at enumeration is quite unnecessary. Any cause capable of producing an irritation, such as heat, cold, wounds, morbid poisons, &c., may excite local inflammation in any tissue or organ.

The predisposition to local inflammation, I am satisfied, may depend upon directly opposite conditions of the system. In the one case, the system being predisposed from a decidedly sthenic, while in the other, from an asthenic condition, irritation being the result of both extremes; and hence a predisposition to inflammation; in the one case of an active, and in the other of a passive character.

The predisposing and exciting causes of inflammation should always be taken into account, not only in forming a diagnosis, but also in arriving at the indications in the treatment; the indications in the treatment often being directly opposite in active and passive inflammation.

Morbid Appearances.—The post-mortem appearances presented in cases in which death has been produced by inflammation, varies with the tissue involved; and also with the period or stage of the inflammation at which death takes place.

The most reliable signs of inflammation are an increase of vascularity, with more or less extravasation of blood, or of coagulable lymph, and also the formation of pus, or various other morbid products, producing a decided change in the structure of the affected part.

If the patient die during the early stage of the inflammation, there is generally found a very minute injection, mainly of the small ramifications of the arterial branches, the whole thickness of the tissue being filled with dots or striæ of a bright red color, and spots of ecchymosis, or infiltration of blood, within the texture, and perhaps effusion on the surface, may generally be found, as evidence of the inflammation.

If the inflammation had progressed considerably before death, there may often be found an exudation of coagulable lymph, pus, ulcers, &c.

If, however, the inflammation had become well advanced when death occurred, there may be found, in many cases, the organization of lymph, and its conversion into new tissues; the complete formation of abscesses, or of purulent sinuses, adhesions, &c.; and the minute vascularity and redness which appears early, will generally have disappeared.

By bearing these facts in mind, and taking into account the termination peculiar to each tissue, correct conclusions may generally be drawn on post mortem examination; not only in relation to the existence or not, of inflammation; but also as to the stage of the inflammation, at which death occurred.

Nature.—Inflammation is a disease, mainly of the capillary system of vessels, consisting in an altered condition of their vitality, and more or less sanguineous congestion; hence, as a general rule, the more abundant the capillaries of a part, the more liable it is to become inflamed.

Now, irritation of a part, whether by a wound or otherwise, either directly weakens the tone of the nerves of the part, so that the small arteries, veins, and capillaries dilate at once, or there is first an increased tone of the nerves, so that the small vessels at first contract, producing a paleness, and then immediately dilate, producing redness.

This irritation of the nerves of a part, is either direct from the injury or else the impression is communicated to the brain, and then, by a reflex action, the nerves of the part become weakened, and congestion takes place.

If the local irritant be slight, and the general system, as well as the small vessels of the part are in a sthenic condition, it is probable, that the local irritation of the part, first produces contraction, and then dilation of the small vessels. But, if the local injury be very considerable, and the system as well as the small arteries, veins and capillaries are in an asthenic state, it is probable that the nerves of these vessels become stunned as it were, and the small vessels are dilated at once, without any previous contraction.

In either case, it is probable that a wound, or injury of a soft part, as a blow, either directly or after an instant of irritated action, stuns or paralyzes the nerves of the part, in consequence of which, the small arteries, veins and capillaries are either immediately or mediately dilated, the blood rushing in to distend them, and this constitutes the congestive stage of an inflammation.

The small vessels during this congestive stage of an inflammation, may become dilated to two or three times their usual diameter, in consequence of which, many small capillaries appear which were before invisible; a result of their admitting red corpuscles, which their calibre did not before allow of.

Now, at the commencement of the congestive stage of an inflammation, the current of the blood is more or less slackened, or it may even retrograde for a moment; but as the dilatation of the vessels becomes complete, the blood in a sthenic state of the system at least, flows more rapidly; a greater quantity passing in a given time, than in a healthy state of the parts. Soon, however, the current becomes again slower, especially in an asthenic state of the system; and there is often complete stagnation in the centre of the inflamed part, while at the circumference the blood may pass still with greater velocity than in health.

The blood thus stagnated is not disposed to coagulate, and the red corpuscles accumulate to such an extent, that they not only occupy the centre of the current, as in health, but fill up the usual colorless space immediately within the walls of the vessels, it being the result evidently of a retention of the corpuscles in the dilated vessels, while the liquor sanguinis is drained off, and passes along. When the stagnation is complete and the accumulation of red corpuscles very considerable, their outlines may no longer be seen, the vessels presenting a crimson hue.

During this dilatation of the small vessels, the liquor sanguinis exudes more or less through invisible pores, being transparent, unless as sometimes happens, the coloring liquid of the corpuscles exude more or less with it; but the red corpuscles are not effused unless the orifices of the capillaries become ruptured, in which case they may be effused into the tissues, and thus give color to the extravasated fluid.

Soon the fibrin of the extravasated fluid coagulates, and then it exhibits organized spherical corpuscles, with numerous interlacing fibrils, **which give** firmness to the mass. These corpuscles resemble those of the chyle, and also the white corpuscles of the blood; and "undergo various changes, being converted into fibro plastic cells, filaments and possibly blood-vessels, when the coagulated fibrin undergoes organization; and degenerating into pus corpuscles, compound granular corpuscles, granules, &c., when **it becomes quite** degraded, **to** be surrendered ultimately to chemical laws."*

In some cases the corpuscles alone are developed in the exuded liquid, on first appearing, in which case, the liquid does not coagulate and is incapable of organization. But in cases in which both corpuscles and fibrils appear, the coagulated lymph soon becomes organized, red lines are seen passing through it in various directions, inosculating and ultimately forming blood-vessels, having connection with those of the surrounding tissues. **In some** cases, during the distension of the small **vessels, the** microscope discovers aneurismal pouches of the small arteries, veins and capillaries, owing probably to a weakening of their coats, caused by the inflammation.

In parts becoming soft from inflammation, the microscope also renders visible granules, and a combination of *granular corpuscles* cohering in a spherical form, **and sometimes** inclosed by a vesicular envelope. **In** other cases "these corpuscles may be seen apparently in the process **of** disintegration, breaking up more or less completely into irregularly clustered and isolated granules."† And if hemorrhage occurs, large corpuscles consisting of several blood corpuscles, surrounded by a **transparent envelope, may** be discovered apparently undergoing degeneration.

Pus, **when it is formed, is a** yellowish or cream-colored fluid, of the consistence of cream and of a sweetish taste, and consists of a serous fluid, in which are floating globular corpuscles about twice the size of the blood corpuscles, being $\frac{1}{285}$ of an inch in diameter. When the pus globules undergo degradation **they** "are resolved into granules and molecules, which ultimately become fluid." Pus also contains, sometimes, "red blood corpuscles, unchanged exudation corpuscles, compound granular corpuscles, numerous isolated granules, fatty matter in molecules or globules, rhomboidal plates of cholesterin, and the debris of the tissues from which the pus proceeds, as cellular fibres, and epithelial or epidermic scales."‡

Theory of Inflammation.—I have already stated that when a cause which may produce inflammation acts upon any part, the nerves of that part become irritated; and that this irritation produces either immedi-

* Wood's Practice of Medicine, vol. i, p. 45.   † Ibid. p. 46.   ‡ Ibid. p 46.

ately or mediately a dilatation of the small arteries, veins and capillaries of the part. If contraction first takes place, as it may in a sthenic condition of the system, there is at first a paleness of the part caused by the blood being in part forced from the small vessels. But if dilatation occurs at once, as it is likely to do, in an asthenic state of the system, no paleness of the part occurs; but we have developed redness, heat, pain and swelling.

Now this dilatation may be, in either case, of an active character, but it appears to me that it is passive, and the result of the debilitated or stunned condition into which the small vessels are thrown by the intruding cause, whatever it may chance to be. However this may be, the blood, as it fills these dilated vessels, is interrupted, for a time, in its onward course. During this interruption to the onward course of the blood, it must necessarily accumulate in the arteries supplying the distended vessels, which accumulation may slightly irritate directly the arteries, and also indirectly through the cerebro-spinal system acting upon the ganglionic nerves. As a consequence of this irritation, direct and indirect, the blood is again sent on through the distended vessels with more or less of an accelerated movement, and as nutrition or deposition of fibrin is suspended in the part, it necessarily accumulates, to some extent, in the blood. This accumulation of fibrin probably causes the corpuscles to cohere, rendering the blood more viscid, in consequence of which, the red corpuscles accumulate in the dilated vessels, thus filling up their increased capacity. In consequence of this clogging of the expanded vessels by the blood corpuscles, the onward movement of the blood is again impeded, and, finally, becomes stagnated in the central portion of the inflamed part, an increased quantity of blood, however, being sent through the vessels towards the circumference of the inflamed part, where the vessels are less clogged if the system is in a sthenic condition; but if the system is in an asthenic state I think it is doubtful.

Thus we have, in my view, the rationale of inflammation fully established, consisting essentially in irritation, debility, and a general deranged condition of the inflamed part, and a more or less irritated condition of the heart and arteries, with either an increased or diminished power of action, constituting active or passive inflammation, according as the system is in a sthenic or asthenic state.

Terminations.—Inflammation having been thus fully established, tends to a termination. There is, however, at this stage, an effusion of serum and lymph, and, perhaps, of blood, which, if the inflammation terminate by resolution, are absorbed, and the part passes on to a healthy state.

If, however, the inflammation continues, the effused fibrin may coagulate, its invisible germs taking the shape of molecules, which unite in rows, forming fibrils, or about a centre, forming granules, and which ultimately develop exudation corpuscles. These fibrils, by uniting, form tissues, while the corpuscles, which become nucleated cells, may undergo various changes, probably elongating and joining end to end in some cases, they form the basis of fibrous tissue, and, perhaps, of blood vessels. But if neither resolution or the organization of the effused fibrin takes place, various degrees of destructive processes may occur, the most frequent of which are suppuration, ulceration, softening, and gangrene, each of which are entitled to a passing notice.

Suppuration may take place in any protracted case of inflammation, especially if from a poverty of the blood, the exuded lymph is of a defective quality. In such cases, the exudation corpuscles lose their vitality, and become pus-corpuscles, and the other ingredients of the coagulated exuded matter, becoming liquid, are converted into granules of pus. And if the pus be not soon evacuated from its cavity in the tissues, its corpuscles may become granules, and these granules, finally, assume a liquid form, which renders it liable to be absorbed.

The tendency to suppuration in an inflammation is supposed to be strongest in cases in which the exuded matter contains the greatest quantity of exudation corpuscles, and the least of the fibrillating material, the fibrils tending, as we have already seen, to coagulation and organization, while the exudation corpuscles, if in excess, tend to pass on to pus-corpuscles.

Ulceration is liable to occur from inflammation in constitutions in which there is a deficiency of vital force, and appears to be the resolution of the tissues into "compound granular corpuscles; or isolated granules, and the ultimate liquefaction of these products, so that they may be absorbed when not eliminated with the pus."*

Softening is another effect of inflammation, and, according to Professor Gross, most frequently occurs in "the brain, the spinal cord, the mucous membrane of the alimentary tube, the spleen, and liver." According to my observation, softening as frequently follows a slow or chronic inflammation, and it appears to be the interstitial cellular tissue of structures, which are mainly at fault, in most cases, at least.

The effused lymph appears to undergo degeneration, and, at the same time, the tissues themselves; the result or product being granules, either diffused or in mass, and "the debris of the softened organs."†

Gangrene, mortification or death of a part, is liable to occur as the result of inflammation, and it is the lowest degree of degradation to which the tissues are liable. It may be a slow or rapid process, varying from a few hours to several months, from the beginning of the inflammation which produces it. When it is about to occur, the temperature of the part as well as its sensibility is lost, and very soon the circulation becomes suspended, and all its other functions.

The textures most liable to gangrene are the mucous, cellular and cutaneous; while those parts in which it seldom or never occurs, according to Prof. Gross, are "the uterus, kidneys, ovaries, supra-renal capsules, the thyroid body, the testicles, pancreas, and salivary glands."

Thus we have traced inflammation through its various processes of resolution, the formation of new tissues and its various destructive terminations. Let us now look at its modifications, depending upon the tissues in which it occurs.

Modifications of Inflammation.—Inflammation may be said to occur under five prominent modifications, corresponding to the five elementary tissues; the cellular, serous, mucous, dermoid, and fibrous.

Inflammation of the *cellular* tissue is characterized by great swelling, throbbing pain, and by the peculiarity of the pus being collected in circumscribed cavities.

---

* Wood's Practice of Medicine, vol. i., p. 49.
† See Gross' Elements of Pathological Anatomy, p. 88.

Inflammation of the *serous* structures is distinguished by acute lancinating pain; little or no swelling, sympathetic reaction of the heart and arteries, by its tendency to terminate in the exudation of coagulable lymph or the secretion of a thin whey-like pus.

It is rapid in its course, and not apt to terminate in gangrene, and is peculiarly liable to form adhesions.

Inflammation of the *mucous* tissue is attended with a stinging burning pain, as we have seen, without much tumefaction of the subjacent cellular structure; the sympathetic fever which it develops is not vehement, and when it terminates in resolution there is an increase of the mucous secretion.

Inflammation of the *skin*, or dermoid structure is attended with a burning pain; is much inclined to spread over the surface, forming small blisters, containing serum, and never forms adhesions, or suppurates in circumscribed cavities.

Inflammation of the *fibrous* structures is attended with intense aching or gnawing pain, and is not disposed to terminate in suppuration or gangrene; it may however, terminate in the exudation of serum, or in the deposition of earthy matter. It is peculiarly liable to change its location from one part to another, and sometimes passes suddenly to internal organs. The sympathetic fever which accompanies acute rheumatism, is generally vigorous; but it seldom proves fatal, unless it passes to some internal part, or organ essential to life.

Though the preceding modifications may not always appear exactly as I have pointed them out, I am confident that the careful observer will find those peculiarities to be the prevailing tendency.

Varieties of Inflammation.—Inflammation may be either acute or chronic, active or passive. *Acute* inflammation is rapid in its course, and violent in its local and symptomatic phenomena. *Chronic* inflammation is generally the result of the acute; and is characterized by a slow progress, and less intensity in all its symptoms.

By *active* inflammation I mean one occurring in a sthenic constitution, and though there be irritation with debility at first, at the seat of the inflammation, and soon an irritated condition of the heart and arteries; when the local inflammation is established and the circulatory system excited, there is with the irritability an increased power of action.

By *passive* inflammation, I mean one occurring in an asthenic or debilitated constitution, in which the debility is a predisposing cause: and though there may be irritation at the point of inflammation, of the small dilated vessels, and also of the circulatory system generally, yet in both there is a diminished power of action.

Treatment of Inflammation.—I do not propose here, to do more than to point out the general indications in the treatment of inflammation, which may serve us as we proceed in our investigation of inflammatory diseases, of the different tissues and organs of the body.

If we take into account the great fact, that a medium action of the circulatory system, not only of the heart and arteries, but also of the capillary vessels is desirable in order to promote resolution of an inflammation, we have reached the great indication in the general and local treatment of all inflammatory affections. For if an inflammation occur

5

in a sthenic or strong and vigorous constitution, and the local and general irritation of the capillary and general circulatory system when the inflammation is fully established, is attended with increased power of action of the heart and arteries, as well as the capillaries of the inflamed part, the action of the circulatory system both general and local may be too strong, and therefore require to be reduced to a medium standard.

In such cases, the deviation from a medium standard should be taken and the general and local action reduced by the most safe, convenient and reliable measures.

If the arterial action be very strong, general and local bleeding may be indicated, with antimony, veratrum, digitalis, &c.; and cold applications to the inflamed parts till the excitement of the circulatory system is reduced to a medium standard. But if the increased action of the circulatory system be not so high above a medium, the warm foot bath, warm sage tea, and perhaps a saline cathartic with small doses of ipecac may be sufficient to bring the circulation down to a medium action.

Thus we have clear indications in all cases of inflammation in which there is increased arterial and capillary action. Let us now inquire into the indications in cases of inflammation occurring in asthenic or debilitated constitutions.

In cases of inflammation occurring in asthenic or debilitated constitutions, there may be, along with the local and general irritated state of the circulatory system, a diminished power of action, not only of the capillary vessels, but also of the heart and arteries, in consequence of which the general and local action of the circulatory system will fall below the medium, and so an indication for tonics, and stimulating applications to the local inflammation may clearly exist. In such cases, to bring up the general circulation, as well as the capillary, to a medium standard, and thus favor resolution, quinine, camphor, and perhaps diffusible stimulants may become necessary, as well as warm or stimulating applications to the inflamed part. Thus we have the great principles to guide us in the treatment of all inflammatory affections.

In relation to the special indications which are liable to arise in inflammation of different organs and structures of the body, it is more proper to leave them to be considered as we take up inflammations of these tissues and organs in their proper place. Enough has, however, been said here to enable us to pass on with our subjects, in the order in which I have already announced them.

I trust that, having thus considered irritation, and having traced it to its terminations by resolution, and also to congestion, and having considered congestion, with its terminations, one of which is in inflammation, and finally, having taken up inflammation, that we shall be well prepared to pass on, in the next chapter, to a consideration of the *pathology, causes* and *phenomena* of fevers, after which we may pass smoothly along through our investigations of the general and local diseases to which the human system is unfortunately liable.

CHAPTER III.

OF FEVER.

SECTION I.—THE PATHOLOGY OF FEVER.

FEVER is the direct or remote effect of some irritant or morbific agent, applied to or acting upon the fluids or solid tissues of the body. Whatever may have been said of the nature, causes, and symptoms of fever, there can be no reasonable doubt but that *irritation* is the immediate cause of the symptoms which are developed in febrile affections.

A morbific agent may act on the system for some time before the system becomes highly irritated and fever developed. Hence it is, that in most fevers there is a forming stage of debility, in which the morbific agent is producing through the blood, or otherwise, a debilitating effect upon the brain and whole nervous system. We find this debilitating agent, whatever it may be, prostrating the strength, confusing the operations of mind, weakening the strength and force of the heart and arteries, deranging the secretion of the liver, kidneys, and glandular system generally, impairing the appetite, constipating the bowels, and rendering irritable the nervous system.

Now, a morbific agent which is capable of producing fever, may produce confusion of the mind by its debilitating or irritating effects upon the brain. If, then, a morbific agent has entered the blood, it soon reaches in its course the brain, and by there producing its peculiar effects, so far deranges that organ as to interfere with the natural development of mind, as we find in the forming stages of most fevers. If that agent be a narcotic, it diminishes action, and renders the circulation through the brain sluggish and tedious, as is the case in most of our miasmatic fevers. Hence the almost universal statement of patients, during the forming stage of fevers, that they feel stupid and sleepy, and have little ambition to do anything. If the morbific agent has been a stimulant, the brain may have been irritated directly, and the circulation through its vessels accelerated for a time. In this case the patient will complain of great irritability, and yet feel weak and prostrated in body and mind.

The debility which almost invariably prevails in the forming stage of fevers, is probably first brought about by the direct action of the morbific agent upon the brain and whole nervous system. And when we take into account the fact that the brain and all the nervous system are the immediate agents by which the different organs derive their ability to act, we shall readily see how the different organs in their turn become involved. By careful observation we find that this disturbed condition of the brain and nervous system is felt immediately by the stomach, probably through the pneumogastric and sympathetic nerves. The appetite becomes indifferent, and by degrees is destroyed. The taste of the mouth becomes changed and unnatural, nausea and a loathing of food follows, and this cutting off of the supply of food or nourishment goes to increase the prostration.

The *liver*, the function of which depends upon a due supply of nervous influence, is either rendered torpid or else is called into increased action, according to the nature of the morbific agent. If its secretion be lessened, the want of bile to mix with the food which is taken, aids in impairing the appetite, and also produces constipation of the bowels. If its secretion be increased by the influence of the morbific agent on the brain, we have an undue amount of bile thrown into the alimentary canal, which may, and generally does, produce a diarrhœa or a morbid condition of the mucous membrane of the stomach and intestines, and this goes still further to impair the appetite and increase the prostration and general debility.

The *kidneys*, too, are soon involved in this sympathetic derangement, and we generally find them either in a morbidly active state, secreting an undue amount of urine, or they become inactive, secreting very little; in either case their derangement goes to increase the morbid condition of the system. For if they secrete too much, they, by robbing the blood of its fluid parts, derange the circulation, and increase the derangement of the various organs, already too much impaired. But if, as generally happens, their secretion be lessened, we have retained in the blood another morbific agent, which, in its turn, acts upon the already deranged brain and nervous system. In fact, so important is this excretion to the well-being of the system, that even life cannot long continue without it; for the urine retained in the blood not only irritates the brain, but poisons, in a greater or less degree, every tissue to which it is carried. And it is very likely that the retained urine, passing through the blood-vessels, capillaries and tissues, produces an irritability, which is carried through their nerves to the brain, and serves, very materially, to derange its function; and that derangement, in turn, being reflected to the various organs, increases the general derangement.

The *skin*, too, is more or less under the influence of the cerebro-spinal and nervous system; and as all this derangement renders the system either languid or irritable, we find the skin either passively exhaling too much of the fluid parts of the blood; or, as is more common, with diminished action, retaining too much.

Now, if the skin is throwing off too much, as we find it in some conditions of prostration, in the forming stage of fevers, the blood, in consequence, is rendered thick, having too little of the watery parts, and as a consequence, the strength of the system is often rapidly impaired, and in the same ratio its various functions become deranged.

But if, as is more common in the forming stage of fevers, the skin becomes torpid, throwing off too little, we have a poison retained in the blood, which going to the brain, increases the morbid condition, and also very likely irritates the heart, arteries, capillaries and veins, or their extreme nerves, and this irritation is sent through their nerves, to the great nervous centre, from which it is reflected to the various organs, the functions of which become still more impaired or deranged. Thus it is that irritating or morbific agents, entering the blood, act upon the brain and the whole nervous system, by either directly debilitating or irritating, and then by direct and reflex action they derange the whole system, in a greater or less degree impairing the functions of its various organs. While it is the direct action of the morbific agent upon the brain which is the cause of the deranged condition which exists during the forming stage of fevers, it is probable, as we have seen, that a part of this influence is carried from the extreme nerves, which supply the inner coats of the arteries, veins, lymphatics and capillaries, to the nervous centre.

Now, the forming stage of fevers is characterized, as we have seen, by debility, which debility and prostration may, and generally does continue for several days; till, though irritable, the system sinks down, the brain and whole nervous system appearing no longer able, by their influence, to carry on the functions of the body: the heart and arteries feel the influence; the blood recedes from the surface of the body and the extreme vessels; congestion of the brain and spinal cord is the consequence; and as the nervous system is impaired, and the circulation sluggish, too little oxygen is received into the blood, by the skin and lungs, to unite with the carbon and keep up animal heat, and there comes a chill. During this chill, or cold stage, the whole system is in an inactive or torpid state, but the brain, probably from undue pressure upon its substance, begins to grow irritable, which irritability increases till it sends forth an influence that compels the heart to renew its activity, and thus reaction is set up.

Now, this command of the brain to the heart and arteries, is carried, I apprehend, through the sympathetic nerve directly to the heart, and through the sympathetic and cerebro-spinal nerves to every part of the system. And with such power and energy is this command obeyed, that the whole system is brought into a state of excitement. The heart contracts powerfully; the pulse beats actively; the blood flows through the heart, arteries, capillaries and veins rapidly, and as a consequence, immense quantities of oxygen are received into the blood by the lungs and capillaries of the skin; animal heat is revived, and generally accumulates to such a degree, that there is a hot skin, dry tongue, and a general feeling of oppressive heat.

This preternatural heat and general excitement continues for a time, and then the heart and arteries, as if exhausted by over action, generally abate their action, and resume their accustomed activity. Or, what is more common, they sink below their accustomed activity, if no local inflammation has taken place during the excitement. The various functions of the body are restored, or approximate more nearly their normal state. And if no local inflammation is set up the skin usually becomes moist, for a time, and the urine free.

Now I have made this increased action of the heart and arteries to depend upon the excitement of the brain and nervous system directly, and such I believe to be the case in the main; but no doubt, the morbific or irritating agent in the blood, may, and does operate to a certain degree, directly upon the extreme nerves of these vessels. And it is probable, that the morbific agent in the blood, directly irritates the different organs as it passes through them, and thus, in part, deranges their functions; but it must be remembered, that if it does, its impression can probably be received only by their nerves, which impression is conveyed by them directly to the brain.

But as the blood cannot be supposed to pass out of the vessels, it is more than probable, that any direct influence of that kind must act upon the extreme nerves, which terminate on the inner surface of the heart, arteries, capillaries and veins. But how far such an influence may go to produce the debility of the forming stage of fever, is not quite certain. I apprehend that the brain receives every impression of that kind immediately, and then transmits the morbid influence to the various tissues and organs, as I have before suggested.

Thus far we have only considered the action of causes on the system through the blood, and producing their effect directly upon the brain and nervous system. Fevers thus produced without any local inflammation, acting as a cause, have been called *idiopathic*. That such fevers occur, there can be no reasonable doubt, for we see them continuing for days, weeks and months, without any local inflammation being developed, at least, so far as we can discover. Such fevers generally have very remarked remisssions or intermissions, if no local inflammation is set up. But if, as generally happens, a local inflammation is set up in some tissue or organ, that local inflammation becomes a constant source of irritation to the brain and the fever becomes more continued.

The frequency of local inflammation occurring in idiopathic fever, is the probable reason why so many have doubted even their ever occurring; and so have regarded all fevers, as the result of local inflammation in some tissue or organ. But when we take into account the various contagious fevers, as well as pure intermittent fever, we cannot reasonably doubt their ever occurring.—But in what way is the poison introduced into the blood to produce idiopathic fevers? It is probable that the lungs is the most frequent medium through which morbific agents are received into the blood, as it is the most direct.

The air is constantly charged, in some localities, with poisons in a greater or less degree. And as the oxygen of the air enters the blood, it is fair to infer, that the noxious agents which the air contains, may, and do enter the blood. For there is only a thin membrane which separates the current of blood in the lungs from the air we breathe; and that membrane expressly adapted to transmitting air. Now if this membrane can take in oxygen and pass out carbonic acid, it can also transmit the noxious gaseous agents which the air contains, and that too, without necessarily producing directly an inflammation of that membrane. And the length of time that certain morbific agents are known to remain in the blood before they develop fever, is sufficient evidence that what we call idiopathic fevers, cannot be located in the lining membrane of the lesser air tubes and cells of the lungs, which, in fact, would render them symptomatic.

The stomach is probably also a medium through which these poisons may be introduced into the blood. For as the air passes freely through the mouth, its noxious principles may become involved in the saliva, and so pass into the stomach and blood, and become the cause of general fever, without necessarily producing inflammation of the mucous membrane of the stomach and intestines.

Certain articles of *food* may contain poisonous principles, which are frequently taken into the blood through the stomach, and gradually develop general fever, without first having produced any local inflammation. Various articles of *drink* also contain poisons which may go with the fluid through the absorbents into the blood, and passing to the brain, sooner or later develop fever without producing first local inflammation.

The skin too, is a medium by which agents may be received into the blood, and go to develop general fever without at first necessarily inflaming the skin.

This absorption of the skin may be from the air, as it is probable that oxygen is constantly entering the capillaries, and carbonic acid being thrown off, as in the lungs; or, the lymphatics of the skin may absorb gaseous or liquid poisons and carry them into the blood.

Thus we have seen, that there are various ways in which morbific agents may be introduced into the blood; and that once in the blood, they may, and do produce general fever without first having produced local inflammation. Let us see now whether these morbific agents pass as foreign substances in the blood, or whether by changing or dissolving the blood they render the blood itself an irritant.

It is possible that the paludal poison which produces ague, may affect directly the blood-corpuscles, but I incline to the opinion, that the symptoms in our paludal fevers which appear to indicate that change is the result of impaired digestion, and temporary derangement of the liver, where it is probable that the blood cells receive their iron. Now, if it be a fact, that the blood-corpuscles are fewer in the blood, after a protracted ague, than in a state of health, I think it is reasonable to infer, that the deficiency may be owing to imperfect digestion, in consequence of which, suitable materials are not furnished for their formation in the mesenteric glands, and especially that they may fail to receive in the liver their due supply of iron; this to me appears the more rational solution of the fact, than to suppose that the blood-cells are dissolved directly by the paludal poison. This appears the more probable when we remember that during the early stages of intermittent fever, the various functions are frequently performed with a degree of regularity that could hardly be expected if the blood-corpuscles had been directly dissolved by the paludal poison.

It is probable then, that the paludal poison which produces our mild intermittent and remittent fevers, may and does pass along with the blood, in something the same way that oil may be mingled with water without a chemical union or decomposition of either; on the other hand, those *morbid poisons* which produce malignant fevers of a contagious character, as for instance, the morbid poison of typhus, may, and probably do unite chemically with the blood, and so far decompose or change it, as to render it unfit to stimulate the brain in a proper manner, so that the various functions of the system are permanently impaired, and consequently no remissions or intermissions occur.

The purpurea which occurs in some malignant fevers, as well as the passive hemorrhages, go to confirm the idea of a decomposed or changed state of the blood; this I conceive to be an important difference between the morbid contagious poison of typhus, and the morbific agent which produces our paludal fevers, and the reason why, in the one class we have intermissions, or marked remissions, while in the other the fever is continued, generally from the first.

Now, as we have seen, it is probable that purely idiopathic fevers may and do occur, and sometimes pass through a regular course; but it is likely that very many cases, which are so considered, are really symptomatic of local inflammation in some tissue or organ which has been overlooked. At any rate, local inflammation is frequently set up in some tissue or organ, which very materially modifies the course, type, symptoms, and danger of fevers.

We must remember that the stomach and intestines sympathize very strongly with the brain, and depend entirely upon the brain for the power to perform their function. And this sympathy is, probably, first through the pneumo-gastric and sympathetic nerves. But we have seen that the circulation through the different organs becomes deranged; and as the blood, too, is not in a natural or healthy state, it is probable that the stomach is unable to secrete from the blood the acetic and hydrochloric acids, and also its nitrogenous principle, for the digestion of what food is taken. In consequence of this, crude articles of food are passed through the alimentary canal in an undigested state, which, acting upon the mucous membrane of the stomach and intestines, sets up an irritation, and we soon have gastro-enteritis, which very much modifies the character of the fever.

And so soon is this inflammation produced, that in many cases it appears to even precede, when it is really a consequence of the general fever.

I am satisfied, from careful observation, that, in many cases, a gastro-enteritis is thus produced, even before the fever is fully established, and becomes so complicated with the first stages of the fever, as to make it even a matter of doubt, which is the cause and which is the effect of the fever.

The liver, too, as we have seen, soon becomes involved in general fever, and either its secretion is materially diminished, or morbidly increased; in either case the effect becomes a source of irritation to the alimentary mucous membrane. If its secretion is diminished, there may not be sufficient bile to unite with the nutritious part of the food which is digested by the stomach; and, in this case, chyle is not formed, or, if it is, it is of an imperfect character, and does, probably, very little towards supporting the decaying strength.

But this deficiency of bile also produces constipation of the bowels, and thus another source of irritation is brought to bear in producing gastro-enteritis. And so if too much bile is secreted, as happens in some cases of fever, the excessive secretion poured into the intestines, produces, as we frequently see, a bilious diarrhœa, and serves, in a greater or less degree, to irritate or inflame the alimentary mucous membrane. Again, the secretion of the liver in the forming stage of general fevers, is probably of an unhealthy character, and though the quantity may be neither too great or too small, yet, being of an acrid character, in consequence of the deranged condition of the organ, when it is poured into the intestines it serves as a direct irritant in producing gastro-enteritis.

Now, as the liver is one of the first organs that becomes involved in the derangement of the system, in the forming stage of fevers, it is probable that it is an important agent in setting up or producing, in the ways I have suggested, gastro-enteritis. And so insidious is this cause in its action, that no wonder that wise men have supposed that the gastro-enteritis was the primary cause of the fever, and that all fevers are essentially symptomatic of local inflammation. I have known many cases of general fever in which, without the most careful inquiry in reference to the early symptoms, I should unhesitatingly have pronounced the fever the consequence, and not the cause of the local disease. Now the causes thus operating so early to produce gastro-enteritis, having once produced the local inflammation of the alimentary mucous membrane, have also furnished another source of irritation to the brain, which sympathetic irritation of the brain modifies very materially the incipient fever, and produces a variety of complications which otherwise might not exist.

The kidneys, too, as we have seen, become involved in the forming stage of idiopathic fevers. And their morbid action, whether it be to secrete too little or too much, has an important bearing upon the mucous membrane of the stomach and intestines, as well as on the brain and nervous system directly. For if they secrete too much from the blood, it robs the mucous membrane of a due amount of moisture, and thus increases indigestion, constipation, irritation, and inflammation of the alimentary mucous membrane, and that, too, at an early stage of the fever. But if, as generally happens, the kidneys secrete too little, the mucous membrane of the stomach and intestines become congested with too much watery blood, and irritation and diarrhœa is frequently the result, even in the first stages of the fever. Again, as we have seen that the parts of the fluid retained in the blood, in consequence of inactivity of the kidneys, are of an irritating character, it is more than probable that, by entering the capillaries of the alimentary mucous membrane, it directly irritates that membrane, and thus produces gastro-enteritis in the forming or any subsequent stage of fevers.

We have seen, too, that in general fever the skin soon becomes involved, either throwing off too much or too little; and this, too, becomes another source of irritation to the mucous membrane.

And that the functions of the skin should become deranged is not strange, when we remember that it depends entirely upon the brain and whole nervous system for the ability to perform its functions in a proper manner. Now if the skin exhale too much of the fluid parts of the blood, it has an effect similar to that produced by a too copious secretion of urine, both generally, and upon the alimentary mucous membrane.

But if, as generally happens, the skin becomes inactive and torpid, and fails to throw off its accustomed secretion, by far more serious consequences follow. For the alimentary mucous membrane, being only a continuation of the skin, sympathizes very strongly with its derangement, and, unless a vicarious discharge is set up by the kidneys, is sure to become congested, and thus irritated and inflamed in a greater or less degree. But if, as generally happens, both the kidneys and the skin are in a torpid state, congestion of the alimentary mucous membrane is

the inevitable result, and that, too, in a very early stage of the general
fever. Now this congestion cannot long continue without developing
irritation and inflammation.

But the mucous membrane sometimes makes an effort to relieve itself,
by pouring out this fluid into the intestines. Hence the diarrhœa which
occurs in the early stages of some fevers, and which may then be regarded as a favorable indication, if it be the result only of congestion. But
the gastro-enteritis which is set up by the various derangements which I
have enumerated, is the cause of that species of diarrhœa which occurs
in the latter stages of fatal cases of idiopathic fevers.

I have been thus particular in expressing my views in relation to the
manner in which general fevers may originate in the system, without a
previous local inflammation. It is a conclusion which I am compelled to
arrive at, after a careful observation of a great variety of cases, during
the past twenty years. I have also attempted to show how general
fever, by deranging the functions of the various organs, may develop,
at an early stage of idiopathic fevers, an inflammation of the alimentary mucous membrane, which inflammation modifies very essentially
almost every stage of general fever, and becomes a fearful source of
fatality, when this local affection is neglected.

And now, if I have established the fact, that idiopathic fevers may
occur, and the manner of their development, as well as the great liability
of subsequent derangement and inflammation of the different tissues and
organs, but especially of the alimentary mucous membrane, I have accomplished the end at which I aimed. And now, having thus laid down the
most essential principles involved in the pathology of idiopathic fevers,
let us examine with patience the principles involved in the pathology of
symptomatic fever; and though some of these principles have been
already anticipated, calling them up in this connection may be the means
of fixing them more permanently in the mind of the student, at least.

Symptomatic Fever.—A symptomatic fever is one which is produced
by a local irritation or inflammation in some tissue or organ, without
any previous general fever. And in this connection, it may be proper
to state that there is not a tissue or organ in which irritation or inflammation may not develop or produce general fever; and though it will
be impossible for us to examine all the local inflammations, in order to
see how they produce fever, we will examine enough to illustrate the
principle at least.

Inflammation of the skin produces general fever. We see this illustrated in scalds or burns, which inflame the skin, and the degree of the
general fever appears to depend very much upon the extent of the surface involved in the inflammation. If the space, or part of the skin
involved in the inflammation be small, the fever which is set up is slight,
and the constitutional derangement not very great. But if the surface
of skin which is involved in the inflammation is great, we have a high
fever, and very general derangement of the system, and especially of
the brain and whole nervous system.

Now, a scald or burn, embracing a large extent of skin, irritates the
extreme nerves of the skin, and this irritation is carried by them to the
brain and spinal cord, and instead of at once developing fever, the brain

appears to be debilitated by it, and great prostration comes on and continues for a time. The skin involved becomes red, hot and inflamed, while the general system appears by this local inflammation to be rendered debilitated, inactive and prostrated. The heart and arteries appear for a time to have lost their accustomed vigor. The circulation becomes sluggish, and in consequence, too little oxygen is received into the blood to unite with its carbon. There is a letting down of vital heat, and frequently a chill more or less severe follows. The brain and spinal cord become congested during the chill, and more or less irritated, and they finally send forth an influence to the heart and arteries which provoke their renewed activity. The blood which had receded from the extreme vessels, and had compressed unduly the vital parts, is sent to the extremities, and into the minute capillaries, which were before contracted. The pulse becomes firm, strong and rapid; the respiration is accelerated; the amount of oxygen consumed is greatly increased; animal heat accumulates, and a general feeling of oppressive heat, irritability and excitement follows, and thus we have a general fever established from a local inflammation. This state of febrile excitement may be continued during the whole period of the local inflammation. But generally, after a longer or shorter period, the febrile excitement abates, and the various organs resume their functions as the local inflammation subsides.

The brain and whole nervous system appear to suffer materially in such cases. Indeed, in no other variety of fever have I ever witnessed such excessive irritability of temper and excitement of the nervous system. This is, doubtless, owing to various causes, not the least of which is the immense number of nerves, the extreme points of which are involved in the local inflammation; they taking the irritation to the brain and spinal cord, and thus producing the general irritability which follows. Another cause of the excessive irritability may be found in the loss of serum which occurs in such cases, and also in the change which takes place in the skin, as well as the various organs of the body.

But by far the greatest source of the nervous irritability in such fevers, I apprehend to be in the sympathetic irritation set up in the alimentary mucous membrane. For if the fever continues unabated until the system sinks down, as some times happens, the local inflammation appears to be translated to the mucous membrane of the stomach and intestines, and thus the patient dies. Now, as we have seen that the mucous membrane is only a continuation of the skin, and strongly sympathizes with it in every inflammation, and as this is the manner of death in such cases, it is reasonable to conclude that during the continuance of the fever, sympathy of the mucous membrane with the suffering skin is a prominent source of the excessive irritation of the nervous system, in this variety of symptomatic fever.

For, though the direct sympathy of the brain with the suffering skin is very great through the cerebro-spinal nerves, yet it must be trifling compared to that between the brain and the alimentary mucous membrane. For, in addition to the cerebro-spinal nerves, they are connected intimately by the sympathetic and pneumogastric.

All the organs of the body suffer in such fevers in a greater or less degree, as in idiopathic fever; but probably not generally to so great an

extent, if we leave out of the account the mucous membranes and the
nervous system, which we have seen suffer frequently immeasurably
more. Thus we have seen how an inflammation of the skin may develop
general fever, and lead on to various complications, and even to death.

Inflammation of the brain, produced by direct injury, develops general
fever. And the first symptoms in such cases do not vary materially
from other causes of symptomatic fever.

Let a blow upon the head fracture the skull and directly irritate the
brain, and we have at first prostration, as in the forming stage of idiopathic fever. The pulse is weak, slow and labored, the face becomes
pale, the extremities become cold, and a general state of prostration
follows. But the receding of the blood from the extremities, and its
accumulation in the larger blood-vessels, finds an undue amount in the
brain; and probably a rush of blood to the capillaries of the brain, at
the point of injury, sets up a local inflammation, even before general
reaction is produced. For pain in the part continues during the whole
period of prostration, which argues, at least, that congestion exists till
the local inflammation is established. And then we have reaction: the
pulse becomes full, strong, and frequent, the extremities become warm,
the face, before pale and sunken, becomes full, flushed and red; in fact,
we have general fever fully established.

Now, in a sympathetic fever from inflammation of the brain, all the
functions of the various organs become deranged, as in other varieties of
general fever, and, of course, similar consequences follow. The skin
becomes hot and dry, probably in consequence of the irritability of its
nerves. And as there is little or no evaporation from the surface to
conduct away the excessive heat, it accumulates and renders the heat of
the system excessively oppressive. The secretion of the liver, kidneys,
and other glands become deranged in a greater or less degree, and thus
we have a complication of morbid actions, as in idiopathic fever. And
the irritation thus set up being carried through the cerebro-spinal and
sympathetic nerves, is immediately felt; and if, as sometimes happens,
the brain has not power to rally, the various functions become suspended,
and life becomes extinct.

But by far the most direct effect of this injury, of a sympathetic character, is felt by the stomach. At the very first, and during the continuance of the prostration, the stomach appears to strongly sympathize, as
in the forming stage of idiopathic fevers, only in a greater degree.
Nausea and vomiting frequently occurs, and is sometimes carried to even
an alarming extent. Here we find the condition of the stomach during
the forming and after stages, very similar to what it is in the different
stages of idiopathic fevers; in both instances loathing food during the
early stage, and entirely refusing it during the stage of febrile excitement. Thus we see how fever is developed by local inflammation of the
brain, and how the different organs become involved, and especially the
functions of the stomach.*

But inflammation of the mucous membrane of the stomach and intestines, or gastro-enteritis, is a cause of sympathetic fever, and probably
by far the most frequent cause. And in children, indigestion from irregular eating, or from intestinal worms, is a very frequent cause. Now,

in either case, no doubt the local irritation of the mucous membrane is carried directly to the brain, through the sympathetic, pneumogastric, and cerebro-spinal nerves, and hence the febrile excitement which follows, and goes on even, sometimes, to a fatal termination. In fact, so strong is the sympathy between the stomach and the brain, that an irritation of the stomach and intestines of children frequently produce not only general fever, but also a local inflammation of the brain itself, which inflammation again produces or increases fever, and often terminates in effusion into the ventricles of the brain, and even in death.

The convulsions of children are generally the result of irritation of the stomach and intestines being transmitted to the brain and spinal cord; the convulsions being the direct result of the congestion, which congestion marks the forming stage of the fever which follows. The large head of the infant, together with the softness of the brain, causing convulsions of the voluntary muscles, while the action of the involuntary muscle, the heart, is only partially interrupted. It is probable that the exercise, during the convulsions of the voluntary muscles, may help to relieve the partially suspended involuntary functions, and bring about reaction, which reaction, when once established, develops general fever. In this way, then, is sympathetic fever produced in children, by irritation or inflammation of the stomach and intestines.

Acrid substances taken into the stomach, as well as crude articles of food, irritate or inflame the alimentary mucous membrane, and develop fever. Now, if an active irritant, such as boiling water, be taken into the stomach, it directly inflames, and gastro-enteritis is produced. Great debility and prostration immediately results, in consequence of the influence being carried to the brain; a chill generally follows, in consequence of the debility, and this debility may end in the complete prostration of the system, suspension of all its functions, and finally in death. But if the brain and whole nervous system retain their vitality, reaction is sooner or later set up by a renewed activity of the circulation through the heart and arteries, and the usual symptoms of fever are developed.

Gastro-enteritis produced by capillary congestion from suppressed perspiration may produce general fever, and this too, from the effects of cold upon the extreme nerves of the skin; in this case, a diarrhœa is generally produced by the congestion, irritation, or inflammation, sometimes before and sometimes after the general fever is established. If the diarrhœa come on before inflammation, and its fever is established, it is generally the direct result of congestion and debility, and being a vicarious discharge may prove salutary, and prevent both the local inflammation and the fever, but if the diarrhœa come on after the fever is established, it is the result of local inflammation, and may be an unfavorable indication. Thus, we see, that while symptomatic fever may be produced by local irritation or inflammation in any and every tissue and organ in the system, gastro-enteritis is probably by far the most frequent cause. As a general rule, symptomatic fevers are more continued than idiopathic, but the general fevers produced by gastro-enteritis are in part an exception to this rule, for in some cases of gastric, or enteric fevers, there are distinct remissions, and regular chills, which may be interrupted and broken up, simply by counter irritation, and a regulated diet.

Thus we have seen, that every possible variety of symptomatic fever may, and do occur, and that though the forming stage of such fevers, is usually shorter than the forming stage of idiopathic fevers, yet the train of symptoms are similar, and for the same reasons, that in both cases the brain is **first** prostrated or weakened, and then roused **by** irritation **to an** excited condition, in which it calls into renewed activity, the depressed circulation, and thus reaction, general **excitement and fever is** the result: and also, that while as a general **rule,** idiopathic fevers are intermittent or remittent, symptomatic fevers **being** the **result of** local disease, are more generally continued ; **and** finally, that while an inflammation of any **organ or tissue** may be the cause **of** general fever, gastro-enteritis is infinitely the most frequent cause, and also, that while idiopathic **fevers may** produce **derangement** in all the tissues and organs, gastro-enteritis is by far the most frequent **complication,** and generally exists in **a greater** or less degree, and sooner or later, in almost every **case of idiopathic** and symptomatic fever.

SECTION II.—THE CAUSES OF FEVER.

The causes of fever are very numerous, and consist of those which are predisposing and those which are exciting. Most of the predisposing causes may, however, under certain circumstances act as exciting causes, and all the exciting causes may, and do act as predisposing causes. I will, however, proceed to consider them under the head of *predisposing* and *exciting* **causes.**

Predisposing causes.—By predisposing causes, **I mean** all those influences which may depress vitality, or in any way increase the liability of fever in any of its forms. Among the predisposing causes are *hereditary predisposition, indigestion,* heat, cold, and *various imprudences,* &c., which we will now proceed to consider, in the order in which **I have** named them.

Hereditary predisposition to fevers, consists in the depression of vitality which the human constitution has suffered, from imprudences and all the depressing influences which have been operating upon it since the creation of Adam. It also includes all the imperfection of organization, as well as consequent inability for the performance of their functions, of all the organs of the body, the result of physical depravity, &c.

Now, as the contagions are mainly generated in the system, it is probable that they were first generated, as well as the predisposition to their impressions, by the imprudences of former ages, among which filthy habits may have been a prominent one. Thus it was that one fever after another arose, as the multiplied deviations from the laws of health created **not** only the predisposition, but also **caused** to be generated in the system the contagion which propagates them.

Now a predisposition to fevers having been in this way once produced, and the contagion which propagates them having been once generated, it is probable that **this** predisposition is transmitted from one generation to another; only requiring the contact, in some way, of the contagion itself **to develop** each peculiar febrile affection. And it should be

remembered that the predisposition has thus become so strong, that most
and probably all the contagious fevers now frequently arise spontaneously or without an exposure to the contagion generated in the system
of a person laboring under the disease, in the precise manner in which
they first arose.

Thus it is no doubt that the human family have acquired a predisposition to the febrile affections, and have caused to be generated in the system
the contagions which produce the contagious fevers. And thus it may
continue to be, till by a rigid observance of the laws of health, the
human constitution shall have been regenerated, and thus the internal
susceptibility to the fevers become destroyed; and the external cause
or poison which propagates them annihilated or rendered harmless.

It is then a lamentable fact, that the accumulated imprudences of the
human family transmitted from one generation to another, have
placed us of the present day in a condition of physical imperfection,
which renders us liable not only to the contagious fevers, but also to a
great variety of non-contagious febrile affections. And though by prudence on our part we may avoid in many cases a frequent attack, yet it
must be confessed that we are predisposed and always liable to an attack.

And now, although we cannot directly destroy the acquired predisposition to fevers, or the contagions or poisons which produce them; we
can, by a due regard to cleanliness, pure air, and careful removal of the
animal excretions, render them milder and less frequent; and when the
laws of health shall have been thoroughly understood and rigidly obeyed
for a reasonable time, I verily believe that the acquired predisposition
to the contagious, as well as to all other fevers, will be permanently destroyed, and the causes which now appear to produce them, may perhaps
be annihilated, or rendered comparatively harmless. Till this very desirable state of things can be brought about, we have got to inquire into
the causes of fever as best we can, and lighten as best we can the burden
of common misery. Let us then proceed to inquire further into the
predisposing causes of fever.

Indigestion, or any derangement in the digestive organs or functions
may predispose to fever. This may be so by cutting off the proper
nourishment; in this way reducing the blood, and through the blood
debilitating the brain and whole nervous system. Indigestion may also
introduce into the blood an impure chyle, which may not only fail to
properly nourish, but may act as a direct irritant to the cerebro-spinal
system; or the crude articles of food which are liable to pass along the
intestines, in cases of indigestion, may by irritating the alimentary mucous membrane, increase the liability to sympathetic fever; or indigestion may predispose to fevers, by producing constipation. It is probable
however, that the constitutional debility that indigestion produces, constitutes the main predisposition to fever which attends it. For no sooner
is the general strength of the system lowered, than the various functions
become impaired, and the brain and whole nervous system grow irritable,
and in a condition to receive impressions from any febrific agent which
it may chance to encounter.

Heat may act as a predisposing cause of fever, especially as it is liable
to prostrate the powers of the system, and by the copious exhalation

which it is liable to produce from the skin, it may rob the blood of its watery parts, and thus hinder its free passage through the capillaries; and thus impairing the functions of the various organs, may predispose to fever. Exposure to undue heat may also cause the liver to secrete an excess of bile, which being poured into the intestines, may act as an irritant, and thus predispose to fever. In these, and probably in various ways an undue exposure to heat may predispose to fever.

Cold, may operate in various ways to predispose to fevers; it is probable however, that it is mainly by checking perspiration and by producing congestion of the brain and spinal cord, that cold becomes more generally a predisposing cause of fever. And it is only when cold becomes excessive, or is long applied, that it may properly be regarded as a predisposing cause. It acts more generally as an exciting cause of fever.

Various imprudences.—In fact every deviation from the laws of health may by debilitating, irritating, and deranging in various ways the organs and functions of the body, predispose to fever. To attempt an enumeration of them, would be like attempting to count the "sands upon the sea shore," for they are "legion." While then I will not attempt an enumeration of them, it may be proper to state, that it is probable that intoxicating liquors, licentiousness in its various forms, and latterly tobacco, with their attendant vices, are the most prominent influences, which undermine the human constitution, depress vitality and thus in ten thousand ways, predispose to fever in its various forms. For it is well known, that any influence which depresses vitality, predisposes to fevers, as well as to all other diseases.

And now in concluding our consideration of the predisposing causes of fevers, it is well to bear in mind, as I have before stated, that they may nearly all under certain circumstances act as exciting causes. And though I have enumerated separately only a few of the causes which predispose to febrile affections, I trust I have said enough to set our minds upon the right track in this matter. Let us then proceed to a consideration of the exciting causes of fever.

Exciting Causes.—By exciting causes of fever, I mean here those imprudences, agents, and influences, which act directly upon the system to produce febrile affections. And if in enumerating them I call up some which I have set down as predisposing causes, it must be remembered that imprudences, agents or influences may act in either way, sometimes as predisposing but under other circumstances as exciting causes.

While then there are really a great variety of exciting causes of fever, such as local irritations and inflammations, retained perspirable matter, retained bile, retained urine, intestinal worms, irregularity in taking food, violent anger, &c., &c.; the action of which in the main are sufficiently clear, I will only consider separately the more prominent exciting causes, and from the manner in which these causes act in producing fever, a clear inference may be drawn in relation to the action of all agents in developing febrile excitement or disease. The more important exciting causes then, to which I will call special attention, are *heat, cold, humidity, electricity, koino-miasmata, idio-miasmata* and *contagions*; and I will take them up in the order in which I have named them.

Heat, though generally a predisposing, may become an exciting cause of fever. This we see illustrated in cases in which there has been a previous exposure to intense cold, the reaction in such cases being sometimes very great. It is probable also that heat may act directly to produce febrile excitement, even in cases in which there has been no previous undue exposure to cold. This may be produced by the direct irritating effects of heat upon the brain and whole nervous system, indirectly producing cardiac and arterial excitement; or the depressing influence of heat, if it be but moderately applied may lead to prostration and congestion of the brain and spinal cord, and the irritation thus set up, may send forth an influence to the heart and arteries which will develop febrile disease. While then heat is generally a predisposing, it may act as an exciting cause of fever.

Cold is a very frequent exciting cause of fever, especially if there had been a previous exposure to undue heat. Cold may, if applied suddenly and of considerable intensity, produce a chill attended with cerebro-spinal congestion, which developing an irritation of the brain and spinal cord, may send forth an influence to the circulatory system which shall develop febrile excitement. It is more generally, however, by checking the perspiration and also the action of the liver, and perhaps kidneys; causing to be retained in the blood urine, bile, or more especially perspirable matter, that cold acts in producing febrile affections. For no sooner are either of these matters retained in the blood, than they irritate the brain, and either directly excite febrile action, or they produce first a depression, and then as a consequence, develop febrile excitement.

Cold may also produce fever by causing to be retained in the system the animal heat, which ordinarily passes off with the evaporation of perspirable matter from the surface of the body; for as the cold checks the perspiration, the animal heat which its evaporation ordinarily carries off, in a latent state, remains, and probably acting as an irritant to the nervous and circulatory systems, is liable to produce febrile excitement; thus in these and various ways, cold becomes an exciting cause of fever.

Humidity too, by checking the perspiration, is liable to become an exciting cause of fever, in the manner I have already suggested. When the respirable matter is retained by exposure to cold, humidity also by letting down the electrical state of the atmosphere may so far rob the nervous system of its electricity, as to cause a general depression of the nervous and circulatory function, which leading to a chill or congestion, and consequent irritation of the brain and spinal cord, may lead on to a febrile action.

Humidity may also by acting in these and various ways, gradually impair the functions of the various organs of the body, and becoming thus deranged, the retained urine, bile and perspirable matters may by acting upon the system, debilitated, produce febrile disease.

Electricity is a powerful excitant to the nervous system, and hence as the atmosphere becomes very dry, and consequently highly charged with electricity, its action upon the nervous system may be too stimulating, and hence, if it produces irritation, it may lead on to febrile excitement. While then humidity with a low electrical state of the atmosphere, is probably a very frequent exciting cause of fever, in persons in a feeble

6

debilitated state, it should be remembered that a high electrical state caused by excessive dryness of the air, may produce febrile affections, in some rare cases, especially in strong constitutions, or in those addicted to over-eating, stimulating drinks, &c.

Now, it should be remembered, that the exciting causes of fever which I have named, as well as numerous other accidental causes which I need not name, produce fevers of a very irregular character, sometimes continuing but for an hour, and again passing on for days or even weeks, sometimes terminating speedily in resolution, while at others passing on and becoming complicated with local inflammations, organic changes, &c. Let us now pass on to a consideration of the remaining causes of our list which produce fevers of a more definite or specific character.

Koino-miasmata, malaria, or the paludal poison, as we have seen in a previous chapter, are generated in marshy places, in which there is decaying vegetable matter with moisture, exposed to a temperature of from 60° to 80° Fahrenheit; and having an affinity for moisture, it ascends with the vapor from such sources, and floating along in the lower current of the air, being apparently heavier than common air, it is taken into the system through the lungs, mouth, and perhaps absorbents of the skin, and produces a variety of febrile affections, of greater or less severity.

It is probable that the paludal poison thus taken into the system, enters the circulation, and passes as a foreign substance with the blood, without any chemical union, or decomposition of the blood. Its action is probably directly upon the brain and whole nervous system; and whatever its composition may be, it doubtless directly diminishes nervous action, or depresses vitality. If the malaria be in a highly concentrated state, and the system be strongly predisposed, there is liable to be developed an intermittent, congestive, or some kindred form of paludal fever in a very short time, the morbid action of the agent producing prostration of the nervous system, with chills, &c., before the different functions of the system have become very materially disturbed.

If, however, koino-miasmata act upon the system in a less concentrated state, and the system is not predisposed to ague or kindred forms of paludal fevers, the agent may, by acting with less violence, produce a gradual impression upon the nervous system, and by thus hindering a proper generation and distribution of the nervous influence, the various functions of the body may become deranged, and with this general derangement, when the system finally sinks down, and there comes a chill, the system does not entirely rally and produce an intermission, as in intermittent fever, but only a remission, and hence a bilious remittent fever is the result.

Finally, if the malaria be still more diluted, and be thus brought to act upon the system in a very moderate degree of concentration, its action upon the nervous system is necessarily very slow, and as it produces only a slight disturbance of the functions of the various organs, we shall get neither intermittent or bilious remittent fevers, but perhaps a bilious diarrhœa, dysentery or intermittent neuralgia, as morbid results of the paludal agent. Hence it is, no doubt, that in localities where but little malaria are generated we have bilious diarrhœa, dysentery and neuralgic affections; where a little more are evolved, bilious remittent fevers; and in localities where the paludal poison is formed in great quantities, we get ague and other kindred paludal diseases.

Now, in relation to the distances to which koino-miasmata may be carried in the air, and yet produce intermittent, remittent and neuralgic affections, there may be some doubt. But from careful observations which I have made in this locality, as well as in Jefferson county, in this State, where I practiced for ten years, I believe that malaria may be conveyed in the air, in a state of concentration sufficient to produce ague, for a distance varying from one to three miles, depending upon the surface, winds, &c.; in a sufficiently concentrated state to produce bilious remittent fevers, it may pass from three to six miles, depending upon the same influences; and finally, that it may be conveyed in a state of concentration sufficient to produce neuralgia, &c., depending upon obstructions, for a distance varying from six to nine miles. I have made this estimate with considerable care, and though it may not be regarded as an absolute rule, I believe it will generally be found, in the main, correct.

While, then, malaria may thus pass in the air to considerable distances when unobstructed, it should be remembered that it will pass but a very short distance over water, being probably absorbed; and that rolling grounds, high fences, if made tight with boards, rows of buildings, and especially forests, may almost entirely obstruct its passage in any part of its course. I have had ample opportunities for verifying these statements in this as well as in other localities; and, as one of them was so very plain and decisive, I will mention it, as illustrating the perfect obstruction which a forest may afford to the passage of malaria.

The land, for a few miles north of Geneva, at the foot of Seneca Lake, in the State of New York, is rather level, and there is, occasionally, a marsh or low wet place, embracing a few acres. On a road which runs east and west, about one hundred rods north of one of these marshes, ague was never known, though it had been settled for years. But on cutting away the forest between the road and marsh, ague speedily made its appearance among the inhabitants along the road, as it was rather thickly inhabited, and it has continued to prevail there, more or less, every season since.

In malarious districts, then, advantage should be taken of all the available circumstances to intercept the poison as far as possible; and if nothing more can be done, a grove of trees may be left or planted between a city, village, or house, and a near marsh; or a high board fence may be made, which would doubtless, in many cases, prove a great protection. In localities, too, where malaria are generated, as the poison is probably brought down by the dews, the inhabitants should not, as a general rule, venture out, either early in the morning or at evening, as the danger from its influence would be much greater at such times.

And while houses in such localities should be built on the highest situations available, the sleeping rooms should be arranged, as is well even in all localities, as near the roof as possible. This gives the advantage of, in part, keeping above the malaria, and also secures the benefit of the heat of the sun during the day, which evidently, in part, destroys the pernicious influence of the agent.

Finally, in malarious districts, where it becomes necessary to live, and it becomes impossible to shut off, or in any way get above the agent, fires should be made in the sleeping rooms towards evening, and thus the malaria may be, in a good degree, destroyed for the time, at least. It is also of the greatest importance, in such localities, that strict cleanliness be observed; for if, in addition to the koino-miasmata, there be generated idio-miasmata, or an animal poison. The combined effects of the two influences upon the human system, I am satisfied, is much greater than of either alone, producing, in some cases, the worst possible varieties of malignant fevers.

Idio-miasmata, or the animal effluvia or poisons which result from a decomposition of the animal secretions and exhalations, as we have seen in a previous chapter, are liable to accumulate in filthy apartments, especially in cold seasons of the year, when they are kept closed; and though they may not generally be conveyed to any very considerable distance, they are liable to be taken into the circulation through the lungs, stomach, and absorbents of the skin; and, by decomposing, or uniting chemically with the blood, are liable to produce fevers of a low typhus character.

Animal poisons thus generated from the decomposition of animal secretions and exhalations, in abodes of filth, may doubtless produce genuine typhus, diphtheria, &c.; and that such effluvia, on entering the blood, unites chemically with it, and probably decomposes it, appears quite certain, when we take into account the low putrid fevers which they evidently produce.

Now, if this supposition in relation to idio-miasmata be correct, their effects are greatly increased by their power of rendering even the blood itself a depressing agent. At any rate, they so completely depress vitality, that instead of getting intermissions or remissions in the typhus and other putrid fevers which they produce, the fevers are of a continued character, as we see in pure typhus; and when we get local inflammations, in such fevers, as we do in diphtheria, the exudations are of a dark putrid character, evidently composed of decomposed albumino-fibrinous matters.

Among the fevers that idio-miasmata may produce, are *typhus, diphtheria, plague;* and probably either alone, or combined with koino-miasmata, this animal effluvia may produce all low putrid fevers of a typhus character, of which it may act both as predisposing and exciting cause.

The distance to which this fearful cause of fevers may be carried in the air, as I have already intimated, is probably not very great; but it is probable that the effluvia may be carried in articles of clothing to considerable distances, and yet retain the power of producing putrid fevers. The best means of preventing, then, the influence of this agent, is to avoid the causes which lead to its formation.

The enforcement of proper sanitary regulations, the breaking up of dens of filth, and securing cleanliness, proper ventilation, &c., would doubtless do much to prevent the generation of idio-miasmata; and as the typhus contagion, which is probably generated in the systems of patients thus accidentally prostrated by the disease, is not very readily communicated for any great distance in pure air, typhus and other kindred fevers, of which this agent is now so frequent a cause, might be rendered comparatively rare diseases.

The combination of *koino* and *idio-miasmata*, as a cause of fevers, is, as I have already intimated, a fearful one; and I make no doubt, from careful observation, but that this combination of causes produces all that class of malignant fevers which cannot be traced to the animal or paludal poisons alone, including yellow fever, and perhaps diphtheria, plague, &c. I am satisfied, too, from careful observation, that the systems of patients prostrated with paludal fevers, if not kept in a cleanly, well-aired room, may occasionally generate the typhus contagion, which shall produce pure typhus in its worst, or at least in a malignant form. And this may sometimes occur even with every precaution in relation to cleanliness; an instance of which fell under my observation a few years since, while practicing in Jefferson county, in this State.

An elderly lady was prostrated by a bilious fever, which reduced her system considerably, during which she was nursed by a lady from ten miles north of her locality, and occasionally visited by a lady whose home was thirty miles east of hers. After her recovery the two ladies went to their homes, one ten miles north and the other thirty miles east, and both beyond the paludal influence which had produced her disease, which consisted of a marsh which from a closure of its drain had been overflowed, and was drying up. Soon after the lady reached her home, ten miles north, she was taken with malignant typhus and died. The other lady, after reaching her home, thirty miles east, was taken with typhus and died. But this is not all: two of her sisters, young women, were taken down with typhus and died.

Now, as there was no typhus fever in either of these localities, it is fair to infer that a typhus contagion was generated in the first patient, laboring under the paludal fever, which, acting upon the systems of the two ladies which attended and visited her, probably in connection with the paludal poison to which they were also exposed, produced in them malignant typhus. In relation to the two sisters of the lady that took the fever and died, they had not been exposed to the paludal poison, as the locality is one of the most healthy in the State.

I will only add, in conclusion, that all these families were cleanly, their houses capacious and well ventilated, and the patients had all the care that wealth could procure or devoted friends could bestow.

But I must pass on to a consideration of the contagions as exciting causes of fevers, which will complete what I have to say, in this place, on the causes of fever.

Contagions, as we have seen in a preceding chapter, consists of those peculiar products of disease, whether in a solid, fluid, or aeriform state; which are capable of producing the same disease in another person, and of propagating the disease of which they are the cause and effect, through any number of unprotected individuals.

Now, the contagions, whether the product of secretion or exhalation, which produce fevers, either in a liquid or solid state, are introduced into the system by contact, generally with the skin: but in the aeriform state they act through the absorbents of the skin, or they may be received into the system with the air through the lungs; or probably by becoming entangled with the saliva in the mouth, they may reach the circulation, and thus produce their effects.

Now, we have little difficulty in relation to the origin of the contagions which produce fevers; when we remember that as with typhus, so with all the contagious fevers, they often arise spontaneously, or from various imprudences, without an exposure to the contagion generated in the system of a patient suffering with the peculiar disease. Thus, then, they all arose originally, and thus they may continue to arise, as well as by a direct contact of the contagion, till the predisposition to them, the imprudences which have led to it, and the contagion which propagates them, shall have been all overcome, by better sanitary regulations, better morals, and in short, by a better observance of the laws of health generally.

The contagions which produce a great variety of fevers, after entering the blood, operate rather slowly in some cases probably to change the blood, but in all to depress vitality, through their influence upon the brain and whole nervous system. The period of incubation then, during which the contagion is operating to derange the system and develop the disease, varies with the variety of fever, from two or three days to several weeks.

After the development of the fever, the contagious secretion or exhalation, begins to be formed; in some fevers however, sooner than in others, and the power of communicating the contagion to other persons continues either from the body or clothes; in many cases through the whole period of convalescence.

The distance to which the contagions may be carried in pure air, is probably not very great; ordinarily not passing beyond a few feet, in a state of concentration sufficient to produce fevers. But it should be remembered, that they may be conveyed in any of their forms, in clothing and other materials, for great distances, and still retain even for a long time, the power of communicating their peculiar disease.

It is always best to avoid an exposure to the contagions, especially if it may be done without any inconvenience; for even though the persons thus exposed, may get the fever at perhaps the best time for himself, yet he is liable to communicate it to another at the worst possible time. And further, there is always the possibility of finally escaping some of them at least.

Finally, the contagions which produce most of our fevers, become harmless, or comparatively so, when brought to act upon persons who have had one attack of the disease which has produced them; the result doubtless of some change wrought in the system by the first attack. Thus, have I completed what I had to say here, on the causes of fever, *predisposing* and *exciting*.

SECTION III.—THE PHENOMENA OF FEVER.

Having in the two preceding sections of this chapter taken a glance at the *pathology* and *causes* of fever, we are now prepared to consider, or even anticipate, the phenomena which will be developed in febrile affections.

Now we find, just as we should have supposed, that as the predisposition to fevers is not alike in any two cases, and as the predisposing and exciting causes differ widely in different localities, and in the same localities at different times, that the phenomena of febrile affections differ widely in different epidemics, and also in the same epidemic in different individuals. And yet, as there is a certain degree of similarity between the constitutions of all mankind, and as there is also a certain degree of likeness in the same causes, even in different epidemics, we find a degree of similarity in the phenomena of fevers developed from like causes. And yet, it must be remembered, that no two cases of fever develop precisely the same symptoms. And that no two attacks of fever, even in the same individual, developed from the same or similar causes, are attended by precisely the same phenomena. And yet the phenomena of fevers, comprising the symptoms, course, type and stages of them, admit of a general description, as a matter of convenience, almost every case, however, being in some degree an exception to the rule. With this qualification let us inquire into the phenomena of fever, comprising the general symptoms, courses, type and stages, as they are most frequently developed.

The *course*, type or order in which the symptoms of fever develop themselves, admit of division into intermittent, remittent and continued. But it must be remembered that while many fevers are remittent, very few are strictly intermittent or continued. Now there has been much speculation as to the reason why some fevers are intermittent, while others are remittent or continued. I think, however, this question admits of an explanation, if we will take into account carefully all the facts in the case, and use the common sense we have, as we would on other subjects.

If we can find a good reason why the fevers produced by marsh or koino-miasmata, are generally of an intermittent character, I think the fact that some of them as well as all the fevers produced by idio-miasmata being remittent or continued will admit of an explanation.—If my supposition be correct, that the paludal poison enters the blood and passes with it through the system as a foreign substance, and without chemical union with the blood, we may reasonably suppose that it may produce temporary prostration of the system and lead on to a chill, in the way I have suggested in a preceding section on the pathology of fever; the febrile reaction being the direct result of the cold stage, the sweating stage and the intermission following as a consequence. This will account for the first chill. And as the paludal poison is still in the blood, and the system debilitated, the system will keep up the intermission for one, two, three or more days according to the degree of its debility, and then another chill will follow. This regularity in the return of the chills being the result, as I apprehend, of the ability of the system to keep up its healthy function for just that length of time.

And this appears more probable when we remember that, if a patient be declining in strength while suffering from ague, the chills often anticipate, while, if the general strength of the patient is improving, the intermissions become longer, the chills at each time occurring a little later.

If this accounts for the chills in pure ague, as well as for the time of

their occurrence, I think we may find a good reason why some paludal
fevers are remittent, and others nearly or quite continued. For we have
only to remember that local inflammation produces general sympathetic
fever, and, also, that local inflammations are liable to occur in intermittents, and we have a solution of this difficulty. For, after the paludal
poison enters the blood, that, or some accidental cause may, and frequently does, set up some local inflammation before the first chill occurs,
or else in the fever which follows the first chill, and, in that case, we
have a symptomatic fever, superadded to the intermittent, which renders
the fever itself more or less remittent, or perhaps nearly continued, according to the extent and intensity of the local inflammation.

This solution of the difficulty appears the more reasonable, when we
remember that a purely intermittent fever often becomes remittent, and
a remittent continued, by the supervention of local inflammation.

And also, conversely, that continued paludal fevers often become remittent, and remittent become intermittent, by the subsidence of local
inflammation. This, in my mind, is a rational solution of the fact that
paludal fevers assume sometimes an intermittent, at others a remittent,
and, occasionally, a nearly continued course. And it also accounts for
the fact that intermittents become remittents, and remittents continued,
and the reverse; and hence, also, the symptoms which arise.

We have, now, only to explain why fevers produced by idio-miasmata
are more continued than those produced by the paludal poison, and we
have a solution of the whole difficulty.

Now if, as I have already suggested in a preceding section, idio-miasmata, on entering the blood, unites chemically with it, and, also, materially changes or decomposes it, we have a clue to the reason why such
fevers are generally of a continued character. For, if the blood, in such
cases, is in a dissolved or materially changed condition, we can readily
see why, after the chill, that ushers in or produces the first febrile reaction, there should be no intermission, or only an imperfect remission,
even though no local inflammation exist. For the first chill would not
occur till this morbid condition of the blood produced sufficient general
prostration of the system to lead to it, and then, after the febrile reaction which the chill produces, the blood being changed, the system could
not be supposed to produce an intermission, if even a remission.

Having thus accounted for the intermittent, remittent, and continued
character of fevers, let us inquire into the phenomena developed in the
symptoms which arise during the progress of fevers through their different
stages.

Now the series of phenomena which occur in the system during a
course of fever, are just what we should naturally suppose when we take
into account the nature of the physical organization, and also the character of the predisposing and exciting causes. And though the predisposition, the predisposing and exciting causes differ, as we have seen, so
that no two cases of fever develop precisely the same symptoms, even
though they occur in the same individual, yet most fevers, left to themselves, develop a train of symptoms, at different periods of their progress, which, for convenience, we may call "stages of the fever."

These stages we may call the "forming, the cold, the hot, the declining, and the convalescent," all of which are attended with symptoms which it may be well for us to consider.

The forming stage of fever includes the period from the time the febrific agent begins to produce its effects on the system till the first chill. This stage may be long or short, depending upon the predisposition of the system, and concentration or activity of the exciting cause. If the predisposing causes which have been operating in the system have been debilitating to a great degree, and the powers of the system thus rendered feeble, a moderately concentrated febrific agent may produce such an impression upon the system, that the forming stage may be of comparatively short duration. While, on the other hand, if the predisposition and predisposing causes have been slight, a comparatively concentrated or active exciting cause may be slow in developing fever.

The symptoms of the forming stage of fevers are what might be expected from a febrific agent in the blood.

There is loss of appetite, a bitter taste in the mouth in the morning, drowsiness, headache, lassitude, wandering pains, a dull heavy pain in the back, disturbed sleep, a feeling of coldness in the morning, and slight thirst at evening, fretfulness, sunken countenance, dry skin, a sluggish, feeble or irritable pulse, loss of energy, and general debility.

Now it is evident, that the febrific agent going into the blood, as we have seen, operates first upon the brain and nervous system, and thus deranges the various functions. The loss of appetite, which occurs early, is no doubt the result of the disturbance of the brain, the impressions being transmitted to the stomach through the sympathetic and pneumogastric nerves. The disturbed sleep occurring during the early part of the forming stage of fevers is no doubt the result of derangement of the brain and whole nervous system, as also, the wandering pains in the limbs, lassitude and general debility; so too of the indigestion, the brain not being able to supply sufficient nervous influence to keep up a healthy action of the stomach and intestines, the liver too, from the same cause not supplying a proper amount of healthy bile.

The dryness of the skin, too, comes probably from cerebro-spinal irritation or derangement, together with the general derangement of the organs and functions of the body; so too the lassitude and general debility; and finally, last of all, the power of the heart and arteries gradually decline; the blood is not sent with its accustomed force to the extremities, and through the capillaries; as a consequence, the extremities become cold, the combustion in the lungs and in the capillaries not being sufficient to produce a due amount of vital heat; this in turn increases the general derangement of the system, and all finally lead on to the chill. The heart and arteries appear to be the last of the organs to be brought into a morbid condition during the forming stage of fevers, and this is not so strange, especially in idiopathic fevers, since the circulation is carried on mainly by the ganglionic nerves.

If, as I have supposed, the febrific agent acts first upon the cerebro-spinal system, and through them and the sympathetic, upon the various organs, it is natural to suppose that their functions should become deranged, one after another, and finally the circulation, the most strictly vital function, carried on by the ganglionic system of nerves, should be the last to fail

This I think, is generally true not only in idiopathic fevers, but also in most symptomatic fevers; for as I have before suggested, the local irritation in symptomatic fever is first transmitted to the brain, and then from the brain through the nerves to the various organs, the functions of which become more or less deranged, and finally, if the shock of the local affection be great, the ganglionic function of the circulation becomes implicated and then there is a chill, as in idiopathic fever. Hence we see that the phenomena or symptoms of the forming stage of idiopathic and symptomatic fevers are similar, and from the same or a similar condition of the various functions, but from the nature of the cause, the period of the forming stage, of symptomatic fevers is much shorter than in the idiopathic.

The cold stage of fever occupies the second place in febrile affections, coming on as we have already seen from the derangement the febrific agent has produced on the brain, nervous system, and various organs of the body; all of which, producing debility, general prostration, and languor, let down the circulation, and produce a chill or cold stage more or less marked.

As the powers of the brain and whole nervous system, and through them the circulation, sink down, the blood does not flow with its accustomed freedom through the heart, arteries, capillaries, and veins, the extremities become shrunken and cold, the surface of the body numb and dry, the countenance pale and sunken, the head feels confused, the tongue is dry, the pulse becomes small, frequent, and feeble, and nausea and vomiting frequently occur.

The symptoms of the cold stage are what might be expected from the debilitated and irritable condition into which the system is thrown during the forming stage. But the stage of congestion of the cerebro-spinal system, in which cold is one of the leading symptoms, is generally of no very long duration. The brain rallies, and through that, as we have seen, the heart and arteries. Reaction occurs. The blood is thrown to the extremities. The capillaries are filled and become active. The surface of the body becomes red. The countenance is flushed. And so the hot stage is developed. The most common symptoms of the hot stage are augmented heat, a full, quick, frequent and vigorous, or else a small and frequent pulse, pain in the head, intolerance of light, dry, hot skin, scanty urine, and wakefulness.

Now all these symptoms result from the excitement and irritability of the cerebro-spinal system, calling into renewed activity the before languid circulation, by which the blood is forced back to the extremities, and into the extreme capillaries, from which it had receded. And while this excited action of the heart and arteries continues, the skin, liver, kidneys, and other organs, become so far over-excited or irritated as to very materially impair their functions for the time, and hence the scanty urine, dry skin, and other deranged secretions.

But this excited action, after continuing for a longer or shorter period, reaches its highest point of febrile activity, and a termination, either fatal or favorable, results. This point we may call the crisis, or period of decision. In rapid, continued fevers, there is but one crisis. And there is only one crisis in a single paroxysm of an intermittent fever.

In remittent fevers, we may also regard them as recurring at each remission during the stage of declension, till the final crisis, when no more febrile paroxysms occur.

At the crisis, if it be favorable, the various organs resume, to a certain extent, their functions. The tongue becomes moist, the urine more copious, the secretion of bile more natural, and the skin becomes moist, or sometimes there is copious perspiration, and then it constitutes the sweating stage.

The period from the time at which the fever begins to abate its violence till convalescence is fully established, may be called the declining stage. The duration of this stage varies, according to the character of the fever, and the length of time the system has been suffering from the febrile affection.

The period which follows between the termination of the fever and complete restoration to health, is properly the stage of *convalescence.* Now the time which intervenes between the termination of the fever and perfect restoration to health, may be longer or shorter, according to the constitution of the patient, the nature of the predisposing and exciting cause, and also the degree of the structural and functional derangement of the various organs.

The symptoms which arise during the convalescence from fevers, are such as are developed by a return of all the organs and functions of the body to a healthy or normal state. The appetite improves, the skin becomes natural, sleep becomes quiet, the countenance appears more lively, there is increase of flesh and strength, the mind becomes cheerful, and, in short, the whole physical, intellectual, and moral man becomes restored to a normal state.

Having completed our consideration of the phenomena furnished by the symptoms which arise during the forming ; the cold, the hot, the critical, the sweating, the declining, and the convalescing stages of fevers, we may now proceed to the consideration of the other phenomena of fevers, and first of the revolution of fevers.

"The time which is occupied by one paroxysm of fever, and its succeeding intermission, or between the periodical exacerbations of fevers not strictly paroxysmal, is called the revolution of a fever." "The revolution of fevers vary in point of duration, some fevers completing their revolution in twenty-four hours, others in forty-eight, while others require seventy-two, and some even ninety-six hours." "The form which a fever assumes, in this respect, is called its *type.* So that a fever which occupies twenty-four hours from the commencement of one paroxysm to another, is said to be of the quotidian type ; while one which revolves every forty-eight hours is of the tertian type ; and when this period is extended to seventy-two hours the fever is of a quartan type ; and a period of ninety-six hours constitutes the quintan type." But by far the most common types are the quotidian, the tertian, and the quartan, the others occurring only occasionally.

Now the cause of this phenomena in fevers, as I have before intimated, is clear to my mind. For, if the system can sustain an intermission only from the close of a paroxysm, on one day, to the same hour of the next, a chill will occur every day ; and if the system be rather declining, the chills will anticipate, but if the general strength is improving the chill will occur a little later each day.

So in those cases in which the chills occur every forty-eight hours; the system retains more vigor, and does not sink sufficiently to produce a chill oftener than every forty-eight hours, the chills in these cases anticipating or postponing according as the vigor of the system is rising or falling. The same rule, I am satisfied, will apply in the quartan and all other types. The same, also, will apply to all remitting fevers; only, as there is generally, and, in my opinion, in every case of remittent fever, a local inflammation, the system can only sustain a remission instead of an intermission. And as the local inflammation is so very liable to change, the remissions are generally less regular in their return and length than the intermissions in pure ague.

In intermittent fevers of the quotidian type; the chill generally comes on in the morning. In tertians, it generally comes on about noon. But in the quartans, they occur more frequently towards evening. At least, such has been the result of my observation.

In remittent fevers, I have generally observed the remissions to occur more frequently in the morning. And this I suppose, is owing in part, to the absence of light, and other causes of excitement, during the night season. This however, is not an invariable rule.

In intermittent fevers, we sometimes have what have been called, "double tertians," the chills occurring every day; but they differ from quotidians, by the paroxysms of alternate days being similar, in relation to the time of occurrence, grade, duration, &c. In such a case, the chills on Monday and Wednesday, may occur at 10 o'clock, in the forenoon; while on Tuesday and Thursday, they may occur at 4 o'clock, in the afternoon; so that, though each day has its chill, on the alternate days, only, do they occur at the same hour.

The double tertians however, assume generally a simple tertian type before they terminate, the weaker paroxysm disappearing first.

Much has been said in relation to critical days in fever, or the days on which a crisis is most likely to occur. From careful observation on this subject, for the past twenty years, I am compelled to believe, that the seventh, fourteenth and twenty-first days, are entitled to consideration in this respect. That is, in all cases of idiopathic fever, depending upon a febrific agent, whether that be *idio* or *koino-miasmata;* the tendency to a crisis is stronger on those, than other days; other things of course being equal. And I suspect that such is the case, to a certain extent, in symptomatic fevers. This if true may not admit of an explanation. But, when we remember that the paroxysms of many fevers, recur with great regularity for a long time, it does not appear so strange, that the system should predispose to a *crisis* on particular days, and such I suspect may be the case.

When a fever has run on, till it has reached its height, and a crisis occurs, that crisis is either favorable or unfavorable. If it be unfavorable there may be a rapid sinking of all the powers of life: the extremities become cold, the skin clammy, the pulse feeble, fluttering, or intermittent, the breathing labored, the countenance ghastly, the eyes dim, the mind wandering, and finally the last breath is exhaled, the heart stops, and the patient is dead.

But if, on the other hand, the crisis be favorable; the various organs of the body resume their functions, more or less rapidly, as the morbid condition subsides. Now, during this period, after a favorable crisis, or at the very crisis, the skin, kidneys and bowels, are liable to a copious or to a preternatural discharge, which as it occurs at or near the crisis, is called a "critical discharge." These critical discharges I think, are only the result of an improved condition of the system; the skin, kidneys, or bowels, thus throwing off from the system that which had been morbidly retained.

A gentle perspiration occurring in this way, if it becomes general over the body, is a favorable indication, as it shows an improved condition of the general powers of the system. So, too a free discharge of urine occurring after a long partial suppression, indicates an improvement in the vital energy, whereby this languid secretion has become restored or improved.

A slight diarrhœa, too, occurring at such a time, if there has been no gastro-enteritis, is a favorable indication. But great care should be exercised in giving an opinion of such a discharge, till the other secretions and the general symptoms have been carefully observed; for too often, a diarrhœa occurring at this stage of a fever, indicates a neglected gastro-enteritis, which would render the prognosis vastly more unfavorable.

This then completes our consideration of the *phenomena* which are furnished by the symptoms which arise during a course of most cases of *idiopathic* and *symptomatic* fevers. And though some of the phenomena here described, may not always occur with the regularity which I have supposed, yet I believe the statements here, are generally correct.

Having now in my first chapter, considered the *nature, causes, symptoms, diagnosis* and *treatment of diseases,* and in the second, taken up *irritation, congestion,* and *inflammation;* and finally in the present, taken a glance at the *pathology, causes,* and *phenomena* of *fever;* we are prepared to pass on from GENERAL to SPECIAL PATHOLOGY and THERAPEUTICS.

CHAPTER IV.

GENERAL FEVERS.

SECTION I.—INTERMITTENT FEVER.

INTERMITTENT fever is essentially a disease of the nervous system. In fact, so markedly is the cerebro-spinal system the seat of this affection, that I seriously question whether it should be regarded as a fever or as an intermittent neuralgia. But as very few cases occur in which there is not some local complication, and as it is a matter of convenience, I have classed it with, and shall proceed to consider it as a fever.

Intermittent fever assumes generally either the quotidian, the tertian, the quartan, or the quintan type. During the forming stage of ague, there is loss of appetite to some extent, the nervous system becoming irritable, the circulation imperfect, the extremities being cold, and more or less pain is felt in the lower portion of the spine. By degrees the circulation becomes still more sluggish, the patient yawns, stretches, and feels more languid, till finally the circulation sinks down, the blood recedes from the extremities and surface of the body, and there comes a chill.

During the cold stage the skin becomes pale, the extremities shrunken, the pulse small and quick, the extremities cold and numb, the fingers contracted, and the whole surface of the body materially shrunken. The breathing becomes hurried and irregular, and there is sometimes a hacking cough. Confusion of the mind is very marked at this stage, probably from the undue pressure upon the brain. In fact, sometimes a complete state of coma occurs, especially in weak, debilitated patients. Vomiting frequently occurs at this period, probably from sympathy of the stomach with the brain. The mouth, too, becomes dry, and usually there is very great thirst. The urine is clear, free, and without sediment, as is usually the case in nervous prostration. Thus it is that the symptoms of the cold stage develop themselves, varying in different cases, more or less, from a period of a few minutes to several hours.

The rationale of the symptoms of the cold stage is plain, when we recollect that the general cause, operating through the blood, produces debility of the brain and whole nervous system, which, sinking down, the circulation becomes languid, the blood recedes from the extremities, they in consequence becoming cold and shrunken, and in fact the whole surface cold and contracted. The internal congestion produces the excitement, restlessness, thirst, &c. Pressure of blood accounts for the difficult breathing, while irritation of the brain is finally set up by the undue pressure, and by it the heart and arteries are called into renewed activity. The blood is now thrown to the extremities, the minute capillaries are injected, the pulse is quick and strong, the respiration is less oppressed but quick, animal heat accumulates, and thus the hot stage is developed.

During the hot stage the skin is hot and dry, the thirst urgent, the pulse full, strong, and frequent, the respiration free and quick. There is usually pain in the head, back and extremities, and the urine is more or less scanty. After this hot stage of reaction has continued for a time, usually longer than the cold stage, and varying from one to six or eight hours, the febrile excitement subsides, and there comes the sweating stage.

The perspiration usually appears first about the head and face or breast, and gradually extends over the whole body. The pulse soon becomes soft, but retains its fullness. The breathing becomes nearly natural. The skin becomes cool, in part no doubt by the evaporation from the surface of the body. The urine, though high colored, generally deposits a pale reddish sediment. And thus, by degrees, the sweating stage passes by. Thus, by the return of the heart and arteries to about their natural degree of activity, the free action of the skin and other secretions, and a degree of natural action of the brain and nervous system, there is brought about a state of perfect apyrexia, or intermission.

During the intermission there is generally, in simple intermittents, no fever, but rather a condition of languor, the system appearing weaker than natural, and easily prostrated or fatigued. There is usually an unnatural sensitiveness to cold, and the countenance is pale, but the appetite is usually quite good. In simple intermittents, the intermissions are nearly perfect, and the various functions are performed with a degree of regularity. But there evidently exists a degree of debility of the cerebro-spinal system.

Complications.—But I have said that simple intermittent fever is quite rare, for we find, if there is no complication at first, they often soon occur, and very much modify the symptoms of every stage, but especially of the intermission. We may have, therefore, the simple, the inflammatory, the congestive, the gastric, and malignant. And as I have already given the general course and symptoms of the simple variety, I will now consider the varieties or complications.

The Inflammatory.—If, as frequently happens, a local inflammation is set up, in some tissue or organ, during the forming stage, or at the first paroxysm of fever, that local inflammation very essentially modifies the symptoms of the different stages, and also the course of the fever; hence they may be called *inflammatory* intermittents. As inflammatory affections occur more frequently in winter and spring, this variety is more frequently met with at this season of the year, and on that account.

Now, as simple intermittents, as we have seen, are carried through all their stages, with only derangement of the blood and cerebro-spinal and nervous system, it is reasonable to suppose that any local inflammation would essentially modify its symptoms, and such we find to be the case. For, in inflammatory intermittents, though the cold stage is about as in the simple, yet the hot stage is generally more or less lengthened, the heat of the surface becoming very intense, and the pulse strong, hard, and full.

The intermissions, as might be expected, are not perfect. The pulse usually remains quick, tense, and accelerated. The thirst is not entirely gone, and the heat of the skin often remains higher than natural. A slight headache is frequently complained of, and sometimes wandering pains in the back and limbs, as well as a short, dry cough. Such are the symptoms arising in ague, with inflammatory complications, varying, however, with the seat and extent of the local inflammation.

The Congestive.—Congestive ague generally occurs in weak and very debilitated persons, in whom the powers of life are incompetent to restore the circulation perfectly, in consequence of which the cold stage is very much lengthened, and is attended with great oppression of the lungs, vertigo, fainting, and sometimes coma. The hot stage comes on slowly, as the heart and arteries appear incompetent to bring up the circulation to hardly its natural standard. The countenance remains pale and sunken. The skin is only moderately warm. The breathing remains somewhat oppressed, and the pulse frequent, small and tense.

The Gastric.—When the miasm which operates through the blood, on the brain and nervous system, is slow in developing fever, and therefore deranges the digestive apparatus, or the stomach, liver and bowels, as is frequently the case in the autumnal intermittents; we have a gastric complication which very materially modifies the course, and symptoms of the fever, and this may be called the gastric variety of intermittents.

In the gastric variety of intermittents, there is a foul tongue, bitter taste in the mouth, pain in the forehead, an icteric hue of the skin and eyes, diarrhœa, the urine loaded with bile, a desire for acid drinks, and great irritability of the nervous system.

The malignant.—Malignant intermittents, are such as occur in hot climates; a familiar examination of which, is the " Chagres fever," which attacked so many Americans a few years since, on the Isthmus, on their way to, and from California. Their malignancy is probably in consequence of the extreme concentration of the miasmata, together with the debilitating effects of excessive solar heat, on such as are unacclimated.

Their peculiarity consists in the excessive nervous prostration and decidedly typhus symptoms which occur. Hemorrhages are very liable to occur from various parts of the body, and such cases often have a rapidly fatal tendency.

It is quite probable, that *idio-miasmata,* combine with the paludal poison, to render such cases malignant, as there is evidence of more or less of a deranged or decomposed condition of the blood.

Irregularities.—Such, is the general course of intermittent fever; and also, the peculiarities of its various complications. But in some cases we have instead of the cold stage, a general numbness, without much coldness. In others, still there will be a diarrhœa, instead of the sweating stage.

In infants, it is not uncommon for convulsions to occur, at the beginning of the cold or hot stage, especially if the child has a large head, or is in any way predisposed to convulsions.

Effects of Ague.—The effects of chills are always more or less pernicious, and ague may produce in the system a variety of diseases, not of an inflammatory character. Apoplexy sometimes results from the undue pressure of blood on the brain, during the cold stage of intermittents. Aneurism of the large arteries, are sometimes no doubt produced during the cold stage of congestive intermittents. In one case that fell under my observation, I have no doubt but that there was an enlargement of the arteries of the brain, produced by congestive intermittent fever, which produced a fatal insanity. Paralysis, neuralgia, and congestion, or other derangement of the spleen, are sometimes the result of intermittent fever.

Intermittent fever, like many other disease, tends, after a longer or shorter time, to a spontaneous termination. I believe however, that the same rule applies to ague, in that respect, as to any other similar affection. If the general circumstances by which the patient is surrounded, are such, as to render the system more feeble, and especially the nervous system; the chills generally anticipate a little, and the fever becomes more remitting in its tendency, and if sufficient complications arise, the fever may become even continued. On the other hand, if the circumstances by which the patient is surrounded, are such as go to improve the general condition of the patient, the chills generally postpone a little; the increasing strength, enabling the system to keep up a healthy action a little longer each day, till finally, the chills cease altogether, and health is restored.

Prognosis.—In simple intermittents, the prognosis is generally favorable. Death may occur, however, from cerebral or pulmonary apoplexy, a fatal case of the latter having occurred in a feeble patient of mine, the present season, during the first chill. In postponing agues, or those in which the chills come on later each time, the prognosis is of course, more favorable than in anticipating, or those in which the chills come on earlier; as this indicates either an improvement or a decline in the general condition of the patient.

I have generally noticed eruptions about the mouth in cases that are assuming a favorable character; at least, I have noticed that this seldom occurs in cases which are becoming more complicated.

In the inflammatory complication the danger is very much in proportion to the parts involved in the inflammation, and its degree of violence. And I am satisfied that these local inflammations, when they exist, are the cause of the irregularity of ague. A diarrhœa occurring and continuing for any great length of time indicates either a congested or inflamed condition of the alimentary mucous membrane, which is always an unpleasant complication.

In the congestive and malignant varieties, delirium, coma, œdema of the feet and legs, and passive hæmorrhages, are always unfavorable symptoms, as well as great prostration during the intermissions. As a rule, however, the prognosis in ague is favorable in temperate climates, the dangerous complications being an exception to the rule. The only fatal case of ague that has occurred in my practice was from apoplexy of the lungs, and this occurred during the first chill.

Causes.—The general cause of intermittent fever is the paludal poison,

7

or koino-miasmata, operating, as we have seen, through the blood upon the brain and nervous system.

The varieties or complications, as we have also seen, are the result of accidental causes, occurring during the forming stage, or after the fever is developed. It is probable, too, that hereditary predisposition favors the various complications which sometimes arise. Thus, the irregularity of the inflammatory variety is caused by local inflammation in some tissue or organ. The cause of the peculiarity in the congestive is hereditary or accidental debility, together with concentration of the febrific agent. The cause of the gastric complication, as I have already hinted, is owing, in part, to the slowness with which the paludal poison sometimes acts to produce its effects on the brain and nervous system, thus giving ample time for gastric and biliary derangement to take place. The cause of the malignant variety of intermittents, as we have already seen, is the combined effects of koino-miasmata and excessive solar heat, and probably, also, the additional cause, in some cases, of idio-miasmata partially changing or decomposing the blood, and thus lessening vital energy.

The length of time the paludal poison may be operating upon the system before it develops fever is exceedingly various. Sometimes the fever occurs in a few days or hours after exposure; but, in other cases, not till several weeks or months, depending, no doubt, much upon the predisposition, and also the concentration of the miasmatic agent.

Gastro-enteritis will produce symptoms similar to ague, even when there has been no exposure to koino-miasmata, very many cases of which have fallen under my observation during the past few years, and one in this village during the past few days. But great caution is necessary in discriminating between such cases and genuine ague produced by the paludal poison.

Proximate cause.—I have no doubt, as I have before intimated, but what the proximate cause of intermittent fever is debility of the brain and whole nervous system, by which the vital energy is, for the time, prostrated, producing the cold stage, while each succeeding stage is the direct result of the one which precedes it. And it appears to me that the regularity of the paroxysms furnishes nothing so very strange, for they only show, as I believe, the length of time the brain and nervous system can carry on or keep up the functions of the system, uninterrupted by a sinking, which sinking produces a chill. Thus, in a quotidian, the brain and nervous system have only strength to carry on, uninterruptedly, the functions of the system, without a sinking, for twenty-four hours, while in tertian they can do it for forty-eight hours, and so on. Hence, while the strength of the system remains stationary, the paroxysms occur at regular periods, and neither anticipate or postpone. But let the strength of the system, by some complication, become lowered, and the paroxysms will anticipate, thus showing that the brain and nervous system is able to keep up the functions uninterrupted for a less time. On the other hand, let the general condition of the system be improving by the subsidence of some general or local debilitating cause, and the paroxysms will postpone, thus showing that the general strength is enabling the brain and nervous system to carry on, uninterruptedly, the functions of the body for a longer time.

Treatment.—The treatment of ague may be considered under two heads: that which is proper during the paroxysms, and that which is proper during the intermissions.

In the simple, or ordinary intermittents, of an uncomplicated character, little or no treatment is generally necessary during the paroxysms. But in the congestive and malignant intermittents, in which the vital energy is deficient, it is well to keep the patient secured from cold air. And sometimes it may be well to give a mild stimulant, to aid the vital powers in preventing fatal congestions; which are liable to occur. If this is not done, the cold stage is very much prolonged; and the reaction, if it does come on, is weak and inefficient, and the subsequent stage also not fairly developed.

In the gastric variety, in which the digestive organs are very much involved, with irritation of the alimentary mucous membrane and acidity of the stomach, if vomiting occurs during the cold or hot stage, sinnapisms should be applied over the stomach, and a little camphor and prepared chalk administered in solution. This may be prepared by rubbing pulv. camphor, one scruple, with prepared chalk, ten grains, and then adding an ounce of water. Of this a teaspoonful may be given every fifteen minutes till the vomiting ceases. If this, however, should fail, gtt. xv of laudanum should be given, which, together with the camphor mixture, will seldom fail of arresting the vomiting in such cases.

In all cases in which the cold stage is very much protracted, and a special tendency to the brain prevails—such as severe congestion—and there is, consequently, imperfect reaction, the head should be elevated, and the feet placed in warm water, and warm, stimulating applications rubbed along the whole length of the spine. That which I prefer, as it is efficient and always at hand, is a decoction of capsicum in vinegar. Two or three red peppers may be broken into half a pint of vinegar, and then, being covered, it should be steeped for a few minutes, till it becomes quite strong, after which it may be applied a little warm, with a soft flannel cloth, along the whole length of the back.

Those who have never used this means of producing, or hastening reaction, in such cases, will be greatly surprised at its effects. It is also exceedingly convenient, as it can be used when the stomach would not retain the least thing that could operate in this way. It probably acts, through the spinal nerves, upon the ganglionic nerves, in consequence of which, they call into renewed activity the languid heart and arteries. It probably also, by stimulating the spinal cord, lessens the determination to the brain, and thus promotes speedy reaction.

Cold drinks may be used during the hot and cold stages of ague, if there is nothing to contra-indicate them; but they should be used with a degree of caution. The drinks during the sweating stage should be warm, and of a mild, unirritating character. Warm toast water, or sage tea, is what I have generally found to agree best. These are the general principles which should guide us in the treatment during the paroxysms of intermittent fever; always, however, waiting for a clear indication before resorting to any of these remedial agents. We now come to consider the treatment proper during the intermission.

Quinine is the remedy for the cure of simple, uncomplicated intermittents, and is so, because it is a tonic which operates probably directly upon the brain and nervous system, giving power and energy to keep up the vigor and action of the various functions of the body. It probably also destroys the paludal poison in the blood, which has produced the disease, and thus renders the cure permanent.

The latter part of the intermission is the time when the quinine is indicated. For then it is that the energy of the system begins to decline, and the ability of the brain and nervous system to carry on the functions of the body begins to be apparent. It is generally best, I think, to commence with the quinine about twelve hours before the chill is expected, and give two grains in solution every three hours, till four doses are taken, and then a dose each hour, till two more are given, which makes twelve grains; the last dose, in that case, would come one hour before the time for the chill. This treatment should be continued till the chills are arrested, which will generally happen by the second or third paroxysm, and sometimes at the first; no more chills occurring.

After the chills are arrested, the quinine should be continued in the same way, on the days the chills would have occurred, dropping the first dose each day till only one dose is given, and that on the hour next preceding the time the chill would have occurred. After continuing the quinine in this way till only one dose is given, it may be omitted, except to give one grain after each meal for seven or eight days.

Or instead of the quinine in grain doses after each meal, the *fluid extract of bark* may be given, in half dram doses, or one-third, if the quinine appears in any way to disagree. By continuing the quinine, or bark, in this way, the tone of the brain and nervous system is sustained, and in my opinion the miasmata in the blood neutralized; and thus the cure rendered permanent, in most cases at least.

I have suggested the treatment for pure ague; but as there are few cases in which the miasmata in the blood has not deranged more or less the gastric and hepatic functions, I think it is generally best to give one dose of blue pills, and follow them in six or eight hours by a half ounce of castor oil, or sulphate of magnesia. In the agues occurring in this vicinity, during the three preceding years, I have given three blue pills at evening, and followed in the morning by the oil, or salts, and then given the quinine, in solution, as above suggested.

I usually take quinine gr. xxx, acid sulphuric aromatic ℨi and add water ℥viii. This makes a convenient solution of the quinine; one table-spoonful being a dose. With this amount, I have generally succeeded in arresting the chills and preventing their return, in the way I have suggested, and then had enough left to give in teaspoonful doses after each meal for several days. But in a few cases, I have had to use more than the thirty grains. I am satisfied from careful observation, that the quinine in solution, is much more certain in its effects, and generally borne better by irritable stomach. I think therefore, that in ague at least it should generally be given in solution, as I have suggested.

In cases of congestive and malignant intermittents, the quinine should I think, be commenced with, in the early part of the intermissions, and continued as I have suggested, to a period, one hour before the chill is expected, and, after the chills are arrested, it should be continued in the same way, as in simple ague, only, the doses given after each meal, should be at first one and a half or two grains, and later at last one grain as first suggested.

In inflammatory intermittents, or those in which some local inflammation exists; the inflammation should first be subdued by bleeding, cupping and blistering if necessary. Blue pills, followed by full doses of the sulphate of magnesia, or a full dose of calomel, in castor oil, may be of very essential service in such cases. Gentle diaphoretics, and the free use of warm sage tea should be allowed in most cases. And, when the local inflammation is subdued, the quinine should be administered, and the treatment continued, as in simple intermittents, guarding of course, the irritated organs or parts.

In the gastric variety, or in cases in which the digestive organs are materially involved, blue pills should always be given, and followed by castor oil; or in children, rhubarb with hydg. cum creta in castor oil, will often be best. When the gastric complications or derangement is subdued, the quinine should be administered, as in simple uncomplicated cases, taking care always to prevent if possible, a return of the gastric derangement, by a well regulated diet and every means in our power.

In those cases of gastro-enteritis in children, which develop symptoms very similar to ague; cupping and blistering over the stomach will generally arrest the chills. It is hardly necessary to say, that in such cases, quinine should not be given.

Such, according to my experience is the best method of fulfilling the indications, which arise in intermittent fever. But, as quinine is not always at hand when ague occurs, at least in some places, it is well to remember, that any nervous tonic rightly administered, may arrest ague; the most convenient of which, according to my observation, are the following:

The bark of the willow, (salix alba), or its active principle, salicin is a valuable substitute for quinine or bark, for the cure of intermittent fever. The bark may be used in substance, or decoction, in the same doses, and prepared in the same way, as the Peruvian bark or cinchona. The dose of the salicin is about four grains, to be given during the intermission, as I have directed for the quinine, it generally requiring about twenty-five grains, between the paroxysms of intermittents. The willow or its active principle the salicin, will often do well, where from some peculiarity, the bark or quinine may not agree. It may also be a convenience when the quinine is not at hand.

The bark of the dogwood, (Cornus Florida), is a very convenient substitute for quinine, in the treatment of intermittents. The bark of the dogwood may be given in substance, in dram doses, at intervals between the paroxysms.—Or, the decoction made by boiling for ten minutes, an ounce of the bark in a pint of water. The dose of this, is one or two fluid ounces, to be administered during the intervals of ague, the same as the quinine or Peruvian bark.

But when the Cornus Florida is to be used, as a tonic, especially in ague, the fluid or solid extract is the most convenient, and should generally be used when it can be obtained, or, what in some cases would be preferable still, is the *cornin*, its most concentrated active principle. Of the fluid extract, the dose is from one to two drams. Of the solid extract from five to ten grains, while the dose of the cornin is only about four grains. In either of these forms the active principle of the dogwood may be very conveniently administered.

The *iron-wood*, is another indigenous remedy, which I do not remember to have seen noticed; but which I am satisfied is little, if at all inferior to the Peruvian bark, as a remedy in intermittent fever. A strong decoction of the bark, or what is better, of the wood, taken at intervals, between the paroxysms of ague, I have known to speedily arrest the chills, and permanently cure the disease. Of a decoction made by boiling four ounces of the rasped wood in a pint and a half of water, down to a pint, doses of from one to two ounces may be administered, during the intermissions of ague, in the manner I have suggested for the quinine or Peruvian bark.

Other indigenous plants or vegetables have more or less efficacy in arresting ague, among which are the *hop*, (Humulus Lupulus), the sage, (Salva Officinalis,) the plantain, (Plantago major,) capsicum and various others of more or less value, which I need not mention. There are also various mineral tonics, which are entitled to more or less consideration, as remedies in the treatment of ague : among which are *arsenic, iron, zinc*, &c.. In obstinate cases of ague, in which the chill comes on very irregularly ; Fowler's solution of arsenic, in ten drop doses three times per day, becomes a very valuable remedial agent. Care should be taken however that the stomach is in a proper condition to bear it; and also, that the remedy be not continued too long.

Of the preparations of *iron*, the *carbonate*, and the *prussiate* according to my observation are the most valuable preparations of this mineral. Of the *prussiate*, from three to six grains may be administered at a dose, and continued during the intermission of intermittents, the same as quinine. The carbonate when used in ague, should be given in doses varying from ten to sixty grains, according to the nature and urgency of the case. In protracted cases, of irregular agues, in which there is great poverty of the blood, the preparations of *iron* become valuable remedial agents.

Of the preparations of *zinc*, the *sulphate* and *oxide*, are the most convenient and reliable, in the treatment of ague. Of the sulphate, two grains may be given three times per day, in protracted cases, where other remedies are contra-indicated on account of some peculiarity of the system, or from a peculiarly nervous condition of the patient. The *oxide* of *zinc* may be given in ague, in cases of extreme nervous excitability, in which other remedies have failed, and the chills have become very irregular. From five to ten grains three times per day may be given with good effect in such cases. I need hardly say, in conclusion, that a clear indication should always be had, before any of these remedies should be prescribed in intermittents.

SECTION II.—BILIOUS REMITTENT FEVER.

Remittent fever is the result of the combined influence of koino-miasmata, and a local irritation or inflammation in some tissue or organ, and generally of gastro-enteritis.

The *symptoms* of the forming stage of mild bilious remittent fever, are very similar to the symptoms of the early stage of intermittent fever. A loss of appetite, bitter taste in the mouth, foul breath, scanty yellowish urine, constipation of the bowels, or diarrhœa, drowsiness, headache, pain in the back, restlessness, and general prostration of the system, are among the symptoms which are developed, during the forming stage of bilious remittent fevers.

After these symptoms have continued for a few days, gradually increasing with the general debility, a slight chill usually occurs, in which though there may be coldness of the extremities, and chilliness along the spine, the general heat of the body appears nearly natural, or accellerated, especially in the irritated or inflamed parts. After this imperfect chill, during which there is evident congestion of the liver, or alimentary mucous membrane, or both, and generally an acceleration of the local irritation of the stomach and intestines, reaction follows, with a good deal of febrile excitement.

As the fever becomes established, the pain in the head, back and limbs, is considerably increased, and sometimes becomes very severe and tedious. The countenance and especially the eyes, assume more or less of a yellowish tinge. The tongue becomes covered with a brownish or yellowish fur, nausea, and generally bilious vomiting occur, during the first twenty-four hours of the febrile excitement. A sense of weight is usually felt in the right hypochondrium, and epigastric regions, and sometimes a dull heavy pain.

The bowels become slightly distended, and are more or less tender to the touch. And the respiration is more or less oppressed and irregular. The *urine* becomes scanty and slightly tinged with bile. The pulse is full and generally frequent, but not very hard and tense. The skin is always dry and hot. These symptoms usually continue till the following morning, when a slight perspiration appears on the superior portions of the body, and sometimes over the whole surface.

The febrile excitement, very considerably abates, but not so as to form a state of opyrexia; the skin remaining preternaturally warm, and the pulse irritable, and frequent. This remission continues from two to four hours; when the febrile excitement rises, with increased activity, and continues for a time, usually twenty-four hours, when another remission occurs. In this way the fever continues with regular revolutions, of exacerbations and remissions, until it either terminates in a crisis, and convalescence, or else assumes a more uniform or continued character.

This is the ordinary course of a bilious remittent fever, in which the miasmatic agent has operated slowly, in producing a debility of the brain and nervous system; and at the same time, has brought about a derangement of the digestive organs, and more or less irritation or inflammation of the alimentary mucous membrane, or some other tissue or organ. This local irritation or inflammation, preventing a perfect intermission, which the paludal poison would otherwise have produced.

Type.—Fevers of this character, usually assume either the double tertian, or quotidian type, and generally the double tertian. For though remissions occur every day, yet they are generally more marked on alternate days, at least such has been the result of my observation. The exacerbations of a remittent of the quotidian type, usually occur, at nine or ten o'clock in the forenoon; while those of the double tertian, occur generally an hour or two later, but this is not invariably the case.

Though as we have seen the remissions usually occur in the morning, yet this is not always the case, for I have known them to occur sometimes during the night; or they may occur at any hour. And in some cases, in which the local inflammation is very marked, the remissions are scarcely perceptible at any hour, the fever being continued in consequence of the leading character of the local inflammation.

Remittents, though mild at the commencement, may assume an aggravated character; if they continue over nine or ten days. The tongue becomes more loaded with a brown fur, and is dry in the middle: the skin assumes a deeper tinge of yellow; debility becomes more conspicuous; the bowels are distended, and tender to external pressure, and frequently restlessness and almost constant delirium occurs.

In localities where there exists, a good deal of humidity in the atmosphere, together with the influence of very warm days and cool nights, remittent fever sometimes assumes a more aggravated character. The cold stage, in such cases is short, but quite marked; the heat during the excitement intense, the thirst urgent, violent pains occurring in the back, and frequently vomiting. The remission is very marked, but short; the next paroxysm is generally more violent than the preceding; the eyes become yellow, nausea and obstinate vomiting occurs, together with a great oppression, and anxiety in the epigastrium. During the second or third remission, a clammy perspiration appears on the surface of the body, and in this way the paroxysms continue to recur, until either a salutary crisis occurs, or death takes place.

If the fever continues beyond the tenth or twelfth day, there is very great prostration, and the fever becomes more continued. The skin too sometimes acquires that stinging heat, called "calor mordax," or else becomes cool and cadaverous to the touch. In this aggravated and protracted state of the fever, the lips become swelled, the tongue brown, or black, the eyes red, the urine dark brown, or entirely suppressed, the alvine evacuations reddish and watery, and generally there is a tympanitic state of the abdomen, and hæmorrhages occasionally occur in the last stage of the disease.

Between the mild and malignant varieties of bilious remittent fevers, there is a vast variety of grades, depending upon the degree of concentration of the miasmatic agent; the violence of the exciting cause, and the degree, nature, and extent of the local inflammation ; in fact no two cases develop precisely the same symptoms.

Now, the general cause operating upon the brain and nervous system, to produce this fever, is *koino-miasmata;* but that alone would have produced as we have seen, intermittent fever. The local inflammation, as I have before stated, so changes the general condition, as to prevent the fever from assuming an intermittent character and renders it remittent ; while the organs involved, and the degree and extent of the inflammation, accounts for the almost endless varieties of grade, or character of this bilious remittent fever.

The parts most frequently involved in inflammation in this form of fever, are the brain, liver, and alimentary mucous membrane. The brain may be the primary seat of the local inflammation, but I am satisfied that generally in the cases in which cerebral inflammation is developed, there is a decided gastric derangement first, or more or less gastro-enteritis preceding it.

The complications then, which occur, in the forming stage of the fever are usually congestion, with irritation or inflammation of the liver, or of the alimentary mucous membrane, and of the two gastro-enteritis is by far the most frequent in temperate climates at least.

Gastric.—Now, in the gastric variety, the nature of the local difficulty is congestion, irritation, or inflammation of the mucous membrane; generally brought about by the combined influence of the paludal poison acting upon the brain and nervous system, deranging digestion; and also the effects of cold operating to check the exhalation from the skin, by which the mucous membrane of the stomach and intestines becomes congested.

The liver too, becomes slightly congested, in consequence of which, its secretion becomes more or less acrid, and this being poured into the intestines, serves no doubt, to irritate and inflame the congested mucous membrane. Gastro-enteritis is thus set up, which producing sympathetic irritation of the brain, serves in a still greater degree, to produce derangement of the various organs, and especially of the liver, stomach, and intestines.

Indigestion, is therefore, among the early symptoms of gastric remittent fever. There is early, a bitter taste in the mouth, the tongue being covered with a yellow mucous fur, which sometimes becomes dry and brown. The appetite is impaired very early, and finally becomes entirely destroyed, as severe vomiting often occurs. The urine is scanty, and tinged with bile. The bowels are tender and distended, and severe pain in the back and head, adds to the already suffering condition of the patient. The tongue becomes more red; the alvine evacuations watery and reddish; and there is also, generally, a desire for cool acidulated drinks. In this way is the gastric variety of bilious remittent fever developed, and its symptoms produced.

Hepatic.—The hepatic complication, in bilious remittent fever, is characterized by intense heat, violent pain in the head, and early delirium, fullness of the right hypochondrium, a clean tongue, and forcible vomiting, of a glairy fluid without bile.

The bowels are confined, during the early stages; but later there is a discharge of dark bile from the bowels; the skin and eyes become yellow, all of which, indicate a deranged condition of the liver.

In the early stages of such cases, there is evidently little or no bile secreted, and later, that which is thrown into the intestines is of an unnatural irritating character. The full hypochondrium indicates congestion, while the yellow skin and excessive vomiting, without the presence of bile, goes to prove that little or no bile is poured into the intestines, at least during the early stages. The dark bilious matter, which passes off later in the disease, goes to prove that some derangement has existed in the functions of the liver, at least.

While there is doubtless considerable congestion, irritation, or functional derangement of the liver, in these cases, I am satisfied that gastro-enteritis exists to some extent, from the very first; having become seated generally, during the forming stage of the fever, or at the first paroxysm. During a practice of ten years, in a locality where bilious remittent fevers, especially prevailed, I do not remember to have treated a case, however great the hepatic derangement, in which there was not, during the course of the fever, evidence of more or less gastro-enteritis. And gastro-enteritis has appeared to me to be developed to some extent during the forming stage, or at least, with the first paroxysm of fever. And, I am also confident, that gastro-enteritis in these cases, is the reason why the fever does not intermit, but assume a remittent character. The local inflammation adding the symptoms of symptomatic, to the paludal fever, and thus rendering the fever remittent.

It has also appeared to me, that the irritation of the brain, when that was a leading symptom, was generally the result of sympathy with the irritated alimentary mucous membrane. And finally, that in those cases in which there was hepatic derangement, that it was produced by a debility of the brain, which deranges its functions; and that the debility or derangement of the brain is generally in part, the effect of gastro-enteritis, and also, in part the effect of the febrific agent.

I have been led to these conclusions by careful observation in every possible variety of bilious remittent fever common to our climate. I was first led to this conviction by noticing the effects of the minutest doses of spirits or any kind of irritating stimulant such patients had taken. And I have been confirmed in the belief by the fact, that in the worst cases of cephalic or hepatic complications, I have usually got sudden and permanent relief by counter-irritants over the stomach and bowels.

I have sometimes in this way changed bilious remittents to intermittents, and in other cases have arrested them altogether, especially if at the time there was not a concentrated miasmatic influence prevailing. Another, and unmistakable evidence, of the universality of gastro-enteritis, in all severe cases of bilious remittent fever, is the fact that in nearly all fatal cases, (that have fallen under my observation at least,) there occurs a diarrhœa of a character indicating gastro-enteritis. The absence of pain is no evidence that gastro-enteritis does not exist. And even absence of abdominal tenderness is no positive evidence. For an inflammation confined entirely to the alimentary mucous membrane, is not necessarily attended with either.

Causes.—The causes of bilious remittent fever as I have already suggested, are koino-miasmata and some local irritation or inflammation, and very generally of the alimentary mucous membrane. Every symptom developed in these fevers are just what we might expect, if such were the case. The paludal poison acting slowly upon the brain and nervous system gradually dibilitates, and acting through the nerves derange the various organs by impairing their functions.

The liver becomes congested, and either too much acrid bile is thrown into the intestines, which serves as an irritant to the alimentary mucous membrane; or if the congestion of the liver be very great, little or no bile is secreted, and hence the food which is taken, passes undigested through the alimentary canal and becomes a source of irritation to its mucous membrane. In either case the digestion becomes impaired, too much bile producing a diarrhœa, and too little or no bile producing constipation, and both acting as direct exciting causes of gastro-enteritis.

If we take into account, too, the fact, that the general cause operates to derange the functions of the skin, and that the secretion of the liver is very much as the action of the skin, we need not be at a loss to account for the bilious derangement that occurs, even when little or no congestion of the liver exists.

It should be remembered, also, that such fevers occur most frequently at a season of the year when the warm days and cool nights, by checking the perspiration, tend to produce congestion, irritation, and inflammation of the mucous membrane of the stomach and intestines.

Another fact should also be remembered, that intestinal worms, or gastro-enteritis from any cause, will sometimes produce remittent fever without the aid of a miasmatic agent. Now all these facts, together with the symptoms which are developed, tend to prove that gastro-enteritis, together with the influence of the paludal poison, is the cause of bilious remittent fever. And this position is further strengthened by the fact, that gastro-enteritis, occurring in intermittent fever, will render the fever remittent. And, also, that removing gastro-enteritis in remittent fever will frequently change them to intermittents.

In malignant cases, it is likely that solar heat, and, perhaps, idiomiasmata, act, in addition to the above causes, in developing the fever and in increasing its malignancy.

Treatment.—The indications in the treatment of bilious remittent fever are very plain, when we take into account the morbid condition or true pathology of the disease. The indications are plainly to equalize the circulation, and prevent local congestions, in order that the functions of the skin and various organs may be called into healthy activity; to remove from the alimentary canal any acrid secretions or irritating substances, which may be operating to produce or increase irritation or inflammation; to counteract local irritation or inflammation, and especially of the alimentary mucous membrane; and, finally, to counteract the febrific agent, and to restore the tone of the brain and nervous system —in all cases, of course, watching with due caution any complications that may arise.

If a patient is seen early in bilious remittent fever, during the forming stage, or soon after the chill, and febrile reaction has occurred, the fever should generally be arrested at once, or in two or three days; at least, such has been my experience during the past few years. At such a stage, the patient should have explained to him his real condition, and should be encouraged to believe, that, with proper care and a little judicious treatment, he may be restored to health without going through a course of fever, always, however, stating that it is not quite certain. A little encouragement of that kind acts like a charm, and sometimes may do considerable towards bringing about that very desirable result.

At such a time, there is a dry skin, a determination of blood to the brain, and general derangement of the whole system, and especially of the circulation. To equalize the circulation, promote perspiration, and relieve the chilliness and heavy pain in the lower portion of the spine, the feet should be placed in warm water, warm sage tea allowed, and, if there is chilliness, the whole length of the back should be rubbed with a warm decoction of capsicum in vinegar. The foot-bath may be continued for two or three evenings, as well as the warm application to the back, if there is chilliness, and it is grateful to the patient. If the patient is a strong man, ten grains of calomel, with twenty grains of rhubarb, may be mixed with half an ounce of castor oil, and given at once, and the oil repeated, every six hours, till it operates.

But if the patient be a female, or a man of slender constitution, or a young person and rather delicate, two or three blue pills should be given, and followed in six hours by half an ounce of castor oil, and this may be repeated every six hours, till it operates. But if the patient be of delicate constitution, or if, from any cause, a mercureal is contra-indicated, the next best cathartic in such cases, is some preparation of the Podophyllum Peltatum (mandrake), of this, one dram of the fluid extract, five grains of the solid extract, or what may be better still, one grain of the podophyllin may be given, and repeated in six hours if necessary. For young children, I generally give either the leptandrin,* or else hydg. cum creta, with a little rhubarb in castor oil, and repeat if necessary.

Sinapisms should be applied over the stomach and bowels, and made to produce a smart irritation, and repeated at evening, for two or three days, if necessary. A mild, nutritious, and digestible diet should be enjoined, and every source of irritation, both of body and mind, should be removed, as far as possible. In this way, bilious remittent fever may generally be arrested, if taken in the forming stage, or early after febrile reaction is established.

If the fever has run on for a day or two, or if some irritating drugs have been taken, we generally have more local irritation or inflammation, and frequently severe vomiting. In that case, the warm foot-bath and stimulating friction along the spine, with sinapisms over the stomach and bowels, will do much to equalize the circulation, promote perspiration, and allay gastric irritation. As soon as the vomiting is thus arrested, a cathartic should be administered. If there is a strong tendency to the brain, and the patient be strong and robust, ten grains of calomel, with twenty grains of rhubarb, may be mixed with half an ounce of castor oil, and administered at once, and the oil repeated in six hours if necessary.

But if the patient is a male or female of slender constitution, I would give three blue pills, and follow in six hours with half an ounce of castor oil, and repeat it if necessary. Or, if the patient be of a very slender and feeble constitution, or a very young child, the Hydg. cum creta, with a little rhubarb, may be mixed and administered with oil, and the oil repeated if necessary. If, however, a mercurial from any cause is contra-indicated, I would give to an adult two grains of podophyllin, and follow with castor oil if necessary. Or, if the patient be a young child, I would give the leptandrin in oil, and repeat it if necessary, remembering that the dose of the leptandrin, for an adult, is two or three grains.

*Active principle of the Leptandra Virginica. Dose of Leptandrin for adults 2 grains; of the fluid extract one dram.

In some cases, in which there is great functional derangement of the digestive organs, without very marked symptoms of inflammation of the stomach or brain, an emetic of ipicac may precede the cathartic. But I am satisfied, from careful observation, that, in a large majority of cases, an emetic is not indicated, and, therefore, would often retard instead of hasten the cure. After getting the operation of a cathartic, I would continue warm sage tea or toast-water, with more or less milk, as tending to promote perspiration and support the declining strength.

To sustain the system, prevent local congestions, equalize the circulation, and arrest the fever, two or three grains of the sulphate of quinine should be given every six hours, and continued till the fever is arrested. During the first two or three days, three or four grains of Dover's powder may be given with the quinine, for the purpose of promoting perspiration, procuring sleep, &c. Later, and after the perspiration becomes profuse, the Dover's powder should be omitted, and one or two grains of pulv. camphor given with the quinine instead, for the purpose of quieting nervous excitability and helping sustain the system.

If the heat of the surface becomes very great, the skin may be sponged with moderately cool water towards evening, but great care should be taken that it be not done unless the skin is hot and dry. If the skin be moist, it may be sponged with tepid water if the condition of the surface requires it, and it is agreeable to the patient.

If much gastric irritation exists, a blister should be applied over the stomach, and, the skin being removed, it should be well dressed with soft wilted leaves, and left to discharge as long as it will. If there arise much irritation of the brain, not of a sympathetic character, a blister should be applied to the back of the neck, and treated in the same manner, that its full effects may be obtained.

In relation to drinks, I am satisfied that, while there is a hope of arresting the fever, warm drinks are decidedly preferable to cold. I generally allow warm crust coffee or toast-water, with nearly an equal quantity of milk, sweetened or not, as the patient may prefer, as it satisfies thirst, favors perspiration, and affords sufficient nourishment, and, is I believe, in the best possible form.

As soon as the stomach will bear it, I allow arrow-root, toast, or a poached egg, in addition to the drink, at regular meal hours. For it is on food that we live. And if it can be taken without injury to the alimentary mucous membrane, in just so far as it is appropriated it goes to sustain the powers of life, which are tottering under the debilitating effects of a poisonous febrific agent.

With this plan of treatment, modified, of course, to fulfill the indications which arise in each particular case, I believe that most bilious remittent fevers may be arrested in five or six days, or, if they are not arrested, will become intermittent, and so require only quinine, given, as I have suggested in simple intermittents, to perfect a cure.

It may appear strange to some that quinine should be given, in two or three grain doses, every six hours when there is a high state of febrile excitement; but I am persuaded, from extensive and careful observation, that its effects are most salutary. And when we take into account the real condition of the system, in such cases, it is rational to suppose that its effects would be salutary.

For we must remember that the system is laboring under a poisonous agent in the blood which has debilitated, and would have produced an ague, except for the local irritation, congestion, or inflammation, which has been set up in some tissue or organ, in consequence of this debility. Now, the quinine, by sustaining the sinking powers of the system, not only lessens the liability to, but also absolutely relieves, by equalizing the circulation, local congestions, irritations, and even inflammations; and just as we should suppose, by arresting the debility of the system which led to them.

If, as will generally happen, the fever be arrested in this way, the quinine should be continued, in diminished doses, during convalescence, with such food, taken at meal hours, as the stomach will bear. And as the stomach begins to bear a reasonable amount of food, I generally reduce the quantity of quinine to one grain after each meal. Or if the quinine is from any cause contra-indicated, I substitute the fluid extract of bark, in doses varying from one-third of a dram to a dram. I have treated, during the past few years, hundreds of cases of bilious remittent fever in this way, with the most satisfactory results; generally arresting the fever in five or six days entirely, or else rendering it intermittent, and then completing a cure with quinine, in the way I have suggested for simple intermittent fever. In some cases, however, if neglected early, or, what is worse, badly treated in the early stages, blisters are required, not only over the stomach but also over the bowels. If this is not done, in such cases, the system may sink into a typhoid state, and be protracted for two or three weeks, or perhaps longer.

In cases which, from neglect or bad treatment, run on to this typhoid state, quinine, with camphor blisters over the stomach and bowels, and a sustaining liquid, mucilaginous, or perfectly digestible and unirritating diet, are clearly indicated. In such a state, two grains of quinine every six hours, either in solution, and administered with mucilage, or else in powder, with camphor, may be indicated. For food, in such cases, arrow-root, cooked in equal parts of milk and water, or else in chicken or mutton broth, does very well for a time; and gradually a poached egg, toast, &c., may be allowed during convalescence.

In the treatment of bilious remittent fevers, after the first cathartic, the bowels should be moved once each day, if necessary, by injections of equal parts of milk and water, with half an ounce of salt, lard, and molasses, administered a little warm. Or if there is not too much prostration, an occasional dose of castor-oil may be indicated. But no active cathartic should be given, generally, after the first dose, and that should always be as mild as may be, and yet fulfill the indication.

In cases in which the stools remain clay-colored after the first cathartic, or if skin and eyes remain yellow, and the bowels constipated, alterative doses of calomel, every six hours for a day or two, may be indicated. In such cases, a grain of calomel may be given with each dose of quinine, till six or eight doses are taken, and then half an ounce of castor-oil administered. Or if a mercurial be contra-indicated, one half a grain of podophyllin, or one grain of the leptandrin may be given instead.

In some rare cases, in which violent congestion, irritation or inflammation of the brain, or other vital part, exists, general or local bleeding may become necessary, and should not be omitted. But as disease is always a condition of *debility*, in one sense, general blood-letting can never be resorted to except at the expense of the general strength, and therefore should not be, except when some violent local congestion, irritation, or inflammation, demand it, to save the part involved from its effects, till such time as the warm foot-bath, friction along the spine, warm drinks, and quinine, shall equalize the circulation, and relieve the suffering part.

I am satisfied, from careful and extensive observation, that the warm foot-bath, with warm sage tea, is often more effectual, when there is time for them to be used, than general bleeding; and they have the advantage of not permanently debilitating, at least to so great a degree. And in all cases in which general bleeding might be resorted to, to prevent fatal congestion, during a chill or cold stage, I am confident that the indication may be much more rapidly and effectually fulfilled, by rubbing the back with a warm decoction of capsicum in vinegar. In local inflammation, in which an immediate relief of the affected organ or part is always indicated, the abstraction of blood, by cups, from over the part affected, or along that part of the spine supplying it with nerves, becomes one of our most valuable remedial agents.

SECTION III.—SIMPLE CONTINUED FEVER.

By simple continued fever, I mean that variety of continued fever which arises from various causes, such as atmospheric vicissitudes, electrical influences, stimulating articles of food and drink, overheating the system by violent exercise, mental excitement, &c., and finally endemical and epidemical influences. Now, as the human system is liable to be thrown into a state of febrile excitement, from these various causes, I have thought proper to consider the morbid condition under the head of Simple Continued Fever, as being the most convenient, and also saving a multitude of names, which by no means increase our knowledge.

It may be thought that the symptoms which are developed by these various causes are not identical, and that, therefore, there should be a variety of names, to suit each class of symptoms. I think, however, that the symptoms developed, by what have been called simple, catarrhal and inflammatory fevers, as a class, correspond as nearly as the symptoms of bilious remittent fevers, or most other febrile affections. I shall therefore include under this head what have been called the simple, the inflammatory, the catarrhal, &c., leaving typhoid, typhus, and yellow fevers and diphtheria, for consideration in the following sections. Hoping that this classification may be found the most convenient, I will proceed to the consideration, in the present section, of simple continued fever, without further apology.

Symptoms.—The symptoms of simple continued fever are just what might be expected, from the nature of the various exciting causes which operate to produce it. There is at first a feeling of languor, continuing for a longer or shorter time, during which the pulse becomes weak and sluggish, the respiration slower than natural, the surface and extremities become more or less cold, the appetite indifferent; there is thirst towards evening; restlessness during the night; headache; sensitiveness to cold; creeping chills along the back, with a dull, heavy pain in the lumbar region; scanty urine; dry skin, with a general feeling of indisposition to perform mental or corporeal exertion.

These symptoms may continue from a few hours to several days, according to the nature and activity of the exciting cause, after which there is more or less sinking down of the circulation, and then comes a chill. The chill may vary from only a slight feeling of chilliness, continuing but a few minutes, to a protracted state of chilliness or severe coldness; depending, of course, very much upon the degree of sinking of the circulation, and the consequent failure of oxidation or combustion in the lungs, and in the capillaries upon the surface of the body. During the chilliness, or cold stage, there is yawning, thirst, headache, restlessness, pain in the back, &c., depending very much upon the severity of the chill, as well as the nature of the exciting cause. After the chilliness, or cold stage, has continued for a time, varying from a few minutes to several hours, or even days, the heart and arteries are called into renewed activity by the irritation which the chill has produced, and by other causes, and then comes on reaction, corresponding, generally, with the severity of the chill.

The surface of the body and extremities becomes warm; the countenance flushed; the eyes red; the pulse active, full, and quick; and there is thirst, restlessness, headache, wandering or darting pains in different parts of the body; the urine becomes scanty and reddish; the skin hot and dry, and thus the febrile reaction becomes fully established, and continues for an indefinite time, varying from a few hours to several days, or even weeks. Thus we may have established simple continued fever, of every possible degree of severity, from the slightest degree of febrile reaction, following a scarcely perceptible chilliness, to the highest grade of febrile excitement, following a severe and sometimes protracted chill, or cold stage.

In the mildest form of simple continued fever depending upon some slight cause, such for instance as mental excitement, there is scarcely any sensible chilliness, the fever is slight and generally passes off in a few minutes, or hours, leaving only a slight feeling of exhaustion, from which the system soon rallies, and the various functions of the body are carried on as usual. But in more severe cases of simple continued fever, such for instance as occur from sudden atmospheric vicissitudes, the febrile reaction may be very considerable, and the fever may continue for several days, or even weeks, developing in its course various local affections, the most frequent of which are of the respiratory organs, such as catarrh, bronchitis, &c.

In the highest grades of simple continued fever, such as occur in full, strong, and robust young people, from active causes, the febrile excitement may run very high, and, unless arrested, may lead to violent inflammations of the brain, lungs, stomach, bowels, or in fact any organ or tissue of the body. And, finally, in the simple continued fever, arising from endemical and epidemical influences, we may have for a time more or less active febrile reaction, but finally a decidedly typhoid tendency of the fever, leading to various local affections of more or less severity.

We must remember, that while all these local inflammatory affections may arise, they are only complications, which are liable to occur during a course of simple continued fever; many cases of which pass on to a favorable termination, without any very special local complications.

Therefore, when these complications do arise, they do not necessarily constitute a new variety of fever, but only local inflammations, supervening upon the simple continued fever. As such, therefore, I shall consider them, and will now proceed to inquire into the causes of simple continued fever.

Causes.—The causes of simple continued fever are almost innumerable. In fact, any general cause capable of producing directly, or after a chill, a febrile condition, might be enumerated as a cause, but I will only mention those causes which are the most prominent, leaving every one to seek out the minor ones, as they may be found to operate in each individual case, as they may occur.

Mental excitement may act as a cause of simple continued fever. This we see illustrated in violent fits of anger; the prostration being marked by paleness, and perhaps slight chilliness, of very short duration; the febrile excitement being by far the most prominent, but generally passing off in a few minutes, or hours, leaving only a feeling of exhaustion, from the overaction of the system. The immediate cause of the paleness of the countenance, trembling of the limbs, coldness, &c., in such cases, is the shock to the brain and nervous system, producing a transient debility, during which the circulation and other functions of the body are more or less interrupted. This slight depression, during which the brain becomes more or less irritable, from undue pressure, is followed, as a consequence, by an excited reaction, which is more or less febrile in its character, as the shock to the cerebro-nervous system has been greater or less.

In many cases the shock to the cerebro-nervous system is followed by almost instantaneous reaction, but in other cases, depending much upon the constitution of course, the shock is followed very slowly by reaction; the brain and nervous system not calling up at once the depressed circulation. Thus it is that mental excitement produces simple continued fever.

Electricity.—If we had always obeyed the laws of health, electricity would probably ever be to us a source of health and comfort. But in our present state of physical degeneracy, it becomes, no doubt, a frequent cause of simple continued fever. Electricity tends to an equilibrium between all bodies; hence, the electricity of the body is very much as that of the atmosphere. Now a damp atmosphere is in a low electrical state, while a dry, or heated atmosphere becomes highly electric. If our systems were in a perfect state of health, and neither too sparely nor fully fed, these natural changes in the electrical state of the atmosphere would regulate the electrical state of our bodies, in a manner most conducive to health and comfort.

But a low electrical state of the atmosphere, making a draft upon the already low electrical state of such subjects as are of a slender, weak, or feeble constitution, still further debilitates; and, by thus deranging the various functions, becomes a predisposing cause of simple continued fever. On the other hand, a highly electrical state of the atmosphere, while it rather helps the weak, feeble, and debilitated subject, acts, or may act, as a direct exciting cause of simple continued fever, in subjects that are overfed, or that are addicted, from any cause, to over stimulation; and

8

thus it is that electricity becomes not only a predisposing, but an exciting cause, of simple continued fever.

Atmospheric vicissitudes, or changes from heat to cold, or from cold to heat, become not only predisposing, but exciting causes of simple continued fever, according to the manner in which the system is exposed to them. If the subject be exposed first to severe cold, and then passes into a heated atmosphere, the effect is at once debilitating; and thus, by deranging the functions of the body, the change becomes a predisposing cause of fever. But if the subject be exposed first to a heated atmosphere, and then passes suddenly into cold, the change, by checking the cutaneous exhalation, becomes an active exciting cause of simple continued fever. It becomes so, in part, by stopping evaporation from the surface of the body, and thereby retaining animal heat; and also, by retaining in the blood, the perspirable matter, which becomes an irritant, directly, to the circulatory system, and also indirectly, by affecting the cerebro-nervous system. Thus it is, that atmospheric vicissitudes, or changes occurring suddenly in the temperature, become not only predisposing, but exciting causes, of continued fever. The fevers thus produced from the merest cold, to the most troublesome influenza, have a strong tendency to produce congestion, irritation, or inflammation of the mucous membrane of the respiratory organs.

Stimulating articles of food, and drink, by throwing into the blood irritating principles, tend to develop simple continued fever. This, we see, illustrated in various articles; in some of which it is the quality; while in others it is doubtless dependent, in part at least, upon the quantity taken.

Violent exercise, by producing languor, and thus deranging the functions of the body, becomes a cause of simple continued fever. It is probable, that over exercise may not only act as a predisposing, but also, in some cases, as an exciting cause, of this variety of fever.

Endemical influences are frequent causes of simple continued fevers. It is a fact well understood, that at certain times, in certain localities, or continually in some other localities, certain influences prevail which produce continued fever. Now the exact nature of these endemical influences, we may not always know positively; yet, by careful observation, in relation to the diseases which occur in different localities, we may form an idea of the endemic tendency, whether it be exciting, debilitating, or malignant.

It is a fact, that each locality has, not only its peculiar endemical influences, which more or less modifies every disease which arises, but also, in certain seasons, a transient endemical influence may prevail, for a time, and then pass away. Now these endemical influences, whether they be vegetable, animal, aqueous, terrestrial or aerial, probably have a powerful influence in producing or developing continued fever. And, although we may not always be able to ascertain even their source, we should remember that they are probably agents, which act upon the system through the blood, affecting either directly the extreme nerves of the circulatory system, and thereby exciting fever; or else, by a reflex action, affecting the circulation through the irritation they set up in the brain and nervous system.

In some localities, this influence appears to be directly stimulant, or irritating; and then it is, no doubt, an exciting cause of fever. In other localities, this influence appears to be debilitating in its tendency, and then it becomes a predisposing cause of fever. While in other localities, still it appears to be decidedly malignant, and therefore may act both as predisposing and exciting causes of continued fever. Transient endemical influences depend, doubtless, upon some transient cause; which, as it vanishes, or spends its force, its influence also passes away.

These endemical causes of disease should not only be thoroughly understood, but their peculiar character should always be borne in mind, in arriving at the indications, in every possible variety of febrile; and, in fact, all other affections. Hence, we see, that endemical influences, though we may not always know their exact nature, are frequent causes of simple continued fever.

Epidemic influences are always prevailing, and either as debilitating, irritating, or malignant agents, are producing, modifying, or materially controlling, not only febrile, but also every possible variety of disease. That epidemic influences depend, in part, upon aqueous, terrestrial, and aerial causes, there can be no reasonable doubt; and yet, the same is true of them, whatever they may be, that is true of electricity; and, in fact, all the natural causes of febrile affections, that they are mainly rendered noxious to us, in consequence of physical degeneracy, brought about by disobedience of the laws of health. It is probable, that filth, gluttony, tobacco, drunkenness, licentiousness, and their kindred vices, have very much to do in producing or modifying epidemical and endemical influences; and also in rendering a community susceptible to these influences, when they are once generated.

Hence it is, that epidemics, sweeping round the globe, become purifiers; not only removing, to a great extent, its cause, but also most of its fit subjects, leaving the better disposed of mankind to struggle on for a time; till the inhabitants of the earth, again, by multiplied imprudence and degradation, become not only the generators of, but also fit subjects for, another mighty epidemic influence, which again sweeping round the world, brushes away the offending cause.

Now whether these epidemic influences are general, passing round the globe, or only affecting a continent, an island, or a state, it matters not; they are, as their name implies, *upon the people;* so that, not only its victims, but also the whole people or race where they prevail, are more or less under their influence.

It is this kind of epidemic influence which is so frequent a cause of simple continued fever, acting either as predisposing or exciting cause, or very likely both. It is, in short, the influence of the sum total of the iniquity of the people, in deviating from the laws of health, falling upon them, not only to produce febrile affections, but, in fact, also producing, modifying, and controlling, to a certain extent, every possible variety of disease.

Its prevailing tendency should always be understood, whether it be irritating, debilitating, or malignant; and its influence carefully estimated, in arriving at the indications, in every possible deviation from the standard of health, whether it be in febrile or any other affection.

It is evident, then, that an influence, so general and searching in its effects, must be a very frequent cause of continued fever.

Other causes might be mentioned of this variety of fever; but I trust enough has been considered to illustrate the principles involved in this variety of febrile affection, and to assist the thinking to search out each particular cause in every case of continued fever which may arise. I will now proceed to consider the nature or pathology of simple continued fever.

Pathology.—In relation to the nature or pathology of simple continued fever, it appears to me there can be no reasonable doubt. In the mildest variety of this continued fever, there is probably only a slight prostration, or cold or chilly stage, followed by slight reaction, which, however, soon passes over, leaving only a feeling of languor.

The cerebro-nervous system probably being at first slightly prostrated, and then more or less irritated, which irritation produces the accelerated action of the heart and arteries, or of the circulatory system. The depression which follows is doubtless from the exhaustion of slight over-action of the system during the febrile state.

In the more active grades of this continued fever, produced by atmospheric vicissitudes, electrical influences, &c., there is probably the same depression of the cerebro-nervous system, followed by an irritated reaction. There is also, probably, in addition to this, more or less retained perspirable matter, which, passing through the heart, arteries, capillaries, veins, &c., irritate more or less directly the inner surface of the whole circulatory system, and, through the extreme nerves of the circulatory system, also irritate the cerebro-spinal system. This irritation is then transmitted through the cerebro-spinal nerves to the sympathetic system, and thus the heart and arteries are kept in an irritated state of excitement, and the fever rendered continued.

In the highest grades of continued fever, such as are produced by sudden changes in the temperature of the atmosphere, in addition to an irritated nervous excitability, and retained perspirable matter, &c., we have various active or passive inflammations supervening, all of which, when they exist, tend to render the original fever continued. And, finally, in that class of continued fever depending upon endemical and epidemical influences, we have the same pathology, and probably, in some epidemics or cases, a morbid febrific agent entering the blood and decomposing, modifying, or materially changing it, and thus rendering the blood itself an irritant to the circulatory system.

Or, in other epidemics or cases, there is probably more or less irritating or poisonous principles entering the blood, and passing with it, as a foreign substance, irritating directly the vascular system, and through its nerves, the cerebro-spinal and ganglionic systems, and hence, by a reflex action, the circulatory system. In either case, the morbid febrific agent becomes a cause of continued fever. In the one case, passing, as a foreign substance through the system, as an irritant, and, in the other case, by changing the blood, and thus not only itself acting as an irritant, but also rendering the blood itself more or less so.

Such I believe to be the true pathology of simple continued fever. And, if it be correct, it explains the cause of the continued character of the fever—at least, till such time as local irritation or inflammation is set up in some tissue or organ, after which, the local complication would render the fever continued, and would also perpetuate it, unless the local inflammation is subdued, far beyond what the original cause would have done.

Prognosis.—The prognosis in simple continued fever is generally favorable, if we except those produced by certain endemical or epidemical influences, which are sometimes of a most fatal and destructive character.

That form of continued fever which sometimes prevails epidemically, under the popular name of "*influenza*," at certain periods, assumes a most malignant character, sweeping off by thousands subjects "fitted for destruction" by their wanton disregard of the laws of health.

All simple continued fevers are liable to the supervention of local inflammations; and, when they do occur, they become a source of danger, according to the parts involved, and also the degree and extent of the local inflammation.

Treatment.—The indications in the treatment of simple continued fever are generally very plain, when we take into account, as we should, the exact deviation from the standard of health. In the mildest form, such, for instance, as arise from a fit of anger, nothing need be done, except to quiet the raging element within, and persuade the subject to avoid a repetition of the cause if possible; but if not, to be sure that, in "being angry he sins not." In more violent cases, such, for instance, as arise from electrical influences, or from atmospheric vicissitudes, there is generally an indication to open the cutaneous exhalent vessels, overcome local congestions, reduce the activity of the circulation, and, finally, to protect the respiratory and alimentary mucous membranes, and subdue local inflammations if they arise.

Now to promote perspiration and reduce the activity of the circulation, the warm foot-bath, with warm sage tea, at evening, are plainly indicated. The feet should be placed in water, at evening, as warm as it can well be borne, and, if it be necessary, the temperature of the bath may be gradually raised by the addition of hot water, till free perspiration is produced, or, if need be, till syncope occurs. The feet should then be wiped dry, and the patient allowed free draughts of warm sage tea, and allowed to go quietly to bed, and, if the fever be in its early stages, the patient may very likely awake in the morning well, or with only a feeling of languor, which a little rest and a gently nourishing diet will soon correct.

But if the case has run on till local congestions, irritations, or inflammations occur, though this treatment would still be indicated, it would become necessary to do still more. If, as is often the case, the respiratory or alimentary mucous membranes are suffering, in addition to what I have already suggested, sinapisms, made of pulverized mustard, three parts, flour, one part, mixed together, and wet with warm vinegar, and spread upon cotton or linen cloth, should be applied over the chest, and, if necessary, over the stomach and bowels. This may be left on from twenty to forty minutes, or till the skin looks quite red.

If the brain suffers materially, cups may be applied to the back of the neck, and two or three ounces of blood taken. If the lungs are involved to any great extent, cups may be applied over the seat of the difficulty and also on the side of the spine corresponding with the affected part. In this way, cases of considerable activity may generally be arrested. It is well however, to use the warm foot bath, and sage tea at evening, for two or three nights, or till all the febrile symptoms pass off.

In cases of continued fever, in which the subject is a full robust young person, and there is an active exciting cause, and a high grade of febrile excitement, general bleeding may be indicated, and when it is, it should be resorted to, early after the establishment of febrile reaction. After bleeding, the warm foot bath, warm drinks, &c., a cathartic of either calomel and castor oil, or of the sulphate of magnesia should be administered, and afterwards repeated if necessary.

To reduce the activity of the circulation, the veratrum viride may be administered, every six hours. The most convenient form for administering this is that of the fluid extract, of which four or six drops may be given every six hours. If there be a strong tendency to the brain, the antimonial or James's powder may be given every six hours, alternating with the veratrum viride, and continued till the fever is subdued. If there be much restlessness or irritability, three or four grains of Dover's powder may be given with the antimonial powders, and continued till quiet is restored.

The drinks in such cases should be of gruel, barley water, or what in many cases would be preferred, crust coffee or toast water. In this way, continued fever of the highest grade may often be subdued in three or four days, and often sooner if no local inflammation supervenes. But if local inflammations do arise, the same treatment should be continued and cupping, blistering, fomentations, &c., resorted to in addition, and continued till both the local inflammation and the general fever are subdued.

In continued fever, endemic to any locality, or which is only transiently so, the character of that local tendency should be carefully studied, as the indications in the treatment depend of course upon the nature of the cause which is operating. If the cause be of an inflammatory or irritative character, nearly the same treatment I have already suggested will be indicated, and should be resorted to, modified of course to suit each particular case. But if, as is often the case, the endemic tendency is depressing or malignant, the warm foot bath, friction along the back, with warm pepper and vinegar, warm drinks, sinapisms, &c., may be indicated, and should not be neglected, being continued if necessary at evening, for two or three days.

A mild cathartic of rhubarb, leptandrin or podophyllin may be given or if indicated hydg. cum creta, or even calomel in castor oil may be indicated. In such cases however, as mild a cathartic as will be efficient, should always be selected.

Having procured the operation of a cathartic, a tonic is generally indicated, and should not be withheld. If only a mild tonic be indicated, the fluid extract of columbo, in doses of from twenty to sixty drops after each meal may be sufficient. But if the prostration be considerable, from half a dram to a dram of the fluid extract of cinchona may be given, every six hours, or three times per day, according to the degree of prostration.

But in cases which are of a congestive or decidedly malignant character, the sulphate of quinine should be given in solution, in doses varying from one to four grains every six hours, and continued till convalescence is fully established, and then in diminished doses during the whole period of convalescence. Or, during convalescence, if the quinine should become very unpleasant to the patient, the fluid extract of bark, or of columbo, may be substituted for the quinine, and given in doses of from one-third of a dram to a dram after each meal. Such, then, are the principles which should guide in the treatment of simple continued fever, depending upon endemic influences.

In simple continued fever of an epidemic character, the tendency of the prevailing epidemic influence should be carefully observed and studied, whether it be irritative, depressing, or malignant. Till the character of the prevailing epidemic influence can be ascertained, a temporizing, expectant treatment should be resorted to, only fulfilling such indications as clearly arise, and waiting with patience for those which are to follow. But when the epidemic influence is fully understood, indications may often be anticipated, and thus much time and suffering saved to the patient.

When the prevailing epidemic, with the endemic influences, are well understood, the indications in the treatment of the fevers which they produce are generally very plain. They are essentially those which I have already suggested as arising in simple continued fever from other causes, such as to equalize the circulation, promote perspiration, &c., and to sustain the sinking powers of the system. The treatment, then, should be much the same as that I have already laid down, such as warm footbaths, warm drinks, and applications along the back, if there is chilliness, mild cathartics, and if, as is generally the case, there be a sinking or a prostrating tendency, tonics, and generally the sulphate of quinine.

There is one important fact or principle to be borne in mind in the treatment of epidemic continued fevers of an asthenic character. It is this. If local inflammations supervene, they generally arise from debility and the consequent imperfection of the circulation, and, therefore, will not admit of general depletion. In such cases, if inflammation arises in any tissue or organ of the body, it is an indication for more quinine, and it should, on no account be withheld, but continued, in full doses, in conjunction with cupping, blistering, fomentations, &c., till both the local inflammation and general fever are entirely subdued.

I speak advisedly from experience on this subject, having observed and treated hundreds of such cases in this way with the most satisfactory results.

This, then, completes what I had to say on the subject of simple continued fever. And I will now only add, that while these are the general principles, it is important to bear in mind that each case will present peculiarities, which must not be overlooked in arriving at the indications for a rational treatment. In every possible variety of simple continued fever, a reasonable amount of nourishment should be allowed, but the greatest possible care should be taken that it be administered in a form suited to each particular case.

In mild cases, in which there is little febrile excitement, a plain, digestible, and nourishing diet may be allowed at meal hours from the very first. But in cases in which there is a high inflammatory fever, there will generally be at first little or no appetite for solid food. In such cases, toast-water, or crust coffee, with a little milk, should be allowed at first, and then, as the febrile excitement gradually passes off, more milk may be added to the coffee, till an appetite for solid food returns, when it should be allowed in a mild form at regular meal hours during the whole period of convalescence. In prostrated or congestive malignant cases, crust coffee, from one-third to one-half milk, should be allowed early; and later, if necessary, beef tea, beef essence, chicken broth, &c., may be given till solid food may be allowed, when it should be given in a plain, digestible, and nourishing form during the whole period of convalescence.

SECTION IV.—ENTERIC CONTINUED FEVER.

By enteric continued fever, I mean that variety of continued fever in which there is congestion, inflammation, and generally more or less ulceration of the alimentary mucous membrane, and especially of the glands of Pyer and the solitay glands situated along the small intestines.

This condition is liable to occur in every possible variety of fever, as I have already suggested; and when it does occur in intermittent, bilious remittent, or simple continued fevers, the symptoms of typhoid, or enteric fever are added to those of the original fever. Or, more properly, the symptoms of the primary fever, or disease, become merged in those of the typhoid or enteric fever. But as typhoid, or enteric continued fever is liable to occur without any other regular fever preceding it, as cause, it is entitled to separate consideration as a fever. I shall therefore treat of it as *enteric continued fever*, preferring that to the term " typhoid fever."

Symptoms.—When bilious remittent, or simple continued fevers, assume a typhoid or enteric character, there is generally more or less tenderness of the bowels, with slight tympanites, and in some cases more or less pain in the bowels. There is often a diarrhœa, more or less troublesome, restlessness, irritability, and more or less prostration of the nervous system; but the most prominent symptoms are stupor, delirium, picking at the bed-clothes, imperfect sight, and dullness of hearing, lying upon the back, and sliding down in the bed; and finally, constant stupor, insensibility, and if not arrested, eventually death.

These symptoms are of course modified by the character of the primary fever, and also by the extent of the gastro-intestinal congestion, inflammation, and especially ulceration, which has supervened. The symptoms which arise, however, in primary enteric continued fever are just what that peculiar condition of the alimentary canal would be likely to develop, especially if we take into the account the nature and effects of the causes operating to produce it. There is generally a slight loss of appetite, more or less weariness, and disinclination to perform mental or corporeal exertion. There is some thirst, especially towards evening, and sometimes a little nausea, with a feeling of heaviness in the stomach and bowels. There is a heavy feeling in the lumbar region, and often more or less pain, and slight creeping chills along the back. There is more or less pain in the head, and especially a dull, lost feeling, which the patient attempts but fails to describe.

After these symptoms have continued increasing for a longer or shorter time, varying from two or three to ten or twelve days, more or less febrile symptoms are developed. These febrile symptoms may be preceded by a marked chill, or they may arise, gradually, without any very sensible cold stage, except the general chilliness which prevails more or less during the whole forming stage of the fever. In either case, there are gradually developed the phenomena of enteric continued fever. There is thirst, dry skin, flushed countenance, headache, loss of appetite, irritability and general prostration of body and mind. The pulse is frequent, full and strong, or small and compressible. The pain in the back and head become more severe. There is more restlessness, and though there is a general feeling of drowsiness, little quiet sleep is obtained.

These symptoms continue on, with either very slight or no remissions, for several days; the pulse becomes more frequent, the skin hot and dry ; the countenance wears an expression of gloom ; the tongue becomes coated, and more or less clammy; the stomach becomes irritable; there is often diarrhœa, with more or less pain in the bowels, slight tenderness on pressure, and generally more or less tympanites, detected by percussion over the abdomen ; and in some cases a slight eruption appears.

But gradually the tongue becomes dark and dry, the abdomen more distended, small vesicles, or sudamina, appear upon the chest, and often on various parts of the body ; there is delirium, obtuseness of hearing, redness of the eyes, and, unless the disease be arrested, the pulse becomes very feeble and frequent , the patient lies on his back, and gets down in bed, picks at the bedclothes, becomes delirious and comatose, has retention of urine, and involuntary fecal discharges ; and finally, if the disease continues, the circulation yields, there is a cold, clammy sweat, the countenance becomes cadaverous, and death closes the scene.

Or if the disease be arrested, the skin and tongue become more moist, the stupor and delirium less, the pulse becomes more regular, the patient becomes more rational, the tympanites less, the appetite returns; and thus the patient becomes slowly convalescent. The duration of this affection may vary, if left to itself, from one to five or six weeks.

Diagnosis.—The diagnosis of enteric continued fever is not generally attended with much difficulty, if we take into the account the most prominent symptoms which arise; as they are peculiar to this form of febrile affection. The diarrhœa is an important diagnostic symptom of enteric continued fever. According to my observation, it generally prevails some time during the forming or subsequent stage of the disease. The discharges may vary from two or three to ten or twelve in twenty-four hours, and they are attended generally by more or less pain.

The tympanitic state of the abdomen, occurring as it does, and gradually increasing as the disease progresses, is an important diagnostic symptom. This symptom usually makes its appearance from the third to the sixth or seventh day, but it often increases till near the termination of the febrile symptoms, and sometimes even later. The slight eruption, which generally makes its appearance about the ninth or tenth day, upon the abdomen and breast, and extending sometimes over other parts of the body, is an important diagnostic symptom when it makes its appearance. The sudamina, or small vesicles, about the size of a split pea, or smaller, which make their appearance about the twelfth or fifteenth day upon the chest, neck, and other parts of the body, are also characteristic of enteric continued fever. The headache, occurring early, and continuing generally till stupor, delirium, coma, and obtuseness of hearing supervene, is more or less characteristic of this enteric affection, and taken together with these symptoms, becomes an important diagnostic symptom.

If we notice then, carefully, the slow and insidious mode of attack, the dejected countenance, the great prostration of strength, the small and frequent pulse, the character of the diarrhœa, the tympanitic state of the abdomen, the slight rose-colored eruption, the sudamina, the headache, the obtuseness of hearing, the delirium, the coma, the getting down in bed, the dry brown tongue ; and, finally, the peculiar gloomy expression of the countenance from the very first, we need not be mistaken in our diagnosis of enteric continued fever.

Nature. — In relation to the nature of enteric continued fever, there has been a good deal of speculation. But it appears to me, that when we take into account the symptoms which are developed from the very first, together with the anatomical appearances, there can be little room for doubt in relation to the nature of the disease. I am satisfied that congestion, irritation, inflammation, and ulceration of the intestinal canal, and especially of the glands of Pyer, and the solitary glands along the small intestines, constitues, essentially, the morbid condition from which most of the symptoms arise.

It is probable, that typhoid symptoms are seldom developed, unless the intestinal disease passes on to ulceration. But the symptoms of the forming stage, are just what we might expect from congestion, irritation, and inflammation of the intestinal canal ; while those of the latter stage are essentially those of ulceration. The headache, stupor, delirium, coma, &c., are nothing more than we might expect from this variety of intestinal disease, the sympathetic derangement, of course, passing from the intestines, or alimentary canal, through the pneumo-gastric and sympathetic nerves, directly to the brain.

I have repeatedly known an incipient typhoid state arrested, and the stupor, headache, delirium, &c., to subside, at once, by the drawing of a blister over the stomach or bowels, in cases in which the subsidence of these symptoms could not be attributed to anything except the arresting of the alimentary, or gastro-intestinal inflammation or ulceration. I have also known the diarrhœa as suddenly arrested by counter-irritation over the stomach and bowels, which could not reasonably be attributed to anything except the counter-irritation subduing the gastro-intestinal inflammation or ulceration. Again, I have known irritating articles of food, evidently by irritating the alimentary canal, to bring on, rapidly, a typhoid state in bilious remittent, and simple continued fevers ; and also, to bring on a return of all the typhoid symptoms in convalescent enteric fevers. It is a fact, also, that drastic cathartics often immediately develop typhoid symptoms, evidently by the irritation or inflammation they produce in the stomach and intestines.

Now with all these facts, and many more which might be mentioned, it appears to me, that there can be no reasonable doubt in relation to the nature of typhoid or enteric fever, especially as all the symptoms from the first, are what might be expected to arise from irritation, inflammation, and ulceration of the intestinal canal; involving, of course, the mucous membrane, with the glands of Pyer, and of the solitary glands, the ulceration sometimes extending **through** the muscular and peritoneal coat of **the intestines.**

It is probable, that the symptoms of the early **stage** may be produced by merely congestion, irritation, and inflammation ; **but** that it requires *ulceration* to develop the stupor, delirium, coma, deafness, &c., which constitute the essential symptoms of confirmed enteric fever But, in **order** to render this matter more certain, it becomes necessary to inquire into the morbid appearances presented on post-mortem examination.

Anatomical characters.—Now it is a well known fact, that in all **marked** cases of enteric continued fever, there is presented, on post-mortem examination, a congested state of the mucous membrane of the small, **and** frequently of the large intestines, and stomach.   There is also evidence of inflammation, more or less general, through the whole alimentary canal, not even excepting the stomach and large intestines ; but generally, more especially along the small intestines.

Patches of ulceration may generally be found along the small intestines, and sometimes in the stomach and large intestines.   These ulcerations may pass only through the mucous coat, or they may perforate the muscular, and even the peritoneal coat, in some cases ; one interesting case of which fell under my observation a few years since.   But the anatomical changes, perhaps the most characteristic of enteric continued fever, are the congested, inflamed, and generally ulcerated condition of the glands of Pyer, situated along the ileum ; and also the small *solitary* **glands** of the ileum, occupying mostly its lower portion. These morbid appearances are, I believe, nearly always found, to a greater or less extent, in all well marked cases of enteric continued fever, and appear to point to the alimentary canal as the seat of this truly formidable disease.

Other morbid appearances are, however, sometimes presented ; such **as** inflammation, or enlargement of the mesenteric glands, enlargement of the spleen, a slight softening of the liver and kidneys, and various other changes of minor importance ; all of which, I believe, may be regarded as the effect, and not the cause, of the disease itself.

Causes. — I have already stated that a typhoid, or enteric fever, may supervene upon a bilious remittent, or simple continued fever, from any cause that will produce ulceration of the alimentary mucous membrane ; involving, of course, the glands of Pyer, and also the solitary glands of the lower portion of the small intestines. Now, this to my mind, accounts for the disease when it occurs as a complication, in more simple febrile affections ; the morbid condition, in such fevers, very much predisposing to this gastro-intestinal ulceration, which being once produced and established, the symptoms of the primary fever are merged in those of enteric continued fever.

But in primary cases of enteric fever, there are generally operating some predisposing, as well as exciting cause of the disease, independent, of course, of any preceding febrile affection. It is probable, then, that any debilitating cause may operate to predispose the system to enteric fever; such as damp sleeping rooms, impure confined air, insufficient clothing, scanty or unwholesome supply of food, impure water, filth, great depression of spirits, taking food at unseasonable hours, the use of tobacco, drunkenness, licentiousness, and various other like causes. By any, and all these causes, the system may be reduced to a condition in which the fluids, and solid tissues of the body, are in a more or less deranged state; and, as the cutaneous exhalation becomes less, the alimentary mucous membrane becomes at first congested, then inflamed; and, finally, more or less ulcerated, developing the symptoms which arise as the disease progresses.

In many cases, the disease appears thus to come on insidiously, without any very marked active exciting cause. But there may be some irritating exciting cause, as I have known in some cases; such as stimulating articles of food or drink, or indigestible food, sudden exposure to cold damp air, wet feet, sudden changes in the electrical state of the atmosphere, &c. These exciting causes, however, when they do operate, probably only hasten on what would otherwise occur in a little longer time. I have known, however, green chestnuts and other such indigestible substances to produce enteric fever, in cases where there did not appear to be any very marked predisposition to the disease. Having now inquired into the symptoms, diagnosis, nature, anatomical characters, and causes of enteric continued fever, I think there should not be room for reasonable doubt in relation to its pathology.

Pathology.—It appears to be plain that the typhoid state, into which other fevers pass, are from the enteric inflammation and ulceration which supervenes; the deranged condition of the fluids and solids of the system constituting the predisposing, and the enteric ulceration the essential condition of the typhoid state. And it also appears to me reasonable, that in primary enteric fever a deranged condition of the solids and fluids of the body, produced by any of the causes I have enumerated, such as filth, bad air, unwholesome food, intemperance in eating and drinking, licentiousness, exposure and want, may act as predisposing and exciting causes of enteric fever. And finally, that in cases in which there is no very marked predisposition, enteric fever may be produced by crude indigestible articles of food, producing directly inflammation and ulceration of the alimentary mucous membrane, involving, of course, the small glands along the small intestines. In this latter case, however, it is probable that the local irritation of the alimentary mucous membrane, very soon, by interrupting digestion, produces the requisite changes in the fluids and solids of the system, and so enteric continued fever is the result.

This, then, brings us to the *essential pathology* of enteric continued fever: a deranged or changed condition of the fluids and solids of the system, and gastro-intestinal ulceration, involving the small solitary glands and glands of Pyer, situated along the small intestines, and chiefly along the lower portion of the ileum. And, while I am confident that this is the real truth of the matter, I am satisfied, from careful observation, that the system, while laboring under typhoid or enteric fever, may, and frequently does generate a virus or morbid poison, which will produce directly genuine typhus fever, and very strongly predispose to enteric continued fever.*

Prognosis.—The prognosis in enteric continued fever is generally not very unfavorable, at least if proper treatment be resorted to in season. In those typhoid or enteric fevers supervening upon bilious remittent or simple continued fevers, the danger depends much upon the general de-**rangement** produced in the system by the primary fever, for I believe the typhoid complication may generally be arrested if attended to immediately ; at least such has been the result of my observation.

In primary enteric continued fever, the danger depends much upon the extent and degree of the abuses of the system which has led to it. If the abuses of the system have been very great, and thus the derangement of the fluids and solids of the system very considerable, it offers a formidable barrier to a resolution of the gastro-intestinal inflammation and ulceration. But in cases in which the exposures have been slight, and thus the integrity of the solids and fluids not very materially impaired, I have generally found the disease to yield to proper and judicious treatment, if timely applied.

Treatment.—In arriving at a rational treatment we have only to take the deviation from the standard of health, and we have the indications, to fulfill which we need only to make the best selection of remedies, and apply them judiciously, and we shall have done our duty.

Now the deviation from health is evidently, as we have seen, a deranged condition of the fluids and solids of the system, such as grow out of imprudence, exposure, filth, &c. Great general debility, and consequent derangement of the functions of the body, and congestion, inflammation, and ulceration of the alimentary canal, but more especially of the minute glands, situated along the lower portion of the small intestines.—The indications, then, are plainly to sustain the sinking powers of the system, arrest the ulceration, subdue the intestinal inflammation, and restore to a healthy state the functions of the body, and especially that of the skin.

When typhoid or enteric fever supervenes upon bilious remittent or simple continued fever, the general indications in the treatment of the primary fever may not be very materially changed, except in so far as the intestinal ulceration is concerned. In such cases, I have generally succeeded in arresting the typhoid symptoms by applying a blister over the stomach, if it is done early, or blistering the stomach and bowels, if neglected for a time. In such cases, as soon as the stupor, delirium, diarrhœa, &c., make their appearance, a blister, the size of the hand, should be applied to the epigastrium, and left on till the skin all raises, when it should be removed, the skin taken off clean, and then the sore should be dressed with soft wilted leaves, and kept running as long as it will, the leaves being changed every six hours.

As soon as the blister begins to discharge, if the patient was not taking it before, the sulphate of quinine should be given in two or three grain doses every six hours, to sustain the sinking powers of the system, and

* See cases reported by me in the Buffalo Medical Journal, Vol. viii., page 354, in the year 1852.

to counteract internal congestions, and especially that of the alimentary mucous membrane. The quinine may be given in powder or solution; in either case it should be taken with mucilage of gum arabic, to prevent any irritating effect upon the alimentary mucous membrane.

As the first blister gets nearly healed, if the tympanitis, diarrhœa, stupor, &c., have not been entirely arrested or corrected, a blister of the same size should be applied to the abdomen, directly below the umbilicus, and made to draw well, and it should be treated as the other, taking care that it be kept moist and well covered with soft wilted leaves. The quinine and mucilage should be continued, and a reasonable amount of such nourishment as the stomach will bear should be given, and, as the blister begins to discharge, the tympanitis, diarrhœa, stupor and delirium will generally entirely subside; at least, such has been the result of my observation. In fact, so uniformly has this been the result of my observation, that I have become accustomed to feel a degree of certainty in such cases that I seldom do in any others.

The only exceptions to a rule has appeared to me to be from perforation, or those extreme cases of intestinal ulceration, in which a wasting diarrhœa has brought the system too low to be sustained by the amount of nourishment and quinine the stomach will bear. So much, then, for the treatment of secondary typhoid, or enteric fever. Let us now proceed to consider the best methods of fulfilling the indications in primary enteric continued fever.

In the treatment of primary cases of enteric continued fever, the debility has generally so far deranged the gastric and hepatic functions, that a cathartic becomes indicated. That which I prefer in such cases, for adults, is two or three blue pills, followed in six hours by half an ounce of castor oil. For children, the hydg. cum creta, mixed with a little rhubarb, or leptandrin, may be given in oil, and the oil repeated in six hours if necessary.

The warm foot bath should be used, morning and evening, for two or three days, and warm sage tea allowed; and, if there is chilliness, with a dull heavy pain in the lumbar region, the back should be bathed with a warm decoction of pepper in vinegar. This application to the back, in such cases, relieves the dull heavy pain in the back, and produces an agreeable glow of warmth, and often sends forth a gentle perspiration, especially with the aid of the warm sage tea; while the warm foot bath not only favors this, but also lessens the tendency to the brain, and also to the alimentary mucous membrane.

To counteract the gastro-intestinal inflammation, and arrest the incipient ulceration, sinapisms should be applied over the whole abdomen, morning and evening, for a day or two, and allowed to irritate, but not to blister. The patient should be directed to take a plain, nourishing, and digestible diet, if the stomach will bear it, but if it will not, warm toast water, with an equal quantity of milk, may be allowed. Or what I prefer, is the crust coffee, one-third, or one-half, milk. This may be made, by toasting brown, a slice of bread, and then pouring upon it a pint of boiling water, and then adding from half a pint to a pint of milk, according to the amount of nourishment required. This may be sweetened, or not, to suit the taste of the patient; and in either case, it is a most pleasant, convenient, and safe method of administering nourishment in such cases, and in fact, in all cases in which nothing but drinks can be borne.

Now I am satisfied, that nearly every case of incipient enteric fever may be arrested in this way, if attended to in season, with perhaps a grain of the sulphate of quinine in solution after each meal, or every six hours, for five or six days. The bathing of the back equalizing the circulation, the foot bath and warm sage tea promoting perspiration, the cathartic improving the gastric and hepatic function, the mustard counteracting and arresting the intestinal inflammation and ulceration; while the nourishment and quinine arrest the sinking tendency, and restore the tone of the nervous system, and with it the functions of the whole body.

But, unfortunately, most cases are neglected during the incipient stage, and as they are, the chances of thus arresting, at once, enteric fever, is of course lessened. In cases, then, that have been neglected, or what is worse, badly treated, that which I have suggested should be resorted to, modified, of course, to fulfill the indications which arise in each case, and much more may become necessary. If, after the operation of the cathartic, the tympanites, diarrhœa, stupor, delirium, &c., increase, a liberal blister should be applied to the epigastrium, for the purpose of arresting the intestinal ulceration, and subduing the inflammation upon which it doubtless depends.

The sulphate of quinine, in two or three grain doses should be given, in mucilage of gum arabic, every six hours, to arrest the sinking tendency, and to prevent congestion, and especially of the alimentary mucous membrane. The crust coffee, one-third or one-half milk, should be allowed freely, so that the patient may get from one to two pints of milk per day, with considerable nourishment from the bread. This affords a pleasant drink, and a reasonable amount of nourishment, and in a very suitable form; without which, the system must sink, for it is on food that we live.

If the diarrhœa is very considerable, two grains of tannin may be given every six hours, alternating with the quinine, till it is in a measure controlled. As the blister to the epigastrium begins to heal, if the tympanites, diarrhœa, stupor, &c., still continue to considerable extent, a blister should be applied to the abdomen, below the umbilicus, and the other treatment continued. In some cases, if the sinking tendency be considerable, one or two grains of camphor may be given with each dose of quinine, as it will add much to its sustaining effects, and often relieves, to some extent, the flatulency, which is apt to be troublesome.

As the blister on the abdomen begins to discharge, the tympanites, and other unpleasant typhoid symptoms, will generally entirely subside; and it will generally only require a continuation of the quinine, in two or three grain doses, in solution, every six hours, with mucilage, and nourishment, to conduct the patient on to convalescence and health. If, however, the case has been protracted, and the strength of the system very much exhausted, arrow-root, cooked in mutton or chicken broth, may become necessary; and when it is, it should be given at meal hours, and continued till toast, a poached egg, and other digestible kinds of nourishing food can be borne.

To prepare the arrow-root in mutton or chicken broth, dissolve a teaspoonful in two ounces of cold water, and while the broth, prepared in the usual way, is boiling, pour the solution of arrow-root into half a pint of it, and let it cook for four or five minutes. Of this, the patient may be allowed as much as he likes, at regular meal hours, the crust coffee and milk being allowed at all hours as a drink.

In this way, I have generally succeeded in arresting severe or neglected cases, in from ten to fifteen days; as convalescence becomes established, diminishing the dose of quinine, usually giving one or two grains after each meal, in solution, and with mucilage of gum arabic. Or if the quinine becomes unpleasant, I sometimes substitute the fluid extract of cinchona, or of columbo, in from one-third of a dram, to dram doses, after each meal, during convalescence.

In some cases, however, in which the patient had been neglected, and badly treated, I have found a lingering tympanites, with more or less diarrhœa, to continue after the other typhoid, or enteric symptoms have subsided. In such cases, I have given the oil of turpentine, in about fifteen drop doses, every six hours, alternating with the quinine. It appears to relieve flatulence, slightly stimulate, and as an alterative, tends, as I believe, to favor the resolution of the gastro-intestinal inflammation and ulceration. I have generally given it in emulsion, with gum arabic, sugar and water, as being the most pleasant, and, also, the best adapted to alleviate the inflamed and ulcerated alimentary mucous membrane.

Perforation of the intestines rarely occurs in enteric fever, if it is properly treated. But as it may occur, it is well to remember, that if it does, and the passage is sufficiently large to let through the contents of the intestines, the patient may suddenly scream, or groan with pain, in some part of the abdomen. In the only case of this kind which has fallen under my observation, every breath was a groan, the countenance grew pale, the pulse weak and fluttering, warm drinks produced an intense feeling of heat in the abdomen, the powers of life gradually yielded, and the patient died in about twenty hours after the perforation occurred. Should symptoms of perforation occur, the case should not be abandoned, the indications should be met as best they may be, by warm fomentations to the abdomen, the free administration of wine whey, &c.; but my firm convictions are that such cases seldom, if ever, recover.

Thus I have completed what I had to say on the subject of enteric continued fever; and while I would confidently urge the treatment which I have suggested, modified, of course, to fulfill the indications which arise in each particular case, I would as earnestly advise to abstain as carefully from the administration of the least thing which is not clearly indicated, and imperatively demanded.

During convalescence from enteric continued fever, no crude indigestible articles of food should be indulged in. The whole surface of the body may be washed with warm rain water, sufficiently often, to keep the skin clean, as it should be, in fact, during the prevalence of the disease. This is of importance, not only to the patient, but also for the attendants; for though I do not believe that enteric fever is directly contagious, I am confident, as I have already stated, that the system may, in enteric fever, generate a virus, which will produce genuine typhus fever, and probably act, as a strong predisposing cause, of enteric continued fever.

SECTION V.—TYPHUS CONTINUED FEVER.

By typhus fever I mean that variety of continued fever which has been treated of under the names of *spotted fever, putrid fever, camp fever, ship fever, jail fever, hospital fever, &c.*

The term *typhus*, from τυφος, stupor, is indicative, perhaps, of the most prominent symptom essential to this febrile affection. But if a suitable term could be agreed upon, which would indicate more directly the real pathology of the disease, it might be preferable. Of the terms suggested, that of "putrid fever," perhaps, might be the least objectionable if it were more generally employed. I shall, therefore, use the term *typhus fever* and *putrid fever* as synonymous, though the terms may indicate conditions not essentially similar or identical.

I believe that a putrid condition of the fluids and solids of the system invariably prevail, to a greater or less degree, in all cases of genuine typhus, and, in fact, that a putrid tendency is an essential characteristic of the disease under whatever form and in whatever locality it may prevail.

Symptoms.—Now, keeping in mind the putrid character or tendency of this febrile affection, the symptoms developed during the different stages are just what might be expected. And especially are they so, when we take into account the fact, that the contagion or morbid poison which produces it, acts directly upon the blood, and, through the blood, upon the brain and nervous system, and, in fact, upon all the fluids and solid tissues of the body.

During the forming stage, while the blood is gradually becoming changed, and the brain and nervous system, as well as all the solid tissues, are becoming more or less affected, there is languor, loss of appetite, an uneasiness at the pit of the stomach, slight giddiness, and nausea, a pale, shrunken countenance, dull, heavy eyes, tremor of the hands, a feeling of weariness, debility, and disinclination to mental or corporeal action.

These premonitory symptoms may continue from two or three to five or six days, at which time the general derangement of the system has become sufficient to develop other and more decided symptoms, such as slight chills, alternating with flushes of heat, entire loss of appetite, pains in the back, loins, and lower extremities, nausea, and sometimes vomiting; a frequent and feeble pulse; the tongue becomes coated with a white fur, there is a confused heavy sensation in the head, and increased disinclination to mental and physical exertion.

The patient takes his bed generally during this stage of invasion. It may last from eight to fourteen hours, during which time there may be frequent chills, with slight flushes of heat, or there may be a continued state of coldness, more or less severe, with great heaviness in the back, and confusion or a lost feeling in the head.

Gradually, as this cold or congestive state continues, the brain, spinal cord, and nervous system become irritable; the impression passes through

9

the sympathetic nerves to the heart and arteries; the circulation is aroused; the face becomes flushed; the heat of the body accumulates or increases; the pulse rises in strength and fullness; the skin becomes dry; there is thirst; the tongue becomes more furred and slimy; the bowels are constipated; the mind more confused; the patient becomes fretful and restless; there is an anxious expression of the countenance; the urine is small in quantity and of a reddish color; the head feels heavy and confused; and generally, during the first two or three days of this stage of reaction, occasional manifestations of delirium occur during the night.

About the second or third day of the reaction, there are frequently slight catarrhal symptoms, such as injected eyes, slight soreness of the throat, painful deglutition, slight difficulty of breathing, and occasionally a slight cough. There is, also, frequently some tenderness in the right hypochondriac region, and more or less pain through the back, loins, and lower extremities, and, in fact, a general soreness felt through the whole body.

The heat of the surface accumulates, as there is little perspiration, producing that peculiar character of heat called *calor mordax*, imparting to the hand a stinging sensation, which may continue even after it is removed from contact with the affected body. The temperature of the surface by the thermometer, according to Professor Wood,* varies from 100° to 109° F. at the height of this disease.

About the close of the third day of the stage of excitement or reaction, there is generally considerable giddiness and sensorial obtuseness, the patient even thus early appearing as if under the influence of some narcotic. The cerebral functions now become more and more disturbed; the pulse becomes frequent and feeble; the respiration hurried; the hearing obtuse; delirium more frequent and considerable; and the general torpor increases.

About the fourth day of the febrile reaction, an eruption makes its appearance, consisting of small red spots, varying from the merest speck to spots one-fourth or half an inch in diameter. This eruption generally occurs upon the neck, breast, and, sometimes, appearing very much like measles, extends over the whole body. The spots are but very slightly raised, and vary in color from a bright red to nearly black, the darker color appearing usually in the more malignant cases.

These spots, though often numerous, are sometimes very few or entirely wanting. And though they generally appear about the fourth day of the fever, they may appear any time from the third to the fifteenth day of the febrile reaction. This eruption is sometimes accompanied with small vesicles or sudamina, and, in some rare cases, the sudamina appear without the eruption. The eruption or the sudamina may continue from four to ten or twelve days, or they may continue till convalescence is established.

One of the most prominent peculiarities of typhus is the almost insurmountable aversion to corporeal and intellectual exertion, which is manifested through nearly the whole course of the disease. The patient

---

* Wood's Practice, vol. i., page 366.

moves slowly and with reluctance, and answers questions with hesitation or fails to answer them at all.

The **stage** of febrile excitement usually continues about six or seven days, when the system, apparently exhausted, sinks into a state of collapse. This collapse may, however, occur any time after the chill or stage of invasion, sometimes even continuing without any marked febrile reaction after the chilly stage, the shock to the brain and nervous system being so great that the circulation does not even rally. When, however, as generally happens, the febrile excitement subsides, after it has continued from four to eight days, the occurrence of collapse is marked or manifested by the subsidence of the previous inflammatory symptoms and the supervention of great prostration.

The pulse becomes feeble and generally more frequent, the tongue becomes brown and eventually black, the teeth and prolabia become incrusted with dark sordes, there is a stunned, confused, and deranged state of the sensorial functions, with more or less constant muttering delirium, and generally total indifference to surrounding objects. There is difficulty of hearing, twitching of the muscles of the face, difficulty of protruding the tongue, recumbence on the back, and gradual sliding towards the foot of the bed, from deficient muscular power, that peculiar biting heat of the skin called *calor mordax*, and finally, in violent cases, dark spots on the surface, hiccough, and generally a more or less tympanitic state of the abdomen.

During the collapse the urine is generally copious, and often foams when voided into a vessel; and, in some cases, there is a tendency to a diarrhœa in the latter stages, the discharges being watery, and highly offensive. Towards the termination of the stage of collapse, when it tends to a fatal termination, coma more or less complete is seldom absent, from which, however, the patient may generally be roused for a few moments. The state of collapse usually continues from seven to nine days, terminating either in slow convalescence or in death.

The occurrence of convalescence is announced by the appearance of a gentle perspiration, and reduction of the acrid heat of the skin, a moist tongue, cleaning along the edges, copious and sedimentous urine, abatement of the delirium, and in some instances a moderate diarrhœa. The phenomena of a favorable crisis may occur any time from the seventh to the thirtieth day, but I think, generally, from the fourteenth to the twenty-first day; allowing one week for the forming stage, one for the stage of febrile reaction, and one for the stage of collapse.

The convalescence is generally tedious, as there remains a general debility and prostration of the whole system, after the total subsidence of the fever. Such are the symptoms of simple typhus continued fever, in its ordinary course, varying, of course, in degree, from the mildest cases, which continue only for a few days, and require little or no treatment, to the most malignant cases, passing through the different stages, and hastening to a fatal termination, or else to a favorable crisis and slow convalescence.

Varieties.—Typhus is liable to assume either an inflammatory or congestive character, both of which it becomes necessary to consider, as the different morbid conditions require correspondingly different modes of treatment. During the continuance of typhus local inflammations are liable to occur, of the brain, lungs, alimentary mucous membrane, of the liver, and peritoneum; and of these, the brain, lungs, and alimentary mucous membrane are the most frequent. When inflammation does occur, it constitutes the inflammatory variety of typhus.

Generally, in these cases, the phlegmasial symptoms do not occur until the second or third day of the stage of excitement, but sometimes earlier. If the brain becomes inflamed there is deep and pulsating pain in the head, flushed countenance, throbbing carotids, redness and sensibility of the eyes, irritability of temper, pains in the extremities, irregular respiration, continued wakefulness, early and uninterrupted delirium, a bloodshot appearance of the eyes, contracted pupils, intolerance of light, agitated countenance, continued moaning, and coma.

When the lungs become inflamed the symptoms of pneumonia are added to those of typhus, only in the pneumonia of typhus the sputa are generally of a rusty appearance, from the very commencement of the inflammation.

The symptoms of enteric inflammation, when it occurs, are tension and tenderness of the abdomen, an anxious and disturbed countenance, a small, quick, and frequent pulse, recumbence on the back, vomiting, desire for cool drinks, difficult deglutition, and great prostration of strength. In such cases, however, the patient seldom complains of much pain in the abdomen, and there is no very great degree of tenderness, unless firm pressure be made, when slight suffering will generally be expressed by the appearance of the countenance.

In the congestive variety of typhus there is a want of reaction after the stage of oppression, or cold stage, the system remaining in a congested and oppressed condition during the whole course of the disease. The vital powers, in such cases, are overwhelmed, and the patient appears to sink, progressively, from the moment the disease commences till the vital action ceases altogether.

In aggravated cases of this kind of putrid fever, there is from the very beginning extreme debility, with deep-seated pain in the head, vertigo, the face is pale, the respiration is oppressed and slow, the pulse is small, feeble, and variable, the skin relaxed and damp, and may be below the natural temperature, the countenance confused and anxious, the eyes watery and red, the bowels torpid at first, but in the latter period of the disease there may be a diarrhœa. The tongue is at first pale and slimy, but it becomes rough, brown, and sometimes nearly black, as the disease progresses. Towards the close, colliquative hemorrhages and involuntary stools generally occur. Sometimes coma is among the first symptoms, and continues to the end of the disease, and I have not unfrequently seen a complete state of insensibility to supervene very soon after the attack of the disease.

Diagnosis.—The diagnosis of typhus, or putrid continued fever, is not usually attended with any great difficulty, if all the symptoms are taken into the account. Typhus may generally be distinguished from enteric or typhoid fever, by careful attention to the following diagnostic symptoms.

Typhus fever is generally less insidious in its approach and attack than typhoid or enteric continued fever, the bowels in typhus also being generally more or less constipated, while in enteric fever there is often a diarrhœa from the commencement. In typhus, the sensorial functions are earlier and more invariably disturbed, and the muscular prostration is greater than is common in typhoid or enteric fever.

Mental depression, a sullen gloom of the countenance, and disinclination to mental and corporeal exertion, are remarkably characteristic of typhus, and not so conspicuously present in enteric fever. In typhus, the patient usually answers questions with reluctance, while in typhoid fever, in most cases, questions are answered more readily. The eruption in typhus generally makes its appearance earlier, is of a darker color, is more abundant, and comes out more generally over the whole body than is common in enteric fever, and besides, in typhus, sudamina are less frequently seen than in enteric fever.

In typhus fever, there is generally a less tympanitic state of the abdomen than in enteric fever, and the average duration of typhus is probably only about two-thirds that of enteric continued fever, typhus also being marked by the three conspicuous stages of invasion, reaction and collapse, which do not mark enteric fever. Typhus fever is very liable to attack old people, while enteric fever more frequently attacks the young. Typhus fever is also highly contagious, while enteric fever is scarcely if at all so; it is probable, however, as I have before intimated, that the system in enteric fever may generate a virus, or morbid poison, which will produce typhus, and very strongly predispose to enteric fever.

It is a fact that in incipient typhus it may sometimes be difficult to distinguish typhus from some cases of simple continued fevers, as well as enteric; but by taking into the account the diagnostic symptoms which I have enumerated, and by keeping in mind the prevailing epidemic and endemic tendencies, the acute observer need seldom if ever be mistaken in the diagnosis of typhus fever.

Anatomical Characters.—The post-mortem appearances, in simple typhus, if we except the appearance of the blood, are not of a degree of uniformity that would justify us in considering the disease one of a local character. In fact, if the patient die early, there is often found no local morbid lesion to which to attribute the dissolution. And in the inflammatory variety of typhus, the morbid lesions of the parts inflamed are the result of an accidental, not an essential complication of typhus, as we have already seen; and the same is also true of the local lesions which may be found in congestive typhus.

But the blood, I believe, is always in pure typhus found in a changed, or abnormal state. Prof. Wood says,* "after death it is found in the veins black, liquid, sometimes resembling molasses, and occasionally mixed with oily globules." And besides, it is a well-known fact that blood drawn in the latter stages of typhus coagulates imperfectly, and sometimes is almost black. So far, then, as the morbid appearances are indicative, the blood is emphatically the seat of typhus fever, as it invariably presents a changed appearance, and as every other morbid appearance, or changed condition, is more or less accidental.

Causes.—Typhus fever may be produced by impurities arising from

* Wood's Practice, vol. 1, p. 370.

crowded apartments, in which filth, and especially animal secretions and exhalations are an ingredient, constituting *idio-miasmata;* or from a *morbid poison*, generated in the system while laboring under typhus, and probably also sometimes in enteric fever, and also from that sum of all human imprudence, epidemic influence.

It is probable that the cause may be one, however, in the three cases, for in crowded apartments in which filth and animal secretions and exhalations are an ingredient, it is probable that its decomposition may produce a *morbid poison*, identical with that generated in, and emanating from, the bodies of patients suffering from typhus fever. And further, as epidemic influences are probably the direct result of the combined influence of the sum total of every species of filth and uncleanness, resulting from human degradation and imprudence, it is more than probable that when typhus is so produced, it is in consequence of decomposed animal matters, as in other cases.

It is rational to conclude, then, that *idio-miasmata*, or a deleterious *morbid poison*, or effluvia, originating from the decomposition of matter, derived from the human body, or emanating from the body of patients suffering from typhus, and perhaps from enteric fever, is the invariable cause of typhus continued fever.

That typhus is strictly a contagious disease, I have not a reasonable doubt, and yet it does not, except in powerful epidemic influences, appear to be conveyed to any very considerable distance by the air, at least in a sufficiently concentrated state to produce the disease. But the typhus contagion may become attached to various articles of clothing, and thus be conveyed to great distances, and also retain its power of infecting for a long time, probably for several months.

Nature.—Having now inquired into the symptoms, cause, &c., of typhus, let us inquire into its nature, if, in fact, it needs an inquiry; for it appears to me there remains little room for doubt. It is probable that idio-miasmata, or the morbid poison of typhus, enters the blood through the mouth, lungs, and perhaps the absorbents of the skin; and that once introduced into the circulation, it unites chemically with the blood, materially decomposing or changing it; thus not only acting upon the inner surface of the circulatory system, but also rendering the blood itself a poison, not only to the circulatory system, but also to the brain and nervous system, and, in fact, to every tissue of the body to which the blood is sent.

Now the blood, in this state, either by directly poisoning, or by a negative property failing to supply or furnish the brain sufficient vital power, disqualifies the brain for generating sufficient nervous influence to keep up a healthy action of the different organs of the body. As a consequence of all this, the powers of life are more or less prostrated even during the forming stage; the circulation yielding at the approach of the chill.

During the chill, the cerebro-spinal system becomes irritated, and as a consequence, an irritated influence is sent through the cerebro-spinal to the ganglionic nerves; and there is set up an irritated reaction of the circulatory system, which continues through the second stage, or that of febrile reaction. This irritated reaction is kept up till the system becomes too much exhausted to sustain even that; and so, after a few days, perhaps a week, there comes a collapse, through which the patient may pass on in another week to dissolution; or else, with proper aid, to permanent but slow convalescence.

In those cases in which local inflammations occur, whether of the brain, lungs, liver, or bowels, it is evidently accidental; of a passive character; and results from the general prostration, and also the deranged condition of the fluids and solid tissues of the system.

In congestive typhus so great is the prostration of the system, in some instances, that there is really no reaction of even an irritated character, the brain and nervous system being apparently incompetent to keep up the circulation, or carry on the functions of the body; and so death occurs without any reaction or local inflammation, at least so far as can be discovered by post-mortem examination. And finally, another strong reason why typhus should be regarded as a disease of the blood, involving a changed or decomposed state of that fluid, is the very general hemorrhagic tendency in this disease, which is not found to prevail to so great an extent in fevers produced by other febrific agents.

Prognosis.—The prognosis in typhus depends very much upon the habits and consequent predisposition of the patient, and also upon the prevailing epidemic influence. In simple typhus, however, the prognosis may be regarded as rather favorable, especially if no complications arise.

But in inflammatory and congestive typhus the danger is greater, in proportion to the parts involved, or the severity of the congestion. A moderate diarrhœa, occurring early in typhus, or a slight bleeding from the nose, the sixth or seventh day, are favorable indications. The most certain sign, however, is to be found in the sensorial functions, for if these are but slightly disturbed during the collapse, it indicates a less changed state of the blood, in consequence of which the case will probably terminate favorably.

The unfavorable symptoms are violent delirium during the stage of excitement, blindness, involuntary flow of tears, paralysis of the tongue, a frequent small, irregular pulse, distortion of the muscles of the face, tenderness of the abdomen, continued motion of the hands, dysenteric stools, aphthæ in the mouth, colliquative hemorrhage, &c.

Treatment.—Taking into account the deviation from the standard of health in putrid or typhus fever, the indications become very plain, not only in the simple, but also in the inflammatory and congestive varieties, and if the indications be promptly fulfilled during the early stages, I believe many cases may be arrested or, at least, very much cut short. We have seen that there are three prominent varieties of typhus: the simple, the inflammatory and the congestive; all of which require modifications in the treatment.

When called to a case of typhus in the early stage, before the febrile reaction is set up, we find the patient prostrated, chilly and irritable, generally with pain in the head, back and limbs, all of which is produced by debility, and especially of the cerebro-spinal and nervous system, involving of course the circulatory system, the blood having ceased to flow freely to the extremities.

Now to arrest the chilliness, equalize the circulation, promote perspiration, and restore as far as may be the functions of the body, the feet should be placed in warm water, the back should be rubbed with a strong decoction of capsicum in vinegar, and warm sage tea allowed as a drink. This will do much to arrest the chilliness, equalize the circulation, and to rally the sinking powers of the system, and will sometimes send forth a gentle perspiration.

If there is much nausea, or a very bad taste in the mouth, an emetic of ipecac may sometimes be given with advantage at first, and then followed by a cathartic. After the emetic, or at first when no emetic is indicated, a cathartic is generally required to affect slightly the gastric and hepatic functions, and to carry from the alimentary canal any accumulation that might become a source of irritation. What I have generally found to do best in such cases are two or three blue pills, followed in six hours by half an ounce of castor oil, and repeated if necessary. Or, if there is a strong tendency to the brain, and constipation, ten grains of calomel, with twenty grains of rhubarb, may be administered in half an ounce of castor oil, and the oil repeated if necessary.

A liberal sinapism should be applied over the stomach and bowels, and repeated morning and evening, as well as the warm foot bath, and stimulating application to the spine, if the exercise can be borne without worrying the patient. By the warm foot bath we lessen the tendency to the brain and promote perspiration; by the stimulating friction along the back the circulation is more or less equalized, while the sinapisms to the abdomen will guard the alimentary canal and other abdominal viscera, and by a general stimulating effect help sustain the sinking powers of the system.

After the operation of a cathartic the sulphate of quina, or quinine,* to restore the tone of the system, to equalize the circulation, and also to counteract the change going on in the blood, is evidently indicated, as well as a mild anodyne and diaphoretic. In such cases I generally use quinine and Dover's powder combined, of each three or four grains every four or six hours, and allow crust-coffee, one-third or one-half milk, to keep up the strength, and as soon as the stomach will bear it, a little toast, or a poached egg, at regular meal hours.

In this way, I believe, a large majority of the cases of typhus that are attended to early may be arrested and convalescing in five or six days, especially if the patients can have the benefit of good fresh air, clean clothing and good nursing. I suspect that the quinine, besides its general tonic effect, operates to arrest or lessen the morbid change going on in the blood, and probably, in some degree, to destroy or neutralize the contagious poison that has entered the blood, and is producing the disease. But if from neglect, or bad treatment at first, the disease has become more seated and complicated, and the febrific agent has had time to produce its worst effects upon the blood, then this course of treatment, though it may not arrest the disease, will always lessen its violence, and render it less tedious and protracted. At least such has been the result of my observation.

If in such cases the stage of excitement has passed by, and the stage

---

* I prefer the term *quinine* to that of *quina*, and shall therefore generally make use of it, as I have done.

of collapse is approaching, the Dover's powder should be discontinued, and the quinine continued with camphor. A convenient mode of administering them is to rub thirty-two grains of camphor, and an equal quantity of quinine, with a scruple of dry prepared chalk, and then add two ounces of water and mix carefully. Of this a teaspoonful, containing two grains of each, may be given every four or six hours.

A blister should be applied over the stomach, and later over the bowels if necessary, in most cases that are not arrested by mild means in the early stages, for by it we secure the safety of the mucous membrane of the alimentary canal, and other abdominal viscera. Even if the abdominal viscera are not involved especially, a blister to the epigastrium has often a salutary effect, allaying the most alarming symptoms of cerebral affections. In cases of that kind in which there was delirium, picking at the bed clothes, and other symptoms of an alarming character, I have known them speedily arrested by a blister to the epigastrium.

After blistering over the stomach, and bowels if necessary, if the symptoms continue alarming a blister may be applied to the back of the neck to allay any local irritation there may be in the brain. After the first cathartic in simple typhus, the bowels should be moved once in every twenty-four hours by castor oil, or by an injection of a pint of milk and water, in which is dissolved a tablespoonful of salt, lard and molasses.

The strength should be sustained by toast water, or crust-coffee one-half milk, taken freely as a drink, at all hours; and during the collapse arrow root may be freely allowed, and in some cases of great prostration wine whey may become necessary. During the stage of febrile excitement, if the heat of the surface be intense, the skin may be sponged with moderately cool water once every twenty-four hours. This, however, should never be done when the skin is cool or moist, moderately warm water being alone suitable in such a state.

In inflammatory typhus the early treatment should be the same, only an emetic should not, I think, generally be given, especially if the brain be involved, as the vomiting might increase the local inflammation. If the brain be the seat of the inflammation, a cathartic of calomel in castor oil should be administered, the warm foot bath, and stimulating friction along the back, should be used; but instead of the Dover's powder and quinine, the antimonial or James's powder should be substituted, in three or four grain doses every four hours, as they lessen the tendency to the brain, while the Dover's would rather increase it, and the quinine might act unfavorably.

A blister should be applied to the back of the neck, and if the inflammation continues after the blister has produced its effects, two grains of calomel may be given with each dose of the antimonial powder, and continued for twenty-four or thirty-six hours, and then half an ounce of castor oil administered, if the calomel has not moved the bowels. In such cases, immediately on the approach of collapse, I would give quinine and camphor with the antimonial, or James's powder. Three grains of quinine and two of camphor may be given with the four grains of James's powder every four hours, and continued through the state of collapse, or as long as the condition may require.

The strength should be sustained by toast-water, at first, and during the collapse, arrow-root cooked in broth, wine whey, and as soon as it may be borne, a poached egg, rice, toast, &c., may be allowed at meal hours.

In cases where there is inflammation of the alimentary mucous membrane, the early treatment should be in every respect as in the simple variety, only a cathartic of an active character should not be given. Twenty grains of rhubarb, with ten grains of the hydg. cum creta, may be mixed with half an ounce of castor oil, and administered; or if the mercurial may be contra-indicated, half a drachm of the fluid extract of the leptandra virginica may be given instead, and repeated in six hours if necessary; or what might be more convenient, one or two grains of the leptandrin, its most concentrated preparation.

The quinine and Dover's powder may be given, as in the simple variety, but all the medicines which become necessary in such cases, should be administered in mucilage of gum arabic, and arrowroot cooked in milk and water allowed for food. This may be prepared by dissolving a teaspoonful of arrow-root in two ounces of cold water, and stirring it into half a pint of milk and water of each equal parts, raised to the boiling point; after which it should be boiled four or five minutes, and seasoned to suit the taste of the patient.

In cases in which the lungs are inflamed, after the early treatment suggested for the simple variety, I would give quinine, Dover's and James's powders combined, in doses of four grains each, every four or six hours, adding a grain of calomel to each dose, for the first day or two if necessary. A blister should be applied over the inflamed lung, and mucilages freely allowed.

In congestive typhus the same early treatment should be applied, and the patient immediately put upon the sulphate of quinine, in six or eight grain doses, every four or six hours, with four grains of Dover's during the first, and two grains of camphor during the latter stages of the disease. The strength should be sustained from the first, and during the whole course of the disease, by broths, beef essence and wine whey, if necessary, and under such treatment the most foreboding cases may sometimes recover. Thus we have the principles which are to guide in the treatment of typhus continued fever, requiring, however, to be modified, to fulfill the indications which arise in each particular case.

SECTION VI.—YELLOW FEVER.

By yellow fever, I mean that variety of febrile affection, regarded by some as remittent, and by others as continued fever, occurring in warm climates, and characterized by a peculiar expression of the countenance, a yellow skin, and by the ejection from the stomach of a dark fluid, which has been called *black vomit*.

This disease seldom prevails, I believe, at a distance of more than 40° north or south of the Equator; being endemic in regions north and south of the Equator to about the 30° or 32°, and occasionally epidemic from that to the 40° or 42°, but by no means equally distributed in those regions.

This disease generally prevails mostly in large towns, situated upon the sea-coast, or along rivers emptying into the sea, where large numbers of human beings are congregated; but seldom prevails, I believe, to any great extent in a scattered population. There, it is thought, that a temperature of about 80° F. continued for nearly, or quite three months, is requisite for the appearance or prevalence of yellow fever; though it is probable that sporadic cases may occur where the temperature never remains to that point for half that time.

Yellow fever generally makes its appearance in localities where it is endemic or epidemic in the latter part of summer and in autumn, but invariably receives a check on the appearance of frosts, and very soon after entirely disappears.

This disease always has some symptoms or features which are uniform or similar, and yet there is scarcely any variety of disease which afflicts the human family, which, in different localities, or in the same locality in different epidemics, presents such varieties of symptoms and severity. And even in the same epidemic, the range of symptoms and severity are very great, varying from the mildest cases, little worse than simple bilious remittent fever, to the congestive, in which vitality is smitten down, as it were, at the very moment of attack, and the patient swept away in two or three days, or perhaps as many hours.

To give the history, with all the peculiarities and varieties of yellow fever, would require a volume. I shall, therefore, not attempt to do more than to give the general character of the disease, as it most generally prevails, leaving the reader to fill up the history and multiplied varieties, as it occurs in different localities, by the accounts furnished by eminent physicians residing in each locality where the disease presents its peculiarities. With these considerations, let us proceed to inquire into the symptoms of yellow fever, as it generally prevails, from a correct understanding of which all its varieties may be readily appreciated and understood.

Symptoms.—In some instances, yellow fever makes its attack with scarcely no premonitory symptoms; but generally there are pains in the back and extremities, and sometimes in the head; the patient feels more or less indisposed for an indefinite time, during which there is a gradual sinking of all the powers of life, involving more or less derangement of all the functions of the body. After these symptoms have continued for an indefinite time, when they do appear, or at the very first when they are not noticed, there is generally slight chilliness, though in many cases no regular chill, and very soon febrile reaction is set up, developing the following symptoms. There is restlessness, thirst, a hot and dry skin, the respiration is hurried, the eyes red and watery, the face flushed, the tongue becomes covered with a white fur, there is more or less soreness of the fauces, and difficulty experienced in swallowing, and sometimes nausea and vomiting immediately on the setting up of febrile action.

Generally, in the course of twenty-four hours from the chilliness, or time of febrile reaction, the patient complains of great oppression, weight, and tension in the epigastrium; there is more or less tenderness on pressure, and vomiting, especially of everything taken into the stomach; the vomiting being often violent and distressing from the flatulence which is apt to attend.

The patient complains of great gastric heat, and therefore has an intense thirst for cold drinks. The bowels are generally constipated, the discharges being also of an unhealthy appearance, and very offensive. There is pain in the forehead, eyes, and sometimes in one side of the head, of a violent and most distressing character, and is only surpassed by the excruciating pains in the back and extremities, which often continue during the period of febrile excitement.

There is more or less mental disturbance at this stage, the patient being exceedingly restless, and more or less delirious, and often wild and furious in his movements; but there may be instead, in congestive cases, drowsiness, stupor, coma, &c.

This stage of febrile reaction may continue, in violent cases, but a few hours; but in mild cases it is sometimes prolonged to three or four days; but the average duration of this febrile stage may be set down at two days, being evidently an irritated reaction of the circulatory system, holding out for a longer or shorter time, according to the violence of the attack, and the natural powers of the system.

As the system becomes exhausted, this irritated febrile action subsides more or less, the skin becoming cool and soft, the pulse more natural, the respiration less hurried, the stomach more quiet, the pain in the back and head subsides, and the general distress of body and disturbance of mind is followed by comparative ease, and more or less cheerfulness.

In mild cases, this subsidence of the fever may be the commencement of convalescence, but not so in severe malignant or congestive cases, for then it only indicates inability on the part of the system to keep up even an irritated action of the circulatory system. The tenderness of the epigastrium is rather increased. In place of the redness of the eyes and face, they have a yellow or orange-colored appearance, which gradually extends itself till it covers the whole body. The urine becomes yellow, the pulse much slower than in health, and in severe cases, there is sometimes a little drowsiness or stupor.

This period of abatement may last for twenty-four hours, when there comes on the stage of debility, prostration or collapse, during which the pulse becomes feeble, quick and irregular; the skin becomes more yellow; the circulation becomes languid, especially in the capillaries; the extremities become purple; the tongue becomes brown and dry, or reddish; sordes gather about the teeth; the stomach becomes irritable, and ejects a darkish liquid, containing darker flakes which finally become black; the blood recedes from the extreme vessels; the urine may appear natural or be nearly suppressed, blood may ooze from the gums, or from the mucous membranes in different parts of the body, or it may be thrown from the stomach, or pass from the bowels, or with the urine; and it is not unfrequently extravasated under the skin, in different parts of the surface of the body

By degrees the pulse becomes almost imperceptible at the wrist, there is frequent sighing, and hiccough, the skin becomes cold and clammy, large quantities of blackish matter are ejected from the stomach, and discharged from the bowels, an offensive odour exhales from the body, the eyes are sunken, there is a low muttering delirium, the countenance

becomes cadaverous, and finally the patient dies, either in violent convulsions, or else passes off rather quietly. In some cases, however, the
patient, instead of passing off thus rapidly, may linger for perhaps two
or three weeks in a typhoid state, and then sink down, or else slowly
convalesce.

When the patient does recover, either from a violent and rapid course,
or from a typhoid and lingering run of fever, the convalescence is slow
and tedious.—Thus we have the general symptoms and course of yellow
fever, exhibiting three distinct stages: the first, that of primary febrile
action, which may generally continue for two days; the second, that of
abatement, lasting about one day; and, finally, the stage of collapse,
which may generally last two days in rapidly fatal cases; or the patient
may pass into a slow typhoid state; or there may instead be slow convalescence.

In some cases, there is no febrile reaction, at least that is perceptible,
the patient being struck down as it were from the first, in a congestive
state; or if there is a slight effort at reaction, it soon gives way, and the
system remains prostrated, and in a congested state, the stomach eject-
ing large quantities of black, or bloody matter, the skin becoming dark
yellow, and the patient rapidly sinking with hemorrhages, and all the
more alarming symptoms of this truly malignant disease. It has been
noticed by some, that there is an offensive breath in these more malignant cases, and in some epidemics, a rash has made its appearance upon
the face, neck, and some parts of the body, during the first stage of the
disease.

It will be noticed, that the most strikingly peculiar, or characteristic
symptoms of yellow fever, are the rapid, irregular, and jerking, or else
slow and feeble pulse, a reddish appearance of the upper part of the
face, and bloodshot appearance of the eyes, the yellow appearance of
the skin, coming on by the third or fourth day; and, finally, the *black
vomit*, which makes its appearance generally during the second or third
stage of the disease.

In relation to the exact nature and composition of *black vomit*, there
has been some doubt; animalculæ and microscopic fungi have, in some
cases, been detected in it; but its principal ingredients have been found
to be disintegrated blood. It is, therefore, rendered probable, that it
results from, or is composed of, dissolved blood; the clear liquid being
the serum, and the insoluble part, the coagulable matter of the blood.

Diagnosis.—In forming a diagnosis in yellow fever, it is necessary to
remember its most prominent characteristic symptoms, the rapid and
feeble, or slow and irregular pulse, the red appearance of the eyes and
face, and the *black vomit*, as well as the severe, and often excruciating
pains in the back and lower extremities. These symptoms, together
with the fact of the prevalence of the disease, and the peculiar symptoms which are developed during the febrile stage, the abatement and
the stage of collapse, will generally enable the acute observer to form a
correct diagnosis in most cases of yellow fever.

Anatomical characters.—The dead body, according to Professor Dickson,* presents various morbid appearances, there being generally more

* Dickson's Elements of Medicine, pp. 273, 274.

or less "ecchymosis or subcutaneous effusion of blood." The blood is also represented as being dark, and frequently incoagulable, "the heart as soft and flabby," sometimes undergoing a fatty degeneracy, the lungs "being firm, dark, and heavy," exhibiting marks of inflammation or congestion. "The brain is usually found more or less altered," its membranes appearing inflamed; and it has been stated, that in some cases, serum has been found in the ventricles.

The liver, I believe, does not present any very constant appearances, it being sometimes "soft and flabby," while in others, it is more or less engorged with darkish appearing blood. Professor Dickson states that he has often found it "in a natural state, so far as could be judged by the naked eye." In six cases examined by him, "one was fawn-colored, very characteristic; a second was fawn-colored on the upper surface, and of a deep purple on the lower side, and around the edges; two were dark chocolate-brown, not far from natural, or olive-colored; the sixth was again of coffee and milk hue, being large and heavy."

The spleen, according to Professor Dickson's observation, is sometimes more or less enlarged, and softened; but often remains nearly or quite natural. The stomach he has always found "reddened and injected on its inner surface," and "sometimes on its outer surface also;" and, occasionally, it was found softened, but he never found ulceration of any portion of it.

The mucous membrane of the œsophagus and duodenum, he found in a similar condition, and in hemorrhagic cases, he found more or less bloody infiltration in every part of the system. Various other morbid appearances may sometimes be found on post-mortem examination of patients dying of yellow fever; but none of them, I believe, are of any very great uniformity or importance.

It will be noticed that I have given the observations of Professor Dickson on the post-mortem appearances of yellow fever patients. This I have been induced to do, from the fact that my position in practice at the North, has not allowed me that amount of observation in such cases that I could desire. Professor Dickson, who has for many years been a practitioner of medicine in Charleston, South Carolina, where yellow fever prevails more or less, has had abundant opportunity for observation; and what is more, has been a close observer, not only in such cases, but also in every variety of disease coming under his observation.

It will be seen, by taking a glance at the general anatomical characters of yellow fever victims, that the most important morbid appearances presented, are the dissolved, or changed appearance of the blood, and the congested and inflamed appearance of the mucous membrane of the œsophagus, stomach, and duodenum; the morbid appearances of the liver, spleen, brain, &c., though very frequent, appear by no means essential morbid appearances in such cases.

Causes.—In relation to the cause of yellow fever there may be some room for doubt, but when we take into account the fact that yellow fever almost always prevails on ships at sea, upon the sea coast, or upon navigable rivers emptying into the sea, we have got at least a clue to the cause which may produce it. And especially have we, when we remember that it almost invariably prevails in these positions, in localities where a large number of human beings are associated, and where the paludal poison prevails to a greater or less extent.

Now, from the fact that yellow fever prevails where the paludal poison is generated, and where a collection of human beings furnish the materials for the generation of idio-miasmata, we have a right to infer that *koino* and idio-miasmata combined produce yellow fever. And though this position may not admit of positive proof, it appears to me that all the circumstances favor this idea.

It may be asked, if the koino and idio-miasmata, with a temperature of 80° F., is all that is requisite, why may not yellow fever prevail away from the sea coast and large rivers emptying into the sea, provided there be the material for generating the requisite animal poison, and also the presence of the paludal poison, as well as a temperature of 80° F.

This question may be one more easily asked than answered. But it should be remembered, that observations long ago made, and confirmed by more recent observation, have rendered it quite certain that a mingling of salt water and fresh along the sea coast favors the generation of the paludal poison, and also greatly increases its virulence.

M. Monfalcon mentions some interesting examples illustrating this fact.* "The extensive pool of Valdec, in the south of France, is quite saline, and only a few rods from it is a large pool of fresh water. Now, when the waters of these two pools rise and run into each other, as they occasionally do, much sickness soon occurs throughout the adjoining parts." "In the vicinity of Lukes, on the south of the Ligurian Appenines, there is a large marshy plain, accessible to the high tides of the ocean. The neighboring districts were almost uninhabitable from the pestilential effluvia which emanated from this marsh, until the waters of the sea were separated from the sweet water of the marsh, by means of sluices and hydraulic works, when it became healthy, and the population increased rapidly." †

Now, if it be a fact that a mingling of salt and fresh water not only favors the generation of the marsh miasmata, but also renders it much more virulent, may it not account for the fact that yellow fever prevails mostly in localities along the sea coast, where there is always more or less mingling of salt and fresh water. And if my position be true, that this peculiarly virulent paludal poison, with an animal poison, are together the cause of yellow fever, then we should suppose it would prevail in such localities upon the sea coast as furnish the most mingling of salt and fresh water, and also the greatest amount of animal secretions, exhalations, &c., and not in localities, even along the coast, where human beings are not assembled, as is really the case.

This position appears to me also strengthened by the fact, that not only a temperature of 80° F. is requisite, but it also requires to be continued for two or three months in order to produce yellow fever, which would be requisite if the decomposition of vegetable matters were taking place at some depth in the water, as it very probably is, the animal filth at the same time undergoing the requisite change.

Another strong reason for supposing that koino and idio-miasmata are,

---

* Histoire Medicale des Marais, Paris, 1828.
† Eberle's Practice, vol. i., p. 53.

together, the cause of yellow fever, may be found in the post-mortem appearances, which answer very well to what we might expect would be found, if an animal and virulent paludal poison produced the disease. For in the body dead from yellow fever we find the changed or dissolved state of the blood, and the general hemorrhagic tendency which we find in typhus fever, and also the marks of congestion, irritation, and inflammation of the alimentary mucous membrane, so common in bilious remittent fever.

And finally, the symptoms and course of yellow fever are what we might reasonably look for, if *koino* and *idio-miasmata* were operating upon the system to produce or develop them, the one passing as a foreign substance with the blood, and producing its effect upon the brain and nervous system, as in bilious remittent and intermittent fevers, and the other, or animal poison, uniting chemically with and changing or decomposing the blood as it probably does in producing pure typhus fevers. This supposition will also account for the fact that some cases, and even some epidemics, partake more of the paludal character, and even run into bilious remittent or intermittent fevers, while other cases, or other epidemics, partake more markedly of the typhus and hemorrhagic character; in the one case the animal, and in the other the paludal poison existing in the greatest degree of concentration.

This view of the cause of yellow fever also accounts for or settles the question as to its being contagious or not, rendering it contagious in just so far as it partakes of the nature of typhus, and not contagious in just so far as it is paludal; and this accounts for the apparent differences in the contagiousness in different cases, and also in different epidemics. Those cases and those epidemics partaking most of the typhus character being more contagious, while those cases and epidemics of a more paludal character being less so.

Besides all this, it was long ago observed, and I am confident, as I have stated elsewhere, that the combination of koino and idio-miasmata, acting upon the system, will produce a most malignant variety of febrile disease, as I have witnessed in our climate.* And if so, may we not suppose that in a hot climate, where there is a mingling of salt and fresh water, and also a large collection of people for furnishing animal matter for decomposition, that the virulent animal and vegetable miasmata thus generated may produce every possible variety of yellow fever.

Nature.—Having inquired into the symptoms, anatomical characters, cause, &c., of yellow fever, we are prepared to investigate its nature, and if we can draw no positive conclusion in relation to its exact nature, it appears to me that we may arrive at a rational conclusion, which may be regarded at least as probable. If, as I have supposed, yellow fever arises from the combined influence of the marsh and animal miasmata, we have an easy way of accounting for the symptoms which arise, as well as an explanation of its nature. For, as we have seen in a former section, if the idio-miasmata be operating upon the system, it probably, on entering the blood, unites chemically with it, changing or decomposing it in a greater or less degree, and this will explain the reason of the great prostration, the typhus and hemorrhagic tendency, and also its contagious characteristics.

* That of Geneva, Central New York.

If the paludal poison be operating upon the system in producing yellow fever, it probably, as we have already seen, on entering the blood, passes with it through the system as a foreign substance, producing its effects directly upon the brain and nervous system; and this will **account** for the apparent **remitting** tendency, for its often running into or partaking strongly **of** the nature of bilious remittent and even intermittent fevers, and it will also account for the apparent similarity, in many respects, **between** the post-mortem appearances in bilious remittent and yellow fevers.

If now, as I have supposed, both the **koino** and idio-miasmata are operating upon the system at the same time—the one decomposing or changing the blood, and producing the prostration, **the** hemorrhagic tendency, the black vomit, and the typhus characteristics, and the other producing the yellow skin, the remittent and other bilious **characteristics** —we have every symptom accounted for, and a rational solution of the nature or true pathology of yellow fever.

That such is the true nature of yellow fever, I have not in my own mind a reasonable doubt.

With these considerations, I shall leave this part of our subject, hoping that the suggestions I have made may receive that degree of consideration on this truly perplexing question that they may deserve—not for my sake, but that the subject may be brought as near to a rational solution as it can well be from the circumstances by which it is surrounded.

Prognosis.—Yellow fever is truly a fatal disease, partaking as it does, in my opinion, of the nature of typhus and bilious remittent fevers. We have, in addition to the dangers usually attending both these febrile affections, a degree of malignancy resulting from this combination, as well as from the peculiar virulence of the febrific agents which produce it. With these considerations, it is not strange that so many should be smitten down by this disease, especially when we remember that every species of imprudence and intemperance practised by the masses of the human family not only act to predispose the system to this disease, but also tend to develop the disease, where it is prevailing, in a way not conducive to a favorable result.

Dr. Fenner estimates that about one-eighth of the cases occurring in New Orleans, in the epidemic of 1847, proved fatal.* Professor Dickson estimates that the proportion of fatal cases in Charleston is about one-fifth or one-sixth, while the average loss in his own practice, in eight epidemics, during a period of about thirty years, was only one in fifteen.†

The unfavorable **symptoms** are violent pains in the forehead, back, and limbs; a frequent and feeble pulse; **a** blood-shot appearance of the eyes; a deep **yellow skin** or mahogany color; short febrile reaction; coma; slow and **difficult** respiration, with hiccough; restlessness; **a morbid appetite;** black vomit; and, finally, a universal **hemorrhagic** tendency and great prostration.

The favorable symptoms are a lengthened febrile reaction; **the occur**rence of gentle perspiration; dark bilious evacuations; cleaning of the tongue; slight vomiting, subsiding early; and a tolerable retaining of

---

* N. Med. and Surg. Journal, v. 206.   † Charleston Med. Journal, xii. 744.

10

the integrity of the system, and the absence of all the unfavorable symptoms enumerated above.

Treatment.—If we take into account the deviation from the standard of health in yellow fever, we shall have the indications for a rational treatment.

It is fair to state, that in yellow fever we have a debilitated condition of the system, produced by a febrific agent or agents, which decompose or change the blood, and also produce a direct effect upon the brain and nervous system of a most debilitating character. And I believe it is also true, that nearly or quite every unfavorable symptom which arises during the continuance of the disease, is either directly or indirectly the result of debility, not excepting even the febrile reaction, which is evidently the result of irritation set up in the cerebro-spinal system during the chilly or cold stage, in congestive cases the system not retaining power enough to get up even an irritated reaction.

This, then, being the deviation from the standard of health, the indications are, plainly, to equalize the circulation; to sustain the sinking powers of the system, and thereby keep up or restore the deranged functions of the body; to counteract local congestions and inflammations; and, finally, to afford the patient sufficient nourishment, fresh air, and every comfort that may in any way be conducive to a restoration to health.

In ordinary cases of yellow fever, then, at the period of attack, if febrile reaction has not been set up, it should be encouraged by rubbing the whole length of the back with a warm infusion of capsicum in vinegar, and the administration of warm drinks. And if there is severe pain in the head and back, the feet should be placed in warm water, rendered stimulating by mustard or capsicum. To aid in this, and prevent early congestion and inflammation, a mustard poultice may also be indicated over the whole abdomen, and, when it is, should never be neglected.

As soon as these indications are fulfilled, if the patient be seen before reaction is set up, or at the very first, if the patient be not seen till reaction is established, a cathartic of calomel, to clear the alimentary canal of any irritating matters it may contain, should be administered. Ten grains may be given at once, and then half an ounce of the sulphate of magnesia given every six hours till the bowels are moved.

Immediately on the operation of the cathartic, a sustaining course of treatment should be resorted to, as it is plainly indicated, and continued as long as treatment may be required—modified, of course, to meet the indications which are furnished in each particular case.

In ordinary cases, to sustain the system, keep up its various functions, &c., four grains of quinine, with an equal quantity of Dover's and James's powders, may be combined, and given every four or six hours, with two or three grains of calomel, if the bowels are rather constipated and the case appears foreboding by a torpid inactive state of the liver. This prescription is well adapted for the febrile stage, the quinine preventing local congestions, and sustaining the system against an approaching collapse, and, as I believe, modifying, or in part neutralizing the febrific agent. The James's powder lessens the cephalic tendency and promotes perspiration, while the Dover's quiets nervous excitability, favors sleep, and also promotes perspiration, and finally, the calomel, when it is indicated, tends to equalize the circulation, promotes a healthy action of the liver and whole glandular system, and lessens the danger of local inflammations.

This treatment may be continued during the febrile stage, with whatever else may be indicated, such as sponging the surface with cool water, a blister to the back of the neck, and sinapisms to the epigastrium, and, if necessary, a blister. Immediately on the approach of collapse the Dover's and antimonial may be discontinued, and the quinine continued in five or six grain doses, with an equal quantity of camphor, with or without the calomel, every six hours, and continued with warm crust-coffee, one-half milk, and, if necessary, mutton or chicken broth, wine whey, &c., till either death or convalescence render them no longer necessary.

To help restore the integrity of the blood, and to exert a styptic effect upon the mucous coat of the stomach, the tincture of muriate of iron has been used, and first, I believe, by the late Dr. Wildman, of Savannah. During the collapse from twenty to thirty drops of the muriated tincture may given in a little water, every six hours, alternating with the quinine. This course of treatment should be continued during the collapse, the quinine and camphor sustaining the sinking powers of the system, while the muriated tincture of iron not only acts as a styptic, but also probably tends to produce a favorable change of the blood, and is generally agreeable to the stomach.

It may be noticed that I have said nothing of the period of abatement in the treatment, regarding the condition immediately on the subsidence of the febrile action as being either convalescence or tending to a collapse, and therefore requiring the increase of quinine with camphor in all cases, except, perhaps, in mild cases, in which it may be the commencement of convalescence, when often the doses of quinine, instead of being increased, may admit of being gradually diminished.

If the vomiting be severe and obstinate, or if there is evidence of gastric inflammation, a blister should be applied to the epigastrium, after sinapisms have done their utmost, and, in some cases, a blister to the abdomen, below the umbilicus, may become necessary. To stop the vomiting when it is commencing, one grain of calomel, with a grain of Dover's powder, may be administered every fifteen minutes, till four or five doses are given, and often with very satisfactory results, as I have found in many cases.

Such appears to be the general character of the indications arising in ordinary cases of yellow fever, and also the best mode of fulfilling them. But cases of a congestive character sometimes occur in which there is no febrile reaction, the vital powers being so far smitten down that the system will not arouse to even an irritated febrile state. In these congestive cases the same early treatment should be resorted to; such as stimulating frictions to the spine, the warm stimulating foot bath, the warm nourishing drinks, cathartic, &c. But, instead of commencing with the quinine in four or five grain doses, with the Dover's and antimonial powders, ten grains of quinine and five grains of camphor may be given from the first every six hours, and alternating with it, from forty to sixty drops of the muriated tincture of iron. And this treatment should be continued with baths, wine whey, and brandy if need be, and sinapisms, blisters, &c., as the case may require.

In such congestive cases, if the quinine and camphor fail to arouse the system, ten grains of capsicum may be given with each dose of the quinine, or what would generally be better, alternating with it at the time of administering the iron. It is most conveniently given in the form of a pill, and may require in some cases to have combined with it a grain or two of opium to render it acceptable to the alimentary mucous membrane.

During the whole course of the disease the patient should get as much nourishment in the form of crust-coffee and milk, broths, wine whey, &c., as the system requires and the stomach will retain. Great care should also be exercised to keep the apartments clean and well aired, and no case should be abandoned as hopeless while one vital spark remains. Thus I have completed what I had to say upon yellow fever, and also the general febrile affections, except plague and diphtheria, which I shall proceed to consider in the following sections.

SECTION VII.—PLAGUE.

By PLAGUE I mean a well-known general febrile affection endemic in Egypt, and other countries bordering on the Levant, and characterized by petechiæ, carbuncles, and swellings of the lymphatic glands of the groin, axillæ, &c. I can only give a general description of plague, embracing the most prominent facts connected with it, as I have never seen a case. The description then which I shall give of this disease will not be from personal observation, but will be such as I have gathered from those who are familiar with the disease.

It appears that plague is endemic in countries bordering on the Mediterranean, and that it has made frequent eruptions into various parts of Europe, having visited at different times France, England and Germany, and having been exceedingly fatal in Paris and London in early times, according to the accounts furnished.

It has been supposed, however, that Egypt is the only country where it is ever engendered, and that when it occurs elsewhere it is as an alien.

In Egypt it is said to appear every autumn, and to prevail more or less till the following June, when its ravages cease and its contagion remains inactive till early autumn when the disease again prevails, being the most fatal about the time of the vernal equinox. It has been supposed to arise in Egypt from putrid emanations exhaling from the decomposition of animal and vegetable substances decomposed in the lakes, formed by the retiring of the waters of the Nile, or in the cemeteries which its inundation often reaches.

Symptoms.—The symptoms which arise in plague, as described, are similar in some respects to those of malignant typhus, the patient being very much prostrated early, with a sense of weight in the head, giddiness, and great depression of spirits. They are silent, making often little complaint, and having little fever they may appear to a careless observer as being only slightly indisposed, and yet they may be smitten down, in some instances, within two or three days. In such rapid cases, however, no buboes or carbuncles make their appearance.*

In other cases, though the first attack may be similar to that I have already described, the disease develops other and more prominent symptoms in a few hours; the eyes become dull, the extremities, and surface of the body become cold, there is drowsiness, and more or less pain in the region of the heart. As the disease advances, the patient loses the power of speech, is covered with a cold clammy sweat, the pulse becomes weak, and quick, there may be more or less delirium, but a comatose disposition generally prevails. Towards the close there is great restlessness. Buboes may appear in those who continue to the fourth day, as well as petechiæ; or else broad livid spots may occur, though the latter generally not till after death.

Such are the ordinary symptoms of plague, as it appears in its most violent forms at the commencement of an epidemic. But at a later period in an epidemic, plague assumes a milder form, though still a very dangerous disease. In this milder, which is by far the most frequent form of plague, the disease commences with more or less coldness, which may amount to shivering, followed by fever, vomiting, and more or less giddiness and pain in the head. The fever increases during the day, and following night; but in the morning, there is usually a very perceptible remission.

Usually in this form of the disease, carbuncles and buboes make their appearance on the first day, and successive eruptions of them may appear throughout the whole course of the disease. During the second day, there is an exacerbation of fever, the patient being afflicted with nausea, distressing headache, severe pain in the swellings, and more or less confusion of mind, with a tendency to coma. The skin becomes hot, the pulse more frequent, and the tongue quite dry; the patient being anxious, restless, and complaining of more or less pain in the region of the heart.

During the following night, all the symptoms become aggravated; the patient either raving incoherently, or else inclined to a profound stupor. Usually on the morning of the third day, there appears a gentle perspiration, with more or less abatement of all the more violent symptoms, and may even be the commencement of convalescence; but it generally only furnishes ground for hoping that a favorable crisis may occur on the fifth day. It is said, that in all cases in which there is no sweating or remission on the third day, great danger is to be apprehended.

The exacerbations may increase in violence up to the fifth day, at which time, the remission and sweating may be more considerable, leaving the patient faint and languid, but in every respect, improved or relieved. After the fifth day, the remissions may become gradually more marked, the exacerbations being lighter each day, and the buboes advancing to suppuration; so that by the eighth or tenth day, little fever remains, except that which is symptomatic of the local inflammations.

In some cases, all these symptoms are greatly aggravated, there being violent headache, vomiting, and purging, the face flushed, and eyes glistening, there is violent delirium or coma, the pulse is full and strong, the tongue dry, and thirst excessive, and the most distressing vomiting

* See Russell's History of the Plague at Aleppo in 1760, 1761, and 1762. Page 96.

occurs, especially during the night. These symptoms may continue with only slight remissions, the exacerbations gradually becoming more violent or distressing, till the tongue falters, the surface of the body becomes cold, and covered with a clammy sweat, the vomiting and diarrhœa continuing, and often hemorrhages occurring from different parts of the body, the buboes appearing on the second or third day, do not suppurate, and in this way the patient, distracted and worn down, dies on the third or fourth day.

In some cases, again, the carbuncles and buboes appear to be the prominent feature of the disease; the patient has little fever, walks about, or even follows his accustomed avocations, as appears to have been the case with many soldiers in the French army in Egypt, during the Syrian Expedition;* and it is stated on good authority, that in some instances, patients suffering from plague, have walked to within a few hours of death.

The bubo of plague appears to be a swelling of the glands of the neck, axillæ, and groin, and sometimes of the parotid, being at first deep-seated, and painful, but gradually advancing towards the surface, and generally proceeding to suppuration. They may terminate, however, in resolution, suppuration, or gangrene; suppuration not being essential even to a favorable termination, but rather a sign of a favorable termination.

Besides the buboes which occur in glandular parts, tumors, which have received the same name, are often found on the head, neck, shoulders, chest, abdomen, or, in fact, on almost any part of the body, even the extremities. These tumors may disappear, or pass on to suppuration, though it is said more slowly than the glandular bubo. Carbuncles may appear in various forms; commencing generally, it is said, as a pustule, but sometimes like a vesication, filled with a yellow or blackish fluid. As they advance, they become hard, painful, and gangrenous, and form eschars, which spread, and destroy the skin, muscles, and sometimes even the tendinous parts. In fatal cases, the eschar, it is said, remains dry, with little or no appearance of being cast off; but in cases in which convalescence is approaching, suppuration takes place around the eschar, by which it is soon cast off, leaving an ulcer of greater or less depth, which gradually heals during convalescence.

Various other cutaneous affections have been described as occurring in plague; such as petechiæ, at first purple, and later of a livid appearance: and an erysipelatous efflorescence of a transient character; narrow streaks of a livid color upon the face, of a frightful appearance; large blue or purple spots occurring in the most dependent parts of the body; the result, probably, of infiltration, caused by the attenuated state of the blood. Thus we have an outline of the symptoms which are developed in plague, in the various forms which it assumes, in regions where it is endemic, or epidemic. Let us now proceed to inquire into its diagnostic symptoms, morbid anatomy, cause, nature, prognosis, and treatment.

Diagnosis. — It appears that there need be little, if any difficulty, in discriminating a case of plague, if the symptoms can be observed during its whole course, as most of the symptoms which are developed, are peculiar to this disease. But at the commencement of an epidemic, when

* See Dictionnaire des Sciences Medicales, vol. xii., p. 77.

the disease is usually very fatal, the patient may die early, or before its most peculiar symptoms are developed; and thus the diagnosis may be attended with some difficulty. In such cases, the following diagnostic symptoms should be carefully noted: the inflamed appearance of the eyes; the swelling of the tongue, and difficulty of articulation; the tottering gait, &c.; all of which, together with the extrinsic circumstances, such as the season of the year, the location, the possibility of exposure, &c., may aid in forming a correct diagnosis in most cases at least.

But if, in addition to the symptoms which I have enumerated above, buboes, carbuncles, petechiæ, &c., make their appearance, the diagnosis may be regarded as certain, and every precaution which prudence can suggest should be taken to prevent a spread of the disease. And even where there is doubt, prudence would dictate that the patient be kept secluded for a sufficient time to allow the disease to develop its real character, for of all diseases plague is one most to be dreaded, on almost every account.

Anatomical Characters.—Various morbid appearances have been described as being found, on post mortem examination of patients dead of plague, the most prominent of which are a changed condition of the blood, dark coagula being found in the heart, large arteries and veins, distension of the gall-bladder, with black or greenish bile; black spots, probably of extravasation, through the lungs, heart, liver, stomach and bowels; hemorrhagic effusion into the cellular tissue, and sometimes into the cavities; an engorged and enlarged condition of the lymphatic glands; and various accidental marks of inflammation, in different parts of the body, and especially in the alimentary mucous membrane.

It is proper to state, however, that in some epidemics, at least, on post-mortem examination, everything is described as being found in a natural state, in some cases; while in others, slight marks of inflammation were found only in the intestines, and that was supposed to have been produced during the last moments of life.[*] It may be seen, then, that the principal morbid appearances, when any are found, are of the blood, and such as are produced by a poisoned, changed, or partially dissolved state of the blood.

Cause.—Great differences of opinion have prevailed in relation to the origin of plague, and also in relation to the manner in which the disease is contracted; whether exclusively from a peculiar state of the atmosphere, or by contagion, or both. From all the facts that I have been able to gather on this subject, I incline to the opinion that the disease, or the morbid poison which produces it, is engendered in Egypt, and perhaps in the lakes left by the subsiding of the waters of the Nile; or, as has been suggested, in cemeteries which are often inundated, and then left exposed to the intense heat of the summer sun.

Or, what is more probable, both these causes may conspire to generate a morbid poison, from the accumulated animal and vegetable matter, which, on being brought to act upon a population filthy and intemperate, in many respects, as the lower classes of the Egyptians are, will readily produce the morbid condition of the system which has been called plague.

---

[*] See Relations Historique de la Peste de Marseille, pp. 447, 448.

When the disease is once engendered, the bodies of patients suffering from it probably generate a specific contagion, which will communicate the disease from one individual to another, in all the modes in which diseases are thus communicated, by contact, by inoculation, through the atmosphere, and by fomites. This, then, accounts for the origin of the disease, which, being once generated, is liable to spread beyond the borders of Egypt, where it is endemic, and thus become a pest and terror in different parts of the world, where filth, intemperance, and various other causes predispose to it.

It may be communicated, probably, to regions not too remote, directly from the lakes or cemeteries where it is generated, by the prevailing wind, and to regions more remote by ships, infected goods, and from the bodies of persons suffering from the disease. It has been a question whether the bodies of the dead retain an infecting power. But I think the weight of testimony is decidedly in the affirmative. This was particularly noticed in the epidemic of Marseilles, in the year 1720, and also in the plague of London, of 1665.

The latent period of this disease has been an important question, on account of quarantine regulations. It appears that the disease may develop itself, after exposure, at any time from a few hours to the tenth day, and perhaps at a later period. But the length of time that contagion may adhere to infected goods, not ventilated, is probably much longer than that. It has been noticed that extremes of temperature modify, or check, the activity of the cause which produces plague. Hence it is, that the abatement of the pestilence takes place along the Mediterranean during the heats of summer, while in the North of Europe, it has been during the cold of winter.

In relation to the causes which predispose to plague, there can be no reasonable doubt, but that filth, intemperance, licentiousness, and in fact every deviation from the laws of health, not only predispose the system to this disease, but also render the air impure, in consequence of which it becomes a more ready medium for communicating the contagion.

Nature.—In relation to the nature of plague, there may be room for doubt; and yet I think it is rational to infer that a poison, probably the result of the decomposition of animal and vegetable matters, is taken into the blood, through the skin and lungs, and effects a marked change in all the fluids and the solid tissues of the body. That this morbid poison is more animal than vegetable, and that it is taken into the system, in part at least, by the absorbents of the skin, appears to me reasonable.

And if so, it accounts in part for the buboes which occur, the lymphatic glands being irritated by the poison which is carried through them by the lymphatics, very much as we see from animal matter taken up by the absorbents, in dissection wounds. But in those rapidly fatal cases of plague, in which no buboes appear, it is probable that a very concentrated poison enters the blood, through the skin and lungs, in sufficient quantity to change at once the blood and solid tissues, and thus by a shock to the brain and nervous system, to destroy life, before the lymphatic glands have time to become inflamed, as we see in some cases of poisoned or dissection wounds.

Prognosis.—Plague has been regarded, and is really one of the most fatal diseases which afflicts the human family; the general prognosis in relation to its spreading, if introduced into a place, is in the highest degree unfavorable, unless the first cases admit of being totally secluded, that a general contamination of the inhabitants may not be produced. And the mortality of cases actually attacked with plague is surpassed, I believe, by few, if any maladies which have ever prevailed since the creation of man. It is estimated that in the plague, as it generally prevails, at least fifty per cent. of those attacked perish. In the plague at Marseilles, in 1720 and 1721, it is supposed that of a population of 80,000, about 40,000 fell victims to the disease, though about 10,000 escaped infection.*

In individual cases, hiccough, cardialgia, diarrhœa, convulsions, and colliquative sweats, are unfavorable symptoms, while the favorable signs are the occurrence of buboes, passing on to regular suppuration, a mild but well-developed febrile reaction, a gentle perspiration occurring at the remissions, and finally, the absence of all the more violent symptoms common to this most malignant disease.

Treatment.—As there is some doubt in relation to the real pathology of plague, there must of necessity be a little uncertainty in relation to the real deviation from the standard of health, as well as to the indications and best mode of fulfilling them. If, however, the morbid condition consists, as I have supposed, of a poisoned state of the whole system, both of the fluids and solids, and that, too, by a poison originating in the decomposition of animal and vegetable matters, but chiefly of animal, or else being generated in the systems of patients suffering from plague; we have a tolerable idea of the deviation from the standard of health, which may suggest for us the indications, and a rational mode of fulfilling them.

To sustain the sinking powers of the system, prevent local congestions, and consequent passive inflammations, to promote perspiration, favor a healthy action of all the functions, and hasten suppuration of the buboes when they occur, are evidently the indications to be fulfilled.

To arouse the system, equalize the circulation, and promote perspiration, it appears to me, an emetic of ipecac, at the very first, might be serviceable, and unless in some way contra-indicated, would be worth a trial. To clear the alimentary canal, and also to help equalize the circulation, if no diarrhœa be present, a mild cathartic would appear to be indicated. And that which would suggest itself to me, would be twenty grains of rhubarb, and ten grains of hydg. cum creta, and followed, if necessary, by a seidletz powder. Immediately on the operation of the cathartic, I would apply sinapisms to the feet; cups, and if necessary, blisters to the back of the neck, and give at once three or four grains of the sulphate of quinine, with an equal quantity of Dover's powder, every six hours, and this, I think, might be continued till the fever subsides.

On the subsidence of the fever, I would discontinue the Dover's, and continue the quinine, with an equal quantity of camphor; in a few days discontinuing the camphor, and gradually lessening the dose of the

* See Traité de la Peste. p. 464.

quinine. The quinine and Dover's early, would sustain the system and promote perspiration, while the quinine with camphor, in the latter stages, would sustain the system, and also quiet nervous excitability. And though the febrile excitement be very great, and local inflammations appear to be forming, I should have no fears from the quinine, for I believe they are generally of a passive character, which would be relieved instead of aggravated by the quinine, as I have witnessed in severe cases of dissection wounds, or poisoning from animal matter.

I would allow the patient early, crust coffee, or toast-water, one half milk; and later, the same, with chicken broth, wine whey, &c., if they were in no way contra-indicated. And as soon as solid food could be borne, a poached egg, toast, mutton, &c., taking care to allow the system as much nourishment as it will bear, during the whole course of the disease. To favor suppuration of the buboes, and separation of the eschars in the carbuncles, a poultice of flax-seed, slippery elm, or bread and milk, appear plainly indicated; and to the ulcers which follow, gently stimulating or soothing applications may be indicated, according to the condition of the patient.

Such, it appears to me, are the indications in the treatment of plague, and also the best mode of fulfilling them in its ordinary form. But in those rapid cases in which there is an early sinking, without the appearance of buboes or carbuncles, I would omit the emetic and cathartic, and give at first quinine and camphor in full doses, and if necessary, wine whey, broths, brandy, &c. During the whole course of the plague, cleanliness, fresh air, and every possible care to prevent the spread of the disease should be carefully attended to.

SECTION VIII.—DIPHTHERIA.

By *diphtheria*, from the Greek διφθερα, 'a membrane,' I mean here that peculiar malignant febrile affection, which, being so generally attended with the formation of fibrino-albuminous membranes, especially in the fauces, larynx, trochea, &c., has been called *sore throat, putrid sore throat, malignant sore throat, &c.* The term is objectionable; but as it indicates one of the most prominent features of the disease, it may be the best term we can get for this truly malignant febrile affection.

I introduce diphtheria in this place because I think it is emphatically a febrile affection, and should thus be classed, and not regarded or treated as a throat, but as a general blood disease, partaking of the nature of typhus and other malignant fevers; the throat complications being the effect of the general condition, and not the primary disease to be eradicated. And though the disease is in some cases attended with a scarlet rash, I do not believe that it is properly an exanthematous fever; especially as enteric and typhus fevers are not generally classed as such, though they are attended in most cases with a slight exanthema. I have, therefore, placed diphtheria with the general fevers, and the last of the list, following *typhus* and *plague*, believing that in so doing, I have it where pathologically it belongs.

Diphtheria may have prevailed more or less extensively in various localities at different periods, but it has been especially prevalent in

Europe and America during the past few years, in some localities assuming a mild, while in others it has assumed a decidedly malignant character. In this respect diphtheria does not differ materially from other malignant fevers, as some cases even of the plague are said to be so mild that soldiers have not left the ranks for even a day while laboring under the disease.

Symptoms.—In *mild* cases of diphtheria there is generally a slight feeling of weariness or languor for two or three days, or perhaps longer, with partial loss of appetite, especially in the morning, a slight headache, a heavy feeling in the lumbar region of the back, and sooner or later slight chilliness, followed by moderate febrile action. Soon after the febrile reaction the patient feels a slight soreness of the fauces, and there is a swelling of one or both tonsils, and very soon the parts inflamed, generally including one or both tonsils, are covered with a fibrino-albuminous membrane, continuous or in patches, and of a gray, brown, or blackish color.

In some cases, especially in children, the symptoms of the forming stage, and even the chill and fever, are scarcely noticed, the soreness of the throat being the first symptom complained of, attended with a slight swelling and soreness of the glands of the neck at the angle of the jaws. In these mild cases the fibrino-albuminous membrane is apt to be of a lightish color, and is often thrown out in patches, instead of a continuous membrane.

In *severe* malignant cases of diphtheria we have, however, a much more violent train of symptoms. The forming stage is attended with a feeling of great prostration, the appetite fails, the tongue becomes coated of a darkish color, the breath becomes offensive, the patient becomes nervous, restless and irritable, and within two or three days from the appearance of these premonitory symptoms, there comes a chill, more or less severe, attended with headache, pain in the back, &c., and followed in some cases by moderate, and in others by considerable febrile reaction.

The fever for two or three days is usually slightly aggravated about the hour the first febrile action was set up after the first chill, but though the pulse may be frequent, and in some cases tolerably full, yet it is not strong, indicating from the very first great nervous prostration, and, in some cases, as in other congestive malignant fevers, little or no reaction takes place, the powers of life steadily sinking from the first. Soon after the first chill, however, whether reaction occur or not, the fauces, and especially the tonsils, become more or less swelled, and in congestive cases the glands of the neck about the angles of the jaws become very much swelled, and often present a frightful appearance.

The fibrino-albuminous exudation, in severe cases, may form a continuous membrane, of a dark gray or brownish color, covering the palate, uvula, tonsils, fauces, and inner surface of the larynx, trachea, and bronchial tubes. In some severe cases, especially of a congestive character, suppuration occurs in one or both tonsils, in which case there may be but slight exudation upon any of the mucous membranes, the matter in such cases having the same dark grayish appearance which the exudation presents, and the same offensive, putrid smell.

In severe cases, unless early arrested, the face and neck become bloated and pale or livid, a bluish streak appears around the mouth, there is a constant cough, with expectoration of detached portions of the exudation, along with a more fluid fibrino-albuminous matter unorganized, a sanious matter runs from the nose, the pulse becomes frequent, quick, and fluttering, the extremities become cold, and the patient dies about the fifth or sixth day of the disease, either from general prostration of vital power, or from retention of carbon in the blood, caused by interruption of the respiration, or shutting off of oxygen by obstruction either in the fauces, larynx, or bronchial tubes. The patient may, however, die within forty-eight hours after the attack from congestion, or from the direct effects of the poison which produces the disease, acting upon the blood and cerebro-nervous system, directly destroying vitality.

Causes.—Now in this, as well as in all putrid fevers, every imprudence and deviation from the laws of health, such as exposure, want, improper food and clothing, low damp apartments, filthiness of every kind, &c., act as predisposing causes of this febrile affection. In addition to all this, it is probable that a damp atmosphere, such as always exists in some localities, from protracted rains, or from warm weather at seasons when it is always wet but usually cold, has very much to do in predisposing to or even producing this disease.

A damp atmosphere at such seasons of the year operates to depress vitality, and thus predispose to this disease in part by checking perspiration, and thus causing to be retained in the blood the perspirable matter. It is probable, however, that it is the low electrical state of such damp air, letting down as it does the electrical condition of the system, and thus deranging all its functions by a failure on the part of the brain and nervous system to generate and distribute sufficient nervous influence or vital force, that damp air becomes more especially a cause of this disease. On the temperate, well fed, well clothed, and cleanly, whose blood is kept up to the standard of health, this low electrical state of the air may scarcely operate unfavorably, but not so with the unfortunate children of want, who are half fed, half clothed, and filthy withal, or with the intemperate, imprudent, licentious, &c., whose blood is always below the standard of health.

I am satisfied, however, that in every case of diphtheria there is an animal poison introduced into the blood, either from the bodies of those suffering with the disease, or else arising from an accumulation, and perhaps decomposition of various animal secretions and exhalations, constituting *idio-miasmata.* I believe, also, that the paludal poison, in malarious districts, is often combined with the animal poison, rendering the disease more malignant, and giving it, in some cases, a slight remittent character.

While then exposure, want, dampness, and a low electrical state of the air, as well as every possible deviation from the laws of health, predispose to diphtheria, I am confident that either a contagion generated in the bodies of patients suffering from the disease, or else an idio-miasmata resulting from animal filth, invariably enters the blood, either through the skin, stomach or lungs, and so far decomposes or changes the blood as to lead to the prostration, chill, &c., and this dissolved state of the blood accounts for the putrid smell, the exudation of the putrid fibrino-albuminous matter, and in fact for every essential feature of the disease.

While then the above cause or causes, if a contagion be generated in the disease, are, as I believe, essential to the development of this febrile affection, *koino-miasmata*, when present in the blood at the same time, I am confident may very materially modify the disease, rendering it more malignant, and rendering the fever in many cases of a slightly remittent character, as it is liable to in other typhus or putrid fevers when it enters the blood with the animal poisons.

I have suggested the possibility of a contagion being generated in the bodies of patients suffering from this disease, and while I am not quite certain that such is the case, I am confident that an exposure to putrid. or even mild cases, renders the person thus exposed much more liable to an attack, and this conclusion is not the result of a limited, but of a careful and extensive observation in a variety of cases too numerous to mention in this place. This fact, however, is not positive evidence of contagion in these cases, for it is probable that idio-miasmata, or a general animal poison, would more generally arise in a degree of concentration sufficient to produce the disease in another in apartments where patients were kept with this or any putrid disease, even though no specific contagion were generated.

Diagnosis.—The diseases with which diphtheria is perhaps most liable to be confounded are typhus fever and scarlatina, from either of which, however, it may generally be distinguished by careful attention to all the symptoms. Diphtheria differs from typhus in the exudations of the fibrino-albuminous matter which occurs on the mucous membrane of the mouth, throat and respiratory tubes, and in there being generally less stupor, and also less mental disturbance. From scarlatina it may be distinguished by greater prostration during the forming stage, the uncertainty of and irregular appearance of the scarlet rash which sometimes appears, the offensive putrid smell, and by the very decided tendency to the formation of the fibrino-albuminous membrane on the surface of the mucous membrane of the mouth, throat, and respiratory tubes. The rash too, when it appears, is liable to be of a darker color than is common in most cases of scarlatina.

Anatomical Characters.—On examination after death from diphtheria, the morbid appearances that are characteristic of the disease are those that are presented by the mucous membrane of the pharynx, larynx, trachea, and bronchial tubes, and also the changed appearance of the inner surface of the heart arteries and veins, as well as the changed appearance of the blood itself.

The mucous membrane of the pharynx, and frequently of the mouth, are either covered or have patches of a fibrino-albuminous membrane of either a lightish gray or dark brown color, if it has not been cast off before death, and if it has, the mucous membrane presents a swelled or congested appearance, as is usual after passive inflammation in a mucous membrane. The tonsils show signs of having been inflamed, and are often found enlarged, and present a dark congested appearance.

The mucous membrane of the larynx and trachea is more or less covered with the fibrino-albuminous membrane, either forming a continuous membrane or else in patches, if it has not been thrown off, as happened in one case in this village, the membrane forming a perfect tube, the exact shape of the larynx and trachea down to its bifurcation. This tubular membrane was coughed up by a girl, I believe about twelve years old, and was preserved by my friend Dr. Jedediah Smith, of this village.

The bronchial tubes are usually found filled with fibrino-albuminous matter in a viscid unorganized state, and in a case which I witnessed a few days since, examined by Dr. C. N. Hewitt, the lungs presented an emphysematous appearance, evidently from the entrance of air into the smaller bronchial tubes and air cells without the possibility of escaping past the viscid putrid matter which they contained. The *heart clot* in this case had the color and general appearance of unorganized fibrino-albuminous matter, and appeared to have been continuous through one of the valves, as if it might have been partially organized, and possibly interrupted its action during the last stages of the disease.

The inner surface of the arteries, veins, and even the lining membrane of the cavities of the heart, are liable to be found covered more or less with the fibrino-albuminous membrane, as was found in a case examined by Dr. Jedediah Smith, a few months since. Other parts and structures are liable to present morbid appearances, but none that I am aware of but what are common in other putrid fevers. The blood is found of a dark color, and presenting an appearance of having been decomposed or materially changed. Evidences of passive inflammation may also be detected in some cases in the lungs, pleura, alimentary mucous membrane, and other structures, as in cases in which death has occurred from other putrid fevers.

Nature.—I need not say much in relation to the nature or pathology of this disease. It appears to me clear, that an animal poison in every case is introduced into the blood, which, on entering it, acts in conjunction with the predisposing influences which I have named, to decompose or materially change the blood; and this changed condition of the blood, together with the direct action of the animal poison upon the cerebro-nervous system, interrupts the due generation and proper distribution of the nervous influence or vital force, and so there comes a chill more or less marked. In cases in which the paludal poison is also acting in conjunction with the animal, the disease is liable to be more malignant, and the fever to be of a slightly remittent character.

Now, the fibrin and albumen of the blood being in a partially dissolved state, it is readily poured out, or allowed to transude from the capillaries of all the tissues, but especially so from the exposed surfaces of the pharynx, larynx, trachea, bronchial tubes, and inner surface of the heart, arteries, and veins, and hence the fibrino-albuminous exudation which occurs, sometimes as an organized, and, in other cases, as inorganized matter in these situations, in this disease. This fibrino-albuminous exudation may be favored by the passive inflammation which is liable to be set up in the parts from which the exudation occurs, but I think that congestion, with perhaps slight irritation of the structures, is all that is requisite for the exudation in its unorganized state, at least, provided the decomposed fibrin and albumen exist in the blood, or if the blood be in a decomposed or materially changed condition, as I believe it always is in this disease.

It may be difficult to account for the special tendency in this disease to this fibrino-albuminous exudation, so especially upon the mucous membrane of the nose, pharynx, larynx, trachea, and bronchial tubes;

and yet, when we remember that the depraved materials exist in the blood in this disease, and that this fever generally prevails in damp weather, and at seasons of the year when catarrhal, throat, and bronchial affections occur, it is not strange that the exudation should take place just as it does. There is nothing more singular in this, than that buboes should occur in *plague*, or eruptions about the mouth in paludal fevers. I do not, however, believe that the fibrino-albuminous exudation is essential to this disease, as I have had many cases developing early all the premonitory symptoms common to this disease, some of which have assumed the character of congestive typhus in the latter stages, while others have assumed a typhus character, being complicated with passive inflammation of the pleura, lungs, or some other organ or structure of the body. These cases, it may be said, were not diphtheria; but they had every appearance of the disease early, and many of them had even the soreness of the fauces, especially those that assumed the typhus character, with pneumonia, pleuritis, &c., the difficulty of the throat disappearing as the pulmonary or pleuritic inflammation became developed. I have also noticed, that in cases of diphtheria in which even abscesses in the tonsils, occur, there is liable to be very little organized fibrino-albuminous membrane on the mucous surfaces, the matter, however, in the abscesses presenting a dark putrid appearance. It may be asked if the pathology of this disease be so near that of pure typhus fever, why there is not the coma and mental disturbance so common in that disease. I answer that it may, and sometimes does occur; but I suspect that in cases in which there is marked irritation or passive inflammation of the fauces, larynx, or trachea, &c., this fact hinders, in part, the cephalic condition which develop those symptoms in pure typhus; and, besides, fatal cases of diphtheria, in which these symptoms would be most liable to occur, generally terminate by the fifth or sixth day, and, therefore, before we should look for the development of these symptoms if they were to occur.

Prognosis.—The prognosis in mild cases of diphtheria, if properly treated in season, may be regarded as generally favorable. In severe malignant cases, however, especially if neglected in the early stages, the prognosis may be regarded as generally unfavorable. The favorable indications are but slight premonitory disturbance; a moderate or imperceptible chill; but moderate fibrino-albuminous exudation, and that of a lightish color; and slight general fever or disturbance of the functions of the body.

The unfavorable indications are great prostration during the forming stage; loss of appetite; a foul and even putrid breath; a severe chill; extensive passive inflammation of the fauces, larynx, trachea, or bronchial tubes; very extensive fibrino-albuminous exudation, either in an organized or very viscid form; and the early development of very decided typhus symptoms.

Treatment.—The indications in the treatment of this disease are evidently to equalize the circulation; to sustain the sinking powers of the system; to restore the blood; and to prevent, as far as may be, suffocation; and, in most cases, especially in paludal localities, to attend to the hepatic or bilious complications.

In most cases, a cathartic of three blue pills, followed by half an ounce of castor oil, or of three grains of podophyllin, or five grains of leptandrin, or, if the patient be a child, of rhubarb, with a little hydg. cum creta, is generally indicated, especially in paludal localities. The warm foot-bath should be used morning and evening, and the patient should be required to drink freely of either sage tea or crust coffee, one half milk, to be taken a little warm, for the purpose of promoting perspiration, and, at the same time, to furnish proper materials for making good blood.

A warm, dry flannel, folded so as to be of two or three thicknesses, should be pinned around the neck, sufficiently wide to extend up to the ears, for the purpose of keeping the surface of the neck warm, and thus promoting perspiration, as well as relieving the congested, irritated, and perhaps inflamed, structure of the fauces; including, generally, the tonsils. Other more stimulating applications, as sinapisms, may be required, if the inflammation assumes in any degree an active appearance; but I believe the dry flannel will generally do best, in the great majority of cases at least.

To sustain the sinking powers of the system, and to arrest the fever, two or three grains of quinine, for an adult, or equivalent doses for children, should be given every four or six hours, and continued during the whole course of the disease; and after the general and local disease is arrested, the dose of quinine should be continued in gradually diminished doses, at least three times per day, during the whole period of convalescence. In malignant cases, it may be best to give, during the progress of the disease, alternating with the doses of quinine, ten or fifteen drops of the tincture of muriate of iron, for the purpose of sustaining the blood as far as may be; but I think the inconvenience of taking it to the patient, may in many cases more than counteract the benefit it is likely to produce, especially if there be great difficulty of swallowing, and especially, also, as the system in such a state appears to be incompetent to appropriate only the simplest forms of nourishment, and hardly to heed remedies, except such as act directly to sustain the sinking powers of the cerebro-nervous system. Great discretion should, therefore, be exercised in prescribing this remedy, as mild cases may not need it, and malignant cases might not heed, or be able to appropriate it.

As a gargle, for the purpose of clearing and cleansing the throat, I prefer a solution of muriate of ammonia, in vinegar and water, for adults, or for children over eight years old; but for children under that age, a solution of common salt in water, as it is less stimulating, and therefore better borne. The solution of the muriate of ammonia may be varied in strength; but two drachms, dissolved in half a pint of vinegar and water, consisting of one part of vinegar to two of water, is a medium strength. The solution of muriate of soda or common salt, for children, may be varied from two to four drachms, to half a pint of water, according to the age of the child. These gargles may be used every four or six hours in the early stages of the disease, but later, only three times per day, after each meal.

Astringent gargles may be indicated in the latter stages of the disease, especially if the inflammation, on subsiding, has left a congested condition of the fauces, with more or less irritation of the mucous membrane. Or in some cases which early assume this appearance, astringent gargles may be used alternately with the cleansing gargles which I have suggested, perhaps from the very first. For adults, an astringent gargle made by dissolving half a drachm of tannin, in eight fluidounces of water, or of equal parts of vinegar and water, if it be desirable to render it more stimulating, will, I believe, generally do best. For children, I prefer a strong sage tea, sweetened with honey or loaf sugar, with the addition of half a drachm each of alum and borax, to a half pint of the tea. These astringent gargles may be used in the early stages of such cases as they may be indicated, every four or six hours, alternating with the cleansing gargles. In cases, however, in which they are only indicated in the latter stages of the disease, these astringent gargles may be used only morning and evening, the cleansing gargles being used after each meal at this stage, if required at all.

The bowels should be kept regular, after the first cathartic, by small doses of castor oil, if necessary, and at such times as the patient does not take sufficient nourishment in his drinks of sage tea, or crust coffee, one-half milk, the proportion of milk should be increased; and besides this, broths, beef essence, and wine whey, may be given in addition; and finally, in cases in which the quinine, together with the nourishment I have suggested, fail to sustain the sinking vital powers, equal parts of whisky, or brandy and water, should be sweetened, and administered as freely as it may be required.

I do not believe that the throat should be touched with anything that can possibly irritate, at any stage of the disease, unless it becomes absolutely necessary. If, however, the patient be likely to suffocate from an accumulation of viscid matter in the fauces, it may be carefully wiped out by a swab, wet in one of the astringent gargles I have suggested. Or if partially detached membranes are interrupting respiration, and they cannot be removed with the swab, as I have suggested, they may be carefully removed by a pair of bent forceps. Or, again, should the tonsils so far fill the fauces, as to render suffocation from this cause inevitable, they should be carefully cut off with Fahenstock's, or some other suitable instrument for the purpose. Finally, as a last resort, should a closure of the larynx render suffocation certain, without a resort to tracheotomy, this last chance should be given to the patient, provided it be a favorable case of suitable age; and provided, also, the patient and friends desire to avail themselves of so uncertain a chance, rather than submit to an inevitable fate.*

Thus have I completed what I had to say on *diphtheria*, as well as on all the *general fevers*. I will, however, suggest, that as soon as the patient acquires an appetite for food, it should be allowed in suitable forms, and at regular meal hours, through the whole period of convalescence.

* Perhaps "tubing the larynx," through the mouth, might do in some cases instead.

CHAPTER V.

EXANTHEMATOUS FEVERS.

SECTION I.—VARIOLA—(*Small-pox.*)

By variola or small-pox I mean that variety of exanthematous contagious febrile affection which is characterized by fever, an eruption, first pimply, but passing on to a vesicle, and finally to a pustule; the duration of the disease varying from two to three weeks.

Much has been said in relation to the origin of small-pox; some being disposed to regard it as a disease of recent origin, or as having originated about the A. D. 550. But it appears to have prevailed in India and China at a period long anterior to this, and it is possible that the disease with which Job was afflicted, near three thousand years ago, may have been small-pox.

Small-pox occurs under a variety of modifications, from the most malignant to the merest varioloid affection, all, however, depending upon the same cause, and being propagated by a specific contagion. Two varieties, however, the *distinct* and the *confluent*, are entitled to consideration, though they are one and the same disease. In that which is usually called the distinct variety, the pustules, are distinct, elevated, distended and circular, and are scattered over the surface of the body, while in the confluent, the pustules are exceedingly numerous, depressed, irregularly circumscribed, and coherent or confluent.

The time which elapses between the reception of the variolous virus and the first manifestation of its influence on the system, is generally from eight to fourteen days. During this period of incubation, no symptoms of indisposition generally manifest themselves, the individual retaining apparently a usual state of health.

DISTINCT VARIETY.—*Symptoms.*—The disease commences with a feeling of languor, weariness, pains in the back and lower extremities, slight chills, with flushes of heat, and pain in the forehead. As the disease progresses, there is usually more or less nausea and vomiting, with thirst, pain in the epigastrium, and more or less soreness in the fauces. When the fever is developed, the skin is hot and dry, the tongue white, with a red point, the bowels torpid, and the urine scanty and high colored.

During the first and second days of the fever the mind often becomes dejected and confused, and towards the end of the third day the tongue acquires a bright red color. And just before the appearance of the eruption there is an unusual tendency to perspiration, generally in adults, and usually drowsiness, and sometimes coma. In children, the eruption is often preceded by convulsions, there being much less tendency in them to free perspiration than in adults. The hands and feet, especially of children, may remain cold during the whole course of the disease.

There is sometimes an increase of the febrile symptoms a short time before the eruption begins to make its appearance, attended in some cases with severe cramps in the limbs, and violent pains in the back. Usually towards the end of the third, or beginning of the fourth day, the eruption begins to make its appearance. It comes out first upon the forehead, face and neck, and soon upon the breast and forearms, and lastly upon the abdomen and lower extremities, so that in about twenty-four hours, or at the close of the fourth, or beginning of the fifth day, the eruption is completed, and the fever has nearly or quite disappeared.

The eruption consists at first of minute red points, which, by the middle of the second day of the eruption, the sixth day of the disease, become slightly elevated, with inflamed bases, but which as yet discharge no serum when punctured; but the cuticle appears slightly distended by a semi-transparent lymph, so that the pimple has become a vesicle.

By the third day of the eruption, the seventh day of the disease, the vesicles are fully formed, being round, with a central depression, which depression is generally perfected by the fourth day of the eruption, the eighth day of the disease, so as to become conspicuous in most of them. When, however, but few pustules exist, they may have only a slight central depression, but they generally assume the umbilicated form. The fluid appears first in the central point of the vesicle, and is of a serous character. But as the vesicles increase in size, they gradually assume a whitish color, and become pustular, being surrounded by a pale red areole. If the pustules are numerous, these areole may run into each other, and give a uniform redness to the interstitial spaces.

The fluid which appears in the central part of the vesicle becomes gradually more abundant as it becomes purulent in character, constituting the vesicle a pustule. This change is fully commenced, as we have seen, at the perfection of the vesicle, on the fourth day of the eruption, and the eighth day of the disease, and marks the commencement of the stage of suppuration.

In the distinct variety, the fever which precedes the eruption always remits, and frequently disappears entirely, as soon as the eruption is completed; but the febrile symptoms reappear when suppuration commences. And as the suppuration goes on the pustules lose their flattened form, are distended with pus, and become spherical. At the commencement of suppuration, about the eighth day of the disease, the face swells considerably, but it may subside in two or three days, and the hands and feet, and much of the surface of the body, in turn become swelled and tender in the same manner.

The period of suppuration is usually attended with more or less soreness of the fauces, and increased secretion of the saliva, the mouth and throat being swelled, and of a bright red color. During the latter period of the suppurative stage, a peculiar odor arises from the patient's body, and this continues till desiccation is completed. There is also frequently more or less drowsiness, and occasionally a diarrhœa towards the completion of the suppurative process.

The secondary, or suppurative fever, which begins about the eighth day of the disease, varies in violence and duration, according to the copiousness of the eruption, and the greater or less activity of the suppuration. In mild cases, the suppurative fever rarely continues more than four days, or to the twelfth day of the disease, as it commences about the eighth. The suppuration, like the eruption, begins on the face, and gradually extends to the extremities, being completed, and the pustules at their maturity at about the twelfth day of the disease.

After the pustules have acquired their full state of development, at about the twelfth day, they sometimes remain stationary for several days, but they generally assume a brownish appearance at the centre of each pustule soon after suppuration is completed, and the liquid oozing, they acquire a rough, deep yellow appearance. Soon after they assume this appearance, the pustules shrink and become gradually drier, darker and harder, till the matter is converted into a brown crust or scab. Desiccation commences on the face, and extends gradually to the body and extremities, so that the scabs begin to fall from the face about the fourteenth day, and are all off the body, and even the extremities, by the twenty-first day of the disease.

Thus we see that the eruption of pimples makes its appearance by the beginning of the fourth day of the disease. The pimples by the seventh day of the disease, contain a little lymph, constituting the pimples vesicles. And by the eighth day of the disease the pocks are fully formed. Suppuration commences about the eighth, and continues to about the twelfth day of the disease, being attended with swelling of the face, and surface generally, and more or less suppurative fever. At about the twelfth day suppuration is completed, the pustules being fully formed, the swelling of the face and surface generally has subsided, and with it the suppurative fever. Desiccation commences about the twelfth day. The scabs begin to fall from the face about the fourteenth, being generally all off by the twenty-first day of the disease, making the duration of the disease from two to three weeks.

CONFLUENT SMALL-POX. — The *symptoms* in confluent small-pox, are generally more violent than in the distinct. The pain in the back and extremities is more severe, the febrile phenomena are more violent, the skin being hot, the thirst urgent, the tongue dry, and covered with a dark brown fur, and the nervous system being much more sensibly affected. In confluent small-pox, there is a diarrhœa before the eruption in many cases, instead of the perspiration which occurs in the distinct; and when this occurs, the secretion of saliva is diminished, instead of being increased.

The appearance of the eruption is more irregular in the confluent, sometimes appearing as early as the second, and in other instances, not till the fifth day after the commencement of the fever, being frequently preceded by a roseolous rash. The pustules of confluent small-pox are more numerous, irregular in shape, often running into each other, and being less elevated than in the distinct variety.

The fever only partially subsides on the appearance of the eruption, the patient being restless, the thirst urgent, and the skin hot and dry. As the disease progresses, the symptoms often become more alarming, and frequently a long train of typhoid symptoms are manifested. There is great prostration of strength, a dry tongue, low muttering delirium, involuntary stools, and flow of urine, picking at the bed-clothes ; and the patient dies about the twelfth day of the disease. Or if the **patient passes** through the suppurative stage, to the period of decline, **various** disorganizing inflammations are liable to occur, such as pleurisy, pneumonia, ophthalmia, &c.; which may set up a violent symptomatic fever, just as the suppurative fever is beginning to decline; and thus **the life** of the patient becomes again materially endangered.

The matter, in the confluent pustules, is of **a dark** brown **color; and** about the twelfth day of the disease, it begins **to escape**, and **hardens on** the surface in brown crusts, which fall off by the **fifteenth or twentieth** day of **the** disease, leaving deep marks, furrows, or pits, according as the disease is more confluent. The eruption of **small-pox** is not confined to the skin, but appears on the mucous **membrane of the tongue, throat,** larynx, and trachea, producing more **or less cough**; and very likely in **other portions of the mucous membranes, as the** mucous membrane is only a continuation **of the skin.**

When small-pox attacks persons in **an asthenic state, it sometimes** manifests a *malignant* or *putrid* character, the patient sometimes sinking from the direct effect of the *morbid poison* during the early stage **of** the disease. In these malignant cases, the prostration is such, that **there is** no efficient reaction; the patient is restless, delirious, or comatose. **The** eruption comes out imperfectly, or suddenly recedes, the perspiration **is** offensive, the face is bloated, and red, or purple, the eyes watery and inflamed; and in some cases, colliquative diarrhœas and hemorrhages occur before death, which often takes place by the seventh day of the disease. In some **of these** malignant cases, the pustules, instead **of** containing purulent matter, are filled with a colorless transparent serum, indicating the greatest degree of danger; **the** secondary fever being generally of a typhoid character, **or else** the patient passing **on**, rapidly, **to** dissolution.

VARIOLOID.—Very mild cases of small-pox are frequently not attended **with** most of the regular phenomena of ordinary small-pox, the attending fever being mild, and scarcely attracting attention. The eruption **often** appears in clusters, or are very scattering **over** the surface, **and occur** at irregular intervals between the **second and fifth** days; nor is it attended with any very marked primary **or secondary** fever. Desiccation may commence earlier than in severe cases, **occurring sometimes** by the eighth or ninth day, and the falling of **the scabs by the tenth or** twelfth.

The **varioloid, however, is** only a mild small-pox, **rendered so by** various causes, such as partial protection by previous **vaccination, slight** hereditary predisposition, cleanliness, atmospheric peculiarities, **and** various other accidental causes; such as prudence, temperance, &c. Such mild cases, it should be remembered, depend upon the **same cause**; the virus generated in the same patient producing in **one, distinct**; in another, confluent; while in a third, only **a mild varioloid disease.**

Such are the ordinary symptoms of small-pox in its various **forms, the** distinct, the confluent, the malignant, and the varioloid. It should be remembered, however, that there is a **vast** variety of modifications in these varieties, and that they run into each other, **or have** no well defined dividing line between them. So that, instead of **attempting to** classify cases of small-pox as they occur, it is far better to forget classification, names, &c., than to fail of getting at the real condition of the patient.

Sequelæ.—Small-pox, from the derangement it produces in the blood and different tissues of the body, is liable to give rise to slow fevers, dropsy, a variety of cutaneous affections, ophthalmia, deafness, phthisis, paralysis, mania, epilepsy, and a variety of general and local affections.

Anatomical Characters.—The principal morbid conditions or alterations found on post-mortem examination are those of the skin and mucous membranes. There are, however, often found other accidental morbid appearances, the result generally of congestion, irritation, or inflammation of different tissues and organs, the most frequent of which are of the brain, lungs, pleura, &c. There are generally found marks of the eruption upon the mucous membrane of the nose, mouth, throat, larynx, trachea, and sometimes of the bronchial tubes, and perhaps in other parts, as upon the genital organs.

The alimentary mucous membrane generally shows signs of congestion, irritation, or inflammation, and it is said that pustules have been found on the mucous membrane of the colon. The seat of the small-pox pustule is in the vascular structure which lies immediately beneath the cuticle; a slight slough of the true skin, however, occurs, which being cast off, gives rise to a depression or pit in the skin.

Cause.—The cause of small-pox is a specific contagious principle, which occurs both under the form of a palpable matter and an imperceptible effluvium, which is generated in the bodies of patients suffering from the disease.   •

It is possible that the disease may now be generated spontaneously, as at first, or be very much favored by certain epidemic influences, which, as I have before said, is the result of the sum total of the imprudence of the human family. But it is probable, that while these influences may now operate, more or less, as predisposing causes, the specific morbid poison generated in the bodies of patients suffering from the disease is the general, if not the universal exciting cause of small-pox.

The susceptibility to the impression of the small-pox contagion appears to vary in different individuals, not only in relation to the liability of becoming affected by it, but also in relation to the degree of violence the disease arising from the contagion assumes.—The variolous contagion possesses the power of destroying the susceptibility of the system to its subsequent operation, so that a second attack of genuine small-pox seldom occurs in the same individual.—The contagious principle may become attached to various articles of clothing, and retained in a sufficient degree of concentration to communicate the disease for a long time. It is probable, also, that the disease may be communicated from the bodies of patients dead of the disease, and also from the bodies of patients suffering from the disease at any time after the establishment of the primary fever.

The period of incubation has been variously estimated, but I believe that the time which elapses between exposure to the poison, and the establishment of the primary fever, may be set down, as I have said, at from eight to fourteen days, at least so far as my observation has extended.

Pathology.—There is an ancient tradition, originating I believe with the Arabian physicians, that small-pox was derived originally from the camel, but it is one of the ways in which the human family have ever been willing to avoid an acknowledgment of the fact of their own self-destruction, either directly or indirectly. As to the precise manner of its origin, or the exact nature of the imprudences which first led to the generation of this disease in the human system, I presume they may never be ascertained, and yet it is as probable that it was the result of imprudence, as it is that the venereal disease resulted from imprudence. The same is also true of all contagious diseases, as I have already stated, in the early part of this work, and in fact of nearly all diseases, whether contagious or non-contagious, there being perhaps not an agent in nature capable of injuring or harming man, so long as he obeys the laws of his being. This is true, as I have before stated, not only in relation to the predisposition to disease, but also of the exciting causes of almost every possible variety of disease.

But in the contagious diseases which we are now about to consider, and of which this is one, and which have arisen one after another from various imprudences, they have only to be originated once. For the virus which is generated in the system during the first attack is sufficient to propagate the disease during all time, or till such time as the human family shall have so far improved their physical condition, by a return to the laws of health, that the susceptibility may be lost. I have made this digression here that we may remember that even the contagious diseases are generally the result, either directly or indirectly, of disobeying the laws of health.

The variolous poison once generated may be received into the system through the lungs by respiration, by the application of the matter to the unbroken skin or mucous membranes, or by the application of the matter to a wound of the skin. In either case the matter entering the blood, and affecting first probably the cerebro-spinal and ganglionic system, letting down the circulation, and producing a chill. Febrile reaction follows, during which the morbid poison being thrown to the extreme capillary vessels, produces in the vascular layer of the skin its specific effect, leading to the eruption, and also affecting to some extent the mucous membranes, especially of the respiratory organs.

With the eruption the general febrile symptoms usually subside, and is only revived by the local irritation set up by the inflammation and suppuration of the pustules. The whole process appears to be an effort on the part of the system to get rid of a morbid poison it has unfortunately imbibed, and when it has done this the system is gradually restored to health, and the susceptibility to the disease appears to have been lost, but failing to eliminate the poison the patient passes on to speedy dissolution.

Diagnosis.—The difficulty of discriminating between small-pox and other febrile affections is confined mainly to the early stage of the disease, at which time the possibility of exposure, the severe pain in the back, and the unaccountable vomiting, should be taken into consideration, and each allowed their due weight in forming what can, at that stage, be only a probable diagnosis. After the appearance of the eruption, the only diseases with which it is likely to be confounded are measles and chicken-pox, but the eruption of measles is generally less prominent than in small-pox, and besides, the eruption of measles does not pass on to a vesicle and pustule as in small-pox. This process, when it transpires, renders the diagnosis plain between measles and small-pox.

Small-pox may generally be distinguished from chicken-pox by the shorter duration of the eruptive fever in chicken-pox, and by the vesicles or pustules of chicken-pox not presenting the umbilicated appearance of the small-pox pustule. By carefully observing all these peculiarities, and taking into account all the circumstances connected with the case, I think a mistake need seldom be made, but yet some mild cases of small-pox appear precisely like severe chicken-pox, or at least such has been the result of my observation in small-pox.

Prognosis.—In distinct small-pox, in patients of good habits, not given to over-eating, or to over-stimulation, the prognosis may be set down as favorable, especially if the patient be not terrified by the name. In confluent small-pox, however, the danger is always considerable, but if there is no hoarseness, and the voice is natural, if the mind is composed, and the patient is quiet at night, and between the age of eight and sixteen years, and temperate in all things, reasonable hopes may be entertained of a favorable result of the case. But if the patient be a young child, or a person over fifty years of age, and the vesicles are flat and the extremities livid, and if there is hoarseness, spitting, and in children moaning and grinding of the teeth, or in adults delirium or great despondency, little hope need be entertained of a recovery.

Malignant small-pox is generally rendered so by some hereditary or accidental depraved condition of the system, in consequence of which the vitality is not sufficient to withstand or eliminate the morbid poison; the patient becomes prostrated, irritable, delirious, and perhaps convulsed. The eruption fails to appear, or the pustules are imperfectly developed, hemorrhages occur, there is great hoarseness, with difficult respiration, and finally involuntary stools, and suppression or involuntary flow of urine, all of which are unfavorable symptoms, and indicate almost certain dissolution.

Patients may die at any period of the disease, but it appears that more cases die during the second week after the appearance of the eruption, and the greatest number on the eighth day—at least, such was the result of Dr. Gregory's observation in 168 fatal cases which fell under his notice in the London small-pox hospital, during the years 1828 and 1829.*

When small-pox attacks women during pregnancy, death may occur at any period—at best, an abortion may be expected, and, very often, both the mother and offspring perish.

Treatment.—As the system laboring under small-pox is evidently making an effort to eliminate the morbid poison it has imbibed, the danger is from a failure to accomplish this, or else from the over-action of the system in this process, in consequence of which, there is too much inflammation of the skin, and, perhaps, of other tissues and organs of the body. The indications, then, are very plain, and may always suggest a rational treatment if we will carefully heed them.

---

* See Cyclopedia of Practical Medicine, vol. iii. p. 166.

In those cases in which the febrile reaction is sufficient, and not too much, as we do not yet know of any measures that will neutralize the poison or arrest the disease, there is but little to be done by the medical attendant. The patient should be kept, however, in a moderately cool and airy room, allowed the free use of cool acidulated drinks, such as lemonade, and be kept on a light and very digestible diet. In such regular mild cases, little more than I have suggested need be done. If, however, a laxative becomes necessary, half an ounce of the sulphate of magnesia may be administered, and, if the skin be dry, two or three grains of the James's powder may be given every four or six hours. This may fulfill all the indications which arise in such cases.

But in violent cases, in which there is too much febrile reaction, much more may be indicated to prevent a too copious eruption, and also to counteract various local inflammations that may arise. In severe cases, if the fever runs high, a cathartic of calomel in castor oil will generally do very well at first, after which, the bowels may be kept gently loose by small doses of the sulphate of magnesia. As a diaphoretic, and to allay the general fever, the nitrate of potash, in five grain doses, may be given every four or six hours, to be given well diluted in cold water; or, if there is a strong cephalic tendency, the James's powder may be given instead in four grain doses every four hours.

Should local inflammations arise of the brain, lungs, or any other part, cups, leeches, &c., may be resorted to, and the local inflammation treated according to the indications, always keeping in mind the general condition of the patient, as well as the character of the inflammation, whether it be of an active or passive character.

In malignant small-pox, occurring, as it does, in persons of a depraved constitution, or in an anæmic state, the system often sinks directly from the effects of the morbid poison, and either no eruption appears, or, if it does appear, it may suddenly recede, and the patient struggles in a low typhus state. In such cases, the system should be sustained by two or three grains of the sulphate of quinine, with an equal quantity of camphor, every four or six hours, and, if necessary, wine-whey or brandy may be given in addition, but such cases, however, often go on to a fatal termination.

After the eruption makes its appearance, in most cases of small-pox, the primary fever in a great degree subsides, requiring generally only gentle laxatives, and perhaps a little James's powder, two or three grains, with or without a grain or two of camphor, and, in malignant cases, a grain or two of quinine. But as suppuration commences, the symptomatic fever arises, being more or less severe, according to the extent of the eruption and the amount of suppuration. At this stage, a combination of James's and Dover's powders, in three or four grain doses of each, every four or six hours, will generally lessen the cephalic tendency, promote perspiration, and quiet general nervous irritability, and, if sinking of the powers of the system occur, the addition of camphor and quinine may be indicated.

Should local inflammations arise during the suppurative fever, cupping, leeching, fomentations, &c., should be resorted to, and, if the inflammation be of a passive character, quinine, with alterative doses of calomel, leptandrin, or podophyllin, may be indicated. The greatest care should be taken to keep the inflamed surface as clean as may be; and the face and eyes may be washed often with cold water or milk and water. And if the mouth and throat be sore, two drachms of the fluid extract of the geranium maculatum may be diluted with an ounce of water, and sweetened with honey or loaf-sugar, and used as a gargle, with or without a little alum or borax.

To prevent the pustules from affecting the globe of the eyes, pieces of folded linen, wet in cold water, may be kept applied to the eyes during the eruptive fever. To prevent pits, which are liable to occur on the face from the small-pox pustule, something may generally be done. Simply opening the pustules when they are completely formed and as suppuration is commencing, and pressing out the matter, and washing the part with tepid milk and water, will always do some good, and should not be neglected if nothing more is attempted. If, however, the pock be opened with a lancet as soon as they become vesicular, and a pointed piece of the nitrate of silver be applied that and the succeeding day, they may not pass on to suppuration, and, consequently, little or no scars will be produced.

The application of the tincture of iodine with a camel hair brush, instead of the caustic, will sometimes prevent suppuration, and, consequently, the formation of pits. Another method, which, in many cases, may be preferable to either, is to apply a fine linen cloth, spread over with the mercurial plaster, prepared according to the United States Dispensatory: of mercury, six ounces; olive oil and resin, of each two ounces; and lead plaster, a pound. This being spread over the fine linen, should be laid over the face as soon as the eruption makes its appearance, and kept on without intermission for four or five days. In either of these ways may the deformity from pitting be greatly lessened, and in some cases almost entirely prevented.

I have said that the patient should be allowed cool drinks, and a light digestible diet, but in many cases no solid food can be taken for a time at least. In such cases crust-coffee, or toast-water, one-half milk, should be allowed till such time as rice and milk, a poached egg and toast may be allowed; or, if the prostration be very great, mutton or chicken broth, wine whey, &c., may be indicated. During the whole course the greatest possible care should be taken to keep the patient and his clothes and bed clothes clean, and to keep the apartments well aired, and during desquamation the whole surface of the body should be sponged or washed with tepid water.

And finally, the greatest possible care should be taken in all cases that no unnecessary exposure be made, and as the patient leaves his sick room, his clothes and bedding should either be destroyed or thoroughly cleansed, and the patient, at least once, thoroughly washed from head to foot, before he goes forth to mingle with his fellow-men. And now, in conclusion, let us remember to keep an eye on the exact condition of the small-pox patient, and be sure that we meddle no more than the case absolutely requires.

SECTION II.—VACCINA—(*Cow-pox.*)

By vaccina I mean that variety of exanthematous disease communicated by inoculation or vaccination, and characterized by the appearance of one or more umbilicated vesicles upon the skin, and serving materially to destroy the susceptibility of the system to the small-pox contagion. The origin of this disease is not quite certain. It has been supposed to have originated in the horse, in the disease called grease, and to have been communicated by farriers to the teat of the cow, and from the cow to the human subject. But it appears to me more probable that it is the result of the variolus disease communicated from human beings to the cow, and producing in them a modified form of the disease which in man has ever been so destructive. The disease having once occurred in the cow, whether from the horse or from human beings, may be communicated to man by application of the matter to a slight wound of the skin, and then is communicable from man to man by vaccination.

It was long ago discovered in Persia, and probably in other parts of the world,[*] that the disease communicated from the cow to man was capable of destroying or lessening the susceptibility of the persons so affected to the small-pox contagion. But it was in one of the dairy counties of England that Dr. Jenner ascertained that persons who had suffered from the vaccine disease could not be brought under the influence of the small-pox contagion, even by inoculation. He also ascertained by experiment that the vaccine disease could be conveyed from one person to another, and that in every case it appeared to destroy or very materially lessen the susceptibility to the small-pox contagion. Dr. Jenner published an account of his discoveries in 1798, and though his views met with violent opposition, vaccination was soon adopted as a prophylactic measure against small-pox in almost every part of the world.

Symptoms.—The symptoms of the disease in the cow are very slight, producing, apparently, it is said, a slight febrile excitement, and on the fourth day minute pocks appear upon the teats, and perhaps about the eyes or nose, which gradually enlarge, assuming the umbilicated appearance, and becoming perfected and beginning to desiccate about the twelfth day of the eruption, the sixteenth of the disease.

To produce the disease in the human subject matter should be taken from the pustulate on the teat of the cow, or from a healthy child, in which the vaccine disease has previously been produced. It is best secured by making slight punctures with the point of a lancet in the pustule, about the seventh day after vaccination. The lymph which exudes, if not immediately used, being preserved between two pieces of glass; or if the scab is used it may be dissolved in a little water, and rubbed down with a knife, and kept in the same manner between glass.

That part of the arm near the insertion of the deltoid muscle is a very convenient place to insert the virus. The insertion I have found most conveniently made with two lancets; with one of which a slight cut is made with the point passing under the cuticle, while with the other some

* See Dict. de Med., xxx. 393.

of the matter may be inserted in a dissolved state with little or no inconvenience.

About the third day after vaccination a slight inflammation arises at the point where the puncture was made, and on the fourth a small pimple appears, encircled by a faint areola. This pimple now gradually enlarges, and on the fifth day assumes a regular circumscribed form, with a flattened surface and a small depression at the centre; the pimple having become a vesicle containing a limpid fluid, mostly around the edges.

From the fifth to the ninth day after vaccination the pock gradually enlarges, being on the sixth day surrounded at the base by a plain narrow red circle; on the seventh the vesicle is tolerably well formed, presenting a shining appearance; on the eighth the areola increases its circumference, till about the ninth or tenth day, at which time the vesicle or pustule is perfected; the pock being from one-third to one-fourth of an inch in diameter, one or two lines deep, slightly umbilicated, and having a small brown scab or scale in the centre.

About the fourth day after vaccination there may be a slight febrile excitement, and about the tenth day there is often slight pain, irritability and restlessness, till the twelfth day, at which time the disease is on the decline. By the twelfth day of the eruption, the fifteenth or sixteenth from vaccination, the scab is well formed, the matter having become purulent and collected in a small cavity on the thirteenth, and the areola having faded on the fourteenth day. There remains then by the fifteenth or sixteenth day little more of the pock than a yellowish brown scab, which gradually dries, becomes more prominent, and falls off during the third or fourth week, leaving an oval scar.

While this is the general course of the vaccine disease, there may be wide variations in the symptoms in different cases, the period of incubation being greatly lengthened in some cases, and the development of the pustule greatly hastened or retarded in others. In some cases there may be no perceptible primary fever, and little or no general irritability, or even local pain, during the suppurative stage, while in others there may be not only a primary fever, but also considerable suppurative fever, with great nervous irritability, and severe pain, inflammation, and much heat and swelling of the arm.

If the vaccine virus be inserted at different points, so as to form pocks at such distance that their areolæ run into each other, the most violent and rapid inflammation with gangrene may be the result, as I have witnessed in several cases, in one of which the patient came near losing his arm, and in fact his life. The virus should never be inserted, then, in more than one point in the arm, at the same time, unless it be done so closely together that but one pock will be produced.

Vaccination may be practised at almost any age after the third month of infancy, and should always be attended to in infancy or early childhood, and should be repeated in after life, once in six or seven years, or on the appearance of any very strong prevailing epidemic influence. This should be done, as it is possible that the system may lose the impression, and therefore become susceptible to the small-pox contagion.

As a general rule, I think it is not advisable to vaccinate females during pregnancy, as serious consequences might result; but in an emergency, where the exposure to small-pox is endangering life, it should be done to avoid a worse calamity. In all cases in which the unprotected have been exposed to small-pox, vaccination should be immediately resorted to, with a hope of preventing, or materially modifying the variolous disease.

Diagnosis.—To distinguish the vaccine disease from various forms of disease produced by vaccination, is of the utmost importance, as without this the patient may not know, till it is too late, that he is unprotected.

No sore, the result of vaccination, should be pronounced as genuine vaccine disease, unless it presents all the essential appearances which I have laid down, as exhibited by the vaccine pustule. There must be an incubation of from two to four days; then a pimple, which passes on to a vesicle by the fifth or sixth day, having an umbilicated surface; by the eighth day a bright areola of greater or less extent should encircle the pock, and go on increasing till the tenth or eleventh day, the vesicle becoming enlarged and turgid at its circumference.

The spurious sores differ from the genuine pock, according to my observation, in these essential particulars they frequently arise in one or two days after vaccination; progress rapidly, being more or less inflamed; are irregular in shape; lack the central depression; scab early, and perhaps re-scab; are disposed to bleed on a slight injury, and, finally, the scar wants those numerous little depressions, so characteristic of the scar following the genuine vaccine pustule or scab. By careful attention to all these peculiarities and differences, a correct diagnosis may generally be formed, so that every person vaccinated may know when he has, and when he has not passed through the vaccine disease.

Prognosis.—The prognosis in the vaccine disease is always favorable, unless violent inflammation occurs in the arm, followed by gangrene, &c., which need not occur, if the virus be inserted in only one point at the same time.

The prognosis, so far as the protection from small-pox is concerned, may be set down as favorable under certain qualifications. That is, the genuine vaccine disease absolutely destroys the susceptibility to small-pox in some cases, and materially lessens or modifies it in every case, so that in case small-pox is contracted, it is of a very mild and benignant character.

Treatment.—In healthy children, and in adults of temperate habits, no special precautions are necessary on resorting to vaccination, only that the patient be of a proper age, and in a good state of health. But if the patient be an adult of imprudent habits, given to overeating, &c., he should be restricted to a moderate diet, and prudence in every respect, and this will generally be sufficient. But in such cases, if a primary or secondary fever arise, with swelling, pain, and violent inflammation of the arm, half an ounce of the sulphate of magnesia should be administered, and a solution of the sulphate of iron, two drachms to the pint of water, applied to the arm early.

This will generally arrest the fever and soothe the local inflammation; but if the inflammation passes on to gangrene or suppuration, as I have seen, then a warm bread and milk poultice, with or without laudanum, should be applied, and the parts freely laid open, for the speedy evacuation of any matter that may collect. Should mortification occur, with general sinking of the powers of the system, the fluid extract of bark, quinine, iron, wine whey, &c., should be resorted to. Such cases, however, are rare, and form only an exception to a rule, as probably not one case in a thousand of the vaccine disease, require any medical treatment or interference whatever.

SECTION III.—VARICELLA—(*Chicken-pox.*)

By varicella, or chicken-pox, I mean that mild exanthematous febrile affection, so common, in which a slight fever is followed by a vesicular eruption, which begins to break, or dry, by the fifth or sixth day.

Symptoms.—In some cases of chicken-pox, there is a slight febrile excitement, preceded, if at all, by a very slight chill, which febrile state may continue from a few hours to twenty-four, at which time a vesicular eruption makes its appearance, first upon the breast, and gradually extends over the body, head, face and extremities. Sometimes there is no sensible febrile excitement preceding the eruption, but instead, headache, drowsiness, a foul tongue, sickness at the stomach, slight heat of the skin, and quickness of the pulse, which continuing perhaps for twenty-four hours, the vesicular eruption makes its appearance as in cases attended with fever.

The pocks may appear at first as small red spots, soon becoming vesicular, but generally, I think, they appear at first as vesicles, being distinct, irregular in shape, though they may be of an irregular circular form, and varying in size from the head of a pin, to that of a split pea. The pocks are filled, on their first appearance, with a clear, and generally inodorous fluid.

On the second or third day the vesicles begin to burst, or are broken by scratching, and on the fourth or fifth day the fluid in those that remain entire, acquires a straw-colored appearance, and gradually dries up, leaving crusts, which crumble away gradually, or fall off in scales by the eighth or ninth day, without leaving pits, or any permanent marks.

The eruption may be numerous, but generally it is distinct and scattering. The pustules sometimes appear in successive crops, so that when some are just appearing, others are in a state of maturity, and others still have dried, and are falling or crumbling away.

Diagnosis.—Chicken-pox is not liable to be confounded with any disease, unless it be mild varioloid affections, from which it generally differs in many very essential particulars. In chicken-pox there is very slight fever, if any, and it lasts generally but twenty-four or thirty-six hours, while in varioloid there is always fever, continuing for two or three days, with headache, and sometimes slight delirium, before the appearance of the eruption.

The vesicles of chicken-pox are filled with a white clear fluid, on the first day of their appearance, while the varioloid pock is much slower in assuming the vesicular character, and has often an umbilicated appearance, and finally becomes pustular. The vesicles of chicken-pox have no central depression, and when punctured they fall to a level with the surrounding skin, while in varioloid there is a hardened base, surmounted with small circular vesicles containing matter, and generally more or less depressed at the centre.

In varicella there is generally no depression or scar left to mark the location of the pocks, while in varioloid there is generally a peculiar pit or scar, with minute depressions, at the points where the pocks were located. And finally, small-pox is a very contagious disease, while varicella is only very slightly so. With all these essential differences, varicella should not generally be confounded with varioloid affections, especially if the case be examined with proper care. Cases sometimes occur, however, in which it is impossible to decide with certainty at first, or till other cases occur.

Cause.—Varicella may be strongly favored, and perhaps originated, by a peculiar epidemic influence, but it is probable that a peculiar contagion is generally, if not invariably, the exciting cause of chicken-pox.

One attack of varicella protects the system against a second, but it is doubtful whether, under any circumstances, the disease can be propagated by inoculation.

Varicella is a disease occurring mostly in children, and is probably distinct, in every respect, from any and every possible variety of varioloid or variolous affection.

Treatment.—Varicella is a disease of so mild a character as to require very little or no medical treatment. If, however, the bowels are constipated, a moderate dose of rhubarb and magnesia, or of leptandrin, may be administered, and the patient should be directed to take a plain vegetable diet for a few days, and not to be exposed to a cold damp air. The skin should be kept clean, and a warm bath or a thorough washing should be resorted to after the disease has run its course.

SECTION IV.—RUBEOLA—(*Measles.*)

By rubeola, or measles, I mean that variety of contagious exanthematous febrile affection characterized by catarrhal symptoms, a peculiar rash occurring about the fourth day, the fever and rash generally beginning to decline about the eighth day of the fever, the fourth day of the rash.

Rubeola is a disease which is highly contagious, and it is probable that it may be communicated by inoculation, but it is apt to prevail epidemically, and peculiar conditions of the atmosphere appear to modify very materially the character of the disease. Measles generally assume a mild form during warm and steady weather, while in cold seasons, and very changeable weather, they are apt to assume a more violent grade or character. The period of incubation may vary from five to fourteen days, but seven days is about the usual period in the majority of cases, according to my observation.

Symptoms.—Slight tenderness and redness of the eyes, with an increased flow of tears, sneezing, cough, and watery discharges from the nostrils, together with slight creeping chills and flushes of heat, are among the first symptoms of this disease. In some cases the catarrhal symptoms precede, while in others they do not appear till the second day of the fever. In all cases, however, they appear, and may be regarded as specific phenomena of the disease. The cough is at first dry, and attended with oppressed breathing, and more or less soreness in the fauces, and often considerable hoarseness. The lymphatic glands along the neck also sometimes become swelled and tender.

About the third day considerable nausea and vomiting is apt to occur. And, if there be considerable fever, slight delirium may take place on the evening of the third day. And in severe cases, slight coma may precede the appearance of the eruption, and in small children convulsions are not uncommon at this period, especially if, as is generally the case, the fever is of a high grade, the pulse being full, hard and quick, and the skin hot and dry.

Generally about the fourth day the eruption makes its appearance in the form of small red spots, apparently papular, at first on the forehead, chin, nose and cheeks, and then successively on the neck, breast, body and extremities. These red spots soon enlarge, and as they increase they run into each other, and form large patches of an irregular or semilunar shape, leaving intermediate spaces, in which the skin retains its natural color.

During the first day of the eruption there may be discovered a small vesicle in the centre of some of the spots. During the second or third day of the eruption, the sixth or seventh of the disease, the eruption on the face is at its highest state of maturity, at which time there is generally heat of the skin and more or less itching. On the next day it begins to subside and fade on the face, while on the rest of the body it remains quite red. On the face the eruption may be felt with the hand, being slightly elevated above the surface of the skin, but on other parts the patches are not so sensibly raised. In some severe cases the face becomes swelled, and in some instances the tumefaction almost closes the eyes.

The fading and disappearance of the eruption proceeds over the body in the same order in which it made its appearance, so that, beginning as it does the seventh or eighth day of the disease, it disappears by the ninth or tenth day from the commencement of the fever upon the back of the hands, where it generally remains longest. About the ninth day the eruption assumes a yellowish appearance, and desquamation commences on the face, which by the eleventh or twelfth day is completed over the whole body.

It should be remembered that the eruption appears about the fourth day of the fever, and that both the fever and the eruption go on increasing till about the eighth day from the commencement of the fever, at which time the eruption begins to fade and the fever to subside, both having disappeared by the tenth, and desquamation being completed by the twelfth day from the commencement of the fever.—The eruption, in this disease, is not confined to the surface of the body, but makes its appearance in the mouth, and probably more or less in the mucous membrane of the trachea, bronchia, and alimentary canal.

The catarrhal symptoms usually subside with the eruption and fever, at which time a slight diarrhœa frequently occurs, which, however, should not be regarded as an unfavorable occurrence. Such is the general course of this disease, but, like all other diseases, it is liable to variations; even the eruption may appear at any time between the first and seventh day of the fever, but on the fourth day more frequently.

The *inflammatory* variety of this disease is characterized by a high grade of fever, the pulse being full, hard, and tense; the skin dry and hot; the cough violent, harsh, and dry; there is headache, with occasional delirium during the night; the eyes are red; and the respiration oppressed and difficult. Pleuritis, pneumonia, bronchitis, cerebral inflammation and gastro-enteritis, are peculiarly liable to occur in this modification of the disease.

Congestive cases of this disease are characterized by the usual phenomena of congestion in other febrile affections, the reaction taking place slowly, and the system remaining in an oppressed condition. The eruption comes out slowly; the extremities are cold; the features have a sunken, anxious appearance; the face is pale; the pulse feeble; the bowels torpid; and, if the internal congestions are not relieved, coma, stupor, or convulsions may occur.

Malignant cases of measles are liable to occur, but, according to my observation, they are exceedingly rare. When measles, however, do assume a malignant character, it is generally from some peculiar epidemic influence, together with a depraved condition of the system in which they occur. The peculiar symptoms of such cases are great prostration; imperfect or irregular appearance of the eruption, it being of a black or purple color; a strong tendency to colliquative diarrhœa and passive hemorrhage; and, finally, all the symptoms which are common to malignant typhus.

Sequellæ.—The diseases most liable to occur after measles, the result of the derangement they produce in the system, are pneumonia, croup, ophthalmia, otitis, arachnitis, bronchitis, phthisis, and a general scrofulous condition of the system.

Diagnosis.—The only diseases with which measles are liable to be confounded are catarrhal fevers, variola, roseola, and scarlatina. A correct diagnosis may, however, be formed by attention to the following diagnostic symptoms.

Measles may be distinguished from catarrhal fever by the peculiar hoarse cough which is generally present in measles, and also by the eruption when it makes its appearance, the diagnosis being presumptive at first, and positive after the appearance of the eruption.

From small-pox measles may be distinguished by the catarrhal symptoms in measles, which do not appear to so great an extent in small-pox; by the eruption in measles being less prominent, and failing to pass on to a vesicular and pustular state; and by the continuance of the fever on the appearance of the eruption in measles; while in small-pox it generally either entirely subsides or materially abates.

From roseola measles may generally be distinguished by the catarrhal symptoms in measles, which are not present in roseola. If, however, roseola be accidentally associated with catarrh, the extrinsic circumstances should be taken into the account, such as the prevalence or

12

not of measles at the time, the liability of exposure, &c., from all of which a correct conclusion may generally be drawn.

From scarlatina measles may generally be distinguished by attention to the following facts:

In measles there are prominent catarrhal symptoms, while in scarlatina no such symptoms occur. In measles the eruption consists of vividly red spots, united into irregular semi-lunar patches, with the intervening skin of a nearly natural appearance, while in scarlatina the redness is more diffused and uniform, consisting of an infinite number of minute red points united together. The eruption of measles, too, generally appears on the fourth day from the commencement of the fever, while in scarlatina the eruption usually comes out on the second day, and sometimes on the first day of the fever.

With all these essential differences, if a correct diagnosis be not arrived at, I think it must be from a want of due attention.

Anatomical Characters.—Patients die of measles either from accidental local inflammations, from congestion, or from a malignancy, owing either to a peculiar epidemic influence, or to an accidental depraved condition of the system. When, therefore, the cause of death is local inflammation, the post-mortem examination reveals only the marks of that inflammation. When death has occurred from congestion, the post-mortem reveals only the general congested state of the internal tissues and organs, and especially of the mucous membranes generally. And when death occurs in malignant cases, the peculiar changed appearance of the blood is found, which is ordinarily found on examination of subjects dead of other malignant affections.

It will be seen, then, that the anatomical characters of measles have no special peculiarities, all depending upon accidental complications which arise, and which are the immediate cause of death.

Causes.—It is probable that certain epidemic influences strongly predispose to and perhaps produce measles, but there is no doubt that a specific contagion is the general exciting cause of this disease. It is asserted, on good authority, that the disease can be produced by inoculation, blood being taken from the eruptive spots, or lymph from the vesicle.

It is evident that the disease must have been originally produced or generated in the human system, like all other contagious diseases, from various systems of imprudence; but the disease once produced, the poison is clearly generated in the bodies of persons suffering from the disease, and in this way it is kept up, prevailing generally more in winter or cold weather.

Most persons take this disease during childhood, but measles may occur at any age. One attack generally destroys the susceptibility to the disease, but, like most other contagious diseases, second attacks occasionally occur.

Prognosis.—Measles generally are not a very dangerous disease unless they assume a highly inflammatory, congestive, or malignant character. The occurrence of inflammation of the brain, larynx, trachea, or bowels, may, in some cases, lead to a fatal termination. Violent congestion, occurring early, with tardy or imperfect appearance of the eruption, renders the prognosis more or less unfavorable. The same is also true in all cases exhibiting marks of malignancy, such as great prostration, black or darkish appearance of the eruption, a disposition to passive hemorrhages, and other malignant or typhus symptoms. Such cases, however, are an exception to the rule, the disease being ordinarily attended with little real danger, unless it occurs in females in the puerperal state, or in the latter months of pregnancy.

Treatment.—As we do not as yet know of any means of arresting this disease, the indications are plainly to favor their regular development, and to meet any accidental complications that may arise, such as inflammations, congestions, and any condition arising from a malignant tendency.

In mild cases that assume a regular form, we can do little that will be of any avail. The patient should be kept warm, and allowed the free use of warm sage tea. If the bowels are confined, a dose of castor oil may be given, and to promote perspiration, one-fourth of a grain of ipecac may be given in solution ; or, what may be more convenient, from five to ten drops of the fluid extract, every four or six hours ; and the patient kept on a mild digestible diet.

But in cases in which there is a high grade of fever, with local inflammations, an active course of treatment may become necessary. If the larynx, trachea, or bronchia be the seat of the inflammation, an emetic of ipecac in the early stages may be of very essential service. A cathartic of the sulphate of magnesia should be administered, and tartar emetic in one-sixth of a grain doses every four or six hours. After the violence of the inflammation has subsided, the hive syrup in one half-drachm doses, every four or six hours should be substituted for the antimony. If the inflammation is active or obstinate, cups, sinapisms, &c., may become necessary, and should not be neglected.

If the brain become the seat of the inflammation, a cathartic of calomel should be administered, and a blister applied to the back of the neck, and four grains of the James's powder given every four hours. Cups to the back of the neck may be indicated in some cases, and in violent cases, general bleeding may be resorted to.

If the alimentary mucous membrane becomes inflamed, a blister should be applied to the epigastrium, and mucilaginous drinks taken, and only arrow-root, or some mild, unirritating nourishment allowed.

In congestive cases, if there is slowness in the appearance of the rash, stimulants and tonics may be indicated. Camphor in such cases, in two grain doses, made into a mixture with chalk and water constitutes a very valuable remedy.

And if the prostration is very great, from two to four grains of quinine may be given with each dose of the camphor mixture, and continued till reaction is fully established, when the dose of the quinine and camphor may be gradually diminished.

In malignant cases after an emetic of ipecac, a moderate dose of hydg. cum creta in castor oil, or podophyllin or leptandrin, may be indicated, after which wine-whey, camphor, and the fluid extract of bark, or quinine should be given, in doses sufficient to sustain, if possible, the sinking powers of the system. The patient should also be allowed chicken or mutton broths, and if necessary, beef essence. And now in conclusion, let us bear in mind, that while mild and regular cases require little or no treatment, in complicated cases we must be prepared to meet promptly any indications that may arise.

SECTION V.—SCARLATINA—(*Scarlet Fever.*)

By scarlatina, or scarlet fever, I mean that peculiar exanthematous febrile affection, characterized by sore throat, a scarlet rash appearing about the second, and declining about the sixth day, and followed by desquamation; the disease being liable to various modifications.

The degrees of modification in the course and symptoms of scarlatina are such, that I shall proceed to consider the disease under three heads: the simple, the anginose, and the malignant; the simple embracing those mild regular cases, in which there is no marked local complications; the anginose, those in which there are local complications, involving the throat; and the malignant, those in which there is a marked typhous, or congestive tendency; all, however, being only modifications of one and the same disease. As a matter of convenience, I will give the general symptoms of mild. uncomplicated cases, including, of course, the simple variety; and then I will give the peculiar symptoms of the anginose and malignant varieties.

Symptoms.—Generally after an indefinite period, varying from one to three or four days, of ordinary premonitory symptoms of febrile disease, the patient has slight chills, followed by flushes of heat, nausea, pain in the back, extremities and head, a hot and dry skin, and generally a frequent quick pulse. As these symptoms progress, and generally within the first forty-eight hours after the fever commences, a scarlet rash or eruption, makes its appearance; first on the face, and then on the neck, trunk, and extremities, spreading finally over the surface of the nostrils, mouth, and fauces; and in some instances, on the tunica albuginæ of the eye.

The rash consists of innumerable red points, which, running into each other, give a diffused blush to the skin, leaving generally no intermediate skin of its exact natural appearance. In some cases, however, the scarlet efflorescence appears only in large irregular blotches, leaving the intermediate portions of skin of its natural appearance. The papillæ of the skin are generally somewhat enlarged, giving a slight roughness to the feel, especially on the breast and extremities.

Generally with the commencement of the fever, there is slight soreness of the fauces, with more or less difficulty of swallowing; and the voice usually becomes thick, and more or less unnatural. The face is slightly swelled, and the tongue is covered with a white fur, through which the enlarged papillæ exhibit their scarlet points, the pulse being frequent, quick, tense, and vigorous.

About the fourth day of the disease, the rash and fever are generally at their complete state of development, and on the fifth day, both begin to decline, and continue to diminish till they are entirely gone, about the seventh or eighth day, at which time, desquamation of the skin commences. Usually while the eruption is declining, the tenderness of the fauces abates, the perspiration becomes more free, the urine becomes copious ; and in some instances, there is a slight diarrhœa. The **fever, as** we have seen, generally abates as the rash declines ; but in some instances, the fever abates considerably, as soon as the rash makes its appearance, but this is not generally the case according to my observation.

In some cases, the fever from beginning is so slight, as scarcely to **attract** attention ; while in others, the arterial excitement is very strong. **Thus we** have the ordinary symptoms of simple scarlatina ; liable, of **course, to various degrees** of febrile excitement, **and also to a** greater or **less extent of the rash,** but affording no tedious **or perplexing complications, as in the** anginose and malignant cases, which **we will now proceed to consider.**

Scarlatina anginosa.—In the anginose variety, **the fever is generally** more violent, and the affection of the fauces very marked. **In the forming stage, there is** generally considerable headache, nausea, **vomiting, and** general muscular prostration. **At** the commencement of **the fever,** and in some instances before the fever commences, there is a feeling **of** stiffness of the neck, and more or less swelling and redness of the palate, tonsils, uvula ; and, in fact, of the whole fauces. The voice **becomes** hoarse, deglutition difficult, and respiration is attended with a **sense of** constriction in the throat. The fever rises rapidly, the pulse is frequent and quick, the thirst is urgent, and the heat of the skin is often more intense than in almost any febrile affection. The tongue, too, becomes dry ; **the** injected, **or** inflamed papillæ, projecting from its **surface.** There is generally considerable pain in the head, with much restlessness, languor, and prostration, through the whole course of the disease.

The eruption does not generally make **its** appearance till the third **day** of the fever, and then is seldom diffused **over** the **whole surface, coming out** in irregular and large patches, on different parts of **the body. It is,** in fact, no uncommon event for the rash to disappear, **for a time, after it has made** its appearance. If the fever declines **on** the **fourth day, the** tonsils and palate seldom become ulcerated, the **local inflammation subsiding** with the fever, without ulceration. But if the fever is **active, and** continues, small ulcers are apt to form about the parts, which may be converted into ash-colored superficial sloughs. As the fever subsides, the sloughs separate and leave red ulcerated surfaces, which generally cicatrize **with** little difficulty. In some cases, however, **the inflammation** extends into the trachea, and the patient dies with **symptoms of acute** bronchitis. Other local inflammations may also **arise, the most frequent** of which, according **to my** observation, are **of the brain and alimentary** mucous membrane.

Scarlatina maligna.—Malignant **scarlatina generally begins like the** preceding, **but soon** assumes a violent **and** dangerous **character.** The **eruption is irregular in** making its appearance, generally coming **out** between the first and fourth days, being pale at first, and acquiring, **in most** cases, a dark livid hue during the progress of the disease ; **and in some** cases, it disappears in parts, and then returns a **day** or two afterwards. The temperature of the skin is variable, and the pulse, active at first, becomes small and frequent in the course of the second day. Delirium, too, **is apt to occur** early, and continue more or less during the whole course of the disease. The cheeks are suffused with a livid flush, the tongue is covered with a dark brown fur, and the breath is generally more or less fetid. Gray colored sloughs appear on the palate and tonsils, which soon acquire a dark color; and if the fever is continued beyond the fourth day, the ulcers are converted into dark or black fetid sloughs, attended sometimes with the most fearful hemorrhage, especially after the supervention of collapse, which not unfrequently occurs by the middle of the second week of the disease, or even earlier.

A fearful case of hemorrhage from the tonsils, uvula and gums, occurred under my observation, a few years since, in the person of a young lady laboring under malignant scarlatina. She was the patient of my friend Dr. Walter Webb, an eminent physician of Jefferson county of this state. I saw the case but once as I was only called in consultation; but the patient, though the hemorrhages occurred often for several days, finally recovered, much to the satisfaction of her persevering and skillful attendant and a large circle of anxious friends.

Malignant scarlatina may assume an inflammatory, or congestive character, or it may partake of the nature of both; depending upon the natural or accidental condition of the patient at the time of attack. Thus we have the symptoms of scarlatina as it occurs in its simple, anginose and malignant forms, but it must be remembered that there is no real definite dividing line between them; the descriptions being designed to answer to all the variations that may occur in this really variable disease.

Sequelæ.—Scarlatina is frequently followed by various troublesome and dangerous dropsical affections, among which anasarca is by far the most frequent, but the serous effusion may be into the cavity of the abdomen, the chest, the pericardium, and even into the ventricles of the brain. Various nervous affections also sometimes occur as sequelæ of the disease, as hysteria, asthma, chorea, epilepsy and various other local and general neuralgic affections.

A scrofulous condition with swellings or abscesses of the cervicle glands, or about the submaxillary or paroted glands, are, according to my observation, a frequent result of scarlatina. Scarlatina also, besides leading directly to dropsical, nervous and scrofulous affections, often produces a general derangement of the system which strongly predisposes to many acute and chronic complicated diseases.

Diagnosis.—The difficulty of distinguishing scarlatina from other febrile affections, is confined mostly to the first period before the appearance of the eruption, during which period the slight redness, and perhaps soreness of the fauces, the frequency of the pulse and the extrinsic circumstances, go to make a probable diagnosis. When the eruption makes its appearance, a positive diagnosis may be formed by a careful observation of all the symptoms.

From measles, it should be remembered, scarlatina may generally be distinguished by the following marked differences. In scarlatina the eruption generally comes out on the second day of the fever, whereas in measles the rash generally appears on the fourth day. The rash in scarlatina, appears like a diffused erythematous blush of the skin with innumerable points intermixed and small papilla dispersed over the cuticle, while the eruption of measles consists of small circular dots of a deeper red in the centre than at the circumference, so that, in running into each other, the skin presents a less uniform flush than in scarlatina. The color of the eruption in scarlatina is scarlet, while in measles it is of a darker red inclining to brown. In scarlatina there are no marked catarrhal symptoms, while in measles catarrhal symptoms are always more or less prominent. And finally in scarlatina the white of eye is of a diffused reddish appearance, while in measles the redness is generally less diffused.

From roseola scarlatina may be distinguished by the redness or soreness of the fauces in scarlatina, while in roseola no such redness or soreness exists. The red points on the tongue, consisting of enlarged papillæ, rising through a white coat is a condition not present in roseola, and, besides, there is generally much more febrile excitement in scarlatina than in roseola.

Anatomical Characters.—In patients that die early of scarlatina there is often nothing found, on examination, to indicate the cause of death, and the morbid appearances found on examination of patients dying later, which reveal the cause of death, are such as result mainly from accidental complications, such as local congestions, inflammations, &c.

There generally remains, however, after death, more or less redness of the fauces, and marks of congestion, irritation, or inflammation of the alimentary mucous membrane, as well as a more or less changed condition of the blood, it being in some cases thick and in others of a watery appearance. The cuticle, in malignant cases, is sometimes found more or less separated from the vascular layer of the skin, and the rash is generally found of a dark livid appearance—in some cases, however, every vestige of the rash disappears at death, so that no trace of it is found.

Cause.—Scarlatina doubtless had its origin in human imprudences of various kinds, and when the disease was once produced, the system, laboring under it, generates a specific contagion which communicates the disease. While, then, a specific contagion, generated in the bodies of patients suffering from the disease, is probably the general cause of this disease, it is possible that the disease is occasionally produced spontaneously by peculiar atmospheric and personal impurities. This is rendered probable by the fact that the disease frequently arises under circumstances in which it is impossible to account for its origin in any other rational way

Most persons contract this disease during childhood, but it may occur at any age, one attack generally destroying the susceptibility to the disease, though second attacks may occur, as in most other contagious, exanthematous febrile affections.

The period of incubation, according to my observation, is about one week, and I suspect the disease may be communicated at any time after the febrile symptoms are developed, but perhaps not till the eruption makes its appearance.

Pathology.—That a morbid poison enters the blood either by the absorbents of the skin, by the lungs, or by the stomach, I believe is universally admitted. Now it is probable that this poison, once in the blood, passes with it through the system, affecting all the tissues of the body, and especially the brain and nervous system. Now, as a consequence of all this derangement, there is more or less debility and consequent letting down of the circulation, and hence the slight chilliness which occurs. Over-pressure upon the brain and spinal cord leads to slight irritation, which, being transmitted to the ganglionic system, reaction of the heart and arteries follows, and the blood is thrown to the extreme capillaries of the vascular structure of the skin, and to the mucous membranes, the morbid poison there producing its specific effect, especially upon the skin and mucous membrane of the fauces.

The capillaries of the skin become congested or enlarged, producing the blush of the skin, and the papillæ of the skin and mucous membrane of the tongue and throat becoming enlarged, makes the minute points and slight elevations on the skin, and especially the minute red points observed on the tongue. Now this deranged condition of the skin and mucous membranes interrupts, to a certain extent, the secretion of mucus, and to a very great extent the cutaneous exhalation, and hence the great liability to dropsical affections after scarlatina, especially as perspiration is interrupted, till desquamation is accomplished.

Prognosis.—The prognosis in simple uncomplicated cases of scarlatina is generally favorable, but in anginose and malignant cases there is always considerable danger, depending, of course, much upon the nature and extent of the complications, and also upon the general condition of the system as well as the nature of the prevailing epidemic influence.

I have known children to die in the early stage evidently from the direct effects of the morbid poison upon the brain and nervous system, no symptom of reaction having appeared, one instance of which fell under my observation during an epidemic which prevailed in Seneca county, in a healthy locality between the Seneca and Cayuga lakes, during the winter of 1855. I saw the case in consultation with my friend, Dr. F. Glauner, an eminent and skillful physician then practising in that locality, and though the treatment had been of the most judicious character, the morbid poison had done its work, and vitality soon became extinct.

The unfavorable symptoms are a tardy and imperfect appearance of the rash, or a sudden retrocession of it; coma or delirium; a livid appearance of the rash and of the fauces; petechiæ, hemorrhages, &c.

The disease is always attended with the greatest danger when it occurs during pregnancy or in the puerperal state.

Treatment.—As we do not know of any means of neutralizing the morbid poison and arresting the disease, the indications are evidently to favor its regular development, to control any undue febrile excitement, and to correct, as far as may be, any complications that may arise.

In mild cases of the simple variety of scarlatina, little treatment is indicated. The patient should be allowed cool drinks, a little toast-water, arrow-root, or other light food, and be kept in a moderately cool room. If the bowels are constipated, a dose of castor oil may be administered, and, as a diaphoretic, ipecac and camphor in one-fourth grain doses each, made into a mixture with prepared chalk and water, may be given every four hours during the continuance of the fever. As the fever subsides, to produce a slight effect upon the kidneys, and to keep the bowels in a moderately relaxed state, from ten to thirty grains of the supertartrate of potash may be given in solution in water after each meal.

In the anginose variety, or in those cases in which the **fauces are** seriously involved, with a high grade of febrile excitement, an emetic of ipecac or antimony should be administered at first, after which, a cathartic of calomel or podophyllin should be given, and a free evacuation of **the bowels** secured. As a diaphoretic, the James's and Dover's powders, of each three grains for an adult, every four or six hours may be given.

As a wash for the throat in such cases, a strong sage tea, with alum and loaf sugar or honey, will generally do well in the early stages, but later, especially **in** putrid cases, Labarraque's solution of the chloride of **soda,** diluted with eight parts of water, will often do better **as a gargle.** Either, **when** indicated, may be applied to the throat, **or used as a gargle,** every four or six hours.

If the skin is very hot and dry, it should be sponged with moderately cool water, once every twenty-four hours, generally towards evening of each day. Great care should be taken, however, not to apply cold water to the surface unless the skin is hot and dry, and no chilliness present, as well in this as in all other febrile affections.

In malignant scarlatina, or in cases in which there is a typhous or congestive tendency, it is generally well to give an emetic of ipecac at first, and follow that with a cathartic of calomel, rhubarb and castor oil, or if a mercurial is from any cause contra-indicated, podophyllin or leptandrin may be given instead. After the operation of a cathartic, and when the sinking or congestive character becomes apparent, camphor and quinine should be given, **in** doses for an adult varying from two to four grains of each, and continued during the sinking or congested state.

In cases of mild scarlatina, the patient **may be** allowed regular meals of plain digestible food, such as toast, rice, milk, a poached egg, &c., during the whole course of the disease; but in anginose and malignant cases the patient should be allowed toast water, or **crust coffee,** one half milk, taken a little cool if preferred, for a drink, **and to sustain the system** till such time as plain digestible articles of food can be allowed or borne. **In** malignant cases, in which there is great prostration, mutton or chicken broths, wine whey, and beef **essence * may be indicated, and should be freely** allowed.

In **malignant** cases, in which there is hemorrhage from the gums, or **any other** parts, in addition to **wine** whey, broths, beef **essence,** and quinine, the tincture of muriate of iron, in ten drop doses, may be given every four or six hours, alternating with the quinine. In fact, I know of no remedy more clearly indicated in malignant hemorrhagic cases **of** scarlatina. It is agreeable to the throat and **stomach, affects** favorably the alimentary mucous membrane, improves **the blood, and serves by** improving the blood and tissues to lessen **the hemorrhagic tendency.**

The room of the scarlatina patient should be kept cool, and also well

---

* Beef essence is best made by cutting **lean beef into fine pieces, and** filling or in part filling with it a quart bottle; set the bottle **into a kettle of water, and boil** two hours; pour off **and** season.

aired, so as to be agreeable to the patient, and also that the morbid poison emanating from the patient's body may become diluted as much as may be, and thus the liability be lessened of communicating the disease to others. Every reasonable precaution should be taken to prevent exposure to scarlatina, but I have little confidence in any prophylactic remedy against this disease, such as mercury, belladonna, &c., but should rely upon prudence and a strict observance of all the laws of health, as a prophylactic against this and all other diseases.

In the dropsy following scarlatina I have generally succeeded with digitalis, squills, and tartar emetic, in moderate doses every four or six hours at first, and then substituting the iodide of potassium in five grain doses three times per day, keeping the bowels gently open with the supertartrate of potash in moderate doses.

SECTION VI.—ERYSIPELAS.

By erysipelas I mean that variety of exanthematous febrile affection, perhaps slightly contagious, characterized by a diffused eruption or inflammation of the skin, of a superficial or phlegmonous, and sometimes malignant character.

Before proceeding to consider this disease, it is well for us to remember the peculiar structure of the skin, and also the fact that the mucous membranes are only a continuation of the skin, lining internal parts communicating with the skin, secreting mucus to lubricate their surfaces, while the skin exhales the perspirable matter. This becomes necessary, as the erysipelatous eruption or inflammation often extends over the surface of the body more or less, and also into the mucous membranes, or it may pass by metastasis directly from the skin to the mucous membranes, and in fact to the serous and other structures of the body.

The eruption of erysipelas may consist merely in a congested state of the minute vessels, constituting the vascular structure of the skin, usually called erythema, or the vascular structure of the skin may become inflamed, constituting what has usually been called simple erysipelas, or not only the skin, but the areolar tissue may become inflamed, constituting the phlegmonous erysipelas; and finally, in either case, if there is a malignant or typhous tendency, the disease has been called malignant erysipelas.

Regarding this disease as an exanthematous febrile affection, I shall proceed to consider its symptoms, under the heads of simple, phlegmonous and malignant erysipelas, as being the most convenient, and including every possible variety of the disease.

Symptoms.—*Simple erysipelas* commences with a general feeling of ill health, such as headache, loss of appetite, nausea, general depression, foul tongue, and a feeling of weight in the epigastrium. After these symptoms have continued for a longer or shorter period, there is more prostration with chills, varying from slight chilliness to a severe chill or cold stage.

During the chilly or cold stage there is often headache, nausea, vomiting, &c., but reaction takes place, and gradually febrile symptoms are developed, the pulse becomes frequent, the skin hot, the tongue coated, there is thirst, nausea, restlessness, and in some cases sore throat, and slight swelling and tenderness of the lymphatic glands of the neck, axilla, or groin, indicating to some extent the point at which the erysipelatous eruption is about to make its appearance.

If the cervical glands are swelled, the eruption will probably appear about the face or scalp; if the axillary, on the arms or hands; but if the glands of the groin, the erysipelatous eruption or inflammation may be expected to appear on the lower extremities; and finally, if there is a general soreness or swelling of the lymphatic glands, the eruption may be expected to be very general over the surface of the body.

At any time after the development of the fever, but generally, I think, about the second or third day, the erysipelatous eruption or inflammation makes its appearance, in the form of an irregular blotch or stain, which may spread over more or less of the contiguous surface. The eruption may be only a congestion of the vascular structure of the skin, or a slight inflammation of the skin, and in either case the redness will disappear on making pressure on the inflamed or reddened part; but if not only the skin, but the deeper tissues become involved, the redness will not disappear on making pressure.

There is generally more or less burning or stinging pain and tumefaction, from the beginning, which may increase, more or less, during the progress of the disease; but in simple cases, the swelling is much less than in the phlegmonous. Usually about the third day of the inflammation or eruption, small vesicles or blisters, of various sizes make their appearance, filled with a yellowish serum. In two or three days these blisters or vesicles break, and discharge a viscid fluid, which sometimes forms crusts or scabs. By the seventh day, from the appearance of the inflammation or eruption, and about the tenth day of the disease, the crusts or scabs have desquamated, a new cuticle having formed.

The slight swelling, redness, and febrile symptoms, progress together, both in increasing and declining; so that the fever, the swelling, and eruption or inflammation, have all subsided by about the tenth day of the disease, when desquamation occurs. It should be remembered, however, that there is not always a direct proportion between the fever and the eruption, or inflammation; in some cases there being a high fever, with only a slight eruption, and the reverse.

In some cases, the eruption extends gradually along the skin, without increasing much in extent, as it disappears from the parts first affected, while it encroaches on the sound skin. The eruption may, in some cases, disappear from its original seat, and make its appearance in some other, and remote part of the body, as I have often seen.

The eruption of erysipelas consists, as I have said, in either a congestion or inflammation of the vascular structure of the skin, and in the simple form, the tissues beneath seldom become inflamed to any great extent; but when the inflammation extends deep into the intermuscular cellular structures, it constitutes phlegmonous erysipelas.

Phlegmonous erysipelas.—In phlegmonous erysipelas, as I have already intimated, there is, in addition to the eruption or inflammation of the skin, more or less inflammation of the intermuscular cellular tissue beneath, the fever is also more violent, the pain much more severe, and the swelling greater, than in simple erysipelas. The disease may progress, decline, and finally disappear in a few days, like simple erysipelas; but it frequently happens, that as the fever diminishes with the local pain and redness, the swelling of the part does not diminish in proportion, **it becomes soft,** and in a few days matter is formed between the **muscles underneath the skin.**

In severe cases the matter may extend along **the cellular tissue in the** direction of the muscles, and destroy the adjacent parts, which are frequently discharged in dirty-looking shreds of mortified substance, mixed with pus. And this discharge may continue for weeks, until it completely exhausts the patient, or till the system rallies, and the discharge finally ceases.

A remarkable case of **this character fell under** my care a few years since while practicing in Jefferson county of this State. The patient, a lady about fifty years old, of delicate health, and a slender constitution; had phlegmonous erysipelas affecting the upper portion of the right arm, and also the right breast, and whole right side of the chest. The fever was of a high **grade at** first, but the suppurative stage was followed by great **prostration, requiring** full doses of quinine and camphor for several days.

The matter formed in the cellular tissue of the right breast, and also in nearly the whole right side of the chest, from near the axilla, almost to the floating ribs, and also in the right arm, from the insertion of **the** deltoid muscle to near the elbow. After the system had sufficiently rallied by the continued use of quinine, camphor, wine whey, broths, &c., I made an incision to **evacuate** the matter from the side, and a pint, or more, of thin brownish appearing pus, with shreds of mortified cellular tissue, was discharged at first.

The matter lay between the outer surface of the ribs, and the muscles of the side, nearly all of which had been dissected up, more or less, by the accumulated matter. The discharge continued for several weeks, and also from the arm ; but by the continued use of the iodide of potassium, iron, and a nourishing diet, **the** discharge ceased, and the muscles formed new adhesions to the surface of the ribs. The cavity formed by the extensive sloughing of the arm gradually healed or filled up, and the lady regained her usual **state of** health.

I have related this case, as it illustrates severe phlegmonous erysipelas, **but we must** remember that most cases of phlegmonous erysipelas terminate by resolution, no matter being formed, as I have before stated.

Malignant erysipelas.—This variety of erysipelas is characterized by a putrid typhoid condition of the system, and often tending rapidly to a fatal termination. It is sometimes the result of poison animal matter taken from the **dead body in** dissecting, or from putrid sores on the living subject ; and in either case, the fever and the local erysipelatous affection are of an asthenic, and highly malignant character.

Malignant erysipelas may, and does sometimes, prevail epidemically ; or in particular localities, in a manner to indicate a contagious tendency ; or at least that some morbid poison exists in the air, which produces or predisposes to it ; while exposure to the disease appears to favor its development, or to act as an exciting cause of the disease. The primary fever is generally preceded by a marked chill, and is at first of a high grade, the skin being hot, the thirst urgent, and there being great restlessness, and general irritability. The eruption may make its appearance at any time after the development of the fever; or in some cases, a slight inflammation may appear to precede it, as in dissection wounds; in either case, it may involve a whole limb, the face, or any other part of the surface of the body, of greater or less extent, and have very much the appearance of **ordinary phlegmonous** erysipelas, except that it is usually of a darker color.

Generally between the first and fourth day, the fever rapidly assumes a typhoid or typhous character, the pulse becomes small, weak, and frequent, the eruption or inflamed surface becomes pale, or of a livid appearance, and great prostration of the nervous and muscular system follows. If the eruption is of the face as the sinking comes on, it frequently passes by metastasis to the membranes of the brain, developing all the symptoms of acute meningitis. If the eruption is on the extremities, as it fades during the sinking, it is frequently translated to the peritoneum, or to the mucous membrane of the alimentary canal, developing the ordinary symptoms of peritonitis, **or mucous** enteritis. If the eruption is on the chest, it may pass to the pleura or to the pericardium, developing symptoms of acute pleuritis, **or of pericarditis,** and generally of the most rapidly fatal character.

In some instances I have known the erysipelatous eruption to pass by metastasis to the meninges of the spinal cord, producing, in the midst of a typhus state, the most violent neuralgic and rheumatic symptoms, proving conclusively to my mind, that rheumatism, in some cases at least, depends upon an irritated, congested or inflamed condition of some portion of the spinal cord or its meninges.

I have noticed that in cases in which the local erysipelatous affection has appeared to pass from the chest to the meninges of the cord, the superior portion appears to be the part involved, and the neuralgic and rheumatic inflammation is developed in one or both arms; but if it pass from the dorsal or lumbar region, it appears to fall on the lower portion of the spine, and in that case the lower limbs become the seat of the neuralgic and rheumatic inflammation.

I am not aware that this peculiarity of the metastasis of erysipelas to the spinal cord, or its meninges, has ever been noticed by any writer, but many very marked cases of this character fell under my observation during an epidemic which prevailed in the range of my practice, in Northern New York, during the winter of 1853 and 1854, a short account of which I afterwards published in the Buffalo Medical Journal.

In malignant erysipelas, if there is no metastasis to internal parts, and the patient survives, the eruption or inflammation may terminate by resolution and desquamation, by about the tenth day of the disease, as occurs in simple erysipelas, and in most cases of phlegmonous; but it often terminates, as in bad cases of phlegmonous, in suppuration or gangrene.

An interesting case of malignant erysipelas, terminating by suppuration and final recovery, fell under my care in this village, during the summer of 1856, in the person of Dr. H. A. Potter, a distinguished and

eminent surgeon of this village, well known in this country and in
Europe, as a successful operating surgeon. The facts and history of
this case I published in the Western Lancet, in the fall of 1856,* from
which, and my note-book, together with my recollection of the case, I
draw the following history and symptoms.

In the early part of July, 1856, Dr. Potter was called on to attend
his brother, who was also an eminent physician and surgeon, residing in
Steuben county, of this State, supposed to have been poisoned by matter
received from an extensive suppurating wound, by a slight scratch in
one of the fingers of the right hand. The general and local symptoms
which were developed in his case, were similar to those usually following dissection wounds, as I understand; there having been at first a
slight irritation of the wound, a chill following, and then a high febrile
reaction, followed by an asthenic malignant erysipelatous eruption or
inflammation of the arm and side, and great tenderness of the axillary
glands, general restlessness and irritability, and finally extensive suppuration in the right axillary region.

He had been bled by his own directions, I believe, before calling his
brother, a circumstance very much to be regretted, as it probably lessened the chances of his recovery, at least it was so regarded by Dr.
Potter when he was first called to attend to the case. Dr. H. A. Potter,
when he was first called to see his brother, regarded his case as a desperate one, and so remained with him up to the time of his death, which
took place in about one week, thus giving him not only the benefit of a
judicious treatment, but also of untiring care.

While Dr. Potter was attending his brother, he opened an extensive
abscess which had formed in the right axillary region, and passed the
fingers of his right hand into the cavity, in order to ascertain its direction, and also otherwise exposed his hands to the matter which was discharged in dressing the parts each day. He was, I believe, attending
another patient with an extensive suppurating wound at the same time,
or during that week.

While attending the funeral of his brother, in Yates county, on the
13th of July, Dr. Potter suffered from a slight uneasiness in a trifling
scratch, received several days before, on the outside of the third finger
of the right hand. In a few hours the irritation had passed along the
lymphatics to the back of the hand, where an asthenic erysipelatous
inflammation was set up after reaction had taken place, from a chill
into which he was at that time passing.

At this time I was called to attend him, he having reached home late
in the evening. I found him restless, irritable and chilly, notwithstanding he had taken nearly a quart of whisky in sweetened milk, during
the preceding eight hours, to counteract the poison, which the Doctor
knew was producing its general, as well as local effects upon his system.

A red streak extended from the slight wound on the finger to the back
of the hand, which was swelled and painful, though no general reaction
had as yet taken place. The axillary glands were also swelled and tender, and the general prostration of the system was very great.

The liquor was continued till something over a quart had been taken, when, as the stomach began to reject it, very little more was given. A warm foot bath and friction along the spine, with a strong infusion of capsicum in vinegar succeeded, with other measures, in overcoming the chilliness and bringing about reaction, so that no more coldness was experienced after three o'clock in the morning.

As reaction took place, the erysipelatous eruption or inflammation extended rapidly up the arm, the whole of which was kept wet with a strong solution of the sulphate of iron. A strong infusion of valerian was given every six hours up to the third day, at which time nearly the whole extremity was swelled to about twice its natural size, and the hand was evidently approaching gangrene, to prevent which, I laid it open freely between the second and third metacarpal bones. A bloody serous discharge from the incision relieved the excessive pain in the hand and arm, and appeared to arrest the deep asthenic erysipelatous or phlegmonous inflammation, and, in fact, the progress of the local affection, as the color of the arm began to fade, and soon the swelling was evidently becoming less.

At this stage, as there was an evident sinking tendency, I gave him the sulphate of quinine, in full doses, with camphor, and applied a bread and milk poultice to the hand. This treatment was continued up to the seventh day, at which time free suppuration took place in the hand and arm; the matter in the hand escaping from the incision made four days before.

On the eighth day of the disease I opened the arm through the extensor muscles, two inches above the wrist, and subsequently at a point nearer the elbow, and still later made another opening on the ulner side of the hand; from all of which points matter continued to discharge till July 30, the seventeenth day of the disease, at which time suppuration was suspended, and the cuts were healing by granulation.

The swelling of the hand and arm was nearly down at this time, but the quinine and camphor was continued till August 4, the twenty-sixth day of the disease, at which time I discontinued my attendance, the hand and arm having healed and only a stiffness remaining. From the thirteenth day of the disease, at which time I made the last opening in the hand, I gave him five grains of the iodide of potassium three times per day, alternating with the quinine, and I believe this was continued for about four weeks or about two weeks later than the quinine, and two or three weeks after the Doctor was about town.

The nourishment during the first eight days of the disease was mainly toast water or crust coffee, one-half milk, and from the eighth to the seventeenth, the period of suppuration, broths, beef essence, &c., were taken in addition, and from the seventeenth day onward, toast, eggs, wild game, fowls, mutton, beef, &c., were freely served up and taken with a good relish.

The hand for a time, entirely useless, gradually by gentle rubbing and careful use was restored to nearly its natural state, and has since been well employed by the Dr. in performing some of the most difficult and successful operations ever performed in this country.*

* See American Journal of Medical Sciences for October 1858, page 571, for cases of Ovariotomy.

I have been particular in relating this case, as it illustrates malignant erysipelas under a controlling and sustaining course of treatment, while the case in Steuben county unfortunately bled, illustrates the rapidly fatal tendency of this disease. In other cases it must be remembered, the patients may die in the early stage, before the erysipelatous eruption makes its appearance or even reaction takes place.

Thus we have the ordinary symptoms of simple, phlegmonous and malignant erysipelas, but we **must remember** that it is an exanthematous febrile affection, liable like all others to great variations, all of which, however, I think may reasonably be brought under these three heads, and especially so, if we have the condition of the patient and not the name of his disease uppermost in our minds.

Diagnosis.—The diagnosis in erysipelas before the eruption makes its appearance, must be to a certain extent, only probable; but, by taking into account, the prevailing epidemic influence, the tenderness and swelling of the lymphatic glands, and especially of the neck, the gastric and cephalic disturbance, together with the character of the chill, and the febrile excitement, this probable diagnosis may be rendered nearly certain. But this strong probability is rendered certain, immediately on the appearance of the eruption, which we have seen, may appear at any time after reaction takes place and fever is developed, but generally about the second or third day of the fever, as the eruption differs from all others to a degree, not allowing of a doubt even.

Anatomical Characters.—In cases of erysipelas that prove fatal very early, from the direct depressing influence of the morbid poison, as in dissection wounds, but little is found to indicate the cause of death, except perhaps, a changed or dissolved state of the blood, and a slightly brownish or purple appearance marking the seat of the erysipelatous eruption. In cases, however, that have progressed, the cuticle is often found more or less detached from the vascular structure of the skin, and the cellular tissue beneath the skin exhibits more or less marks of inflammation, according as the disease had been more or less phlegmonous in its character.

In cases in which there has been metastasis of the eruption to internal organs, as the brain, lungs, liver, &c., marks of inflammation, with collections of pus or serum, are often found to a greater or less extent, occupying the organ or part to which the erysipelatous eruption or inflammation had been translated.

Causes.—The exciting causes of erysipelas are various; such as wounds, or any irritant affecting the skin, but general or constitutional causes operate to predispose the system to this disease, and often develop the disease without any apparent local exciting cause. Taking food irregularly, or that which is indigestible or unwholesome, by deteriorating the blood, and irritating directly the alimentary mucous membrane, is a frequent cause of simple erysipelas. And when we remember, that the alimentary mucous membrane, is a continuation of the skin, and that there exists a strong sympathy between them, we need not wonder that a mucous gastro-enteric irritation, together with a deteriorated state of the blood, and the fluids of the body should develop the erysipelatous eruption or inflammation.

Phlegmonous erysipelas may be produced by the same causes, but, generally I think koino-miasmata operating on systems predisposed by habits of imprudence, acts as an exciting, and probably also as a predisposing cause of this variety of erysipelas. The paludal poison by operating through the blood on the brain and nervous systems produces debility, prostration, and generally a chill; and, if now the skin becomes irritated by filth, a cold damp air, or from any other cause, the alimentary mucous membrane being irritated also, the one or the other will become the prominent seat of irritation.

If now, as febrile reaction occurs, the local irritation be concentrated upon the alimentary mucous membrane, gastro-enteritis is the result, with perhaps bilious remittent fever; but if from any predisposition or accidental cause, the irritation be transmitted to the skin, we have the erysipelatous eruption or inflammation instead, complicating the bilious derangement, and of course, the attending fever modified according to the nature or extent of the local erysipelatous affection. And as in such cases, not only the skin, but the adjacent cellular tissues are in a relaxed state, the irritation and inflammation is very liable to extend from the skin to the intermuscular cellular tissue, producing phlegmonous erysipelas, and if the intermuscular cellular tissue continue long in an inflamed condition it may terminate in suppuration, and in extensive wasting purulent discharge, as we have already seen.

Now in the production of malignant erysipelas, all the causes which I have enumerated as predisposing to, or exciting, or developing the simple and phlegmonous varieties may operate to a certain extent in producing the malignant, but I am satisfied that malignant erysipelas, whether it occur from dissection wounds, or in hospitals, or from an epidemic influence is invariably produced by an animal or vegetable poison, or by a combination of the two.

It is probable that a concentrated koino-miasmata may operate alone to produce the disease, if the patient be filthy, and confined in bad air, so that the system itself becomes the generator of the idio-miasmata necessary to produce a dissolved state of the blood, and also the typhus condition, as well as the malignant character of the local erysipelatous affection. But while the paludal poison may thus operate to produce malignant erysipelas, it is probably combined in most cases, with poison, animal matter in hospitals, in epidemic influences, and in sporadic cases which occur; and in very many cases, as in dissection wounds, or poison animal matter taken from wounds on the living subject, in hospitals, and in some epidemic or endemic influences, in filthy localities where human beings are congregated; poison animal matter is alone the exciting cause of this disease.

Thus we have the principal causes of simple, phlegmonous, and malignant erysipelas, and some of the reasons why each assume their peculiar characteristics.

Pathology.—In relation to the pathology, or nature of erysipelas, it appears to me there is room for no reasonable doubt. If we take into account the system in a state of health, the causes which operate, and the symptoms which are developed, as well as the anatomical characters

presented in fatal cases, I think we must regard erysipelas as an exanthematous febrile affection, liable, of course. to wide variations.

The causes, whether filth, unwholesome food, marsh, or idio miasmata,
or even poison animal matter taken directly from the dead or living subject, enter the blood either through the stomach, lungs, or skin, being
taken up by the absorbents, more or less, in either case, passes into the
blood; affecting, to some extent, the lymphatics, and especially the
lymphatic glands, causing in them more or less irritation and swelling.

The morbid poison, in either case, once in the circulation, passes with
the blood to every part of the system; affecting, more or less, all the
tissues and organs, but especially the brain and nervous system, producing derangement, prostration, and finally by letting down the circulation, a chill, more or less, marked.

During the chill, or cold stage, the cerebro-spinal system becomes
irritated; and this irritation being transmitted to the ganglionic system
produces an irritated action of the heart and arteries, and hence, febrile
excitement follows, and the blood with whatever *morbid poison* it may
contain is thrown to the extreme capillaries in the vascular structure of
the skin; and hence, sooner or later, the erysipelatous eruption or inflammation is developed.

If the morbid poison be only such as are received from unwholesome
food or drink, and the predisposition be slight, simple erysipelas may be
the result; but if, in addition to these slight causes, the paludal poison be
also added, phlegmonous erysipelas may be produced if the patient is
strongly predisposed; and finally, if there be added poison animal matter
in any form, or if only that alone be operating, as in the cases I have related, we may have malignant erysipelas of a most rapidly fatal tendency.
In the simple and phlegmonous cases, if no animal poison be operating,
it is probable that the blood is not materially changed; but in the
malignant, and in all cases in which poison animal matter operates as a
cause, the blood evidently becomes more or less dissolved or changed,
and hence the typhus symptoms and rapidly fatal tendency of such cases.

The translation of the eruption to internal parts occurs generally in
cases in which the general powers of the system are sinking, the metastasis being to such internal part as happens to be most predisposed, and
is probably the result of a letting-down of the circulation; the blood,
together with the morbid poison receding from the vascular structure of
the skin and surface of the body, and being concentrated upon internal
parts, and affecting that organ or tissue most predisposed.

Prognosis.—The prognosis in simple erysipelas is generally favorable
unless it is translated to some internal tissue or organ, as the meninges
of the brain or spinal cord, the pleura, pericardium, peritoneum, or to
the various mucous membranes; in which case, the danger is always increased and is greater or less according to the internal parts thus
involved.   In phlegmonous erysipelas the danger of metastasis is, I
believe, less than in simple cases, but the general febrile excitement and
derangement of the system is usually much greater, and there is always
danger of suppuration and a protracted wasting discharge which renders
the prognosis more unfavorable in the phlegmonous than in simple cases.
In malignant erysipelas the prognosis is decidedly unfavorable, whether

it occur accidentally, sporadically, or epidemically; the cause generally as we have seen being the same, and the disease tending, unless early arrested, to a typhus state, and to a fatal termination.

In middle-aged persons of good habits, the system may endure even malignant erysipelas, if the general strength be sustained, as we see in the case of Dr. H. A. Potter, but if by accident the patient be bled, or not sustained in the early stages, as was the case with his brother in Steuben county, the result is almost invariably unfavorable, as it was in his case. In very young children, or in persons of advanced age, or of filthy intemperate habits, the prognosis is always more unfavorable than in middle-aged, temperate, and more cleanly patients, other things being equal.

Treatment.—If we take into account the condition of the patient, and the exact deviation from the standard of health, the indications in the treatment are very plain.

In mild cases of simple erysipelas, two or three blue pills may be given and followed in six hours by half an ounce of the sulphate of magnesia, to act gently upon the liver, and to evacuate any irritating matters that may be lodged in the alimentary canal. Or if a mercurial is from any cause contra-indicated, one or two grains of podophyllin, or a drachm of the fluid extract of mandrake, may be given, and followed in six hours by the sulphate of magnesia, or castor-oil.

If a diaphoretic be indicated, three grains of Dover's with an equal quantity of James's powder, may be given every six hours, while the febrile state continues, and the patient should be kept quiet, and allowed a plain, digestible, and moderately nourishing diet. In most mild cases this is all that will be indicated, unless the eruption threatens to spread to parts in which it might be very inconvenient, or dangerous, as to the scalp. In that case, the extension of the cutaneous disease may generally be arrested by surrounding the eruption, or making a line on the side where it is desirable to arrest its progress, with either the nitrate of silver in substance, or a strong solution of it; or what in most cases is better, making a line an inch or more broad with the tincture of iodine, with a small brush. The line should be made just at the edge of the eruption, so as to extend about one-third of its width over the inflamed part, and may be applied each day for as long a time as may be necessary.

If a local application becomes necessary to the eruption, an ointment made by mixing two drachms of the sulphate of iron, finely pulverized, with an ounce of lard, will do very well; but if the surface is very hot, and the patient complains of severe burning or itching in the part, from two to four drachms of the sulphate of iron may be dissolved in half a pint of water, and cloths wet in the solution kept applied to the inflamed surface, instead of the ointment.

Mild cases will generally pass on favorably with little more than I have suggested, but violent cases, of a bilious character and a strong phlegmonous tendency, may require active and persevering treatment. In such cases an emetic of ipecac will occasionally be of service at first, in equalizing the circulation and ridding the stomach of any bilious matter that may have accumulated; and also strongly tending to promote perspiration. **After** the emetic, when it is indicated, and at first, when an emetic is **not** required, a cathartic of calomel, rhubarb, and castor-oil should be administered, and a free operation secured; after which, the bowels should be **kept moderately** loose, by teaspoonful doses of the sulphate of magnesia, administered once or twice every twenty-four hours.

As a diaphoretic, four grain doses of the Dover's and an equal quantity of the James's powder may be combined and given every four or six hours, and the sulphate of iron applied as before suggested, and if there is a good deal of gastric irritation, a blister may be indicated to the epigastrium. If, as sometimes happens, suppuration occurs, and the fever assumes a typhoid character, the sulphate of quinine, in from two to four grain doses, should be given, with the Dover's and James's powders, and the system supported by wine-whey, broths, beef-essence, &c., and later, by other varieties of plain, digestible and nourishing food.

If matter collects, it should be evacuated at the most convenient point, and a slight compress applied, to secure an adhesion between the different parts of the cavity which has been formed. In this state, if the patient be inclined to scrofula, five grains of the iodide of potassium should be given three times per day, and continued till after suppuration is suspended, when the syrup of the iodide of iron may be substituted, in ten drop doses, and continued during convalescence, if necessary, together with a good nourishing diet.

In malignant erysipelas, the prostration during the first or chilly stage is very marked, and the patient may die without any febrile reaction, from the immediate effects of the morbid poison upon the system, as we see in some cases of poison from reptiles, and from the dead and living human subject. In such cases, if the patient is seen early, to arrest the sinking tendency, diffusible stimulants may be indicated. Half a pint of good whisky may be mixed with an equal quantity of milk, and sweetened, and a gill administered every two or three hours, till it is all taken. This insures reaction, and keeps up the sinking powers, till more permanent tonics can be brought to bear in sustaining the system.

If, however, in malignant erysipelas, the patient is not seen thus early, and reaction has taken place, the diffusible stimulus should be omitted, and a moderate dose of calomel, hydg. cum creta, or leptandrin should be given, in half an ounce of castor oil; and the bowels kept open, if necessary, by a teaspoonful occasionally.

Immediately after the operation of the cathartic, to sustain the system, keep up the functions of the body, and promote perspiration; quinine, Dover's, and James's powders, of each three or four grains should be given, every four or six hours, during the first two or three days; when, if typhoid symptoms make their appearance, the Dover's should be omitted, and two or three grains of camphor given with the quinine and antimonial powder, instead; and this should be continued during the whole course of the disease, the quinine being increased if necessary. By commencing with the quinine thus early, the distressing sinking which would otherwise occur is frequently prevented, the severity of the asthenic erysipelatous inflammation lessened, and also the liability of metastasis greatly diminished.

In all cases of metastasis of the eruption of erysipelas, to internal parts, the local inflammation thus produced, should be treated as ordinary acute inflammation of the part; taking into account, of course, the general condition of the patient.

In phlegmonous cases, if resolution of the local inflammation cannot be secured, and suppuration is tardy, gangrene being threatened, the parts involved may be freely laid open, and a poultice applied, in hope of securing suppuration.

Should suppuration take place in malignant cases, the matter should be evacuated early, the patient sustained by wine-whey, broths, &c., as already directed; and iodide of potassium given, alternating with the quinine, during the suppurative stage; but later, the tincture of muriate of iron, in ten drop doses may be given instead; and, in fact, this preparation of iron may be indicated in the early stages of malignant cases, and when it is, should be given alternating with the quinine. Thus we have the general principles, to guide us, in the treatment of every possible variety of erysipelas that may occur.

SECTION VII.—PURPURA.

By purpura, I mean that peculiar exanthematous affection, sometimes febrile, and in other cases apparently not; characterized by spots or patches of a livid hue on the skin, varying in size from the merest speck, only one or two lines in diameter to several inches.

The eruption consists in a subcutaneous hemorrhage, and may appear in malignant typhus, in malignant erysipelas, and various other malignant diseases, in which the blood is in a more or less dissolved or changed state.

Purpura may, however, occur uncomplicated with any other disease, the eruption in some cases being the first intimation of a diseased state; but not generally I think, as the ordinary premonitory symptoms of some of the exanthematous febrile affections sometimes precede the eruption, and generally more or less symptoms of deranged health. The disease might be described under two heads, the simple, and the hemorrhagic; the difference, however, being merely in degree rather than in kind; and, as there is always at least a hemorrhagic tendency, with more or less subcutaneous hemorrhage, I shall make no division of the disease.

Symptoms.—The premonitory symptoms of purpura, when closely observed, are just what might be expected, if from some cause the blood had become suddenly or gradually deteriorated, and the solids materially relaxed or changed. The petechiæ or eruption may make its appearance without any marked symptoms preceding or indicating it. But generally there is for a time an imperfect or irregular appetite, indigestion, slight stupor, and a general languid tendency, restlessness, and irritability, indisposition to mental and corporeal exertion, muscular debility, a sallow or pale complexion of the countenance, pain in the back, limbs and extremities, headache, and sometimes disposition to fainting on taking active exercise.

After these symptoms have continued, or some of them, for a longer or shorter period, a slight coldness of the extremities may be experienced, and in some cases slight chilliness, or a peculiar sensitiveness to cold may be noticed. After this depressed or chilly stage has continued for an indefinite time, varying from a few minutes to several days, a slight effort at reaction may be noticed, which may in some cases amount to a febrile state; the face having a hectic flush, the pulse being frequent, the thirst more or less urgent, and there being a general state of restlessness and irritability.

After these symptoms have been developed, when they occur, and at first, when no premonitory or febrile symptoms are noticed, the eruption makes its appearance, the spots generally appearing first upon the lower extremities, and then on the arms, trunk, and perhaps over most of the body. The size of the petechiæ, or spots, varies from the merest speck, to several inches in diameter, being irregular in shape, and appearing like an ordinary bruise. In young persons, in whom there is often slight febrile symptoms, the color of the spots may be at first of a reddish color, but they soon become purple, as they are at first in older patients, and such as have no sensible febrile symptoms. The spots gradually fade, and may disappear in from five to eight days; fresh spots having, however, made their appearance, so that some are appearing while others are fading, till the disease disappears altogether.

In cases of a strongly hemorrhagic character, small blisters, containing blood, often mingle with the dark spots upon the skin, and may also be found on the tongue, and perhaps other parts of the mucous membrane of the mouth and throat, and they probably exist, more or less, upon all the mucous membranes. In these cases hemorrhages are liable to occur, from the nostrils, mouth, conjunctiva, stomach, bowels, urinary passages, and vagina. and sometimes of an alarming character, but generally not very copious, it being mostly of a passive nature. Blood is also sometimes poured into the brain, lungs, or other vital organs, seriously endangering the life of the patient. The blood in such cases generally coagulates imperfectly, and is evidently in a reduced and half dissolved state. The pulse is weak, the digestive organs more or less deranged, and great debility prevails throughout the whole course of the disease.

In some cases of purpura, there are slight elevations of the cuticle, similar to those of urticaria; the eruption, however, may disappear as in ordinary cases, fresh spots and elevations appearing while the first are subsiding; the disease continuing three or four weeks, or till the blood is restored, as in all other cases of purpura.

Diagnosis.—The only disease with which purpura is liable to be confounded is scurvy, from which it may be distinguished by the following essential differences:

In purpura there is not that tenderness, softness, and swelling of the gums which occur in scurvy. The eruption in purpura is at first of a more reddish cast than in scurvy, and besides, purpura usually occurs in autumn, when fruits and vegetables are freely used, while scurvy generally occurs in the latter part of winter, when little fruit and few vegetables are to be had, and generally very soon disappears if a supply of fruits and vegetables can be obtained.

Anatomical Characters.—The blood is usually found in a dissolved or fluid state, and very generally effused into the different organs and tissues of the body. It is extravasated blood immediately beneath the cuticle, or else in the tissue of the true skin, which constitutes the eruption or purple spots in this disease.

Petechiæ, or purple spots, similar to those of the skin, are found upon the mucous membranes, and especially of the alimentary canal, and in many cases upon the pericardium, pleura, and peritoneum; and blood is found extravasated, more or less, into all the organs and tissues of the body.

Causes.—Purpura may be produced by any cause capable of deteriorating the blood sufficiently to produce the hemorrhagic condition; among the most frequent of which are the putrid fevers, habits of filth, impure air, unwholesome articles of food, such as putrid meats, &c., great despondency, and frequent fits of anger, intemperance of every kind, and finally, various drugs, long continued, which prevent the formation, or diminish the proportion of fibrin in the blood.

Pathology.—The nature of this disease is a little obscure, and yet I believe if its symptoms be closely scanned they will be found to indicate a pathology that will not totally exclude it from among the exanthematous febrile affections. The impure air, putrid articles of food, habits of filth, &c., which operate to produce this disease, may generate in the system itself a morbid poison, of an animal character, which helps materially to deteriorate, dissolve, or deprave the blood.

The system gradually becomes deranged, and sinks down, as in the forming stage of fevers: but such is the state of the blood, and consequent relaxed, and debilitated condition of all the tissues, even the muscles, involving of course the heart, that in aged and feeble patients, no sensible febrile reaction occurs, and in the middle aged and more robust, only a feeble, and slight febrile reaction is produced. During the effort at reaction, blood is thrown into the vascular structure of the skin; where, in consequence of its dissolved, or watery state, and the relaxed condition of the solid tissues, it is extravasated; producing the eruption, or purple spots, which appear: the same condition also occurring to some extent, in all the tissues of the body.

If the patient be middle aged, and the reaction considerable, the eruption is free, but the internal hemorrhagic tendency is slight. But if the patient is advanced in life, and the effort at reaction slight or imperceptible, the eruption may be slight; but the internal hemorrhagic tendency is much more considerable; being liable to occur from all the mucous surfaces, as well as into all the tissues and organs of the body. In those cases of purpura, in which there are slight elevations of the cuticle, it is evidently owing to an enlarged or congested state of the vessels of the skin at that point, the blood not being entirely extravasated. Such, it appears to me, may be the true pathology of this disease.

Prognosis.—The prognosis in purpura is generally favorable in young or middle aged persons, if the internal hemorrhagic tendency is not very considerable; but in patients of advanced age, and of filthy intemperate habits; and in cases in which there is little or no apparent reaction; and a strong internal hemorrhagic tendency, the prognosis is always more or less unfavorable.

Treatment.—When we take into account the deteriorated condition of the blood; the relaxed condition of the solid tissues; the consequent hemorrhagic tendency, and the general deranged condition of the system, the indications in the treatment of purpura are very plain. Habits of filth, intemperance, &c. should at once be corrected, and the patient restricted to a plain, digestible, and nourishing diet; taken with regularity and if the bowels are constipated, a full dose of rhubarb or leptandrin may be exhibited at first, and the bowels kept regulated if necessary, by a pill of aloes and rhubarb, of each one and a half grains, taken at evening, if the bowels have not been moved during the day.

To improve the appetite, favor digestion, and help sustain the system, and keep up the circulation, the infusion, tincture, or fluid extract of bark should be given in full doses, three times per day, before each meal. To change the blood, lessen the hemorrhagic tendency, and help sustain the sinking powers of the system, the tincture of muriate of iron should be given in ten drop doses after each meal, in a little water, and continued till the integrity of the blood, and tone of the solid tissues are fully restored. Should hemorrhages occur, two grains of tannin, or from half a drachm to a drachm of the fluid extract of geranium meculatum, should be given every six hours, till the hemorrhage is arrested.

SECTION VIII.—GLANDERS—(*Equinia.*)

By glanders I mean that malignant exanthematous febrile affection contracted by human beings from the horse suffering with glanders, and characterized by a muco-purulent and sometimes bloody discharge from the nostrils, a peculiar pustular eruption, and by tumors occurring in different parts of the body, of either a purulent, bloody, or gangrenous character.

This disease has been recognized in the horse from the time of Hippocrates, but until a comparatively late period it had not been discovered or described as occurring in the human subject, and even now, the disease in the human subject is probably of very rare occurrence; no well marked case having fallen under my observation.

Glanders occurs in the human subject under an acute and chronic form, the symptoms of which, for the sake of convenience, I will give under two heads, that of acute and chronic:

Symptoms of the acute.—Acute glanders usually commences with the ordinary premonitory symptoms of the exanthematous fevers, such as, depression, weariness, languor, pain in the back, and sooner or later, slight chilliness, followed by febrile reaction; the pulse becomes frequent; the skin hot and dry; the thirst urgent, and the tongue more or less coated. Usually, about one week after the exposure, and among the first symptoms which occur are neuralgic or rheumatic pains, affecting the joints of the limbs, and sometimes darting through different parts of the body. Portions of skin over the seat of these local pains may become red or assume an erysipelatous appearance, turning of a darkish color, and containing vesicles, or terminating in patches of a gangrenous character.

Usually, in about one week from the commencement of the disease, being about the fourteenth day from the time of exposure, a pustular eruption makes its appearance on different parts of the body, but generally most abundant on the face and limbs.

The pustules vary in size, from that of a pea to two or three times that size, and have a slight resemblance to the vaccine vesicle, being sometimes umbilicated, and having a red areola, being filled with a purulent fluid along with coagulable lymph. Generally, along with the eruption, small shining tumors appear on different parts of the body, which soon turn brown, and may crack and discharge an acrid thin serous fluid, or they may mortify or form connections with abscesses in the intermuscular structures.

The pustular eruption in this disease appears in successive crops, so that some are appearing while others are maturing at the same stage of the disease. Neither is the eruption confined to the surface of the body, but appears on the mucous membrane of the nose, mouth, fauces, larynx, and it is said, upon the alimentary mucous membrane. In some cases, at the very commencement of the disease, but generally after the eruption makes its appearance, a muco-purulent fluid begins to be discharged from the nose; being at first yellowish, but later having a bloody appearance and an extremely offensive smell, and excoriating the parts with which it comes in contact. Some of this matter passes along the posterior nares to the fauces; which, with the mouth, larynx, lips, eyelids, forehead, and in fact the whole face becomes red, inflamed, and often very much swelled, closing the eyes and materially distorting or mutilating the features.

As these severe local symptoms are developed, the skin becomes hot, the tongue more dry, the respiration difficult, the pulse frequent and feeble; and there is generally thirst and an offensive colliquative diarrhœa. Finally a fetid odor exhales from the body, the mind becomes unsteady, coma or delerium occurring; and by the end of the second or during the third week the patient sinks into extreme prostration, through which, with involuntary discharges, the miserable sufferer passes on to speedy dissolution.

The acute form of the disease does not always present precisely these symptoms, being liable, of course, like all other diseases, to variations. In some cases, the nasal symptoms predominate from the first, while in others the nasal discharge may not occur till the latter stage of the disease; the carbunculous and eruptive tendency being by far the most prominent feature.

Chronic Glanders.—The symptoms of chronic glanders differ considerably from those developed in acute cases.

In the mildest cases no febrile symptoms are noticed at first, the discharge of an offensive, purulent, or senious matter from the nose being the first indication of the disease. Such cases may pass on for months and finally recover, or acute symptoms may occur, with malignant tumors, or abscesses, and a fatal termination be the result.

Generally, however, in chronic cases, slight neuralgic or rheumatic symptoms may occur, with more or less redness of the surface at the seat of the pain, and perhaps slight febrile symptoms may be developed, and continue for a longer or shorter time. At an indefinite period of the disease the nasal discharge may make its appearance, and, together with suppurating tumours, in different parts of the body, may pass on for months, and perhaps recover, but more probably, go on to a fatal termination.

Diagnosis.—The only disease with which glanders would be likely to be confounded, is acute or chronic catarrh, of the schniderian membrane, occurring in its worst form, and being complicated, as it often is, with chronic neuralgic or rheumatic symptoms.

In glanders, however, the nasal discharge is of a muco-purulent, and perhaps bloody character, while in simple catarrh of the schniderian membrane, the discharge is less acrid, wants the offensive odor, and is generally of a thin mucous character.

In the neuralgic or rheumatic localities in glanders, more or less swelling and redness appear, which is not generally the case in simple catarrh; and, finally, the presence or absence of the eruption, the abscesses, and the febrile symptoms, together with the general condition of the patient, and the possibility of exposure, will generally render the diagnosis clear, between glanders and simple catarrh of the schniderian membrane.

Anatomical Characters.—On post-mortem examination, besides the external lesions, there is found numerous small white pustules, interspersed with patches of ulceration on the mucous membrane lining the nasal cavities. The septum nasi is always more or less ulcerated, and often perforated, and the nostrils and frontal sinuses contain a viscid mucus.

On laying open the gangrenous tumours, the muscles are of a dark color, and appear to be more or less decomposed. They emit an offensive odor, and contain more or less purulent matter, either in small specks, or infiltrated through their structures.

White pustular eruptions, similar to those in the nasal cavities, are generally found in the soft palate and fauces, and sometimes in the alimentary mucous membrane. Large abscesses are found between the muscles, and in some cases pus is found in the articulations.

Cause.—Glanders, in the human subject, arises from a morbid poison, generated probably in the horse, and communicated either by contact of the poison with an abraded surface, or else through the stomach or lungs. When the matter is introduced into the system, through an abrasion of the skin, it affects the lymphatics very much like the animal poison received in dissection from the human subject, and there is a striking similarity in many respects between the diseases which the two poisons produce.

While glanders is generally taken from the horse, it is probable that the disease may be communicated from man to man, and that the predisposition to the disease is much stronger in persons of filthy intemperate habits than in the strictly temperate and cleanly, as is the case with almost all contagious and noncontagious diseases.

Nature.—That the morbid poison which produces this disease operates upon the system, in many respects, like the poison received from the dead human subject, is evident from the symptoms which are developed. In glanders, however, the poison may be received through the stomach, and probably the respiratory passages, and this may account in part for the nasal affection in this disease, and also for that of the mouth, fauces, &c., which do not occur from dissection wounds.

Prognosis.—The prognosis in glanders, as we have already seen, is always unfavorable, and yet it is possible that in mild cases of the disease patients sometimes recover.

Treatment.—From the morbid condition in acute glanders, it appears that an emetic at first, followed by a mercurial cathartic, may be indicated, and afterwards quinine, and perhaps the iodide of potassium.

In chronic cases, in which the nasal discharge is the prominent symptom, it appears to me that the iodide of potassium, in five grain doses, three times per day, continued for several weeks, and then, followed by the syrup of the iodide of iron, for a long time, might be of service. As a local application to the schniderian membrane, I would use a solution of corrosive sublimate of the strength of half a grain to the ounce of water, snuffed up the nose once each day.

SECTION IX.—DENGNE—(*Dandy Fever.*)

By dengne, or dandy fever, has been designated a peculiar exanthematous febrile affection, characterized by a high fever, violent rheumatic or neuralgic pains in different parts of the body, and by a rash or an eruption of either a papular, vesicular, or pustular character.

This disease appears to have prevailed epidemically, in all its visitations, having appeared in Philadelphia in 1780 ; in Calcutta in 1824, and in the West Indies in 1827, from whence it extended to New Orleans, Charleston and Savannah, in the summer and autumn of 1828. The term *dengne* is supposed to be, according to Prof. Wood* "a Spanish corruption of the word dandy ; the name of *dandy fever* having been jocosely conferred on the disease by the negroes of St. Thomas, from the stiff carriage of those affected with it."

Since 1828, the disease appears to have prevailed, according to Prof. Dickson† "in Mobile in 1844, and in Natches in 1848," and also in Savannah, and finally more or less along the "southern seacoast," till 1850, at which time, an epidemic of considerable severity again visited Charleston, an interesting account of which is furnished by Prof. Dickson, in his "Elements of Medicine."

As dengne has never prevailed within the range of my practice, I shall attempt only a general description of the disease, leaving the reader to search out the minutia in the able accounts which have been furnished by eminent physicians, practicing in localities where the disease has prevailed.

Symptoms.—The symptoms of this disease, according to Prof. Dickson's account, are very similar in the first stages, to those occurring in cases of epidemic rheumatic fevers, which occasionally prevail in our climate ; a short account of an epidemic of which I furnished for the Buffalo Medical Journal of 1855. The disease commences with pain, more or less stiffness of the neck, back, and loins, and slight swelling of the muscles of the limbs, and of more or less of the joints. There is also slight intolerance of light, restlessness, and more or less chilliness, followed by fever,

* See Wood's Practice, vol. 1, page 458.
† See Dickson's Practice of Medicine, page 734.

headache, redness of the eyes; a full, frequent, and quick pulse ; a hot and dry skin, with a more or less urgent thirst. The fever usually declines on the second or third day, leaving only slight swellings of the lymphatic glands of the neck, axilla and groin, which had taken place during the initial fever, and slight rheumatic symptoms, with general debility.

After an interval of two or three days, there is a return of the fever, and an increase of the neuralgic, or rheumatic pains, the tongue becomes coated of a yellowish color, the stomach irritable, and the patient dejected and fretful.   Vomiting also occurs, with great lassitude and restlessness.   On the fifth or sixth day of the disease, an eruption makes its appearance, attended with more or less relief of all the distressing symptoms.   The eruption is sometimes continuous as in scarlatina ; in other cases it is in darkish patches, as in measles, while in other cases still, it is "either papular, vesicular, or pustular," and in some cases, a mixture of all these forms, or it may present an erysipelatous appearance.

The rash or eruption makes its appearance first on the face, and then gradually on the body and extremities, and when fully developed, is attended with an itching or burning sensation, and sometimes with a secondary fever, and a return or increase of the rheumatic pains.

The eruption disappears after two or three days ; the color of the skin gradually fading, with slight desquamation of the cuticle.   But the affection of the joints may continue more or less troublesome for several weeks ; the patients remaining weak, or depressed in body and mind.

The duration of the disease is about eight days, and it attacks equally all ages, and in some localities where it has prevailed, almost the whole population have been affected by it; but generally, nearly or quite all have recovered from it.   The inflammation of the lymphatic glands of the neck, axilla, and groin, continues in some cases after convalescence is established, and in some cases the tumors have suppurated.

Diagnosis.—As dengue always prevails epidemically, little difficulty attends its diagnosis; except, perhaps in the first cases which occur, which might be mistaken for measles, scarlatina, or yellow fever.

From measles it may be distinguished by the want of catarrhal symptoms, which invariably attend that disease.   From scarlatina it differs in the time of the appearance of the eruption, it generally coming out later in dengue than in scarlatina ; and from both scarlatina and measles it differs widely in its neuralgic or rheumatic symptoms, which do not prevail in those diseases.   From yellow fever it differs in not presenting the yellow appearance of the skin, which is an attendant on that disease, and also in the fact that yellow fever is often a very fatal disease, while dengue is seldom or never so.

Cause.—Dengue is an epidemic, and probably contagious disease, but the exact nature of the cause is as yet entirely unknown.   An epidemic of this disease differs from many others, in nearly the whole population being affected by it; in this respect partaking of the peculiarity of influenza.   The period of incubation is supposed to vary from one or two, to ten days.

Pathology.—From the symptoms which are developed in this affection, and from their similarity in many respects to epidemic rheumatic fevers, which have occurred under my observation, I am disposed to regard dengne as an exanthematous febrile affection, of a rheumatic character.

The morbid poison, whatever it may be, on entering the blood may affect at an early stage the cerebro-spinal system or their meninges, either directly or by the congestion the forming or chilly stage produces; and this may account for the rheumatic character of the disease.   As reaction is fully established, the blood, with its morbid influence is carried to the minute capillaries in the vascular structure of the skin, where irritation and an eruption is the result.   This eruption on the skin may act as a counter irritant to the cerebro-spinal system or their meninges; and hence the great palliation of the symptoms which occur as the eruption makes its appearance.

The swelling and irritation of the lymphatic glands probably arises from the morbid febrific agent in its passage through them, as it is carried along the lymphatics, as more or less of it would be in this disease.

Prognosis.—The prognosis in dengne, it appears, is always favorable, so far as recovery is concerned; as in the few fatal cases which have occurred, it appears to have been from accidental complications.   The prognosis, in so far as it relates to the probability of an attack, when an epidemic is prevailing, may be set down as unfavorable, as nearly the whole population are likely to be affected by it.

Treatment.—From the morbid condition, in this disease, the indications appear to be very plain, as I have found them in epidemic rheumatic fever, to equalize the circulation, promote perspiration, and to sustain the system; meeting, of course, any complications which may arise.

To equalize the circulation and promote perspiration, the warm foot bath and warm sage tea are clearly indicated.   The warm foot bath might be used morning and evening for two or three days, and the warm sage tea used freely as a drink at all hours, with or without milk, as the patient might be able or not to take solid food at regular meals.   If the bowels are constipated, a mild cathartic of rhubarb or leptandrin, in castor oil, might be indicated, but not otherwise I think.

To sustain the system, quiet nervous irritability and promote perspiration, it appears to me that quinine, Dover's and James's powders are clearly indicated.   Four or five grains of Dover's, with two or three grains each of quinine and James's powder may be combined and given every six hours, as I have done, with the most happy effects in our epidemic rheumatic fevers.

Should the rheumatic symptoms predominate or require any attention, I think the iodide of potassium, in five grain doses, three times per day, alternating with the quinine, or given before each meal, after the quinine is no longer indicated, would be found to speedily and permanently remove the rheumatic symptoms and materially favor convalescence.

I offer these suggestions in relation to the treatment of dengne with more confidence, having treated on this plan very many cases of a rheumatic fever, which prevailed epidemically, in the range of my practice in the winter of 1853 and 1854.   In that epidemic there was a chill, febrile reaction, an erysipelatous eruption, and more or less rheumatic complications.

CHAPTER VI.

GENERAL INFLAMMATORY DISEASES.

SECTION I.—ACUTE RHEUMATISM.

By rheumatism I mean a peculiar general inflammatory disease, liable to affect any organ or tissue of the body, but more generally locating in the fibrous and muscular structures, from which, however, it is very liable to pass by metastasis, or extension, to other tissues, and to internal organs.

As the fibrous and muscular tissues, or structures, are the parts most frequently involved in, and peculiarly predisposed to rheumatic inflammation, it follows, that when rheumatism becomes seated in any organ which is composed in part of muscular or fibrous tissue, these structures become the immediate seat of the disease, but not necessarily the exclusive seat.

Rheumatism in all its forms is generally immediately preceded by neuralgic pain and irritation, which neuralgic pain or irritation I have generally been able to trace to an irritation, or else congestion, of that part of the brain or spinal cord supplying nerves to the part. This peculiar irritation of the brain, or spine, which produces acute rheumatic inflammation, I believe may be either irritation, inflammation, or congestion, but probably in most cases there is congestion. Now this congestion may be the result of any cause which deranges the flow of the nervous influence and disturbs the circulation, such as retained perspirable matter, bilious derangement, or any other irritating cause, acting through the blood on the cerebro-spinal and nervous system.

Symptoms.—There is generally a dull pain in the head or back, or more frequently of both, for a time, after which there follows chills, more or less severe, alternating with flushes of heat, with general lassitude, loss of appetite and depression of spirits. At this stage there are darting pains in the principal nerves which supply the limbs or part, which are to become the seat of the rheumatic inflammation. And I have generally noticed that the rheumatic inflammation locates, in most cases, in the limbs or parts which receive their nerves from that part of the brain or spinal cord in which the congestion, irritation, or pain, is first felt. If the early pain, congestion or heaviness is in that portion of the spine which furnish the brachial nerves, the neuralgic pain and the succeeding rheumatic inflammation will generally be, according to my observation, in the superior extremities or arms.

If, however, the early irritation, congestion or heaviness, be lower in the spine, as at the point which gives off nerves to supply the lower extremities, the neuralgic pains and the succeeding rheumatic inflammation will generally be found, I think, to occur in the lower extremities. And the same is true, I am satisfied by careful observation, in all other points, in the brain and spinal cord. And in cases in which there is no sensible heaviness even, experienced at any point in the brain, or along the spine, I am confident from the effects of cupping, in such cases, that there is more or less congestion acting as the direct cause of the neuralgic pain, and subsequent rheumatic inflammation.

After the chill, during which the local and general cerebro-spinal congestion is greatly increased, and the neuralgic pain severe, there is developed a general febrile excitement, and very soon a local rheumatic inflammation. In some cases the febrile reaction becomes completely established before the local rheumatic inflammation is set up; but in other instances the local inflammation is developed before the general febrile reaction takes place.

Now I suspect that the neuralgic pain in the limb, or part, produces an irritation in the capillaries of all the tissues, but more especially in the fibrous or muscular tissues involved, on account perhaps of the firmness of their structure, and hence the rheumatic inflammation which follows. Perhaps another reason why the fibrous or muscular structures are more liable to rheumatic inflammation than any other tissues, is the fact of their being more abundantly supplied with nerves of sensation and motion, and also the fact that the nerves in dense, firm tissues become vastly more irritated, from slight congestion of the capillaries, than in looser tissues, which might tend to develop rheumatic inflammation. This appears the more probable when we reflect that the looser the tissue, the less liable it is to become the seat of rheumatism; for the firm serous membranes are vastly more liable to rheumatic inflammation than the less firm and loose mucous or cellular structures.

The parts affected by rheumatic inflammation are generally swelled, and extremely painful, the slightest pressure or motion causing the greatest degree of suffering. And this is not so strange when we remember the firmness and density of the structures involved. Sometimes, if the patient be kept perfectly at rest, some abatement of the severe gnawing pain is experienced during the day; but at night, as the exhalations from the skin grow less, by the influence of cool damp air, intense exacerbations take place.

Whether the fever precede or follow the development of the local inflammation, it generally acquires additional violence, as soon as the local affection is fully established. The pulse becomes full, frequent and vigorous; the skin hot and dry; the tongue coated with a white fur, changing to a brown color as the disease advances; the thirst is urgent; the bowels constipated; and the urine scanty, and of a reddish color. In very severe cases of rheumatism, headache, and slight delirium attend during the exacerbations.

Rheumatic inflammation may pass by metastasis, or extend to internal and vital parts; the most frequent point, perhaps, is to the heart and its lining and investing membranes. If the pericardium or endocardium become involved, there is oppression in the region of the heart; hurried and difficult breathing, palpitation, frequent pulse, and a disturbed and anxious expression of the countenance.

If the muscular structure of the heart itself become involved, as happoned in one fatal case under my care, there is in addition to the symptoms enumerated above, the most distressing pain in the heart, apparently increased by its own action. so that its pulsations or contractions at times will almost cease, the countenance becoming livid, the extremities cold; and the patient for a time enduring the most distressing faintings, sinking, &c., passes on to speedy dissolution. In this case, I am satisfied that death occurred from an inability on the part of the heart muscle to carry on the circulation, in part perhaps by the morbid changes in its structure, and also in part from the violent pain its own action produced, both operating together to suspend its function entirely.

Anatomical Characters.—The blood contains an unusual amount or proportion of fibrin in this disease, sometimes amounting to nearly or quite ten parts to the one thousand, the healthy standard being about three. In relation to the quantity of uric acid in the blood, it is doubtful if more exists, as a general rule, than in health.

On post-mortem examination, the synovial membranes are found thickened and red, and generally an increased amount of liquid is found in their cavities, and very rarely pus, false membranes, &c. Signs of inflammation are exhibited in some cases, by softening of the fibro-cartilages, and of the muscles, and signs of inflammation are sometimes presented by the lining membrane of the arteries. In cases in which the heart is involved, various morbid appearances are presented, according to the time the disease has progressed. In one case, I found the pericardium very much thickened, and firmly united to the heart, and the valves thickened, and rendered nearly incapable of acting, from appearances at least.

Causes.—If, as I have supposed, the causes of rheumatism act through the blood on the cerebro-spinal system and their nerves, any cause capable of producing debility, congestion, or irritation of the brain, or spinal cord, may produce rheumatism. Among these causes, retained perspirable matter, being an irritant, may produce rheumatic inflammation, or that condition of the system which develops it. Hence cold, with moisture, or atmospheric vicissitudes, is one of the most frequent causes of this disease. This accounts for its occurring most frequently in damp and variable seasons of the year, and also for its being aggravated during the night time when it does exist.

The paludal poison, by the debility or prostration it produces in the cerebro-spinal system, and the functional derangement which follows, may produce acute rheumatism, many marked cases of which have fallen under my observation during the past few years, and some complicating severe bilious fevers. The sudden suppression of a gonorrhœa, by throwing into the blood a poisonous principle, may so far affect the brain and nervous system as to develop acute rheumatism.

It is probable also, that very many other morbid agents, may, by acting through the blood on the cerebro-spinal system, to produce acute rheumatism. It is probable also, that in a highly sthenic constitution, a high electrical state of the atsmosphere may act as an exciting cause of acute rheumatism, while in debilitated, or asthenic constitutions, it is probable that an opposite, or low electrical state of the atmosphere, may produce this disease, by taking off the already too scanty supply of electricity from the system.

Now I am satisfied that any one, or all these causes, may act to produce the debility, congestion, or irritation of the brain or spinal cord, which is necessary for the development of acute rheumatism; while the precise seat of the inflammation, will depend upon the natural or accidental point of irritation, in the cerebro-spinal system, as I have before suggested. If the brain, from hereditary or accidental causes, is more irritable than any point in the spine; there will be rheumatism of the fibrous structures of the brain, or its meninges, or of some muscular or fibrous structure about the head or face, which receive their nerves directly from the brain. But if, as generally happens, some point of the spine is from natural or accidental causes most irritable; then as I have before suggested, the neuralgic irritation and inflammation will be developed in that part of the system, which the irritated portion of the spine supplies with nerves, and for the reason, as I have before suggested, that the primary neuralgic irritation produces an irritation, congestion and rheumatic inflammation of the part or parts so supplied.

In cases in which rheumatism passes from one point to another, rapidly, I am satisfied from careful observation, that it is in consequence of a change of the point of local irritation, in the cerebro-spinal system. That when it passes from the brain or heart to an extremity, it is from increased vital power, in consequence of which, the point of local congestion, or irritation in the brain or spine, is carried further from the nervous centre. But that when rheumatism passes from an extremity, to the heart or brain, it is from diminished vital energy, or power, in consequence of which, the point of irritation or congestion in the cerebro-spinal system, approaches the nervous centre.

Such, I believe, are the causes direct and remote of rheumatic inflammation, and also the principles involved in its development.

And I will only add, that my view of the increased proportion of fibrin in the blood; and of uric acid, when it is in excess, is this; that both are the effects, and not the cause of the disease. The increased proportion of fibrin, accumulating in the blood, I believe may be accounted for from a partial suspension of nutrition, and perhaps slight wasting in the muscular structures, during their diseased state, in consequence of which, the fibrin supplied by the food to nourish them, is retained, and accumulates in the blood in a greater or less degree, according to the extent and duration of the disease of the muscular structures.

The presence of uric acid, when it exists in excess in the blood, if such be a fact, I think may be accounted for from the derangement of the perspiratory and secretory organs generally; but I doubt, if an excess of uric acid, will very generally be found to exist in this disease. It is possible, also, that the derangement which I believe invariably exists in the cerebro-spinal system in this disease, may so far affect the ganglionic system of nerves, as to account in part, for the excess of fibrin in the blood, and also for the excess of uric acid, if such be a fact.

Diagnosis.—Formerly rheumatism and gout were considered as the same affection; but there are sufficient marks of distinction to entitle them to a separate consideration. The principal distinguishing features between the two affections are, the periodical recurrence of gout, after it has once invaded the system; whereas rheumatism does not possess this

14

tendency so markedly, patients frequently passing without a second attack. Rheumatism is not so generally produced like gout, by indulgence in the free use of vinous drinks, and high seasoned, and stimulating articles of food. It is also less hereditary; gout being often transmitted from parent to offspring, while such is rarely the case with rheumatism. Rheumatism too, is apt to occur in persons of debilitated relaxed conditions, while the reverse generally obtains with gout.

Prognosis.—Although rheumatism is an exceedingly painful affection, it is not generally dangerous, so long as the inflammation is confined to external parts. But when it extends, or is translated to internal structures, or organs; as the heart, meninges of the brain, stomach, or lungs; the most serious consequences may be apprehended. Such cases may, however, from careful and judicious treatment, sometimes recover.

Treatment.—If my view of the nature of rheumatism is correct, we have as the deviation from the standard of health, together with a too high or too low electrical state, generally retained perspirable matter, with perhaps the paludal poison, and various other morbid agents. There is also an irritation or congestion of some point of the brain or spinal cord, of either an active or passive character, and more or less functional derangement of the different organs, and especially of the secretory, and the local inflammation.

The indications then are very plain; to counteract any local irritation or congestion of either an active or passive character there may be in the cerebro-spinal system, to remove from the alimentary canal any bilious or irritating matters, to call the cutaneous exhalents into action, to restore the secretion and functions of the glandular system, and to subdue the local rheumatic inflammation.

Cups should at once be applied over that part of the spine which supplies nerves to the inflamed part, for the purpose of removing any irritation or undue pressure that may exist, wet or dry according as the patient is in a sthenic or asthenic state. If the inflammation is in the meninges of the brain, or in the scalp, or about the face, the cups should be applied to the back of the neck, and if the patient be in a sthenic state, from two to four ounces of blood should be taken. If the rheumatic inflammation is in the arms or upper extremities, the cups should be applied on each side of the spine, between the shoulders, over the seat of the bronchial plexis of nerves. If the disease be in the lower extremities, the cups should be applied to the lumbar and sacral regions, in order to relieve the irritation of that portion of the spine.

Having thus applied the cups on each side of the spine at the point supplying the inflamed parts with nerves, and taking from two to six ounces of blood, according to the general strength of the patient, and the degree of febrile reaction, a cathartic should be administered to clear the alimentary canal. Two or three blue pills, or ten grains of calomel should be administered, and followed by a full dose of the sulphate of magnesia, and a free evacuation of the bowels secured. If from any cause, a mercurial be contra-indicated, from half a drachm, to a drachm of the fluid extract of mandrake, or one or two grains of podophyllin may be given instead, and followed by castor oil, or half an ounce of the sulphate of magnesia if necessary.

To promote perspiration, act gently upon the kidneys and bowels, and to reduce the general febrile excitement, the nitrate of potash, in from twenty to thirty grain doses, every four or six hours, dissolved in half a pint of warm crust coffee or gruel, is clearly indicated; the gruel or crust coffee serving as drink and nourishment. If there is much nervous irritability and the stomach is in a tolerable state, ten drops of the fluid extract of colchicum, or twenty or thirty drops of the wine may be given with each dose of the nitrate. The nitrate, with or without the colchicum should be continued during the active stage of the general fever, and the local rheumatic inflammation.

As an application to the inflamed part, from two to four drachms of the muriate of ammonia may be dissolved in half a pint of vinegar and water and applied cool or warm as is most agreeable to the patient, two or three times per day. After the general fever and the severity of the local inflammation is subdued, there generally remains more or less tenderness, thickening, and lameness of the parts involved. To remove this and to restore the action of the lymphatic and glandular system, the iodide of potassium, in ten grain doses, every six hours, should be given and the nitrate of potash discontinued. The iodide of potassium should be continued in this way till the soreness, lameness, and thickening of the parts are removed, when the dose should be diminished to five grains three times per day, and thus continued for one or two weeks to remove the effects of the disease.

In bilious rheumatism, I have generally given about two grains of quinine and four or five of Dover's powder every six hours, at first, alternating with the nitrate for the purpose of arresting the fever, of which the rheumatism appears to be a complication in such cases; the Dover's acting as an anodyne and diaphoretic. In some cases, attended to early, I have succeeded with the cups, a cathartic, and warm sage tea; but if the local rheumatic inflammation has become established the other treatment is rendered necessary. If the rheumatic inflammation pass suddenly to the heart, lungs, or brain, it is generally in consequence of great prostration, and it is usually followed by a decided sinking tendency. In such cases, the sulphate of quinine and Dover's powder, in five grain doses each, should be given every six hours, alternating with the iodide of potassium, and blisters applied over the inflamed part.

Such, I believe, are the principles which should guide us in the treatment of acute rheumatism; and having pursued such a course of treatment for several years, I have no hesitation in recommending it; to be modified, of course, to meet the indications which arise in each particular case. If the indications be thus early met, I am satisfied that few cases need continue for many days, and none go on to a protracted chronic state; at least, such has been the result of my observation. But if acute rheumatism be neglected or mal-treated in its early stages, it may pass into a chronic form and continue for many weeks.

The gruel or crust coffee in which the nitrate of potash is given, affords sufficient nourishment while that is continued, but when that is no longer indicated, toast and other light food should be allowed at regular meal hours.

SECTION II.—CHRONIC RHEUMATISM.

By chronic rheumatism, I mean that peculiar chronic rheumatic inflammation, affecting most frequently the joints, but which may occur in
the fibrous, synovial, muscular, or in fact any tissue or organ of the
body. Chronic rheumatism is very generally the result of neglected or
mal-treated acute rheumatism, but it may arise in the system in the
same manner, and from the same causes as acute rheumatism, no active
inflammatory fever being developed, and the local rheumatic inflammation being of a passive or chronic character.

When chronic rheumatism is not a consequence of the acute disease, it
arises from the same causes, and is developed in the same manner as the
acute variety, only there is frequently a distinct tenderness of the spine,
at the point where the nerves come off which supply the affected part.
The neuralgic irritation which is set up is generally not so violent, and
does not so readily develop the rheumatic inflammation; and when the
inflammation does arise, it is of a slow chronic or passive character.

Symptoms.—At first there may be a dull, heavy feeling in the head,
or along the spine, with slight drowsiness, or lassitude, and a general
feeling of coldness or chilliness. Sooner or later slight febrile reaction
may take place, with slight increase in the frequency of the pulse; more
or less thirst; a dry skin, and perhaps scanty high-colored urine. Soon
the dull, heavy feeling in the head, or along the spine, appears to be
concentrated at some particular point in the brain or along the spine,
which point, if it be in the spine, becomes often more or less tender, and
immediately the darting neuralgic pains are felt in the limb or part
which this irritated point supplies with nerves.

After this neuralgic pain has continued for a longer or shorter period,
the patient being more or less restless or uneasy, especially at night, a
slight swelling, tenderness, and perhaps redness appear in the joints,
limbs, or parts in which the neuralgic irritation had been set up, and
this, if neglected, may continue for days, weeks, months, or even years,
being increased or lessened according to the exposure or imprudence of
the patient, and also modified more or less by the dampness or electrical
state of the atmosphere. It may continue confined to some particular
part for a long time, with only a slight swelling, and lameness of the
part, or it sometimes assumes an erratic character, passing from one
part of the system to another, and attended with considerable pain, but
no great tenderness or swelling.

In cases in which it remains stationary for a long time, there is generally a decided irritation, and often tenderness at the point of the spine
supplying nerves to the affected limb or part; but in the erratic cases,
I have generally detected only a transient irritation, with slight pain,
but little or no tenderness of the spine, to retain the local inflammation
in one part, so that it shifts, as the point of greatest irritability in the
cerebro-spinal system passes from one point to another.

In erratic cases, when the general strength is declining, I have generally noticed the local disease to pass to more important or vital parts,
indicating that the point of irritation in the cerebro-spinal system is
approaching the nervous centre. But if the general strength is im-

proving, and the cause being removed, I have generally noticed that the rheumatic inflammation passes from important or vital parts towards the extremities, or less vital parts, thus indicating that the point of irritation in the cerebro-spinal system is passing from the brain or nervous centre.

Thus we have the symptoms, and as I believe the principles involved in chronic rheumatism, which are very similar to those of the acute, only the disease is less violent, and of a more passive or chronic character, depending upon the peculiarly weak or debilitated condition of the patient, and perhaps a less concentrated exciting cause.

Diagnosis.—The only diseases with which chronic rheumatism is liable to be confounded, is neuralgia and gout, from both of which it may generally be distinguished by attending carefully to the following diagnostic symptoms. I regard the early pain, which precedes the inflammation, as purely neuralgic. But when the rheumatic inflammation takes place, the pain becomes more continuous, is less lancinating, and more dull, and there is generally tenderness, swelling, and sometimes redness, which rarely occurs in neuralgia, unaccompanied with rheumatic inflammation.

Chronic rheumatism may be distinguished from gout, by the fact of its being less hereditary, by its occurring in weak, relaxed, and debilitated constitutions, while gout generally occurs in persons of luxurious and idle habits. Besides, rheumatism is less periodical than gout, less liable to return, is attended with less derangement of the digestive organs, and while the blood in rheumatism contains only a trace of uric acid, it exists in an abnormal proportion in gout.

Causes.—A debilitated, relaxed and depraved condition of the system, together with an irritable condition of the cerebro-spinal system, constitutes, I am confident, a predisposition to chronic rheumatism; while among the exciting causes, retained perspirable matter, a low electrical state, and various irritations of the brain and spine, operate as exciting causes.

In persons strongly predisposed to chronic rheumatism, by a depraved, relaxed and debilitated condition of the system, the various functions of the body are more or less imperfectly performed, and especially the perspiratory function. In such persons the skin is stimulated to carry on its functions while the weather is fine and the air dry, and well charged with electricity, but as the weather becomes changeable, the air damp, and consequently in a low electrical state, the system parts with so much of its electricity that the different functions of the body become more or less impaired, or interrupted, and especially that of the skin. As a consequence of this, the perspirable matter is retained, and passing with the blood to the brain and spinal cord, produces a debilitating or irritating effect, which effect falls, or is concentrated upon the most irritable point or points of the brain and spinal cord, and hence a neuralgic irritation, and finally rheumatic inflammation is developed, of a slow or chronic character, and attended with little if any general febrile excitement, in some cases at least.

Pathology.—After what I have said of acute and chronic rheumatism, but little need be added to explain my views of the pathology of this disease in all its forms. I am satisfied that retained perspirable matter, or something that will produce an equivalent effect, passes with the blood to the brain and spinal cord, producing a poisonous effect, debilitating perhaps in all cases, at first, but in sthenic constitutions soon producing irritation, followed by more or less active congestion of that portion of the brain or spinal cord which, from hereditary or accidental causes, is most predisposed to take on irritation or congestion.

In consequence of this congestion or irritation of some point of the cerebro-spinal system, a neuralgic irritation is set up in the limb or part which the irritated or congested portion of the brain or spine supplies with nerves. This neuralgic irritation irritates or debilitates the capillaries of the fibrous, muscular, or other tissues to which the irritated nerves are distributed, and as a consequence we have a local rheumatic inflammation developed, of a greater or less extent, according to the degree and extent of the neuralgic irritation, and the accidental predisposition to the disease. There is also an irritation transmitted from the cerebro-spinal to the ganglionic system, in consequence of which we have an irritable condition of the heart and arteries, which in sthenic constitutions is attended with increased power of action, and so a general febrile excitement is set up, either before or after the establishment of the local rheumatic inflammation.

In debilitated, relaxed, and depraved constitutions we have in the same manner set up a passive irritation of the brain or spine, and also an irritation transmitted thence not only to the capillaries of the part where the local disease is to be developed, but also to the heart and arteries ; but in both, as the system is in an asthenic state, the irritations are attended with diminished power of action in the capillaries, as well as in the heart and arteries ; hence we have either a passive or chronic inflammation as the result.

The agency of electricity in the production of this disease, I have already explained, but it should be remembered that electricity should only be supposed to act as a cause of rheumatism in constitutions which are either above or below a medium, or in a sthenic or asthenic state ; in the sthenic producing an influence in developing acute inflammatory rheumatism, while in the asthenic it favors the development of passive or chronic rheumatism.

The presence of an undue proportion of fibrin in the blood in rheumatism, I have accounted for in the preceding section, as generally an effect and not a cause of the disease ; supposing that it may arise from a partial suspension of nutrition in the muscular structures, and perhaps also from a slight wasting of the muscular structures involved, thus causing to be retained or accumulated in the blood an abnormal proportion of fibrin.

Prognosis.—The prognosis of chronic rheumatism is favorable so far as a fatal termination is concerned, unless the heart, or some other vital organ or part becomes involved ; in which case the disease is always attended with more or less danger.

The prognosis in chronic rheumatism, so far as an ultimate and permanent recovery is concerned, depends upon the time the disease has continued, and also upon the hereditary and accidental predisposition to the disease.

Treatment.—The indications in the treatment of chronic rheumatism are to relieve any local irritation there may be in the brain, or spinal cord ; to promote perspiration ; to restore the healthy action of the glandular system, and to correct any deranged, or depraved condition of the system that may exist. If there is much spinal tenderness; dry cupping, blistering, or tartar emetic pustulation should be resorted to at once, on each side of the spine, at the tender point.

Having removed the local irritation of the brain or spine, if it exists ; if the inflammation is sub-acute, half an ounce of the sulphate of magnesia should be administered at once, and then ten drops of the fluid extract, or thirty or forty drops of the wine of colchicum should be administered every four or six hours, till the general febrile excitement is checked, and an impression produced on the local rheumatic inflammation. As soon as a decided impression is produced, the wine of colchicum should be given, in twenty or thirty drop doses, or the fluid extract, in from eight to ten drop doses, every six hours, and alternating with this, the iodide of potassium, in ten grain doses should be given every six hours, till the local disease is subdued ; after which, the iodide should be continued in five grain doses, three times per day, for a few days, to remove the effects of the disease.

In cases of chronic rheumatism, which depend mainly on the influence of cold, a torpid condition of the skin, and a general debilitated and relaxed condition of the system, the gum guiac, in twenty or thirty grain doses, three times per day, is a valuable remedy. The pulverized gum may be mixed with gum arabic and sugar, and then by the addition of water, a convenient mixture may be formed. If such cases prove obstinate, twenty-five drops of the fluid extract of the blue cohosh * may be given, alternating with the guiac, four times per day, and continued till the disease, and its effects are removed as far as may be.

In all those cases of chronic rheumatism, in which the abuse of mercury is a cause, or in which there is a syphilitic taint, with a marked derangement of the skin and glandular system ; the iodide of potassium should be given in ten grain doses, three times per day, with full doses of the compound decoction of sarsaparilla, and continued till the disease is removed. This treatment in some cases, may require to be continued for several weeks, or even months ; but if persevered in for a reasonable time, it will generally effect a cure, at least such has been the result of my observation.

By a strict observance of these principles, varying of course to meet the indications which arise in each particular case, carefully fulfilling the indications by the proper remedies, the most obstinate and protracted cases of chronic rheumatism may be greatly relieved, and very many permanently cured.

Rheumatic patients should be directed to keep the skin clean, and the feet dry ; to wear flannel next the skin, during the cold, damp, and variable weather, and to observe with the greatest care, those rules of propriety in all things, most conducive to health, and its preservation, always preserving an even and cheerful temper of mind.

* Leoutice Thalictroides.

SECTION III.—GOUT.

By gout I mean that peculiar general inflammatory disease occurring in either an acute or chronic form, and affecting most frequently the smaller joints, but liable, like rheumatism, to be translated to internal and vital parts. In acute cases of gout the disease is attended with more or less febrile excitement; but in cases which become chronic, or which are chronic from the commencement, there is little or no general fever attending the disease.

Symptoms.—An attack of acute gout is generally preceded by more or less derangement of the digestive organs; the tongue is foul and redder than natural; there is loss of appetite, sour stomach, and perhaps vomiting, distress in the stomach and drowsiness after eating, and generally depression of spirits and restlessness during the night. The feet are cold and perhaps soon distressingly hot; a pricking, darting or numb sensation is felt in the legs and feet, especially in the foot that is about to be attacked; and some hours previous to the paroxysm there is slight shivering alternating with flushes of heat.

After more or less of these symptoms have continued for a time, the paroxysm develops itself generally between one and three o'clock in the morning. The patient awakes with a violent throbbing pain, generally in the ball of one of the great toes, but sometimes at the heel, instep or ankle. This pain goes on increasing accompanied with a sensation of burning heat, weight and stiffness of the part, and severe shooting pains in the limb. The severe pain is usually attended at first with shivering, but soon the joint begins to swell, and more or less fever is developed with great restlessness. As the local inflammation is developed, the affected joint becomes swelled, hot, red and tender; so that the weight of the bedclothes can hardly be borne. The superficial veins about the affected part and for some distance up the limb become turgid, and all the symptoms of the local disease continue from six to twenty-four hours; after which the violence of the pain subsides, but the swelling rather increases and becomes more or less edematous.

Nearly or quite simultaneously with the local affection, the febrile symptoms are developed, the chilliness being followed by an increased heat of the skin, disgust of food, a furred tongue and scanty high-colored urine. The fever continues with the local affection, unabated for a period varying, as I have stated, from six to twenty-four hours. Generally I think the violence of the local pain, as well as the general fever, begin to subside about midnight, following the morning of the attack, so that by two or three o'clock in the morning the patient may fall asleep, after twenty-four hours of severe suffering.

The following evening the pain and fever are renewed and continue through the night to abate again in the morning; and thus the disease passes on, with remissions during the day and exacerbations at night, each paroxysm being less severe than the preceding one, till at last the attack terminates perhaps after having continued a week or ten days, especially in the first attacks of the disease. After the pain and fever have subsided, or during their subsidence, there is often slight looseness of the bowels, a gentle perspiration and a copious discharge of urine. There is slight itching in the part which has been inflamed, the skin falls off as the swelling subsides, and the patient finally recovered from the attack, feels in every respect much better than before the attack.

The first attacks of gout as we have seen, seldom last but a few days, or affect more than one joint, but when the disease has gone on for some time, the inflammation when declining in one foot, suddenly attacks the other, and sometimes the fingers, wrists, knees; and then the pain, shivering, fevers, swelling and redness, and all the symptoms recommence. At the end of three or four days the pain is again relieved, but the attack does not end here, as a similar fit supervenes affecting the same or other joints, accompanied with the same series of symptoms, and continuing perhaps for the same length of time. Hence to complete an attack of gout, three or four fits may occur, each taking from three to five days to run its course; about fifteen days, perhaps, being the average duration of an attack, but it may continue much longer than that.

After gout has affected a person for several years, the attacks are not only lengthened, but they occur more frequently At first, the attacks may occur no oftener than once a year; later, they may come on every six months, then every three months, and finally, as the paroxysms occur with much greater frequency, and are of much longer duration, the patient is scarcely over one attack, before another occurs, and so the patient is laboring under the disease, as it were, almost constantly, with only imperfect remissions. But as the disease progresses, and becomes more continuous, the attacks are less violent, there being generally less pain and redness, though the swelling may be quite as much or more. The joints do not now completely recover between the paroxysms as at first, but remain more or less stiff and incapable of motion.

The grade of the inflammation is lower and the general fever less vigorous than in the early attacks of the disease, corresponding more or less with the diminished vigor of the system. At this stage of the disease, it is very liable to affect internal or vital parts, as the brain, heart, lungs, stomach, bowels, liver and kidneys, and when it does, the symptoms of inflammation and functional derangement of the internal part which becomes affected, are developed with more or less intensity. If the stomach becomes involved, we have the ordinary symptoms of gastritis; if the bowels, there is often a diarrhœa or severe colic; if the heart, there is precordial oppression, violent pains, dyspnœa, syncope, &c.; if the lungs, difficult breathing, lividness of the face, a feeble pulse, &c.; if the liver or kidneys, there is developed symptoms of inflammation or functional derangement of those glands, and finally, if the brain becomes involved, there is more or less stupor, and in some cases, palsy or apoplexy.

As gout thus assumes a chronic character, it may appear at each paroxysm in some one of these internal vital organs, or it may appear externally, in some of the small joints, and be translated to the internal part in consequence of the debilitated condition of the system. In either case there is danger of organic visceral disease, which may lead to a fatal termination, such as of the brain, heart, lungs, liver or kidneys; if these internal parts become often seriously affected.

Chronic gout may be the result of the acute form of the disease, as we have seen, or it may assume a chronic form from the first invasion of the disease; especially in persons of a feeble constitution, and of indolent intemperate habits. The disease may be regarded as chronic when with the moderate local affection there is produced little or no general fever. In chronic gout the paroxysms are frequent, but less severe than in the acute form of the disease. There is little pain except on motion ; but slight increase of heat ; and, instead of redness, more or less of a purplish hue. If there is swelling it is of an edematous character, unless it be from an increase of the synovial fluid of the joints, as is sometimes the case.

The paroxysms occur very irregularly, sometimes coming on after an interval of a month, and at other times they may run into each other—a new attack appearing before the preceding has subsided. In some cases the disease remains fixed in particular joints, while in others it may pass from joint to joint, or it may be translated to internal or vital parts.

In fixed cases of chronic gout the joints often become nearly or quite useless, in consequence of a thickening of the ligaments, with the addition, in some cases, of a deposition of an earthy matter like chalk, either into the cavity of the joint itself or into the cellular tissue about the joint. The uric acid and urate of soda, of which the concretions chiefly consist, is probably first deposited in a semi-liquid state, and afterwards becomes dry by absorption of the fluid parts, leaving the chalk stones, varying in size from that of a mustard seed to a walnut, and sometimes much larger than that; producing, in some cases, swelling, ulceration, and even mortification of the surrounding tissues.

The characteristics then of chronic gout, are very little or no fever, an edematous swelling of the inflamed parts of a purplish hue, the patient having a sallow complexion, with a more or less flaccid state of the flesh, and a general appearance of being worn down by protracted suffering and disease, and finally either a deposition of uric acid in or about the joints, or else a more or less copious discharge of it with the urine, especially as a paroxysm is about terminating.

Thus we have the general symptoms of acute and chronic gout, under which heads I have included all cases of gout in which there is either general fever and local inflammation, or inflammation without fever. There is however a gouty diathesis, in which various neuralgic affections are developed, with more or less functional derangement of the parts, in the treatment of which the gouty diathesis should always be taken into the account ; but I doubt if such neuralgic affections should be regarded as gout, any more than neuralgia should be confounded with rheumatism.

Various conditions, too, have, from the resemblance of the symptoms developed to those of rheumatism and gout, been denominated rheumatic gout ; but I believe that the condition partakes, in every case, much more decidedly of either the one or the other disease, and that the morbid condition should therefore be regarded as either rheumatism or gout, unless per chance both diseases be found to exist in a patient at the same time, which very possibly may sometimes be the case.

Anatomical Characters.—In acute gout the patient generally dies from an extension or translation of the disease to some internal or vital part, in which case the ordinary signs of inflammation of the affected part or parts are presented on post-mortem examination. In cases in which the patient has suffered from chronic gout evidences of inflammation are exhibited in the synovial membrane of the affected joints, as well as in the interarticular cartilages. The bones, too, near the joints are found injected with blood, perhaps softened, and, in some cases, eroded, and the cartilages covering the articulating ends of the bones may be found more or less absorbed. The ligaments are found injected, and sometimes thickened; and the muscles in the vicinity are either contracted or more or less wasted. And finally, layers of chalky matter are sometimes found, either without the joint in the cells of the areolar tissue, of the periosteum, ligaments, fascia, or muscles, or else within the synovial membrane, or between it and the adjoining cartilages.

Diagnosis.—Gout is not likely to be confounded with any other disease, except rheumatism; from which it may generally be distinguished by attention to the following differences: Gout is much more hereditary than rheumatism, and it very seldom attacks children, which is not so generally the case with rheumatism. Gout is found most frequently affecting the idle, luxurious, and wealthy; while the reverse of this is true of rheumatism.

Gout too is very prone to attack the small joints, is more paroxysmal, the fever is less continuous, the pain is more violent, the inflamed parts are of a brighter red, or else of a purplish hue, there is greater swelling and a more edematous tendency; and finally there is desquamation following the subsidence of the inflammation, which is not the case in rheumatism. If all these peculiar differences are carefully observed, and the history of chronic cases taken into the account, I think little or no difficulty need be experienced in arriving at a correct diagnosis in genuine cases of gout.

Causes.—The predisposition to gout is no doubt often hereditary, being transmitted through several succeeding generations, even though the disease may not be developed in each; in many cases, thus, "the iniquities of the fathers are visited upon the children to the third and fourth generation." It is probable also that the gouty diathesis may be generated in those not hereditarily predisposed, by the use of an undue proportion of animal food, indulgence in alcoholic liquors, gluttony and an indolent habit of life, together with various other imprudences. When the gouty diathesis exists, whether it be hereditary or acquired by imprudent habits, very slight causes, such as a full meal, the use of alcoholic liquors in any form, exposure to cold, venereal excesses, anger, indigestible food, acids, injuries; and various other like causes may develop the disease.

Gout is most liable to make its first appearance between the twentieth and fortieth years of age, and it most frequently occurs about the time of the vernal and autumnal equinoxes. It is probable also that gout is more liable to occur in damp variable climates, as it is more common in temperate latitudes where the weather is variable, than at the extreme north or in the tropical regions.

Pathology.—In relation to the nature or pathology of gout, there may be grounds for doubt. But when we take into account the facts connected with the disease, so far as we are capable of ascertaining them, it appears to me that we have at least a clue to what may be regarded as the probable pathology of this peculiar disease. When we remember that the gouty diathesis may be produced by the use of alcoholic drinks, or by an undue proportion of animal food, the latter of which favors the formation of uric acid in the system, and both if used in excess, not only impair digestion, but also render the nervous system more or less irritable, involving of course the brain and spinal cord; it appears to me that we have at least a starting point from which we may proceed to a probable pathology.

If now we have this **peculiar** irritable **condition** of the nervous system, whether it be hereditary or produced by imprudence in eating and drinking, as I have already suggested, it appears to me that the presence of uric acid in the blood, if the quantity be considerable, may, by passing with the blood through the system, not only affect the brain and spinal cord to produce the violent neuralgic or darting pains, but also directly to assist in developing the gouty inflammation. The uric acid, as it accumulates in the system, may produce a debilitating effect upon the brain and spinal cord, and lead on to the chills which occur, and also to the violent, **darting, neuralgic pains which** immediately precede the local inflammation.

It is possible that the local inflammation may be the result of the neuralgic pain, set up in the part, affecting the capillaries about the joints, or parts which become the seat of the disease. But I suspect, that in addition to the peculiar diathesis, and the neuralgia which is set up, as I have suggested, that the development of the inflammation, is aided at least, by the uric acid passing through the capillaries of the tissues of the small joints, or parts which become inflamed.

In acute gout then, we have the irritable condition of the nervous system, the presence of uric acid in the blood, irritating, or debilitating the cerebro-spinal system, which leads on to a chill, during which, there is more or less congestion of the brain and spinal cord; **which congestion produces** an irritation of the spinal cord, or brain, in addition to that already set up. Now as a consequence of all this, the ganglionic system becomes involved, producing an increased action of the heart and arteries, developing general fever. At the same time, the neuralgic pain in the small joints, or parts to become affected, develops, by the help of the irritant in the blood, the local inflammation.

Now in acute cases, occurring in plethoric overfed constitutions, the irritation set up in the capillaries of the joints, as well as the irritation set up in the heart and arteries, are attended with increased power of action, so that there is developed an active inflammation, as well as a vigorous general fever. The peculiar appearance of the inflamed part, as well as the peculiarity of the fever, may be owing to the peculiar irritable condition of the system, constituting, as I believe, the gouty diathesis; and also in part to the excess of uric acid, or urate of soda in the blood.

The disease remits during the day, and has exacerbations during the night, perhaps in part, owing to the dampness of the night air, and its consequent effect upon the cutaneous exhalation, and perhaps in part also to changes in the electrical state of the air. After the paroxysm has continued for a few days, the kidneys, either by the natural effort of the system, or by the effects of remedies, drain off the excess of uric acid in the blood, and so the paroxysm, with all its general and local symptoms subsides, and leaves the patient feeling in every respect better than before the attack, when the morbid product was accumulating in the blood.

As the uric acid accumulates in the blood, time after time, the paroxysms are produced in the same manner; only the irritation of the brain and spinal cord becomes more permanent. The general irritability of the system is increased, but vital power is diminished; and as a consequence of all this, the paroxysms occur more frequently. But as both the capillary, and general circulatory systems become irritated, they have gradually a diminished power of action; and hence the passive character of the general fever, and local inflammation, as the disease becomes chronic. And finally as the brain and spinal cord become constantly irritable, the neuralgic pains become almost constant. The heart and arteries are also constantly irritated; but as they have diminished power of action, there is finally no general fever developed. And as the local irritation of the capillary vessels is also attended with diminished power of action, the local inflammation assumes a slow passive character; and thus we have accounted for the phenomena of both acute and chronic gout, in all its forms.

In the same manner we may account for chronic cases, that never had assumed the acute form.

But in those cases of neuralgia which occur from a gouty diathesis; that is from an irritable condition of the cerebro-spinal system, in which various neuralgic symptoms are developed, with more or less functional derangement of the different organs, without any general fever, or local inflammation, I prefer to consider the morbid condition as neuralgia, depending upon the gouty diathesis, as a cause. I shall therefore treat of those neuralgic affections, which have been called nervous gout, under the head of neuralgia, where it appears to me, they more properly belong.

Now, if my view of the pathology of gout be correct, we have a ready solution of the reason why gout so often passes suddenly from the small joints of an extremity, to some internal or vital part, and the reverse.

For, as we have seen in rheumatism, it is only necessary that the general powers of the system decline, for the point of greatest irritability in the cerebro-spinal system to approach the nervous centre; and hence the point of the neuralgic irritation, and inflammation is changed to the internal or more vital part. And on the other hand, if gout attack an internal or vital part, in a feeble patient, if the vigor of the system be improved, the point of greatest irritation in the cerebro-spinal system may recede from the nervous centre, and as a consequence, the neuralgia pass from the vital part to an extremity, and with it, the local inflammation.

This view of the pathology of gout, it appears to me, receives confirmation in the fact that the gouty diathesis often produces various neuralgic affections, and functional derangement of the different organs of the body, which appear to point directly to the brain and spinal cord as the seat of the difficulty. Thus I have explained what appears to me to be the nature or pathology of gout, and while it must be confessed that the subject is involved in what may be regarded as a reasonable doubt; I know of no symptom developed in this disease, which may not be rationally accounted for, allowing this to be the true explanation of its pathology.

Prognosis.—The prognosis in gout so far as a fatal termination is concerned, is not very unfavorable, unless it passes by extension or translation to some internal or vital part, in which case the danger is always considerable. In protracted cases too, in which digestion becomes seriously impaired, the nervous system excessively irritable, and the powers of life very much reduced, with general functional derangement of the different organs of the body, the patient is liable to be cut down by some acute disease, or else to be taken off by some form of dropsical affection.

If the inflammation tends to pass from the internal vital parts towards the extremities, it indicates increased vigor of the system, and the prognosis becomes more favorable. But if the gouty inflammation tends to pass from the extremities to the internal vital parts, it indicates diminished vital power, and the prognosis is always rendered more unfavorable.

Treatment.—The deviation from the standard of health in gout, consists evidently in a peculiar irritable condition of the system, constituting the gouty diathesis, in a deranged condition of the digestive organs, caused generally by abuse in eating too much animal food, and drinking too much alcoholic liquors, and in an accumulation of an undue amount of uric acid in the blood, together with the local inflammation, and the general febrile excitement. The indications then are very plain, both during a paroxysm, and also between the attacks, or in the intervals.

The habit of using alcoholic drinks in any form should be avoided by persons of a gouty diathesis, and should not be allowed except in cases in which the habit has been of long duration, and even in such cases, it may generally be broken off by degrees, and the condition of the patient very materially improved. In persons that have not the gouty diathesis, the use of alcoholic liquors should be avoided, lest the diathesis be formed, as well as various other morbid conditions of the system produced.

As the use of an undue proportion of animal food, favors the formation of uric acid in the system, gouty patients should always use due caution, that no more than a reasonable amount of animal food be taken; and that food always be taken at regular meal hours. If, however, the gouty diathesis be hereditary, or acquired by intemperance in eating, drinking, &c., and with this predisposition, a too free use of animal food, with indolence and other causes which operate to produce the disease, be indulged in, uric acid accumulates in the blood, and a paroxysm of gout is the result.

In such a case, as the digestive functions are always, impaired, a cathartic of calomel or blue pill, followed by half an ounce of the sulphate of magnesia, is generally indicated, or if from any cause a mercurial be contra-indicated, from half a drachm, to a drachm of the fluid extract of mandrake, or one or two grains of podophyllin may be given instead, and followed by the sulphate of magnesia if necessary.

Having thus cleared the alimentary canal, and gently stimulated the liver and glandular system, it is desirable to bring the skin and kidneys into a state of renewed activity, and also to favor the secretion by the kidneys, of the excess of uric acid in the blood. To act thus upon the skin and kidneys, no remedy appears better adapted in acute cases, than the colchicum. From ten to twelve drops of the fluid extract, or from thirty to forty drops of the wine may be given every four or six hours, and continued till the paroxysm has abated, when the dose may be diminished one-third, and continued till the fever and local inflammation have entirely subsided.

If an anodyne becomes necessary to secure rest, fifteen drops of the fluid extract of hyoscyamus may be given at evening, and repeated if necessary once during the night. Or if the bowels are in a relaxed state, Dover's powder may be given instead. If there is much acidity of the stomach, half an ounce of the bicarbonate of potash may be dissolved in a pint of water, and an ounce of the solution given three times a day, at meal hours, and continued till the acidity is corrected, or, if the acidity be slight, half an ounce of the solution may be given instead. If the colchicum fails to keep the bowels sufficiently loose, small doses of magnesia, or of the sulphate of magnesia may be given occasionally during the continuance of the paroxysm, that one or two movements may be secured every twenty-four hours.

While the fever continues, the diet should consist mainly of light farinaceous substances, as sago, arrow-root, &c., with a little chicken broth and toasted bread, and as a drink, toast water may be allowed.

The patient should take exercise as soon as the paroxysm is over, and he will generally find some relief from friction of the part with the hand, or an application of the tincture of camphor if swelling remains after the inflammation has entirely subsided.

In acute gout, if the disease occur in or extend to the stomach, heart, brain, or some other internal or vital part, prompt measures should be taken to arrest the inflammation. If it be not a translation of the disease, the result of a sinking tendency, general bleeding may be indicated, and should be resorted to, together with cups, blisters, and other measures ordinarily indicated in inflammation of the part.

In chronic gout the patient should be restricted to a regulated diet; should avoid the use of all kinds of alcoholic liquors; should wear warm clothing; take sufficient exercise; preserve a mild and even temper of mind; and, in short, should conform most rigidly in every respect to the laws of health. If there is acidity of the stomach, it may be corrected by a weak solution of the bicarbonate of potash, taken after each meal for a few days; or if that be unpleasant, the prepared chalk may be taken instead. If the bowels are confined, a pill of equal parts of aloes and rhubarb should be taken at evening, and continued till the habit is corrected.

If, however, in spite of every precaution a paroxysm occurs, ten drops of the fluid extract, or thirty drops of the wine of colchicum should be given three times a day, after each meal, and continued till the paroxysm is over, and then diminished one-third and continued for a week or two. If from any cause the colchicum disagrees, twenty or thirty grains of gum guiac may be given instead, made into an emulsion with sugar, gum arabic and water taken after each meal, and continued, if necessary, for several weeks. In those cases of chronic gout in which the joints become permanently swelled, stiff and lame, or almost useless, five grains of the iodide of potassium may be given three times a day before each meal and continued for several weeks or even months, often with the most happy effects. It should be given well diluted in water, or what in many cases will do better, with the compound decoction of sarsaparilla.

In cases of chronic gout, if from great debility the disease be suddenly translated to some internal vital part, warm stimulating applications should be made to the original seat of the disease, and cups, wet or dry, applied over the inflamed part, and the system sustained by two or three grains each of quinine and camphor every six hours. Blisters also over the inflamed part may be of essential service in such cases.

In all protracted cases of chronic gout, in which, with the indigestion, great poverty of the blood occurs, with a dropsical tendency, a reasonable amount of animal food may be allowed, and every possible care taken to improve the digestion, and to sustain the sinking powers of the system. And if dropsy occur, it should be treated on general principles, taking into account, of course, the general condition of the patient.

In all cases of neuralgia, occurring in patients of a gouty diathesis, the treatment must be as in neuralgia from other causes, taking into account, of course, the gouty diathesis.

CHAPTER VII.

DISEASES OF THE NERVOUS SYSTEM.

SECTION I.—CEPHALALGIA—(*Headache.*)

By the nervous system I mean the brain, spinal cord, cerebro-spinal nerves, and the sympathetic or ganglionic system. The cerebro-spinal system is composed, it will be remembered, of white fibrous tubular matter, and a gray vesicular substance. The gray vesicular substance occupies the surface of the cerebral mass, but the central portion of the spinal cord, while the cerebro-spinal nerves are composed entirely of white fibrous tubular material.

Nervous matter consists of albumen, phosphorus, and various fatty substances. The sympathetic nerves, extending from the brain along each side of the spine, are of a grayish color and consist of numerous ganglia, connected by branches, and receiving branches from the cerebro-spinal nerves.

The brain is supplied with blood by the carotid and vertebral arteries, is surrounded by a fibrous, serous, and vascular membrane, and, together with the nervous system generally, is the medium through which mind controls the body, and by which all the functions of the body, both voluntary and involuntary are sustained. The cerebral hemispheres appear to be the more immediate seat of the mind, and especially their external, gray, or vesicular portion appears to be devoted to intellection; while the function of the spinal cord appears to be physical, and the nerves "mere conductors of sensations, volitions, and impulses direct and reflex, between the nervous centres and their extremities."

It is probable that the nervous centres generate a nervous power or influence, which being sent forth through the nerves exercises a controlling influence upon the various functions of the different organs of the body; the sympathetic system influencing especially the circulation, digestion, absorption, nutrition, and secretion; while the cerebro-spinal nerves influence more especially the voluntary functions. The brain, then, the great centre of the nervous system, receiving near one-fifth of all the blood which passes through the heart, invested with its membranes, of a fibrous, serous, and vascular character, is liable to a great variety of inflammatory, organic, and functional derangements, many of which are attended with cephalalgia or headache.

Cephalalgia.—Cephalalgia is a symptom, not a disease. And I have introduced it here for the purpose of inquiring into the morbid conditions which it indicates, or of which it is a symptom. Headache may arise from disease of the brain, or its membranes; or it may be sympathetic, from a local disease in any tissue or organ of the body.

If headache arise from disease of the brain or its membranes, its character will point out, to a degree of certainty, the seat and character of the cerebral disease. Thus, if the dura mater be inflamed the pain is of a dull, aching, or gnawing character. If the cephalic pain be acute, sharp, and lancinating, it indicates inflammation of the arachnoid membrane; while, if it is of a throbbing character, and neither very acute or dull, the disease is probably more especially confined to the pia mater or vascular membrane of the brain. And finally, if it partake of the characters of the three kinds of pain, it is probable that the three membranes are involved, and perhaps also more or less of the cerebral mass. If the headache be confined to one side of the head, or to one temple, it is generally of a neuralgic character; and especially may it be known to be neuralgic if it intermits—the paroxysms coming on at regular hours each day, or on alternate days.

When headache is sympathetic of gastric derangement, I have generally noticed that it occurs in the forehead or temples; and it is often attended with a feeling of languor, drowsiness, and more or less dizziness, especially if there be acidity of the stomach as a cause. If a protracted pain be complained of, in the top or back portion of the head, there is probably either irritation at some point along the spinal cord, or the patient is suffering from some derangement of the genital organs, and most likely, from excessive sexual indulgence or solitary pollution.

Thus we have some of the important indications which cephalalgia furnishes of local and sympathetic diseases. It may also arise from a congestive state of the brain, in which case it is of a dull, heavy, aching character; or it may arise from a directly opposite or anemic condition.

15

in which case there is rather a feeling of lightness than fullness attending the cephalic pain.

Causes.—Any influence capable of producing either disease of the brain or its membranes directly, or which may derange any portion of the system, either organically or functionally, may be a cause of headache. Hence, every imprudence may operate either directly or indirectly; but among the many, those which most frequently operate as causes of cephalalgia, are tobacco, drunkenness, gluttony, onanism, sexual excesses, self-pollution, irregular eating, fits of anger, constipation, and various occupations which require a stooping posture for an unreasonable time.

Pathology.—The great point to be born in mind, in the pathology of cephalalgia, is that it depends upon irritation of some portion of the nervous tissues, and that it may depend upon directly opposite conditions; in the one case the nervous irritation, depending upon congestion or an undue amount of blood, while in the other condition there is an anemic state, or too little blood supplied to the nervous matter. In either case I believe the function of the nervous matter involved is equally deranged, but the indications in the treatment are, of course, directly opposite; in the one case requiring stimulation, while in the other abstraction of blood or depletion may be necessary.

Treatment.—As cephalalgia is a symptom, it should not of course ever be treated as a disease. But it should be examined as a means of ascertaining the morbid condition upon which the pain depends. And the great points to be settled in the examination are, the exact seat and nature of the morbid condition upon which it depends, and also the causes which are operating to produce the morbid derangement, whether it be local or general, functional or organic.

When the morbid condition is ascertained, and the cause which has produced or is producing it is sought out, it should be abandoned; whether it be drunkenness, tobacco, licentiousness, gluttony or any other cause, and then the morbid condition corrected as far as may be, by temperance, prudence, &c., and also by such other remedial measures as may be indicated, always keeping in mind the general condition as controlling the indication for depletion or stimulation.

SECTION II.—MENINGITIS.

By meningitis I mean inflammation of any one or all the membranes of the brain. It may, therefore, be of either the dura mater, the pia mater or the arachnoid membrane, or of all of them at the same time.

The *dura mater*, it will be remembered, is a fibrous structure lining the interior of the cranium. It also sends processes from its internal surface to support and protect the brain; while from its exterior, processes are sent off to form sheaths for the nerves as they quit the skull. Its external surface is rough, while its internal surface is smooth and lined by the arachnoid serous membrane.

The *pia mater* is a vascular membrane, being composed of innumerable vessels held together by a thin layer of areolar or cellular tissue. It is the inner membrane of the brain, the whole surface of which it invests, dipping into its sulci between the convolutions and forming a fold in its interior, the *velum interpositum*. It also forms folds in the third and fourth ventricles and along the longitudinal fissure of the spinal cord.

The *arachnoid* membrane is of a serous character, enveloping the brain and being reflected upon the inner surface of the dura mater, and like other serous membranes is a shut sack. The arachnoid lies between the dura mater and the pia mater, the external, covering the convexity of the hemispheres. The interior arachnoid is continuous with the exterior, and was formerly supposed to penetrate into the ventricles and line their internal cavities; but it is probable that the serous membrane which lines the ventricles is not continuous with the arachnoid membrane, the arachnoid being reflected inwards upon the venæ Galeni for a short distance, and then returning to the dura mater of the tentorium.

Thus we have a glance at the different membranes of the brain, the inflammation of which we will now proceed to consider. And though they may all be more or less involved in every inflammation of either, I am satisfied that the inflammation is generally more especially located in some one of the membranes, and shall, therefore, give the symptoms peculiar to each.—The dura mater, as we have seen, is a fibrous structure, and though it may be involved in every variety of meningeal or membranous inflammation, that to which it is by the peculiarity of its structure especially liable, is of a rheumatic character.

Symptoms of Meningitis.—I will take up first the symptoms of inflammation of the dura mater, and then of the pia mater, and finally of the arachnoid membrane, and then consider the symptoms attending an inflammation of all the membranes occurring at the same time.

The symptoms of rheumatic inflammation of the dura mater do not differ very materially in the early stages from those of rheumatism affecting other fibrous structures. It generally occurs at seasons of the year when rheumatism prevails, and attacks persons of a rheumatic diathesis, and in whom this membrane of the brain is more predisposed than any other part of the fibrous structures to take on rheumatic inflammation. It may also be the result of metastasis of rheumatism from other and less important parts. When not the result of metastasis, its progress may be slow, the patient complaining of a dull, heavy, aching pain in the head, which gradually increases from day to day, till it becomes very severe.

The appetite by degrees is entirely lost. The patient is at first restless and irritable, and finally becomes wakeful, passing his night without any rest, complaining more and more of a constant maddening pain, which renders him at times, entirely uncontrollable. The face is sometimes flushed, and at others of a very pale appearance. The eyes are slightly red, and there is intolerence to light. The pulse is at first quite frequent, but later it becomes very slow, full and heavy, frequently making no more than forty pulsations to the minute. In severe cases, there is during the latter stages, almost constant delirium, as is also the case when it is the result of metastasis, from other and remote parts.

If, as sometimes happens in this inflammation, the other membranes become involved, the symptoms are materially modified. There is set up a high state of febrile action, the eyes become red, the face becomes flushed, the pulse frequent, full and quick, and there is a peculiar uneasiness experienced along the course of the spine, with general restlessness and irritability. Delirium becomes almost constant, and acquires a degree of violence resembling the most furious and ungovernable mania.

In this aggravated state, the face becomes turgid, the eyes wild and furious, the carotid beats strongly, vision is imperfect, and the whole system is in a state of restlessness and agitation. Such are the symptoms of rheumatic inflammation of the dura mater, and also their modifications when the other membranes of the brain becomes seriously involved in the inflammation.

The *pia mater*, as we have seen, is a cellulo-vascular membrane dipping into the amfractuosities of the brain, and also extending into its internal cavities. Being thus exceedingly vascular, and its vessels being connected by a loose cellular tissue, this membrane is peculiarly liable to become inflamed. The pia mater is liable to acute active inflammation, in full plethoric habits, and especially during the continuance of a high grade of febrile excitement, the irritation of its small vessels, as well as of the heart and arteries being in such cases attended with increased power of action.

During low typhus fevers, passive inflammation of this membrane is liable to occur ; the capillaries and small blood vessels becoming weak, and more or less congested ; and when the distension has produced a sufficient degree of irritation of the small vessels, inflammation is the result, of a more or less passive character ; the irritation of the capillaries of this membrane, as well as the irritability of the heart and arteries, being attended with diminished power of action. And as the pia mater is supplied with branches of the sympathetic nerve, it is probable that a good deal of the sympathetic irritation of the brain, especially from irritation of the alimentary mucous membrane, is received by the pia mater, and then communicated to the other cerebral tissues, in all febrile affections.

Inflammation of this membrane, whether active or passive, is generally at first attended with a feeling of fullness, and pain in the head ; which symptoms are very soon followed by nausea, and sometimes vomiting. As the pain and febrile reaction increase, the eyes becomes red ; the face flushed ; the pulse frequent, and there follows a general state of excitement and irritability.

Delirium is generally an early attendant, and in many instances soon acquires a degree of violence, resembling the most furious mania.

The face becomes turgid ; the eyes wild and furious ; the carotids may beat strongly ; vision is imperfect, and the whole system is in a state of restless agitation.

During the early part of the disease, the sense of hearing may be painfully acute ; but in the latter stages, deafness often occurs. Respiration, though hurried at first, becomes slow and laborious in the latter stages ; and in many cases deglutition is performed with more or less difficulty. The various organs become deranged ; the liver frequently pouring out a copious secretion of acrid bile. And not unfrequently hemorrhages from the bowels occur towards the termination of fatal cases.

The *arachnoid membrane*, as we have seen, is of a serous character, lying between the pia mater and dura mater; and its inflammation is characterized by the symptoms that attend inflammation of other serous membranes. As we have already seen, inflammation of serous membranes tend strongly to a termination, by effusion of serum; hence the frequency of hydrocephalus, as a consequence of inflammation of the arachnoid membrane. And though it is probable that this membrane is seldom inflamed, without involving the dura and pia mater, more or less, yet it is frequently more especially the seat of meningitis, especially in children.

The commencement of this disease is marked by very great irritation, rather than by symptoms of inflammation, the approach being very gradual, especially during early childhood. The patient is irritable in temper, wakeful, and evinces a repugnance to bright light, on account of the sensitive state of the retina. The pupils are contracted; the disposition fretful and variable; young children often crying in a sharp and spiteful manner, for very trifling causes. Often when sleeping they may start suddenly, and exhibit symptoms of terror, or great fright; and finally they exhibit the most decided and violent dissatisfaction, at everything which occurs around them.

This state of irritability continues for a time, and then in some cases gradually subsides, without passing into the inflammatory state; the child gradually returning to its ordinary condition of health. If, however, some accidental cause supervenes; such as dentition, cold, or intestinal irritation; the irritation of the arachnoid membrane is increased; its capillaries become irritated, and finally either an active or passive inflammation in set up, according as the system is in a sthenic or asthenic state.

The patient complains now of pains in the head; the restlessness and irritability of temper increase; the pulse becomes irritated, quick, tense and active; the countenance is expressive of discontent and suffering; the eyebrows are knit, and the eyes half closed, in consequence of intolerance to light, and the bowels are either torpid or relaxed, with a greenish appearance of the stools. As the disease progresses, the cephalic pain becomes more severe, and it may be general, effecting the whole head, or it may be mainly in the forehead; children frequently manifesting this, by applying the hands to that part. Vomiting occurs, and the patient is unable to sleep, tossing about in bed, and throwing the arms in every direction, and frequently groaning, as if suffering from the most intolerable distress.

Towards the close of the inflammatory stage, when effusion is about taking place, there is frequent sighing, such as I have seldom noticed in any other disease, and delirium is a more or less constant attendant. The skin, during the continuance of the inflammation, is generally warm and dry, and the tongue may remain nearly clean, unless there be marked gastric disturbance, in which case it may be covered with a thick brown fur.

After an indefinite period, a new train of symptoms supervene, which mark the stage of cerebral oppression. The delirium becomes more continuous, the countenance exhibits an aspect of surprise and stupor, the pupils are either dilated or contracted, the conjunctiva becomes reddish, the eyes turn up under the upper lids, constant somnolency supervenes, and the patient becomes entirely inattentive to surrounding objects. The mind becomes torpid, and incapable of any attention, and the drowsiness increases till a complete state of coma occurs. In some instances the coma occurs suddenly, with paralysis of a limb or one side. In other cases of inflammation of the arachnoid membrane, only slight febrile excitement is developed, the drowsiness being almost the first manifestation of the disease. In such cases the inflammation is of a passive character, and has passed through its stages without developing much pain, or any considerable degree of febrile excitement.

Generally before the occurrence of paralysis, one hand will be firmly clenched, and the thumb turned in upon the palm. The upper eyelids too may become paralyzed, and strabismus may occur. When effusion takes place, if it occurs, all the more violent symptoms may abate, and friends may be deceived into the belief that the patient will recover. But sooner or later convulsions supervene, or the patient passes into a state of fatal coma or stupor. During this stage the pulse becomes slow, and often irregular, and both seeing and hearing are sometimes totally lost.

I have spoken of arachnitis as terminating in effusion of serum, and it is most probable that such is generally the case, especially if, as generally happens, the serous membrane lining the ventricles becomes involved in the inflammation. It should be remembered also that while the arachnoid membrane is the more special seat of the inflammation in such cases, it is generally by no means the exclusive seat, the other membranes, and especially the pia mater, becoming more or less involved in the inflammation.

Thus we have at a glance the more special and important symptoms developed in inflammation of the fibrous, vascular and serous membranes of the brain. Let us now, bearing these peculiarities in mind, inquire into the symptoms which are developed when all these membranes become alike inflamed, at the same time, or so nearly so, that a distinction may not be made between them. The ordinary symptoms of general meningitis are restlessness, loss of appetite, depression of spirits, vertigo, headache, thirst, and as febrile excitement with the local inflammation are fully developed; there is violent headache, the face and eyes become red, there is a wildness of expression, giddiness, intolerance to light and sound, and the pupils are sometimes almost closed.

Later, the patient becomes restless, his movements are spasmodic, delirium more or less violent occurs, the pulse is frequent, hard, and perhaps irregular, the respiration difficult, the skin hot, but perhaps moist, and the tongue is often clammy, or covered with a lightish colored fur. Vomiting too generally attends from the commencement, and the bowels are generally, though not invariably, in a constipated state.

At a later period still, the delirium yields to stupor, and soon the patient becomes decidedly comatose. The pupils become dilated, the patient neither sees or hears, the surface becomes less sensitive, the muscles become rigid, the movements become convulsive, there is picking at the bed-clothes, the pulse becomes slow and intermittent, the respiration difficult, and finally with partial paralysis, the pulse becomes feeble, the skin cold and clammy, the features cadaverous, and the patient usually dies either in convulsions, or in a state of profound coma or insensibility. Such are the ordinary symptoms of general meningitis. But it must be remembered that in many cases of this character, the inflammation not only involves the meninges, but also extends, more or less, to the cerebral substance.

Meningitis may assume a chronic form, or the disease may be chronic from the first, in which case the most prominent symptom is that of insanity. The patient is more or less excited, and is apt to express great eagerness in whatever theme may be most prominent in the mind. He may be exalted by some vain, ambitious notion, or possessed by some monomaniacal propensity, being restless, and talking with very unnatural gesticulations. Or the patient may be in an opposite dejected, or gloomy mood, being tormented with foreboding apprehensions of some imaginary evil. Or the patient may pass suddenly, from very slight causes, out of the deepest gloom into a violent passion of pride, vanity, or ambition.

For a time the functions of the body may be kept up with a tolerable degree of regularity, but sooner or later there is a hesitancy of speech, or difficulty in pronouncing words; there is an imperfect control, and more or less rigidity of the voluntary muscles, paralysis occurs, and finally the sensation of the limbs, as well as motion, is lost. Soon the patient loses the power of speech, the sphincters give way, and, finally, in a state of idiocy, the patient dies, either paralytic or in the most violent convulsions.

Such are the ordinary symptoms of chronic meningitis. It is liable, however, to great variations, like almost every other form of disease.—In relation to the duration of meningitis there is a liability of great variations. Generally, I think, acute cases may continue from one to three weeks, while chronic cases may continue for several months, or even years.

Thus I have completed what I had to say on the symptoms of meningitis, not only as it occurs in each of the membranes of the brain separately, but also as it occurs in all the membranes simultaneously, both in its acute and chronic forms.

Anatomical Characters.—The dura mater may be thickened, if it has suffered from rheumatic inflammation, and in cases of simple inflammation, the result of injury, or which has extended from the other membranes, it may be reddened, ulcerated, or gangrenous, and in some cases covered with a layer of lymphs or pus, either on its external or internal surface.

If the pia mater has been the main seat of the inflammation, it is found reddened, infiltrated with serum, or pus, and adhering with more or less firmness to the brain. I have known this adhesion so strong that portions of the cortical substance have been torn off, as the pia mater was removed from the surface of the brain.

In cases in which the arachnoid has been the principal membrane inflamed, it exhibits signs of sanguineous congestion, is more or less opaque, and perhaps thickened, and more or less covered with pus on both surfaces. There may be found, in some cases, layers of coagulable lymph on one or both surfaces of the arachnoid, but the most invariable morbid appearance presented, if the arachnoid has been highly inflamed, is the presence of fluid, of a serous, purulent, or bloody character, in the arachnoid cavity, and in the ventricles, if, as generally happens, their lining serous membrane has been involved in the inflammation. The quantity of the fluid thus found in the ventricles and general arachnoid cavity may vary from a few ounces to half a pint or more. But it should be remembered that while this is a general, it is not an invariable morbid condition following arachnitis.

If all the membranes of the brain have been inflamed simultaneously, the dura mater may be found thickened, reddened, ulcerated, or gangrenous, and the pia mater reddened, infiltrated with serum, or pus, and adhering with more or less firmness to the brain, and the arachnoid congested, opaque, thickened, and perhaps covered with pus or lymph on one or both surfaces; and, finally, more or less bloody, serous, or purulent fluid may generally be found in the arachnoid cavity, as well as in the ventricles of the brain.

In cases of chronic meningitis, in which insanity is a prominent symptom, the arachnoid membrane on the surface of the brain, and on the inner surface of the hemispheres, is opaque, and thickened so as to resemble the dura mater, being quite as thick, and having in some cases false membranes adhering to its cerebral and parietal surfaces. There may sometimes be found adhesions between the opposite surfaces of the arachnoid. The pia mater, too, is generally found red, thickened, and more or less filled with a bloody or serous fluid, and it sometimes adheres very strongly to the cerebral mass over which it is spread. In the ventricles, the pia mater is not only thickened, but its surface is often found rough, feeling much as if covered with sand; and this is also sometimes the case with this membrane on the surface of the brain.

There is generally found, in these cases of chronic meningitis, in which insanity is a prominent symptom, an abnormal quantity of serum, not only in the ventricles of the brain, but also in the arachnoid cavity, amounting, in some cases, to half a pint or more. The serous membranes generally are apt to show signs of inflammation, and the disease is often complicated during its course with gastro-enteric inflammation. In some cases of both acute and chronic meningitis, the brain presents a nearly natural appearance, but generally it shows marks of having participated in the inflammation, to some extent at least. The brain may, however, appear swelled, reddened, congested with blood, and perhaps softened; in acute cases, the cerebellum appearing to have participated in the disease; while in chronic cases the cerebellum seldom, if ever, appears to have been involved in the inflammation.

Diagnosis.—The diseases with which meningitis is most liable to be confounded, are tuberculous meningitis, sympathetic irritation of the brain, occurring in enteric and other fevers, and that peculiar anæmic state of the brain liable to occur in debilitated conditions of the system, and which we see illustrated in delirium tremens. By careful attention, however, to the following differences, a correct diagnosis may generally be formed.

Simple meningitis differs from tubercular meningitis, in occurring in very young children, while the tuberculous seldom occurs before the second or third year, and sometimes much later than that. Simple meningitis is often an acute disease, occurring in children of strong or sthenic constitutions; while tuberculous meningitis is slower in its progress, is attended with less acute pain and vomiting, and always occurs in children or patients of a tuberculous or scrofulous diathesis.

Meningitis differs from sympathetic irritation of the brain, occurring in many fevers; by the pain in the head and vomiting being more constant; and by the want of development in sympathetic cases, of several important symptoms, such as sighing, clenching of the hands, convulsions, paralysis, &c.; all of which differences, together with the history of the case, may render the diagnosis probable at first, and in the latter or more advanced stages perfectly certain.

From an anæmic condition of the brain, such as occurs in debilitated conditions, as in delirium tremens, meningitis may be distinguished by the history of the case, and by the following differences: In meningitis, the face is flushed, the skin often hot and dry, and the patient apparently quite strong, resolute and fearless; while in anæmic cases in which the brain is suffering from a want of proper stimulus, the face is pale, the surface cool, the pulse feeble, and the patient, instead of appearing strong, resolute and fearless, is feeble, tremulous and timid, being often tormented with the most ridiculous apprehensions of danger.

In discriminating between chronic meningitis and chronic anæmic cases attended with cephalic and general nervous irritability, the history of the case, together with the constitutional predisposition, and all the symptoms which are developed, will generally point to a correct diagnosis which may be of very great importance, as the treatment proper in the two conditions might be widely different.

Causes.—The predisposition to meningitis may depend upon directly opposite conditions of the system; in active acute cases arising from a sthenic, while in passive and chronic cases it may be the result of an asthenic or feeble condition of the system. A rheumatic or gouty diathesis also probably strongly predisposes to certain forms of meningitis. Meningitis probably occurs most frequently between the ages of ten and forty years, though it may occur in children much younger or in persons of a more advanced age.

Overeating, or the free use of alcoholic liquors, or any other excesses, such as great mental excitements, may, by temporarily elevating the system above a medium of excitability, act as causes of acute active meningitis. It may also be produced by injuries, especially in children, and it doubtless sometimes arises from sympathy with intestinal irritation, and also from dentition.

Chronic meningitis may arise from any cause which depresses and permanently debilitates the system, such as indigestion from irregular eating, and self-pollution in children and young persons, venereal excesses, protracted drunkenness, licentiousness, and the use of tobacco in adults; and, in short, everything which weakens the blood, and relaxes and irritates the solid tissues of the body. Chronic meningitis may thus arise, or it may be the result of the acute form of the disease.

Active cases may occur during any inflammatory grade of fever, and

especially of the exanthematous fevers, such as scarlatina, erysipelas,
&c. In such cases, the irritation set up in the minute capillaries of the
meninges, as well as that of the general circulatory system being attended
with increased power of action.

Passive meningitis may also occur during any of the low forms of the
exanthematous or other fevers, the irritability or irritation of the meningeal capillaries, as well as that of the heart and arteries, being attended
with diminished power of action.

Treatment.—The indications in the treatment of meningitis are very
plain, whether one or all the membranes are involved. In rheumatic inflammation of the dura mater, involving more or less, of course, the other
membranes, if it be acute, general bleeding may be required, after which
the head should be elevated, and the feet placed in warm water, if the
patient can bear it, morning and evening. A full dose of calomel, or if
that be contra-indicated, of podophyllin should be given at once, and
followed, if necessary, in six hours by the sulphate of magnesia, and a
free operation secured. After general bleeding, when that is indicated,
or at first, when it is not, three or four ounces of blood may be taken
from the back of the neck, by cups, and afterwards a blister applied.

After the operation of the cathartic a diaphoretic of Dover's and
James's powders, of each three or four grains, may be given every six
hours, and the patient immediately placed on the use of the iodide of
potassium, in full doses. Ten grains of the iodide may be given, well
diluted, every six hours, alternating with the antimonial and Dover's
powder, and continued till the disease is entirely removed; which may
take, according to my experience, from one to two weeks, or perhaps
longer.

In cases in which the rheumatic meningitis is the result of metastasis
of rheumatism, from some other part, occurring from great debility of
the patient, the general bleeding should be omitted, and perhaps the wet
cupping, and two or three grains of quinine given every six hours, with
the antimonial and Dover's powders. In other respects the treatment
should be the same. Toast water should be allowed at first, and soon a
plain, digestible diet should be given, and continued through the whole
course of the disease, if the stomach will bear it.

In this way I have succeeded in arresting rheumatic inflammation of
the meninges of the brain, involving no doubt, in some instances the
fibrous structure of the brain itself. The iodide of potassium should be
continued, however, in diminished doses, for one or two weeks after the
disease is arrested, for the purpose of removing the effects of the
disease.

Inflammation of the pia mater, of an active character, coming on
during an inflammatory grade of fever, or developing a high degree of
febrile excitement, may require active treatment. General bleeding may
be required in strong plethoric cases, to lessen the violence of the arterial action, and to relieve the irritated capillaries of the suffering membrane. After bleeding and cupping, if necessary, ten grains of calomel
should be given, and followed in five or six hours by a full dose of the
sulphate of magnesia.

After the operation of the cathartic, to reduce arterial action, and promote perspiration, one-fourth of a grain of tartar emetic, or four drops of the fluid extract of the veratrum viridi, may be given every three hours, and continued till the general fever and local inflammation are subdued.

The warm foot-bath should be used, and sinapisms applied to the feet, and, if necessary, a liberal blister applied to the back of the neck. Toast water should be the nourishment at first, and later arrow-root, and finally toast may be allowed. Light and every other exciting agent should be excluded from the sick chamber, during the continuance of the febrile excitement.

In cases of *passive* inflammation of the pia mater, occurring in low fevers, or in very debilitated patients, depletion is not indicated. In such conditions, an elevated position of the head, blisters to the back of the neck, and sinapisms to the extremities, with quinine and camphor internally, and proper nourishment, constitutes the course of treatment clearly indicated.

In inflammation of the *arachnoid* membrane, the treatment to be effectual must be early applied. During the first stage of irritability attention should be directed to the cause of the irritation, and, if possible, it should be removed, whether it be from intestinal worms, indigestion, dentition, or any other cause. But if inflammation becomes established, general or local bleeding may be necessary, in active cases, after which a cathartic of calomel or podophyllin should be administered, and then small doses of calomel with James's powder may be given, and continued every four hours, during the active stage of the disease.

The head should be kept high and cool, and the feet warm, and blisters applied back of the ears, or to the back of the neck, and if the small doses of calomel does not keep the bowels loose, small doses of castor oil should be administered, once every twenty-four hours. After the mercurial impression is produced the calomel should be discontinued, and full doses of the iodide of potassium given every four or six hours, alternating with the James's powder, and continued during the whole course of the disease, for the purpose of preventing effusion into the ventricles of the brain, or into the arachnoid cavity.

Blisters should be applied to the temples, back of the ears, or to the back of the neck; and the patient should be sustained early by toast water, with, perhaps, a little arrow-root or crust coffee. Thus we have the treatment proper for inflammation of each of the membranes of the brain, whether it occurs in an active or passive form, the principles of which may guide us in the treatment of meningitis, involving all the membranes simultaneously, both active and passive.

In acute general meningitis, then, our main reliance must be on general or local bleeding, the warm foot-bath, sinapisms to the extremities, cathartics, alteratives, arterial sedatives, diaphoretics, blisters, low diet, &c.; keeping, of course, an eye upon the general condition of the patient, whether the system be in a sthenic or asthenic state.

In general chronic meningitis, in which insanity is a prominent symptom, a careful removal or avoidance of the cause, or causes, which have been operating, a regulated diet, and strict observance of all the laws of health, together with tonics, if the patient be anæmic, and dry cupping, or blisters to the temples or back of the neck, &c.. involve the principles upon which our main reliance is to be placed. Thus we have the principles which are to guide us in the treatment of every variety of meningitis.

SECTION III.—CEREBRITIS.

By cerebritis I mean inflammation of the cerebral substance, which, if only partial, may be exclusive of meningitis, but which if quite general is apt to involve to some extent the meninges of the brain.—The brain includes properly those parts of the nervous system, exclusive of the nerves, which are contained within the cranium, embracing the cerebrum, cerebellum and medulla oblongata. The cerebrum, it will be remembered, consists of two lateral hemispheres, which are divided inferiorly into an anterior, middle and posterior lobe; beneath which posterior lobe is the cerebellum lodged in the posterior fossa of the base of the cranium.

The medulla oblongata appears like an enlarged portion of the spinal cord. It is slightly conical in shape, about an inch in length, and extends from the pons varolii to the upper border of the atlas, being the tract of communication between the brain and spinal cord. It is probable that its anterior portions convey motor influences, while its posterior convey sensations. And besides, the medulla oblongata acts as a nervous centre, as respiration and deglutition depend upon it.

We have already seen that the brain has certain cavities, the lateral, third and fourth ventricles, and that its structure consists of an external, gray, vesicular substance, and an internal, white, fibrous, tubular matter, and that it is supplied with blood by the carotid and vertebral arteries. Thus constituted the brain is liable to partial or general inflammation of its substance, either complicated with, or exclusive of, inflammation of its membranes.

Symptoms.—Inflammation of the brain may, in some cases, pass on to destruction of its substance, or at least some parts of it, before any very marked symptoms are developed. Generally, however. there is at first a fixed and violent pain in the head, which may continue for several weeks or even months. There is also vertigo obtuseness of the mental functions, confusion of the ideas, loss of memory, hesitancy in answering questions or partial loss of speech; and the patient not unfrequently becomes dejected and difficult or else indifferent to surrounding objects.

There may be numbness of the extremities, perverted vision, and in some cases total blindness occurs. The hearing also is apt to become either dull or acute, the epigastrium tender, the bowels constipated, and the appetite variable. The pulse may be full but it is liable to great variations, being often slow, irregular and exceedingly variable. Thus the symptoms are gradually developed during the first stage, the headache continuing often with unabated violence till a new train of symptoms are developed during the second stage of the disease.

The symptoms developed thus far, are probably the result of congestion, irritation, and slow inflammation: but as softening of the cerebral substance takes place, there is rigidity, and in some cases continuous spasm of the flexor muscles of some part of the body, as of the limbs, or perhaps of a single muscle, or possibly symptoms of paralysis of sensation or motion, or both may occur.

In one fatal case of this disease which fell under my care, during the second stage, the patient had suffered from intolerable pain in the head for several months, during the first stage of the disease. And, though with the softening process, or second stage, there was only a hesitancy of speech, with inability to connect words to form a sentence, in attempting to answer questions, or express his ideas; he was seized suddenly with convulsions of the flexor muscles of the limbs, and died after two or three days of the most profound coma. In some cases, however, the second stage is merged in one of paralysis; the limbs which were before flexed with violent contractions of their muscles, now become perfectly paralyzed, losing all sensation, and power of voluntary motion.

In such cases, the affected portion of the brain has generally passed into a disorganized state, being either in the last stage of softening, or thoroughly dissolved into pus. The patient finally loses the senses of sight, hearing, &c., and by degrees, the different functions of the body, become suspended, till at last he dies, having been reduced to the extreme of physical and mental weakness.

Thus we have the symptoms of cerebritis, as they are generally developed : but it must be remembered, that the disease is liable to great variations. In some cases, if the inflammation be quite general, and violent, convulsions may occur in a few days, from the first attack, or after a short period of headache, with slight febrile excitement, coma may supervene, and the patient may die in one or two weeks, without having had any considerable convulsions.

Generally, however, the disease is of a chronic character; being marked as I have said, by three distinct stages, of variable duration. During the first stage, it should be remembered, while the inflammatory process is going on, violent pain in the head is the most prominent symptom. During the second stage, while the process of softening, or suppuration is going on, the pain in the head may subside; and instead, the patient be tormented with irregular action of the muscles, and more or less paralysis, or partial loss of voluntary motion, and sensation, of different parts of the body. And finally, during the third stage, when the brain has become completely disorganized, there is stupor, paralysis, and a gradual suspension of the voluntary and vital functions, till at last the patient dies, after having suffered for an indefinite period, varying from a few weeks, to several months, or perhaps years.

Anatomical Characters.—In cases of cerebritis, which are rapid in their course, and quite general, congestion of the diseased portion of the brain is generally found, and often the meninges of the brain show signs of having been more or less involved in the inflammation. In cases in which the brain alone has been inflamed, its substance is found changed at the point of disease, which may be any portion of the cerebral mass ; but I believe it is most frequently in the external gray vesicular substance. The character of the morbid change, varies also with the stage of the disease, at which death occurs ; and also somewhat with its location, whether it be in the vesicular, or white tubular portion.

If the patient dies during the early stage, numerous red points may be seen, and the brain may be found of a red, violet or brown color, and it is generally softer, though it may be harder, than in a state of health. In advanced stages of the disease, pus is found infiltrated into the brain, or what is more probable, the substance of the diseased brain is found converted into pus. At least, the structure of the brain, at the diseased point, has the appearance of being dissolved into pus, by which it is colored of a brown or greyish color, the mass being apparently a mixture of blood, pus, and dissolved cerebral substance. Finally, if the patient has passed on to the third stage of the disease, no trace of blood or even of the cerebral substance may be found, nothing but pus occupying the diseased part.

In some cases, the cerebral substance is found softened, without any appearance of pus. Especially may this be the case, if the disease has been located in the medullary portion of the brain, in which case the softened cerebral substance is sometimes found of a milk white color, and without any apparent mixture of pus or blood.

It is well to bear in mind then, that if death occurs during the early stage, the diseased part presents a congested appearance. At a later period there is found more or less softening, with purulent infiltration. And, finally, if the patient passes into the third stage, clear pus is generally found, if the disease be of the cortical substance, though white softening may be found, if the disease has been of the medullary substance of the brain. Induration may, however, be found in some rare cases, and so may gangrene; the diseased part in the latter case, presenting a livid softened appearance, having a fetid odor, and being mingled with a very offensive greenish liquid.

Thus we have the principal morbid appearances presented in the different stages of cerebritis, but it must be remembered that obstruction of the cerebral arteries sometimes leads to softening of the brain, in cases in which no cerebral inflammation has occurred. In such cases, however, no granules or granular corpuscles are found in the softened mass, while if the softening be the result of inflammation, they may be detected by the microscope, "resulting from the exudation of the inflammatory process."

Diagnosis.—Cerebritis may be confounded with meningitis, hysteria, and apoplexy; from which, however, it may be distinguished by careful attention to the following differences.

From meningitis this disease differs by being attended with less fever, less acute delirium, or general convulsions; by being attended with more rigidity and tonic spasm in the latter stages of the disease, especially of the flexor muscles, and finally by the much greater duration, as cerebritis may continue for several weeks, months, or even years.

Cerebritis differs from hysteria in the violent protracted pain in the head, which does not necessarily occur in hysteria, in the distressing and often protracted vomiting, and also in not being so generally attended with alternate laughing and crying, as is apt to be the case in hysteria.

From apoplexy, cerebritis may be distinguished by the history of the case, by the pain, vomiting, and rigidity of the flexor muscles in cerebritis, and by its generally occurring in more anæmic patients than apoplexy

Causes.—The causes of cerebritis are various, such as imprudence in eating, drunkenness, the suppression of some habitual sanguinous discharge, the depressing mental emotions, the use of tobacco, and also indulgence in vinous and other fermented liquors. It is probable, however, that masturbation, and venereal excesses, constitute by far the most frequent causes of cerebritis. By these and like causes, the substance of the brain becomes involved in a slow passive inflammation, which may terminate in softening, suppuration, and other varieties of morbid conditions, affecting in some cases a very limited amount of the cerebral mass, while in others, a whole hemisphere may become involved.

Prognosis.—Cerebritis is always attended with considerable danger. If, however, the cause can be ascertained and removed, and the patient can be brought to conform, rigidly, to the laws of health, and can have the benefit of a careful judicious treatment during the early stages, the patient may very much improve, and perhaps entirely recover. If, however, the disease is very extensive, and has passed on to the third stage of softening, or suppurative disorganization, the case may be expected to terminate fatally; the only hope in such a case being that the disorganized part may become inclosed in a cyst, and thus lessen its serious effects, till finally it may be absorbed.

Treatment.—In acute cases of cerebritis, if the inflammation be quite general, or slight bleeding from the arna, and cupping about the temples or back of the neck may be indicated, after which a cathartic of calomel or podophyllin should be given, and a reasonable operation secured. The warm foot-bath should be used morning and evening, if the patient can bear it, and gentle diaphoretics, such as ipecac, antimonials, sage tea, &c., given, taking care, however, not to increase the nausea, if that be already troublesome. If, however, the case becomes chronic, as it often does, the indications are the same as those which exist in cases chronic from the first, the treatment of which we will now proceed to consider.

In all chronic cases of cerebritis, in which the inflammation is generally of a passive character, the depressing or debilitating cause should be sought out and removed, if possible; whether it be tobacco, drunkenness, self-pollution, venereal excesses, indulgence in fermented liquors, or any other cause. Cups, wet or dry, according to the degree of debility, should be applied to the temples or back of the neck, and repeated if necessary, once or twice a week at first, but later, instead, blisters should be applied to the temples, back of the ears, or back of the neck; and this should be kept up during the whole course of the disease. The bowels should be kept gently loose during the early stage of the disease, by an occasional blue pill, or if that be contra-indicated, small doses of podophyllin or leptandrin. In the more advanced stage, however, a pill of aloes and rhubarb, of each one and a half grains, I believe will generally do best.

The patient should be restricted to a plain, digestible, and moderately nourishing diet, to be taken with strict regularity; or if the patient be decidedly anæmic, a good, nourishing diet should be allowed, and the citrate, carbonate, or syrup of the iodide of iron given and continued in moderate doses for a long time. The patient should be directed to take a reasonable amount of exercise; to keep the head well elevated and cool and the feet warm and dry; but violent exercise, either of body or mind, should be strictly prohibited, and a fit of anger should on no account be indulged in.

In very protracted cases, in which a small cyst may have formed, containing pus, the absorption of which is desirable, five grains of the iodide of potassium may be given three times per day, before each meal, and continued for a long time. In cases in which the disease has been removed, if the brain remains in a torpid, inactive state, indicated by derangement or an impaired state of the functions of the body, some preparation of the nux vomica may be indicated; in such cases, a grain of the muriate of strychnia may be dissolved in six or eight ounces of water, and a teaspoonful given three times per day till the tone of the nervous system is restored.

SECTION IV.—TUBERCULOUS MENINGITIS.

By tuberculous meningitis I mean inflammation of the membranes of the brain, involving, to some extent, the cerebral substance, and being attended with tubercles in the pia mater, and always occurring in patients of a scrofulous or tuberculous diathesis. This disease may occur at any age, but it most frequently occurs during childhood; and from the fact of its terminating so often in the effusion of serum into the ventricles, it has been called acute hydrocephalus; but in some cases of this disease no serous effusion takes place into the ventricles.

Symptoms.—In scrofulous children, or adults, in which this disease is being developed, there may be a protracted stage of premonitory symptoms, such as irritability, slight fever, paleness, emaciation, and various dyspeptic symptoms. There may be also occasional headache, with vomiting coming on irregularly, or at regular hours each, or every other day; and these symptoms may be gradually aggravated till inflammatory symptoms are developed.

After these premonitory symptoms have continued for an indefinite period, when they occur, or are noticed, and at first when they are not visible, the patient is seized with pain in the head, and vomiting, which may occur two or three times during every twenty-four hours for the first three or four days before the patient takes his bed. Slight chills may precede the febrile excitement. In either case, when the disease becomes established, the pain in the head becomes almost constant, being, however, greatly aggravated at times. The pain is most frequently felt in the forehead or temples, and is indicated in young children by a frequent application of the hand to the fore part of the head. The vomiting is apt to occur with the aggravation of the pain, and not unfrequently with the food more or less bile is ejected. This symptom may subside in three or four days, or it may continue much longer than that.

There is often pain in the abdomen, and constipation at first, or if the bowels are loose, the stools are generally of a dark greenish color, according to my observation. The pulse may be slightly excited, but the face is apt to be pale, except during the paroxysms of severe pain. There is little thirst, and the tongue may be moist, and only slightly furred. The eyes are sensitive to light, the pupils either contracted or dilated, and the child becomes peevish, fretful, and perhaps at times slightly delirious, even during this first stage of the disease.

As the disease passes into the second stage, more decided symptoms supervene. The pulse may become slower, but it is very irregular, and often intermittent. The respiration becomes irregular, is slower than in the first stage, and I have generally noticed frequent and deep sighing. By degrees the skin becomes cool, and partial sweating occurs, the patient becomes drowsy, often moaning, or sending forth sharp screams, or else is delirious, being wild, and sinking in two or three days into profound coma.

One hand is apt to become clenched, with the thumb pressed into the palm; the eyes turned up, or else either in or out, and the pupils dilated; vision is impaired or lost, and the patient evidently without taste, swallows, if able, anything placed in the mouth. Thus the patient may continue for a week or more, with perhaps periods of marked remissions, till a third stage, indicating more profound cerebral lesion supervenes.

In this third stage of the disease, there is more or less tonic spasm of some of the flexor muscles, generally of one of the extremities; the jaws may be closed, the head turned back, and not unfrequently there are violent convulsive movements. Finally, paralysis takes place in some one of the extremities, the surface becomes insensible, sight and hearing are lost, profound coma supervenes, the pulse becomes more irregular and frequent, the eyes become dim, the surface cold, and the patient finally dies either in a state of profound coma, or in violent convulsions.

Thus we have the ordinary course and symptoms of tuberculous meningitis, but it must be remembered, that like every other disease, it is liable to great variations, according to the general condition of the system, and the character of the meningeal inflammation. These circumstances also materially affect the duration of the disease, which may vary from five or six days to six or eight weeks. Generally, however, I think the patient may survive two or three weeks, after the supervention of the violent headache, vomiting, &c.

Anatomical Characters.—There is found, on post-mortem examination in these cases, minute, gray, white, or yellowish granules, or tubercles, varying from the merest speck to the size of a mustard seed, and being dispersed over the surface of the cerebrum and cerebellum, in the substance of the pia mater. In some cases, however, they may be confined to a small spot, or to one of the hemispheres; but I believe they are more generally diffused on the lateral portions, base, anfractuosities, and fissures of the brain, filling up more or less the loose tissue of the pia mater. In some cases the tubercles are of much larger size, or several of them may be aggregated together, including the pia mater in their substance.

The arachnoid membrane on its free surface may appear nearly natural, but it is sometimes slightly thickened, and occasionally it contains a little serum in its cavity. Marks of inflammation are almost always exhibited by the pia mater, as it is thickened, and infiltrated more or less with a turbid fluid, or a concrete matter is deposited in its tissue, consisting, probably, of coagulable lymph or concrete pus. This concrete

16

matter may generally be found either in patches, or in lines close along the blood-vessels, and being interspersed more or less with many tubercles.

The brain itself may appear flattened, and the cortical substance reddened, and perhaps softened, while the medullary portion may be covered with minute red specks, and the same appearances are presented if it be cut through. In some cases no more than a natural quantity of serum is found in the ventricles, but generally from an ounce to half a pint or more of serum may be found in the ventricles, the result, probably, of inflammation of their lining membrane, its appearance being modified, and its quantity increased by the softening of the brain which occurs, as the medullary substance bounding the ventricles is often found softened to a creamy consistence. Tubercles are generally also found in other parts of the body, thus rendering it certain that this disease is the result of a general tuberculous condition of the system.

Diagnosis.—Tuberculous meningitis differs from simple meningitis in the following particulars, by which it may be distinguished. Tuberculous meningitis always occurs in patients of a tuberculous or scrofulous diathesis, and seldom occurs in children under two years of age, while simple meningitis may occur in early infancy, and in patients of a strong vigorous constitution from various exciting causes.

In simple meningitis, the disease is liable to be more acute and sudden in its attack, and to be attended with more fever, heat, thirst, delirium, vomiting, and to be more continuous and rapid in its course than the tuberculous disease. Besides, tuberculous meningitis may pass on from three to six weeks, before a fatal termination takes place, while simple meningitis often terminates by the seventh or eighth day, though it may continue longer.

From certain conditions of congestion and irritation of the brain, occurring in enteric and other fevers, tuberculous meningitis may be distinguished by the history of the case, by the absence in the febrile affection of the violent paroxysmal pain, and vomiting, of the clenching of the hand, and deep sighing, and finally, by the want of the spasms and paralytic symptoms, which attend tuberculous meningitis. By careful attention to all these differences, I believe a correct diagnosis may generally be formed in tuberculous meningitis.

Causes.—It is probable that this disease is often hereditary, as tuberculous diseases are apt to be. Children born of tuberculous parents, with large heads and short necks, are peculiarly liable to become affected with meningeal tuberculous disease as the brain or its membranes in such cases is most prone to become the seat of the tuberculous deposit. Various causes may, however operate to produce this disease, in children or adults, in whom no decided hereditary predisposition to the disease existed.

In children, this disease may be produced by exposure to damp apartments, by insufficient clothing, by irregularity in taking food, and by any and every cause capable of impairing digestion, reducing the blood, and at the same time tending to congestion and irritation of the brain or its meninges. In adults, tuberculous meningitis may be produced by the causes enumerated above, together with the use of tobacco, drunkenness, masturbation, licentiousness, and every other cause capable of impairing the blood and of increasing cephalic congestion and irritation.

In children or adults in which there is the scrofulous or tuberculous diathesis, it is probable that slight causes may develop the disease, such as falls, dentition, the healing of sores of the scalp, intestinal irritation from irregular eating, and worms, in children and adults; in addition to all these causes, great mental excitement, exposure to solar heat, various occupations, fits of anger, &c.

Pathology.—It is evident, that in every case of this disease, there is a scrofulous or tuberculous condition of the system, and generally tuberculous deposits, not only in the pia mater, but in other tissues of the body. Now it is probable that in many cases, tubercles are deposited in the pia mater, from slow congestion and irritation of the meninges of the brain, and as the tubercles accumulate they increase the irritation, and finally develop, with some slight accidental cause, meningeal inflammation. In other cases, slight inflammation may exist from the first, and only be increased by the tubercles, till by the addition of some accidental exciting cause or without it, meningeal inflammation is fully developed, and being complicated with the tuberculous disease, passes on often in spite of remedies, to a fatal termination.

It is probable that the serous membrane lining the ventricles becomes inflamed, as well as the medullary substance constituting their walls, in all those cases of tuberculous meningitis, in which serum is found in their cavities and the surrounding cerebral substance is found softened. But it is possible that in such cases the serum in the ventricles, may by acting upon the medullary substance constituting their walls, produce or increase the softening of the cerebral substance. In cases in which no more than a natural quantity of serum is found in the ventricles, and in the general arachnoid cavity, the presence of the tuberculous matter together with the congested and inflamed condition of the brain and its meninges, account for the symptoms which are developed.

Prognosis.—The prognosis in tuberculous meningitis is generally unfavorable, especially if the disease has progressed considerably, and any considerable amount of tuberculous matter has been deposited in the pia mater. It is possible, however, that if the inflammation be slight and only a few scattering tubercles have been deposited, that the disease may be arrested so that no further tuberculous deposit may take place if the inflammation be subdued, and the condition of the blood improved as soon as may be. The scattering tubercles in such a case, remaining in a latent state, and failing to produce any very considerable local or general disturbance for a long time at least.

A case fell under my care a few weeks since, in this village, in which I have reason to believe there was incipient tuberculous meningitis. The patient a child two years old, of a scrofulous diathesis, had all the early symptoms of this disease; which, however, appeared to be arrested, for the time at least, by cathartics, iodide of potassium, blisters, &c. As the more alarming symptoms subsided, a copious eruption of mattery pimples appeared upon the scalp and the child, though feeble, exhibits now no marked symptoms of the disease.

Treatment.—In cases which inherit a scrofulous constitution, or in which there is an acquired predisposition to this disease, every possible care should be taken to prevent its development. The child should be

kept in comfortably warm and dry apartments : should be allowed food
only at regular hours, and that of a plain, digestible and nourishing
character, and should be allowed pure, fresh, dry air, as far as may be.
If, however, the disease develops itself, more active measures should be
resorted to.   On the first appearance of the symptoms, a cathartic of
calomel, hydg. cum creta, podophyllin or leptandrin, should be administered, and a free operation secured.   The feet should be set in warm
water morning and evening, and as much irritation kept up to the bottom
of the feet by mustard applied fresh morning and evening, as the patient
can bear, and this should be kept up, I am satisfied, during the whole course
of the disease.   Great care should be taken to keep the feet warm, and
the head elevated and cool.

As the blood is always in a reduced state in these cases, and the inflammation, consequently of a more or less passive character, I believe
that general bleeding is seldom, if ever indicated in this disease ; cups,
however, or leeches, may be applied to the temples, or to the back of the
neck, and repeated, if necessary, during the early stage of the disease ;
and later, blisters should be applied back of the ears, and to the back of the
neck, and repeated at intervals while the disease continues.   After the
operation of the first cathartic, moderate doses of James's powder may
be given, every six hours, for the first few days, to promote perspiration,
and lessen the cephalic tendency.

To correct the scrofulous condition ; arrest the tuberculous deposit,
and prevent the effusion of serum into the ventricles, and general arachnoid cavity, the iodide of potassium, in full doses, should be given
immediately after the operation of the first cathartic, and continued
every six hours, during the whole course of the disease.   The strength
of the patient should be sustained from the first, by crust coffee, one
third, or one half milk, as a drink ; and if the stomach becomes able to
retain it, mild digestible food should be allowed at meal hours ; at least
after the inflammation is in a measure subdued.

The James's powder should be given, alternating with the iodide of
potassium, till the meningeal inflammation subsides ; after which, if the
patient is decidedly anæmic, small doses of the citrate of iron may be
given instead, every six hours, alternating with the iodide of potassium,
and continued, as well as the iodide, for a long time.   It is doubtful
whether any advanced case of tuberculous meningitis is ever permanently
cured : but I believe in this way, the disease may often be arrested, so
that no more tubercles may be deposited ; and thus a check, at least, put
upon the development of the disease for a time, if not permanently.

SECTION V.—SPINAL MENINGITIS.

By spinal meningitis, I mean inflammation of the membranes of the
spinal cord, in which the medullary substance of the cord, is liable to be,
though not necessarily involved.   The membranes of the spinal cord, it
will be remembered, are continuous with those of the brain ; consisting
of the dura mater, pia mater, and arachnoid membranes.   Their structure
is identical with the cephalic membranes, except that the pia mater is
less vascular, and is of a more fibrous structure, than the pia mater of
the brain.

The dura mater, like that of the brain, is the outer membrane, being united firmly to the margin of the occipital foramen superiorly, and to the coccyx inferiorly; but in the rest of its extent, it is only connected by a very loose areolar tissue, to the walls of the spinal cord. The pia mater immediately invests the cord, being a continuation of the pia mater of the brain. It sends off duplicatures into the anterior and posterior longitudinal fissures, and also a process on each side of the cord, its entire length, the ligamentum dentatum, which separates the anterior from the posterior roots of the spinal nerves.

The arachnoid membrane of the cord is continuous with that of the brain, and encloses the cord very loosely. It lies between the dura mater and pia mater, forming the outer serous covering of the pia mater, passing off with the other membranes to form a sheath for the nerves, and then being reflected on the inner surface of the dura mater, thus giving it a serous lining, and constituting the arachnoid cavity of the cord a shut sac, similar to that of the brain.

The arachnoid cavity of the cord, like that of the brain, is occupied by a serous fluid, which expands the arachnoid, and keeps up a gentle pressure, which yields with facility to the various movements of the cord.

Now it must be remembered, that all these membranes pass off together, to form sheaths for the spinal nerves. And further, that the pia mater is supplied by a number of plexuses, derived from the sympathetic system of nerves. Now the spinal meninges, thus constituted and arranged, are liable to become inflamed, the simple inflammation of which we will now proceed to consider.

Symptoms.—Spinal meningitis may be either acute or chronic; the symptoms, of course, varying with the intensity of the inflammation. The symptoms vary, also, according to the membrane chiefly involved in the inflammation. If the dura mater or the pia mater be the seat of the inflammation, the pain may be of a dull heavy character, as the pia mater of the cord is of a fibrous character, as well as the dura mater. If the arachnoid membrane lining the dura mater, or that covering the pia mater, become materially involved, or be the special seat of the inflammation, the pain may be of a sharp lancinating character.

In spinal meningitis, then, involving as it does, generally, more or less, all the membranes, we have pain of either a dull, aching, gnawing, or else of a sharp lancinating character along the spine, and extending into the limbs, or parts supplied with nerves from the spinal cord. This pain is generally increased by motion, and is often attended with tetanic contraction, and violent spasms of the muscles supplied with nerves from the diseased track, or point of the spine; and there is also a feeling, either in the neck, back, or abdomen, according to the seat of the meningeal inflammation, as if they were girt by a tight bandage. Besides, there is generally constipation, and often retention of urine, with a feeling of suffocation, and chills, of more or less severity.

After the chill or chills, more or less fever is developed, the pulse being hard and frequent, but perhaps small, the skin hot, but perhaps profusely moist, the respiration tedious, &c. Finally, if the disease progresses, the pulse becomes irregular and feeble, drowsiness, delirium, and coma occur, with paralysis, and involuntary discharges, and at last the patient dies, at a period varying from one to two weeks from the commencement of the disease.

In chronic spinal meningitis, then, there is generally a dull pain in some portion of the spine, with more or less deranged sensations in the extremities, as well as functional derangement of the thoracic, abdominal, or pelvic viscera.   After an indefinite time tonic spasms occur, the limbs are flexed, the head is drawn to one side or backward, and, unless an acute attack supervenes, there may be paralysis, exhaustion, and a gradual failure of all the functions of the body, till the patient, worn down by wakefulness, irritability, and physical prostration, dies in a state of profound coma, having suffered, perhaps, for weeks, months, or even years.

Anatomical Characters.—The membranes of the cord are found thickened, the dura mater presenting a deeper color than in health, the arachnoid is more or less opaque, and the pia mater injected, red, and swelled. Turbid serum is found in the lower portion of the arachnoid cavity, and more or less lymph or pus is found between the arachnoid and pia mater, and also between the pia mater and the spinal cord.

Diagnosis.—Spinal meningitis may be confounded with myelitis, or with rheumatic inflammation of the spinal ligaments, or the surrounding muscles.   In spinal meningitis, tonic spasms of the muscles, with flexion of the limbs, is a prominent symptom ; while in myelitis, paralysis, following inflammation, is a leading feature of the disease.

From rheumatism of the spinal ligaments, spinal meningitis may be distinguished by the history of the case, and by the fact that in meningitis pressure on the spinous processes causes most pain, while in rheumatism of the ligaments, most pain is produced by making pressure along the side of the spine.

Causes.—Spinal meningitis may be produced by direct blows or violent straining, by venereal excesses, by atmospheric vicissitudes, by alcoholic liquors, and by the metastasis of rheumatism, erysipelas, &c.

Prognosis.—Spinal meningitis though a dangerous, is by no means always necessarily fatal.   According to my observation, most cases, if attended to properly in season, may be arrested, though I believe one attack strongly predisposes to subsequent attacks of the disease.

Treatment.—In acute spinal meningitis, occurring in strong plethoric patients, general bleeding may be indicated, at first, as well as an active cathartic of calomel or podophyllin, after which, cups should be applied along each side of the spine at the inflamed point, and more or less blood taken.

If the disease becomes chronic, and in all cases which have not assumed an acute form, cups, wet or dry, should be applied on each side of the spine, at the inflamed point, and repeated every day or two at first, and then blisters should be applied, an inch wide, on each side, along the inflamed part, and repeated at intervals of a few days till the inflammation is subdued.

Mild cathartics may be given, such as the blue pill, followed by the sulphate of magnesia, or perhaps the podophyllin or leptandrin if the mercurial be contra-indicated.

In all cases of a rheumatic character, or in which there is danger of effusion into the arachnoid cavity, the iodide of potassium in full doses should be given three times per day, and continued till all symptoms of the disease are removed; in conjunction with the remedies suggested above.

SECTION VI.—MYELITIS.

By myelitis I mean inflammation of the substance of the spinal cord. The spinal cord, it will be remembered, is a continuation of the brain, extending from the brain, and including the medulla oblongata, from the pons varolii to the first or second lumbar vertebra. It passes along in the midst of the vertebral canal, which it by no means fills, and exhibits, besides the medulla oblongata, two enlargements corresponding with the origin of the nerves for the upper and lower extremities.

The spinal cord presents an anterior and a posterior longitudinal fissure, the latter of which terminates in the gray substance of the interior of the cord, while at the bottom of the anterior there is a layer of the white structure connecting the two lateral portions of the cord. Two other lines are observed on each side of the medulla, the anterior and posterior lateral sulci, corresponding with the anterior and posterior roots of the spinal nerves.

The anterior lateral sulcus is a mere trace, marked by the attachment of the filaments of the anterior roots. The posterior lateral sulcus is more plain, being a narrow grayish line derived from the gray substance of the interior of the cord.

Now the anterior lateral columns, comprehending all that part of the cord situated between the anterior longitudinal fissure and the posterior lateral sulcus are columns of motion, while the posterior columns, situated between the posterior lateral sulci and the posterior longitudinal fissure are columns of sensation.

If a transverse section of the spinal cord be made, it appears to be composed of two hollow cylinders of white matter, placed side by side, and connected by a narrow white commissure; each cylinder is filled with gray substance, which is also connected by a commissure of the same matter. The form of the gray substance is that of two curved lines joined by a transverse band; the extremities of the curved lines corresponding with the sulci of origin of the anterior and posterior roots of the spinal nerves. The anterior extremities of the curved lines being larger than the posterior do not quite reach the surface at the point of origin of the anterior roots of the nerves; but the posterior extremities of the curved lines of gray matter reach the surface forming a narrow gray line; the sulcus lateralis posterior, the point of origin of the posterior roots of the spinal nerves. The white substance of the spinal cord is composed of parallel fibres which are collected into laminæ and extend the entire length of the cord.

"The spinal cord transmits impressions from the periphery to the brain and conversely enables the brain to bring into action the motor nerves." If the cord be divided, it interrupts voluntary motion and sensation in those parts supplied by nerves from that portion of the cord below the point of division, the functions of the parts supplied with nerves from above the point of division remaining unimpaired.

It is probable that the transmission of influences upward and downward, is to a considerable degree accomplished, through the vesicular substance, which has the power of transmitting influences, not only in a longitudinal but in a transverse direction. But the exterior white fibrous structure possesses a like function, the anterior columns being motor, and the posterior sensory, at least, in a longitudinal, if not in a transverse direction.

There are nine pairs of cranial nerves, and thirty-one pairs of spinal nerves. The spinal nerves, as we have seen, arise by two roots, an anterior motor and a posterior sensitive; the anterior proceeding from a narrow white line, the anterior lateral sulcus, while the posterior proceeds from the posterior lateral sulcus, the narrow gray stria, formed by the posterior extremities of the curved lines of the gray substance of the cord.

The posterior roots form ganglions in the intervertebral foramina, after which the two roots unite, constituting a spinal nerve; which, escaping through the intervertebral foramen, divides into an anterior branch, for the supply of the anterior aspect of the body, and a posterior branch, for the posterior aspect. Thus constituted and arranged, the substance of the spinal cord is liable to become inflamed, and to develop a peculiar train of symptoms, in either an acute or chronic form, both of which we will now proceed to consider.

Symptoms.—In acute cases considerable pain may be experienced in the inflamed portion of the cord, which is liable to be greatly increased by motion and pressure, and convulsions or loss of sensation and voluntary motion generally attends, as a consequence of the disease.

If the superior portion of the cord be inflamed, there is liable to be convulsions of the muscles of the head and face, inarticulate speech, loss of voice, difficult swallowing, spasmodic breathing, and irregular action of the heart. If the inflammation be a little lower, there is constriction of the chest, vomiting, pain in the bowels, and a sensation as if a cord were drawn round the abdomen. Finally, if the lower portion of the cord be the seat of the inflammation, there may be retention or incontinence of urine, constipation, tenesmus, involuntary stools, &c.

The voluntary muscles supplied by the inflamed portion of the cord may become convulsed or paralyzed, or more frequently they may be first convulsed, and then paralytic, and sensation may be either lost or retained, according as the anterior or posterior portion of the cord becomes involved in the disease. If, however, both the anterior and posterior nerves are affected, there is loss of sensation and voluntary motion. As the muscles thus lose their power, the organic functions become impaired or deranged, as of the heart, stomach, liver, kidneys and genital organs. Febrile symptoms, according to my observation, are seldom developed to any considerable degree, the respiration and pulse being slower than in health.

The duration of the disease varies according to the seat of the inflammation. If it be in the superior portion of the cord, above the origin of the respiratory nerves, the patient may die in two or three days; but if the same amount of disease be situated in the lower part of the spine, the patient may survive for a much longer time, as the vital functions are not so directly interrupted. Acute myelitis may thus pass on and become chronic, or the disease may be chronic from the first; in which case the most prominent symptoms are slight uneasiness of the spine, and more or less deranged sensations of the limbs, with fatigue attending exertion.

By degrees paralytic symptoms are developed, the patient becomes wakeful and tremulous, the limbs become weak, and the gait tottering; the different parts of the body become paralyzed, the surface becomes pale, the pulse slow, and the limbs more or less swelled; and, finally, death takes place, the result of a gradual suspension of the voluntary and vital functions.

Anatomical Characters.—Inflammation of the substance of the spinal cord may lead to softening, induration or suppuration, as is the case with inflammation of the cerebral matter; but softening is, I believe, by far the most frequent result. The softening may consist in only a slight diminution of consistence, or it may be complete; the softened mass having either a reddish, yellowish, or nearly natural color; and it may occupy only a portion of the diameter of the cord, or it may involve its whole thickness.

Diagnosis.—Myelitis may be confounded with spinal meningitis, or rheumatism of the spinal ligaments or muscles, from both of which diseases, however, it may be distinguished by the following diagnostic symptoms.

In spinal meningitis the prominent symptoms are severe pain at the point of inflammation, extending to the muscles or organs supplied by the inflamed part with nerves, and also tonic spasms or convulsions of the muscles so supplied, while in myelitis there is less pain in the back or in the muscles supplied by nerves from the inflamed portion of the spine, and more derangement in the motions, sensations, and functions of the muscles and organs, and finally more paralysis, or loss of sensation or voluntary motion or both, gradually increasing till there is a total suspension of the voluntary and vital functions in fatal cases.

From rheumatism of the spinal ligaments or muscles, myelitis may be distinguished by making pressure along the spinous processes, and by the side of the spine. If more tenderness is experienced by the side of the spine, than along the spinous processes, the disease is rheumatism, or irritation of the spinal ligaments or muscles; but if most pain is experienced from pressure along the spinous processes, the disease is either myelitis or spinal meningitis; and to ascertain which of these affections it may be, we have only to take their diagnostic symptoms into account and the diagnosis is clear.

Causes.—Myelitis may be produced by injuries, caries, tubercles, &c., but according to my observation, masturbation, venereal excesses, drunkenness, and the use of tobacco, are by far the most frequent causes of this most distressing affection.

I have noticed that cases produced by tobacco are most liable to commence at or near the lower portion of the cord, while cases produced by masturbation, or venereal excesses, may commence in the superior portion of the cord, perhaps extending from the brain. This, however, though it be a general rule, may not be an invariable one. I do not, however, remember an exception, among the cases which have fallen under my observation.

Pathology.—There is no disease, the nature of which is more clear than myelitis, if we keep in mind the structure and functions of the spinal cord, and remember the office of the cerebro-spinal and sympathetic system of nerves, the former controlling sensations and voluntary motions, and the latter presiding over the involuntary or vital functions. During the inflammatory stage of the spinal cord, its power of conducting influences, both to and from the brain is more or less interrupted, both sensory and motor, and hence the derangement of the sensations and voluntary motions which occur in the early stage of the disease; the sensory or motor being most affected, according as the anterior or posterior roots of the nerves become most involved in the disease.

As the inflammation of the cord passes on to softening, or disorganization, sensation, or voluntary motion, or both, are lost to the part supplied with nerves from the diseased portion of the spine; and as the disease progresses, not only sensation and involuntary motion are lost, but by degrees the vital functions become interrupted, and though mainly carried on through the agency of the sympathetic system, their functions finally become suspended, and thus the patient dies.

Prognosis.—Inflammation of the substance of the spinal cord is a very dangerous disease, and yet I believe that by a judicious treatment in the early stage of the inflammation, it may often be arrested. But if the disease passes on to disorganization of the cord, a recovery need not be expected. In such cases, however, if the disease be below the origin of the respiratory nerves, the disease may assume a chronic form, and the patient live on in a wretched state for weeks, months, or even years, as happened in one finally fatal case that fell under my care.

Treatment.—In acute cases of myelitis, if the patient be strong and plethoric, and the inflammation active, general bleeding should be resorted to, after which, blood should be taken by cups from each side of the spine, at the diseased point, and a cathartic of calomel or podophyllin administered. If, however, the disease assumes a chronic form, and in all chronic cases, cups, wet or dry, along each side of the diseased spine, an occasional cathartic, and finally blistering, pustulation with tartar emetic ointment, &c., together with rest, quiet, and a plain digestible diet, are the remedies upon which we are mainly to depend. In very chronic, passive cases, occurring in anæmic patients, from masturbation, venereal excesses, &c., iron, and strychnine, as tonics for the blood and nervous system, I have often found to do well.

SECTION VII.—CEREBRO-SPINAL MENINGITIS.

By cerebro-spinal meningitis, I mean a peculiar epidemic, malignant disease, characterized by inflammation of the meninges of the brain and spinal cord, and assuming generally a congestive or typhous character.

This disease more frequently attacks young persons, and especially males, and has within the last twenty-five years prevailed extensively in different parts of Europe, and especially in France, where it has prevailed most extensively in garrisoned towns. In 1848, the disease made its appearance in this country, epidemics of it having since prevailed in different parts of the United States, till the winter of 1857, at which time the disease prevailed extensively in central and western New York, but most malignantly in Chemung and Onondaga counties, where the disease had a rapidly fatal tendency. During the winter and spring of 1857, a few cases of cerebro-spinal meningitis occurred in this village and vicinity, and a few scattering ones since, one fatal case of which fell under my care.

Wherever this disease has prevailed epidemically, the mortality has been very great, not less in some cases, than seventy or eighty per cent, and in some instances even more than that. It is evident that this disease has its origin in a malarious agent of an uncommon degree of concentration, either of an animal or vegetable derivation, or what is more probable of both, and this appears to have been the opinion of the most acute observers, in the localities where the disease has prevailed, both in this country and in Europe. The disease appears most frequently in low moist localities, and during warm weather in February, March, or April, or else in the warm, wet weather which is liable to occur in our climate during November and December.

From the general character of this disease, there would be no impropriety in placing it among the malignant fevers, but as the cerebro-spinal meningeal inflammation is its leading characteristic, I have retained the consideration of it for this place, the symptoms of which we will now proceed to consider. We should, however, bear in mind what I have before explained in relation to the fibrous, vascular and serous membranes fo the brain and spinal cord, their relation to each other, and to the cerebro-spinal substance, and also that the membranes of the cord are continuous with those of the brain, and identical in structure, except that the pia mater of the cord is less vascular and more fibrous than that of the brain.

Symptoms.—In many cases of this disease there is at first, slight disturbance of the stomach and bowels, with perhaps nausea, vomiting and diarrhœa, and soon slight chills and more or less pain in the head and along the spine. As the cerebro-spinal pain increases, there is in a few hours, restlessness, and an anxious expression of the countenance, delirium, a cold skin, and often a frequent irregular, and intermittent pulse. As the disease progresses, the muscles become spasmodically contracted, the head is drawn back, or to one side, and in some cases the whole body is rendered stiff or rigid, or in some way distorted, in which condition it may remain for days, or even weeks, and finally coma and paralysis supervene, and the patient dies, unless a favorable change takes place.

Such are the ordinary symptoms of a mild form of the disease. But in the more severe cases, there is at first a slight feeling of indisposition, continuing for an indefinite time, and then there comes a severe chill, or congestive stage, with coldness of the surface, irregular pulse, and coma more or less profound. The patient sometimes sinks, with perhaps violent vomiting and purging, into a state of collapse, very much like that of cholera; in which state death may take place in a few hours. Or, if reaction becomes established, the pulse is strong, frequent and irregular, the skin hot and dry, the thirst urgent, and there is severe pain in the head and along the spine; and, finally, there is general spasms of the muscles, vomiting of a greenish liquid; the head is drawn backward, or to one side, and there is also delirium, petechia, and great restlessness, with excessive irritability, till profound coma and death puts an end to the extreme suffering. The tongue, during the early stage of the disease, may remain nearly natural, or be covered with a white or yellowish fur, but during the latter stages, the tongue may become dry and reddish, or be covered with a brownish coat, and instead of diarrhœa there is in many cases obstinate constipation from the very first.

A prominent symptom of this disease are the petechial spots which appear on the second day, or later; first upon the face, neck, and breast, and then extending gradually over the surface of the body and extremities, more or less general. The spots vary in size from the merest speck to near half an inch in diameter, and in color from a bright red to a purple, or dark brown. In some cases, there are but a few scattering scarlet spots, while in others, one-fourth or one-sixth of the whole surface may be covered with spots of a dark purple color. This eruption may continue till death in fatal cases, but in cases that pass on favorably, it gradually fades, till it disappears about the seventh day from the appearance of the eruption, the ninth day of the disease. It should be remembered, however, that the eruption is not an invariable symptom in this disease.

The duration of this disease varies from a few hours to two or three weeks, the average of fatal cases being probably about nine or ten days.

Anatomical Characters.—The post-mortem appearances in this disease correspond, very nearly, with those presented after death from cerebral and spinal meningitis, the substance of the brain and spinal cord remaining often in a nearly natural state.

The meninges of the brain and spinal cord are found more or less congested, and generally there is found an effusion of serum, pus, or else a layer of yellowish or greenish lymph between the arachnoid and pia mater, or else between the pia mater and brain, rather scanty on the hemispheres, abundant at the base of the brain, and extending and investing the whole length of the spinal cord, and sometimes even giving a coat to the roots of the spinal nerves.

Diagnosis.—Little difficulty need be experienced in distinguishing this disease, if we remember the localities, and seasons of the year in which it is most liable to occur, the slight premonitory symptoms, the chill, congestive stage, or collapse, the violent pain in the head and along the spine, the rigidity and spasms of the muscles, more or less general, the nausea, vomiting, and either diarrhœa or obstinate constipation; and, finally, the petechia, delirium, coma, &c.

If all the extrinsic circumstances, the history of the case, together with the preceding characteristic symptoms, be taken into the account, as well as the irregular pulse and the strong tendency of the head to be drawn backwards, or to one side, I think no one, even though he had never seen a case, need be mistaken in forming a diagnosis in this disease.

Cause.—It is probable that the epidemic influence which produces cerebro-spinal meningitis is either of an animal or vegetable origin, and very likely both. The rapidly fatal tendency of this disease, in its worst form, indicates a concentrated morbid poison as the cause; and the seasons of the year, the localities, and the damp, open weather in which it makes its appearance, all appear to indicate a concentrated combination of koino and idio-miasmata as the cause of this disease.

It is possible that the paludal poison alone may produce this disease, but as it occurs generally in open winter weather, when animal poisons are very liable to be generated, and more or less of the paludal, and as the symptoms developed are such as might be expected to arise from the combined influences of the two causes upon the system, I think we have a right to infer, at least, that the two influences combine in producing cerebro-spinal meningitis. It is probable also that a low electrical state of the atmosphere, which exists during its humid state, when this disease makes its appearance, strongly predisposes to the disease, and perhaps also favors the cerebro-spinal inflammation, of a passive or congestive character, which is the prominent feature of the disease.

For, if the system part with its only reasonable amount of electricity, to restore an equilibrium between itself and a humid atmosphere, vital action must be in a measure reduced. If now, while in this reduced state of vital power, the combined influence of koino and idio-miasmata be be brought to act upon the system, it is not strange that it should strike at the very foundation of voluntary and organic life, and a congestion or inflammation of the cerebro-spinal membranes be the result.

Pathology.—Having already intimated my views of the nature of this disease, but little further need be said. That a low electrical state of the system, caused by a humid condition of the atmosphere, strongly predisposes the system to this form of disease, in the manner I have suggested, I have no reasonable doubt. That a morbid animal or vegetable poison, or probably both, enters the blood, through the skin, stomach, and lungs, I think there can be no reasonable doubt.

Now, if the animal and vegetable poisons both reach the blood in this way, as I believe they do, the animal may be supposed to produce its usual effects in materially changing or decomposing the blood, while the vegetable only produces a debilitating effect upon the cerebro-spinal and nervous system, and thus both leading, the vegetable directly and the animal indirectly, to a chill or congestive stage, during which a passive inflammation is set up in the cerebro-spinal meninges, and hence the symptoms which follow.

If the predisposition be slight, and the vegetable posion be in excess, the blood may be but slightly changed, and then reaction is soon established; the eruption has a bright red appearance; the convulsions may be slight; and, with proper remedial measures, the cerebro-spinal meningeal inflammation may be subdued, and the patient may finally recover. But if, on the other hand, the predisposition be strong, and the animal poison in excess, the blood may undergo great change, reaction may not be established after the chill or congestive stage, passive inflammation, with severe congestion of the cerebro-spinal meninges takes place, and finally, with vomiting, and perhaps purging, dark petechia, convulsions, and coma, the patient passes on to speedy dissolution. Such, I am convinced, is the true pathology of this most distressing, malignant disease.

Prognosis.—The prognosis in cerebro-spinal meningitis is generally unfavorable, but with proper remedial measures, timely applied, many cases may recover.

The unfavorable symptoms are a protracted cold stage, great prostration, obstinate vomiting, violent pain in the head and along the spine, continuous delirium, obstinate general convulsions, or profound coma, and a dark or livid appearance of the eruption. While the favorable symptoms are a slight chill, followed by moderate febrile excitement, a bright scarlet appearance of the eruption, and an absence of all the unfavorable symptoms enumerated above.

Treatment.—When we take into account the general debilitated condition of the system in this disease, and remember that the cerebrospinal meningeal inflammation must be of a passive character, produced by and kept up in a great degree at least by congestion, the indications in the treatment become very plain. At the commencement of the disease a full dose of calomel should be administered in castor oil, and the oil repeated in six hours if necessary. The feet should be placed in warm water, and then sinapisms applied to the bottom of the feet and also over the abdomen, or to the epigastrium, and warm sage tea allowed for drink.

Immediately on the operation of the cathartic, the sulphate of quinine should be given in two or three grain doses, with an equal quantity of James's powder every four or six hours. The quinine should be continued during the continuance of the disease, but two or three grains of camphor should be substituted for the antimonial as soon as the febrile excitement is subdued. Cups may be applied, wet or dry, in the early stage, and later blisters may be of service along the spine, and the patient should be sustained by crust coffee, one-half milk, till such time as solid food may be allowed.

In many cases the operation of the first cathartic will consist of a lightish watery substance, containing little or no bile; the vomiting of a greenish watery bilious fluid being continued. In such cases, a grain of calomel should be added to each of the powders of quinine and the antimonial, and continued till the vomiting ceases and the alimentary evacuations become bilious; at which time it may be discontinued.

Such is the general plan of treatment indicated in ordinary cases of cerebro-spinal meningitis, and which, if timely applied, may prove effectual, in many cases, at least. But in those decidedly malignant cases in which the patient is at once smitten down into a state of collapse, with vomiting, purging, and extreme prostration, in addition to the warm foot bath, sinapisms to the extremities and epigastrium, warm ginger or capsicum tea should be administered internally, with hot brandy sling, wine whey, &c., till reaction is established, when the case should be treated in all respects according to the plan I have laid down above for ordinary cases of the disease. Care should be taken, however, that a reasonable amount of proper nourishment be allowed from the very first.

SECTION VIII.—APOPLEXY.

By apoplexy, from ἀποπληττειν, to "strike with violence," I mean a peculiar disease of the nervous system, characterized by sudden diminution or loss of sensation, voluntary motion, and consciousness; the respiration and circulation continuing; the disease depending on pressure upon the brain at some point within the cranium.

The nervous system, embracing the brain, spinal cord, the cerebro-spinal and sympathetic nerves, together generate and distribute an influence by which the different functions of the body are performed; the cerebro-spinal system presiding over the voluntary functions, while the sympathetic control the involuntary or vital functions.

The nervous system is also the medium through which mind operates upon the physical organization, in a state of health, controlling the voluntary functions.

It is through the brain and nervous system that the mind is enabled to control, in a state of health, the voluntary muscles or those parts of the system placed under the influence of the will, while the involuntary muscles or those that are not subject to the dictates of the will, as the heart, have a power of action peculiar to themselves, or are mainly under the control of the sympathetic nerves. But the involuntary muscles, although they have wisely been placed beyond the control of a depraved will, are notwithstanding, dependant upon a healthy action of the cerebro-spinal system for the perfect performance of their functions. The cerebro-spinal system may, however, be very much deranged, and yet the functions of digestion, absorption, nutrition, secretion and circulation, which are mainly under the influence of the sympathetic nerves, be carried on with a comparative degree of regularity.

Now, if any part of the brain or spinal cord becomes the seat of congestion, or local disease, to an extent sufficient to arrest the flow of nervous influence, the voluntary muscles which are thus deprived of their supply of nervous influence are either convulsed or paralyzed, and in either case are not subject to the control of the will. If, now, a severe pressure be made in the middle of the spine, so as to compress the spinal cord to a degree sufficient to arrest the flow of the nervous influence, there will be paralysis of the lower limbs.

If, however, pressure be made only on one side of the cord, the flow of nervous influence is arrested only on one side, and paralysis of only one limb will be the result.

The paralysis thus produced may be of motion, or of sensation, or of both. If the nervous influence be interrupted only in the anterior roots of the spinal nerves, the power of voluntary motion will be lost, but sensation will remain; while if the injury cause only an interruption of the nervous influence in the posterior roots, sensation will be lost to the paralyzed part, but the power of voluntary motion will remain. If, however, the injury cause an interruption of the nervous influence in both the anterior and posterior roots of the spinal nerves, we have loss of both sensation and voluntary motion.

Now, let the whole brain become very much congested, and it at once stops the flow, and perhaps the generation of nervous influence, so that none passes down the spinal cord, and to its own nerves, and there is, for the time, a suspension of the power of motion and sensation, in all the voluntary muscles of the body, and the involuntary functions, or those under the control of the sympathetic nerves, are carried on very imperfectly. That portion of the system ordinarily under the influence of the will, heeds no longer its dictates, and appears to be in a passive state; animal heat being kept up by the continued action of the involuntary muscles, the heart and arteries, though generally in a more or less imperfect manner.

Now this condition, which I have here described, constitutes apoplexy. But if the injury of the brain be still greater, so as entirely to suspend its functions, the sympathetic nerves have their powers suspended also; and with them the functions of the involuntary muscle, the heart, and instead of apoplexy, immediate death is the result. Hence we see that there is not a tissue or organ of the body that can continue its functions if the flow of nervous influence to the part be cut off; and further, that the brain is the point whence this influence proceeds, not only to the cerebro-spinal, but to the sympathetic nerves. Thus we have the real condition in apoplexy, which will account for the symptoms of the disease, which we will now proceed to consider.

Symptoms.—As we have already seen, apoplexy consists of a suspension of the voluntary functions, while the involuntary are carried on in an irregular or imperfect manner; the attack coming on, in some cases, without any indications of its approach. Generally, however, various premonitory symptoms, indicative of cerebral disturbance, precede the attack.

The symptoms which most frequently precede and indicate an apoplectic attack, are deep-seated pain in the head, especially on stooping, or turning the head suddenly around; a turgid state of the veins of the head; throbbing of the temporal arteries; ringing in the ears; inability to articulate distinctly; dimness of sight; obtuseness of hearing; flashes of light before the eyes; bleeding at the nose; drowsiness; confusion of ideas; disturbed sleep; loss of memory; irregular spasmodic contraction of the muscles of the face; and occasionally pains in the pit of the stomach, and nausea. Of these symptoms, however, vertigo; ringing in the ears; dimness of sight; pain and heaviness in the head, are by far the most frequent precursors of this disease.

Now, it is evident that these symptoms come from a greater or less degree of congestion of the cerebral vessels, and as the degree of cerebral congestion differs in different cases, these symptoms may continue for a longer or shorter time before an apoplectic attack, or before the cephalic congestion produces the apoplectic state. When the cephalic congestion or pressure becomes sufficiently strong, whether any premonitory symptoms have been noticed or not, the patient falls, and sinks at once into a state of profound stupor, resembling deep and profound sleep, from which it is impossible to arouse him, even in the slightest degree.

If now, as we have seen, the cephalic congestion be very great, and the flow of nervous influence suspended, both in the voluntary and involuntary muscles and organs, the apoplexy terminates in death. If, however, only congestion of the cephalic vessels takes place, and the vital or involuntary functions continue, and no effusion takes place into the brain, as the congestion of the brain is overcome by the circulation becoming equalized; sensation, voluntary motion and consciousness again return, and the whole system regains its natural vigor, with only a strong predisposition to another attack.

If however, during the congestion, there is effusion of blood or serum into one side of the brain, it frequently so far interrupts the flow of nervous influence to one side of the cerebro-spinal system, and generally to the opposite to that on which the effusion occurs; that a paralysis of one side of the body is the result, either of sensation, or voluntary motion, or as generally happens of both. The circulation is kept up in such paralyzed parts by the involuntary, or vital function of the heart; though in consequence of the suspension of sensation and motion in the paralyzed part, the circulation is quite languid, and the paralyzed part is, in consequence, not so warm as the opposite side.

In other cases, as the patient emerges from the apoplectic state, instead of paralysis, there is a high febrile reaction, which will sometimes continue for several days, depending no doubt, upon irritation, or inflammation of the brain, or its meninges, produced by the severe congestion during the apoplectic state; the local irritation developing a general febrile excitement.

Such are the ordinary symptoms, and terminations of apoplexy; liable, of course to variations, like all other affections.

Diagnosis.—The diagnosis of apoplexy, is not in general, attended with much difficulty. When a loss of consciousness of the sensorial function, and voluntary motion come on, with a more or less active state of the pulse, and full respiration, the case must be regarded as apoplexy. The conditions with which apoplexy is most liable to be confounded, are syncope, asphyxia, and deep intoxication.

From syncope and asphyxia, apoplexy may be distinguished by the almost imperceptible pulse, and respiration in these two affections: while in apoplexy, the pulse is quite full, and the respiration more or less free. It is, however, quite difficult to distinguish apoplexy from intoxication. But by a careful observation in reference to the habits of the patient; the smell of his breath; and the general relaxed condition of his muscles, a correct diagnosis may be formed in such cases. There is also frequently more or less drawing of the mouth to one side in apoplexy, which seldom, if ever occurs in intoxication.

Anatomical Characters.—In some very rare cases, death may occur from apoplexy, and yet no well marked lesion of the brain may be found on post-mortem examination; while in other cases, only marks of congestion remain. In some cases again, an effusion of serum is found in the ventricles of the brain, or else perhaps into the arachnoid cavity; the result, probably, of a sudden extravasation of the fluid, without any marked inflammation, or even congestion.

While other conditions may be found, effusion of blood at some point within the cranium, is the most frequent lesion. It may be found between the dura mater, and the cranium, between the membranes, in the arachnoid cavity, or in the ventricles; or finally what is more frequent in the substance of the brain itself. When blood is found extravasated, it may be infiltrated into the cerebral substance; but it is more frequently found in a cavity, which it has formed, varying in quantity from less than a teaspoonful, to a half a pint, or more, and occupying any part of the

17

cerebrum, or cerebellum ; and generally occupying the side of the brain opposite the paralyzed side of the body.

The effused blood, after a few days, forms a darkish coagulum, which gradually becomes of a yellowish appearance, and is finally entirely absorbed, leaving a cavity with smooth sides, which, if small, may be found in contact, and perhaps connected with small interlacing filaments. Blood effused into natural cavities may coagulate, lose its redness, become organized, and finally form a false membrane, or else it may become encysted in a well formed serous sac.

Prognosis.—Apoplexy is a dangerous disease. If there is a complete suspension of the sensorial functions, with great irregularity of the vital or involuntary functions, attended with difficult breathing, and an irregular intermittent pulse, the vital functions are becoming suspended, and the patient will probably die. If, however, there be total suspension of the animal or voluntary functions, and still the involuntary, or vital functions are carried on with a degree of regularity, attended with a moderately full pulse, and a regular respiration, the patient may, and very likely will recover.

The duration of an apoplectic attack varies from a few minutes to several days, by which time it generally terminates in resolution, paralysis, or in febrile excitement. In those cases which terminate in a paralysis of one side there is generally an effusion of blood or serum, either of which are liable to continue, and so prevent a perfect recovery of the affected side. But as the congestion of the engorged vessels of the brain is overcome, there may be a sensible improvement, and it is possible that the clot of effused blood may be absorbed, and thus a healthy action of the affected side be restored.

Causes.—In relation to the causes of apoplexy, very much depends upon the peculiar conformation of the body. For the brain, being situated but a little distance from the heart, is supplied, by the carotid and vertebral arteries, with about one-sixth of the blood in the system.

If now a person has a large head and short neck, thus bringing the head still nearer the heart, it will constitute a strong predisposition to apoplexy. The blood, in such cases, is carried with stronger impulse into the cerebral vessels, and a congested state of the arteries of the brain, by producing pressure upon the sinuses or large veins, hinders a free return of the venous blood, and thus produces congestion of the cephalic veins.

Age, too, constitutes another predisposing cause of apoplexy, as most cases occur after the fortieth year. This, however, is not an invariable rule. Whatever tends to plethora also predisposes to apoplexy, such as a full, nourishing diet, the habitual use of stimulating drinks, and an indolent, sedentary course of life. All the preceding causes may predispose to apoplexy, by tending to active congestion of the brain ; but there are a class of predisposing causes, which, by debilitating, tend to produce a passive congestion of the brain, and so predispose to a passive form of this disease.

Now among the depressing causes, the use of tobacco, venereal excesses, frequent and long-continued warm bathing, and drunkenness or habitual intemperance, are probably the most frequent. Various organic affections, such as aneurism of the aorta, hypertrophy of the heart, bronchocele, or other tumors about the neck, increase the liability to apoplexy. Softening of the brain, involving, as it generally does, a more or less changed condition of the cerebral vessels, also strongly predisposes to apoplexy, generally of a passive character; one interesting case of which fell under my observation the last season, in a patient not otherwise predisposed to apoplexy.

Distention of the stomach by immoderate eating, the intemperate use of spirituous liquors, violent exertion in lifting, or anything capable of producing a determination of blood to the head, either active or passive, may act as exciting causes of this disease, in those who are natually or accidentally predisposed to it.

Pathology.—Congestion of the brain, either active or passive, is the immediate cause of the suspension of sensation and voluntary motion, and consequently of all the symptoms which are developed in this disease. If, as we have seen, the cerebral congestion be very great, involving to a considerable extent the medulla oblongata, respiration may be suspended, and death takes place. But if the pressure be less, and the medulla oblongata becomes but slightly compressed, respiration and all the vital functions may continue, more or less perfectly, and the case terminate, either by resolution and restoration to health, or else in bloody or serous effusion, leaving a paralysis of one-half of the body.

Now it is evident that, in each of these cases, congestion of the brain is the immediate cause of death when that occurs, of serous or bloody effusion when that occurs, with its attendant paralysis; while in those cases in which there is not a suspension of the vital functions, or effusion into the brain, a removal of the congestion restores the powers of sensation and motion in the voluntary functions, and health is restored.

Now this congestion of the brain may be either active or passive, as we have already seen. The congestion is active, when in consequence of a large head, a short neck, and a powerful heart, together with great plethora, the blood is forced with violence into the arteries of the brain, producing a congestion, or distention of all their minute ramifications. Now by this active congestion of the cephalic arteries, they are necessarily enlarged, and consequently press upon the cerebral mass, which being confined within the bony cavity has little space to expand. In consequence of this it probably tends to force the brain to close, more completely, the different passages from the cavity of the cranium, and in this way impedes the free return or passage of the venous blood from the large sinuses, through their constricted passages or channels from the cavity of the cranium; and hence the congestion of the cerebral substance is greatly increased. In this way, I am satisfied, does active congestion produce apoplexy; very many cases of which have fallen under my observation, during the past few years.

But we have seen that passive congestion of the brain may cause apoplexy, very many cases of which have fallen under my observation, and I am inclined to the opinion that a larger proportion of apoplectic cases are the result of passive congestion than was formerly supposed. It is very common to see the most debilitated patients in malignant cholera dying apoplectic in a few hours from the attack, from purely passive congestion of the brain.

In cases of great debility, the blood does not circulate freely in the extremities, and in consequence more accumulates in the larger vessels near the heart, and especially in the vessels of the brain. This produces a passive congestion of the brain, and as the brain in debilitated anæmic patients can hardly be supposed to have the ordinary powers of endurance, the flow of nervous influence is perhaps more readily arrested or suspended, and thus there is a suspension of the voluntary and perhaps vital functions, according to the degree of the passive congestion, and consequent interruption of the vital force.

Thus we see that while apoplexy, probably in most cases, depends upon an active congestion of the brain, very many cases depend upon an opposite and purely passive congestion, brought about by an anæmic debilitated condition in the way I have suggested.

Treatment.—A patient in a fit of apoplexy should be taken immediately to a moderately cool apartment, into pure fresh air, and be placed with the head considerably elevated; and have everything removed from about the neck capable of preventing the free return of blood from the head. If the patient is plethoric and the congestion of the brain is of an active character, blood should be drawn from the arm, and afterwards cups applied to the back of the neck, and three or four ounces of blood taken, and repeated if necessary. This general and local bleeding will, in such cases, very much lessen the danger of meningeal inflammation, and also tend to lessen the danger of rupture of the cephalic vessels or the effusion of blood into the substance of the brain, or of blood or serum into the ventricles.

The feet should then be placed in warm water to lessen the cephalic tendency, and also cold applications made to the head, the head being kept a little elevated. If the patient can swallow, two drachms of the fluid extract of senna may be given with half an ounce of the sulphate of magnesia, and the latter repeated every six hours till a free operation is secured; or, if there is evidence of bilious derangement, ten or fifteen grains of calomel, or two or three grains of podophyllin may be given in half an ounce of castor oil, and the oil repeated every six hours till a free operation is secured.

If, however, the patient be unable to swallow readily, a drop or two of croton oil may be placed on the back part of the tongue, and repeated every two or three hours, and if it does not reach the stomach in sufficient quantity to operate upon the bowels, a strong infusion of senna should be given as an injection, with half a pint of milk and water, and perhaps half an ounce each of salt, lard and molasses, and repeated at intervals of six hours, till a free operation is secured.

In cases of apoplexy, depending upon congestion of a passive character, general bleeding should not be resorted to. The patient in such a case should be placed with the head elevated, and the feet set in warm water, and have moderately cool water applied to the head. Dry cups may be applied to the temples, and two or three ounces of blood taken by cups from the back of the neck, and dry cupping repeated at intervals during the continuance of the apoplectic state; sinapisms too should be applied to the feet and lower limbs, and continued as much of the time as the patient can bear it, without making sores.

To assist in equalizing the circulation, the spine should be rubbed morning and evening with a strong infusion of capsicum in vinegar. A cathartic of calomel or podophyllin should be administered in castor oil at first, and later, a pill of aloes given each day to secure a regular action of the bowels, and also to lessen the determination of blood to the head. Blisters to the back of the neck may be of very essential service after cupping has accomplished what it may do.

In decidedly anæmic cases in which debility has been the entire cause of the attack, in addition to what I have suggested, two or three grains of the sulphate of quinine, with or without camphor, may be given every six hours, to promote a better circulation and thereby lesssen the cephalic congestion. In apoplexy from active congestion, toast water with a little milk should be allowed; but in passive cases the patient should be sustained with a reasonable amount of nourishing food from the very first. All persons that have suffered from one attack of apoplexy, or who are in any way predisposed, should observe most rigidly all the laws of health, but especially in relation to eating, drinking, exercise, &c.

SECTION IX.—PARALYSIS.

Paralysis consists in a loss of sensation, or motion, or both in certain parts of the body; generally embracing one side of the body the lower limbs or some particular part, without coma or loss of consciousness. It may be the effects of apoplexy, but it may also come on gradually without any apoplectic attack.

Hemiplegia, or that variety which affects one side of the body, is generally the effect of an apoplectic stroke, more or less blood or serum being effused into the cavities or substance of the brain. It is probable also that paraplegia, or paralysis of the lower limbs, may be the result of serous or bloody effusion into the cavities or substance of the brain; but generally I think it is the result of injury or pressure in the spinal cord, in the dorsal or lumbar region. In local paralysis or that which occurs in different parts of the body, of small extent, pressure on the nerve or nerves which supply the part, either at their roots or at some point before they reach the paralyzed part, is most frequently the cause according to my observation.

Besides hemiplegia, paraplegia and local paralysis, we have a shaking palsy consisting in a constant tremulousness, commencing generally with the head and upper extremities, and finally extending to the whole body, the result generally of either excessive mental labor, intemperance or licentiousness in some form, or else of extreme old age. Besides we have what has been called rheumatic, hysteric, and lead palsy, but they are really only varities of local paralysis, which may conveniently be considered with local paralysis produced by other causes; the mere fact of their being produced by rheumatism, hysteria and lead, not necessarily entitling the conditions to separate names.

In hemiplegia and paraplegia, the result of serous or bloody effusion into the cavities or substance of the brain, there is usually more or less mental weakness; and the same may be true of other forms of palsy, but not necessarily.

Paralysis may be either transient or permanent, depending upon the causes, or condition upon which it depends. If it depend upon sanguineous or serous effusion within the cranium, it will generally remain, with little or no permanent improvement. But if it depends upon spinal congestion, or irritation, it may frequently be removed, or it may recover spontaneously. And, finally, if paralysis depend upon a slight general softening of the brain, or congestion of the brain, it may in some cases be restored, if the imprudence which has led to it can be corrected.

Paralysis may consist only in a loss of motion or of sensation, or it may consist in the loss of sensation and motion, depending upon the nerves involved, whether anterior or posterior, or both; the rationale of which I have given in the preceding section. The principal varieties of paralysis, then, are hemiplegia, affecting one side of the body, paraplegia, affecting the lower extremities, and paralysis partiales, or local paralysis, involving only parts of a limited extent, including paralysis from rheumatism, hysteria, and lead, as well as all other varieties of local palsy.

Local paralysis may affect only a particular muscle, or limb, or it may involve an entire organ, as the liver, kidneys, or alimentary canal, partially or entirely suspending their functions. In fact, there is not a tissue or organ in the body, but an injury of its nerves may produce a paralysis of the part, and thus impair or destroy both sensation and motion, and interrupt or suspend the function of the organ or part. With these preliminary considerations, let us proceed to consider the different varieties of paralysis, and the treatment proper for each

HEMIPLEGIA.

Hemiplegia is a paralysis affecting one half of the body, as we have already seen; depending sometimes upon an apoplectic attack, and at other times coming on slowly or suddenly, without apoplexy or any other very marked symptoms of its approach.

Symptoms.—Generally, when hemiplegia makes it appearance without an apoplectic fit, more or less of the ordinary premonitory symptoms of apoplexy precede the paralytic attack, indicating strong sanguineous congestion of the cephalic vessels, such as a flushed face, distension of the veins of the head and neck, vertigo, a sense of fullness and pain in the head, ringing in the ears, drowsiness, impaired articulation of words, or loss of speech, confusion of the mind, loss of memory, and a marked change of the habitual disposition. In some cases that come on gradually, there is slight distortion of the mouth before the hemiplegia supervenes, and by careful observation the respiration will be discovered more perfect on one side of the chest than the other.

The involuntary or vital functions are seldom very materially affected in hemiplegia; especially in that which comes on without an apoplectic attack at first. If hemiplegia be the result mainly of congestion, it will gradually pass off, as the congestion is overcome, and the natural use of the affected side will be restored. But if the hemiplegia be the result of effusion of blood or serum within the cranium, it will probably be only relieved as the congestion is overcome; and partial or entire paralysis of the side will generally continue during life. It is possible, however, that in some rare instances, by a marked change in the constitution, the clot may be partially removed or changed, so as to allow of a partial or entire restoration of the functions of the affected side.

In those cases of hemiplegia which arise from softening of a portion of the brain, there may be a decided improvement in the paralysis, if the cause of the morbid change of the brain can be removed before the disease has progressed too far. In that variety of paralysis which occurs in the insane, being in some cases hemiplegia, in others paraplegia, and in others still, general; the result, almost invariably, of softening of the gray substance of the brain, with either softening or induration of the white matter, together with chronic meningitis; there is little hope of a permanent improvement after the disease has progressed to any considerable extent. There is generally hesitancy of speech, a tremulous tongue, an unsteady gait in walking, difficulty in voiding urine, and, finally, along with paralysis there may be contraction of the muscles, especially of the extremities; involuntary discharges may occur; respiration becomes difficult, the heart acts feebly, and the patient dies, having been reduced to idiocy during a distressing illness, which may have lasted from a few weeks to several years.

Treatment.—The treatment of hemiplegia, coming on with apoplexy, has already been given with the treatment of apoplexy, but we have yet to consider the treatment of hemiplegia not produced by an apoplectic fit, and that will also apply to those cases which follow apoplexy.

During a paralytic shock, the patient should be allowed good air, in a moderately cool apartment, and have the head well elevated and kept cool, and the feet placed in warm water. If the patient is plethoric, or of a sthenic constitution, blood should be drawn from the arm, and afterwards a few ounces of blood taken from the back of the neck by cups. If, however, the patient be feeble, blood should only be taken by cups; and in very debilitated cases, only dry cups should be applied.

A cathartic of calomel or podophyllin should be administered at first, and the bowels kept moderately loose in plethoric patients, by the neutral salts, but in feeble patients, by a pill of aloes, taken at evening each day. To lessen the cephalic tendency, and promote perspiration, the James's powder may be given, in three or four grain doses, every six hours, for the first few days.

Dry cups, blisters, or pustulations, with tartar emetic ointment, should be kept up to the back of the neck, during the first few days or weeks, if it continue so long. If, however, after continuing this treatment for a reasonable time, the condition of the paralyzed side continues stationary, we have a right to infer that it is the result of effusion within the cranium; further medical treatment should be abandoned, and the case left to the efforts of nature, the patient being directed to observe most rigidly, the laws of health, and especially to avoid all those causes which tend to produce cephalic congestion.

In the paralysis occurring with insanity, whether it be hemiplegia, paraplegia or general, the cause of the disease should be sought out and removed, if possible, and then the case treated on general principles, while any grounds of hope remain in the case. Generally in such cases, gentle cathartics, or laxatives, a regulated diet, tonics of iron, and perhaps strychnine, good air and quiet, together with blisters back of the ears, or to the back of the neck, constitute our main reliance. If this treatment be resorted to early, a recovery is possible; but if the case has progressed to any considerable extent, the patient will ultimately die, in most cases at least.

PARAPLEGIA.

By paraplegia, I mean paralysis affecting the lower limbs, and lower part of the body. Paraplegia may sometimes be the result of an affection of the brain, as of softening, or effusion of blood or serum, but generally it arises from some injury, irritation or congestion of the spine, or else from effusion of blood or serum at some point along the dorsal or lumbar portion of the spine.

Symptoms.—Paraplegia generally comes on gradually, unless it be the result of direct mechanical compression of the spinal cord. At first the patient, after having suffered from slight uneasiness of the back for a time, feels an awkwardness in the motion of the lower extremities. Early this is slight, and experienced mostly on first rising to walk, or when the patient becomes very much fatigued. This awkwardness, however, increases, till the patient is unable to walk without the aid of a cane.

As the disease advances, the bladder becomes more or less paralyzed, so that the urine is expelled with difficulty, the stream becoming more and more feeble, till at length it dribbles away involuntarily. The bowels, too, are more or less paralyzed, and consequently very much constipated, frequently being entirely inactive, without the use of a stimulating cathartic or injection. If now, in such a case, the kidneys become paralyzed, as happened in one case that fell under my observation, the secretion of urine is suspended, and the patient very soon dies.

In paraplegia, the power of motion in the lower part of the body and limbs may be lost, and the sensation remain nearly natural. In other cases, however, sensation and motion are both lost at the same time, or the one first, and then gradually the other.

Causes.—The causes of paraplegia are various. In those cases that depend upon cephalic derangement, such as congestion of the brain, the effusion of blood or serum, or softening of the brain, if it is not the effects of an apoplectic stroke, it is brought on by the same causes which produce apoplexy; the immediate cause of the paraplegia being pressure upon some portion of the brain, or a morbid change of its structure, such as softening, induration, &c.

Any direct mechanical injury of the spinal cord, by which its substance becomes congested, or compressed, may produce paraplegia; such as blows, fracturing or depressing the spinous processes of the vertebræ, &c. Or an injury of the spine may, by producing irritation of the cord, cause so much congestion of its blood vessels, as to interrupt the flow of nervous influence, and thus produce paraplegia. An inflammation of the spinal cord, or its meninges, may terminate in softening of its substance, or in effusion of serum, &c., into the arachnoid cavity, or elsewhere, which may produce paraplegia, by interrupting the flow of the nervous influence.

Again, a poisonous agent, as tobacco, may, by debilitating the brain and nervous system, so far impair the functions of the brain, that there will not be a sufficient amount of nervous influence generated in the brain to supply the nervous system, and what is generated will not receive sufficient vital force to be sent to the lower extremities, and consequently paraplegia will be the result.

Many cases of this character have fallen under my observation during the past few years. Dr. Harvey Jewett, an eminent physician of Canandaigua, in this State, has also noticed, it appears, the debilitating effects of tobacco upon the lower portion of the spinal cord, several interesting cases of which he has reported in the Buffalo Medical Journal, for May, 1853.*

In one marked case of this character that fell under my observation during the summer of 1853, there was at first weakness and tremor of the lower limbs, then paralysis of the bladder and lower portion of the alimentary canal, and finally of the kidneys, and on the suspension of their functions the patient died, evidently the result of the excessive and protracted use of tobacco.

Masturbation or venereal excesses may produce paraplegia in very nearly the same manner as tobacco, by impairing the vital functions of the brain, and also impairing the ability of the brain to produce a proper distribution of the nervous influence to remote parts, and especially to the lower extremities, in consequence of which paraplegia is the result. These causes may also produce paraplegia, in part at least, by the local irritation which is set up, and perhaps softening at the origin of the spermatic nerves, near the junction of the dorsal and lumbar regions of the spine.

I am satisfied that tobacco, licentiousness, and drunkenness, with various other excesses and abuses of the system, are the most frequent causes of that peculiar tremulous condition, either partial or general, which has been called shaking palsy. It may, however, occur from age, and general debility, from various depressing agents, accidentally brought to bear upon the system. In conclusion then, I am compelled to believe, from careful observation, that paraplegia occurs most frequently in children, and young persons from masturbation; in middle aged persons from venereal excesses, onanism and various other abuses, while in advanced life, it is frequently the result of tobacco, and perhaps in some cases of intoxicating liquors.

Prognosis.—The prognosis in all those cases of paraplegia which arise from congestion and irritation only of the spinal cord, may be regarded as favorable, while all those cases which are the result of sanguineous or serous effusion, either into the brain or along the spinal cord, may be regarded as unfavorable.

In those cases of paraplegia produced by masturbation, sexual excesses, tobacco, drunkenness, and other abuses of the system, the prognosis is generally unfavorable; but cases may recover, if the cause can be removed at an early stage of the disease, and the patient can be induced to conform strictly to the laws of health.

Treatment.—In those cases of paraplegia, depending upon a cephalic cause, the same treatment is proper that I have suggested in the treatment of hemiplegia, and it will generally be applied with only a partial

* See Buffalo Medical Journal. Vol. 8, No. xii., p. 721.

relief to the patient. Sometimes, however, a more favorable result follows. In those cases too, which depend upon the continued use of tobacco, and other abuses of the system, and especially sexual abuse, nothing very favorable can generally be accomplished, except to remove the cause. In some cases, however, of this kind, in young patients, by enjoining the most strict and rigid observance of the laws of health, and pursuing a tonic course of treatment, the patient will sometimes recover, after some of the most unpromising symptoms have occurred.

Such cases are generally attended with great debility, both of the blood and nervous, or cerebro-spinal system. Iron, and generally the citrate, in small doses, continued for a long time, will do all for the blood that can be reasonably expected in such cases. As a tonic for the cerebro-spinal and nervous system, strychnia in small doses, continued for some time, will generally do well. A grain of the muriate of strychnia may be dissolved in six or eight ounces of water, and a teaspoonful given three times per day, before each meal, and continued for a long time, in cases in which there is no marked local irritation of the brain or spinal cord.

In those cases of paraplegia, depending upon irritation, with more or less congestion of the spine, in the dorsal, or lumbar region ; cups at first, wet or dry, on each side of the spine, and then blisters, or pustulation, with tartar emetic ointment, continued for a reasonable time, will sometimes remove the disease ; one terrible case of which, I succeeded in removing, after nearly the whole system had become partially paralyzed. In this case, the child had a fall on the door-sill, which injured the middle portion of the spine, and produced a passive inflammation, which extending to the brain, so far interrupted its function, as to produce decided idiotic symptoms.

After having remained in this deplorable state for about twelve months, with the whole body partially paralyzed, this apparently hopeless case was cured by blistering, successively, the whole length of the spine, by a little at a time, and giving for several months, the citrate of iron in two grain doses, three times per day, after each meal. I succeeded in regulating the bowels in this case, with a solution consisting of rhubarb, and the sulphate of magnesia, of each four drachms, in eight ounces of water ; a teaspoonful of which was given every morning, till the constipation was overcome. Thus in cases of this kind, by the aid of cups, blisters, tonics, and laxatives, success may attend in apparently hopeless cases.

In cases of paraplegia depending upon effusion into the arachnoid cavity of the spine, I believe there is no great certainty of relief from medical treatment ; one fatal case of which fell under my care a few years since, terminating probably in softening of the cord itself. The only reasonable hope of relief in such cases, is from the long continued use of the iodide of potassium, internally, and the application of blisters at first along each side of the spine, and then the continued application to the spine of iodine ointment, with a hope of effecting absorption of the effused fluid, which may possibly sometimes be accomplished in this way.

In that peculiar tremulous condition of the system, which has been called shaking palsy, the cause should be sought out, and, if possible removed, and then by a prudent observance of the laws of health, proper exercise, and tonics for the blood and nervous system, such as iron, strychnia, &c., an improvement may follow in some cases; but this is by no means certain. Such then are the principles which are to guide us in the treatment of paraplegia, which having considered, let us proceed to the consideration of partial paralysis.

PARALYSIS PARTIALIS.

By partial or local paralysis, I mean that variety in which the morbid condition extends only to a limited extent. It may be of sensation, or motion, or both; and it may involve a single muscle, a whole limb, or any organ of the body. When it involves a whole limb, as a leg or arm, it generally depends upon an irritation, congestion, or some other morbid condition of the corresponding side of the spine, at the point where the nerves come off to supply the limb.

If the paralysis extend only to a particular muscle, or set of muscles, as of the face, it is generally from pressure on a nerve at some point away from the brain or spine; often in its passage through a foramen in a bone, by which the flow of nervous influence is interrupted, and a partial, or local paralysis is produced. The most frequent seat of local, or partial paralysis, involving only a small part of the body, is that which effects one limb, one arm, or one side of the face; but no tissue or organ of the body, is exempt from this variety of paralysis.

Causes.—The immediate cause of paralysis of a lower limb is generally, as we have seen, pressure on the corresponding side of the spine, or spinal nerves, in the lumbar or sacral region. The cause of paralysis of one arm is pressure on the corresponding side of the spine at the point between the shoulders, where the brachial plexus of nerves are given off. While paralysis of the muscles of one side of the face, generally depends upon pressure on the facial nerve, in its passage through the stylomastoid foramen, as it emerges to be distributed to the muscles of the face.

As to the remote causes, the same operate, both of a local and general character, that I have mentioned as producing paraplegia; such as injuries, with serous or other effusions along the spinal cord, or softening from inflammation, &c. Rheumatism, hysteria, and some other diseases, occasionally appear to act as causes of local paralysis, or at least are attended with paralysis; generally, however, of a transient character. Local paralysis occurring in rheumatism, or in rheumatic conditions, probably depends upon thickening of the nervous sheaths, if it be of only a limited extent; but when paraplegia occurs in such cases, it depends upon congestion of the spinal cord, or else thickening of its meninges, passing on to an extent sufficient to produce paralysis.

In cases of local paralysis, caused by hysteria, or occurring in hysteric patients, there is generally diminished sensibility, with temporary loss of voluntary motion, depending, I am inclined to believe, in part upon cerebro-spinal congestion, and also perhaps in part upon a bad distribution of the nervous influence, which is apt to be a prevailing tendency in hysteric conditions. Finally, it is probable that lead, in its various forms, is a very frequent cause of local paralysis; affecting generally, at first, the extensor muscles of the hands, and perhaps extending more or less

to all the muscles, which become flabby, the whole surface often presenting a pale leaden hue.

In some cases, however, I have known lead to produce paraplegia, and
in other cases still paralysis of one limb; in every case falling under my
observation, being attended with neuralgic pains, of a more or less
tedious character, either in the affected limb or in other parts of the
body. The lead probably acts in such cases not only upon the brain
and spinal cord, but also upon the tissues of the muscle, or muscles, involved in the paralysis.

Treatment.—In the treatment of partial or local paralysis, cups,
blisters, and other counter-irritants, over the seat of the local irritation
or congestion, may be indicated, to take off the pressure on the nerve or
nerves involved. After the local irritation or congestion is subdued, and
the undue pressure on the nerves removed, if the paralysis continue, the
muriate of strychnia should be given in moderate doses till the paralyzed
part is restored.

In local or general paralysis, depending upon lead, I have generally
succeeded with the iodide of potassium, in five grain doses three times
per day, continued for several weeks, and the application of iodine ointment over the seat of local irritation, if any exist, along the course of the
spine.

SECTION X.—EPILEPSY.

By epilepsy, from the Greek επιλαμβανω, "I seize upon," I mean a disease of the nervous system, manifested by convulsions, recurring at irregular periods in paroxysms, accompanied by a temporary loss of consciousness, sense, and voluntary motion, and usually terminating in
somnolency.

The brain appears to be the special seat of epilepsy, and it may be
either organic or functional, or symptomatic of irritation or derangement in other parts, as of the stomach, bowels, genital organs, &c., but
the whole nervous system is always more or less involved in this disease.

Symptoms.—The epileptic attack may come on suddenly, without any
symptoms of its approach. Generally, however, certain symptoms precede the occurrence of the paroxysm, such as a disturbed or distressed
feeling in the head, confused state of the mind, giddiness, dimness of
sight, ringing in the ears, flashes of light before the eyes, distention of
the veins of the neck, an anxious feeling in the precordial region, starting during sleep, loss of distinct articulation, temporary deafness and
drowsiness.

Just before an attack, some patients are gloomy or peevish, while
others are exceedingly lively and cheerful; depending, no doubt, upon
the effects which a rush of blood to the head produces in different individuals. For, whatever may be the remote cause of epilepsy, the immediate cause of the epileptic fit is a rush of blood to the brain; which,
suspending the generation of nervous influence, or interrupting its distribution, produces, doubtless, the peculiar sensation which many epileptics experience, of cool air upon the body, beginning often in the lower
limbs and passing to the head, as the flow of nervous influence is gradually withdrawn or withheld.

After the premonitory symptoms have continued for an indefinite time, when they occur, with or without the sensation referred to above, the *aura epileptica*, the epileptic seizure occurs: when, if the patient be sitting or standing, he suddenly falls down in a state of insensibility, and immediately becomes **convulsed** more or less violently. In many instances the convulsive action of the muscles, especially of the face, are frightfully violent. The whole frame, too, is frightfully **agitated**; the eyes roll about, the lips and eyelids are convulsed, **the tongue is** spasmodically thrust from the mouth, which, with the foaming **of the mouth**, gives the countenance a very wild appearance.

In some instances the teeth are firmly pressed **together, at other** times the jaws are widely distended; the thumbs are generally firmly pressed in upon the palms of the hands, but this is not invariably the **case.** The spasms are generally of the clonic kind, but sometimes the **muscles** remain for a time rigidly contracted, the body being bent either backward or to one side, as in tetanus. The face is occasionally pale, but generally livid, with a turgid state of the veins of the neck. The heart usually palpitates **tumultuously**; the pulse is irregular, contracted and weak, and the **respiration difficult and in violent** cases sonorous.

After an **indefinite period, varying** generally from five to thirty minutes, either **suddenly or gradually, these** spasmodic symptoms abate; the respiration becomes full, and more regular; the countenance **more composed, and the patient finally** falls into a state of stupor or deep sleep, out of which he wakes with a feeling of languor, confusion and torpor of mind, which state may continue declining for several hours, and in some cases for even days. The countenance generally exhibits a more or less vacant or stupid expression, and the eyes are apt to be dull and wandering. During the somnolent state, the patient generally perspires freely, especially about the head, neck, and breast; sometimes, however, the perspiration is confined to one side.

In mild cases of epilepsy the attack comes on suddenly, and after a few minutes of partial convulsions of the muscles of the face and neck, is quickly subdued, and the patient becomes conscious. The average duration of an epileptic paroxysm is probably about fifteen minutes, but it may continue for several hours. Usually but one fit occurs at a time, but occasionally several succeed each other in rapid succession, with only short intervals between; a remarkable instance of which fell under my observation a few years since.

The interval between the paroxysms varies from a few hours to several days, weeks or even months, generally growing more frequent the longer the disease has continued. In some cases the paroxysms occur with great regularity for a long time, and this, I believe, is most apt to be the case in females, if there is suppression of the mensis. In such cases, I have known the paroxysms to occur once in four weeks, or thereabouts, for a great length of time. In most cases, however, the paroxysms come on at irregular intervals, varying from a few days to several weeks, depending, in some degree, upon the state of the atmosphere, and very much upon the accidental occupation of the patient, and the degree of physical and mental excitement.

In some rare instances, epilepsy terminates spontaneously, but generally the fits continue to occur more frequently during the life of the patient, or it may terminate in apoplexy and death, as happened in one case under my care about two years since.

In those cases that commence early in life, and continue for many years, there is often more or less mental weakness exhibited, approaching sometimes to idiocy, several marked cases of which have fallen under my observation, one of which is under my care at the present time.

It has been thought by some that in those cases in which idiotic symptoms make their appearance, there is no hope of improvement or permanent recovery; but this is not invariably the case, for in one patient in which the disease commenced in early childhood, and continued on to the twelfth year, with a decidedly idiotic expression of the countenance, I succeeded in entirely arresting the disease, and the boy, now grown to be a man, is smart, bright and intelligent, and enjoying good health. Generally, however, in those cases which exhibit much mental weakness, the prospect of recovery, or even permanent improvement, I believe is very slight.

It should be remembered, however, that while the tendency of epilepsy is generally towards imbecility, that it is apparently not invariably so; some patients passing on even to advanced age without any material apparent impairment of the mental faculties, as was the case with Cæsar, Mahomet, and Napoleon, and as has been noticed in other epileptics of less distinction.

Anatomical Characters.—On the examination of the brain of those who have died of epilepsy, during a paroxysm, the brain and its meninges are found greatly congested, the cineritious substance being of a deep red, and the white substance of a more or less reddish appearance.

This appearance, however, is probably the result of the recent congestion, as there is generally no morbid lesion indicative of permanent disease anterior to the fatal congestion. This is not always true, however, in very protracted cases, in which there has been marked mental weakness, paralytic symptoms, &c. In these cases there may be found marks of meningeal inflammation, more or less induration or softening of the cerebrum and cerebellum, a general injection of the cerebral vessels, thickening and adhesion of the membranes to the substance of the brain, and more or less effusion, of blood or serum, into the arachnoid cavity, or into the ventricles of the brain; the result, probably, rather than the cause of the protracted epileptic disease.

The cerebellum is often found in epileptics of a softer consistence than in health, and there is frequently some thickening or internal projection of the cranium, the result of external violence, which has evidently produced the disease; one interesting case of which I saw in Philadelphia, during the winter of 1845. This patient had received an injury, some years previous, which had depressed the skull, I believe near the lambdoidal suture, a little upon one side of the head; the symptoms following the depression not being very great, the bone was not disturbed, and the patient became epileptic, the paroxysms occurring quite regularly once in about three weeks. Professor Mütter removed the depressed portion of the bone, and there was, I believe, no return of the epileptic paroxyms.

It is probable that tubercles, abscesses, and thickening of the membranes of the brain, which are sometimes found in epileptics, may have preceded, and in some cases produced, the epileptic disease.

Causes.—There appears to be a hereditary predisposition to epilepsy in some individuals and families, but in what that predisposition consists, is not quite certain. But as the immediate cause of the epileptic paroxysm is pressure of blood upon the brain, and apparently more especially of the cerebellum, it appears to me quite possible that among the predisposing physical conditions, a preternatural development of the vertebral arteries, by which they convey an undue amount of blood to the cerebellum, is one of not the least importance.

Hereditary peculiarity of organization, by which the tissues, and especially the brain, are rendered more soft, and consequently more susceptible to slight impressions, and especially to a congested state of the cephalic vessels, may predispose to epilepsy. This appears the more probable, when we remember that children are more liable to epilepsy than adults, in whom the brain has acquired more firmness and power of endurance. It appears that more cases of epilepsy occur at or before puberty than subsequently to that period. And it is probable that the nervous irritability and cephalic tendency, which occurs in many at about the age of puberty, may at least strongly predispose to epilepsy, if it does not act as an exciting cause of the disease.

It is probable that the habit of masturbation so common in both sexes about that age, does much to render irritable the nervous system, and to change more or less the substance of the brain itself, and thus acts as a strong predisposing or exciting cause of epilepsy.

Irregular eating, which is so common a habit with children, by producing dyspepsia and gastro-enteritis, and affecting the brain sympathetically, very often acts as a cause of epilepsy in those who are, or are not, hereditarily predisposed. In one case of epilepsy, in Jefferson county, N. Y., beginning in a child at about three years of age and continuing to the twelfth year, going on constantly from bad to worse, so that at last he averaged about three fits per night, I succeeded in permanently arresting the disease with very mild measures, after I had corrected his habit of irregular eating.

In this case the boy from early childhood had been in the habit of taking food at all hours of the day, and eating but very little at meal time; which habit had so far impaired his digestion, and produced and kept up gastric irritation, affecting the brain sympathetically, that it had not only produced, but kept up the epileptic paroxysms; and no measures had any influence in arresting the disease, till this habit of irregular eating was corrected.

Another case, a girl twelve years old, residing in Madison county of this state, fell under my care about four years since; in which irregular eating was probably the cause of the disease. In this case I corrected this habit of irregular eating; and with very mild measures succeeded in effecting a cure in about twelve months, though she took tonics for about six months after the epileptic paroxysms were arrested. Other cases of epilepsy have fallen under my observation, in which I am confident that irregular eating had been either the predisposing or exciting cause of the disease; one of which, of an epileptic character, I now distinctly remember from the state of Michigan.

Among the exciting causes of epilepsy, some act directly upon the brain, while others act on distant parts, as the stomach, bowels, liver, kidneys, genital organs, &c., the impression being conducted to the brain through the medium of the nerves. The causes which act directly upon the brain, are, as we have seen, depression of some portion of the bones of the cranium; thickening or tubercles of the membranes of the brain, tumors or some morbid change in the cerebral substance, and bloody or serous effusion into the ventricles of the brain. The disease is said to have been produced directly by seeing patients in the epileptic paroxysms, and also by persons feigning the disease.

Among the exciting causes of epilepsy, which affect the brain sympathetically, are indigestion, intestinal worms, suppression of the menses, the repulsion or sudden drying up of eruptions on the scalp, the habitual use of tobacco and intoxicating liquors, irritation from biliary and urinary calculi, atmospheric vicissitudes; and in fact any cause capable of producing irritability of the nervous system, and a determination of blood to the brain, and especially to the cerebellum, may be either direct or indirect causes of epilepsy.

Pathology.—I am satisfied that epilepsy consists in a morbid excitability of the whole nervous system, including, of course, the brain; and that each paroxysm consists in an increased excitement, or irritation of the whole nervous system, but especially of the brain itself, in consequence of which it becomes congested, and hence all the symptoms of the disease are developed. The cause of this morbid excitability of the nervous system may be, as we have seen, in the brain, or it may be in some remote part; only a slight increase of the excitability to an irritation being sufficient, in either case, to produce enough cephalic congestion to bring on a paroxysm, more or less violent.

Immediately before the epileptic attack there is vascular turgescence of the brain, including the cerebellum, but not sufficient to produce apoplexy; hence we have a suspension of the mental functions, as sensation, perception, consciousness, &c., while the motor functions are deranged, but not abolished, as is the case in apoplexy. Thus it is probable that we may have epilepsy without any organic derangement of the brain, its functions only being deranged by the irritation and congestion alone, caused, as we have seen, by a general irritable condition of the whole nervous system, produced by causes acting upon parts remote from the brain.

But while epilepsy may be produced in this way, it should be remembered that organic disease of the brain, or its meninges, such as inflammation, tumors, thickened membranes, exostosis, depressed bone, serous or bloody effusion, &c., may produce sufficient derangement, in the cortical substance, to suspend the mental functions, and also sufficient in the white to derange the motor function, and thus epilepsy may be produced by organic disease of the brain or its meninges.

Now that epilepsy should be paroxysmal is not strange, whether it be produced by organic or functional disease of the brain, for nearly or quite the same degree of excitability of the brain may exist in either case; only requiring this excitability to be increased to irritation, by some exciting cause, to bring on sufficient congestion to develop a paroxysm. Slight accidental causes may develop a paroxysm, if the morbid excitability of the brain be considerable; but if the morbid excitability of the brain be slight, a more considerable exciting cause is required, and so the paroxysms are liable to occur less frequently.

Thus it is that the paroxysms come on so irregularly, except in cases in which the exciting causes occur at regular periods, as is the case in females laboring under menstrual irregularities, the paroxysms coming on, in such cases, according to my observation, when there is an effort in the system to bring on the menstrual discharge, when, failing in this, the brain becomes irritated and congested, and an epileptic paroxysm is the result. Thus we have, in my opinion, the true pathology of this perplexing, and most distressing paroxysmal disease; all the symptoms and peculiarities of which may thus be accounted for.

Diagnosis.—Epilepsy may be distinguished from apoplexy, hysteria, and various accidental convulsons. by the following differences. In epilepsy the muscles of the face are contracted, as well as those of other parts of the body, while in apoplexy the muscles are relaxed and motionless. During the latter stage of an epileptic fit, when coma has supervened, the face is pale, the pulse is feeble, and there is foaming at the mouth; while in apoplexy the face may be flushed, the pulse full and strong, and, in addition, there is generally stertorous breathing, which is not the case in epilepsy.

In hysteria the face is not so much distorted, there is not the foaming at the mouth, and the fit does not terminate in a heavy sleep, as is the case in epilepsy. In hysteria, too, there is more generally involuntary laughing or weeping, and, generally, a continuation of some degree of consciousness, which is not the case in genuine epilepsy.

In accidental convulsions, occurring from dentition, intestinal irritation, from the puerperal state, &c., there is less frothing at the mouth, less stertorous breathing, and a more sudden return of mental activity, after the cessation of the convulsions, than is the case in epilepsy.

Feigned cases differ from the real disease, in every essential particular; so that in order to detect feigned cases, it is only necessary to notice all the symptoms which are developed, and to take into account all the extrinsic circumstances, to avoid deception, and imposition in such cases.

Prognosis.—The prognosis is unfavorable in all those cases in which there is organic disease of the brain, or its investing membranes, or bones; at least so far as a permanent recovery is concerned. But in those cases of epilepsy, in which the cause of the general nervous excitability is in some part of the system remote from the brain, the brain being irritated sympathetically, there may be reasonable hope of a permanent recovery, if the imprudence which has produced this irritable condition can be removed, or corrected, and the patient subjected to a judicious treatment.

It has been supposed by some, that those cases of epilepsy which occur in early childhood, are necessarily exceedingly obstinate, and seldom if ever yield to remedial measures; but such has not been the result of my observation in epileptics; the only cases permanently cured, occurring in my practice, having been under fifteen years of age. As a general rule, the prognosis is rendered more unfavorable the longer the disease

has continued, and especially if decided paralytic, or idiotic symptoms
have supervened.

Treatment.—During an epileptic paroxysm, the patient should be
placed upon a bed, with the head a little elevated, and all reasonable
measures taken to prevent the patient from injuring himself, and others,
till the paroxysm passes off. If apoplectic symptoms occur, blood should
be drawn from the arm, or taken by cups from the temples, or back of
the neck, and cold applications made to the head, with sinapisms to the
extremities, warm foot baths, &c.

In very debilitated patients, if the pulse become feeble during the
paroxysm, and there appears to be danger from apnœa, ammonia, or
camphor should be held near the nostrils, sinapisms applied to the extremities, and the whole length of the spine rubbed with a strong infusion
of capsicum in vinegar, applied a little warm. While then it is possible
that circumstances may occur, during the paroxysm of epilepsy, rendering some such interference as I have suggested necessary; it should be
remembered that symptoms very rarely arise, which render any interference necessary, except to prevent the patient from injuring himself and
others.

The indications in the treatment during the intervals, for the purpose
of permanently eradicating the epileptic disease, are generally very plain
if the real condition of the patient can be ascertained, and also the cause,
or causes which have been operating to produce it. In all those cases in
which the primary irritation is about the brain, either intrusion of bone
upon the brain, thickening, or tubercles of its meninges, or some morbid
change of the brain itself, or effusion into its ventricles; the exact condition should be ascertained as near as may be, and also the causes which
have led to it.

The cause in every case, which may have been operating to produce
the cephalic irritation should be removed, whether it be the result of an
accident, as depression of bone, or of imprudence of any kind, such as
the use of tobacco, intoxicating liquors, masturbation, excessive venery,
&c., which have produced a morbid change in the brain, without necessarily having first produced local disease in other parts of the system.

If the epilepsy has been produced by depression of a portion of bone,
it should be removed with the trephine, which will give the patient a
tolerable chance of a permanent cure.

If there is evidence of inflammation, or thickening of the membranes
of the brain, or spinal cord; cups, wet or dry, blisters, or pustulation
with tartar emetic ointment should be persevered in, to the back of the
neck, or along the spine, according to the seat of the disease, till it is removed as far as may be. This course of counter-irritation, with an
occasional saline cathartic, a plain diet taken with regularity, and a strict
avoidence of every cause capable of producing, or increasing the general
excitability of the nervous system, and especially irritation of the brain,
will generally best fulfill the indications in epilepsy, from cerebral, or
spinal meningitis.

In scrofulous cases, in which there is danger of tuberculous deposits,
or of serous effusion into the arachnoid cavity, or into the ventricles of
the brain; in addition to what I have suggested, the iodide of potassium

or the syrup of the iodide of iron should be given, in moderate doses, and continued for a long time, with, or without cod liver oil. This course of treatment in epilepsy, from such conditions, may not effect a cure : **but** I am satisfied that it is clearly indicated, and that it may palliate **the** disease, and perhaps prevent an early fatal termination.

In cases in which there is evidence of **softening** of the brain, from masturbation, excessive venery, &c., the patient should be prohibited from further abuse of himself; should avoid all **manner of** excitement; **should be placed upon** a plain, digestible, and **nourishing** diet, to be **taken with** regularity, and should take for a long time some preparation **of** iron for the blood, and the oxide of zinc as **a tonic** for the nervous **system.** In such cases, three or four grains **of the carbonate of iron,** with an equal quantity of the oxide of zinc, **may be given three times** per day, after each meal, and continued for as **long a time as the patient** continues to improve **from its use.**

In the treatment of cases of epilepsy in which the **local** irritation **is** situated at some point distant from the brain, the brain being **irritated** sympathetically, the remote causes should be sought out and **removed,** and then the local irritation or derangements treated on general **princi-ples, whether it be in the liver, kidneys, womb,** or alimentary canal. **If the local irritation or derangement, of which** the cephalic irritation is **symptomatic, can be removed in such cases, the** sympathetic irritation **of the brain will generally subside, and hence the** epileptic paroxysms **will no longer occur.**

In cases of epilepsy in which the **cephalic irritation or congestion is symptomatic of disease of the liver or kidneys, of an** inflammatory or **functional character, the treatment should be directed to** the affected organ, **and counter-irritants, alteratives, diuretics, &c.,** used according to the nature of the morbid condition, **till** the healthy **hepatic or** renal functions be restored. When this is accomplished, the patient may require gentle vegetable or mineral tonics for a few weeks, **and** perhaps months, to correct the fluids and solid tissues of **the** body, **and** to break up the tendency to the disease, which the epileptic paroxysms have **pro-duced.**

If **vegetable tonics** be required, the fluid **extracts of cinchona or of** colombo will generally do best. From thirty to sixty drops of the fluid extracts of either may be given in such cases, before each meal; and if the blood and nervous system require tonics, from three **to five grains** each of the carbonate of iron and oxide of zinc may be given **three times** per day, after each **meal,** and continued **as** long as the patient improves under their use. With such **a** plan of treatment, a regulated diet, **and** a careful avoidance of **all** the causes calculated to perpetuate **the disease,** many epileptics may be greatly benefited, and some cases permanently cured.

In cases **in which the womb is the** primary seat of irritation, **with** menstrual irregularities, the treatment should be directed to **that organ,** and **by a** judicious treatment, adapted to each particular case, the **irrita-**tion subdued, and the menstrual secretion restored. When this is accomplished, if the epileptic disease continues, a tonic course of treat-ment similar to that suggested above, which shall exactly fulfill the indications which arise, should be continued while the patient continues to improve.

If the alimentary canal be the seat of primary irritation, it may be from indigestion or from verminous irritation. If from the latter, anthelmintics suited to the case should be administered, and the offending cause removed, after which the case should be treated according to the principles I have already laid down, with vegetable or mineral tonics, or both, according to the indications which arise.

If the gastro-intestinal irritation be from indigestion, it is generally, according to my observation, the result of imprudence in eating and drinking; and if the patient be a child, from eating between meals, and perhaps if an adult. In such cases the patient should be allowed a plain, digestible and nourishing diet, to be taken invariably at regular meal hours, and on no account anything between meals. To improve the digestion, allay the gastro-intestinal irritation, and to improve the blood and nervous system, a pill composed of rhubarb, extract of conium, carbonate of iron and oxide of zinc, of each one grain, with a little ginger, may be given three times per day, after each meal, and continued till a permanent cure is effected. With this pill alone, and a regulated diet, as well as regular hours of taking food, I have succeeded in permanently curing two epileptics of this character, in one of which the fits had continued for ten years, and in the other, a little less than that time.

Thus it may be seen, we have the principles which are to guide us in the treatment of epilepsy, the grounds for every remedial measure being purely rational, thus leaving no occasion for empirical remedies, as there really is none in any disease, with which the human family are afflicted.

There is no disease with which I am acquainted, in which so much can be done by way of correcting various imprudences, such as irregular eating and drinking, masturbation, excessive venery, the use of tobacco, drunkenness, licentiousness, &c., as in epilepsy. In fact, no prescription should ever be made in this or any other disease, without first getting at the imprudence which has led to it, and then demanding and enforcing its abandonment, as well as a strict observance of the laws of health and propriety. By taking this course, we go to work for the patient with his co-operation and assistance, but neglecting this essential preliminary measure, we may prescribe with the wisdom of an Esculapius, but the counteracting influence of the patient's imprudence will foil all our efforts, and be sure in the end to bring down upon our guilty heads the deserved anathemas of our patients and their friends.

SECTION XI.—CATALEPSY.

By catalepsy, from the Greek καταληψις, "a seizure," I mean that peculiar disease of the nervous system characterized by a temporary suspension of consciousness, sensorial power, and volition; the body remaining in the precise position it was when the attack came on, without coma, muscular rigidity, or spasm; the involuntary functions being carried on with little or no interruption. Catalepsy is a paroxysmal disease, each attack being the immediate effect of irritation and congestion of the brain, sufficient to suspend, for the time, the influence of the mind upon the body, through the medium of the brain and nervous system, but not sufficient to interrupt the generation and flow of nervous influence, by which the voluntary **muscles receive their** tone.

Hence in catalepsy the voluntary **muscles** are not paralyzed as in apoplexy, or convulsed as in epilepsy; **but being** shut out from the influence of the mind **or will,** and receiving **at least** a partial supply of **nervous influence, their tone is** retained, and **they** consequently remain during a paroxysm in the precise position **they were at the** time of attack, unless they are moved by some other **person, and then the** part so moved remains in the new position in which **it is placed.**

Symptoms.—The attack very generally comes on without any warning of its approach, but sometimes slight premonitory **symptoms occur,** such as vertigo, cephalalgia, a flushed face, forgetfulness, pain in the bowels, yawning, depressed spirits, and sometimes other symptoms indicative of slight cephalic congestion. When an attack occurs, **every part of the** body remains in precisely **the same position** it **was at the moment of seizure.** If the attack **comes on while the patient is in the act of doing** any thing, as eating or drinking, the hand remains in the precise position it was in till the paroxysm passes over.

In a case of a boy, nine years old, that came under my care a few years since, from Lewis county, N. Y., I have known the paroxysm to come on at meal time, and while the patient was in the act of passing food to the mouth, the mouth also being open. Not a motion would be made, either with the hand or mouth, till the paroxysm passed off, when the food would be carried **to the** mouth, as though nothing had happened. On another occasion he ran up a board, one end of which lay on the top of a fence, and when at the height, being seized with a paroxysm, he remained poised upon the extreme end of the board over the fence, till the paroxysm passed over, when he jumped to the ground as though nothing had happened.

The eyes, during a **paroxysm, are** generally **open, fixed, and slightly turned up,** but sometimes they are spasmodically **closed.** The extraordinary peculiarity of this affection consists in the tendency of a limb or part to remain in the precise position in which it is placed, **even though** it be the most awkward, as that of a hand high above the **head. In** complete catalepsy the sensorial functions **are** entirely **suspended, and** the patient on recovering, remembers nothing of his own **sensations, or** of what has been doing about him during the paroxysm.

If a cataleptic attack comes on while the patient is conversing, and during or interrupting a half pronounced word, as soon as the paroxysm is over, **the word** will generally be finished, and the conversation continued as though **nothing had** happened. In very mild cases, a slight degree of sensorial power remains, and the patient retains an **indistinct** recollection of what had taken place during the paroxysm; but **in very** severe cases, there is not only a suspension of the sensorial power, **but** the involuntary functions of respiration and circulation are partially interrupted or quite imperfectly performed.

The duration of a cataleptic paroxysm varies from a few minutes **to** several days, but in the majority of cases, it does not exceed two minutes, at least such has been the result of my observation. There is generally no regularity in the return of the cataleptic attacks. **They may** come on several times in a day, or only once in several days, weeks, or even months; but I think in most cases they recur either every day or every few days.

Severe cases of catalepsy sometimes terminate in epilepsy, an increase of the cephalic congestion not only interrupting the mental functions, but also so far interrupting or deranging the generation, flow, and distribution of nervous influence, as to produce convulsions of the voluntary muscles, as an increase of the cephalic congestion in epilepsy may produce apoplexy.

Diagnosis.—The diagnosis of catalepsy is attended with little or no difficulty, as there is no other disease in which the different parts of the body retain their exact position during an unconscious state.

Causes.—The causes of catalepsy are various, such as irregular eating and drinking, intestinal worms, masturbation and sexual excesses, the use of tobacco, drunkenness, licentiousness, &c., in all respects similar to those which produce epilepsy. But catalepsy occurs most frequently in young females, coming on apparently in consequence of a nonappearance of the menses, or else from their partial or total suppression after they have been established. It may occur, however, in either sex, and at any age, from intense mental application, violent anger, protracted grief, hatred, terror, and many other like causes. Finally, every variety of imprudence or abuse capable of producing epilepsy may act as a cause of catalepsy; but next to menstrual irregularities, I believe the disease is most frequently produced by gastro-intestinal irritation, from irregular eating, or from taking, at all hours, crude, unwholesome, and indigestible articles of food.

Pathology.—In relation to the pathology of catalepsy, it appears to consist in an irritable condition of the whole nervous system, including, of course, the brain; the paroxysms being brought on by any cause capable of producing a slight congestion of the brain. This congestion may be either in a sthenic or an asthenic patient; in the one case, being of an active, while in the other, it is of a passive character.

In catalepsy occurring in plethoric or sthenic patients, the paroxysms are usually more violent and protracted than in cases in which there is slight debility, and the paroxysms brought on by passive congestion of the brain. The reason why epilepsy or apoplexy is not produced instead of catalepsy, is because of the less intensity of the cephalic congestion, leaving the brain the ability to generate and convey to the muscles sufficient nervous power or influence to keep up the tone of the muscles, while the power of the mind, or will, over the physical organization is for the time entirely suspended.

In those severe cases of catalepsy which terminate in epilepsy, the cephalic congestion becomes sufficient to produce convulsions, while in very slight cases of catalepsy, the degree of the cephalic congestion is not sufficient to completely destroy or suspend consciousness, so that the patient may have an imperfect recollection of what occurs during the paroxysm. It is possible, too, that in catalepsy the congestion of the brain may not be so much of the cerebellum as in epilepsy, and that may be a reason, in part, for the absence of convulsions in patients with catalepsy.

In those cases of catalepsy which occur from irritation of the stomach and intestines, the brain becomes affected to irritation through the **sympathetic nerves**, and, consequently, more or less congestion of the brain occurs, **which** produces or develops the cataleptic paroxysm. The same is also true **in** all sympathetic **cases**, no matter at what point away **from the brain the** primary local **irritation** is located. There appears, then, to be nothing **very** mysterious **in** this affection, **or its** symptoms, when we take into **account** the nature **of the disease, and the** various functions of the brain and nervous system.

Prognosis.—The prognosis in **catalepsy is not generally very** unfavorable **so far as** a fatal termination is concerned, **but the patient** may pass on to an epileptic state, and this finally to apoplexy, **and thus the** patient **may** perish. Many **cases** however, terminate spontaneously, while a much greater number **recover** by the removal of the **cause or imprudence** which has produced them, together with a strict observance **of the laws** of health, and a careful judicious **course of medical treatment.**

Treatment.—The treatment of **catalepsy should be on strictly rational** principles, and does not differ in its indications materially **from those of** epilepsy. In every **case the** cause should **be** sought out and removed if possible, **and then the** attention and treatment should be directed to the general condition of the patient, **and** to the primary seat **of the local** irritation, which either directly **or** sympathetically irritates the **brain** and produces the **cephalic congestion, upon** which the paroxysms depend. If the patient **be plethoric, general and local bleeding** may be indicated, with an occasional saline cathartic, but if the patient be anæmic, a tonic course of treatment, with alœtic purgatives from the first is clearly indicated.

If the irritation which **produces the attack be primarily of the** brain, or spinal cord, after a **general bleeding** and a **saline cathartic, if** indicated, **cups,** wet **or dry, over the seat of the** cerebro-spinal irritation, repeated at intervals of a few days at first, **will generally do well, and** later, blisters or pustulation with tartar emetic **ointment to the back of the** neck, or along the spine, **may** be indicated, **and should not be neglected.**

Having thus subdued the cerebro-spinal **irritation the** cataleptic paroxysms may disappear, **but** it generally becomes **necessary,** to place the patient on a restricted, plain, digestible diet, to **insist upon a** reasonable amount of **exercise, to avoid** any undue physical **or mental excitement,** to keep the **bowels regulated** by an occasional **saline cathartic, and to** enforce a strict observance of the laws of health **in every respect.**

In cases of catalepsy **in which** the irritation and **congestion of the** brain is sympathetic of a primary irritation in some **part of the system** remote from the **brain, as** in the liver, kidneys, **uterus, or alimentary** canal, after removing or correcting the imprudence which has **led to it,** the general condition of the patient being taken into **account, the** treatment should be directed to the **organ** or part primarily **affected.**

If the liver or kidneys be the seat of primary **disease or derangement,** cups, counter-irritants, alteratives, and diuretics **may be indicated, and** if so, should be judiciously persevered in till the **hepatic or renal disease** be subdued and a healthy action restored, and **then the case should be** submitted to that course of treatment best calculated **to restore a** healthy state of the nervous system, which being accomplished, the cataleptic disease may disappear

In cases of catalepsy in which there are menstrual irregularities, operating as the primary derangement, that function should be restored by a judicious course of treatment, and then the case treated if necessary, on general principles, till the blood and nervous system are restored to a healthy state.

If the disease depends upon gastro-intestinal irritation from worms or from imprudence in eating, these causes should be corrected; the worms being removed by remedies the least liable to produce or increase the irritation of the alimentary mucous membrane, and then the case treated with tonics, &c., if necessary, till the paroxysms cease.

In those cases which occur from indigestion, the result of irregular eating, the patient should be placed upon a plain, digestible and nourishing diet, to be taken at regular meal hours, and the indigestion and gastro-intestinal irritation corrected as well as the state of the blood and nervous system, by small doses of rhubarb, extract of conium, carbonate of iron and oxide of zinc; a pill of one grain of each, with a little ginger, after each meal, as I suggested for epilepsy, will, I believe, best fulfill the indications in such cases. In cases of catalepsy from general nervous irritability, without any local disease, the result generally of masturbation, sexual excesses, the use of tobacco, drunkenness, licentiousness, &c.; after removing the cause, the pill I have suggested above may do very well, or small doses of the citrate of iron, one or two grains, after each meal, and $\frac{1}{15}$th or $\frac{1}{60}$th of a grain of strychnia before each meal, continued for a long time, may be indicated.

SECTION XII.—CHOREA—(*St. Vitus's Dance.*)

By chorea, from the Greek χορεια, "a dance," I mean that peculiar disease of the nervous system, in which, without loss of consciousness, the voluntary muscles are rendered, in a greater or less degree, uncontrollable by the dictates of the will. In slight cases, the influence of the will over the voluntary muscles is only very slightly impaired, and hence the irregular action of the muscles is not very marked; but in very severe cases the nervous derangement is so great that the voluntary muscles are rendered almost independent of the influence of the will, at least for a part of the time.

Symptoms.—Chorea generally makes its appearance very gradually, under a variety of premonitory symptoms, varying in duration from a few days to several months, and marked by a deranged state of the digestive organs and nervous system. The most common of these premonitory symptoms are flatulent pains in the stomach and bowels, variable appetite, constipation, with a tumid and hard abdomen, vertigo, anxiety and oppression in the præcordial region, tremors of the extremities, oppression in the chest, palpitation, fullness of the head, confusion of mind, itching in the nose, cold feet, and a general nervous and irritable condition of the system.

After these premonitory symptoms have continued for a longer or shorter time, irregular spasmodic contractions are observed in some of the voluntary muscles; frequently of the face, or one of the extremities. These spasmodic or imperfect voluntary motions of the part or parts are at first slight and may only occur from physical or mental excitement. Gradually, however, these involuntary, or imperfect unsteady voluntary movements become stronger and more constant, till at times almost every muscle of the body is in a state of continued involuntary action. In some cases, this irregular muscular action is confined to one side of the body, or to a particular part; but generally it extends more or less to all the voluntary muscles.

From the imperfect command of the will over the voluntary muscles, the patient when he attempts to walk has a starting, hobbling, and irregular gait, with an awkward dragging, in some cases, of one of the legs. In some very severe cases, the empire of volition over the voluntary muscles is so completely lost that progression and even the erect posture are rendered utterly impossible. The hands and arms are usually in constant motion, the patient being unable to direct them, and frequent ineffectual efforts are made before the hands can be brought to the desired point. In attempting to carry food to the mouth, the hand is carried in almost every direction before it finally succeeds in reaching the mouth, the head being sometimes thrown from side to side, or backward and forward, and the mouth suddenly widely opened and then forcibly closed.

Respiration is often anxious and irregular, the voice being altered, and articulation indistinct; and in one case that fell under my care, the voice was entirely lost for a time. In severe protracted cases, the mind appears materially disturbed, and the authority and commands of the will are almost entirely disregarded; the whole muscular system being thrown into a state of insubordination, its action being irregular, lawless, and not conducive to the general welfare of the system. Chorea sometimes assumes a very extraordinary character; the patient being seized with paroxysms of violent dancing, with various antic contortions of the body, and appears to have an almost irresistible propensity to leap upon chairs, tables, and to exhibit other like performances.

During sleep, while volition is in a state of temporary suspension, all the spasmodic motions which characterize this affection cease entirely. Thus showing that the irregular action of the voluntary muscles, is owing to an effort of the will to command them; but on account of the deranged condition of the nervous system, this command is imperfectly obeyed, and thus the irregular motions of the limbs, and other parts may be accounted for. Chorea is sometimes a paroxysmal affection, the irregular motions coming on by turns. But in most cases that have fallen under my observation, the irregular muscular motions have been continuous, during the waking hours.

Anatomical Characters.—In many cases in which death has occurred from chorea, or in which the patient has died from some accidental cause, the brain, spinal cord, and nerves, have been found in an apparently healthy state. It is therefore probable, that in those cases in which there has been found a serous effusion into the ventricles of the brain, and into the meningeal cavity of the spinal cord, injection of the meninges of the brain, or softening of the brain or spinal cord; that the conditions are accidental, and not a necessary morbid appearance in this disease.

Diagnosis.—The only conditions with which chorea is liable to be confounded, are mild cases of epilepsy, and certain forms of hysteria and paralysis; but when we remember that in chorea there is neither, coma, delirium, or rigid spasms, but only such symptoms as arise from an imperfect control of the will over the voluntary muscles, the diagnosis may be rendered clear.

Causes.—According to my observation chorea usually attacks children between the eighth and twelfth years of age; but it may occur earlier than this, and in some rare cases it may occur in persons of even advanced age. Chorea most frequently occurs in patients of a nervous temperament; thus indicating that its development depends upon a deranged condition of the nervous system, by which the nerves supplying the voluntary muscles, become imperfect conductors of the dictates of the will.

Any cause then, capable of rendering the nervous system weak and irritable, may act as an exciting cause of chorea. Among the most frequent, are mental emotions, especially terror, fear, and religious enthusiasm, gastro-intestinal irritation from crude articles of food, worms, &c., and the accumulation of irritating substances in the intestines. But there are other causes which often appear to produce chorea, such as repulsion of cutaneous eruptions, the suppression of habitual discharges, over-excited sexual propensities, and the ruinous habit of taking food at irregular hours, or between meals.

Pathology.—The real nature of this disease appears to consist in a weak, deranged, and irritable condition of the whole nervous system, including, of course, the brain, and especially the cerebellum, as well as the spinal cord; but not necessarily connected with any local lesion, organic disease, or even congestion. The disease, then, is functional, and it is probable that the nerves are involved as well as the brain, and it appears to me in a greater degree, for the brain appears capable of generating the nervous power, and also of receiving from the mind the dictates of the will; but as this impression of the will is sent forth on the nerves, which supply the voluntary muscles, it is very much like sending a locomotive on a deranged track—confusion and irregular muscular action being the result.

Prognosis.—The prognosis in chorea is generally favorable, as most cases, under proper treatment, finally recover. Only one fatal case has fallen under my observation, and that I suspect was at first neglected, which may account for its fatal termination.

Treatment.—The indications in the treatment of chorea are plainly to remove the cause, to overcome any local or general cause which may be irritating the nervous system, and then to restore the blood, and also the tone of the nervous system. The cause should be sought out at once, and whether it be irregular eating, masturbation, or any other imprudence, it should be corrected or removed, and the patient made to take a plain, digestible, and nourishing diet, with regularity; and to conform rigidly, in every respect, to the laws of health and propriety.

In those cases that have been produced by mental emotions, the greatest care should be exercised to remove or avoid that source of nervous excitement, after which the patient should be put upon a tonic course of treatment. That which I have found to do best, as a tonic for the blood, for the nervous system, and to improve digestion, is a combination of carbonate of iron, oxide of zinc, and rhubarb, given in powder, after each meal, so that five grains of the iron, three grains of the zinc, and two grains of rhubarb shall be given at a dose.

In very debilitated cases, if a nervous stimulant be indicated, half a drachm of the fluid extract of the cimicifuga racemose may be given three times per day, before each meal, for a time, with very happy effects; or instead of this a grain or two of ginger may be given with each dose of the iron, zinc, and rhubarb, after each meal.

In those cases in which the nervous irritability or derangement has been brought about by gastro-intestinal irritation, from worms or retained fecal matter, a cathartic of calomel and castor oil, or oil of turpentine if it be worms, should be administered. And if slight constipation continues, a pill of a grain of rhubarb and two grains of aloes should be continued, after dinner each day, and the combination of iron, zinc, and rhubarb, with or without a little ginger, given as suggested above, and continued till a cure is effected.

In cases in which there is suppresion of the menses, causing or complicating the nervous irritability, after using proper cathartics, a pill of the sulphate of iron, extract of gentain, and rhubarb, of each a grain, with half a grain of ginger, taken after each meal, will often best fulfill the indications. Pursuing the principles I have here laid down, I have generally succeeded in the treatment of this disease.

SECTION XIII.—INSANITY.

By insanity I mean that peculiar disease of the nervous system, including of course the brain, in which the mind fails to communicate and receive correct impressions; the disease being purely physical and not of the mind, which being a spirit is not liable to disease.

Now the mind, being a spirit, is capable of existing independent of the body, of thinking, acting and reasoning within itself, as it will after the death and decomposition of our bodies; but our bodies are the instruments of our minds in our present state, the nervous system, including the brain, being the only part of the physical organization through which the mind can act. It is my opinion that the whole nervous system is properly the organ of the mind; but it is probable that the cortical or gray matter of the brain is the more essential seat or organ with which the mind immediately communicates, receiving from it and communicating to it all impressions sent forth, or received by the mind though the physical organization.

It is probable that the impressions so sent forth by the mind through the gray matter of the brain, is communicated to different parts of the body through the white tubular matter of the brain, spinal cord and nerves; while impressions from without are received through the senses, and communicated through the nerves, spinal cord, and white tubular matter of the brain to the gray matter of the hemispheres, and through this gray matter directly to the mind itself. It is probable that the cerebellum so far influences impressions sent forth by the mind to the voluntary muscles, as to insure harmony in all the minute muscular motions, such as are necessary in standing, walking, running, &c.

If now we bear in mind that the intellectual principle is a spirit, possessing powers, faculties and properties of its own; and that our bodies are mere instruments of our minds furnished by the Creator for the purpose of enabling mind to communicate with mind, and with the material world; the phenomena of insanity is really no mystery. In fact, when we remember, the functions of each part of the nervous system, and that it is the only medium through which the mind receives and communicates impressions, all the phenomena of insanity are just what might be expected to arise; varying of course with the seat, nature, and extent of the disease or derangement of the nervous system.

In a healthy state of the nervous system, the mind receives and communicates ideas or impressions in a correct, plain and rational manner, so that the mind is not deceived through the senses from external objects, and the ideas which the mind sends forth to other minds, are communicated correctly, and hence we call the person rational or sane. But if from any considerable cause the brain or any part of the nervous system becomes sufficiently deranged, the channel of communication becomes more or less interrupted, and the ideas, though they may be rational in themselves, are imperfectly conveyed through the shattered or deranged brain and nervous system, and so we call the patient insane.

Now the degree of insanity will depend upon the degree of physical derangement of the brain and nervous system, but especially of the cortical or gray matter of the brain. If from meningeal inflammation the cortical matter of the brain becomes more or less involved, there may be an entire interruption to the communication of ideas to and from the mind, in a rational manner at least, and this constitutes mania.

In cases in which there is only a limited degree of cephalic derangement, of either an inflammatory or congestive character, involving perhaps only a limited portion of the gray matter of the brain; or if general which to me appears the more probable condition in such cases, the congestion and derangement being brought on only by a particular subject, we have insanity upon one subject, or monomania.

If now, instead of an inflamed or actively congested condition of the meninges and gray matter of the hemispheres, we have an anæmic, relaxed, debilitated, and perhaps irritable state of this cineritious portion of the cerebral matter, with perhaps more or less passive congestion, there may be a multitude of ideas or impressions communicated to and from the mind, but it is so imperfectly done, that there is a want of coherence or connection between them, and this constitutes dementia.

When from various derangements of the nervous system, the mind has long been deceived by false impressions or impressions communicated to it through deranged organs of sense, the mind fails by degrees to appreciate those principles of utility, order, or morality, by which it should be governed, and this constitutes moral insanity.

And finally, if from protracted inflammation of the brain or its meninges, or from any other cause, sufficient disorganization or morbid changes occur in the gray matter of the brain, to utterly destroy its capacity for communicating impressions to and from the mind, the mind though sane and rational within itself, is cut off from communicating with other minds, and with the material world, and this constitutes idiocy

Hence we have, as the result of different degrees and varieties of derangement of the nervous system, involving to a greater or less extent, probably in every case, the gray matter of the hemispheres of the brain, mania, monomania, dementia, moral insanity, and idiotism; the symptoms of each of which we will now proceed to consider.

MANIA.

Symptoms.—The symptoms of mania differ widely in different cases; they are generally, however, so obvious and well understood that no more than a general description need be given of the peculiar symptoms of each variety. In mania there is generally a rapid succession of incoherent ideas and violent excitement of the passions, expressed by great agitation, loud vociferations, singing and fury, just as might be expected when we remember that the mind is deceived by impressions received through the senses, and that the ideas or impressions sent forth by the mind are communicated through a deranged track or channel of communication.

Mania is usually preceded by a marked change in the habits, tastes, attachments and passions of the patient. The patient is generally either animated, irritable, jealous or wayward. The patient may be eccentric in his conversation, viscious in his disposition, may sleep little or be harrassed with dreams, and he often forms extravagant plans either to increase his fortunes, or for the public good. Sometimes he engages in ruinous speculations, squanders his means in extravagant amusements, or in unnecessary articles of clothing, or allows himself to be cheated by bad bargains, &c.

While this train of premonitory mental symptoms are being developed, the patient is often afflicted with constipation of the bowels, irregular appetite, vertigo, cephalagia, a sense of throbbing in the head ,a wild expression of the eyes, and a general feeling of restlessness, uneasiness and irritability; indicating slight congestion and irritation of the brain, which condition is evidently producing the deranged mental phenomena. But as the cephalic congestion, irritation, and meningeal or cerebral inflammation becomes fully developed, the expression of the countenance is wild and often ferocious; the eyes are prominent, sparkling, and in constant motion; the patient sings, whistles, walks to and fro with rapidity, or stands still with his hands and eyes directed towards some real or imaginary object; and in some cases the patient fails to sleep for many successive nights.

The pulse may be full, strong and frequent, but it is often after a time very irregular, unnatural, and sometimes intermittent. The skin is at first hot and dry, but later it may become cool, and perhaps covered with a profuse perspiration. The sensorial organs are apt to be excitable, the appetite variable, the bowels constipated, and the urine high-colored and small in quantity. During this inflammatory or highly irritated state of the brain and its meninges, the mind forms erroneous perceptions of the impressions of external objects on the senses, or the senses convey erroneous impressions to the mind, and impressions sent forth from the mind are imperfectly communicated in consequence of derangement in the channel or medium of communication.

In consequence of this derangement in the communication of impressions, the patient may mistake friends for strangers; may take the kindest words or treatment as the highest abuse, and may even suppose himself conversing with persons that are silent in his presence, or even with imaginary beings. On the other hand, ideas or impressions of the patient, perhaps rational in themselves, are so imperfectly communicated to those around him, that they by no means express the feelings or sentiment of the mind of the insane; the patient often expressing curses and violence when really the mind intended no such thing.

In consequence of this deception of the patient through his senses and also from the inability of the mind to communicate its real impressions, the empire of volition appears at times entirely suspended, and the patient is really no longer master of his own determinations; he may injure himself, or inflict injuries upon others; tear his clothes; leap out of windows, and commit various acts of fury.

Such are the ordinary symptoms of acute mania, as it has developed itself in most cases that have fallen under my observation; but as the irritation or inflammation of the cerebral hemisphere, or their meninges gradually subsides, or is subdued, the mania may cease, or it may assume a chronic form. When acute mania thus assumes a chronic form, or in cases which are chronic from the first, instead of the highly excitable condition of the acute state, the physical symptoms are those of chronic inflammation of the brain or its meninges, or of some morbid change involving the gray matter of the hemispheres.

The mania in this chronic state may be continued, remittent, or even intermittent; but in either case, as the cephalic disease assumes a more passive or chronic form, the patient has a sullen or downcast look; he is apt to neglect his dress and person, and in many cases that have fallen under my observation, there has been an apparent indifference to surrounding objects.

Thus we have the most common symptoms of both acute and chronic mania, as they have been developed in cases that have fallen under my observation. But it must be remembered, that as no two cases have exactly the same degree of physical derangement, upon which the mania depends, there must of necessity be a variety of symptoms developed in different cases, and also in the same cases in different stages of the disease.

MONOMANIA.

Monomania consists, as we have seen, in an insanity upon one subject, while on all other subjects the patient may appear quite rational.

The *symptoms* of monomania differ widely in different cases, as it includes hypochondria, fanaticism, melancholy, misanthropy, and, as I believe, what has been called insane impulses, depending generally upon the degree and extent of the physical derangement, and also in part upon the topics which have been uppermost in the patients mind previous to the insanity.

In *hypochondria* there is generally preceding the mental disturbance more or less physical derangement for some length of time, such as indigestion, flatulence, constipation, a furred tongue, a dry skin, a sallow complexion, cold feet, disturbed sleep, and especially is the patient apt to complain of pains in the chest or abdomen under the false ribs, which symptom has given the name to this variety of insanity.

After these physical ills have continued for a time, the nervous system becomes very sensibly involved, and probably some portion, or even all the gray matter of the hemispheres of the brain especially so; and then the mind being deceived by every symptom being magnified through the shattered nervous system exhibits monomanical symptoms.

The morbid condition of the gray matter of the brain may be partial in such cases, and thus the insanity upon one subject be accounted for; but it appears to me more probable that it is general, and that it is irritation from congestion of the cephalic vessel; the congestion and its attendant irritation being produced whenever the mind calls up the favorite topic of health, and thus the impressions to and from the mind are magnified or deranged, and hence the monomania on that subject. The patient, under such circumstances, is apt to suppose himself afflicted with consumption, cancer, intestinal worms, liver complaint, syphilis, and a host of other distressing, loathsome and dangerous affections: and what is more, in many cases no course of reasoning appears capable of reaching his mind to correct the delusion.

Fanaticism is another form of monomania, most liable to be developed, I believe, from religious or political excitement; the physical derangement consisting in an excitable, irritable condition of the nervous system, involving, of course, the cerebral hemispheres. In cases of this kind, the mind is apt to dwell upon its peculiar theme, whether it be of a religious, political or other character, till the concentrated influence of the mind overtaxes the nervous system, and therefore rational ideas are not communicated through the brain and nervous system; but when the mind is called off upon other subjects, ideas are communicated quite rationally.

If *religion* be the theme upon which the patient has become a monomaniac, the patient is apt to believe that he has received some direct revelation, or that he is commissioned of God to perform some important or miraculous mission not within the sphere of mortals. If politics be the theme, the patient is apt to suppose himself fitted for some important office, to which he believes he has every prospect of being promoted. One monomaniac, that fell under my observation during the administration of a late President, supposed that he was destined for the presidential chair, which, however, was to be brought about by the help of a large army. While laying his plans, and explaining them, he mischievously remarked to me, that he thought the President, calling him by name, would regret having taken the chair.

Melancholy is another form of monomania, in which the nervous system is generally reduced to a weak, irritable, and deranged condition; the result generally of some debilitating agent or pernicious habit, or else of protracted grief; the nervous system in either case being unstrung, whenever the prominent theme of misery is called up by the mind, but serving quite well as a conductor of impressions if the mind be called off upon other subjects. In some persons the depression and gloom become so great that the patient has a strong suicidal tendency, in some cases desiring to destroy himself to get rid of the burden of life, while in others he may regard it as a duty, which has perhaps been revealed to him, and with which he thinks it would be an unpardonable sin not to comply. In some cases, again, melancholy leads the patient to destroy his family, either to better their condition or to save them from some real or imaginary impending disgrace, or calamity, which is supposed to be pending.

Misanthropy is another form of monomania, in which the derangement of the nervous system is often brought on by a train of unfortunate circumstances, in which the patient has in reality, or in imagination, been unkindly treated by his fellow men. The effect of this kind of monomania in some cases is to make the patient a silent, sullen hater of his fellow men; the patient regarding it as a righteous indignation; while in other cases the most revengeful disposition to homicide is developed. In either case, the condition is a fearful one, and may lead to the most fearful consequences, unless reasonable precautions and proper restraints be resorted to, to prevent the patient from injuring himself or his fellow creatures, from whose hands he often believes he has received such high-handed and unmerited abuse.

Insane impulse is the last form of monomania which I will mention, as entitled to a separate consideration. This variety of monomania appears to be the result of a blind impulse, coming on suddenly, to commit some crime without a motive, which the patient in his rational moments would by no means be guilty of. It most frequently occurs in persons of strong feelings and sympathies, and is the result of learning or witnessing some important serious or other event, such as an execution, a suicide, &c.

I believe it may be accounted for in the same way that other cases of monomania must be. The person witnesses or hears of a case of suicide, or of an execution, or some other like event occurs, which, at the time, so far excites the sympathies as to produce a temporary rush of blood to or from the brain, during which state the person becomes temporarily insane. During this temporary state of insanity the patient may very possibly commit a homicide or suicide, according as the one or the other theme was occupying the mind, as the cause of the temporary monomania or insanity; and yet the patient may not be in the least responsible for the deed so committed.

Great care should be exercised however in all legal proceedings, that this fact be not taken advantage of, to screen from justice the wretch who willfully and maliciously dyes his hand in the blood of his fellow man. Having thus completed our consideration of the symptoms of mania and monomania, let us proceed to inquire into those dementia.

DEMENTIA.

Dementia which consists, as we have seen, in that peculiar debilitated condition of the nervous system, in which there are a multitude of impressions or ideas conducted to and from the mind, in a disconnected or incoherent manner, may follow mania or monomania; or it may be an original affection, the result of inflammation of the brain, and its meninges.

In cases of mania and monomania in which dementia supervenes, or when it occurs as an original affection, the inflammation of the brain and its meninges generally assumes a passive form, and, sooner or later, there is apt to occur a hesitancy of speech, weakness of the lower limbs, and more or less paralytic symptoms. As these physical symptoms are developed, a multitude of ideas may be communicated to and from the mind; but as the tone of the brain and nervous system is deranged, there is an association of unrelated perceptions—the ideas appearing to be collected together without any degree of order or harmony.

There may be in such cases profound silence for a long time, as happened in one case that fell under my care; but there is apt to be great volubility of speech, with more or less bodily gestures, performed in convulsive rapidity. In dementia the mind appears fleeting and changeable, the patient being mild, malicious, generous, and miserly, all in the course of a few hours; depending upon the weak, irritable, and deranged condition of the brain and nervous system.

Dementia is apt to come on gradually, and may exhibit a variety of grades, from the slightest mental disturbance to a state approaching imbecility; the patient being either silent or indisposed to speak, or else attempting or disposed to communicate his ideas, without the ability of doing it in a quiet and rational manner.

In many cases the patient, unable to get off his ideas by speech, calls into aid the voluntary muscles, and hence the bodily gestures which are so often exhibited by patients in this condition. Thus we have the most common peculiar symptoms of dementia; but it must be remembered that they are liable to great variations in different cases, and in different stages of the disease in the same case, like all other affections.

MORAL INSANITY.

Moral insanity consists, according to my view, in an inability to discover those principles of order or morality which should govern or control every intelligent rational being. Hence a person morally insane may commit many acts of violence against his best friends, or fail to exercise that discretion and prudence which is always due from one person to another, whether friends or foes.

This want of discretion, or common prudence, may be very trifling, or it may be exhibited in falsehood, theft, violence, or even murder; requiring the greatest possible prudence on the part of friends or attendants to exercise a judicious restraint. The patient may appear physically well. But it will generally be found that such patients have led idle, vicious, and profligate lives; disregarding in almost every respect the laws of health, which will account for the moral weakness or insanity.

Now, to make moral insanity perfectly plain, it is only necessary to remember that the human mind is endowed by the Creator with an intuitive consciousness of self, and also an intuitive notion of God, while a knowledge of the material world is acquired through the bodily senses. Now, by the knowledge thus acquired, through the senses, the mind improves its knowledge of self, as well as its notions of God; and by the sum of this knowledge, the mind learns its obligation to self, and its fellow beings, to God, and to all things; and hence it may act in manner conducive to the greatest good of all.

19

Let now any human being pursue a reckless course of life, abusing his bodily senses and nervous system generally by various imprudencies, such as the **use of tobacco**, gluttony, licentiousness, drunkenness, &c., and the mind forms very imperfect notions of the material world, and **consequently of** self and of God. Now, as a consequence **of this want of the essential** elements of knowledge, the **person so situated forms no** just conceptions of his obligations to himself, to his fellows, and to God; **and hence** the reckless disregard of the interests, welfare, **and safety, of** even near relatives, exhibited by patients morally insane.

In this light, a person intoxicated is morally insane. His sense of obligation being while in that state deranged at least, and hence the acts of violence which are so often committed by persons in that deplorable condition, in some cases being of the most serious and fatal character. I believe that persons morally insane should have wholesome restraint exercised over them. But legally, as in drunkenness, I believe it should be no palliation for **crime** actually committed. Thus we **have the** symptoms peculiar to moral insanity, as well as the cause and manner of their development, **there remaining only** those of idiotism for our consideration.

IDIOTISM.

Idiotism consists, as we have seen, in such a hereditary or acquired condition of the brain and nervous system, that there is an utter inability on the part of the mind to receive and communicate impressions or ideas, however rational they may be in themselves. It may be congenital, or it may supervene upon other forms of insanity. It either case, it consists in a deranged condition of the nervous system, involving probably in every case the gray matter of the hemispheres **of the** brain, **either** directly or indirectly. Whatever be the nature of the morbid change in the brain in different cases of idiocy, the function of the nervous system, including the brain is destroyed, at least so far as its being the organ of the mind is concerned.

As there are no mental symptoms in utter idiocy, we have only to **remember the vacant stare,** the expressionless countenance, and the **automatic** movements, and **the** symptoms are all told. And sad indeed as is the picture, it would be doubly sadder if it were not for the fact, that the disease is physical, not of the mind, in congenital cases at least. For it is probable that the mind in such cases may be rational within itself, thinking, acting and reasoning by God-given powers. **But as its** instrument is useless, it has to think, act and reason within itself, without the ability of receiving or communicating a single idea.

Thus then we have the peculiar symptoms of mania, mono-mania, dementia, moral insanity and idiotism; as they have generally occurred **in** cases that have fallen under my observation and treatment. It should **be** remembered, however, that no two cases of any variety of insanity exhibit precisely the same symptoms, if we except perhaps utter idiocy, which really has no mental symptoms, and only antomactic physical symptoms.

Anatomical Characters.—The appearances presented on post-mortem examination of the bodies of persons that have died insane vary according to the nature of the insanity, and the period of the disease at which death has taken place. There may be various accidental morbid lesions of different parts of the system, as of the lungs, liver, alimentary canal, kidneys, &c., but by no means necessary conditions in insanity. But in addition to the accidental lesions, when they exist, and in fact in all cases in which death has occurred from insanity, or from accidental causes during insanity, the gray matter or cortical substance of the brain, together with the membranes of the brain exhibit signs of disease and generally of inflammation. In those cases of insanity which are complicated with more or less extensive paralysis, in addition to the signs of inflammation of the gray matter and membranes of the brain, there is generally a preternatural whiteness and induration of the medullary, tubular or white substance of the brain.

The following then are the most frequent morbid appearances presented in connection with insanity. The cranium may be thickened, and either hardened or softened, and in some cases more or less irregular in shape. Of the membranes, the dura mater is apt to be thickened and adherent, the arachnoid thickened and opaque with false membranes adherent to it, within its cavity; and the pia mater swelled, red, infiltrated with blood or serum, and more or less strongly adherent to the brain in most cases at least.

The cortical substance of the brain is reddened, sometimes generally, but in other cases in points; its blood vessels are enlarged, and while it may be slightly indurated on its surface where it adheres to the pia mater, the portions beneath are apt to be softer than in health. In cases in which paralysis has occurred, the white or medullary portion of the brain is almost invariably found indurated.

In protracted cases the bulk of the brain is often diminished, and minute cavities are found in the cortical substance, the result probably of former extravasation or else of softening. In chronic protracted cases the specific gravity of the brain may be diminished; there may be serous or sanguineous effusion into the cavity of the arachnoid membrane, into the tissue of the pia matter, or into minute cavities in different parts of the encephalon, or finally into the ventricles of the brain in various quantities.

Finally in conclusion, it is well to bear in mind, that among the morbid appearances thus presented, marks of disease, and generally of inflammation of the gray matter of the hemispheres, are a general, if not an invariable morbid condition in fatal cases of insanity; and that in cases in which more or less paralytic symptoms have been developed, there is in addition, a preternatural whiteness, and induration of the white, madulary, or tubular matter of the brain.

Causes.—The causes of insanity, both predisposing and exciting, may be either physical or moral. The physical causes operate in most cases to derange the different organs, and their functions; and thus either directly, or indirectly the brain and nervous system, while the moral causes consist in an overaction of the mind, directly upon the brain and nervous system, the different organs and functions of the body becoming affected secondarily.

Among the *predisposing* causes of insanity, both physical and moral, the most frequent are hereditary tendency, time of life, constitution, condition, education, celibacy, profession, and previous insanity.

Hereditary predisposition to insanity consists in a peculiarity of the physical organization, and especially of the brain and nervous system, by which the brain and its meninges are exceedingly liable to take on either active or passive inflammation, which condition, when it is produced by accidental causes, is liable to develop various forms of insanity.

Time of life too, appears to have an influence in predisposing to insanity, as few cases occur in children, and it is not very common in advanced life, most cases probably occurring between the twentieth and fortieth years of age. Various causes may contribute to render insanity more frequent at this age, among which the embarrassments, and perplexities attendant upon the commencement of the business transactions of life, may not be the least.

A nervous temperament, whether inherited, or acquired from various imprudences in early life, is a strong predisposing cause of insanity, especially if in addition the various passions predominate, such as pride, ambition, &c.

Wealth and extreme *poverty* predispose to insanity; the former by leading to idleness, and various excesses, and the latter by exposing to hardships, privations, distress, worriment, and unmerited neglect.

Education, if attempted too early, or neglected, or what is worse, if it be of an improper vicious character, predisposes to insanity. If attempted too early, the brain is rendered irritable, and hence only slight causes may lead to inflammation of the brain, and its meninges, and so insanity in some form may be the result. If it be neglected altogether, or if it be of a vicious improper character, the mind fails to form correct notions of its obligations, and hence a strong predisposition to insanity, and especially moral insanity is the result, which only requires the continuation of the unfavorable beginning to develop the disease in mature years.

Celibacy in both sexes, appears to predispose to insanity, perhaps in part from physical causes; but more frequently I suspect from moral cause; the unmarried, as a general rule, being more dependent upon strangers for many of the necessaries and comforts of life; and besides, the cares and perplexities of life, though perhaps fewer are more apt to be shared alone in most cases.

Profession predisposes to insanity; those whose business require a constant exercise of the mind upon the brain, such as artists, poets, politicians, and professional men, being by far more strongly predisposed to insanity, than those who pursue a tranquil course of life. Finally, one attack of insanity by the change that is apt to take place in the brain, and its meninges, strongly predisposes to another, for some time at least, and probably during the whole subsequent period of life, in most cases.

The *exciting causes* of insanity are very numerous, embracing every imprudence, whether physical or moral, capable of irritating the brain, and of producing inflammation of the brain or its meninges, either of an active or passive character.

Among the *moral causes* of insanity, or those in which there is an overaction of the mind, producing the cephalic inflammation and nervous irritability, are severe study, perplexing investigations, political or religious excitement, fits of anger, remorse, erotic exuberance, connubial disturbance, fear, despair, loss of friends, abuse of confidence, public disgrace, violated chastity, destitution, and finally, the witnessing or hearing of executions, murders, suicides, &c.

Any *physical cause* capable of producing either an active or passive inflammation of the brain or its meninges, directly or by sympathy, may be an **exciting cause of** insanity. But the most frequent causes thus operating, are direct injuries, protracted watchings, irregular eating, masturbation, onanism and sexual excesses, the use of tobacco, drunkenness, licentiousness, &c. Of these causes, the use of tobacco, masturbation, protracted drunkenness, and licentiousness, **together with** depressing moral causes, tend to produce dementia, while those causes calculated to excite a particular passion, **as** ambition, **revenge, &c., tend to develop** monomania.

Of the other causes which I have enumerated, **both physical and** moral, **some tend to produce** or develop one form of **insanity and** others another, according to the nature and extent of the inflammation which they set up in the brain and its meninges, **as** well as **the general disturbance** they produce in the nervous system generally.

The puerperal state has been supposed to predispose to, or act **as a** cause of insanity. **And it is not** strange that it should, when we remember the great change the female system undergoes from the period **of** conception to **the** termination of lactation, especially if we take into account the unnatural and imprudent sexual indulgences which are too often resorted to in this delicate state of the female constitution.

Pathology.—I have already anticipated much in relation to the nature or pathology of insanity, so that but little remains to be said. I will therefore only arrange and give a synopsis of my views of the pathology of insanity, and dismiss this branch of our subject.

It is my opinion, that the mind comes directly from God, "who is the Father of our spirits," that it possesses properties, powers **and** faculties of its own, and that it is capable of thinking, acting, and reasoning within itself, independent of the body, by its own God-given powers. And further, that it is immaterial or spirit, and not like the physical organization liable to disease. That the body is not a necessary appendage of the **mind, but a mere** instrument, which **in our present** degenerate state, serves only for a **limited** period of **time. And** finally, that the mind being a spirit is immortal and will exist after the death and decomposition of **our** bodies, and that it is liable only to moral **disease.**

Now if this be true, we have a rational solution of all the phenomena of every variety of insanity. For as the body is **the instrument of the** mind, and the nervous system including the brain, **the medium through** which the mind receives and transmits its impressions **or ideas, any** derangement of the brain and nervous system, sufficient to hinder **a correct** communication of impressions through the senses to the mind, **or from** the mind through the brain and nerves to other **minds is what we call** insanity.

Now we know that impressions from **without are** received through the senses, being transmitted along the nerves, spinal cord, and white tubular matter of the brain, to the gray matter of the hemispheres, and through this gray or cortical matter directly to the mind itself. **Conversely** when the mind sends forth impressions or ideas, it passes them directly to the gray matter of the hemispheres, and thence along the white tubular matter of the brain, **spinal** cord, and nerves, to the voluntary muscles designed **to** be called into action, whether for locomotion or for the communication of **ideas to** other minds, through the **organs of speech, the** cerebellum harmonizing these motions. Now so **long as this channel or** medium of communication, between the **mind at one extreme, and the** material world **and** other minds at the **other** extreme remains perfect, impressions are communicated in a correct and rational manner; but when it is more or less interrupted we have the different varieties of insanity, varying from the slightest mania to utter idiocy.

If the derangement in this **medium of communication consists in active** irritation, or inflammation of the gray **matter of the brain, or its men-** inges, involving the brain, we have that variety of interruption to the communication of **ideas,** which we call mania.

If we have temporary **congestion and irritation of the brain** from over action of the mind upon some particular theme, whenever the mind calls up that **theme, and a remission when the mind is called off on other sub-** jects, we have that variety **of interruption to the communication of ideas,** which constitutes monomania.

But if we have a debilitated condition of the nervous system, with irritation, **and perhaps passive inflammation** of the brain, or its men- inges, we have **that derangement in the** communicating medium, which allows the transmission of **a multitude of ideas, but in so imperfect, dis-** connected a manner, as to constitute dementia.

Again, when by long abuse of **the system the organs of sense have** become so much deranged as to communicate, for a long time, false im- pressions to the mind, **the mind so** situated fails to acquire correct notions of its obligations, and hence becomes reckless of those principles of order or morality which should guide all rational intelligent beings, thus con- stituting moral insanity.

Finally, when from inherited **congenital, or acquired** imperfection of this medium of **communication, there is an entire** incapacity for the transmission of **ideas, even in the most imperfect** manner, utter idiotism is the result.

Thus we have, **in my view, the true** pathology of insanity: a deranged condition of the nervous system being an invariable condition, while de- bility, irritation, congestion, **or an** active or passive inflammation of the brain, and especially of the gray matter of the hemispheres is a general, if not an invariable morbid condition in every variety, except perhaps in moral insanity.

Diagnosis.—The diagnosis of insanity is not in general attended with any considerable difficulty, if the history and symptoms of the case be carefully learned, and all the extrinsic circumstances be taken into account. In ordinary acute and chronic meningitis there is generally sufficient irritation, or inflammation of the cortical or gray matter of the brain to interrupt, for the time, the correct communication of impressions or ideas, developing symptoms of mania, or dementia, according to the nature or extent of the cephalic inflammation.

The same is also true in the delirium of fever, and in delirium tremens, though the irritation in such cases of the nervous system may be from passive inflammation, or mere debility of the brain, as is evidently the case in the latter affection. So, too, a wicked profligate course of life, from a willful and malicious disposition, may develop symptoms similar to moral insanity, though not identical with it, as in such cases the injuries are apt to be inflicted upon real enemies, while in moral insanity they may fall upon intimate friends, and without the least provocation.

So, too, common eccentricity of character may simulate, in many respects, monomania; but it differs in this, that such peculiarities are apt to be more or less congenital, while monomania is generally acquired from a worriment or over-action of the mind upon the subject of the insanity. Finally, it may be an important point to decide upon the insanity of a patient in doubtful cases, for legal purposes, to establish the validity of wills, or to guide in the trial and decision of cases in which there is the charge of capital or other high crimes.

Now, it is my opinion that every condition of a patient whether it be the result of acute or chronic meningitis, or of inflammation, irritation, or debility of the brain arising during fever, or any other general condition of the system, whether temporary or permanent, should be regarded and treated legally as insanity so long as the brain and nervous system generally are not in a condition to transmit or receive impressions or ideas in a correct or rational manner.

And finally, that all those cases of gross immorality in which capital or other crimes are committed upon friends, without any provocation, or any real or imaginary cause, should be regarded as moral insanity, and that such persons should be subjected to proper restraint. But that such a state should be legally no palliation for crimes actually committed any more than drunkenness is, which is really a temporary moral insanity.

Prognosis.—The prognosis in insanity varies with the hereditary predisposition, and also the nature and extent of the physical derangement, as well as the duration of the disease. In patients in which there is a strong hereditary predisposition to insanity, if the disease be developed early in life, it may often be arrested for a time; but it is exceedingly liable to return, sooner or later, and at last to become permanent.

In cases of insanity from accidental causes, and not hereditary, if proper treatment be resorted to early, the cause having been removed, I believe reasonable hopes may be entertained of a recovery in most cases. Among the cases of insanity that have fallen under my observation and care, most have recovered, if the cause has been early removed, and the patient properly treated, before a chronic state had supervened. But among cases that have fallen under my care, in which there was a strong hereditary predisposition, and the patient advanced in life, I have had one most deplorable fatal case, and another in which there was only a partial recovery. Excepting these two cases, I have generally succeeded, sooner or later, in my cases of insanity, if I except also cases of moral insanity, in which I have generally failed.

Finally, from all the observations that I have been able to make on this subject, I believe that nearly or quite nine-tenths of all the cases of insanity not strongly hereditary, that receive proper treatment during the early stage of the disease, may reasonably be expected to recover. In this estimate, however, I do not include cases of moral insanity and idiocy; the **former** of which, under proper moral, and the latter under proper physical and moral treatment, may possibly improve ; but I seriously doubt if a perfect, permanent recovery is often effected in such cases.

Treatment.—In the treatment of insanity the cause should be sought out, and if possible removed, whether it be physical or moral; after which, the patient should be treated on strictly rational principles, physical and moral. The patient should be placed in a situation either at home or in some approved institution for the insane, where, without unreasonable restraint there will be a reasonable security that he will inflict no injury upon himself or any one else.

Every cause of physical or mental excitement should be carefully removed and the patient kept cleanly, allowed good air, and should have his food served up with regularity and of a character exactly suited to his condition. These preliminaries having been arranged, if any overaction of the mind from worriment, excitement, or any other mental excess is found to have been operating to produce the derangement of the physical organization, that source of worriment or excitement should be carefully avoided, and the mind drawn off, as much as may be, upon other subjects suited to his state.

If the physical derangement has been produced by any imprudence, such as masturbation, onanism, sexual excesses, the use of tobacco, drunkenness, licentiousness, or any other varieties of imprudence, they should be prohibited, and the patient made to conform rigidly to the laws of health and propriety. When all this has been carefully and prudently done, the exact deviation from the standard of health of the physical organization should be ascertained, and then the indications fullfiled in every respect, and the treatment persevered in till the physical derangement is corrected, at which time the brain and nervous system will resume their function of communicating impressions or ideas to and from the mind, in a correct or rational manner.

If there be acute or chronic inflammation of the brain, a cathartic of calomel or podophyllin may be indicated at first, and later, a pill of aloes and rhubarb, two grains of aloes and one grain of rhubarb may be given at evening each day, to secure a regular action of the bowels, and to lessen the determination to the brain. Cups, wet or dry, to the temples or back of the neck occasionally at first, and later, blisters or pustulation with tartar emetic ointment are generally indicated, to be varied of course, according to the nature and extent of the cephalic inflammation. The warm foot-bath should be used at evening for a time, and the head kept elevated, and if nothing contra-indicate, it may be showered each morning with cold **water.**

With this course of treatment, the inflammation of the brain or its meninges may generally be subdued, if resorted to in season, and thus in a reasonable time the insanity may disappear. At least such has been the result of my observation in such cases.

If, however, the cephalic inflammation be of a passive character, the result of a debilitated anaemic condition of the system, the drastic cathartics should be omitted, and only dry cups, blisters, &c., applied to the temples, back of the neck, or back of the ears.

The warm foot-bath may be used, as well as moderately cool water to the head by showering in the morning. And besides the pill of aloes and rhubarb at evening, or after dinner each day, some preparation of iron should be administered and continued after each meal till a healthy state of the blood is restored. If only the simple tonic effects of iron are indicated, the citrate in solution, or the iron by hydrogen in powder three times per day may do best. If there be a dropsical tendency, ten or twelve drops of the tincture of muriate of iron is a preferable form, to be administered in a little water. But if there be a scrofulous tendency, the syrup of the iodide of iron, administered in a little water before each meal will generally do best.

Such, according to my view, are the general indications in the treatment of the inflammation of the brain or its meninges, which is so apt to be a condition in almost every variety of insanity. If other portions of the system be in a healthy state, little treatment further than I have suggested may be required. But if the cephalic inflammation be symptomatic of, or complicated with disease of other organs, as the liver, alimentary canal, kidneys or genital organs, such disease should, if possible, be subdued, and the general health and tone of the system restored.

If the liver or kidneys are in a torpid inactive state, alteratives or diuretics may be indicated, and should be resorted to in conjunction with the other treatment. If there is indigestion with debility of the stomach, or irritation of the alimentary mucous membrane, antacids, tonics, and perhaps counter-irritants may be required. Finally, if the patient be a female, and the insanity has supervened upon suppression of the menses, or any menstrual irregularity, a healthy state of that function should be restored by such remedies as appear to be indicated in the case.

Such I believe are the principles which should guide us in the treatment of all ordinary cases of insanity. But cases may occur in which only proper air, exercise, and diet, with perhaps tonics, is all that may be required. This, in fact, may often be the case in puerperal cases, as I have succeeded with a moderate dose of blue pill, followed by castor-oil, and the subsequent use of moderate doses of the sulphate of quinine, continued for a few days.

In cases of moral insanity, I believe that the general health and tone of the system should, if possible, be restored, by judicious medical treatment, and then the patient brought under the most approved moral and religious influence. " Religion," says Hufeland,* " is the best thing in man ; it is that which constitutes him such an one ; it is the essential part of his life ; therefore it is religion, when all other means fail, that the patient's thoughts and actions must be referred to. On this account, going to church and conversation with a sensible clergymen is advised."

Now I can give no better advice in cases of moral insanity, after restoring the health and tone of the physical organization, than to repeat the suggestions quoted above from one of the noblest minds that ever graced the medical or any other profession. And now, in conclusion, I

---

* See Hufeland's Enchiridion Medicum. p. 202.

will only suggest, that in cases of partial idiocy, if any treatment be attempted, it should be of a physical and moral character, and such as is adapted to each particular case.

SECTION XIV.—MANIA A POTU—(*Delirium Tremens.*)

By delirium tremens, or mania a potu, I mean that peculiar affection of the nervous system, the result, generally, of the use of alcoholic liquors, opium, or tobacco, and characterized by muscular tremors, want of sleep, and delirious hallucinations. It occurs in habitual drunkards, and in such as are addicted to the use of opium; and it also occurs, occasionally, from the excessive use of tobacco.

Now, by the long continued use of these poisons, the system becomes accustomed to their stimulating effects, the various functions of the body become deranged, and the tone, strength and vigor of the brain and whole nervous system become impaired, in consequence of their debilitating effects. In consequence of the debility thus produced, the stimulus has to be gradually increased, in order to keep up the impression, and while this is done, no delirium ordinarily occurs. But if, in consequence of sickness, an injuriy, or from any other cause, the accustomed stimulus be removed, or suspended, or even greatly diminished, the system sinks down, or appears as it really is. The nerves become unstrung and the brain irritable, in consequence of which muscular tremors, wakefulness and delirium, with all the symptoms of this affection, make their appearance. While, then, delirium tremens is thus the result of the debilitating effects of alcoholic drinks, opium and tobacco, the delirium and other symptoms of the disease do not ordinarily make their appearance till the system is allowed by their suspension to feel its real condition, by a removal of the intoxication.

Symptoms.—On the removal of the intoxicating agent, whether it be alcohol, opium or tobacco, there is generally exhibited more or less lassitude and general indisposition. There is a feeling of distress in the epigastrium, nausea and vomiting, giddiness, confusion in the head, want of sleep, and an anxious expression of the countenance, with tremor of the hands.

After these premonitory symptoms have continued for a day or two, the patient exhibits symptoms of suspicion and alarm, the eyes are fixed upon some real or imaginary object, or cast about with rapid scrutinizing glances, the patient becomes restless and irritable, he walks to and fro by day, and is unable to sleep during the night. The tremor increases, and the patient becomes loquacious and more or less deranged, and though he may claim to feel well, he is apt to be tormented with various disgusting, alarming and ludicrous imaginary appearances. He imagines that he sees dogs, snakes, cats, mice, and various other animals in his room; and lice, bugs, or other disgusting vermin crawling over his bed, and on his clothes; or that various persons of a fearful character and appearance have entered his room for the purpose of robbing or killing him.

In one case that fell under my care a few years since, the man, at home, supposed himself among strangers, and surrounded by what he called a host of "land-pirates," and he appeared to be in constant and most terrible fear of becoming a victim to their cruelty. His wife, he supposed, was his land-lady, and he indulged in many curious speculations in relation to her hospitality, as well as her good looks; and he appeared to hope that possibly she might befriend him. But still he had constant suspicion as to her character, and a constant unconquerable dread of the host of "land-pirates," by which he imagined himself surrounded and fearfully beset.

Patients, to avoid such and other horrible allusions, will often call loudly for assistance, and becoming greatly agitated, will often threaten, and sometimes rave violently. The mind and body are in a continued state of action. They will walk hurriedly about the room, or run to the window and call to some imaginary person in the street, or start suddenly with horror, terror, and agitation, from the presence of frightful and disgusting apparitions, and insist perhaps that they are well and confined unjustly, and that they should be permitted to attend to some important and pressing business transaction.

In most cases that have fallen under my observation, if the patient be not too plainly contradicted, or too sternly opposed, they have been controlled without any very coercive measures; but in some cases, when the disease rises to a higher grade of violence, the patient becomes furiously delirious, and being unable to recognize even his nearest relatives and friends is restrained with difficulty.

One severe case of delirium tremens that fell under my observation a few years since, exhibited most of these symptoms. In this case, the patient was so annoyed by imaginary, disgusting, and fearful objects that he emptied the straw from his bed into the street, and set it on fire to get rid of his fearful intruders, all the time brushing them from his person and trying to prevent their escape from the flames, which, however, he found not very effectual, especially in destroying several small and very saucy looking evil spirits, or "little devils" as he called them, which he thought "bore the fire very well."

He insisted that his head was covered with lice and various kinds of bugs, and fearful insects, and so certain was he of this, that he procured a tub of water, and placing himself over it, he labored for hours to detach them by means of his hands, combs, brushes, &c., and at the same time defending himself against the serpents, evil spirits, and numerous other fearful imaginary enemies, which he represented as acting in the most saucy, impudent, and provoking manner, and aiming all their spite, venom, and impudent tricks at him. This case exhibited various other symptoms, which I need not mention, but recovered after having been delirious for about one week. According to my observation, persons suffering from delirium tremens do not appear to be susceptible to much bodily pain, or even to suffer much from bodily injuries, so long as the delirium continues.

The pulse in this disease varies in different cases, being sometimes hard, full and frequent, but generally soft, full, and quick, without much strength, or any considerable tension. The skin is generally moist and cool, the tongue covered with a white fur, the bowels torpid and along with slight thirst for cool drinks, there is a disinclination to take food, during the whole course of the disease.

The duration and degree of violence of this affection vary in different cases, sometimes only slight tremor of the hands. with transient delirium, sensorial illusions and wakefulness, continue for two or three days, and then pass off.

In other cases the wakefulness, tremor of the hands, general restlessness and agitation continue for five or six days, with delirium and annoying apparitions at night, while during the day, the patient remains tolerably quiet. But in severe cases, the symptoms continue day and night, with but slight remissions, for one or two weeks, or even longer, if not arrested by proper treatment. If the disease be left to itself, or maltreated there is danger of protracted or even permanent insanity.

Diagnosis.—The diagnosis of this disease is not generally attended with any considerable difficulty, if all the symptoms, as well as the extrinsic circumstances be carefully taken into account. From the dilirium of fever, from meningitis, and from ordinary insanity, this disease may be distinguished by the following differences.

From the delirium of fever it differs in wanting the general febrile symptoms, in its fantastic hallucinations, and in the general tremor, fear, and constant apprehensions of various real or imaginary objects. From meningitis it differs in wanting the local and general inflammatory symptoms, by the want of convulsions and coma, and by the imaginary and fearful apparitions and muscular tremors in this disease. From ordinary insanity this disease differs by being generally more rapid in its approach; by the patient being more constantly tormented by apprehensions of danger, and finally in the fact that delirium tremens is either the direct or indirect result of alcohol, opium or tobacco, while ordinary insanity may be produced by various other causes.

Anatomical Characters.—No appearances are presented on the postmortem examination of patients dead of this disease, which account for the symptoms which are developed. In some cases, more or less serum has been found in the cavity of the arachnoid, in the ventricles of the brain, and infiltrated into the cerebral substance, and also signs of inflammation of the brain or its meninges; all of which, however, appear to be the result of accidental complications, as they are entirely wanting in other cases.

Various other accidental appearances are occasionally presented in different parts of the body, the most frequent of which appear to be signs of pneumonia, pleurisy, gastritis, enteritis, and inflammation, fatty degeneration, &c., of the liver or kidneys; the result generally of the intemperance which has led to the disease, but not an invariable morbid condition in this disease.

Causes.—Alcoholic drinks are probably the most frequent cause of delirium tremens, while opium and tobacco are occasional causes.

Alcohol being a stimulant and irritant, without containing any material nutritious properties, its long continued use creates an irritation of the mucous membrane of the stomach; which being transmitted by the sympathetic and pneumogastric nerves to the brain, produces a temporary stimulation, but a permanent debilitating effect upon the brain and the whole nervous system. It is probable also that a portion of the alcohol enters the circulation, and so passes directly to the brain, producing its temporary irritating and permanent debilitating effects. Now as the tone and strength of the brain and whole nervous system become gradually lowered; more of the alcoholic stimulant is required to keep up sufficient intoxication, to prevent the individual from knowing or feeling the real debility which is being produced, and this is why drunkards increase the quantity of liquor each succeeding year.

If now, in such a case, from any cause, the alcoholic irritant be withheld, the removal of the irritant and its intoxicating effects allows the system to sink down to its real condition, very much in the same manner that a person irritated and intoxicated, by anger, gradually assumes his real state of mind and strength of body, after the fit of anger passes off. Now, as the system in such a case assumes its real level, the brain and whole nervous system which have long been accustomed to this irritated condition, become completely unstrung, and thus the symptoms which follow.

The brain and nervous system are no longer in a condition to transmit impressions to or from the mind. As a consequence of this, every impression received through the organs of sense, many of which are imaginary, coming to the mind through this shattered and unstrung nervous system, is converted into an object of terror, which keeps the mind in constant dread of their annoyance; while, on the other hand, the commands of the will, or impressions from the mind sent forth upon the brain and nerves to the voluntary muscles are very imperfectly communicated, whether for locomotion or for articulate sounds or speech, hence the insanity in this disease.

Finally, as the brain is debilitated in such a state, sufficient nervous influence is not generated and distributed to the nerves to keep them in proper tone, and hence the muscular tremors which occur in this distressing affection.

Opium, we have seen, is a cause of delirium tremens, and though it does not so markedly irritate the mucous membrane of the stomach, it deranges the secretion of the liver and more or less of the whole glandular system and produces the same temporary stimulating and permanent debilitating effects as are produced by alcohol. After its continued and protracted use the brain and nervous system become so accustomed to its impressions that its removal produces the same train of symptoms which are developed in the affection when produced by alcohol, and for the same reasons.

Tobacco is an occasional cause of delirium tremens; and it is not strange, when we remember that it contains nicotin, nicotianina, and an empyreumatic oil; all of which are most virulent and deadly poisons. Its long continued use, either by chewing, snuffing or smoking, so far poisons the blood and debilitates the brain and whole nervous system, that when from any cause it is suddenly withdrawn and the intoxication which it produces passes off, the patient becomes conscious of his real condition, and the brain and whole nervous system being unstrung and excessively irritable, tremor, wakefulness, delirium and all the symptoms of delirium tremens from the effects of alcohol is liable to be the result.

It is possible that tobacco may produce this disease by its direct debilitating effects while the system is under its influence, but according to my observation, the horror, wakefulness, muscular tremors and delirium occur when, from a removal of the drug, its intoxicating effects have passed off, and the patient feels his real debility.

Pathology.—From what we have already seen, the pathology of this disease may readily be inferred. That it depends entirely upon the debilitating effects of alcohol, opium, or tobacco, upon the general system, and especially upon the brain and nervous system, generally, it appears to me, that there is no reasonable room for doubt, when their direct or indirect effects are taken into account. That the sudden removal of these debilitating poisonous irritants and their attendant intoxication should so far let down the system to its real condition that the brain and whole nervous system should be so irritable, weak and tremulous or debilitated as to develop all the symptoms which arise, is not at all strange, and that, too, without producing, necessarily, any congestion or inflammation of the brain or any other tissue of the body.

Delirium tremens then consists, essentially, in a weakened, debilitated and irritable condition of the general system, and essentially of the brain and whole nervous system; in consequence of which, the brain is unable to generate and supply to the nerves a sufficient amount of vital nervous power to enable them, with the brain, to act correctly as the organ of the mind, or to enable the various organs of the body to perform with due regularity their accustomed functions; hence the tremor, wakefulness, delirium, &c., which occur in this disease.

Prognosis.—The prognosis in delirium tremens, so far as the recovery from an attack is concerned, may be regarded as generally favorable, unless the patient dies from exhaustion, or passive congestion, or inflammation supervene. But when we take into account the terrible condition of the system which develops this affection, and its causes, and that almost the only hope of a permanent relief of the horror which attends such a condition, is to be found in the poison which has produced it, or a kindred one, the prognosis so far as the ultimate result of the case is concerned, is always unfavorable.

Treatment.—The indications of treatment are plainly to keep the patient under proper restraint, to stimulate to a condition compatible with rest, and quiet the brain and nervous system, to afford a due amount of proper nourishment, and, finally, to leave the patient in the most favorable condition possible for the permanent abandonment of the ruinous habit which has led to the disease, whether it be from the use of alcohol, opium or tobacco.

I believe sufficient restraint should be exercised over the patient to prevent him from injuring himself or others, but that further than this should never be attempted, as it generally tends to increase the irritability, according to my observation. In mild cases of this affection, whether it be the result of liquor, opium, or tobacco, an attempt should be made to arrest the disease, without resorting to the poison which has been the cause of it. After having arranged the preliminaries in relation to proper restraint, the patient should not be too abruptly opposed or disputed in relation to his imaginary dangers, and should be persuaded to take either a strong infusion of the tincture, or else the fluid extract of valerian or hops. Of these, two or three fluid drachms of the tincture, or one or two of the fluid extract of either, may be given every three or four hours at first, and later every six hours, while the restlessness and tremor continue; the patient being persuaded, if possible, to take a little milk in crust water, or some other mild digestible and nourishing food.

If symptoms of passive congestion of the brain supervene, the warm foot-bath may be resorted to, and dry cups applied to the back of the neck, but this I think will very rarely be necessary. With this mild course of treatment, some cases may recover in a few days, and if they do, due caution should be taken to prevent the patient from returning to the ruinous habit of drinking, opium eating, or tobacco using, which may have been the cause of the disease.

In more severe cases, or in mild cases in which this mild course fails to arrest the disease, five grains of Dover's powder, with three grains of the James's powder, may be combined, and given every four or six hours. Or if the antimonial be contra-indicated, thirty or forty drops of the tincture or fluid extract of opium may be given every four or six hours, instead of the antimonial and Dover's powder, and continued till rest, quiet and sleep are procured. If, however, as sometimes happens, this plan of treatment also fails, good brandy should be given in wine glassfull doses every six hours, and alternating with it one or two fluid drachms of the tincture of asafetida, and continued till rest, quiet and sleep is procured, when the brandy should be withheld, but the asafetida continued till the disease and its effects have passed off, when that should be gradually discontinued also.

The patient should be allowed crust coffee, one-half milk at first, and later, a poached egg, toast, and other varieties of plain, digestible and nourishing food, to be taken at regular meal hours; and after the disease has passed off, and as the stimulants are discontinued, moderate doses of an infusion of columbo, or twenty or thirty drops of the fluid extract may be given after each meal, and continued for a time, to improve the appetite and favor digestion.

In those cases of delirium tremens produced by opium or tobacco, the same course of treatment should be pursued, and the patient cured, if possible, without a resort to the poison which has produced the disease. If, however, it becomes necessary to resort to the opium or tobacco temporarily, they should be discontinued gradually, as the delirium passes off, and finally abandoned altogether. If opium has been the cause, it may thus be abandoned permanently, as is the case with alcoholic drinks; but to abandon tobacco is generally, according to my observation, attended with more difficulty.

There is, however, at this hour, a respectable gentleman of this village in my office, who suffered with all the essential symptoms of delirium tremens twelve years ago, from the use of tobacco, and who, having abandoned its use, since that time is in the enjoyment of tolerable health, though he assures me that his system has not entirely rallied from its pernicious effects.

In those cases of delirium which occasionally occur during the continued use of alcohol, opium or tobacco, which have some of the symptoms of delirium tremens, I believe the symptoms are from inflammation of an active or passive character of the brain or its meninges, and the disease not strictly delirium tremens.

SECTION XV.—ECLAMPSIA.—(*Convulsions.*)

By eclampsia, or convulsions, I mean here that condition of the nervous system, but especially of the brain and spinal cord, in which, without constituting an essential part of any recognized disease, there are clonic spasms of the muscles with irregular motions of the limbs and various parts of the body, and attended generally with temporary unconsciousness.

Now this condition being the result of accidental causes, acting directly upon the brain or else upon other parts, as the alimentary canal, and affecting the brain sympathetically, is liable to occur at any period of life. But as the head of children is larger in proportion to the rest of the body than in the adult, and as the brain is probably softer; accidental convulsions are much more liable to occur in infants or children than in adults.

These accidental convulsions, however, are not necessarily confined to infancy or even childhood, but they may occur at any age from various causes and conditions; and especially in females during pregnancy, or in the puerperal state. Accidental convulsions may be preceded by slight indisposition, or they may come on suddenly, without any special warning of their approach.

Symptoms.—Accidental convulsions are most liable to occur in children or adults of unusually large heads and short necks, other things being equal. The spasms may effect all the voluntary muscles, more or less, or they may be of one half of the body or even of a single part, as a limb or the muscles of the face, as I have known in some cases. In some cases there is only a single attack, lasting from five to fifteen minutes, while in other cases there may be several following in more or less rapid succession, depending in some degree upon the nature and permanency of the accidental cause, and also the means made use of to arrest them.

If the convulsions depend upon irritation of the brain, with vascular congestion, the face will often appear swelled and the face and lips livid, or of a bluish appearance, the jugulars will be distended, the surface hot, and the pulse frequent, but often very irregular; but if, as is sometimes the case, the convulsions are the result of nervous irritation with depression, the face may be pale, and want that swelled or congested appearance, the skin may be cool and the pulse though frequent, is generally feebler than in health, according to my observation at least.

During the continuance of the convulsions, which may last only for a few minutes, or in other cases, occurring at irregular intervals for several days, or even longer, the abdomen is apt to be tympanitic, involuntary evacuations may occur, and as an attack of the convulsions subsides, there is a comatose tendency, the patient being more or less disposed to sleep for a time, in most cases. In cases, however, in which the convulsions have been produced by irritation and depression, without much, if any, congestion of the brain, the patient especially if an infant, appears quite natural, immediately after the cessation of an attack of convulsions.

Accidental convulsions are not generally attended with any very considerable danger, but there occurs sometimes serious cerebral or spinal lesion, leaving strabismus, local paralysis, or perhaps some serious mental disturbance, or death may occur from an interruption of a due innervation of the heart or lungs, or perhaps from other causes. In cases of death occurring thus in accidental convulsions, the post-mortem may reveal nothing to account for the sad result, but generally there are found signs of congestion or inflammation of the brain or spinal marrow, or else tumors, effusions or softening, which may account for the convulsions at least, and generally for the fatal termination.

Causes.—A large head and a short neck, with an irritable condition of the nervous system, and especially of the brain and spinal cord, strongly predisposes to convulsion, in both infants and adults. And while accidental convulsions may occur from various exciting causes in any one, it should be remembered that with this physical peculiarity, it may occur from slight causes, such as in the other cases might produce little or no perceptible disturbance.

The causes most liable to produce convulsions either directly or indirectly in adults are imprudence in eating and drinking, the use of tobacco, excessive venery, fear, excessive anger, direct injuries, protracted grief, and finally, pregnancy or the puerperal state. Of these causes, some may act by producing direct irritation, and either an active or passive congestion of the brain or spinal cord, while others may act by affecting the brain or spine sympathetically, while others still, may act either mechanically, or else by producing an anæmic condition, and consequently an irritability of the brain, and whole nervous system.

The frequency of convulsions during pregnancy or in the puerperal state, may be accounted for in part I think, from the changes which the system undergoes from the period of conception to the close of lactation. These changes rendering the whole nervous system more or less irritable, especially if, as is too often the case, sexual intercourse be indulged in, with various other imprudences. It is possible also that convulsions may be produced during the latter months of pregnancy, from the slight mechanical obstructions to the free passage of blood along the abdominal vessels, especially in anæmic debilitated females. Convulsions may also occur immediately after the delivery or during labor, and in many cases I am satisfied from worriment and general nervous irritability. Convulsions in *infants* may be the result of imprudence on the part of the mother, such as taking improper food, indulging in fits of anger, allowing the infant to nurse at irregular, unseasonable hours, or of feeding young infants unnecessarily with various kinds of solid food, and other like imprudences.

During childhood, various causes may operate to produce accidental convulsions, such as dentition, direct injuries, intestinal worms, &c., but none, according to my observation so frequent, as the indigestion, irritation, and derangement of the system, brought about by irregular eating, or taking food between meals. This habit is sure sooner or later to destroy or impair the natural appetite, and to create instead, a morbid one, so that the child may eat little at one meal, and far too much at another, and as indigestion with gastro-intestinal irritation is sure to be set up, and as a consequence, more or less cerebro-spinal irritation, occasional convulsions are very apt to occur.

20

The direct cause of the convulsions in such cases, may be active congestion of the brain and spinal cord, or it may be from passive congestion from the debility which the indigestion produces in such cases, or it may be from the irritability of the cerebro-spinal, and whole nervous system, which according to my observation is an invariable result of protracted irregular eating, or taking of food between meals. Now I am satisfied, from careful observation, that a large majority of the accidental convulsions occurring in children, and generally supposed to be the effects of intestinal worms, are really the result of irregular eating, the gastro-intestinal irritation, together with the debility, which it produces, causing sufficient cerebro-spinal and general nervous derangement, to produce the convulsions which occur.

Diagnosis.—It may be difficult, in many cases, to decide whether convulsions are the result of some accidental cause, or whether they may not be the commencement of some specific disease, as hysteria, epilepsy, &c. If, however, the habits of the patient, as well as all the circumstances connected with the attack, be carefully taken into account, a probable diagnosis may be formed between purely accidental cases and those which are to attend some permanent specific disease.

Practically, it is of vital importance to ascertain the direct and remote cause of the convulsions. The direct cause, that proper treatment may be had during the convulsons; and the remote cause, that it may if possible be removed, and thus a return of the convulsions be prevented. If the convulsions be the result of active congestion of the brain and spinal cord, the face is generally flushed, the pulse strong, and the heat of the surface above a healthy standard. But if the convulsions be the result of passive congestion, or of general nervous irritation from debility, the face may be pale; the pulse is more feeble than in health, and on the subsidence of the spasms there is not generally so marked comatose symptoms as occur in cases from active congestion.

Prognosis.—In those cases of accidental convulsions in which there is no considerable hereditary predisposition to cephalic congestion, or cerebro-spinal irritation, the prognosis is generally favorable, if the cause which is operating can be ascertained and removed. But in cases in which there is a strong inherited predisposition, or in which convulsions are produced by very slight causes, serious apprehensions may reasonably be had as to their final result. Cases of this character may terminate fatally, during a fit of convulsions; or some serious cephalic or nervous affection may follow, as the result of the repeated cerebro-spinal congestions.

Treatment.—The treatment of accidental convulsions consists in that which is proper during a convulsive attack, and that which may be necessary during the intervals, to prevent a return of the spasmodic affection.

During an attack of convulsions, if it be from active cerebro-spinal congestion, the patient, whether an adult or an infant, should be placed with the head elevated, and the feet in warm water; should be allowed fresh air; and if there is much heat and dryness of the skin, and the head is hot, cloths wet in cool water may be applied to the head. Immediately on removing the feet from the warm water, they should be wrapped up, and kept warm, and sinapisms not too stimulating should be applied and continued to the bottom of the feet, while the cephalic tendency continues.

If the patient be plethoric, and the convulsions are obstinate, blood may be taken from the arm, or cups or leeches applied to the temples, or back of the neck. And if symptoms of cerebro-spinal inflammation continue after the convulsions subside, a cathartic of calomel, or podophyllin may be administered, and followed, if necessary, by a full dose of the sulphate of magnesia, and blisters applied back of the ears, or to the back of the neck; and all the indications fulfilled, till the inflammation is subdued.

In cases of convulsions from passive congestion, the warm foot bath may be used, and an infusion of capsicum along the spine, as well as sinapisms to the feet, or extremities; but general bleeding should not be resorted to, or even wet cupping, as a general rule. But if symptoms of passive inflammation continue after the convulsions subside, mild cathartics, such as the blue pill, or a pill of aloes and rhubarb may be indicated, as well as blisters back of the ears, or to the back of the neck; and thus the inflammation may be subdued.

In cases of convulsions from debility, with irritation of the whole nervous system; but especially of the brain, and spinal cord, such as are liable to occur during labor, or after delivery; in cases in which there is no inflammation, or even active congestion; every cause of irritation should be removed, and along with sinapisms to the extremities, elevation of the head, and fresh air, the tincture of asafetida, with the infusion, ammoniated tincture, or fluid extract of valerian may be administered, to diminish nervous irritability and to promote a proper distribution of the nervous influence; upon the derangement of which, the convulsions in such cases probably depend. To prevent a return of each paroxysm; cloths wet in cold water, should be applied suddenly to the face and neck, the shock of which, I believe may generally arrest them, as it did in one puerperal case that fell under my care. Such I believe is the most safe and convenient method of fulfilling the indications, during the spasms, in the various forms of accidental convulsive affections.

Now, immediately on the subsidence of the convulsions, the remote cause should be sought out, if possible, and removed; and the condition of the system, which is the immediate cause of the convulsions corrected, so far as it may be. If the patient be an infant, and the cause irregular nursing, or improper feeding, the imprudence should be corrected. If the convulsions arise from the irritation of dentition, the gums may be carefully cut, and thus that source of irritation removed. If intestinal worms be the cause, a full dose of calomel may be administered, at first, in castor oil; and if necessary, small doses of the oil of turpentine administered at meal hours, for two or three weeks, or some other vermifuge given, if that be contra-indicated.

If indigestion and gastro-intestinal irritation, from irregular eating, or taking food between meals be the cause, the child or adult, whoever it may be, should be allowed a reasonable amount of plain, digestible, and nourishing food at regular meal hours, but not one morsel between meals. If, with the indigestion there be a diarrhœa and acidity of the stomach, small doses of the compound tincture of rhubarb with a little prepared chalk, taken after each meal, may be sufficient to correct the difficulty.

If the indigestion be attended with constipation, and the patient be an infant or a very young child, a drachm each, of the sulphate of magnesia, and rhubarb should be added to half a pint of water, and a teaspoonful given each morning, and if necessary, at evening, till the constipation is corrected. If the patient be an adult, a pill of aloes and rhubarb, of each a grain and a half, may be given after dinner each day, and continued till the constipation is overcome.

If the use of tobacco, masturbation, onanism, or excessive venery, be the cause of the convulsions, the imprudence should be corrected, and the patient made to conform rigidly to the laws of health and propriety. After having corrected the imprudence, if the anæmic debilitated state, in such cases require a tonic, the citrate, carbonate or ferrocyanuret of iron should be administered in moderate doses, till the strength of the blood and the tone of the nervous system are restored.

In puerperal cases, every possible cause of irritation should be removed, as well as imprudence corrected, and along with a regulated diet, good air, and a reasonable amount of exercise, absolute freedom from worriment should be enjoined. Such I believe are the principles which should guide us in the treatment and prevention of the various forms of accidental convulsions.

SECTION XVI.—HYSTERIA.

By hysteria, from the Greek ὑστερα, "uterus," I mean that peculiar affection of the nervous system, characterized by uterine or other primary local irritations and more or less sympathetic irritation and congestion of some portion of the spinal cord or brain, with irregular distribution of the nervous influence, and in some cases convulsions.

When the primary uterine or other local irritation is not very considerable, and the sympathetic irritation of the brain or spinal cord is quite general, and there is no special local point of irritation along the spine, or in the brain, the hysteria, according to my observation, generally develops itself in a slow and irregular manner, constituting chronic hysteria. When, however, the primary uterine or other local irritation is considerable, and the sympathetic irritation of the brain and spinal cord is very decided, and especially of the superior portion of the spinal cord, with more or less active congestion of these parts, the disease generally develops itself in paroxysms of convulsive hysteria.

Finally, when the primary uterine, gastric, or other irritation is very great, and the sympathetic irritation of the cerebro-spinal system falls mainly upon the brain, producing at times severe cephalic congestion, the hysteric paroxysm is often characterized by coma, insensibility, or stupor. Hence, we have three prominent modifications, or varieties of hysteria, depending, according to my observation, upon the degree and extent of the primary uterine, gastric, or other irritations, and also upon the nature, seat, and degree of the cerebro-spinal irritation and congestion. Let us now proceed to examine the symptoms of the different modifications of hysteria, and first of the chronic.

Symptoms.—Chronic hysteria, as we have seen, is that variety in which the sympathetic irritation affects the cerebro-spinal and ganglionic systems without fixing especially upon any particular part of the brain or spinal cord, in consequence of which, the patient is almost constantly afflicted with various hysteric symptoms, without suffering from any very severe paroxysms. The symptoms of chronic hysteria are just what might be expected from such a condition of the nervous system. The temper is variable, the patient being sometimes animated, and at others peevish and gloomy. The patient may pass rapidly from laughing to crying; from gaiety to melancholy; from despondency to hope; and all for the most trifling causes.

The sympathetic irritability of the cerebro-spinal and ganglionic systems develops often various distressing sensations in the head, chest, abdomen and pelvis, such as a sense of weight or bearing-down in the region of the uterus; flatulency, colic, pains or rumbling in the bowels; a feeling of emptiness or tension in the pit of the stomach; variable appetite, slow digestion, eructations, &c.; palpitations, weakness and syncope; ringing in the ears; confusion of the mind and pains in various parts of the body, and finally, the "globus hystericus," or the sensation of a ball rising in the throat to the top of the sternum, causing oppressed and hurried respiration, and often a feeling of impending suffocation.

Such, according to my observation, are the usual symptoms of chronic hysteria; and the changeable character of the symptoms, or the rapid change of location of the morbid symptoms, depends, I believe, upon the general sympathetic irritation of the nervous system. The slight local irritation of the brain or spinal cord, changing from one point to another, develops the symptoms which arise in the head, chest, abdomen, pelvis and other parts.

Paroxysmal or *convulsive* hysteria is developed, as we have seen, in consequence of sympathetic irritation of the cerebro-spinal system, and especially of the superior portion of the spinal cord; the congestion which attends producing the paroxysms of convulsive hysteria. Now, if this idea be correct, the symptoms which are developed in convulsive hysteria are just what might be expected.

Various premonitory symptoms may occur, such as a very lively or else an opposite, sullen, or gloomy mood; but generally, I think the paroxysms come on quite suddenly. Sometimes the fit consists in violent and convulsive laughing, alternating with crying and screaming, or there is rapid and incoherent talking, singing, suffocative spasms of the throat; a wild and furious expression of the countenance; raving; gnashing of the teeth; tearing out the hair; biting; beating the breast with the hand, &c. Sometimes these symptoms subside without the congestion or disturbance of the nervous influence being sufficient to develop convulsions; but generally, in such cases, convulsions of considerable violence speedily supervene.

The spasms, when they occur, usually partake of the tonic character, the body being rigidly bent backwards, or variously contorted; the breast is projected forward and the head drawn backward; the face is swelled; the tongue is either protruded or the jaws firmly closed; the eyes are rolling, prominent, and red; the fists are clenched; the arms may be spasmodically thrown about; the abdominal muscles contracted, and the whole muscular system is often thrown into such violent spasms that it is sometimes very difficult to keep the patient on the bed. When the paroxysm ceases, the patient is sometimes left in a stupid or somnolent state, which passes off in an hour or two, very much like the somnolency of epilepsy.

This modification of hysteria does not usually occur in weak or debilitated patients, as is the case with the chronic variety, but more frequently in sanguineous, plethoric, and robust females of strong passions, and occurs, usually, according to my observation, in consequence of some disagreeable mental emotion, or from a sudden suppression of the menses. In some cases of this variety of hysteric convulsions, the heart and arteries appear to be prominently affected; the heart palpitates violently, the pulse beats tumultuously, the carotids throb, the face becomes flushed and turgid, there is headache, with slight delirium, and a more or less hurried and anxious respiration.

When the point of local sympathetic irritation is in that portion of the spinal cord which supplies, most directly, nerves for the respiratory organs, the hysteric paroxysm may assume the character of violent asthma. I have known the spasm of the larynx so great, in such cases, that the patient would appear to be suffocating, when the spasms would yield, and the patient would again revive. Such, according to my observation, are the symptoms of the acute, convulsive, hysteric paroxysm.

The symptoms of that modification of hysteria which is developed in consequence of a sympathetic irritation affecting the cerebro-spinal system, and producing a congestion of the brain, are very similar to those of apoplexy, and for similar reasons. In this modification of hysteria there may be slight premonitory symptoms, but the patient usually, without any considerable spasms, sinks into a state of insensibility. The patient in such a state generally lies on the back, with the limbs relaxed, and perhaps extended; the eyes are closed; the breathing is slow and laborious, but scarcely audible; the pulse is slow; the countenance pale; the extremities cold, and generally the power of swallowing is lost, and the sensorial functions for the time appear suspended. Occasionally, while the patient lies in this state, a long inspiration is made, and I have known patients, as partial sensibility returns, to tear their clothes from the bosom, or to press upon it with both hands firmly.

The duration of this variety of hysteric paroxysm may vary from a few minutes to several hours, and I have known patients to remain in this state for a whole day, and then rather suddenly to awake, as from a deep sleep. As the patient thus awakes, I have seen them raise up to a sitting posture, and look around, apparently with surprise, and very soon recover the entire possession of their corporeal and mental powers. This variety of the hysteric paroxysm may occur from various causes, but I have most frequently noticed it in young females, from suppression of the menses.

Anatomical Characters.—The post-mortem examination of hysteric patients, that have died of some acute disease, reveals no morbid lesions which appear at all connected with the hysterical affection. This negative testimony, together with the fact that patients seldom or never die of pure hysteria, is sufficient evidence of the purely functional character of this disease.

Causes.—Any inherited or acquired undue excitability of the nervous system predisposes to hysteria. Girls of a delicate habit of body, with precocious intellects, an animated disposition, and an early sexual development, are generally most predisposed to hysterical affections. In such females, slight causes may produce gastric, uterine, or some other local primary irritation, and then either general or local irritation of a sympathetic character is set up in the nervous system, involving the spinal cord, and perhaps the brain, and thus some form of hysteria is developed.

I believe the most frequent exciting causes of hysteria are constipation of the bowels, suppression of the menses, and the frequent excitation of voluptuous feelings, by improper reading, conversation, and the working of an unchaste and too active imagination.

Pathology.—In relation to the nature of hysteria, but little more than I have already suggested need be said. The disease occurs almost exclusively in females; and generally between the commencement, and final cessation of the uterine functions, which is a strong argument in favor of the uterus, as the seat of primary irritation in this disease. It is probable, however, that other parts, as the alimentary canal, may be the seat of primary irritation, or derangement in hysteria.

It is probable that there is generally a primary local irritation of the womb or alimentary canal, or some other part; from which a sympathetic irritation is communicated to some portion of the spinal cord, or brain, through the ganglionic and cerebro-spinal nerves. If, as we have seen, there is no special point of irritation along the spine, or in the brain— we have the symptoms of what I have called chronic hysteria. If there be some special point of sympathetic irritation along the superior part of the spine—we have the hysteric convulsions. But if the sympathetic irritation, and congestion involve the brain, as well as the spinal cord, we may have hysteric insensibility.

Now it is probable, that the direct cause of the symptoms which are developed in every form of hysteria, is a derangement in the generation, or bad distribution of the nervous influence; in consequence of which more or less of the functions of the body are either deranged, or temporarily suspended, during the hysterical condition.

It will be seen that I have supposed both a primary and sympathetic irritation in every form of hysteria; and I believe that such is generally the case. But it is probable that an anæmic condition, with excessive nervous irritability, involving the cerebro-spinal, and ganglionic system of nerves; as well as the brain and spinal cord may lead to sufficient derangement in the generation, and distribution of the nervous influence, to develop the different forms of hysteria, without any special primary, or sympathetic local irritation, acting as a cause. If now my views of the pathology, or nature of hysteria be correct—we have a solution of all the symptoms which are developed in every possible modification of the disease.

Diagnosis.—Hysteria may generally be distinguished by careful observation of all the symptoms; especially if the character, history, and general condition of the patient be taken into the account. The only disease with which it is very liable to be confounded, is epilepsy; from which it differs in several essential particulars.

The hysteric paroxysm wants the "aura epileptica," the sudden cry at the seizure of the fit; the distortion of the features; the foaming at the mouth; and finally the deep coma which attends, and follows the epileptic paroxysm. Besides, hysteria has the alternate laughing and crying; the "globus hystericus," and the healthy expression of countenance between the paroxysms, which is not common in epilepsy.

The functional derangements which occur in the different organs of the body in hysteria, and which are apt to simulate inflammatory, and other serious or troublesome affections, may generally be distinguished from those diseases, by a careful observation of all the symptoms in the case.

Prognosis.—The prognosis in hysteria, so far as a fatal termination is concerned, may be regarded as favorable. But hysteria is an affection, which, if once fully developed, is very liable to continue, or to return, at least, at intervals for many years; often till after the cessation of the menstrual function, even though it commence with the first dawning of menstruation.

Treatment.—During a paroxysm of hysteria the patient should be placed in as convenient and comfortable a position as may be, and should be kept from injuring herself, or any one else, with as little absolute restraint as may be sufficient for that end. The feet should be placed in warm water, if necessary, and the head kept elevated and moderately cool, and, if the cephalic congestion be very considerable, a few ounces of blood should be taken from the back of the neck by cups. Or if some point in the spine appears to be the seat of the greatest sympathetic irritation and congestion, cups should be applied on each side of the spine, at the point of greatest irritation and congestion.

Sinapisms should be applied to the epigastrium, and to the lower extremities, and a brisk irritation produced. In this way the hysteric paroxysm may generally be overcome, after which the cause should be sought out and removed, as far as may be, and the treatment should be directed to the removal of the primary local irritation, whether in the uterus or elsewhere, and thus an effort made to correct the general nervous irritability, which generally exists in such cases.

Having inquired into the remote cause, as well as the general habits of the patient, every imprudence should be corrected, and the patient made to conform, rigidly, to the laws of health and propriety. In this way alone may the source of the primary irritation be cut off, and tolerable hope of its removal be entertained. Having arranged these essential preliminaries, attention should be directed to the removal of the primary local irritation.

If constipation of the bowels be the cause, the diet should be of a nature calculated to overcome that difficulty, as far as may be, and a pill of aloes and rhubarb be given after dinner, each day; or, if there be hepatic derangement, a drachm of the fluid extract, or a wineglassful of the infusion of taraxicum, made from two ounces of the dandelion root to a pint of boiling water, may be given after each meal, and continued till the hepatic secretion is restored, and the constipation overcome.

If suppression of the menses appears to be the cause, a full dose of the sulphate of magnesia, or calomel and castor oil, should be given at first, and then a pill, of equal parts of aloes and rhubarb, should be directed after dinner, each day, till the bowels are regulated. After regulating the diet, and correcting the constipation, if there be heat, pain, or uneasiness in the region of the womb, cups may be applied to the sacrum, occasionally, and at the period when the menses should appear; the warm foot-bath should be used at evening, and sinapisms applied to the sacrum, on retiring to bed at night. If the menses do not appear from this course of treatment, Dewee's Tincture of guaiac should be given, in teaspoonful doses, after each meal; and if the patient be anæmic, ten drops of the syrup of the iodide of iron, or else tincture of the chloride, should be given before each meal, and this treatment continued till the menses are restored, when the hysteric symptoms will generally subside.

In *chronic hysteria* and all other varieties occurring in girls or young women, and not dependent upon constipation or suppression of the menses, the cause which is keeping up the primary, sympathetic, or general nervous irritability, should be ascertained, and removed if possible. If there be uterine irritation from any imprudence, without menstrual irregularities, the imprudence should be corrected, and its consequences explained to the patient, as well as the local uterine irritation subdued by dry cups to the sacrum, and injections of cool water into the vagina, if necessary.

In hysteria occurring in anæmic females about the time of their final cessation of the menses, a gentle tonic course is often indicated. A pill of two grains of the ferrocyanuret of iron, with one-fourth or one-half a grain of aloes, I have found of very essential service, given after each meal, and continued for a reasonable time.

In all cases of hysteria in which it becomes necessary to arrest spasmodic actions, and at the same time to allay nervous irritability, ten drops each of the tincture of stramonium, tincture of opium, and sulphuric ether, may be administered, every six hours, till the spasms and excessive nervous irritability are subdued. In hysteria occurring in debilitated females, attended with irregular spasms, a teaspoonful of the tincture of asafetida, or a pill of three or four grains of the gum, may be given every four or six hours, and continued till the spasms are subdued.

A regulated diet, regular habits, and an even temper of mind should be enjoined in all females suffering from hysteria, or who are predisposed to this affection. By attending thus rigidly to the laws of health and propriety, and doing just so much by way of medicines as is indicated and necessary, and nothing more; most nervous hysterical patients may be greatly benefitted, and many permanently cured.

SECTION XVII.—SPINAL IRRITATION.

By spinal irritation I mean here that variety of spinal affection in which there is more or less tenderness, on pressure, upon the spinous processes along some portion of the spine.

Symptoms.—Spinal irritation is generally attended with more or less numbness, stiffness, slight pain, and more or less acute tenderness of that portion of the spine involved in the irritation. But the most prominent symptoms of this affection are remote from the spine, at the extremity of, or along the course of the nerves proceeding from that portion of the spine involved in the irritation.

If the irritation be in the superior portion of the spine, there is often pain, spasms, and a distressing cough, from derangement in the nerves of the larynx. If the irritation be a little lower in the cervical portion of the spine, the brachial plexus of nerves becomes involved, so that there is often numbness, pain, or perhaps partial paralysis of one or both arms, and palpitation of the heart; asthma; and perhaps sufficient irritation or derangement produced in the pulmonary nerves to produce a most distressing spasmodic cough.

If the irritation be in the dorsal region of the spine, there is apt to be pains along the intercostal muscles, in the region of the spleen, stomach and liver, and very often functional derangement of the stomach, liver, kidneys, and alimentary canal. Among the functional derangements of the abdominal viscera which I have noticed, from spinal irritation, are nausea, vomiting, constipation of the bowels, a torpid state of the liver, spasm of the bile duct, and a tympanitic state of the abdomen. If the spinal irritation be along the lower portion of the dorsal, or in the lumbar or sacral region, there may be neuralgia or functional derangement of the kidneys, bladder, rectum, womb, testicles, and of the lower extremities; all of which parts I have repeatedly found more or less deranged from spinal irritation.

Now the patients may suffer some pain and uneasiness in the spine at the point of irritation, but more generally the pain or functional derangement is complained of along the course of the nerves supplied by the irritated portion of the spine, or at their extremities. The patient, however, may suffer acute pain in the spine if they sit back against a chair, or if firm pressure be made along the spinous processes, and generally the pain or derangement along the course of the nerves, or at their extremity, is increased by this pressure if firmly made.

In cases in which the irritation is in the superior portion of the spine, involving the nerves of the respiratory organs or heart, if firm pressure be made it will often produce cough or palpitation or both. If there be neuralgia of one or both arms, pressure will often increase the pain, and the same is true of other neuralgic or functional derangements of the abdominal and pelvic viscera, as well as of the lower extremities.

Diagnosis.—There is little difficulty in distinguishing spinal irritation as there is always tenderness on pressure, if firmly made along the spinous processes of the affected part. This will always enable the careful observer to distinguish spinal irritation, proper, from transient neuralgic spinal pains which are generally sympathetic of irritation in other parts, as of the womb, testicles, bladder, kidneys, alimentary canal, &c.

Causes.—I am satisfied, from careful observation, that in most cases of spinal irritation in which there is tenderness on pressure, that some direct injury, such as wrenching, overlifting, falls, bruises, and other accidents, is generally the cause of the slow irritation which is set up. It is probable, however, that some cases in which there is tenderness on pressure, are the result of sympathetic irritation, from some remote disease in the stomach, liver, intestines, kidneys, womb, bladder, or other parts.

When such local irritation of the spine exists, it is increased or aggravated by very many causes, such as concussions, lifting, a low electrical state of the atmosphere, mental excitement, and various other causes, which alone might **not be sufficient to produce spinal** tenderness if it did not already exist.

Nature.—I believe that spinal irritation, **with tenderness,** consists in a **chronic** inflammation of the spinal ligaments, **of either a** simple or rheumatic character; the irritation, if not the inflammation, extending **more or less to the spinal cord** or its meninges, and **involving the** roots **of the spinal** nerves, as well as their envelope derived **from the** membranes **of the cord.** It is probable that in severe protracted **cases there is congestion or chronic** thickening **of** the ligaments of the vertebræ, **of the** spinal membranes, and probably of the origin of the nervous envelopes or neurilemma, which **disturbs, by the** pressure which is made upon the spinal cord and its nerves, more or less, the flow of nervous influence, not only along the cord, but also along the nerves to the muscles and other parts which **they** supply; **and** hence the neuralgia, and functional derangements which we have seen so often occur.

Prognosis.—The prognosis, **so far** as a serious result is concerned, is generally favorable, as most cases improve from proper treatment. But according to my observation, it is very liable to return, from various imprudences, **after it has been** materially **improved,** or even apparently **cured, by a proper course of** treatment.

Treatment.—To arrive at the **indications of treatment of** spinal irritation, it is necessary to ascertain whether it is entirely **the** result of some local injury of the part; or whether it may not have been produced, **or** greatly increased, sympathetically, by some organic, or other disease of the thoracic, abdominal, or pelvic viscera. If then the disease be sympathetic, **or if** it originated in injury of the spine, and has produced important sympathetic derangement of the different organs, or functions **of the** body; the first indications are to correct, as far as may be these deranged organs, or their functions, and then to subdue, if possible, the spinal irritation, if any remains. If the patient be of a strong vigorous constitution, a little blood may be taken by cups on each side of the **spine, at the** irritated point. But if the patient be **anæmic, dry cups only should be** applied for a few times; and then blisters, tartar **emetic pustulation, or** issue may be made use of, and continued till the **irritation is subdued.**

SECTION XVIII.—TETANUS—(*Locked Jaw.*)

By tetanus, from τείνω "I stretch," I mean that peculiar deranged condition of the nervous system; especially of the superior portion of the spinal cord, including the medulla oblongata; attended with tonic spasm of the voluntary muscles, without coma, or any essential disturbance of sensation, or of the mental faculties. This affection differs from many of the convulsive affections, in there being generally no sensorial, and intellectual derangement. This, however, is not an invariable rule; for

I have seen cases in which there was more or less mental disturbance from the first.

That variety of tetanus which affects the muscles of the throat and jaws, has been called *trismus;* when the extensor muscles of the trunk, and lower extremities are mainly implicated, the body being bent violently backward, it is called *opisthotonos;* when the body is curved forward, it is termed emprosthotonos; and finally, when it is curved laterally, it is called *pleurosthotonos;* all, however, being mere modifications of the same disease.

Tetanus has been again divided into *idiopathic* and *symptomatic:* the idiopathic including all cases which arise, without any other known disease condition; while the symptomatic is the result of some previous affection, involving generally, more or less, the nervous system. Cases of the symptomatic variety, which originate in wounds, or direct injuries, have been denominated *traumatic tetanus.*

This division, however, is of little consequence, as the real condition is the same in all, and the division into traumatic, including all cases the result of wounds or direct injuries; and idiopathic, including all other cases, may be sufficiently definite for our present purpose; as it appears to be the division more generally understood.

Tetanus generally approaches gradually, so that several days may elapse between the first manifestation of its invasion, and its complete development. Sometimes, however, its approach is rapid, its symptoms being fully developed in a few hours after the injury of which it is symptomatic is received.

Symptoms.—The most common early symptoms of tetanus, are stiffness of the back of the neck, spasmodic sensations in the muscles of the larynx, by which the voice is slightly changed, difficulty in swallowing, rigidity of the jaws, with pain in opening the mouth, and an uneasiness in the epigastrium, with pains darting from the pit of the stomach towards the spine. Soon the muscles of the neck and jaws become more stiff, the jaws being firmly closed. Deglutition is now performed with pain, and is apt to produce spasm. As the disease advances, the pain and retraction, with the darting pains, return at the epigastrium every ten or fifteen minutes, in violent paroxysms; and are generally followed by spasmodic retraction of the head, and soon by a rigid contraction of nearly all the voluntary muscles of the body.

There is a permanent spasm of the muscles, but as these paroxysms return, the muscles of the throat and chest are painfully contracted, the arms and legs may be forcibly extended, the shoulders drawn forward, the abdominal muscles retracted against the viscera, and the whole body thrown into a most painful and unyielding tonic spasm. These paroxysms at the commencement usually last but a few minutes, but in the latter period of the disease, the spasms remit but slightly, the muscular contractions being general and extremely violent.

During the progress of the disease, the countenance becomes distorted, copious sweats break out, the pulse becomes quick and irregular, the jaws immoveably fixed, the respiration laborious, the sphincters rigidly contracted; and towards a fatal termination, slight delirium occurs, and finally, a severe spasm terminates the scene, or the muscles become relaxed, and the patient dies in an apparent apoplectic state.

The voluntary muscles supplied by the spinal nerves are the ones first affected, but before a fatal termination, the involuntary muscles, controlled more especially by the ganglionic nerves may become affected, and thus by interrupting the involuntary functions, as the circulation, respiration, &c., be the cause of death. In some cases, the mind continues clear to the last, but in most cases that have fallen under my observation, there has been more or less mental disturbance during some period of the disease.

Tetanus may terminate fatally by the fifth or sixth day, or recovery may take place in a week or less. But the disease often continues for several weeks, and thus assumes a chronic form; and even after recovery there is apt to be a degree of stiffness of the muscles, and nervous irritability for several months.

Trismus nascentium or that variety of tetanus which occurs in infants soon after birth, supposed by some to arise from the irritation of cutting the cord; by others, from displacement of the occipital bone, but probably the result of filth, exposure, and various other imprudences, is of an alarming fatal character, calling loudly for cleanliness, fresh air, and a proper management of infants, as prophylactic measures against the disease. The disease occurs most frequently in hot climates, being very destructive in the West Indies, and among certain classes, especially the blacks in some of our Southern States. The attack generally occurs during the first or second week after birth, the disease being attended with spasmodic closure of the jaws, and the various other tetanic symptoms.

Anatomical Characters.—In some fatal cases of tetanus no traces of disease have been found, while in others, signs of congestion, or sanguineous injection of the cerebro-spinal meninges, and of the roots of the spinal nerves have been discovered, and sometimes increased vascularity of the brain and spinal cord, with or without serous effusion into the cavities. Blood has been found within the spinal sheath, and in some cases evidences of inflammation of the meninges of the brain or spine, and also softening of the brain or spinal marrow.

In traumatic cases, the nerves involved in the wound have been found variously lacerated, thickened, softened, &c. And the cervical and semilunar ganglia are sometimes more or less injected. Various other appearances are occasionally presented, such as tubercles, cerebral abscesses, or cartilaginous or bony formations in some portion of the spinal meninges, &c.

Diagnosis.—The only diseases with which tetanus is liable to be confounded, are hydrophobia, and convulsive hysteria, from both of which it may be readily distinguished by careful attention to all the symptoms. The symptoms of tetanus, including the permanent tonic spasm of the voluntary muscles, with paroxysms of greater violence, the peculiar pain at the epigastrium, the expression of the countenance, together with the absence of coma or marked delirium, will always enable the careful observer to distinguish tetanus from hysteria, hydrophobia, and in fact all other affections. It is of the greatest importance, however, to distinguish between inflammatory and non-inflammatory cases of tetanus, in arriving at the indications in the treatment, which may generally be done by noticing carefully any symptoms of cerebro-spinal inflammation which may arise, whether of the cerebro-spinal substance or their meninges.

Causes.—There appears to be a hereditary predisposition to tetanus in some constitutions, but in what that predisposition consists is not quite certain. I believe that exposure to filth, bad air, and various other imprudencies, which render the nervous system irritable, also predispose to this affection, as we have seen in the infantile cases occurring among the blacks in some portions of our Southern States.

It is probable, also, that males are more predisposed to tetanus than females, and that other things being equal, the disease is more liable to occur in hot climates than in temperate or cold climates, and, finally, that while the disease may occur at any age, it is vastly more liable to occur in the young and middle aged than in advanced life.

The *exciting causes* may act generally upon the nervous system, as sudden extremes of temperature, fear, &c., producing the idiopathic variety, or they may act locally, as wounds, bruises, &c., producing the traumatic. Of local injuries, those which bruise or injure a nerve, appear most liable to produce tetanus, especially if the wound has partially divided a nerve, or is of a punctured character.

In one successful case of traumatic tetanus which fell under my care, the injury was from the thorn of a plum tree, which penerated the heel. In another case which fell under my observation, the injury was of the nerves in the palm of the hand, produced by pressing it too firmly against the point of a heavy weight, which the patient had been lifting. This case was treated successfully by my friend and former partner and preceptor, Dr. Wm. V. V. Rose, of Watertown, N. Y. In traumatic cases, the interval between the reception of the injury and the occurrence of the tetanic symptoms may vary from a few minutes to two or three weeks, but I think it seldom occurs later than that.

Pathology.—Whether tetanus be the result of a general or local cause, irritation, congestion, or inflammation of the spinal cord, including the medulla oblongata, or of the spinal and perhaps cerebral meninges, invariably exist to a greater or less degree and extent in every case, and in some cases probably the white tubular, and perhaps even the gray matter of the brain become more or less involved. Now the degree and extent of this irritation, congestion, or inflammation of the spinal cord and brain, as well as of the cerebro-spinal meninges, makes the differences which occur in different cases of tetanus, as well as the differences which occur in different stages of the same case.

If there be only irritation with slight congestion of the spinal cord or its meninges, the tetanic symptoms are often mild and the case may terminate favorably. But if the spinal cord or its meninges becomes severely irritated, congested or inflamed, the tetanic symptoms are usually more severe, and the violent paroxysms of spasms more frequent, only slight remissions occurring early, while in the latter stages the violent spasm becomes continuous. In those cases in which the spinal irritation, congestion or inflammation extends either early, or during the latter stages of the disease, to the substance of the brain, there may be coma and more or less mental disturbance, especially if the gray matter of the brain becomes implicated; and the patient may die either from the involuntary muscles becoming involved, from exhaustion, or else in an apoplectic state.

Now in every case, whether there be only irritation of the spinal cord, or irritation, congestion, inflammation, &c., there is an interruption to, or derangement in the flow of nervous influence along the spinal cord and its nerves; in consequence of which the voluntary muscles are thrown into spasms. If, however, the brain itself becomes seriously involved, the power to act, or supply of nervous influence to the ganglionic system is interrupted or cut off, and hence the suspension of the vital functions and speedy dissolution which follows in such cases.

Prognosis.—The prognosis in tetanus is generally unfavorable, and especially is this the case in traumatic cases. In many cases of idiopathic tetanus, a recovery may reasonably be expected if the patient receive proper treatment, and with the same favorable circumstances, some traumatic cases may recover, at least such has been the result of my observation. The favorable symptoms are, mildness of the attack, relaxation between the paroxysms, and a tolerable performance of the respiratory and circulatory functions; while the unfavorable symptoms are the opposite of the above, and an evidence of an extension of the spasms to the involuntary muscles, indicated by a derangement in or interruption of more or less of the involuntary functions, as of the respiration, circulation, &c.

Treatment.—As tetanus is a disease in which the spine or its meninges are always either irritated, congested or inflamed, the indications to be fulfilled are usually very plain. In mild cases of traumatic or idiopathic tetanus, there may be only irritation with perhaps slight congestion of either an active or passive character, depending upon the general condition of the system at the time of attack. In cases of this character if it be the result of local injury, warm soothing fomentations or poultices should be applied to the wound, and suppuration promoted as far as possible. In addition to this in traumatic cases, and in mild idiopathic cases, the feet should if possible be placed in warm water, and a cathartic of the sulphate of magnesia, podophyllin or colomel in castor oil, should be administered and a free operation secured.

After the operation of a cathartic five grains of Dover's powder may be given, every four or six hours, and if the patient is weak, anæmic, or debilitated, three or four grains of the sulphate of quinine may be given with each Dover's powder, for the purpose of equalizing the circulation, and overcoming the cerebro-spinal congestion and irritation. If, however, the patient is of a strong and vigorous constitution, and the spinal congestion is of an active character, instead of the quinine, three or four grains of the James's powder may be given with the Dover's, and dry cups applied along each side of the spine, and thus the spinal congestion and irritation be overcome, as far as may be.

Thus I believe the indications may be fulfilled, in mild traumatic or idiopathic tetanus, in which there is only irritation with slight congestion of the spinal cord, active medication in such cases, in my opinion, aggravating rather than palliating the disease. But in severe cases of tetanus, whether traumatic or idiopathic, there may be not only irritation and congestion, but there may be a high state of inflammation of the spinal marrow, or its meninges, so that the case demands early and more active treatment, as it tends rapidly to a fatal termination—the inflammation extending to the brain or its meninges.

In such cases, after dividing the wounded nerves, if necessary, applying poultices to the wounds, &c., if the tetanus be traumatic; general bleeding should be resorted to, if the patient is strong, or of a full plethoric habit. After general bleeding, when it is indicated, and in all such cases in which general bleeding is not necessary, cups should be applied to each side of the spine, along its whole length, and from three to six ounces of blood taken at first. A cathartic of calomel and castor oil should be given, and its operation secured by repeated doses of oil, if necessary; or if calomel be contraindicated, a full dose of podophyllin, or senna, may be given, and followed by the sulphate of magnesia, if necessary.

After the operation of a cathartic, sixty drops of laudanum, or ten drops of the fluid extract of the cannabis indica, may be given every four hours; and with each dose of laudanum, or of the hemp, from one-fourth to half a grain of tartar emetic may be given. Alternating with the anodyne, two or three grains of calomel may be given, and continued till slight ptyalism is produced.

After the bleeding, cupping, and operation of a cathartic, if the tetanic symptoms continue unabated, vesication should be produced along each side of the spine, its whole length, either with a strong solution of corrosive sublimate, Granvill's lotion, or a solution of caustic potash, one drachm to an ounce of water. This solution of potash is the most convenient, in such cases, if it happens to be at hand; and it may be applied with a sponge, drawn quickly along each side of the spine, from the neck to the sacrum.

If the cupping and vesication be resorted to early, in these inflammatory cases of tetanus, before any effusion or other serious organic change occurs, the tetanic symptoms may subside, and the patient recover. It is well in such cases, after the tetanic symptoms subside, to give, for a week or two, five grains of the iodide of potassium, three times per day, to prevent any effusion into the cerebro-spinal arachnoid cavity, or into the ventricles of the brain; as there may be a tendency to such a result.

The patient should be sustained by a reasonable amount of proper nourishment, such as toast water, or crust coffee with milk, toast, rice, &c.; and if the powers of the system sink in this disease, broths, wine-whey, quinine, brandy, &c., may be indicated, and should be administered to fulfill the indications as they arise. The patient should be kept quiet, till the disease entirely passes off, and then great care should be taken to so regulate the habits of the patient, as to avoid any excitement of the body and mind, and that too for a long time.

By thus fulfilling the indications as they arise, in each case without doing anything more, I am satisfied that many mild cases of tetanus, and some severe ones may recover, at least such has been the result of my observation in this disease. In localities where tetanus is liable to occur in new-born infants, prophylactic measures are all important, such as keeping the child clean, dry, and properly nourished, and allowing it good, dry, fresh air.

SECTION XIX.—HYDROPHOBIA—(*Canine Rabies.*)

By hydrophobia, from, ύδωρ "water" and φοβος "dread," I mean that disease involving the nervous system, the result of the entrance into the system of a specific virus from rabid animals, communicated generally by the bite of the animal so affected. It is probable that the hydrophobic virus is contained exclusively in the saliva of the rabid animals, hence it is by the bite of the animal that the disease is almost invariably communicated to the human subject. **This disease** may, however, be **communicated by bringing the poison saliva in contact** with an excoriated **surface, or with a mucous membrane.**

In the dog, wolf, cat, or some other animals, hydrophobia **has** at some period been originated spontaneously, and it is probable, I think, that the disease **is now** developed in these animals from general causes, independent of the direct contact of the virus. It is possible that the disease sometimes occurs spontaneously in the human species, from want, hunger, **and** an undue degree of anger or mental excitement. When this disease **has** appeared to occur spontaneously in the animals most liable to it, it **has** followed the eating of unwholesome putrid food, and great rage, and **the** spontaneous symptoms in the human species, approaching most nearly to this **disease,** I believe have arisen from want, exposure and putrid food, **or excessive** hunger approximating starvation.

The period which intervenes between the reception of the contagion of rabies, and the development of the disease, may vary from a few days to several weeks, months, and perhaps years, but it appears probable, that the most frequent period of incubation is from one to three months.

Symptoms.—At an indefinite period **after the** reception of the virus, **the wound** is apt to assume a livid appearance, and it may discharge a **thin, ichorous matter, before** the accession **of the** hydrophobic symptoms. **There is also more or less** pain in the wound, which may pass along the nerves of the part towards their origin. During this period there is muscular prostration, flushes of heat, followed by chilliness, nausea, vomiting, thirst, constipation, and generally most of the symptoms which **usually precede the** development of **febrile disease.**

These symptoms may continue for five **or six** days, when **the** patient feels a stricture about the throat, and finds a difficulty in attempting to swallow liquids; **which** difficulty increases till it becomes impossible to swallow water, the distress being so great on attempting it that the very thought of water becomes intolerable to the patient. The disease being now fully established, the patient becomes nervous, irritable and excited; there is a copious secretion of saliva, stricture about the throat, embarrassed and interrupted **breathing, an** apparent necessity for fresh air, and the whole frame is in a state of tremor, agitation and almost convulsions.

If the patient lies **down** to rest, he starts up again **with** apparent anguish of feeling: the countenance expressing anxiety, terror, or deep despair. An attempt to swallow liquids, or even the thought or sound of water poured from one vessel to another, and various other slight causes will throw the patient into painful agitation, with, perhaps, convulsive spasms of the most distressing character. Acute pains are felt

21

in the epigastrium, back of the neck, and at other points along the spine, and the patient is apt to spit from the mouth, almost convulsively, a viscid mucus which collects in the fauces.

Remissions of partial quiet occur between the paroxysms, but during the exacerbations the countenance is wild and furious, the eyes are red and projecting, the muscles of the face, throat, chest, and perhaps extremities are thrown into spasms; the arms are thrown about, and the fists clenched; there is foaming at the mouth, and sometimes a disposition exhibited to bite whatever comes within reach of the patient. In violent cases, furious and maniacal raving occurs during the paroxysms, but during the intervals, in cases of less violence, consciousness is generally retained and the patient appears quite rational, often appearing fearful that he may be allowed to inflict some injury upon his attendants during the insanity of his paroxysms.

During the early stage of the disease, the tongue is furred, and there may be nausea and vomiting, but the pulse is sufficiently strong, and the skin warm or nearly natural; but as the disease advances, the pulse becomes weak and feeble, the skin becomes cool and finally covered with an offensive viscid sweat; and at last the patient dies, either quietly, or in violent convulsions. Violent cases usually terminate fatally by the fifth day; but it is possible that some cases may linger much longer than that, perhaps to the ninth or tenth day.

Anatomical Characters.—In some cases, the post-mortem reveals nothing which might be supposed to throw any special light upon the disease, but, generally, I believe, the brain and spinal cord, or their meninges, are found to have been in a more or less congested state, and in some rare cases there may be found slight softening of the brain or spinal cord at some point.

The mouth, tongue, fauces, œsophagus, and trachea, are found either reddened or else very pale, and the papillæ of the tongue are often enlarged, especially on its back part. The mucous membrane of the stomach often exhibits signs of inflammation; the lungs are found engorged; the blood is found changed; and the body tends rapidly to putrefaction after death.

Diagnosis.—Tetanus and certain forms of hysteria are the only diseases with which hydrophobia is liable to be confounded, and the discrimination in most cases is attended with little or no difficulty. Hydrophobia differs from tetanus in the dread of water, in its spasms being clonic instead of tonic, in the motion of the jaws being more free, and, finally, in there being a more copious secretion of saliva, and more derangement of the circulation, digestion, and mental faculties in hydrophobia than in tetanus.

Peculiar forms of hysteria, and other nervous affections, which may resemble hydrophobia, may generally be distinguished by the absence of an exposure to the poison of rabies, and by the dissimilarity in the essential symptoms of the two affections. In hysteria there is the alternate laughing and crying, "globus hystericus," &c.; while in rabies there is the dread of water, copious secretion of saliva, &c., which do not attend hysteric or other nervous affections.

Causes.—Hydrophobia had its origin, at some period, either in the human species or in the lower animals, and when once originated, is readily communicated from one animal to another, and also from animals to the human species, **and perhaps** from **man** to man, if the saliva be brought to act upon an abraded **surface, or a** mucous membrane.

The animals that have been known **to** communicate this disease from one to another, and to the human species, **are** the dog, cat, fox, wolf, jackal, and various other carniverous animals; and generally by the bite; the saliva, which contains the poison, **being thus** brought in contact with abraded or lacerated parts. It is very **probable,** however, that the saliva of all the **lower** animals laboring under **this disease,** and even of man, if brought to act upon an abraded surface, **or a** mucous membrane, may communicate the disease. But as the herbiverous animals **and the** human species are less given to **biting** than the carniverous animals, **the disease is** very rarely **communicated by them, by biting** at least.

While it is true that this disease **is now generally communicated by** the virus in the saliva of **rabid** animals, **by a bite or otherwise, I believe** that genuine hydrophobia **may arise spontaneously in the lower animals,** or even in the human **species; though it is probable that** such an event may very rarely occur. **That it may now so occur in** the one or the other appears to me **the more** probable, **when we** remember that it has **so arisen at some** period; and I can see no reason why the same liability may **not now exist, and** the same accidental **causes** be brought to bear, in developing the disease. Among the accidental influences which it appears to me may have an agency in producing hydrophobia independently of the direct contact of the saliva of a rabid animal, are extreme madness, want and exposure, with hunger approaching starvation, or **else the use of filthy,** unwholesome, **and putrid articles** of food, such as **the human species occasionally, and dogs and other carniverous animals very often partake of.**

It is sometimes a matter of importance to decide on the existence **of** hydrophobia in the lower animals, and especially in the dog, that proper precautions may be taken, if necessary. This may generally be done by noticing carefully the appearance and motions of the **animal.**

The animal has at first an altered suspicious look; is restless, and **exhibits** a disposition to take up, and even swallow bits of **straw, hair, wood,** &c., and not unfrequently vomits. As the disease advances, he is apt to become exceedingly irritable, quarreling with other dogs, and flying furiously at strangers, resisting correction; being rather enraged instead of terrified, by threatened chastisement. A copious secretion **of** saliva flows from the mouth; **there is extreme thirst;** labored breathing**; a** sort **of** howling bark; he is constantly **in motion;** has visual illusions**; and** finally the animal becomes paralytic, totters and falls; and dies **by the** fifth or sixth day, either quietly, or in more or less violent **convulsions.**

The period of incubation in the **dog,** varies from **three or four days, to** as many weeks, being shorter than in **the human** species.

Nature.—It appears to me probable, that hydrophobia is the direct result of a specific animal poison, derived **either** from the saliva of a rabid animal, or from various unwholesome putrid articles of food; the disease being called into activity from a latent state, perhaps by various accidental causes, among which fear, privation, and extreme anger or rage, are probably the most frequent.

Now that a specific poison should thus be taken into the system, from the saliva of a rabid animal, or be generated in the system from want, privation, and putrid articles of food; and there remain for a time in a latent state; and finally, by some exciting cause, such as violent passions, or irritation of any kind, produce sufficient derangement in the nervous system, including the brain and spinal cord, to develop all the symptoms of hydrophobia, is not so very strange. If then, there is a specific morbid poison in the system in this disease, I can see no reason why it should not produce exactly the effects it does upon the cerebro-spinal, and nervous system; and that there should be an effort on the part of the system to eliminate the morbid poison, which has produced the disease, by the salivary glands, is to me no more strange than that the poison producing mumps should fall mainly upon the parotid glands, and produce a specific inflammation.

Now the fact that the blood is often found in an altered condition in hydrophobia, and that the bodies of patients dead of this disease very soon become offensive, and tend rapidly to putrefaction, favor the idea of a poison in the blood, which is producing the disease; and that an effort on the part of the system to eliminate the poison through the salivary glands, may render the saliva poisonous; and hence the propagation of the disease by biting, or contact of the saliva.

Prognosis.—I believe that the prognosis in genuine hydrophobia, may be set down as invariably unfavorable. It is possible, however, that very many cases of nervous derangement, produced by fear, from the bite of a rabid animal, may recover after many symptoms analogous to hydrophobia have been developed, when none of the virus has really entered the system. And it is possible that the disease itself, may be palliated, and the patient partially recover, and live on for several years; as happened in one case, bitten by a rabid cat, that has fallen under my observation.

Treatment.—Immediately on receiving a bite by a rabid animal, or on receiving a portion of the saliva upon a raw surface, the part should be washed clean, and then removed by a free use of the knife; carefully avoiding to leave any of the affected part. A cup may then be applied over the wound, in order to draw out a little blood, and with it any poison that may have reached it. After removing the cup, a stick of lunar caustic may be thoroughly applied to the cut surface; and then, if necessary a bread and milk poultice applied till the surface is thrown off, after which simple cerate may be all the dressings necessary, during the healing of the wound.

If, however, the wound from the bite be of a deep lacerated character, and in a situation in which it cannot be readily removed; if it be in a finger or toe, the limb should be amputated, as a less calamity than risking the effects of the poison and the chances of hydrophobia. But if the bite be in a situation in which it does not appear judicious to amputate, the wound should be carefully washed with warm water, as much of its surface removed as may be with the knife, caustic thoroughly applied, and the wound kept open by blisters, poultices, &c., to favor as far as may be the elimination of the poison. If, however, in such cases, at any future period, hydrophobic symptoms make their appearance, the limb should be amputated at once, as affording, perhaps, the only chance of saving the life of the patient.

At any period within two years after the reception of the poison, the treatment I have here suggested may be resorted to, in case it was neglected at first. If the patient is of a good constitution, nothing further need be done, except that he should be made to conform rigidly to the laws of health in every respect. But if the patient be feeble, or debilitated, two ounces of the cinchona bark may be put into a quart of good port wine, and the patient directed to take a wine-glassful after each meal, for a time, to sustain the strength and prevent, if possible, the ordinary effects of this morbid agent upon the system. In these debilitated cases in which the wine is contraindicated, or does not agree, one or two grains of the sulphate of quinine may be given instead, and continued for a time, care being taken to keep the mind at ease, and the various functions of the body as well performed as possible. If, however, after every reasonable precaution has been taken, the disease makes its appearance, the patient should be kept from injuring himself, or any one else, and treated, if at all, upon general principles; fulfilling, as far as may be, the indications as they arise. The feet of the patient may be placed in warm water, a mild cathartic administered, and cups, wet or dry, applied to the back of the neck, or along the spine; the patient quieted, as much as possible, by moderate doses of the fluid extract of hops or valerian; and, finally, if the strength fails, quinine, broths, wine-whey, &c., may be given, as long as the patient is able to swallow them, that our duty in the case may be fully done.

SECTION XX.—NEURALGIA.

By neuralgia, from νευρον, a "nerve," and αλγος "pain," I mean that derangement of the nervous system in which there is at some point in its structure, acute nervous pain, without inflammation or any apparent structural change at the seat of the pain. Neuralgia may occur in any part of the system, as well in parts supplied by the ganglionic system of nerves, as in those supplied by the cerebro-spinal, as there are numerous branches connecting the two systems throughout their whole extent.

Symptoms.—The pain in neuralgia is always of a sharp, acute, darting character, passing rapidly from the seat of the irritation along the course of the affected nerve. The pain may come on and be continuous, or it may remit at irregular periods, or it may even intermit, the paroxysms coming on at regular periods, in such cases being generally quotidian or tertian.

During the paroxysms of neuralgia the pain is severe, with frequent transitory shocks of darting pain, so extremely agonizing as sometimes to cause a temporary loss of reason and consciousness. At such times the irritated parts are sometimes tender to the touch, a very slight touch often producing more suffering than firm pressure. During the continuance of neuralgia, there may be more or less congestion at the seat of the pain, and slight tenderness may remain even after the neuralgia has passed off. In violent cases of neuralgia, the muscles of the affected parts are liable to become affected with spasms, and when it occurs in the face, there may be a copious secretion of saliva, and flow of tears during the paroxysms.

Strictly periodical neuralgia I believe is generally of malarious origin, partaking very much of the nature of intermittent fever, except that there is no febrile reaction as in ague. But cases of neuralgia not of malarious origin, may only slightly remit, and some cases of a chronic character, continue almost uninterruptedly, for weeks, months, or even years, producing the most severe and protracted suffering. There is not an organ or tissue of the body in which neuralgia may not be developed, but some parts are more liable to become the seat of it than others, and among the most frequent are the following :

The *head* is very liable to a neuralgic affection, and in some cases being confined merely to the scalp, while in others, the brain is its special, if not, exclusive seat. Its symptoms differ from those of inflammation, in being confined to a particular part, and also in the absence of general febrile excitement. This variety of neuralgia generally affects the whole of one side of the head, but it is sometimes confined to the temples of one or both sides.

The *face* is exceedingly liable to become affected with neuralgia, and when it is, one side only of the face may suffer, all the parts supplied by the portio dura generally being more or less involved. In a case of this character that fell under my care a few years since, which was evidently the effects of the continued use of tobacco, the paroxysms which came on at evening, were attended with so much pain, that the patient became uncontrollable, groaning, screaming, and raving alternately, and getting little or no sleep for several nights.

The *optic nerves* is sometimes the seat of severe and protracted neuralgia. It may affect both nerves, but it is often confined to one side, and is attended with deep, sharp and acute pain, which produces more or less intolerance of light.

Neuralgia sometimes affects one or both of the *arms*, and it may be confined to one or all the brachial nerves, depending in the one case upon irritation of the spine at the origin of the brachial nerves, while in the other case, it depends upon some accidental disturbance in the nerve at some point between the seat of the pain and the brachial plexus.

The *intercostal nerves* are exceedingly liable to become the seat of neuralgia, and generally of a severe and darting character, sometimes very much resembling the pleuriatic pain. This variety of neuralgia depends upon congestion or irritation at or near the origin of the intercostal nerves.

The *breasts* of females sometimes become the seat of neuralgia, either one or both at the same time. In cases of this kind I have generally found a tenderness of the spine at the point which supplies them with nerves, and the neuralgic pain more or less increased by pressure on each or one side of the spine.

The *abdominal muscles* are occasionally, though not very frequently, the seat of neuralgia ; the dorsal portion of the spine in such cases being more or less congested, irritated or deranged as a cause. I have noticed this variety of neuralgia most frequently during irregular miasmatic fevers, in some cases of which it has been very severe.

One or both the *lower extremities* are sometimes the seat of severe and obstinate neuralgia. It generally follows along the ischiatic nerve, or it may affect mainly the crural or tibial nerves. This affection generally depends upon congestion or irritation in the lower portion of the spinal cord, in the lumbar or sacral nerves, or else in some disturbance of the nerve between the seat of the pain and the origin of the nerve.

Besides these external seats of neuralgia occurring in parts supplied almost exclusively with nerves from the brain and spinal cord, neuralgia is liable to occur in all the internal organs of the body, from the larynx to the rectum; affecting more or less those vital parts which, though mainly under the influence of the ganglionic system, are nevertheless more or less supplied by the cerebro-spinal nerves. Of the internal parts, liable to become the seat of neuralgia, the larynx, lungs, heart, stomach, liver, kidneys, womb, testicles, bladder and rectum, are, according to my observation, by far the most frequent seats.

When neuralgia effects the *larynx* or *lungs*, the pain is not from the nature of the structures involved, very acute, but there is very great distress, and an intolerable and almost constant disposition to cough.

When neuralgia affects the *heart* it is attended with the most excruciating pain and very great distress, affecting more or less the various functions of the body, and especially the circulation.*

When the *stomach* becomes the seat of neuralgia, there is generally a most unpleasant pain experienced, coming on often in paroxysms, and in very chronic cases, an hour or so after each meal.

The *liver* is quite often the seat of neuralgia, being attended with severe sharp, darting pains in the right hypochondrium, and more or less pain in the right or left shoulder. Neuralgia of the *liver* may be distinguished from an inflamed condition, by the want of febrile excitement in the neuralgic affection, and also by the slight functional derangement which occurs in such cases.

The *kidneys* are exceedingly liable to neuralgia, affecting sometimes one and at others both kidneys, and appearing in many respects very much like nephrites. It may be distinguished, however, from inflammation by the intermittent character of the pain, want of tenderness, and also of a general febrile excitement in the nephralgic or neuralgic affection.

The *uterus* is occasionally the seat of neuralgia; the pain being of a darting, lancinating character, and the affections being attended in some cases with slight hysteric symptoms, and more or less sympathetic irritation of the whole system.

The *testicles* are liable to neuralgia, the affection being often attended with a dull heavy pain in the spine, at or near the origin of the spermatic nerves. In one case of this character, that came under may care a few years since, the result of masturbation and excessive venery, the disease had been of several years standing, and was of a most distressing character.

Neuralgia may affect the *bladder*, and when it does, it is apt to produce considerable local and general disturbance; but it may be distin-

* See neuralgia of the heart or angina pectoris, in another part of this work.

guished from cystitis, by the absence of a general febrile excitement, in this affection.

The *rectum* is liable to a neuralgic affection of a most distressing character, in some cases being remittent or intermittent; but often of a chronic and continued character. Cases of purely intermittent neuralgia of the rectum are generally, I believe, of malarious origin.

It is probable that in neuralgia of all the internal vital organs of the body, there is in addition to an irritation of their spinal nerves, either at the spine, or at some point between the seat of the pain and their origin, also derangement with irritation, either primary or secondary, of the ganglionic or sympathetic nerves, by which these vital organs are also more especially supplied.

The *spine* is liable to a species of neuralgic irritation, which may locate at any point; but which more frequently passes from one point to another, along the spine, varying its seat, as I believe, according to the general strength of the patient. If the strength is declining, I have noticed that the point of local irritation is apt to pass towards the brain. But if the general strength is improving, the local point of neuralgic irritation will generally pass from the brain towards the lower portion of the spine, and finally disappear altogether.

As this local neuralgic irritation passes along the spine, opposite the different organs, their functions are apt to be more or less affected. If the brain be the seat of this neuralgic irritation, there is generally an undue degree of mental excitement; if it be in the cervical portion of the spine, there is more or less derangement in the functions of the larynx, heart, and lungs; if it be in the dorsal region, the stomach, liver, or intestines have their functions more or less affected; and finally, if the irritation occupies that portion of the spine opposite the kidneys, these glands become deranged in their functions, generally secreting an undue quantity of clear transparent urine. The same is also true in relation to the uterus, testicles, bladder, and rectum; and in fact of all the internal organs, as well as external parts of the body.

This neuralgic irritation of the spine sometimes occupies the same location for a long time, while in other cases it will pass rapidly from one extreme of the spinal cord to the other, in the course of a few hours; seldom occupying two distinct points of the spine at the same time, at least in any great degree.

Anatomical Characters.—Dissections have afforded no very positive evidence as to the nature of this disease. Generally, however, signs of congestion, irritation, or inflammation, have been detected in the cranium, spinal canal, or in the nervous trunk, at some point between the seat of the neuralgia, and the origin of the nerve.

Diagnosis.—Neuralgia may be distinguished from inflammation by the sharp and darting character of the pain; by its intermitting tendency; by the absence of any special tenderness to firm pressure; and also by the absence of any general febrile excitement. And besides, in neuralgia, the pain follows the track of some prominent nerve or nerves, which is not so remarkably the case in inflammation.

Causes.—I believe that *koino-miasmata* and atmospheric vicissitudes are among the most frequent causes of neuralgia. It is probable that the paludal poison operates through the blood, upon the brain and nervous system, to produce neuralgia, in the same manner that it does to produce intermittent fever, though generally in a less degree; while atmospheric vicissitudes act by interrupting the cutaneous exhalation, as in producing catarrhal fevers, though perhaps in a less degree.

It is probable also that the electrical state of the atmosphere has an important agency in producing neuralgia, a damp or low electrical state of the air developing the disease in debilitated, anæmic patients, while an opposite state may be an exciting cause in plethoric, or sthenic constitutions. Gastro-enteritis is also a frequent cause of neuralgia, and sometimes of intermittent neuralgia, especially if the system has at the same time been exposed to a malarious influence, many marked cases of which have fallen under my care during the past few years. Local injuries are also frequent causes of neuralgia, especially if important nerves are involved in the wound or injured part. Indigestible articles of food, taking food at irregular hours, masturbation, excessive venery, onanism, and the use of tobacco and intoxicating liquors, are very frequent causes of neuralgia.

A weak or depraved state of the blood, in consequence of which the circulation is illy performed, and therefore local congestion produced, is probably the condition of the system most compatible with neuralgic irritation. In this condition a very slight exciting cause will develop the affection. This condition, I believe, is the immediate cause of that neuralgic irritation which affects the brain and spinal cord, when no other exciting cause happens to be operating.

Now, while the above are some of the most prominent causes of neuralgia, I believe that the immediate cause is pressure, congestion, irritation, or inflammation, either of the brain or spinal cord, or their meninges, or else at some point along the nerve, between the seat of the pain and the origin of the nerve.

Nature.—As either pressure, congestion, irritation, or inflammation, at some point in the brain, spinal cord, or nerves, or else in their investing membranes, very generally exists as the direct cause of the neuralgia, it appears to me probable that the pain is from a derangement in the distribution or flow of the nervous influence, produced by its interruption, either at some point along the nerve, or in the spinal cord or brain. Further than this, I am not disposed to offer any opinion as to the nature of this affection.

Treatment.—The indications in the treatment of neuralgia are to correct the derangement of the system, to subdue any local congestion or irritation that may exist in the brain or spinal cord, or at any point in the nerve involved, to equalize the circulation, and to restore the strength and tone of the blood and nervous system. The direct and remote cause should also be sought out, and if possible removed or corrected, as well as all the habits of the patient, that might serve to deprave or weaken the blood, or in any way injure the tone of the nervous system. The state of the digestive organs should be inquired into, and if there is indigestion, or constipation of the bowels, the condition should be corrected, by antacids, laxatives, vegetable tonics, &c., and a properly regulated diet directed.

If a malarious influence has been operating to produce an intermittent neuralgia, two or three blue pills or two grains of podophyllin may be given, and followed by half an ounce of castor oil or of the sulphate of magnesia, and if there is much gastric irritation, a blister should be applied to the epigastrium; after which, the patient should be treated in all respects as for intermittent fever; twelve grains of the sulphate of quinine may be given in divided doses during the six hours next preceding the expected paroxysm. Two-grain doses may be given every hour till four doses have been taken, but four grains may be given at the fifth dose, which should be administered at one hour anterior to the expected paroxysm. As soon as the paroxysms are arrested, the medicine may be continued, dropping the first dose each day till but one dose is given, and then the quinine should be continued in one or two grain doses after each meal, for a time, to prevent a return of the disease.

If, however, the neuralgia be *continued*, or only slightly remittent, and continues after removing the cause, correcting the digestion, &c., some permanent local cause should be suspected along the nerve, in the spine, or perhaps in the brain itself. If, on examination, a local irritation be found at either of these points, cups should be applied and repeated if necessary; and later, should the disease continue, blisters may be of service, and perhaps pustulation with tartaric emetic ointment.

If, on examination, no local irritation be found operating as cause of the pain, and requiring cups, blisters, &c., and the patient be anæmic, congestion from debility, in some portion of the brain or spinal cord, with the irritation of the nervous system, which an anæmic state produces, are evidently the conditions which require to be corrected. In such anæmic cases of continued neuralgia, twenty or thirty grains of the subcarbonate of iron may be given at first, four times per day, till the neuralgia subsides, and then it may be continued in three or four grain doses, after each meal, till the integrity of the blood is fully restored. Or the iron by hydrogen, or the ferrocyanuret may be given instead of the subcarbonate, in case that should in any way disagree. As soon as the blood is thus corrected, and the tone of the nervous system restored, the circulation will be equalized, the cerebro-spinal congestion and nervous irritation overcome, and thus the neuralgia be permanently subdued.

Anodynes, either externally or internally, may become necessary to secure rest, but further than that I do not believe they are generally indicated, as they are liable to derange the various functions of the body, and permanently increase, rather than correct, nervous irritability. During the violence of the neuralgia, fifteen drops of the fluid extract of hyoscyamus, or ten drops of the fluid extract of the cannabis indica may be given at evening, to secure rest; and in very severe cases, in which the pain becomes a greater calamity than the remedy, similar doses of one of these anodynes may be given every six hours, till the violence of the pain subsides, when it should be gradually discontinued.

Externally, the tincture of opium may sometimes be of very essential service, especially in neuralgia of the spine; in such cases, thirty or forty drops of the laudanum may be rubbed along the painful portion at evening, with very happy effect. It may also be applied in neuralgia of other parts, should an external anodyne become necessary. **By thus** removing the cause and correcting the general and local derangements of the system upon which the neuralgia depends; not only the neuralgia is overcome, but the integrity of the system is restored.

SECTION XXI.—AMAUROSIS.

By amaurosis, from αμαυρος, "obscure," I mean an affection of the optic nerve, or its expansion, the **retina, in which it becomes** incompetent to receive or convey to the brain clear visual impressions of objects.

This impaired condition, or loss of sensorial function **of the optic nerve and** its expansion, **may** depend upon organic disease, or merely on functional torpor, or paralysis of these parts, without any structural lesion.

Among the organic conditions which give rise to this affection, are extravasation of blood, structural lesion, and deposition of lymph upon the surface of the retina; morbid growths, dropsy, and atrophy within the eye, and also sanguineous or serous effusion, tumors, &c., pressing upon the optic nerve within the head. The loss of functional **power of** the retina and **optic nerve** may depend **either** upon vascular turgescence of the retina, or sheath of the optic nerve, **or** on deficient arterial circulation in these parts, or it may be the result of an idiopathic paralysis or loss of the sensorial power of the **retina** and its nerve.

Symptoms.—Functional amaurosis usually comes on gradually, the patient at first complaining of some weakness of sight, which goes on gradually increasing till almost or quite total blindness is the result. There is usually more or less pain in the head and temples, which diminishes as the dimness of vision increases, and frequently ceasing altogether, when the amaurosis is complete. In cases, however, in which the pain continues and increases, **we may suspect the existence** of some organic affection as the cause. The eye in amaurosis has generally nearly its natural appearance, only the pupil is usually dilated, and immovable, and sometimes irregular in shape.

Diagnosis.—Amaurosis may generally be distinguished from incipient cataract by a careful attention to all the symptoms. The dilated or possibly contracted, and generally fixed or immovable condition of the pupil may serve in most cases to characterize amaurosis. In imperfect amaurosis, vision is frequently increased or diminished under different states of the circulation, while in incipient cataract no such change occurs.

In amaurosis the sight or pupil of the eye retains its natural **dark** appearance, while in cataract it very soon acquires a lightish **milky** appearance. And besides, in cataract, the patient is apt to see more distinctly just after sunset, on account of a dilatation of **the** pupil which takes place as the light is withdrawn, while such is not the case **in** amaurosis.

Causes.—When amaurosis is not the result of organic structural disease of the optic apparatus, it arises, in most instances, from pressure **on some** portion of the visual nervous texture. Even in those cases **that** arise from excessive losses of blood, it is probable that a passive **congestion** of the retina or optic nerve is the immediate cause of the disease. Amaurosis may also sometimes depend upon **mere** functional torpor, from previous over-excitement of the retina and optic nerve, with diminished vitality or sensibility of the nerves.

Among the exciting causes of amaurosis are metastasis of other affections, suppression of the catamenial or hæmorrhoidal discharges, the healing up of old ulcers, and the long continued use of tobacco and intoxicating liquors.

Prognosis.—The prognosis in all cases depending upon organic lesion, may be regarded as unfavorable; but when the amaurosis is incomplete, and not attended with protracted pain in the head or eyes, and the pupil retains its natural shining appearance, a cure may reasonably be expected.

Treatment.—The imprudence or cause which has been operating to produce the disease should be sought out, and removed if possible, and the general and local derangement corrected as far as may be, by a judicious course of treatment. At first, a cathartic of calomel and castor oil may be administered, and also a little blood taken from the temples and back of the neck, by cups; after which a pill of aloes and rhubarb may be given each day, after dinner, to correct the digestion, and regulate the bowels. As an alterative, a pill composed of two grains of the extract of conium, and two grains of blue mass, may be given, morning and evening, and continued for one or two weeks, depending upon the severity of the case; dry cups being applied to the temples and back of the neck each day, or every two days. After having continued the blue mass and conium as long as is expedient, it should be discontinued, and the iodide of potassium given in five grain doses three times per day, and continued for a long time. Blisters should now be applied to the temples, back of the ears, and back of the neck, and repeated; the patient being allowed a proper diet, and if necessary, iron to restore the blood, and finally strychnine, in small doses, if a nervous tonic should be required.

SECTION XXII.—SINGULTUS—(*Hiccough.*)

By singultus or hiccough, I mean that disturbance of the *phrenic nerve*, attended with sudden involuntary contractions of the diaphragm, with simultaneous contraction of the glottis, producing a peculiar laryngeal sound.

The phrenic nerve it will be remembered is formed from the third, fourth, and fifth cervical nerves, with a branch from the sympathetic. It descends to the root of the neck, enters the chest, passes through the middle mediastinum in front of the root of the lung to the diaphragm, to which it is distributed, some of its filaments reaching the abdomen through the openings for the œsophagus and vena cavâ, and communicating with the phrenic, solar, and hepatic plexus.

Symptoms.—The phenomena of hiccough consists in a sudden and rapid inspiration, suddenly followed by expiration, the movements being accompanied by a noise, not attending common respiration. These convulsive movements succeed each other at intervals of a few seconds, and are attended with more or less uneasiness at the præcordia, sometimes amounting to actual pain.

A paroxysm of hiccough may last from a few minutes to hours, days, weeks, or even months, but I have seldom known a case to continue more than two or three days, and often no more than a few hours, or even minutes. Generally in cases that are protracted, it is not entirely continuous, slight remissions or even intermissions occurring every few hours. At least such has been the result of my observation, in cases of hiccough.

Hiccough being purely a nervous affection, is more annoying than dangerous. It may, however, be indicative, in some cases, of a condition of the system, fraught with more or less danger. It is liable, like many other nervous affections, to assume an intermittent form, the paroxysms occurring at regular hours each, or every other day, &c.; the disturbance of the phrenic nerve, in such cases, being probably the result of some general derangement of the nervous system of a periodical character.

Causes.—In order to appreciate the causes which operate in producing hiccough, it is necessary to remember that the direct cause of the spasm of the diaphragm is probably a disturbance of the phrenic nerve, which we have seen receives a branch from the sympathetic; and also communicates freely with the phrenic, solar, and hepatic plexus of nerves; thus bringing the diaphragm into close sympathetic relation with the stomach, and all the abdominal and pelvic viscera.

Now, any cause capable of producing, either directly, or by sympathy, a certain degree of disturbance, or derangement of the phrenic nerve, will produce spasmodic contraction of the diaphragm, which, with the contraction of the glottis, develop the phenomena of hiccough. We find then, as we should suppose, that the most frequent exciting causes are the swallowing of improperly masticated food, in a dry state; over distension of the stomach; flatulence; acidity of the stomach; the use of alcoholic drinks; congestion, irritation, or inflammation of the spinal marrow; and finally mental emotions, and general nervous derangement from any cause.

Treatment.—The indications are evidently to first quiet the undue excitability of the phrenic nerve, and then to correct the local, or general conditions of the system, upon which the phrenic disturbance depends.

If the hiccough be from swallowing improperly masticated food in a dry state, a draught of cold water, or a little sugar or simple syrup, will generally allay the irritation and hiccough at once. If it depend upon spinal irritation, a dry cup on each side of the spine, in the cervical portion, or opposite the diaphragm, or to the epigastrium, may stop the spasms at once. A full inspiration, and then holding the breath for a short time, may arrest the spasm in slight cases. And finally, a sudden surprise, frightful or agreeable, will very often arrest hiccough.

In the worst case I ever saw, in which all the usual remedies had been judiciously applied in vain, by his medical attendant, I succeeded in arresting it, by taking the light from the sick room, and giving as my reason, to the patient and his attendants, that if left in the dark, he could not see to hiccough. The patient had been very sick, and his nervous system was in a high state of excitability; and yet, ridiculous as was the idea of being unable to see to hiccough, the impression it made upon the nervous system, through the mind, so far affected the phrenic nerve, as to suspend the spasms of the diaphragm, and the patient speedily recovered.

In obstinate cases, in which though the paroxysms may be arrested, they are liable to return, from very slight causes; attention should be directed to the local or general conditions of the system, of which the hiccough is symptomatic. If there be acidity of the stomach, indigestion, or constipation; a teaspoonful of magnesia may be given each morning; and if necessary, a pill of aloes and rhubarb at evening, till the gastro-intestinal derangement is corrected; after which, if the patient is anæmic, iron with mild vegetable tonics, &c.,may be necessary, till the tone of the system is restored.

CHAPTER VIII.

DISEASES OF THE DIGESTIVE SYSTEM.

SECTION I.—STOMATITIS—(*Sore Mouth.*)

BY the digestive system, I mean the alimentary canal and its accessory organs and parts concerned in digestion. The digestive tube, it will be remembered, is lined throughout its whole extent with a mucous membrane, which is a continuation of the skin. Exterior to the mucous membrane is a muscular coat, while that portion of the canal forming the stomach, small and large intestines, has an external serous coat, which extends itself and lines the cavity of the abdomen. Its accessory organs, the salivary glands, the liver and pancreas, are glandular structures having ducts emptying into the canal, while the spleen, though without a duct, is probably concerned in the digestive process.

Having thus called to mind the anatomy of the digestive system, let us before we proceed to the consideration of stomatitis, remember that digestion commences with the mastication and mixing of the salivary secretion in the mouth; that the food is still further dissolved by the gastric secretion in the stomach; that the bile and pancreatic secretion changes it still further; and finally that the nutritious portion is taken up by the absorbents along the intestines, carried by the lacteals through the mesenteric glands which produce another change, and then goes to the thoracic duct through which it passes into the blood.

Having thus taken a glance at the anatomy and physiology of the digestive system, the diseases of which will occupy the present chapter, let us now proceed to the consideration of inflammation of the mucous membrane of the mouth or *stomatitis*, the subject of our present section. By *stomatitis*, from στομα, "the mouth," and *itis* inflammation, I mean here inflammation of the mucous membrane of the mouth, in whatever form it may occur.

Symptoms.—*Simple* or common inflammation of the mucous membrane of the mouth is characterized by redness, heat and tenderness, if the tissues beneath become involved. Blisters may form, and even ulcerations occur, involving more or less of the mucous membrane, especially in certain depraved conditions of the system. There is generally a copious secretion of saliva, the sense of taste is impaired, mastication is difficult, the tongue is covered with a white fur, or else red and glossy, the gums may swell, and perhaps ulcerate, and in severe cases a slight symptomatic fever may be developed.

Thrush is another form of inflammation of the mouth occurring at all ages, but more frequently in early infancy; and presents, in addition to the ordinary symptoms of mucous inflammation, patches of curd-like matter, appearing first on the inside of the under lip, and extending gradually to the inside of the cheeks, roof of the mouth, tongue, and perhaps fauces. This curd-like exudation may become thick, assume a darkish color and fall off, its place being supplied by a new crop; and this may occur repeatedly, the mouth becoming hot, and the voice more or less changed in most cases. In some cases the disease extends to the stomach and intestines, the patient being afflicted with acidity, colic pains, diarrhœa, vomiting, &c., and frequently with more or less febrile disturbance.

Follicular inflammation is another form of stomatitis, characterized by red eminences, especially upon the tongue and palate; which inflamed eminences are apt to pass on to ulceration, being distinct or confluent, according to the number of granules involved.

Aphtha is still another form of stomatitis, and consists of small ulcers, the result generally of a vesicular eruption. When the ulcers are fully formed they are whitish, painful, and surrounded by more or less redness and inflammation, and if they are confluent, more or less constitutional disturbance attends.

Ulcerative Stomatitis consists of ulcers, which may occur in any part of the mouth, having a white or darkish surface, and being surrounded by more or less inflammation and swelling. The mouth becomes swelled, the ulcers are painful, the breath becomes offensive, there is a copious secretion of saliva, and the disease is attended with more or less febrile excitement.

Nursing Sore Mouth is an ulcerative inflammation of the mouth, occurring in the latter months of pregnancy, or in women while suckling. It is attended with a loss of taste, heat in the mouth, tenderness, a smooth, red tongue, a copious secretion of saliva, diarrhœa and emaciation, and unless arrested it may terminate fatally.

Gangrenous Inflammation of the mouth is most frequently met with in children, but it may occur at any age if the system be in a certain morbid or depraved state. There is, in some cases, a slight inflammation, and soon a grayish eschar, either on the gums or at some point in the mouth, attended with some swelling, if it be upon the inside of the cheek. The patient becomes languid and weak; the slough spreads; the breath becomes fetid; the alveolar processes become necrosed; the teeth fall out, and thus the disease passes on, till portions of the cheek, palate, and upper jaw are destroyed, and the patient finally dies from extreme exhaustion.

Mercurial Stomatitis commences with a slight redness and swelling of the gums, a metallic taste, and increase of the salivary secretion; the gums become tender; there is stiffness of the jaws; the palate becomes swelled: the breath is offensive; ulceration occurs; the mouth is opened with difficulty; hemorrhage may occur; and in bad cases, sloughing may take place, and the bones of the jaw be laid bare.

Such, then, are the symptoms peculiar to the *simple* and other varieties of *stomatitis*, the causes of which we will now proceed to consider.

Causes.—The different varieties of stomatitis, as we have seen, may arise from various causes: the *simple* from direct irritants; *thrush* from impure air, unwholesome diet, &c.; the *follicular* from the exanthematous fevers, phthisis, or inflammation of the abdominal viscera; *aphtha* from gastro-intestinal irritation or disturbance; the *ulcerative* from want of cleanliness, bad air, deficient or improper food; the *nursing* from the puerperal state; the *gangrenous* from a malarious influence, insufficient or unwholesome food, crowded apartments, &c.; and the *mercurial* from the improper use of that mineral, especially in scrofulous, debilitated and depraved constitutions.

Now, while these various causes operate to produce different forms of stomatitis, it is probable that if we except direct irritants, and perhaps the use of mercurials, the other causes may operate in producing any form, to which there is in the system, either a hereditary or acquired predisposition. Now, the predisposition to stomatitis, consists, as I believe, in a debilitated, depraved, and deranged condition of the system, and especially of the digestive apparatus and its functions; in consequence of which, the mucous membrane of the mouth takes on some one of the diseased conditions above described.

Treatment.—On examining a case of stomatitis, not only the condition of the mouth should be ascertained, but also of the digestive system, and in fact the whole body. When this has been done, the direct and remote causes should be sought out, and removed or corrected, as far as may be. If the exciting cause be a direct irritant it should be removed; if from nursing, the child should be weaned; if from the use of a mercurial, it should be suspended; and finally, if it be from filth, bad air, deficient or unwholesome food, &c., these unfavorable circumstances should be removed, and the patient be kept clean, allowed good air, and a proper amount of plain, digestible, and nourishing food, to be taken with regularity. Having thus removed or suspended the cause or causes that have been operating, attention should be directed to correcting the local disease of the mouth, and also the general condition of the system, upon which it depends.

If the general disturbance consists merely in indigestion, with constipation or diarrhœa, with acidity, &c., the acidity should be corrected by small doses of prepared chalk, if there be diarrhœa, but with magnesia, if there be constipation. If, however, there be debility of the digestive organs, with poverty of the blood, small doses of the compound tincture of rhubarb, or of the fluid extract, may be given, with the infusion, or fluid extract of columbo, before each meal, to improve the digestion; and to restore the blood, small doses of the ammoniated citrate of iron may be given after each meal, and continued for a time. If, however, there be a scrofulous condition, the syrup of the iodide of iron may be given instead of the citrate, and, if the stomach will tolerate it, small doses of the cod liver oil, an hour after each meal, may help to supply nourishment for the system, without much labor of the digestive organs.

If there is a general depraved condition of the system, and especially of the glandular system, or a syphilitic taint, or mercury in the system, the iodide of potassium may be indicated, and should be given, instead of, or with the vegetable bitter, before each meal, and continue for a long time. If other general morbid conditions exist, they should be corrected by proper remedies, judiciously applied; and while this general treatment is being pursued, various local applications may be resorted to, for the purpose of correcting the local morbid condition of the mouth.

In *simple* cases of sore mouth, a strong sage tea with alum, one or two drachms to half a pint, sweetened with honey or loaf sugar, may generally do very well, as a wash.

In *thrush*, a strong decoction of the root of the geranium moculatum, with one or two drachms of borax, made sweet with honey or loaf sugar, and applied six or eight times per day, is an excellent application. Or if the fluid extract of the cranesbill be at hand, an ounce of it may be added to six ounces of water, and then the wash prepared with the borax and honey, or sugar, as directed above; and applied several times per day.

In *aphtha*, in the early stage, mucilage of gum arabic, with a little laudanum, may be indicated, till the general inflammation of the mouth subsides, leaving only the ulcers, which may then be touched, carefully, with the nitrate of silver, in substance, taking care not to injure the surrounding parts. In case, however, the ulcers are very numerous, instead of applying caustic, a saturated solution of alum, in simple syrup; or water, acidulated with muriatic acid, and added to an equal part of honey, may be used as an application to the ulcerated surface.

In *ulcerative stomatitis*, the alum in simple syrup, or the dilute muriatic acid, with honey, as directed in aphtha, should be applied at first, and if after continuing the application for a reasonable time, points of ulceration remain, the lunar caustic in substance, or in strong solution, may be applied to complete the cure.

In cases of *nursing sore-mouth* a decoction of the petals of the red rose, made from two drachms of the petals, and half a pint of water, to which two drachms of borax, and two fluid drachms of laudanum should be added, will generally do well as a local application. In cases, however, in which this fails, a solution of the nitrate of silver, two grains to the ounce of water may be used instead, till a cure is effected.

In *gangrenous stomatitis*, as soon as the white or grayish surface appears, the solid nitrate of silver may be applied, or, what may perhaps answer, the tincture of the chloride of iron. After the separation of the sloughs, a solution of alum, the decoction of peruvian bark, or the tincture of myrrh may be applied, to promote the healing process. In mild cases of this affection, a solution of the chloride of soda (Labarraque's), diluted with eight parts of water, is a very valuable local application.

In *mercurial stomatitis*, a strong decoction of the water-pepper, or smart-weed (Polygonum punctatum), used as a wash, especially if there be general inflammation and swelling of the mouth, is a valuable remedy. Instead of the decoction, an ounce of the fluid extract may be diluted with seven ounces of water, and this may be used till the height of the inflammation is subdued, and then it may be continued with alum till a cure be effected. Thus have I completed what I had to say, on stomatitis, in its various forms.

SECTION II.—GLOSSITIS.

By glossitis, from γλωττα, "tongue,' and *itis*, "inflammation," I mean here inflammation of the substance of the tongue, involving its muscular tissue. The tongue, it will be remembered, is composed of muscular fibres, distributed in layers, some longitudinal, others transverse, and others still oblique and vertical, between which there is considerable cellular tissue and adipose substance. The tongue, thus constituted, is liable to become inflamed, and when it does, from the nature of its structure, it is apt to be rapid in its course, and to be attended with considerable swelling.

It generally begins with a throbbing pain in the tongue, which is soon followed by redness and swelling. In a few hours, the whole tongue becomes involved in the inflammation, and a high state of general febrile excitement is developed. The swelling of the tongue increases rapidly, till it nearly fills the mouth, and is sometimes thrust out between the teeth. The respiration becomes extremely difficult, sometimes threatening even suffocation, and deglutition becomes difficult, if not impossible, in bad cases at least.

The tongue is generally of a darkish red color, and dry, though it may be moist, and covered with a reddish-yellow, or whitish fur. As the general febrile excitement becomes developed, the pulse is quick and strong, and the skin hot and dry; but if the breathing becomes greatly embarrassed, the pulse has less strength, and cold sweats may occur. The inflammation may terminate in resolution, or it may pass on to suppuration or gangrene. If suppuration takes place, one or several abscesses may form. If gangrene, only portions of the tongue may slough, and so a tolerable reparation of the loss takes place after the disease subsides. In some cases a permanent induration remains after the subsidence of the inflammation.

Causes.—Inflammation of the tongue may occur by extension of the inflammation in tonsilitis, or it may arise in the exanthematous fevers, or from a mercurial course.   •

Glossitis may also occur from atmospheric vicissitudes, but I believe it is most frequently produced by local irritating causes, such as acrid substances taken into the mouth, wounds, bruises, or the sting or bite of venomous insects, &c.

Treatment.—Immediately on the appearance of inflammation of the tongue, the patient should have his feet placed in warm water, and if the patient be of a sthenic constitution, blood may be taken from the arm, and then cups applied to the back of the neck, and leeches along the margin of the lower jaw, if they are at hand. If leeches are not to be had, and the inflammation and swelling are very great, incisions may be made into the tongue from the base to the tip, on each side of the median line, care being taken not to wound the ranular arteries.

The warm foot-bath may be used, morning and evening, and immediately after the general bleeding, if the patient can swallow, a teaspoonful of the fluid extract of senna should be administered, with half an ounce of the sulphate of magnesia, and the dose repeated in four or six hours if necessary. In case the patient is unable to swallow, the bowels should be kept loose by enemata, and should suffocation be threatened, the larynx or trachea should be laid open at once.

SECTION III.—PHARYNGITIS—(*Sore-throat.*)

By pharyngitis, from φαρυγξ, "pharynx," and *itis*, "inflammation", I mean inflammation of the mucous membrane of the pharynx; whether simple or ulcerative; acute or chronic.

The pharynx is the musculo-membranous canal, situated between the base of the cranium and the œsophagus, and in front of the vertebral column. It is formed externally by a muscular coat, and internally by a mucous membrane which is continuous above with the schneiderian membrane; in the middle, with that of the mouth, and below with the mucous membrane of the œsophagus.

The pharynx is a common origin of the digestive and respiratory passages, giving passage to the air in respiration, and to the food in deglutition, and is more or less liable to become inflamed. Inflammation of the mucous membrane of the pharynx may be acute or chronic; simple or ulcerative. It may involve only the superior portion of the pharynx, or it may extend down to the œsophagus.

Symptoms.—*Simple inflammation* of the mucous membrane of the pharynx is first noticed by a feeling of dryness in the throat, and more or less pain in swallowing. Soon the fauces assume a red color, become swelled, and finally, present white patches of lymph, which is thrown out by the inflamed follicles. The dryness and heat of the throat increase, and the pain becomes intolerable, if an attempt be made to swallow. The voice becomes hoarse, a viscid mucus collects in the throat, and slight febrile excitement follows, with loss of appetite, headache, a dry skin, frequent pulse, &c. The inflammation generally terminates by resolution in a few days, but it may gradually extend to the schneiderian membrane, eustachian tube, or larynx, and the inflammation assume a chronic form.

When inflammation of the pharyngeal mucous membrane assumes a chronic form, it may continue for a long time, being attended with heat, redness, &c.; the membrane being thickened, and the mucous glandules enlarged, presenting eminences in different parts of the pharynx. A muco-purulent secretion collects in the fauces, ulceration sometimes takes place either superficial or involving the tissues beneath the mucous membrane, especially if there be a depraved condition of the system; and finally, unless the disease be arrested, the uvula becomes enlarged, the hearing indistinct, and the patient is tormented with a distressing laryngeal cough.

Pseudo-membranous inflammation of the pharynx sometimes occurs, being characterized by the exudation under the epithelium of fibrinous matter. This form of pharyngitis commences with the ordinary symptoms of mild sore throat. Soon, however, the fibrinous exudation commences, exhibiting, irregular patches, of a whitish or gray appearance, extending, more or less over the pharynx. This exudation under the epithelium is of the character of false membrane, presenting under the microscope, according to Professor Wood, "interlacing fibrils, molecular granules, epithelial cells in different stages, and often pus or blood corpuscles."* In severe cases of this affection, the exudation may extend till it covers the whole pharynx; and it may extend into the larynx, and even along the trachea to the bronchia, obstructing respiration, and leading to the most disastrous results. This exudation may finally be absorbed, or it may soften and become mingled with the fluids of the mouth, the surface being covered with a viscid puriform mucus mingled with more or less blood, which keeps up a constant hawking and spitting, with more or less cough.

This variety of pharyngitis is evidently connected with a deranged condition of the system, and especially with an abnormal state of the blood. It may be mild, terminating favorably in two or three weeks; or it may extend, as we have seen, assume a malignant form, and lead to the most serious results.

An *ulceration* or *gangrene* may occur in any form of inflammation of the mucous membrane of the pharynx, or either of these conditions may occur very early, so as to constitute a primary feature of the pharyngeal affection.

When *ulceration* occurs, either as a primary or secondary affection, there is a sharp, pricking sensation, in addition to the ordinary symptoms of sore throat; and on looking into the fauces, one or more oval spots of ulceration, at first white, but later excavated and red, are discovered. If the derangement of the system upon which the ulceration depends, be corrected, these excavated points may speedily fill up, and cicatrize. But if the derangement continue, the ulceration may not only continue, but spread more or less extensively.

When *gangrene* occurs in simple or pseudo-membranous pharyngitis, or as an apparent original affection, it may involve only the mucous membrane, or it may extend to the structure beneath; more or less extensive sloughs being formed. I believe that in all primary or secondary gangrenous affections of this character, there is a pre-existing state of the system, and especially of the blood, which leads to the mortification which occurs.

The symptoms attending this gangrenous condition, including the general inflammatory symptoms, the gray or darkish slough, and the irregular depressions which they leave together with the fetid breath, are sufficiently indicative of the destructive processes which are going on.

Diagnosis.—There is no difficulty in distinguishing simple pharyngitis, or even the pseudo-membranous, if the parts be carefully examined. When ulceration or gangrene occurs in either variety, or as original affections, a careful observation of all the symptoms, will enable the

* See Wood's Practice of Medicine, vol. i. p. 551.

acute observer to get at the real condition, which should always be the point aimed at in making an examination in any case.

Causes.—The most frequent causes of simple mucous pharyngitis are, according to my observation, damp air, atmospheric vicissitudes, furs worn about the neck, the use of tobacco, drinking hot tea or coffee, acrid eructations from the stomach, gastro-intestinal irritation, wet feet, and finally living or sleeping in low damp apartments.

Now any one or all these causes may operate to produce pseudo-membranous pharyngitis in patients whose constitution, and especially abnormal state of the blood, predispose to the fibrinous exudation which occurs in such cases. And if there is, in addition to this condition, the debility arising from hunger, unwholesome food, and various abuses of the system, we may have ulceration or even gangrene supervening or even occurring apparently, as the primary affection.

Treatment.—In simple pharyngitis, it may be sufficient to remove the cause which has been operating, whatever it may be; to give a full dose of the sulphate of magnesia, to direct the warm foot bath at evening for two or three nights, and to have the throat gargled, three or four times per day with a strong sage tea made sweet with honey or sugar, and containing in solution alum, one or two drachms to the half pint.

If, however, the inflammation assume a chronic form, or be so from the first, the general condition of the system should be inquired into, and especially that of the digestive system; and the general health corrected by a regulated diet, and by such remedies as are clearly indicated, at least as far as it may be. If there be acidity of the stomach with constipation, a teaspoonful of magnesia may be given each morning, but if diarrhœa attend, a little prepared chalk may be given after each meal and a pill of aloes and rhubarb at evening, each day, till the constipation is corrected.

While this general condition is being corrected in the early stages, astringent applications may be of service. If there be a general relaxed condition of the mucous membrane, with only a moderate degree of tenderness, a strong solution of tannin, a drachm to half a pint of water will do very well as a gargle after each meal. Or if the patient be a child, an ounce of the fluid extract of cranesbill may be added to eight ounces of water and sweetened with honey or loaf sugar, and used instead.

As the relaxed condition of the pharyngeal mucous membrane is thus corrected, if there remains some tenderness, a fine powder made of equal parts of alum and loaf sugar may be dropped into the throat each night on retiring to bed. The head may be held back, and two or three pinches of the powder dropped carefully into the fauces, and as it gradually dissolves it passes down, and thus comes into contact with the mucous surface in the lower portion of the pharynx.

After the general condition of the system is corrected, if the astringent applications have failed in effecting a cure of the local difficulty, a strong solution of the iodide of potassium, or nitrate of silver, should be applied to the fauces, with a camel's-hair pencil, once each day, and continued till the local morbid condition is corrected. If there be only soreness, without much thickening of the mucous membrane, the nitrate of silver may do best, of the strength of thirty or forty grains to the ounce of water, or even stronger. But in cases in which there is thickening of the mucous membrane, with an edematous state of the tissues beneath, the iodide of potassium will generally do best, of the strength of three or four drachms to the ounce of water.

In cases of acute pharyngitis, in which there is considerable swelling, without very much tenderness of the throat; and in chronic cases, in which there is a viscid secretion, and the astringent and caustic applications are not indicated, or appear to disagree, a solution of the muriate of ammonia, in vinegar and water, will do very well. Two drachms of the muriate of ammonia may be dissolved in half a pint of vinegar and water, four fluid ounces of each, and used as a gargle, morning and evening.

In *pseudo-membranous* pharyngitis the general condition should be corrected, depletion being resorted to, with mercurials, if the system be in a sthenic condition; and tonics, with the iodide of potassium, if the system be in a debilitated or asthenic state.

While the general condition is being thus corrected, and the extension of the disease to the larynx carefully guarded against, fifteen or twenty grains of the sulphate of zinc to a fluid ounce of water may be applied, twice each day, to the pseudo-membranous patches, by means of a camel's-hair pencil. In case the zinc should prove inefficient, either the undiluted muriatic acid, or solid nitrate of silver may be used instead, and continued till a cure is effected; mucilaginous gargles, sweetened with honey or sugar, being used between the caustic applications.

In cases in which ulceration or gangrene occur, the general condition should be corrected by the iodide of potassium, iron, quinine, a regulated diet, &c., and the ulcers touched daily with a solution of the nitrate of silver, twenty grains to the fluid ounce, till their surface becomes red, when they will generally heal, as will gangrenous depressions, as the general condition is corrected, and the sloughs removed.

SECTION IV.—TONSILLITIS—(*Quinsy.*)

By tonsillitis, I mean here inflammation of the substance of the tonsils, which it will be remembered, are two glandular organs, situated between the anterior, and posterior pillar of the soft palate, on each side of the fauces. They are composed of an assemblage of mucous follicles, which open through the mucous membrane, which lines their inner surface, and pour out a transparent viscid mucus, which lubricates the fauces, and favors deglutition.

The tonsils, thus situated, are quite liable to become inflamed, either one or both; but more generally both, I believe, and the inflammation is of a phlegmonous character. In our climate, quinsy is very common, and though in general not a dangerous disease, may under certain circumstances, assume a malignant character.

Symptoms.—The disease generally begins with slight chills, succeeded by a high grade of febrile excitement, attended with an uneasy feeling in the throat, and some uneasiness in swallowing. Generally in a few hours, the uneasiness in the throat or tonsils becomes very marked, and deglutition more painful, till at last it becomes extremely difficult or impossible. On examination of the throat, one or both tonsils are found swelled, and generally the whole fauces somewhat red and tumefied. The tongue is also sometimes slightly swelled, and covered generally with a thick layer of transparent mucus.

The face is red; the carotids beat strongly; respiration is difficult; the hearing obtuse; the pulse frequent, hard and full; and generally the voice is indistinct. The pain shoots into the ears, if the patient attempts to speak, the mouth being opened with great difficulty. A thick ropy mucus adheres to the inflamed parts, which very much increases the difficulty of respiration.

When both tonsils become inflamed, they frequently nearly fill the throat; the uvula and soft palate being generally more or less swelled. The external part of the throat, in the region of the tonsils, is generally more or less swelled, and tender to the touch. This inflammation generally terminates in resolution or suppuration; and not unfrequently will abscesses form in a few days, especially if neglected during the early stages. These abscesses usually point internally; but sometimes they point externally, under the angle of the jaw. Frequent attacks of this inflammation, generally produce more or less permanent enlargement, or induration of the tonsils.

Causes.—Some individuals appear strongly predisposed to tonsillitis; but other things being equal, persons of a strumous diathesis, are most liable to this affection. The disease is much more liable to occur in damp changeable weather, than when the air is dry, either warm or cold. Among the exciting causes, sudden exposure to cold, or standing long on cold wet ground, or ice, by checking the perspiration, will frequently produce tonsillitis, in those who are predisposed.

Treatment.—At the very commencement of this affection, warm drinks of sage tea, and a warm pediluvium will be of essential service; and sometimes by restoring the perspiration, may arrest the disease. But if, as generally happens, the fever and local inflammation continue, cups should be applied to the back of the neck, and a brisk cathartic administered. Half an ounce of the sulphate of magnesia, with a drachm of the fluid extract of senna may be administered at once, and repeated in six hours, if necessary.

To allay the general febrile excitement, promote perspiration, &c., one-eighth of a grain of tartar emetic, or one-fourth of a grain of ipecac, or five drops of the fluid extract, with four or five drops of the fluid extract of the veratrum viridi may be administered every four hours, the warm foot bath being used morning and evening. As a gargle, half an ounce of the muriate of ammonia, dissolved in half a pint of vinegar and water, four fluid ounces of each, to which two drachms of laudanum may be added, will do well in all stages of the inflammation, and may be used every four or six hours. Sinapisms in the early stage, with fomentations of hops wet in warm vinegar, may be of essential service, if applied over the region of the tonsils, but later, blisters may become necessary.

If however, as sometimes happens, abscesses form, an early evacuation of the matter should be sought, by a free opening in the most prominent part, care being taken not to reach the carotid artery with the instrument. A mild unstimulating diet should be directed for several days, and every possible exciting cause of this affection should be carefully avoided. In cases of tonsillitis that assume a malignant or typhoid character, quinine, wine-whey, and animal broths may become necessary, and thus the system be sustained till the local affection subsides.

SECTION V.—PAROTITIS—(*Mumps.*)

By parotitis, I mean inflammation of the parotid gland. This gland named from its position, παρα, "about", and ους "the ear," is the largest of the salivary glands, and situated in front of the external ear. It is composed of many separate lobes with excretory ducts, which unite to form Stenon canal. This canal passes forward horizontally and terminates in the mouth, opposite the second upper molar teeth. The parotid glands secrete saliva, which is poured through this duct into the mouth, to moisten the food as it is masticated.

The parotid glands thus situated, are liable to a specific inflammatory affection, capable of being propagated by contagion.

Symptoms.—Parotitis usually commences with slight febrile symptoms, a stiffness of the jaws, and a slight pain and swelling in one or both parotid glands. The swelling generally gradually increases till the end of the third or fourth day from the beginning of the disease, at which time the glands are usually considerably swelled, and tender to the touch. The skin generally remains nearly of its natural color over the inflamed part, but sometimes it is slightly red.

Mastication and deglutition are generally attended with considerable pain. And though there is only a mild fever it is often attended with considerable nervous irritability and restlessness. From about the fourth day of the disease, the swelling, tenderness, and fever gradually decline, disappearing entirely by about the seventh day, and being succeeded by perspiration about the face, and sometimes over the whole body.

After the swelling of the parotids begin to decline, and the inflammation to abate, the breast in females, and the testicles in males, are liable to become painful and swelled. Or metastasis to the brain, or its meninges, may suddenly occur, and even endanger life. In cases in which the testicles become inflamed, the affection becomes very tedious, and I have known the most violent acute gastritis to occur from a translation of this disease. Children and young persons are more liable to parotitis than persons of more advanced age, and it generally occurs but once in the same individual, resembling, in this respect, other contagious diseases.

Causes.—It is probable that a specific contagion is the most frequent cause of mumps, but I suspect that it may originate from endemic or epidemic influences, independent of contagion, as the disease arises, in some instances, under circumstances favoring this supposition. In cases, however, in which mumps do thus arise, if such be a fact, it appears to be as contagious as when contracted in the usual way.

Treatment.—Mild cases of parotitis require little more than for the parts to be kept warm, and for the patient to avoid cool or damp air, and to take a plain digestible diet. If there be constipation of the bowels, a seidlitz powder may be given, or a drachm of the fluid extract of rhubarb, with half an ounce of the sulphate of magnesia.

If, however, the inflammatory symptoms **run** high, ten grains of **calomel**, or two grains of podophyllin may be given, in half an ounce of castor oil, and the oil repeated in five or six hours, if necessary. Tartar emetic may be given in one-eighth of a grain doses, with five drops of the fluid extract of ipecac, every four or six hours, and if the arterial action be very considerable, three or four drops of the fluid extract of the veratrum viridi may be added to each dose.

When the inflammation is translated to the breast, or testicles, the same general treatment should be continued, and **sinapisms, or** if necessary, a blister applied to the parotids. A warm hop poultice should be applied to the inflamed testicle or breast. Should the translation of the inflammation be to the stomach or brain, a blister should be applied to the parotid, and **a warm** hop poultice be kept to the sides of the **face.** If the brain be the seat of the inflammation which **is set up, the head** should be elevated, and the feet placed in **warm water, and cups, and** then blisters, applied to the back of the neck. The tartar emetic, ipecac and veratum should be continued, and, if necessary, alterative doses of calomel given every four or six hours, till the cephalic inflammation is subdued.

Should the translation of the inflammation be to the stomach, as happened in one case under my care, after applying a blister to the parotids, as I have suggested, cups, wet or dry, may be applied on each side of the spine, opposite the stomach, or to the epigastrium, and a warm hop poultice applied **to the** epigastrium. The violent vomiting **may be** arrested, in such cases, by giving a grain of **Dover's powder,** with **one** fourth of a grain of calomel every fifteen minutes, after which, **ten** drops of the fluid extract of hyoscyamus may **be** given, with mucilages, every six hours, but on no account should blisters, or even mustard be applied to the epigastrium in such cases.

SECTION VI.—ŒSOPHAGITIS.

By œsophagitis, I mean inflammation of the œsophagus, which is that portion of the alimentary canal extending from the pharynx to the stomach, passing behind the trachea, along the posterior mediastinum, and into the abdomen through the œsophageal opening in the diaphragm.

The œsophagus, it will be remembered, from οω, "I carry," and φαγω, "I eat," conveys the food from the mouth or pharynx to the stomach, for which purpose it is provided with longitudinal and annular muscular fibres, its lining mucous membrane being continuous above, with that of the pharynx, and below with the mucous membrane of the stomach. The œsophagus, thus constituted, is liable to become inflamed; the inflammation, in some cases, being confined to its mucous lining membrane, but in others, extending to its muscular structure.

Symptoms.—Inflammation of the œsophagus is attended with heat and pain, more or less aggravated by swallowing, at some point along the tube, but generally referred either to its upper or lower extremity.

There may be pain between the shoulders, and perhaps tenderness on pressure, with more or less difficulty in swallowing. There is generally little or no fever, but there may be hiccough, and perhaps nausea and vomiting.

The inflammation generally terminates by resolution, but suppuration may take place, small abscesses forming in the submucous cellular tissue. Ulcers may form, or a pseudo-membrane may extend from the pharynx, along more or less of the œsophagus. Œsophagitis is generally of an acute character, but it may assume a chronic form.

Causes.—This disease may be an extension of inflammation, either from the pharynx or stomach; or it may be the result of retrocession of various cutaneous affections; but generally it is the result of mechanical violence, or of some acrid, corrosive, or hot substance swallowed.

Treatment.—The indications in the treatment of this disease, may best be fulfilled, by warm pediluvia, morning and evening; cupping to the back of the neck, and leeches along its side, over the œsophagus; saline cathartics; a liquid farinaceous diet; and should the disease become chronic, blisters may be applied, and mercury or iodide of potassium given internally.

SECTION VII.—ACUTE GASTRITIS.

By acute gastritis, I mean here an acute inflammation of the stomach, involving essentially its mucous membrane; but extending, in many cases, to the muscular, and even peritoneal coats. The stomach, it will be remembered, is an expansion of the alimentary canal, situated beneath the diaphragm, between the liver and spleen; occupying the epigastrium, and a portion of the left hypochondrium. The stomach has a mucous, muscular, and serous coat, and is continuous above with the œsophagus, and below with the duodenum; and is that portion of the alimentary canal, in which the food is converted into chyme.

The stomach is liable to become inflamed, in all its coats; but its mucous membrane, which is a continuation of the skin, passing inward to line the alimentary canal, is especially so. This membrane in the stomach, is well supplied with arteries, veins, and nerves; but its structure being loose, its inflammation is not necessarily attended with any considerable pain, and what there is, may be of a biting, stinging character.

Symptoms.—The symptoms of acute inflammation of the mucous membrane of the stomach, involving more or less, its submucous cellular tissue, and in some cases its muscular coat, and perhaps peritoneal, are just what we should suppose from the nature of the structures involved.

Acute gastritis generally commences with violent vomiting, and sometimes purging, attended with a burning stinging pain in the stomach. The vomiting is generally increased by the swallowing of warm liquids; but cool drinks may produce a transient alleviation. If the inflammation be severe and general, extending to the submucous cellular tissue, and muscular coat of the stomach, the respiration becomes anxious and difficult, as the diaphragm does not descend to produce free expansion of the lungs.

There is in such cases, tenderness on pressure, an increase of pain on making a deep inspiration, and also in the act of vomiting. The substances thrown from the stomach, consist first of the food, and later of mucus, bile, and in some instances a tinge of blood. The thirst is urgent, and the tongue is either red, dry, and smooth, without fur; or else it is covered in the middle and posterior part, with a white fur, its tip and edges being red; the papillæ being prominent through the coating, and of a deep red appearance.

The brain sympathizes with the stomach, **producing in** some cases, **more** or less delirium, and there is great depression of spirits and pros-**tration of strength.** The pulse, at first moderately full, becomes contracted, quick, and tense, and finally so small as scarcely to **be felt.** A short, sympathetic cough generally attends, and the voice **may** become altered, and sometimes nearly extinct. If the inflammation be confined to the stomach, the bowels are constipated; but **if** they participate in the inflammation, there is a diarrhœa and perhaps dysenteric discharges.

If the disease tends to a favorable termination, the vomiting gradually abates, the tongue becomes moist, the pulse fuller and less frequent, the skin becomes moist and soft, and finally, convalescence is fully established. **But** should the disease tend to a fatal termination, the tongue becomes more dry, the skin cold and pale, the pulse feeble and thread-like, the patient becomes restless and delirious, dark matter is ejected from the stomach, the countenance becomes cadaverous, and finally the extremities become cold, and the patient dies from extreme exhaustion. Or if perforation of the stomach takes place, the pain increases and becomes diffused; **the whole abdomen** becomes painful and tender, and each breath is attended with a groan; the patient passing on speedily to dissolution.

Gastritis may assume a much milder form, **being attended with little** pain or tenderness, and continuing for several **weeks, while violent cases may terminate in one or two days, especially if** some irritating poison **has been the** cause of the inflammation. **In cases of acute** gastritis, in **which there is** little or no pain or tenderness, and only sympathetic disturbance of the brain and other portions of the body, darting pains in the chest, &c., it is probable that the inflammation is confined almost exclusively to the mucous membrane; and hence the absence of pain, and in fact, of most other symptoms which are usually developed in gastritis.

Anatomical Characters.—On post-mortem examination of the stomach, it is found contracted, and its mucous membrane more or less wrinkled. In violent cases, where death has occurred suddenly, in the early stages, there is generally discoloration, with but slight ulceration of the **mucous** membrane. But in cases that prove fatal at a later stage of the disease, there is ecchymosis and ulceration, and some portion of the membrane is in a softened and broken down state, and appears of a red, yellowish, or dark brown color.

Diagnosis.—Acute gastritis may generally **be** distinguished **from cramp** and flatulent pains, by attention to the following facts: In gastritis, the pulse is small, tense and quick; while in spasm or flatulent pains, there is seldom vomiting, and warm drinks produce no unpleasant effects. The pain is continuous in gastritis, unless it be temporarily alleviated by cool drinks, while in spasm the pain frequently intermits for several minutes. In gastritis, the patient lies on his back, and moves as little as possible; while in cramp, he walks about with the body bent forward, or throws himself about on the bed.

In gastritis the skin is hot and dry, and the pain of a burning, stinging character, while in spasm the skin remains nearly natural, and the pain is sharp, cutting and extremely severe. Finally pressure over the stomach in gastritis is attended with some degree of soreness, and an increase of pain, while in cramp, pressure generally affords some relief.

Causes.—Various causes may operate in producing acute gastritis, among the most frequent of which, are acrid substances received into the stomach, cold water taken whem the body is heated, over-distension of the stomach by indigestible food, stimulating drinks, suppression of habitual sanguineous discharges, metastasis of parotitis, erysipelas, or rheumatism, and finally, mechanical injuries of the epigastrium. It is possible also, that *koino* and *idio-miasmata* by entering the stomach with the saliva, or by passing to its mucous membrane through the blood, may produce acute gastritis, but this is not quite certain.

Prognosis.—A gradual subsidence of the pain and vomiting, a sedimentous urine, a gentle moisture of the skin, and a more developed pulse, indicate a favorable termination. If, however, the vomiting continue with difficult respiration, and hiccough, the pulse becomes smaller, more frequent, and corded, and finally, the pain suddenly subsides, the extremities become cold, with dimness of sight, and delirium, a fatal termination is inevitable.

Treatment.—In the incipient stage of acute gastritis, warm pediluvia, with warm stimulating friction along the spine, and sinapisms to the epigastrium, together with cool mucilaginous drinks, may at once arrest the disease. If, however, the patient has been neglected, or we do not succeed in arresting the disease by these mild measures, cups or leeches should be applied to the epigastrium, and cups on each side of the spine, opposite the stomach, and from two to four ounces of blood taken, and after the cupping, a blister may be applied to the epigastrium.

The bowels should be moved by an injection of castor oil, in flaxseed tea, or in equal parts of milk and water. And to allay the nausea, stop the pain and arrest the vomiting, one fourth of a grain of calomel, with a grain of Dover's powder should be given every hour, till these unpleasant symptoms subside, and cool mucilages freely allowed as the only drink or nourishment during the acute inflammatory stage.

In cases in which some active poison has been taken, the whites of half a dozen eggs may be swallowed, and the whole contents of the stomach thrown off by a full dose of ipecac, administered in mucilage of gum-arabic. After free vomiting has been produced in such cases, proper antidotes should be administered for neutralizing the poison, should any still remain, and if no diarrhœa attend, small doses of calcined magnesia, or castor oil, may be given to remove any of the poison which may have accumulated along the intestines.

Having thus cleared the stomach by the emetic, and the intestines by a laxative, and neutralized as far as may be, the poison in the system, as well as counteracted its affects, the case should be treated according to the principles I have already laid down, modified of course to meet any peculiarity the case may present. In the early stage of acute gastritis, no food except mucilage should be allowed, a little later, arrow-root and rice may be taken, and gradually other mild varieties of digestible food.

SECTION VIII.—CHRONIC GASTRITIS.

By chronic gastritis I here mean to include all cases of inflammation of the stomach, not of an acute character; though many cases of chronic gastritis are the result of the acute form of the disease.

When we take into account the great variety of abuse to which the stomach is liable, from irritating and unwholesome articles of food and drink, and its variety of sympathetic relations, we need not wonder that this is one of the most frequent of the phlegmasical affections. The slow and insidious progress of this grade of gastric inflammation, especially during its early stages, is probably the reason why there is so frequently a misapprehension as to its true character.

Symptoms.—In chronic gastritis, not the result of the acute form of the disease, the affection generally approaches very slowly and insidiously, so that it has often progressed considerable before it receives particular attention. The early symptoms of chronic gastritis are those that usually characterize indigestion, such as acidity, flatulence, a sense of oppression after eating, eructations, and transient pain in the region of the stomach.

The patient may feel comfortable when the stomach is empty, only there is apt to be a languid dissatisfied craving or appetite for food. As the disease progresses the epigastrium becomes slightly distended, and tender to the touch, and nausea, and sometimes vomiting occurs an hour or two after eating. The gastric distress gradually becomes more troublesome, especially after eating, the pain in the stomach being confined to a circumscribed spot, and of a lancinating character. The patient frequently complains of feeling as if a heavy substance were lodged in the stomach, and pressing against the diaphragm.

In some cases there is a thick ropy substance raised from the stomach, while in others there is thrown up an acrid, sour, watery, liquid, especially after eating. In the latter stages of the disease the appetite generally fails; and in very aggravated cases the patients even loathe food.

In the early stages of the disease the bowels are apt to be constipated, but later there is generally a diarrhœa, at least a portion of the time. The patient becomes dejected, impatient, acquires an irritable temper, and is frequently even indisposed to give an account of his sufferings. The tongue is generally red and granulated, or covered with small red points, or there may be a streak of brown fur along the middle, the edges being red and clean.

In protracted cases of the disease emaciation goes on rapidly, the adipose structures becoming sometimes almost entirely absorbed. The skin becomes dry, and of a yellowish-brown color, and is drawn tightly over the muscles, giving the surface a rough or rigid appearance. The pulse may be nearly natural, or quick and tense, in the early stages; but later, it is apt to become irregular, contracted, hard, and frequent· There is generally much prostration, and great indisposition to bodily and mental exertion.

The disease may continue for months, till the system becomes worn down and debilitated, and the patient dies from exhaustion; or a sudden perforation of the stomach or douodenum may occur, and the patient die of acute general peritoneal inflammation, as has happened in one or two cases that have fallen under my observation.

When perforation occurs, the patient may scream out with an acute pain in the region of the stomach, which, however, gradually extends through the whole abdomen. Every breath is attended with a groan, warm drinks produce a general sensation of heat through the abdomen, and after a slight arterial reaction, the pulse and general system sink, and the patient dies, according to my observation, within twenty-four hours after the perforation.

Anatomical Characters.—The inflamed portions of the mucous membrane exhibit either a red, gray, brown or blackish color. Or sometimes minute black points are seen, giving the surface a dark or grayish appearance. In some cases white patches are found, and the mucous coat may be thickened in spots or to a considerable extent, and it may present a granulated appearance, either from an enlargement of the gastric glands, or from an exudation into the tissue of the membrane.

The membrane may be indurated, softened or ulcerated, the ulcers penetrating, in some cases, the muscular and even the peritoneal coat. If perforation has taken place, the edges of the orifice may be rough; but in one case which I examined, it was circular, perfectly smooth, and the size of a half-dime, except in the serous membrane in which it was irregular.

In cases in which perforation has taken place during life, and the contents of the stomach have escaped into the cavity of the abdomen, the peritoneum presents signs of having been very generally inflamed.

Diagnosis.—Chronic gastritis may be confounded with dyspepsia, gastralgia and cancer of the stomach, from all of which affections, however, it may generally be distinguished by careful attention to the following differences. From dyspepsia it differs in there being more pain at some particular point; more frequent vomiting after taking food, more redness of the tongue, and finally more weakness and general emaciation, than attends in simple dyspepsia.

From gastralgia chronic gastritis differs by the pain being more continuous and of a stinging character; the tongue is more dry, and either coated or red, the appetite is more invariably bad, and in gastritis there is tenderness on pressure, and generally a frequent pulse, hot or dry skin, a sallow countenance and general emaciation, which symptoms do not generally exist to so great a degree in simple dyspepsia unattended with gastritis

From cancer of the stomach, chronic gastritis differs in there being no tumor in the region of the stomach; in the matter vomited being generally of a less viscid or ropy appearance, in the dry, red, or coated appearance of the tongue, and finally in the absence of the cancerous diathesis, and the more rapid termination of the inflammatory than of the cancerous affection.

Causes.—Chronic **gastritis is** sometimes the result of the acute form of the disease, but **it is generally** produced by irritating substances acting **on the mucous coat** of the stomach; such as heating or indigestible articles of food, acrid **medicinal substances, as** capsicum, &c.; mental **despondency,** alcoholic **liquors, and repelled** cutaneous eruptions. And **besides, chronic** gastritis may **arise from congestion of** the portal **circulation, or it may** attend phthisis, with which affection it is a very **frequent** complication **in its** latter stages, and sometimes even from the commencement.

Prognosis.—The prognosis in chronic gastritis is generally favorable if proper treatment be resorted **to** in season. But if the case be neglected, and the imprudence which has led to it is continued, ulceration, and **finally perforation may occur, and thus the** case terminate fatally.

Treatment.—**The first** indication in the treatment of chronic gastritis is to remove the cause which has been operating to produce it, and to regulate the diet, allowing just that kind **of** food that is proper, and nothing more. Early in the disease, mucilages, with arrow-root, boiled in milk, **or milk** and water, may **be the** only food allowed. Later, however, when the inflammation is in a measure subdued, a poached egg, boiled milk, and gradually plain digestible varieties of food may be **allowed, to** be **taken** with strict regularity.

Cups or leeches should be applied to the epigastrium, and repeated, if **necessary, every two or** three days, at first. After the cupping or leeching **has been continued as long as is consistent with** the general condition of the system, a blister should be applied to the epigastrium, and as it heals another should be applied, and this should be continued till the inflammation **is subdued,** the blisters **being repeated at reasonable** intervals.

Having removed the cause, regulated the **diet, and applied** cups or leeches, and thus in **a** measure subdued the inflammation, a little may be done by way of medicine internally, if necessary. If the bowels are constipated, a teaspoonful of calcined magnesia may be given each morning, till they are regulated, and this will also correct the acidity, which is often so troublesome in such cases. If, however, there is diarrhœa, the acidity may be corrected by small doses of prepared chalk, **or limewater,** taken before each meal.

As an astringent, anodyne, and **tonic, a pill** made of one grain, each, **of** the sulphate of iron, extract of gentain, and extract of conium or hyoscyamus may be given after each meal, and continued with a regulated diet till the disease is subdued. If, however, evidence of ulceration should supervene, the pills may be prepared with one fourth of a grain of the nitrate of silver, instead of the sulphate of iron, and given instead.

SECTION IX.—CANCER **OF THE STOMACH.**

By cancer of the stomach, I mean that variety of malignant disease, attended with the formation of a scirrhous livid tumor, which finally ulcerates. The stomach is liable to this cancerous affection, which may locate in any portion of its structure, but generally in either the cardiac or pyloric extremity, and of the cases that have fallen under my observation, the pylorus has generally been the part involved. Cancer of the stomach may, however, be of a scirrhus, encephaloid or colloid character, and it **may** pass on rapidly, and terminate in **a few** months, or it may continue to progress steadily for several **years,** and then terminate **fatally.**

Symptoms.—The early **symptoms of cancer of** the stomach are generally very similar to those **of chronic gastritis;** there is uneasiness in the stomach after eating, **variable** appetite, and sooner or later nausea **and** vomiting. In cancer **of** the stomach, however, in addition to the pain in the region of the stomach, the patient complains of lancinating pains extending to the spine and frequently along the spine, and this is often, according to my observation, much more distressing than the pain experienced in the region **of** the stomach.

In a case of cancer of the pyloric **extremity of the stomach which had** been progressing for many years, that came under my **care the** past season, the patient, a lady of about forty, was constantly annoyed by the most distressing pain along the spine, and though she had an indifferent appetite, with some acidity of the stomach, and at last vomiting of a darkish glairy fluid; the prominent symptoms in the case were the pain along the spine, loss of appetite, and emaciation towards the fatal termination **of the disease.**

Generally, however, there is vomiting early in the disease, of a glairy mucus; later it becomes sour, and finally bloody, giving it the appearance of coffee-grounds or soot and water. This appearance of the matter vomited, the loss of appetite, the pain extending to and along the spine, together with the peculiar cathetic countenance, and the great emaciation which occurs, with perhaps a tumor in the region of the stomach, **are** the most prominent symptoms **of this disease.**

Anatomical Characters.—The stomach may be found contracted or very much enlarged, especially if the pylorus is the seat of the disease; I have seen the stomach dilated to more than twice its ordinary capacity, and adhering to the parietes of the abdomen.

The cancerous disease may occupy any portion of the stomach, but it is more frequently found at the cardiac or pyloric orifice, and sometimes extends even beyond the stomach, along the duodenum or œsophagus.

If ulceration **has not** taken **place, the** coats of the stomach are whitish, indurated, semi-cartilaginous, and more or less thickened. The thickening may be very slight, but I have seen it an inch or more, so as nearly to close the pyloric orifice. The disease may be confined to the mucous membrane, but it **may** involve the submucous cellular tissue, and even the muscular and peritoneal coats, as has been the case in most examinations **that** I have made of this affection.

Instead of mere thickening and induration **of the** coats of the stomach, there **may be** roundish masses projecting **either** from the mucous membrane **into the** stomach, or from the peritoneal coat into the cavity of the abdomen. And this tumor may be white and indurated, constituting scirrhus; soft or brain-like, constituting medullary cancer; or it may consist of cells filled with a gelatinous deposit, separated by fibrous partitions constituting colloid cancer.

If the patient has died at a very advanced stage of the disease, ulcers are found in different stages of their progress; the peritoneal coat in such cases having often contracted adhesions with the pancreas, spleen, liver, and in one case that I examined, with the abdominal parietes. The disease, in such cases, penetrates these parts in its destructive progress; and in one case that I examined, the pancreas was completely destroyed, being converted into an ulcerated mass.

Diagnosis.—Cancer of the stomach may generally be distinguished from chronic ulcerative gastritis, by the cachectic countenance, the tumor in the epigastrium, the pain along the spine, the steady, but perhaps slow progress of the disease, the obstinate and repeated vomitings, the steady and great emaciation, sometimes leaving little more than the skin and bones, and by the evidence of a hereditary tendency to the disease.

Some of these symptoms, it is true, exist in chronic ulcerative gastritis, but generally not in as marked a degree, at least according to my observation, in these affections. In cases in which the disease extends to the pancreas, I have noticed that the appetite sometimes becomes entirely indifferent, the patient eating one thing as readily as another.

Causes.—There is evidently a hereditary predisposition to cancerous affections in some constitutions, but in what that predisposition consists is not quite certain. It is probable, however, that the various imprudences of parents are visited upon their children in this predisposition, as well as in all others which are inherited.

The exciting causes are various irritants, such as excesses in eating, indigestible articles of food, alcoholic liquors, stimulant medicines, such as capsicum, &c., and besides, the disease is doubtless produced, in some cases, by depressing mental emotions, and perhaps by excessive venery, and various other kindred abuses of the system.

Prognosis.—The prognosis in cancer of the stomach is always unfavorable, but by a rigid observance of the laws of health, and palliative measures, some cases may live on perhaps for several years, with the certainty, however, of its producing a fatal result, unless the patient should be cut down by some other cause. Cancer of the stomach may however pass on rapidly to a fatal termination.

Treatment.—The treatment of cancer of the stomach must be strictly palliative, and that which is of most importance is a proper regulation of the diet. The patient should be directed to take a plain digestible, and nourishing diet, with strict regularity. And among the articles most suited for such patients are bread and milk. The drinks should be water, or milk and water, in which, if there is acidity of the stomach, small quantities of lime-water may be taken, or of the bi-carbonate of potassa, or prepared chalk. If the symptoms of gastric irritation be aggravated from any cause, cups, leeches, or blisters may be applied to the epigastrium ; and in the very last stages of the disease, anodynes of conium or hyoscyamus may be required.

23

SECTION X.—PERITONEAL ENTERITIS.

By peritoneal enteritis, I mean inflammation affecting mainly the peritoneal coat of the intestines, but extending more or less, in most cases, to the muscular coat. The peritoneum, it must be remembered, is a serous membrane, which, after lining the abdominal cavity, extends over the intestines, forming their external or serous coat. Now this outer or serous coat of the intestines is liable to be the chief seat of intestinal inflammation; the muscular coat is generally, however, more or less involved.

Symptoms.—Peritoneal enteritis usually commences with a feeling of uneasiness in some part of the abdomen, which after a longer or shorter period terminates in a fixed burning pain, generally in the umbilical region, and finally becomes gradually diffused throughout the whole abdomen, or most of it at least. If the inflammation is confined to the serous and muscular coats of the intestines, there is generally obstinate constipation. Nausea and vomiting generally attend, and sometimes even stercoraceous matter is thrown up.

The tongue is generally dry, and may be covered with a white fur, having red edges, but in some cases there is a streak of brown fur along the middle. The thirst is usually urgent, the urine scanty, and high colored, and sometimes voided with considerable difficulty. The skin is hot and dry on the body, but it may be moist on the forehead and on the hands. The pulse is usually frequent and tense, and the respiration is more or less disturbed, each inspiration adding to the local abdominal suffering. The patient lies on the back with the knees drawn up, to avoid pain from the pressure of the abdominal muscles.

Sometimes this disease commences with chills, followed rapidly with febrile excitement, and finally with a state of collapse. If collapse supervene, the extremities become cold, the pulse weak and undulating, the countenance cadaverous, the abdomen tense, and finally a sort of passive vomiting occurs, the contents of the stomach being apparently forced up by the distended intestines.

Peritoneal enteritis is usually rapid in its course, and may terminate in gangrene, in which case the pain subsides, the pulse sinks, the countenance becomes pale, the extremities cold, the surface is covered with a clammy sweat, and with perhaps hiccough and convulsions the patient finally dies. The acute form of this disease seldom continues more than one week, without terminating in resolution or death.

Peritoneal enteritis may, however, assume a *chronic form*, in which case all essential symptoms of acute cases are developed, with less intensity, and the disease may pass on for weeks, or even months; the muscular coat of the intestines probably always suffering more or less in such cases.

Diagnosis.—Peritoneal enteritis, if it be in the arch of the colon, may be mistaken for pleuritis, or hepatitis, but it may be distinguished by attention to the following essential differences. In pleurisy the pulse is full, hard, and active, while in enteritis it may be contracted, quick, and frequent. In pleurisy the respiration is carried on mainly by the abdominal muscles, while in enteritis the chest expands freely, and the abdominal muscles are comparatively quiet. In pleurisy pain is felt **by** pressure in the intercostal spaces, while in enteritis, pain is produced **by** pressure on the abdomen.

Peritoneal enteritis may be distinguished from simple peritonitis, by the constipation and vomiting which attend the intestinal inflammation, which symptom is not a necessary attendant in simple peritonitis.

From spasmodic pain, enteritis may be distinguished by attention to the following differences. In enteritis the patient lies quiet on the back, moving as little as possible, while in colic he throws **himself** about continually. In enteritis the pain is increased by pressure, while in colic it is often a relief. In enteritis the pain is continuous, **while** in colic it often intermits for **a time.** In enteritis the skin is hot and dry, and **there** is thirst, **while in colic there is no** thirst, and the skin is **nearly natural or** moist.

Finally, in discriminating between peritoneal and mucous enteritis, **we** should **remember** that in peritoneal enteritis the bowels are **constipated,** while in mucous enteritis there is generally a diarrhœa.

Anatomical Characters.—Resolution is the only favorable termination in peritoneal enteritis, and when it occurs, it may be attended with **a** moderate diarrhœa. There is generally found on post-mortem examination of patients dead of this disease, a gangrenous appearance; but it is possible that this disease sometimes terminates fatally in the early stages, by the shock to the general system. In some cases, if the inflammation has been considerable, coagulable lymph is thrown out and folds of the intestines have thus formed adhesions, presenting an irregular mass, more or less sero-purulent fluid being generally found in such cases in the cavity of the abdomen.

Causes.—Various causes operate to produce peritoneal enteritis, among the most frequent of which are fecal accumulations in the bowels, mechanical injuries, hernia, drastic purgatives, sudden suppression of the perspiration from cold, metastasis of external inflammations, as **ery**sipelas, rheumatism, &c.

Prognosis.—In **cases in which the vomiting is moderate, and the pulse** not contracted and obscure, there may be a good prospect of **recovery** with proper treatment. But in cases in which there is frequent **and** obstinate vomiting with a contracted and obscure pulse, considerable danger may be reasonably apprehended of a fatal termination **of the** case.

Treatment.—At the very commencement of peritoneal **enteritis, the** inflammation may sometimes be arrested by placing the feet in warm water, applying cups on each side of the spine opposite the abdominal pain, and cups or leeches to the abdomen. After the cupping or leeching, a warm mustard poultice should be applied over the abdomen, and kept on till it has produced a good degree of irritation. Twenty drops of laudanum, or fifteen drops of the fluid extract of hyoscyamus with half a grain of camphor and a little prepared chalk, may be given every four or six hours to quiet pain and allay vomiting. The spine may be rubbed morning and evening with a strong infusion of capsicum in vinegar. Two or three blue pills may be given, and followed in five or six hours with a full dose of castor oil. With this mild course of treatment

I am satisfied that many cases of this disease may be arrested if resorted to early. But if the case has been neglected or mal-treated, it may require more active measures.

In such cases general bleeding, warm pediluvia, cupping on each side of the spine, and cupping or leeching the bowels, and finally sinapisms over the abdomen, and blisters if the case is obstinate may be required. After the bleeding and cupping, a full dose of calomel in half an ounce of castor oil may be administered, and its operation aided by an injection if necessary. Full doses of opium or hyoscyamus may be given every four or six hours, and this gradually diminished in quantity and frequency, as the inflammation and pain subside. A liberal blister followed by a warm hop poultice over the bowels will **generally be of very essential service in such cases.**

Should symptoms of gangrene occur, the case should not be at once abandoned as hopeless. Stimulants, properly administered, may possibly arrest the alarming symptoms in such cases. Camphor, the sulphate of quinine, wine-whey, carbonate of ammonia, or even brandy may be administered; and if they do not avert the fatal tendency, the attendant may have the satisfaction of having done what he could.

Mucilages and toast-water may be allowed during the course of the disease, if the stomach will retain them. During convalescence the patient should be directed to take nothing but the most digestible and unirritating diet, even for several weeks, lest there be a return of the disease.

In chronic peritoneal enteritis, the warm foot-bath at evening, cups along the spine, with cups or leeches to the abdomen, and finally blisters, mild laxatives, or enemata, with a light, plain, digestible diet, taken with regularity, constitute the treatment proper in such cases. After an attack of enteritis, the bowels are peculiarly liable to distension by wind, to relieve which, and to give tone, a weak infusion of columbo, or from one fourth to half a drachm of the fluid extract may be given, three times per day, immediately after each meal, and continued for a time.

SECTION XI.—MUCOUS ENTERITIS.

By mucous enteritis, I mean here inflammation of the mucous membrane of the small intestines, including the duodenum, jejunum, and ileum; and also of the large intestines, if not attended with griping pains in the lower portion of the abdomen, and mucous or bloody evacuations. But that variety of inflammation of the large intestines, involving mainly the colon and rectum, and attended with tenesmus and mucus or bloody evacuations, I shall consider in the following section.—The length of the intestines, it should be remembered, is about thirty feet, of which the small intestines constitute about twenty-five, and the large, the remaining five feet. The large intestines may become involved, to some extent, with the small intestines, in simple mucous enteritis, without developing the ordinary symptoms of dysentery, or griping pains, with mucous or bloody evacuations. While then, by *simple mucous enteritis*, I mean mainly inflammation of the mucous membrane of the small intestines; it must be remembered that I include under this head simple mucous inflammation of the large intestines, not attended with griping pains, and mucous or bloody evacuations. It should be remembered also that while the mucous membrane of the intestines, and especially of the small intestines, is the principal seat of this disease, the inflammation may extend to the muscular, and even to the serous coat of the intestines.—Let us now proceed to the consideration of inflammation of the mucous membrane of the intestines, commencing with those that are developed, if the duodenum be the seat of the inflammation.

Symptoms.—If the duodenum be the seat of the inflammation, there may be pain in the vicinity of the pylorus, and **more or** less pain in the back, along the lower portion of the dorsal region. There is also many **of the** symptoms common **to gastritis**, together with a yellowness of the skin, and generally a yellow appearance of the urine. If the inflammation be of a chronic character, food **taken is apt to produce pain an hour or two** after eating, and the same **is true in acute** cases **if food be taken.** Finally, if the duodenum alone **be** the seat of the inflammation **there is** generally **little or no** diarrhœa, and there may be constipation.

If, however, the inflammation **extends** along the mucous **membrane of** the jejunum and ileum, and perhaps the large intestines, the following symptoms may be expected. In the *acute* form of mucous enteritis, there is at first uneasiness, followed by griping pains, which gradually increase, and very soon there is more or less tenderness on pressure. The pain may occupy any portion of the abdomen, but the region of the umbilicus is its most frequent seat.

Soon after mucous enteritis becomes established there is along with the griping pains, in most cases a diarrhœa, especially if the inflammation extends to the lower portion of the small intestines. The diarrhœa may be continuous, or it may cease for a **little time, and** then suddenly return from very slight causes. The discharges are generally of a liquid character, being an increased serous exhalation, with **or** without bile, and perhaps some fecal matter. There **may** be tympanitis, **and** in some cases if the muscular and serous coats become involved, there may be constipation of the bowels instead of diarrhœa.

More or less febrile symptoms are developed, either before or after the symptoms which are developed by the local inflammation, **and the febrile** symptoms may be either continued, remittent, or even almost intermittent, the paroxysms being preceded by slight chilliness in many cases. The pulse is excited, the skin quite dry, and the urine scanty, and headache is an occasional, though not an invariable attendant. The disease may terminate favorably in a few days, or the pain and tympanitis may increase, the tongue become red and dry, the pulse frequent, and the patient may recover after a lingering illness, or the inflammation may extend to the muscular and serous coats, or perhaps ulceration and perforation occur, and the case terminate unfavorably.

Such are the ordinary symptoms of acute mucous enteritis, but the disease may become chronic, or it may assume a chronic form from the first, in which case the same train of symptoms are developed, with less intensity, if we except perhaps the diarrhœa, which is often more considerable and obstinate than in acute cases.

Chronic mucous enteritis is generally attended with serous evacuations, more or less frequent, which, however, in the latter stages may be mingled with more or less pus. There is generally some pain, and more or less tenderness; the pain being aggravated by a movement of the bowels, and sometimes at a period of a few hours after each meal. The appetite may be variable, or craving, the skin dry, the tongue furred, the pulse frequent; there is more or less emaciation; and, finally, the patient is irritable and gloomy, magnifying every unfavorable symptom, and appearing blind to every indication of a favorable termination. Chronic mucous enteritis may terminate in a few days, or it may run on for weeks, months, or even years, with occasional remissions and exacerbations, till at last, irritable, worn down and emaciated, the patient is relieved by death from his protracted sufferings.

Anatomical Characters.—In acute mucous entritis, the mucous membrane is found reddened, brown or of a livid appearance. And this redness may be uniform, or in patches. The follicles may be enlarged, and perhaps ulcerated on their points, and surrounded by more or less redness; or several of these may run into each other, forming an irregular ulceration.

Ulcers may also exist in any portion of the mucous membrane. Or they may extend to the submucous, and even through the muscular coat of the intestine, and sometimes the serous coat may be found ruptured, the whole peritoneum in such cases presenting signs of inflammation. In some cases portions of the bowels are found gangrenous, their contents having escaped into the peritoneal cavity.

In cases of mucous enteritis that become chronic, or that are chronic from the first, in addition to the appearances presented in acute cases, the mucous membrane appears covered with prominent ulcerated follicles, and the intervening ulcers are more numerous, generally, than in acute cases. And in very protracted chronic cases, in scrofulous patients, the mesenteric glands may be found enlarged, and either suppurating or indurated.

Diagnosis.—Mucous enteritis may be distinguished from peritoneal enteritis by attention to the following differences. In peritoneal enteritis there is great tenderness and intense pain, of a sharp, lancinating character, as well as protracted vomiting and obstinate constipation; while in mucous enteritis the pain is less acute, and the tenderness very much less, and little or no vomiting.

In colic the pain is more severe, but may be relieved on pressure, and there is constipation of the bowels, with little or no fever, by which it may be distinguished from mucous enteritis.

Causes.—Among the causes of mucous enteritis are atmospheric vicissitudes, the translation of cutaneous eruptions, the suppression of accustomed discharges, and the retrocession of gout, rheumatism, &c. It may also be produced by crude articles of food, drastic medicines, and various irritating or poisonous substances swallowed. Besides, mucous enteritis may be produced by acrid bile, and other secretions, by intestinal worms, and it may be a result of scalds, burns, &c., or it may arise during the continuance of febrile and other affections.

Treatment.—At the very commencement of acute mucous enteritis, if the disease be active, and the patient of a strong, vigorous constitution, blood may be taken from the arm, if necessary; and if there be constipation, a cathartic of calomel and castor-oil may be administered. Immediately after the general bleeding, when it is necessary, and at first when it is not, cups should be applied along each side of the spine, and two or three ounces of blood taken, and cups or leeches applied to the abdomen.

The warm foot-bath should be used morning and evening, warm fomentations of hops applied over the abdomen, and if there is pain, with diarrhœa, fifteen or twenty drops of laudanum may be given, with a little prepared chalk, every four or six hours, and this may be continued till the disease is arrested. The patient should be nourished during the acute stage of the disease, by mucilages, or crust coffee, with a little milk, and as convalescence approaches, the return to the use of solid food should be cautious and gradual, and every possible care should be taken to prevent a relapse.

Should the inflammation assume a chronic form, and in all chronic cases after the cupping along the spine, and cupping or leeching of the abdomen, blisters over the stomach and bowels, repeated if necessary, will be of very essential service and should not be neglected. If there be diarrhœa, Dover's powder or laudanum, with a little prepared chalk, and if necessary, tannin may be given every four or six hours and continued till the diarrhœa is corrected.

In very chronic cases, a pill of one grain each of the sulphate of iron, extract of gentian and extract of conium may be given after each meal, and if symptoms of ulceration occur, one-fourth of a grain of the nitrate of silver may be substituted in the pill for the sulphate of iron, and given after each meal, till an impression is produced upon the diseased membrane.

In chronic mucous enteritis, a plain, digestible, and unirritating diet only should be allowed; and among the articles most likely to be acceptable to the patient are arrow-root, rice, bread and milk, &c. The food should be taken with regularity, and every precaution should be taken that nothing be done to aggravate or perpetuate the disease.

In those cases of mucous enteritis which occur in bilious and other fevers, and which are indicated by a dry tongue, tympanitis, and perhaps diarrhœa, a blister should be applied to the epigastrium, and, if necessary, over the bowels, in addition to the other treatment which may be indicated in the case. By thus allowing only proper varieties of food, and fulfilling, as they arise, every indication, and doing nothing more, most cases of acute or chronic mucous enteritis may be palliated, and generally permanently cured.

SECTION XII.—DYSENTERY—(*Bloody Flux.*)

By dysentery, I mean that variety of inflammation of the mucous membrane of the large intestines, and extending sometimes along the small intestines, attended with griping pains in the lower portion of the abdomen, mucous or bloody evacuations and tenesmus. Inflammation of the mucous membrane of the large intestines including the caecum, colon, and rectum, may exist from various causes, in connection with that of the small intestines, or even independent of it, and not be attended with tenesmus and mucous or bloody discharges; in which case the disease falls under the head of *mucous enteritis*, which I have considered in the preceding section.

Let us now proceed to the consideration of that variety of inflammation of the mucous membrane of the intestines, and especially of the large intestines, attended with pain in the lower portion of the abdomen, mucous or bloody evacuations and tenesmus, constituting dysentery or bloody flux. Dysentery may be either benign or malignant, and it may assume an acute or chronic form. I shall treat here only of the simple or benign variety, leaving the consideration of malignant dysentery for the following section.

Symptoms.—Simple dysentery generally commences with slight chills, a loss of appetite, a bad taste in the mouth, and a slightly depressed pulse; soon the chills alternate with flushes of heat; there is thirst; a dry skin; pains in the lower portion of the abdomen, and either constipation or diarrhœa. Or the disease may come on suddenly, with chilliness, griping pains, and mucous bloody stools, without any marked premonitory symptoms, and especially is this the case when it arises from sudden exposure to cold, or from irritants acting directly upon the intestinal mucous membrane.

In some cases febrile symptoms are developed before the dysenteric discharges appear, while in others the bloody mucous stools are, as we have seen, among the first symptoms which are developed. During the whole course of this disease, there are generally no fecal discharges, the stools consisting entirely of intestinal mucus, mixed with more or less blood. Tenesmus too is one of the most constant symptoms in this disease, and its degree of severity generally indicates more or less the degree and extent of the intestinal inflammation. In some cases the discharges are almost entirely of intestinal mucus, but generally it is mingled with more or less blood.

In some cases there is only a slight fever, while in others there is a high state of febrile excitement. And in the latter stages of unsubdued cases, a colliquative diarrhœa may occur and various other unfavorable symptoms may supervene. The pulse becomes small, corded and very frequent, the countenance contracted, the abdomen tender, the skin harsh, and considerable prostration may attend the disease. The skin is generally harsh and dry through the whole course of the disease, the urine is scanty and high colored, and the hepatic functions are often more or less deranged.

Cases of simple dysentery generally terminate favorably in seven or eight days with proper treatment. But if cases be neglected or maltreated, they may pass on to a chronic form and continue perhaps for weeks, months, or even years; the inflammation in such cases involving to a greater or less extent, the mucous membrane of the small intestines.

Chronic dysentery is characterized by frequent small evacuations, consisting mostly of mucus mingled sometimes with blood, and more or less purulent matter with bilious or feculent discharges. The symptoms are, however, in the main similar to those of the acute form, being developed with less intensity; and if the disease be mainly confined to the lower portion of the intestines, the constitutional symptoms may be slight, even though the disease continue for a long time. But very protracted cases, especially if the patient be of a scrofulous constitution, are apt to be attended with emaciation, a sallow and shrunken countenance, general debility, and unless arrested, finally lead on to dropsy, consumption, or some other fatal affection.

Dysentery, whether acute or chronic, is liable to various modifications, depending upon the condition of the patient, at the time of attack, and also upon the accidental circumstances of locality, comfort or want, &c. All these circumstances should be taken into account in the examination and treatment of patients suffering from dysentery as well as in all other affections.

Diagnosis.—The presence of pain in the lower portion of the abdomen, of tenesmus, and of mucous or bloody evacuations, together with the other symptoms which I have enumerated, are sufficient to characterize a case of dysentery. But to distinguish the benign from the malignant variety of the disease, it becomes necessary to take into account the extrinsic circumstances, such as the condition of the patient, the epidemic or endemic influence, &c.

Causes.—Various causes operate to produce this disease, but cold damp air, after the body has been heated, by a hot summer or autumnal sun, is by far the most frequent cause, in temperate climates. During the months of August and September, the days in our latitude are usually warm, and the exhalation from the skin free; but as night approaches, the cool damp air irritates the extreme nerves of the skin, the exhalent tubes are closed, the small capillaries contracted, and an internal congestion produced. This internal congestion injects the minute capillaries of the mucous membrane of the alimentary canal, and produces irritation. It is possible also that the blood, containing the retained perspirable matter, acts as a direct irritant to the mucous membrane, and also the sanguiferous and nervous system, exciting the circulation, and deranging the functions of the various organs. But the alimentary mucous membrane is the part in which the local congestion and inflammation appears most prominent, and hence the pouring out of a bloody mucus, attended with tenesmus, mainly from the large intestines.

It is probable that the liver too, being in a congested state, obstructs the free passage of the blood through the portal vessels, and thus very essentially increases the congestion and inflammation of the alimentary mucous membrane. Now the cause being general, it is reasonable to suppose that the mucous membrane of the whole alimentary canal should suffer; but in dysentery, or those cases in which there is tenesmus, with mucous or bloody discharge, the disease appears to develop itself, as we have seen, most prominently in the colon and rectum. In this way, it is, I believe, that cold, with dampness, produce in our climate the simple cases of dysentery, so common in August and September.

It is possible, also, that the paludal poison, when it exists in the air, serves to predispose the system to this variety of disease; as more cases occur in miasmatic districts than where this agent does not exist. It is probable, too, that crude and indigestible articles of food, and especially unripe fruit, coming as it does at the season when the days are warm, and the nights cool and damp, tend strongly to produce this disease, in those who are any way predisposed.

Chronic dysentery is generally the result of the acute form of the disease; but it may arise from an extension of the inflammation, in hemorrhoidal, and other general or local affections, operating directly or through the general system.

Anatomical Characters.—The whole alimentary mucous membrane may be found, on dissection, to present a more or less congested, and perhaps inflamed appearance, which however becomes more marked, as the large intestines are approached; while in the colon, and perhaps rectum, more or less ulceration is generally found.

In that portion of the large intestines nearest the ileum the ulceration is apt to be superficial; but if the examination be continued along the colon and rectum, the ulcerations are apt to be more extensive, and deeper, if we except the extreme portion of the rectum, in which there are frequently no ulcerations to be found.

In cases of dysentery in which death has taken place early, the alimentary mucous membrane is found of a deep red color; and either soft and pulpy, or of a granular appearance. The liver is almost always functionally deranged in this complaint; but in mild sporadic cases, organic derangement very seldom occurs. In chronic cases, however, occurring in patients of scrofulous constitutions, abscesses and other organic changes of the liver may be found, together with extensive ulcerations of the alimentary mucous membrane, and especially of the large intestines.

Prognosis.—The prognosis in simple uncomplicated cases of dysentery is favorable, if proper treatment be had. If, however, from some accidental complication, from neglect, or from mal-treatment, colliquative discharges occur, with tympanitis, cold extremities, a clammy sweat, and a frequent feeble pulse, with delirium, a fatal termination may be expected.

Treatment.—The indications to be fulfilled in the treatment of dysentery, are to equalize the circulation, to allay the irritation, and subdue the inflammation of the alimentary mucous membrane, and to restore a healthy action of the skin and liver. Hence, in the common simple dysentery, produced by atmospheric vicissitudes, and other causes, if the patient be seen early, the disease may generally be arrested by very mild measures, unless some complication arise.

The feet should be placed in warm water, in order to call the blood to the extremities, and promote perspiration, which this simple measure will very frequently do. Rubbing the back with a warm infusion of capsicum in vinegar, will also aid very materially, in equalizing the circulation, and restoring the action of the skin, if resorted to early, while chilliness prevails.

A large mustard poultice should be laid over the stomach and bowels, covering the whole abdomen; and reaching back towards the spine, and it should be allowed to produce a thorough irritation; and this should be repeated, every six or eight hours, till the disease is arrested. If there has been a bitter taste in the mouth, and a poor appetite, for a few days, and the digestion is bad, a cathartic of calomel, or hydg. cum creta, with rhubarb and castor oil may be administered, and a free motion of the bowels produced.

After the operation of the cathartic, when it is indicated; and at first, when it is not indicated; five grains of **Dover's** powder should be given, every four or six hours, and the patient should drink freely of warm sage tea, to promote perspiration. Or, if little or no fever attend, fifteen or twenty drops of laudanum may be given instead **of the** Dover's powder. If this does not arrest the disease, the dose of the laudanum, or Dover's powder may **be** increased, and **two** grains of tannin **given,** either with, or alternating with the anodyne; and this should be **continued till** the disease is arrested.

In cases not attended **with** much previous gastric derangement, a cathartic should not be given; but a grain of calomel may be given with the Dover's powder or anodyne, till three or four grains have been given. This, however, may generally be omitted, and should not be **resorted to,** unless clearly indicated.

In obstinate **cases, which** pass **on to a** *chronic* state, or in **cases which** through neglect or mal-treatment become chronic, a blister should **be** applied over the stomach, and if necessary to the abdomen, and if obstinate, cups or leeches may be resorted to in case blisters prove insufficient.

In chronic cases which are obstinate, ten drops of the oil of turpentine, or ten drops of the balsam of copaiva may be given, in emulsion with sugar, gum arabic and water, three or four times a day, and continued **for** a long time. If, however, these measures fail, small doses of the nitrate of silver **may** be given instead, and **continued for** a reasonable **time.**

Injections **of fifteen** drops **of** laudanum, in **a fluid ounce** of liquid starch, may be of very essential service in obstinate acute cases, while in chronic cases they may be of great service, used morning and evening, with a few grains of tannin **or** the sulphate of zinc.

Toast water, with a little milk, may be allowed in acute cases, as well as arrow-root cooked in milk, or equal parts of milk and water, till the disease is arrested, and then a plain digestible diet should **be directed** for several days, and the patient should be warmly clad, and should avoid exposure to cool damp air for a long time. In chronic cases of this affection, a plain, digestible, and nourishing diet should be allowed, and the patient should wear flannel next the skin, and if possible, sleep in flannel sheets at night.

SECTION XIII.—MALIGNANT DYSENTERY.

By malignant dysentery, I mean that peculiar miasmatic affection in which the mucous membrane of the alimentary canal, and especially of the large intestines become the principal seat of the local inflammation, the disease being attended with serous or mucous and bloody discharges, pain, tenesmus, &c.

It differs from simple dysentery, **in** being brought about by a miasmatic agent, acting through the brain and nervous system, developing a mucous inflammatory affection, of a malignant character, affecting mainly the large, but more or less the small intestines, and being attended with great prostration and a typhous tendency.

Symptoms.—There is generally a forming stage, during which the miasmatic agent is producing its effects upon the brain and nervous system, the various functions of the body becoming more or less deranged. There is a bitter taste in the mouth, loss of appetite, headache, restlessness during the night, and general irritability of the nervous system. Soon there is chilliness, attended with headache, thirst, difficult breathing, violent pains in the head, back and limbs, and severe pain in the abdomen.

After the chills, reaction is set up, generally with increased pain in the bowels, and dysenteric discharges, with violent tenesmus, attended with a sense of weakness and general prostration of the powers of the system. The discharges may be, at first, of a bloody mucous character, but frequently become in a little time of a reddish watery, and finally of a dark putrid appearance, looking very much like dissolved blood, and the discharges are attended with severe tenesmus, and marked fainting, sinking symptoms.

The general fever which is developed may be of an active grade, but it generally inclines to a typhoid character, very soon after the bloody serous discharges commence. And unless the disease be arrested, the discharges increase in frequency, the extremities become cold, the breathing oppressed, the abdomen exceedingly tender, the thirst urgent, the tongue dark and dry, the pulse small, tense and thread-like, and before the fatal termination, there may be a dark, putrid, or bloody serous fluid thrown from the stomach, very similar to that passed from the bowels.

Such are the ordinary symptoms of malignant dysentery, but in some cases the patient sinks rapidly from the very first, the discharges being putrid, the countenance cadaverous, livid spots appearing, and finally, delirium, stupor, and death being the result.

Anatomical Characters.—The post-mortem in cases of malignant dysentery reveals signs of inflammation of the alimentary mucous membrane, involving sometimes the stomach and small intestines, but more especially the large intestines. The mucous membrane presents a dark livid appearance, or it may be mortified, or it may present more or less extensive patches of ulceration, in the large intestines, and perhaps along the small intestines, even to the stomach. The mesenteric glands are sometimes found enlarged and softened, and abscesses or some other organic change of the liver may be found in some cases.

Diagnosis.—There is generally no difficulty in distinguishing malignant dysentery, after the first few cases, as it generally prevails epidemically, or in some special locality in which filth, bad air, and other unfavorable circumstances serve to render it so. It may be distinguished from simple dysentery by the marked premonitory symptoms of debility, prostration, and irritability of the nervous system, and the general derangement of the stomach, liver, and other organs of the body, and also by the watery, bloody, or putrid appearance of the discharges, as well as the rapidly fatal tendency of the disease.

Causes.—Koino and idio-miasmata are together the cause of malignant dysentery, or either of the causes may operate separately, to produce the disease. The febrific agent or agents, appear to act upon the brain and nervous system to produce prostration, and also probably directly upon the mucous membrane of the alimentary canal, and in something the same manner that it does in yellow, and other putrid fevers, and when *idio-miasmata* is the agent or one of the agents, I suspect that the blood becomes more or less dissolved, as in putrid fevers. It is probable also that prevailing, as it is apt to, at a season of the year when the days are warm and the nights cool, that atmospheric vicissitudes has an agency in producing malignant, as well as simple dysentery.

Pathology.—As we have seen, the general cause or causes, operate to prostrate the powers of the brain and nervous system, the blood becomes more or less dissolved, the various functions of the system deranged, the liver congested, and the portal circulation obstructed, and as a result, the pain, tenesmus, putrid discharges, and typhous symptoms are developed, which are followed, if the disease be not arrested, by a cadaverous countenance, delirium, and death.

Prognosis.—The prognosis in malignant dysentery is generally unfavorable, but if attended to properly at once, some cases may terminate favorably. The favorable symptoms are a subsidence of the pain, tenesmus, putrid discharges, &c., while the unfavorable are an aggravation of all the symptoms, with a cadaverous countenance, cold extremities, and delirium, together with a rapid sinking tendency.

Treatment.—Owing to the great prostration of the general powers of the system, cathartics, or anything of that character, are entirely inadmissible in malignant dysentery. This may be too, in part, owing to the excessively irritable and inflamed condition of the alimentary mucous membrane, for the mucous membrane of the stomach and small intestines are often involved, as well as that of the colon and rectum, though generally in a less degree.

If the patient be seen at the very commencement of the disease, twenty drops of laudanum, with a teaspoonful of brandy, mixed with as much loaf sugar, may be administered, at once, and repeated every three or four hours; sinapisms being thoroughly applied over the whole abdomen. If this treatment be resorted to early, the disease may possibly be arrested at once, but if the case be neglected, or if it be improperly treated at first, the disease may continue from ten to fourteen days, unless the case sooner terminates fatally.

In cases which are not thus early arrested, and the chilly stage is followed by febrile reaction, four grains of Dover's powder, with two grains of the sulphate of quinine, may be given every six hours, and continued till typhoid symptoms supervene, when two grains each of camphor and tannin may be given, in addition to, and alternating with the quinine and Dover's powder, every six hours. This course of treatment may be continued throughout the disease, being modified of course to fulfill the indications as they arise.

If the tenesmus be very distressing, and the discharges frequent, injections of ten drops of laudanum, with an ounce of fluid starch, may be of very essential service, used three or four times per day. As much irritation should be kept up over the stomach and bowels, with mustard, as can be borne, without producing vesication; and in cases that are obstinate, and attended with tenderness, tympanitis, &c., even blisters may sometimes be indicated. Warm toast water, with milk, may be allowed at first, and later arrow-root cooked in milk and water, or mutton or chicken broth, may be allowed, till the patient is restored.

SECTION XIV.—CANCER OF THE INTESTINES.

By cancer of the intestines, I mean that variety of malignant disease of the intestines, attended usually with a livid scirrhous tumor, which ulcerates, or may ulcerate. Every part of the intestines are liable to a cancerous affection, but the duodenum, cæcum, sigmoid flexure of the colon and rectum are the parts in which this affection most frequently occurs.

Symptoms.—The early symptoms of cancer of the intestines are more or less pain or uneasiness in the part affected, and also pain in that portion of the spine opposite the intestinal disease. As the calibre of the intestine becomes materially diminished, there is apt to be constipation, and if the deodenum be the seat of the disease, more or less frequent vomiting is liable to occur.

Sharp lancinating pains are liable to occur, and to be referred to some part of the abdomen, and to extend to, and more or less along the spine. Especially are these pains liable to occur at a particular period after eating. If the duodenum be the seat of the disease, the pain generally occurs, attended perhaps with vomiting, about three hours after eating; but if the disease be seated lower along the intestines, the period becomes a little longer, till we reach the rectum, at which point the pain may be acute, only at or near a movement of the bowels.

When ulceration occurs, there may be a diarrhœa, with bloody or sanious evacuations, which increase, and finally become very offensive; and if the disease be situated in the colon or rectum, there is apt to be the most distressing tenesmus. After ulceration commences, emaciation progresses rapidly, the countenance is cachetic, the appetite fails, the pains become distressing, and, finally, the patient exhausted of flesh, strength and courage, dies, after a train of the most distressing sufferings.

Anatomical Characters.—The appearances presented on post-mortem examination are similar to those of the disease when it affects the stomach. The parietes of the bowels are apt to be thickened, rendering the calibre very small, or there may be tumors projecting into their cavity, and perhaps ulceration, of greater or less extent, may be found. Not unfrequently the disease is found to have extended to the kidneys, liver, or pancreas. In one case I found the pancreas a complete ulcerated mass.

Diagnosis.—If the disease be of the rectum, it may be detected by passing the finger, which will encounter a hard resisting mass, perhaps nearly closing the passage. If it be of the duodenum, it may be detected by the pain, of a lancinating character, which occurs about three hours after eating, together with the vomiting. And if the disease be situated at any intervening point, along the intestines, it may generally be distinguished by the countenance, the pain, and the constipation, together with all the other symptoms which are developed.

Treatment.—During the early stage of the disease, cups, leeches, or blisters may be applied over the seat of the disease, and cups along the spine opposite. Later, conium, hyoscyamus, or stramonium, may be indicated, to allay pain and produce relaxation of the intestine, either taken internally, or used by injection, if the disease be of the rectum. The diet should be mush or corn-bread, and milk.

SECTION XV.—ACUTE PERITONITIS.

By acute peritonitis, I mean acute inflammation of the peritoneum, but especially of that portion lining the abdomen, and reflected over the abdominal and pelvic viscera, except that which belongs exclusively to the serous coat of the intestines, which I have considered in a previous section, under the head of peritoneal enteritis. It must be remembered, however, that in general acute peritonitis, that portion of the membrane reflected over the intestines is very liable to become involved.

It must be borne in mind, that the peritoneum is a serous membrane, which lines the cavity of the abdomen, and is then reflected over the abdominal viscera, and also descends into the pelvis in front of the rectum, from whence it extends over the posterior surface of the bladder, and also of the vagina and uterus in the female.

The peritoneum thus constituted, with its various folds and reflections, is liable to become inflamed, the symptoms of which we will now proceed to consider.

Symptoms.—Acute peritonitis is generally preceded by a feeling of lassitude, pain in the back and limbs, and slight creeping chills, alternating with flushes of heat. There may also be headache, and a feeling of uneasiness in the epigastrium, during the first stage of the disease. Immediately succeeding the chills, or else after febrile reaction is established, pain commences, perhaps in a small space at first, but it soon extends throughout the whole abdominal cavity. In some cases, however, the pain is not general but local, and either stationary or wandering; and in some rare cases, only a slight uneasiness is felt in the abdomen.

While pain is thus a general attendant on acute peritonitis, tenderness is, I believe, an invariable symptom if firm pressure be made on different parts of the abdomen. The patient lies on his back with the knees drawn up and the shoulders elevated, in order to take off the tension of the abdominal muscles, and also the pressure of the bed clothes.

The bowels may become constipated, especially if that portion of the peritoneum forming the outer coat of the intestines becomes involved in the inflammation. The pulse is generally frequent, tense, contracted and sharp, but sometimes it is round and full. The tongue may be moist and covered with a white fur, the edges sometimes becoming red in the progress of the disease.

In some instances the stomach sympathizes strongly with the abdominal affection, and frequent nausea and vomiting may occur, especially if the intestinal peritoneum becomes involved. The face is generally pale and exhibits an expression of anxiety, the patient is wakeful, and delirium is apt to occur towards the termination of fatal cases. The abdomen may remain flat or contracted, but it generally becomes tense and elastic during the course of the disease.

The respiration is oppressed, each inspiration producing an aggravation of the pain. In consequence of this, the respiration is carried on mainly by the expansion of the chest with little or no aid from the abdominal muscles. The secretion of urine is generally more or less suppressed; and when that part of the peritoneum covering a portion of the bladder becomes involved, it is voided with difficulty, and a good deal of pain is experienced in that region. When that portion of the peritoneum adhering to and covering the inferior surface of the diaphragm, becomes involved in the inflammation, hiccough is a very constant attendant. When peritonitis occurs in the puerperal state, the lochia generally cease to flow, the secretion of milk is diminished, and the general powers of the system sink much earlier than when it occurs in other conditions of the system.

Acute peritonitis may be connected with a typhoid condition of the system, the inflammation being of a passive character, the pulse being feeble, the tongue dark and dry, and finally along with great debility there may be a hemorrhagic tendency, with delirium, coma, &c.

Acute peritoneal inflammation is generally rapid in its course, seldom continuing more than a week without terminating in resolution or death, or else passing into a chronic state. Violent cases may terminate fatally in two or three days, or even earlier than that. If, however, the inflammation assumes a subacute grade, the patient may continue on for several weeks.

This variety of inflammation is liable to terminate in gangrene, and when it does, the extremities become cold and clammy, the countenance pale and contracted, and finally, slight wandering delirium occurs before a fatal termination.

Diagnosis.—There is little difficulty in distinguishing acute peritonitis if proper attention be paid to its peculiar symptoms. The supine position, the pale countenance, the tenderness on pressure, and the continuous pain, together with the drawing up of the limbs are sufficient to distinguish this disease from *colic, spasmodic* and neuralgic affections.

From peritoneal enteritis, general inflammation of the peritoneum may be distinguished by the absence of constipation, which attends that disease. But if the peritoneal coat of the intestines become involved in general acute peritonitis, constipation may attend.

Anatomical Characters.—The post mortem appearances vary with the time at which death takes place. If the patient dies early, there may be only a redness of the peritoneum. If the case continue longer, before terminating fatally, there is apt to be a fibrinous exudation on the surface of the peritoneum, of a white or greenish-yellow color, perhaps organized, forming a false membrane, connecting more or less opposite or adjoining folds of the peritoneum.

There is generally more or less liquid found in the cavity of the abdomen, of either a colorless, whey-like, milky, sero-purulent, or bloody character, mixed sometimes with pus, or blood, and containing floating fibrinous flake. Dark spots may be found, either the result of gangrene, or else of effused blood, occupying the submucous cellular tissue.

In cases in which the peritonitis is the result of perforation of the intestines, an offensive gas escapes on laying open the abdomen, and I have found the cavity containing a dark fetid liquid, with more or less fecal matter, the peritoneum presenting a dark injected appearance in most cases at least. In puerperal cases, the liquid found in the abdominal cavity is of a serous, milky, or bloody character.

Causes.—Acute peritonitis may be the result of mechanical injury, violent exertion, stricture of the colon, hernia, the extravasation of blood, urine, or bile into the peritoneal cavity, the action of cold on the surface of the body, causing suppression of the perspiration, cold and wet feet, taking cold water when the body is in a free perspiration, the suppression of hemorrhoidal discharges, or of the menses, and metastasis of erysipelas, rheumatism, and other external inflammations. Acute peritonitis may also occur from an extension of inflammation of the womb, bladder, liver, stomach or intestines, or the disease may prevail epidemically, in which case it is apt to assume a malignant typhoid character.

Treatment.—If the patient be seen very early, there is a possibility of arresting the disease at once. The feet should be placed in warm water, and if there is chilliness, the back may be rubbed with an infusion of capsicum in vinegar. A large sinapism should be applied over the whole abdomen, and allowed to irritate a little short of vesication. By thus equalizing the circulation, promoting perspiration, and producing thorough counter-irritation, the internal congestion, irritation, and inflammation may often be arrested, and convalescence established.

But we are not always so fortunate as to see patients thus early, and if we do, we may not always succeed so readily. If, then, this course of treatment fail, and in all cases in which acute peritonitis is fully established, other and more active measures may become necessary. In cases of acute peritonitis, of an active character, in strong and vigorous constitutions, general bleeding may be indicated, and when it is, should not be neglected.

Immediately after general bleeding, when it is indicated, and at first, when it is not, cups should be applied along each side of the spine, and cups or leeches to the abdomen, and from four to six ounces of blood taken.

After the cupping or leeching, sinapisms should be applied over the abdomen, and a cathartic of calomel and castor oil administered; and if the pain be severe, a full dose of opium may be given with the cathartic. About two grains of calomel, with five grains of Dover's powder may be given every four hours, and with every other powder half an ounce of castor oil may be administered, till a free movement of the bowels is produced. A large blister should be applied to the abdomen, and then it should be kept covered with a warm hop poultice, moistened with vinegar, being laid over the blister at first, and then over the dressings, after vesication is produced.

After getting the effects of the sinapisms, cupping, blistering, and cathartic; if the disease has in a good degree subsided, the calomel may be omitted, and five grain doses of Dover's powder continued every six

24

hours. If, however, the disease continues, the calomel should be continued with the Dover's powder, and the blistering kept up over the abdomen, till the inflammation is subdued.

In cases of this character, which pass on to a state of collapse; and in all cases of acute peritonitis of a passive character, attended with a typhoid condition of the system; two grains each of the sulphate of quinine, and camphor, with five grains of Dover's powder, with or without two grains of calomel, may be given, and continued with dry cupping, blistering, &c., till the inflammation is subdued. And in case there be very great prostration, wine-whey, brandy, &c. may be given in addition, and continued till reaction is fully established.

The nourishment in acute peritonitis should consist of toast water, with a little milk, rice-water, mucilages, &c.; to which may be added, in case of collapse, or in typhoid cases, mutton, or chicken broth, arrow-root, a poached egg, &c., and wine-whey if necessary. During convalescence from this disease, the greatest care should be taken to avoid a sudden exposure to cold, lest by a sudden check of the cutaneous exhalation, a relapse be produced.

SECTION XVI.—CHRONIC PERITONITIS.

By chronic peritonitis, I mean a slow inflammation of the peritoneum, whether of a simple, or tuberculous character. Chronic inflammation of the peritoneum is quite a common occurrence; but unless it be the result of the acute disease, it may come on in so insidious a manner, as not to attract special attention, until incurable structural changes have taken place, or there is effusion into the cavity of the abdomen. The disease may be simple or tuberculous, each variety having some symptoms in common, as well as others, which are peculiar to each.

Symptoms.—In simple chronic peritonitis, there is slight pain in the abdomen, with more or less tenderness, experienced on pressure, as well as in coughing, sneezing, or from any sudden jar of the body; referred generally to the umbilical region. The abdominal pain, however, is not severe, and the inflammation may pass on to disorganization of the peritoneum, without having been attended with any considerable pain.

The abdomen may become tympanitic, and the patient may complain of tightness across the lower part of the abdomen, after exercise; and indurated spots, which are more or less tender, may sometimes be found, consisting probably of folds of the intestines, which have contracted adhesions. The bowels are apt to be torpid in this disease, especially if the peritoneal coat of the intestines becomes involved in the inflammation. The pulse may remain nearly natural; or it may become accelerated, quick and contracted towards evening, in the early stages; and almost continually so, in advanced periods of the disease.

The appetite may be quite good, but in some cases occasional vomiting occurs; the face is generally pale, with an expression of the countenance indicative of languor and ill health. There is apt to be a slight febrile exacerbation towards evening, with oppressed breathing and sometimes slight cough when the patient lies on the back. The feet, too, are apt to become edematous, and the urine scanty, about the time that effusion takes place into the abdomen.

This disease may terminate in fatal disorganization in a few months, but cases occur in which it continues in a slow and insidious manner, for many months before the patient is worn down by the irritation it produces. Whether this disease be protracted or short in its duration, the inflammation generally terminates in effusion into the cavity of the abdomen, this being frequently the manner in which ascites is produced.

In cases of chronic peritonitis depending upon tubercles in the peritoneum, the symptoms are very similar to those **developed** in simple peritonitis; but the disease is apt to be steady in its progress and obstinate, being but slightly, if at all, affected by remedial measures. In this form of the disease, the external lymphatic glands are apt to be enlarged, and also the mesenteric glands presenting small tumors on a close examination of the abdomen. Tuberculous peritonitis occurs in patients of a scrofulous diathesis, and is apt to be complicated with tubercles in the lungs, and in various parts of the system, and especially with tuberculous ulceration of the bowels and an obstinate diarrhœa.

Anatomical Characters.—More or less fluid is generally found in the cavity of the abdomen, usually of a lightish whey-like color, but sometimes of a yellowish or reddish appearance, containing more or less pus or blood.

Different portions of the peritoneum are often found adhering, as well as folds of the intestines, by false membrane; or the intestines may be found adhering to the parieties of the abdomen. The false membrane may be thick and of a gray or dark red color, or it may be in small spots, appearing more or less thickly over the surface of the peritoneum.

In tuberculous cases, in addition to the false membrane, the tuberculous deposition may be found in distinct granulations or in agglomerated masses, and in various degrees of development. The mesenteric glands are apt to be found enlarged and more or less indurated, as well as the external lymphatic glands in various parts of the body.

Causes.—Chronic peritonitis is frequently the result of the acute form of the disease, but it may occur from an extension of the inflammation in mucous enteritis, from congestion of the liver, interrupting the portal circulation, or from blows inflicted on the abdomen; or it may occur from suppression of the perspiration, or from metastatis of erysipelas and other external inflammatory affections. Tuberculous cases arise from a hereditary predisposition as well as various abuses of the system, such as impair digestion, and tend to produce a general scrofulous or tuberculous condition of the system.

Treatment.—After ascertaining and removing the cause, as far as may be, blisters should be applied over the abdomen, or pustulation produced with tartar emetic ointment, according to the condition of the patient, and continued till the tenderness subsides. After having used the counter-irritants for a reasonable time, strong iodine ointment should be applied twice each day, so as to use at least an ounce per week; and the patient should be directed to wear a flannel next the skin, and so applied as to cover the whole abdomen, in order to promote perspiration.

The patient should also be directed to wear a flannel wrapper, to sleep in flannel sheets, and to take extreme care to keep the feet warm and dry. To promote a healthy action of the glandular and lymphatic systems and to prevent effusion, with its consequences, five grain doses of the iodide of potassium may be given in simple syrup, three times per day and continued for a long time. Should an anodyne become necessary to quiet restlessness and procure sleep; the solid or fluid extract of hyoscyamus, or conium in moderate doses will generally do best. The bowels should be kept gently relaxed by small doses of cream of tartar, administered occasionally during the whole course of the disease; a teaspoonful may be given at evening if the bowels have not moved during the day.

If, as generally happens, the blood becomes weak from defective chymification during the latter stages of this disease, some preparation of iron is clearly indicated. The syrup of the iodide of iron may be given in ten drop doses, three times per day, as a tonic and alterative, and continued for a long time. Or, in case that should disagree, the citrate of iron may be given instead, in two or three grain doses after each meal.

In tuberculous cases, in addition to the iodide of potassium and some preparation of iron, the cod-liver oil may be indicated, and should be given in moderate doses an hour after each meal.

The food during the early stages of chronic peritonitis should be mild and unirritating and easy of digestion; but during the latter stages, bread and milk, with meats, may be freely allowed. Great regularity should be observed in taking food, and every precaution taken to guard the general health, that the system may be able to bear up under this extensive, insidious and complicated variety of chronic peritoneal inflammation. By a proper and judicious treatment, faithfully applied, I believe that even tuberculous cases may be retarded in their progress, and many simple cases permanently cured.

SECTION XVII.—ACUTE HEPATITIS.

By acute hepatitis I mean acute inflammation of the liver; which, it will be remembered, is a large conglomerate gland appended to the alimentary canal, and occupying the right hypochondrium.

The liver is placed obliquely in the abdomen; its convex surface being upwards and forwards, and its concave downwards and backwards. It is in relation superiorly and posteriorly with the diaphragm, and inferiorly with the stomach, duodenum, colon, and right kidney; its free border corresponding with the lower margin of the ribs.

The liver is supplied with arterial blood by the hepatic artery, and it also receives, through the vena porta, the blood from the chylopoietic viscera taken up by the gastric, splenic, superior and inferior mesenteric veins.

The liver probably separates impurities from the blood, which passes through it along the vena porta, on its way to the general venous circulation, and it also secretes the bile, which is conveyed along its excretory duct, and emptied into the duodenum, along with the pancreatic fluid; this fluid being essential to chylification. The blood introduced into the liver by the hepatic artery and vena porta, is taken up by the hepatic veins, and carried by them to the vena cava; its impurities, as well as the bile, having been drained off, and returned to the alimentary canal.

The liver, thus situated and constituted, with its firmness of structure, and great vascularity, is liable to severe congestions, and inflammations; the acute form of which we will now proceed to consider.

Symptoms.—Acute hepatitis generally makes its attack suddenly, and with considerable violence, especially in marshy or miasmatic localities. The disease generally commences with chills, and a pain, either acute, or dull and heavy, in the right hypochondrium; accompanied with a sensation of tightness across the abdomen, some difficulty in breathing, and a disinclination to the recumbent posture; the patient generally feeling easiest while sitting, slightly inclined forwards, and to the right side.

Sometimes the attack is less impetuous; the patient complaining of a feeling of tightness in the right hypochondrium and epigastric regions, with very slight febrile symptoms, for some time, before anything serious is apprehended. More or less fever, however, is developed in all cases; and sometimes this precedes any well marked symptoms of the local inflammatory affection.

The pain attending acute inflammation of the liver, frequently extends to the breast, clavicle, and shoulder of the right or left side. And in some cases, these sympathetic pains are more severe, than those in the liver itself. If the inflammation be confined to the substance of the liver, the pain is of a dull, heavy, aching character; but if the inflammation extend to its *peritoneal coat*, the pain becomes acute, and sometimes very severe.

If the right lobe of the liver be the seat of the inflammation, the pain and tenderness is in the right side, and the patient complains of pain in the right shoulder. If, however, the left lobe be the part inflamed, there is pain and tenderness in the epigastrium; and more or less pain is felt in the left shoulder. When the inflammation occupies the lower portion of the right lobe, there is apt to be more or less intestinal disturbance; but if the superior portion of the lobe be the seat of the inflammation, there is generally considerable pulmonary derangement. Finally, in cases in which the inflammation is of the left lobe, there is usually considerable gastric disturbance, with nausea and bilious vomiting.

There is generally in this disease, a yellow appearance of the eyes, and also of the skin, of the chest, neck and face. And the urine is generally charged with more or less bile, being of a deep yellowish color. The thirst too is urgent, and the skin hot and dry. The pulse is usually full, active, and firm; but if the inflammation extend to the peritoneal coat of the intestines or stomach, it may become small, tense, and quick.

The tongue is either smooth and glossy, or else it is coated with a white or yellowish fur, and in this case the mouth has a dirty bitter taste. The bowels are apt to be constipated, but a looseness with dysenteric or serous discharges may attend from the very first, produced probably by congestion of the liver interrupting the portal circulation. The brain is apt to sympathize with the inflamed organ, often giving rise to slight mental disturbance.

Acute hepatitis rarely continues beyond the seventh day without terminating either in resolution or suppuration. When suppuration takes place the pain may become less, there is a sense of weight and throbbing in the region of the liver, with rigors, night sweats, sinking, anxiety, oppression, and a cold clammy skin.

Sometimes large abscesses form in the liver as the inflammation subsides, and if the liver has formed adhesions with the parietes of the abdomen, the abscess may point externally, and the matter be safely discharged, and recovery follow. The matter may point in the superior part of the liver, which, having formed adhesions with the diaphragm, the matter enters the thorax. In this case it may pass by ulceration into the bronchial tubes and be expectorated, but care should be taken not to mistake the bronchial secretion sometimes set up from sympathy, or from an extension of the inflammation in hepatitis, for a purulent expectoration from the liver.

In other cases the liver forms adhesions with the intestines, and the abscess bursts into the alimentary canal, and is discharged by stool, a final recovery being the result, or the hepatic abscess may burst into the cavity of the abdomen, and the case thus terminate fatally.—In cases of hepatic abscess in which pure pus is discharged, it has generally collected between the glandular substance and its peritoneal covering; but if the matter discharged be of a dark gray color, it has probably come from the interior of the liver, or from the glandular portion of the organ.

Acute hepatitis may also terminate in gangrene, but in fatal cases, the substance of the liver may become softened and of a darkish color.

Anatomical Characters.—In cases of hepatitis, in which death has taken place early, if the peritoneal coat has been involved in the inflammation, it is found red, more vascular, and perhaps thickened, and sometimes covered with coagulable lymph, advancing more or less towards organizatian. If the parenchyma has been inflamed, it is found congested with blood, enlarged, and softer than in health, and on being cut, there is apt to be an oozing of blood, which appears to have occupied small cavities in the substance of the gland.

If the patient has died at an advanced stage of the disease, abscesses are found occupying the substance or surface of the liver. If the abscess occupies the substance of the gland, the tissue about the abscess is found more vascular than at other points, and the abscess contains a sero-purulent fluid, mixed with the softened matter of the gland.

In some cases but one abscess is found, while in others they are quite numerous, and in some rare instances, nearly the whole gland is converted into a sero-purulent mass, contained in its investing peritoneal membrane. The pus may appear quite natural, if it only occupies the surface of the gland, but otherwise it is apt to be of a reddish, green, or dark appearance, being mixed with bile, and perhaps more or less of the disorganized tissue of the gland.

Diagnosis.—Hepatitis may be confounded with pleurisy, pneumonia, or gastritis, from each of which, however, it may be distinguished by careful attention to the following differences.

In pleurisy the cough and oppression in the chest are more severe than in hepatitis; and besides, in hepatitis the patient generally rests on the affected side, while in pleuritis, the patient lies on the well side. In hepatitis the tenderness is in the epigastrium, or right hypochondrium, while in pleurisy the tenderness is felt in the intercostal spaces.

From pneumonia, hepatitis may be distinguished by attention to the manner in which the respiration is performed. It being carried on in pneumonia mainly by the abdominal muscles, while in hepatitis it is performed almost entirely by the intercostal muscles.

From gastritis, this disease may be distinguished by the full, hard pulse of hepatitis, while in gastritis, it is generally more or less contracted and weak. And besides in gastritis, there is a good deal of prostration from the first, and obstinate vomiting on taking anything into the stomach, while in hepatitis the vomiting is not so constant, and there is not so much prostration at the commencement of the disease.

From the passage of gall stones, hepatitis may be distinguished by the fever, by the continuous pain, and by the position of the patient, in hepatitis the patient inclining a little forward and to the right side, while in the passage of biliary concretions, the patient is easiest when the body is bent forward on the pelvis.

Thus may hepatitis be distinguished from the various affections with which it has some symptoms in common.

Causes.—The causes of acute hepatitis are various, but it is probable that the paludal poison, heat, or atmospheric viscissitudes, together with the use of intoxicating liquors, and tobacco, and other species of intemperance, are by far the most frequent causes.

We have seen that koino-miasmata, has a debilitating affect upon the brain and nervous system, and that when the brain and nervous system become thus prostrated, the functions of the different organs become impaired, and none more sensibly than that of the liver. When, therefore, this agent acts upon the system through the blood, the liver becomes torpid and congested, and this congestion irritating the capillaries of the arteries, veins, and portal vessels, provokes inflammation.

Atmospheric vicissitudes produce hepatitis, in part probably, by the sympathy that exists between the action of the skin and liver. And hence, during the autumnal season, or when we have hot days and cool nights, the perspiration becomes suddenly checked, during the night, and the action of the liver more or less torpid. As a consequence of all this, the portal vessels become congested, and their capillaries in the liver, which congestion provokes irritation and inflammation, and this is probably the most frequent cause of hepatitis. It is probable, however, that the paludal poison and heat, or atmospheric vicissitudes, combine in most cases, to produce acute hepatitis.

It is probable that tobacco produces hepatitis, by its debilitating effects, in a manner similar to the paludal poison. But intoxicating liquors, as they are probably in part taken up by the portal vessels, and carried through the liver into the circulation, may act as a direct irritant to the capillaries of the liver, and thus produce hepatatic inflammation.

Besides the causes already enumerated, it is probable that violent exercise, injuries of the right hypochondrium, wounds of the head, the passages of biliary calculi, the abuse of mercury, the translation of gout or rheumatism, the suppression of hemorrhoids, terror, mental despondency, and various other causes may produce acute hepatitis.

Pathology.—It appears probable that direct irritants in the blood, act primarily upon the portal veins, and their capillaries, in consequence of which they become congested and inflamed, and very soon all the minute arteries, veins, and capillaries of the liver, and with them its whole structure.

In hepatitis, produced by agents which act indirectly through the system, the congestion, irritation, and inflammation, whether active or passive, may commence in any of the capillaries of the organ, and extend, soon involving its whole structure, always producing, however, congestion of the portal vessels.

Treatment.—If this disease be attended to in its incipient stage, great hope may be entertained of arresting the inflammation. The feet should be placed in warm water, and if there be chilliness, warm stimulating friction may be made along the spine. After this, a few ounces of blood should be taken from the right side of the **spine, opposite the liver, and also** from the right hypochondrium.

A full dose of calomel **and castor oil should** be administered, and the patient directed **to drink freely of warm sage tea.** Or, in case the mercurial is contra-indicated, **two** or three grains of podophyllin may be given with the **oil, instead of the** calomel. In this way incipient hepatitis may sometimes be **arrested : the** warm foot-bath taking off the undue pressure on the brain; the warm friction along the spine equalizing the circulation; while the cupping relieves any irritation that may have **been set up in the** spine or liver; and the cathartic relieves the congestion of the portal vessels.

But unfortunately we do not always see such cases thus early, or, if we **do, we may not** always succeed so readily in arresting the inflammation. In cases that have passed on till the inflammation is fully established, **and a** high febrile reaction is set up, the same measures, except warm friction to the spine, should be resorted to, as helps; but we may **be under the** necessity of resorting to general bleeding, and other measures, **to prevent the most** fearful consequences.

After getting **the effects** of the warm foot-bath, general bleeding, cupping, and a cathartic; in such cases, a large blister should be applied over the right hypochondrium, and kept discharging. If the inflammation persists, three grains each of calomel, Dover's and James's powders **may** be given, every four or six hours; and this treatment may be continued till **an impression is** made on the disease, or ptyalism is produced, **the bowels being moved two or** three times every twenty-four hours, by **small doses of the sulphate** of magnesia.

If **an impression be made on** the disease, or ptyalism is produced, the calomel should be omitted, **and** the other treatment continued, modified, of course, by the indications as they arise. Should suppuration take place, and the matter approach **the** surface, a poultice should be applied, and the matter evacuated, as **soon** as it is in a fit condition. In such cases, after matter has formed, the iodide of potassium should be administered, in five grain doses, three times per day, before eating, and continued for **a** long time.

If an anodyne be indicated to quiet restlessness and irritability, the solid or fluid extract of conium will generally do best, or the solid or fluid extract of conium and taraxacum may be combined, and given in moderate doses after each meal, and continued for a time as an alterative and anodyne. If the digestion is bad, and the stomach distended with wind, a mild tonic bitter of columbo in cold water, or half a drachm of the fluid extract may be administered before or after each meal for a time.

During the debility which follows extensive suppuration, the syrup of the iodide of iron may be given in ten drop doses three times per day. During the inflammatory stage of the disease, the nourishment may consist of crust coffee with a little milk, but later, a plain, digestible and nourishing diet may be allowed.

SECTION XVIII.—CHRONIC HEPATITIS.

By chronic hepatitis I mean a slow or protracted inflammation of the liver. Chronic inflammation of the liver is generally of its glandular structure; and when not the result of the acute form, is usually rather insidious in its attack.

Symptoms.—Chronic hepatitis commences usually with symptoms of dyspepsia. The patient complains of irregular appetite and acidity of the stomach, as well as of colic pains, nausea, and a sense of fullness in the region of the stomach. In some instances there is pain and perhaps tenderness in the epigastrium or right hypochondrium, and pain in the right or left shoulder; but very often no pain is experienced in the region of the liver, unless firm pressure be made on the surface. In most cases firm pressure produces an uneasiness, and by careful examination the liver may be found enlarged, and projecting below the short ribs.

If the superior portion of the right lobe be the part involved, the lungs may suffer considerably, and there may be more or less cough. When the inflammation is confined to the lower portion of the right lobe, there is apt to be intestinal disturbance, with perhaps a diarrhœa or dysenteric discharges. Finally, when the left lobe of the liver is the part inflamed, the stomach generally suffers more or less; and nausea with distressing vomiting may occur. The pain, if the inflammation be of the right lobe, may extend to the right shoulder, but if the hepatic inflammation be of the left lobe, it is generally referred to the left shoulder. In most cases of chronic hepatitis the eyes, as well as the skin of the face, neck and chest, become tinged of a yellowish hue.

The bowels may be constipated with occasional diarrhœa; the discharges being dysenteric or else of a dark muddy appearance. The urine is generally highly tinged with bile. The tongue is white and dry, the taste bitter; and in advanced stages of the disease, the gums are apt to assume a firm hardened appearance. The skin in this disease is dry and harsh, and as it progresses the patient loses flesh, slight febrile exacerbations come on towards evening with heat in the palms of the hands and soles of the feet; and unless the disease be arrested, it may terminate in suppuration, emaciation and death.

Anatomical Characters.—Various morbid appearances are presented on the examination of subjects dead of this disease; among which are induration, softening, suppuration, &c.

Induration is probably the result of the exudation of coagulable lymph into its tissue; and it may be partial, or it may extend throughout the whole gland. The indurated liver may be either enlarged or contracted, and it is generally of a yellowish color.

While induration is occasionally found, partial or general softening is a much more frequent morbid condition, following this disease. In cases of chronic hepatitis, in which the gland is thus found softened, the portion involved is generally of a dark reddish color, and it may be only slightly softened, or it may be reduced to a mere pulp, or a portion of it at least.

Besides softening and induration, as morbid conditions in this disease, abscesses are frequently found; in some cases large and single, but in others, smaller and quite numerous. Various other morbid appearances may be presented, but softening, induration, or abscesses, are the appearances most frequently presented on the post-mortem of subjects dead of this disease.

Causes.—Chronic inflammation of the liver is frequently the result of the acute form of the disease. Or it may arise from a slow operation of the causes which produce acute hepatitis; such as heat, or atmospheric vicissitudes, the paludal poison, intoxicating liquors, tobacco, and various abuses of the system, such as masturbation, excessive venery, &c. It is also sometimes the result of direct injury, or the metastasis of gout or rheumatism, or it may arise from the suppression of the hemorrhoidal or other accustomed discharges.

Treatment.—The indications in the treatment of chronic hepatitis may depend very much on the constitution of the patient. The cause should be sought out and removed, if possible, at first, in all cases. And if a mercurial is not from any cause contra-indicated, two grains of the blue mass may be given, morning and evening, with an equal quantity of the extract of conium, as an alterative and anodyne, and continued for a few days; the bowels being kept regular by drachm doses of the fluid extract of taraxacum, after each meal. If from delicacy of constitution, or from any other cause, a mercurial be contra-indicated, one-third of a drachm of the fluid extract of the Leptandra Virginica, or else of the podophyllum peltatum, with ten drops of the fluid extract of the conium maculatum, may be given after each meal, for a time, as an alterative, anodyne, &c.

Cups should be applied along the right side of the spine, opposite the liver, and also over the epigastrium and right hypochondrium, and more or less blood taken at first. Dry cups may then be repeated every day or two, or blisters may be applied, or pustulation produced by tartar emetic ointment, till the inflammation be subdued.

After having continued the mercurial for a reasonable time, in cases in which it is indicated, it should be omitted, and five grain doses of the iodide of potassium given three times per day, with the fluid extract of taraxacum, the podophyllin, or leptandrin, being given as an alterative, if an alterative be still required. In cases in which a mercurial is not used, the iodide of potassium may be commenced with as soon as the pain and tenderness subside, and it may be given in connection with the fluid extracts of the leptandra or mandrake, conium and taraxacum, or such of them as may be indicated, till the disease is subdued. Should a bitter tonic be indicated, a cold infusion of columbo, or moderate doses of the fluid extract, may be given for a time after each meal.

The nitromuriatic acid foot-bath may be of service in obstinate cases of chronic hepatitis. Equal parts of the nitric and muriatic acids may be combined with an equal quantity of water, and of this an ounce **may** be put into a gallon of warm water, and the feet and legs kept in it **for** twenty or thirty minutes, before going to bed each night, for two or three weeks. By thus removing the cause, and keeping up an alterative course, with counter-irritation, and fulfilling the indications as they arise, with safe remedies judiciously applied, many troublesome cases of chronic hepatitis may be either palliated or permanently cured.

The diet, during the early stages of this disease, must be of a plain, unirritating character; one of the best articles of which is wheat or Indian bread, and milk. Later, the diet may be highly nourishing, but digestible, and may consist of bread and milk, meats, &c., taken with regularity.

The patient should wear flannel next the skin, and sleep in flannel sheets; and he should keep in a temperature that is agreeable, and by no means be exposed to dampness, or suffer from chilliness or cold. Such, then, are the principles which are to guide us in the treatment of chronic hepatitis; subject, however, to modifications, from the variety and degree of the hepatic inflammation.

SECTION XIX.—SPLENITIS.

By splenitis I mean **inflammation of the spleen**, which is an oblong, flattened organ, of a bluish red color, situated in the left hypochondrium. The spleen is in relation superiorly and on its external surface with the diaphragm; by its concave surface, with the stomach, pancreas, and left kidney; and by its lower end with the transverse colon. It is invested by the peritoneum, and by a yellowish elastic tissue, which penetrates the organ, forming sheaths for its vessels.

The spleen is supplied with blood by the splenic artery, a division of the cœliac axis, and is a very vascular organ, even its lymphatics being very numerous and of a large size. The veins of the spleen pour their blood into the splenic vein, which is one of the principal trunks of the portal vein; its lymphatics terminate in the lumbar glands, and it is supplied with nerves by the splenic plexus. It is probable that the spleen serves, as Professor Draper has suggested:* "As a receptacle for any excess of blood" in the portal vessels; and that it also aids in "the dissolution of the disorganizing blood-cells, preparatory to the action of the liver, in which hæmatin is to be converted into the coloring matter of the bile."

The spleen thus constituted and situated, is liable to become inflamed; the symptoms of which we will now proceed to consider:

Symptoms.—Inflammation of the spleen generally commences with chills or coldness, followed by febrile excitement, more or less marked, and it is attended with pain and tenderness to external pressure in the left side, immediately under the false ribs. The left hypochondrium is generally fuller than in health, and there may be pain in the left and perhaps the right shoulders.

The white of the eyes are apt to assume a bluish appearance, and the skin may assume a yellowish tinge. In consequence of the relation of

* See Draper's Physiology, pages 211 and 212.

the spleen to the stomach, liver, lungs, left kidney and colon, more or less derangement in the functions of some or all these organs or parts is lable to occur in **splenitis.**

An oppressive sensation is apt to be experienced in the stomach, with dyspeptic **symptoms,** and perhaps nausea and vomiting. By an extension of the **irritation** to the lungs, through the diaphragm, there may be produced troublesome cough, and hiccough may also **attend,** with dyspnœa, palpitation of the heart, &c. The liver or kidneys may become functionally deranged, and the bowels may become constipated; or there may be a looseness of the bowels, with serous, bloody, **or** dysenteric discharges.

Splenitis may be of an **active, or** passive character; and it may assume an **acute or** chronic form. The acute active variety **of** splenitis generally occurs in sthenic constitutions, **while the** passive and chronic varieties are apt to occur in the asthenic, or in patients worn down by some miasmatic, or other variety of disease, deranging the portal circulation.

Chronic splenitis is attended with a feeling of uneasiness, **and** fullness in the left hypochondrium, and sometimes with pain and tenderness on pressure. There is also apt to be pain in the left shoulder, and on examination, the spleen is found considerably enlarged; forming a more or less prominent fullness in the left hypochondrium.

In chronic cases, the functions of the stomach, liver, lungs, heart, kidney, and intestines, are as liable to suffer as in the acute form of the disease, though generally in a less degree. The disease seldom continues in its acute form, without terminating in resolution, or suppuration, longer than one or two weeks; but it may assume a chronic form, or it may be so from the first, in which case it may continue on for months, or even years.

Anatomical Characters.—If the patient has died early, the organ is found enlarged, and of a darkish color, being also more or less softened. If the patient dies later; pus, mingled with blood, is found, either infiltrated, or collected in cavities; sometimes occupying most of the organ, or filling completely its investing capsule.

In *chronic* cases, the spleen is generally found enlarged, and either hypertrophied, indurated or congested; or else softened with purulent or bloody infiltration. Its proper coat is often found cartilaginous or osseous, and its peritoneal investing membrane, adhering to the stomach, intestines, or kidney.

Diagnosis.—Splenitis may be distinguished from hepatitis, by the pain, tenderness, and fullness in the left side, by the bluish appearance of the eyes, and by the absence of the marked jaundiced appearance, so common in hepatitis.

Causes.—Splenitis may be produced by violent exercise, by injuries, by mental emotion, and by the metastasis of other disease, or the suppression of some accustomed discharge. Splenitis too is often produced by miasmatic, and other affections, which interrupt the portal circulation.

Chronic splenitis may be the result of the acute **form of** the disease. But it most frequently occurs from miasmatic influences, or from the chills or congestions which attend intermittent, remittent, or congestive fevers.

Treatment.—In acute inflammation of the spleen, cups should be applied to the left hypochondrium, and along the left side of the spine, opposite the organ; and from four to six, or eight ounces of blood taken. A mercurial or saline cathartic should be administered; the warm foot bath resorted to, and the cupping repeated, if necessary, or blisters applied, till the inflammation be subdued.

In chronic cases cupping over the left hypochondrium, and along the spine, without taking much if any blood, should be resorted to at first, and then blisters may be applied to the left hypochondrium, or pustulation produced with tartar emetic ointment. A blue pill may be given at evening, for a few days, and a teaspoonful of the sulphate of magnesia in the morning, and then five grain doses of the iodide of potassium may be given three times per day, before each meal, and continued till the fullness entirely disappears. After counter-irritants have been used for a reasonable time, iodine ointment may be freely applied. The patient should wear flannel next the skin, should take a plain, digestible, and nourishing diet, and if the splenitis be the result of or complicated with other affections, such affections should be judiciously removed.

SECTION XX.—DYSPEPSIA—(*Indigestion.*)

By dyspepsia, from δυς, "with difficulty," and πικτω, "I concoct," I mean difficult, disturbed, or imperfect digestion, from whatever state or condition it may arise; the indigestion being the predominant symptom. In order to appreciate dyspepsia, or indigestion, let us take a glance at the processes by which healthy digestion is performed, and then we may be able to appreciate indigestion, in its various forms.

It is necessary for healthy digestion, that the food be of a proper quality, that it be taken at regular hours, and that it be well masticated, and moistened with saliva, before it passes into the stomach. In the stomach it must meet with sufficient gastric juice to thoroughly dissolve it, and the stomach must possess sufficient muscular power to carry it into the intestines, after it is converted into chyme. In the duodenum the chyme must meet with sufficient bile and pancreatic juice, to dissolve the fatty portions of the chyme, that it may be mixed with the albumen, and form a milky substance, or chyle, of a character fit to be taken up by the absorbents of the small intestines. The mucous membrane of the alimentary canal, and especially of the small intestines, must be in a healthy state, so that the lacteals may be able to take up the chyle, and the muscular coat of the intestines must also possess tone sufficient to carry along, and expel from the bowels, the residual matter of the food.

And in order for the chyle, thus prepared and taken up by the intestinal lymphatics, to be taken into the system and appropriated to its use, the mesenteric glands, through which the lacteals pass, must be in a condition to aid in converting the fat and albumen of the chyle into fibrin, which in a healthy state passes to the thoracic duct, and thence to the general circulation, and to every part of the system. Such then is healthy digestion.

Now any deviation from this natural process of digestion is indigestion, however slight or trifling it may be, and if continued for any length of time, is generally connected with a weak, relaxed, and debilitated condition of the mucous membrane, and muscular coat of the stomach, or else with an irritated, excited, or inflamed condition of the mucous and muscular coat of the stomach and intestines.

Let us now bear in mind that dyspepsia may depend upon directly opposite conditions of the alimentary canal; in the one case the mucous and muscular coats of the stomach and intestines, being in a weak, relaxed and debilitated condition, while in the other, they are in a morbidly irritated, excitable or inflamed state. With these considerations, let us proceed to inquire into the symptoms which are developed in dyspepsia, in its various forms.

Symptoms.—In that variety of indigestion in which there is a debilitated and relaxed condition of the stomach and intestines, with no marked irritation, or inflammation, the appetite is generally poor, and often entirely destroyed. The patient is troubled with flatulence, acid eructations, and colic pains; the mind is depressed and languid; the bowels are usually constipated, and if the disease be protracted there is almost always a marked degree of despondency and general emaciation.

In cases of indigestion in which there is more or less irritation of the mucous membrane of the stomach and intestines, and a morbidly excitable condition of their muscular coat, the appetite is frequently morbidly increased, but no sooner is the food in the stomach, than the irritated organ, by an excited morbid action, hurries it along into the intestines in a partially digested state. Immediately after eating, the pain in the stomach subsides for an hour or so, when it usually returns and continues, more or less severe, for two or three hours; then it gradually subsides, and gives place to a gnawing or morbid sensation of hunger, which goes on increasing till it becomes exceedingly distressing, or till food is taken, when there is again a temporary abatement.

The patient has frequently a sour stomach, with eructations; there is yellowness of the skin, and sometimes of the eyes; there is distress in taking very cool or very warm drinks, and not unfrequently there is an appetite for some particular kind of food for a time, and then no relish for any. In this way the patient passes on, with tedious days and restless nights, with great physical and mental irritability, till the most intimate friends can hardly be tolerated, and existence itself becomes almost a dread. Such are the most prominent symptoms of dyspepsia, attended with nervous irritability, affecting especially the stomach, but generally more or less the whole system, if the disease continues for a long time.

Indigestion, attended with inflammation of the mucous membrane of the stomach, is generally attended with an irregular appetite, sour stomach, nausea, tenderness to external pressure, a yellow tinge of the skin, a red and dryish tongue, thirst for cool drinks, irregularity of the bowels, and more or less general nervous irritability. There is distress or pain in the stomach immediately after eating, or taking warm drinks, which, however, gradually subsides as the food passes from the stomach. Any very stimulating food or drink greatly aggravates the distress, and increases the morbid thirst.

The pulse is small, tense and quick, the skin harsh and dry to the feel; there is pain in the head, and frequently along the spine, and nausea is

produced if firm pressure be made on the epigastrium. The bowels are
either constipated or very much relaxed; the food frequently passing in
an undigested state. The urine is scanty, and of a yellowish appearance, indicating more or less functional derangement of the liver, or obstruction to the free passage of bile into the intestines; perhaps from an
extension of the inflammation along the duodenum to the mouth of the
bile duct.

Thus, then, we have the symptoms developed in dyspepsia in its three
prominent forms. But it must be remembered that while every case of
dyspepsia may come under one of these heads, there is an almost endless
variety of degrees in each. And also, that the three varieties may be
developed in the same case, beginning with the relaxed or debilitated,
and ending with the inflammatory, or the reverse; especially if the imprudence producing it be continued, and the case neglected.

Causes.—The causes of dyspepsia are various, but the most frequent
are irregularity and imprudence in taking food, improper exercise, &c.
The stomach in a healthy state will call for the kind and quantity of
food that the system requires; so that in a state of health if its calls
were always properly heeded, it would be a safe guide, as to the quality
and quantity of food to be taken. Dyspepsia is often contracted in
early life, and frequently in early childhood, or even infancy, and the
following is, according to my observation, the manner in which it is
generally brought about.

Infants are allowed to nurse at irregular hours, or when they cry, in
order to pacify them, and at the very time when the stomach is least able
to digest its food. By this imprudence and irregularity many infants get
dyspepsia, attended with sour stomach, vomiting, &c.

After children begin to take solid food, they are often allowed to eat
a *piece*, to allay the morbid appetite which this imprudence has produced,
or to please the taste, or to allay anger, when the system is not in need
of food, or the stomach in a condition to digest it.

In this way, a habit of eating at irregular hours, or between meals is
contracted, and being continued till the child grows to maturity, tobacco
chewing or smoking is taken up to cure the dyspepsia. Now the system
once under the intoxicating effects of this narcotic poison, the natural
appetite is lost, and the mucous and muscular coats of the stomach become narcotized, debilitated, and incompetent to digest food enough to
supply the actual wants of the system, and that little is but poorly done.
In this way is that variety of dyspepsia, attended with a debilitated condition of the mucous and muscular coats of the stomach, very frequently
brought about.

If now, as generally happens in such a state, the system becomes debilitated, and the appetite morbidly craving, the system calls for more
food than the stomach can digest. An extra amount of food is taken,
but the stomach fails to digest it properly. This undigested mass acts
as an irritant to the mucous membrane, and the muscular coat also becomes irritated, in consequence of which the undigested mass is carried
into the intestines, and acts as an irritant to the alimentary mucous
membrane. Thus it is that dyspepsia, marked by irritability of the stomach, is very often produced.

If now, as frequently happens, this manner of taking food, and this imperfect process of digestion continue, with, as occasionally happens, the use of intoxicating liquors, the mucous membrane of the stomach becomes congested and inflamed, and the muscular coat highly irritated, and this constitutes the third variety of dyspepsia, or that in which there is inflammation of the mucous membrane of the stomach, and a morbidly irritable condition of the muscular coat.

There are various other causes of the different varieties of dyspepsia, such as want of, or improper exercise, eating unwholesome food, eating too fast, drinking too much hot tea or coffee, drinking too much at meals, or moistening the food in the mouth with drinks, instead of saliva, intense mental application immediately before or after eating, and various other like imprudences.

Treatment.—As we now proceed to the consideration of the treatment of dyspepsia, let us first examine those rules or regulations which are adapted to all cases of indigestion, and then we will inquire into the peculiar indications in each variety.

All dyspeptics, as well as every body else, should be directed to take only wholesome, digestible food, and that at regular hours, three times per day; and no dyspeptic or other person should be allowed, on any account, to take food between meals, even though the stomach can bear but little food at a time, except in the case of infants, and persons reduced by disease, in which case nourishing drinks may be allowed.

This regulation is absolutely necessary, not only for dyspeptics, but for those in health, in order to give the stomach time to rest after digesting, or attempting to the preceding meal. If this rule be rigidly adhered to, very much will be accomplished by that alone, to prevent and cure dyspepsia.

Food should be taken slow, and be well masticated, and the patient should be directed to allow the food to be moistened in the mouth by the saliva, and not to drink too much during meals. If this rule be rigidly adhered to, the food will go into the stomach in a fit condition for the action of the gastric juice. And by avoiding too much drink, the gastric juice will not be too much diluted, to dissolve the food which is taken.

The dyspeptic should be directed to take enough exercise, and if possible at some employment calculated to occupy both the body and the mind. But active exercise should be suspended for at least one hour before and after eating, in order to give the stomach the full benefit of the energies of the system during the effort at digestion. If the appetite be morbid, and therefore not a guide as to the quality, kind or quantity of food to be taken, the judgment of the patient, together with the advice of his physician, must be the dictator in these respects, till such time as the appetite becomes natural, and its calls an infallible guide when it should be rigidly adhered to.

Clean cold water should be recommended for all dyspeptics who have not become the slaves of tea and coffee; being the most natural and wholesome drink for man, as it is for the whole animal creation. But in cases in which tea and coffee cannot be suspended by the dyspeptic, on account of long continued use, they should be taken with moderation and not too hot. Tobacco and intoxicating liquors are poisons and should not be used by any one ; the dyspeptic, therefore, who has been made so by their use, should be directed to abandon them, as fast as it can be done, without producing delirium tremens.

If these rules were obeyed, we should get few cases of dyspepsia, and if rigidly enforced with the dyspeptic, it would entirely cure very many cases and greatly relieve those it might fail to cure.   It appears to me utter folly to prescribe medicines for indigestion or any other disease, while a violation of these and other laws of health have not only produced, but are keeping up the morbid condition.   But having corrected the habits of the dyspeptic, various remedies may be of service in correcting the morbid conditions which the imprudence has produced.

In dyspepsia attended with a weak and debilitated condition of the stomach, a good nourishing diet of animal food may be allowed ; and the tone of the stomach may be improved by moderate doses of the cold infusion, or the fluid extract of columbo, taken three times per day after each meal ; and if the blood is weak, wo grains of the ammoniated citrate of iron may be taken in solution, at or before eating, three times per day.

In that variety of dyspepsia in which the stomach is in an irritable, but not in an inflamed condition, the food should be mild and nourishing, but not very stimulating.   To allay the nervous irritability of the stomach and correct acidity in such cases, a pill of two grains each of the extract of hyoscyamus and the subnitrate of bismuth may be given for a time after eating, morning, noon and night.   If there be a weak state of the blood, the ammoniated citrate of iron may be given in two grain doses in solution, three times per day and continued for a time.

In cases of dyspepsia attended with inflammation of the mucous coat of the stomach, the food should be of the most mild, digestible and unstimulating character, with very little, if any, animal food.   The drinks too should be cool, and in no case of a stimulating character. To allay irritation and constringe the enlarged capillaries of the inflamed mucous membrane, a pill composed of the sulphate of iron and extract of conium, of each one grain, may be given after eating three times per day.   Cups or blisters should be applied over the stomach, or what in many cases may do better, pustulation may be produced by tartar emetic ointment.

In cases of dyspepsia attended with acidity and looseness of the bowels, a little lime water or prepared chalk may be taken each morning for a time ; if there be acidity, with constipation, a teaspoonful of magnesia may be given instead ; but if there be constipation without acidity, a pill of aloes and rhubarb may be given after dinner each day.

SECTION XXI.—DIARRHŒA.

By diarrhœa, from δια, "through," and ρεω, "to flow," I mean an affection of the intestines or alimentary canal, attended with frequent, and generally more or less copious liquid olvine evacuations of a feculent character.

The immediate condition upon which diarrhœa depends is generally either relaxation, congestion, irritation or inflammation of the alimentary

25

mucous membrane; in consequence of which there is poured into the alimentary canal a serous mucus or other fluid, which, mixing with the fecal matter produces the liquid evacuations in their various forms.

Symptoms.—In cases of diarrhœa depending upon a relaxed condition of the alimentary mucous membrane, the discharges are generally quite frequent though attended with little or no pain; there is apt to be emaciation, and if the disease be protracted, the discharges may become very copious and the prostration very great.

In diarrhœa from congestion of the alimentary mucous membrane, there is apt to be a feeling of fullness in the abdomen, especially before an evacuation, and sometimes more or less pain, and the evacuations are apt to be copious and may occur with more or less distressing vomiting.

Diarrhœa depending upon irritation of the mucous and muscular coats of the intestines, is attended with more or less pain and an increased peristaltic action of the intestines, the discharges, in many cases, being small but very frequent. The discharges may be attended with slight sinking, but more or less febrile excitement is liable to attend this variety of the disease.

In cases of diarrhœa depending upon inflammation of the alimentary mucous membrane, with irritation of the muscular coat, there is tenderness, pain and increased peristaltic action of the intestines, with frequent discharges, though in many cases not very copious. The pain and tenesmus are often very considerable, and there is apt to be more or less fever, with thirst, a dry, hot skin, &c.

Thus we have the symptoms common to diarrhœa from relaxation, congestion, irritation, and inflammation of the alimentary mucous membrane. There may be, however, an obstinate diarrhœa depending upon hepatic derangement, or upon ulceration, or some other organic disease of the alimentary canal. In such cases, the symptoms vary according to the seat and nature of the disease upon which the diarrhœa depends.

If the liver be the cause, there is apt to be a jaundiced appearance. If ulceration or other organic disease of the intestines be the cause, there is liable to be pain, tenderness, emaciation, debility, and finally death, unless the disease be arrested.

Causes.—The causes of diarrhœa are exceedingly numerous, and consist of those that operate directly upon the alimentary mucous membrane, and those which operate through the system, in various ways.

All the causes which lead to indigestion, such as indigestible or unwholesome food, irregular eating, the use of intoxicating liquors, &c., may, by producing dyspepsia, lead to diarrhœa. The causes operating directly in this way, generally produce more or less irritation or inflammation of the alimentary mucous membrane, but if the causes have operated for a long time, they are not only direct, but indirect agents, in producing this disease; and may produce diarrhœa from the debility or relaxation which they have caused.

Of the indirect agents, any cause capable of producing a general relaxed, congested, irritated, or inflamed condition of the alimentary mucous membrane, may produce a diarrhœa: among which are great fatigue, want of proper nourishment, the use of tobacco, intoxicating liquors, and various other abuses of the system, all of which, by producing debility of the system, and relaxation of the alimentary mucous membrane, may become the cause of diarrhœa.

Cold is a very prominent cause of diarrhœa, as it checks the exhalation from the skin, and thus retains in the blood an irritating fluid, which, passing to the alimentary mucous membrane, produces congestion, irritation, and perhaps inflammation, and so a diarrhœa.

Congestion of the liver, from various causes, by interrupting the portal circulation, increases the congestion of the alimentary mucous membrane, and so becomes a cause of diarrhœa. There are many other causes of diarrhœa, or conditions of the system in which a diarrhœa occurs, it being sometimes colliquative, and at others critical, indicating a favorable change in the primary febrile or other affection.

Prognosis.—In cases of diarrhœa, not depending upon great debility, or inflammation, ulceration, or some other organic disease, the prognosis is generally quite favorable, and the disease may be expected to yield in a reasonable time, to proper remedial measures.

In cases, however, in which there is great debility, and causes operating of a serious character, which cannot be removed, the diarrhœa may pass on for weeks, months, or even years, in spite of remedial measures.

Treatment.—In the treatment of diarrhœa the cause should be sought out and removed, the patient should be placed on a properly regulated diet, and then just so much direct medical treatment should be resorted to as is absolutely necessary, and no more. If the diarrhœa depends upon indigestion, the dyspepsia should be corrected by such measures as may be indicated, when the diarrhœa will generally subside.

If it arise from exposure to cold, by which the exhalation from the skin is checked, the patient should be warmly clad with flannel next the skin, be directed to sleep in flannel sheets, and to take an unstimulating and digestible diet. If this fails to correct the disease, five grains of Dover's powder may be given, every six hours, till the diarrhœa is arrested.

In cases of diarrhœa depending upon hepatic congestion, and consequent congestion of the portal vessels, a blue pill may be given morning and evening, till the hepatic derangement is corrected, when the diarrhœa will generally subside, or may be corrected by very mild anodynes or astringents, such as three grains of Dover's powder, with two of tannin, taken four times per day.

In diarrhœa depending upon great debility of the system, and relaxation of the alimentary mucous membrane, the tincture of the chloride of iron, in ten drop doses, three times per day, is a valuable remedy for the local affection, and also for the general strength of the patient. Or, in case this should disagree, the elixir of vitriol may be given, in twenty drop doses, in two fluid ounces of water, four times per day, instead of the iron.

In cases of diarrhœa depending upon congestion of the alimentary mucous membrane, the warm foot-bath, sinapisms over the abdomen, warm clothing, and a properly regulated diet will do very much towards arresting the disease. If, however, the diarrhœa continues, the fluid extract of blackberry, in drachm doses, may be given three times per day, to constringe the alimentary mucous membrane.

In cases of diarrhœa depending upon irritation of the alimentary
mucous membrane, twenty drops of the elixir of vitriol, with ten drops
of laudanum, may be given four times per day, in a little water, and
continued with sinapisms, warm clothing, and proper food, till the
diarrhœa is corrected. Finally, in all cases of diarrhœa depending upon
inflammation of the alimentary mucous membrane, cups, blisters, or pustulation with tartar emetic ointment should be resorted to; the patient
should be properly fed and clothed, and if this fail to effect a cure, three
grains of Dover's powder, with two of tannin may be given, four times
per day, till the disease be arrested.

SECTION XXII.—CHOLERA MORBUS.

By cholera morbus, or sporadic cholera, I mean that peculiar affection
of the alimentary canal, characterized by vomiting and purging, and
attended frequently with more or less cramps or spasms of the abdominal muscles, and also of the extremities, and in severe cases, with considerable prostration.

Symptoms.—Cholera morbus generally makes its attack suddenly, and
is very apt, in our climate, to come on during the cool hours of the night,
at the season of the year when the days are warm and the nights cool.
The attack is sometimes preceded, for a day or two, by a feeling of indisposition, with impaired appetite, headache, &c. But as often, no premonitory symptoms are noticed, the attack being sudden and unexpected.

The disease commences with a feeling of oppression in the stomach
and bowels, with perhaps chilliness, and a feeling of great distension,
with violent pain in the umbilical region. Soon there is nausea and
griping pains, with obstinate vomiting and copious liquid discharges from
the bowels.

The matter discharged by vomiting, and by stool, consists, at first, of
the contents of the stomach and intestines, with a serous fluid. In some
cases, however, the discharges are mingled with bile, especially the
matter vomited. The pulse becomes weak and frequent, the skin cool
and moist, the strength is prostrated, and along with cramps, there is
apt to be prostration, and there is sometimes wandering of the mind, or
slight delirium.

If the discharges continue, they may lead on to a fatal termination.
Or reaction may take place, followed by febrile action, with a hot and
dry skin; and all the symptoms which usually attend gastro-enteritis.
The discharges may become scanty; the epigastrium tender; the abdomen
distended, and the fever, unless the disease be arrested, is apt to assume
a typhous character, and the patient, with great prostration, and muttering delirium, passes on to dissolution, or to a slow and tedious convalescence.

Causes.—Atmospheric vicissitudes, or sudden checking of the cutaneous
exhalation, by cool night air, is probably the most frequent cause of this
disease. This sudden retention of the perspirable matter, renders the
blood more dilute or watery, and ready to part with its serum; and also
accounts for the chilliness, prostration, and congestion, which occurs by
its debilitating effects upon the brain and nervous system.

The paludal poison also, by acting upon the brain and nervous system, through the blood, is a frequent cause of cholera morbus; producing its effects probably in nearly the same manner, as retained perspirable matter, especially at seasons of the year favorable to the occurrence of this disease. Besides cold and the paludal poison, which may act as predisposing, or exciting causes of cholera morbus, there are various other exciting causes; among the most frequent of which, are violent fits of anger, or strong mental emotions, sailing upon a rough sea; the drinking of too much cold water in hot weather, and finally, the eating of green vegetables, such as cucumbers, melons, unripe fruits, &c.

Diagnosis.—Little difficulty need be experienced in distinguishing a case of sporadic cholera, or cholera morbus, from malignant cholera; or in fact from any other disease. It is only necessary to take into account the season of the year, the epidemic tendency, the mode of attack, the character of the discharges, and all the peculiar symptoms of this disease, and the diagnosis may be clear in all cases.

Nature.—When we remember that retained perspirable matter, and the paludal poison, together with the eating of crude green vegetables, are the most frequent causes of cholera morbus; there need, I believe, be little doubt, in relation to the pathology of the disease.

The retained perspirable matter, as well as the marsh miasmata, evidently directly prostrate the cerebro-spinal system, and both directly, and through this general prostration, produce congestion of the liver, of the portal vessels, and of the alimentary mucous membrane. If now green vegetables have been taken, to disturb still further, the alimentary mucous membrane; we have not only congestion of the liver, and portal vessels; but also congestion, with irritation of the alimentary mucous membrane.

Now, in this state of the system, the blood readily parts with its serum, which in the congested state of the alimentary mucous membrane, is poured into the stomach and intestines, and mixing with the fecal matter, is thrown off by vomiting and purging. In cases in which green vegetables have been taken, in addition, there is apt to be a strong tendency to gastro-enteritis, as the vomiting and purging subside, probably from the gastro-intestinal irritation which they produce.

Treatment.—The indications in the treatment of cholera morbus, are plainly to equalize the circulation, to remove any offending matters from the alimentary canal, to counteract gastro-intestinal irritation, and to arrest the vomiting and purging as well as to correct the hepatic and other functions of the body.

If the patient is chilly and prostrated, the whole length of the back should be rubbed with a warm decoction of capsicum in vinegar, to equalize the circulation and produce warmth. A large sinapism should be applied all over the stomach and bowels, and allowed to irritate a little short of vesication. And it should be reapplied every six or eight hours while the disease continues.

To stop the vomiting, one-fourth of a grain of calomel, with a grain of Dover's powder may be given every fifteen minutes, till it be arrested after which, if there is evidence of irritating matters in the alimentary canal, a full dose of calomel may be given, and followed if necessary in five hours, with half an ounce of castor oil. In cases, however, in which no offending matter is suspected in the intestines, a cathartic need not be given, but after arresting the vomiting, if the diarrhœa continues, five grains of Dover's powder, with one or two grains of calomel may be given every four or six hours, till the hepatic function is restored and the bowels corrected,

Or, if the calomel is not indicated, there being only a looseness of the bowels, with increased peristaltic action instead of the calomel and Dover's powder, fifteen drops of laudanum, with a grain of camphor may be given, the camphor and laudanum being suspended in water by means of a little prepared chalk. While the nausea continues, chicken tea, or crust coffee, with a little milk may be allowed ; but later, a poached egg, toast, and other plain and digestible varieties of nourishment may be allowed. In case gastro-enteritis follows an attack of cholera morbus, blisters and other measures proper in the treatment of mucous gastro-enteritis should be persevered in till the inflammation is overcome.

SECTION XXIII.—MALIGNANT CHOLERA.

By malignant cholera, I mean that peculiar epidemic malignant variety of disease characterized by vomiting and purging, and attended with cramps, or spasmodic contractions of the abdominal muscles, and some-times of the extremities, and in severe cases with great prostration and a state of collapse.

Malignant cholera appears to have been long known in the east, but it did not attract the attention of the medical profession to any con-siderable extent, till it appeared in Hindostan, in 1817, from whence it spread over Asia, and subsequently over Europe and America, and in fact, almost every part of the globe.

It appears to have been truly "the pestilence that stalks abroad at noon-day," sweeping off, in its onward march, that portion of the human family fitted for destruction, generally by a reckless disregard of the laws of health; and taking, occasionally, an unfortunate victim whose greatest misfortune consisted in being surrounded by the filthy, reckless and debased.

Symptoms.—Malignant cholera may be preceded by a slight diarrhœa, for a time, but it more frequently comes on suddenly, with a sense of tension in the epigastrium, and cramps of the stomach and intestines, with exceedingly distressing nausea. Very soon severe and obstinate vomiting and purging commence and continue, with but very short intervals, till the system is exhausted or a collapse occurs unless the dis-ease be arrested. During the intervals between the attacks of vomiting, the patient is harassed with continual nausea, and a feeling of great distress in the epigastrium.

The patient is usually cold or more or less chilly at the commence-ment of the disease, and in some instances this coldness continues and becomes so considerable that I have found even the tongue feeling cold to the touch. In these cold cases the face and whole surface of the body has a shrunken, livid, or blue appearance, and the cramps are often very severe. No bile is thrown from the stomach or passed by the bowels, and after the first two or three discharges, no fecal matter ; the matter vomited and discharged by the bowels being liquid and of a rice-water appearance.

Sometimes very early in this disease, and at other times later, the discharges become less frequent and copious, and collapse supervenes, attended with great prostration and apparent insensibility; the evacuations being unheeded by the patient, and little or no regard paid to surrounding objects. The pulse becomes weak and scarcely perceptible, the skin cold, livid, and covered with a clammy sweat; there is great thirst; the voice fails; the countenance is cadaverous, the face being shrunken, and the eyes hollow; respiration is slow, and in some cases the patient is only aroused from apparent insensibility by the cramps which occur, affecting mainly the bowels, abdominal muscles and the inferior extremities.

Malignant cholera is thus one of the most rapidly fatal diseases, death being liable to take place at any period, from five or six hours, to one, two, or three days from the time of attack. If a collapse occurs, the case is always attended with the greatest danger, but if the discharges become less copious, and assume a bilious or feculent character, the case will generally terminate favorably. The convalescence may be rapid in some cases which are arrested early, but in obstinate protracted cases, as the vomiting and purging cease, the skin may become hot and dry, the tongue clean and red; delirium may occur; and there may be a general irritated febrile reaction, with symptoms of gastro-enteritis, the fever assuming a typhoid character. Thus, then, we have in malignant cholera, in some cases, a premonitory stage, during which there may be languor, nausea, and diarrhœa. There is also the active stage of the disease, attended with nausea, vomiting, purging, cramps, &c. There may be also the stage of collapse, marked by extreme prostration, coldness, thirst, clammy sweats, and partial or entire insensibility to surrounding objects. And, finally, there is the stage of convalescence which may be rapid, or it may be tedious, developing at first, an irritative febrile reaction, with symptoms of gastro-enteritis, the fever having a typhoid tendency.

Such then are the general symptoms of malignant cholera, but it must be remembered that the symptoms vary widely in different cases, all, however, exhibiting more or less malignity.

Anatomical Characters.—In cases in which death takes place very early, no marked morbid changes may be discovered on examination of the body. In cases, however, in which death has taken place during a state of collapse, the arteries contain but little blood, but the veins and right side of the heart are gorged with dark blood, imperfectly coagulated, and nearly all parts of the body exhibit more or less signs of venous congestion.

The lungs, however, contain little blood, the spleen is contracted, the peritoneum and other serous membranes are dry, the small intestines are injected and of a rose color, the mucous membrane may appear nearly natural, or it may be reddened, with enlargement of the intestinal follicles, and there may be small semi-transparent vesicles along the mucous membrane of the small intestines. The alimentary mucous membrane is sometimes found coated with a whitish matter identical with that contained in the evacuations, and the intestines are found distended with the same rice colored liquid of which the discharges consist, mixed, in some cases, with effused blood. A glairy mucus is sometimes found adhering to the mucous membrane of the stomach, but this tissue may retain nearly a natural appearance. It appears from microscopic examinations, that the whitish matter adhering to the alimentary mucous membrane consists of the epithelium partially detached from the basement membrane, and that the white deposit in the evacuations consists of "disintegrated epithelium of the alimentary mucous membrane."*

In cases in which death has taken place after reaction, the venous injection is scarcely, if at all apparent, the blood is less dark and viscid, and the contents of the intestines instead of presenting the rice water appearance, is of a more or less bilious or bloody appearance. The alimentary mucous membrane presents signs of inflammation, being red, and sometimes softened, the mucous follicles presenting signs of incipient ulceration.

Such are the ordinary post-mortem appearances presented in subjects dead of this most malignant disease, the differences depending upon the time at which death occurs, as well as the condition of collapse or reaction.

Diagnosis.—To distinguish malignant from sporadic cholera, it is only necessary to take into account the epidemic influence, the rice-water discharges, the shrinking, coldness, and blueness of the extremities, and surface of the body, the great violence of the spasms, together with the rapid sinking tendency, and the collapse when it occurs, to render the diagnosis clear.

Causes.—The predisposition to cholera arises, I am satisfied, from the various imprudences to which the human family are addicted; the most prominent of which are irregularity in eating, taking improper or unwholesome food, the use of tobacco and intoxicating liquors, masturbation and excessive venery, filthy habits of every kind, mental despondency, low and vicious habits of life, damp and filthy apartments, and finally, every deviation from the laws of health to which depraved mortals are addicted.

It is probable that all these imprudences with others of a kindred character, not only impair vitality and thus predispose those guilty of them to malignant cholera, but it is probable that the sum total of human imprudences, produces or causes to be generated the epidemic influence, whether it be animal, vegetable or inorganic, which is the general exciting cause of this disease.

Now with the predisposition and a general exciting cause thus produced, it is easy to see that a damp atmosphere with a low electrical state, together with various imprudences in eating and drinking, may act also as occasional exciting causes of malignant cholera, as is well known to be the fact. But in relation to the natural agents, such as electricity, planitary influences, &c., they cannot probably be operative in producing this or any other disease only in so far as the human system is depraved, or has been let down from its original state of physical perfection, by imprudences which have been operating since the fall of man.

Nature.—In order to appreciate the nature of malignant cholera, it is

* See Wood's Practice of Medicine, Vol. i, p. 722.

necessary to remember that it generally occurs in persons predisposed by imprudence of some kind, and that a poison epidemic influence of either an organic or inorganic nature, is received into the system through the skin, stomach or lungs, which acts as the exciting cause of the disease. It is probable that this poisonous agent passes with the blood to every part of the system, producing a direct depressing effect upon the cerebro-spinal and ganglionic systems, thus deranging the functions of all the organs of the body.

It is possible also that this poison produces a direct change in the blood, in consequence of which it more readily parts with its serum, and it may possibly irritate directly the alimentary mucous membrane; be this as it may, there is very soon cerebro-spinal congestion, as well as congestion of the alimentary mucous membrane. Now the cerebro-spinal congestion with that of the alimentary mucous membrane, together with the state of the blood, account for the coldness, nausea, vomiting, purging, cramps, and other symptoms which are developed.

The collapse may occur from the combined influence of the poison, the loss of the serum of the blood, and from the degranement in the pulmonary circulation, preventing a due decarbonization of the blood. In cases in which there is an irritated reaction, with a fever of a typhoid character, following a collapse, there is evidently a gastro-enteritis which accounts for the symptoms which are developed, which gastro-enteritis may be the result of the disease, or of the remedies which have been resorted to for its cure.

Prognosis.—Malignant cholera is always a dangerous disease. But if the case receive early and proper attention, and the predisposition is not strong, or the surroundings unfavorable, a recovery may reasonably be expected. If, however, the patient is strongly predisposed, if the case be neglected or imprudently managed, or if collapse has supervened, with coldness, lividity, &c., a fatal termination may be apprehended. The ratio of deaths in some localities has been one-half, or even more than that; but, judging from my own observation, I should think that of all the cases of malignant cholera which occur, a large majority recover; but in some localities and epidemics it may be otherwise.

Treatment.—The indications in the treatment of malignant cholera are to equalize the circulation, to arrest the vomiting and purging, to counteract local irritation, and to restore the functions of the liver, alimentary canal, and other organs.

If a case of cholera be seen early, it may sometimes be arrested by very mild measures. The patient should have the feet placed in warm water, and active friction should be made along the spine, with a strong decoction of capsicum in vinegar, used a little warm, with a flannel cloth. A mustard poultice should be applied, large enough to cover the epigastrium and whole abdomen, and allowed to irritate a little short of vesication. Ten drops of laudanum, with the same quantity of the tincture of camphor, may be administered every four hours, with a teaspoonful each of brandy and loaf sugar, and the patient may be allowed to drink warm crust coffee, one-third milk, or chicken tea.

The warm foot-bath favors the circulation in the extremities, the warm friction along the spine equalizes the circulation, the sinapisms counter act the local irritation and congestion of the alimentary mucous membrane, while the small doses of laudanum and camphor, with the brandy and sugar, tend to allay the nausea, arrest the purging, and to quiet general restlessness and irritability. With this plain and simple course of treatment, many cases of malignant cholera may be arrested, if early and judiciously applied.

In case, however, the vomiting continues obstinate, the laudanum, camphor and brandy being rejected, this prescription may be discontinued, and a grain of Dover's powder, with one-fourth of a grain of calomel, may be given every half hour, till with the other measures the vomiting is arrested. After the vomiting is arrested, or as soon as the stomach will retain it, four grains of Dover's powder, with two grains each of calomel and pulverized camphor may be given every four hours, with a little prepared chalk, and continued till the discharges become bilious, when the calomel should be discontinued, and two grains of tannin given with the Dover's and camphor, instead, and continued till the disease is arrested.

In case the rice-water discharges continue obstinate, fifteen or twenty drops of laudanum may be given by injection, with half an ounce of liquid starch, and repeated at intervals of four hours, alternating with the powders, if they appear to produce relief. Should the disease continue in spite of the remedies I have suggested, and great prostration, with collapse supervene, from two to four grains of the sulphate of quinine may be given every four hours, with, or alternating with, the Dover's, tannin and camphor, and continued till death or reaction takes place. Finally, should reaction take place, and symptoms of gastro-enteritis supervene, with an irritated febrile reaction, the fever assuming a typhoid character, the quinine and camphor may be continued, with or without the Dover's, and a blister applied to the epigastrium, and if necessary to the abdomen, and thus the gastro-enteritis subdued, and the typhoid tendency counteracted, till convalescence is fully established.

During the continuance of the disease, the patient should be nourished by crust coffee, one-third, or one half milk, or chicken tea, or mutton broth may be allowed; and if necessary, wine-whey may be resorted to, in case of great prostration or collapse. During convalescence, a plain digestible, and nourishing diet may be allowed, consisting in part of animal food, to be taken with strict regularity. And should the appetite be indifferent, twenty drops of the fluid extract of columbo, or small doses of the cold infusion may be given, after each meal for a time, till the tone of the stomach is restored.

SECTION XXIV.—CHOLERA INFANTUM.

By cholera infantum, I mean that peculiar disease of infants, and very young children, occurring during the warm season, and more especially in cities and large towns, characterized by vomiting, purging, and great emaciation, and sometimes assuming a rapidly fatal character.

Symptoms.—Cholera infantum may come on suddenly, with scarcely no premonitory symptoms, or it may commence in a slow and insidious manner, with occasional vomiting; a little diarrhœa, and perhaps symptoms of slight congestion, or inflammation of the alimentary mucous membrane. The brain appears to suffer materially in some cases; being either congested, irritated, or inflamed, either in its substance, or its meninges. The liver appears to be deranged in the early stages of this disease, as very little, or no bile is visible in the evacuations, the discharges consisting generally of a colorless liquid. Later, however, the discharges may assume a yellow, brown, or green appearance.

In acute cases, in which no premonitory symptoms are apparent, the disease commences with vomiting and purging; the extremities become cold; the respiration difficult; the face pale, and the eyes sunken; the pulse almost imperceptible; the vomiting and purging are almost incessant, and the little sufferer exhibits symptoms of spasmodic pains in the stomach and bowels, and unless the disease be arrested, great prostration ensues, with pale and shrunken features, coma, insensibility, and finally death takes place, at a period varying from six to twenty-four hours from the attack.

In other cases, however, the disease comes on more slowly; the child suffering for a few days, with occasional vomiting, and slight diarrhœa, and perhaps with an irregular fever, apparent headache, &c. After these symptoms have continued for a time, active vomiting and purging is apt to occur during the night, with most of the symptoms of an acute attack of the disease.

Along with the vomiting and purging, in such cases, the extremities become cold; the pulse small and frequent; the tongue is furred, and if the vomiting, and excessive purging be arrested, there is often a diarrhœa remaining; which may continue on for several days, weeks, or even months, unless arrested. In such cases, the child suffers from an irregular fever; the flesh becomes soft; emaciation goes on rapidly; the face becomes pale, and the eyes sunken; the abdomen becomes either distended or sunken; the mouth becomes apthous; the child becomes restless, and moans almost incessantly; and unless the disease be arrested, the little sufferer may die, either in a comatose state, or else in violent convulsions.

The duration of this disease in its acute form as we have seen, may vary from six to twenty-four hours, but a diarrhœa with all the usual symptoms of gastro-enteritis may continue on for days or even weeks, unless arrested by proper remedial measures, or terminated by the return of cool weather, or some other favorable circumstance.

Anatomical Characters.—In cases in which death occurs early, the liver is found congested, and the mucous membrane of the alimentary canal is either congested, as is found in some cases, or else it presents unusual paleness with little or no congestion.

In cases in which death has occurred at a later stage, signs of inflammation are presented by the alimentary mucous membrane. And in chronic cases, the mucous follicles are enlarged, and ulcerated not only in the small, but in the large intestines. The brain is sometimes found congested with serous effusion into its ventricles, and in some chronic cases that have exhibited symptoms of hydrocephalus, the brain is found more or less softened.

Causes.—As cholera infantum occurs most frequently at the season of the year when the days are warm and the nights cool; it appears probable that cold, following moderate heat is one of the most frequent causes of this affection. This is not strange when we remember how sensitive the infant is, to any impressions of an exciting irritating or depressing character. The sudden check to the perspiration during the night, produces congestion of the cerebro-spinal system, of the liver, and of the alimentary mucous membrane, and hence, if the child be predisposed from impure city air, or from any other cause, a rapid and, perhaps fatal attack of the disease is produced. Cases occurring from this cause are generally rapid in their course, and may terminate in death or convalescence within a few hours.

While cold is the most frequent exciting cause of the acute form of this disease, there are various causes which operate either directly or indirectly in producing the more chronic form of this affection, among the most frequent of which are impure city air, damp and filthy sleeping rooms, dentition, improper food, or irregularity in taking it, and *koino* or *idio miasmata.* Any or all these causes may predispose the infant or child to this affection, or they may not only predispose, but act as exciting causes, developing generally the chronic form of the disease, with cephalic, hepatic, or severe gastro-intestinal complications.

Nature.—In the acute form of the disease, there is evidently depressed vitality from some predisposing cause, and a weak state of the blood. This condition, together with the effects of retained perspirable matter, miasmata, or some other depressing agent, leads to congestion of the cerebro-spinal system, liver, portal vessels, and alimentary mucous membrane, and hence the vomiting, purging, and other symptoms which are developed in this disease. But in chronic cases which are preceded and followed by diarrhœa, there is evidently in addition to the congestions; irritation or inflammation of the alimentary mucous membrane, affecting especially the mucous follicles; the cephalic, and other complications being probably accidental in most cases at least.

Treatment.—The predisposing and exciting cause or causes of this affection, should be sought out and removed as far as possible, and the case treated according to the condition of the patient at the time of attack. In those cases which occur from a sudden check of the perspiration, from cold, the feet should be placed in warm water, and the back of the child rubbed with a warm flannel, to help equalize the circulation. Mustard moistened with vinegar and spread on a thin cloth, the cloth being folded over it, should be applied over the stomach and bowels, and also to the bottom of the feet, and kept on as long as the child can bear it without producing vesication.

To allay the vomiting in such cases, ten drops of paregoric may be given, with a grain of prepared chalk, every three or four hours, the child being allowed a little chicken tea, if it can be retained by the stomach. If, however, the case continues obstinate, and the discharges contain no bile, the paregoric should be suspended, and half a grain of Dover's powder, with a grain of calomel, given every three hours, till the discharges assume a bilious character, when the calomel should be omitted, and the Dover's powder continued, with one sixth of a grain each of camphor and tannin, every four hours, till the discharges are arrested, when they should be gradually left off, and the child fed on chicken tea or arrow-root, cooked in milk or chicken broth.

The child should be dressed in flannel, and should lay in flannel sheets, and should the case become chronic, and symptoms of gastroenteritis supervene, a blister may be applied to the epigastrium, and also to the abdomen, if necessary, and thus this complication be subdued. In those cases which appear to be of a malarious origin, a small dose of calomel, or mercury with chalk, may be given with a little rhubarb and castor oil, and then the case treated in every respect according to the principles already laid down, adding the sulphate of quinine to the treatment in all cases in which it is clearly indicated.

In those cases depending upon unwholesome food, and damp impure air, the diet should be corrected; the child, if possible, removed to more suitable quarters, a flannel should be made to fit snugly to the whole body, and the feet and legs should be protected by woollen stockings, to keep the surface warm and promote perspiration. Dover's powder, with the sulphate of quinine, in half grain doses each, may be given with a little prepared chalk at first, and later, if a diarrhœa follow, the child may be fed on milk and water with a little arrow-root, in which has been boiled the root of the *geranium maculatum*, so as to render it considerably astringent.

In those cases in which there is much irritation from dentition the gums should be cut, and if the brain or its membranes are irritated or inflamed, small blisters may be applied back of the ears.

To prevent this disease, and to aid in its cure when it has occurred, all children should be kept clean, dry, and warm, and allowed good pure dry air, and proper wholesome food, which should be taken with regularity. For infants, the mother's milk is the most natural and proper food, and it should be allowed in all cases in which it is possible to do so. But in case this cannot be had, or that of a healthy wet nurse, fresh cow's milk, diluted with water, may be substituted. For older children, arrow-root cooked in chicken broth, or milk and water, will generally be found the most suitable food in this variety of disease.

SECTION XXV.—FLATULENT COLIC.

By flatulent colic, I mean that variety arising from an accumulation of air in the alimentary canal, characterized by pain in the bowels, of a paroxysmal character, to some extent, and unattended with inflammation. This variety of colic is apt to attack persons of weak stomachs, or those whose intestines, or their muscular coat, is in a weak and debilitated or relaxed state. Various articles of food are liable to provoke an attack of flatulent colic in those who are predisposed, among the most frequent of which are cucumbers, unripe fruit, fresh, warm bread, &c. In some cases, the colic pains are felt in about an hour after taking food, but frequently the food passes into the intestines, before the disease is developed.

Symptoms.—The patient usually experiences a sense of distension and uneasiness in the pit of the stomach, which is soon followed by a distressing pain in the abdomen, referred in most cases to the umbilical region. The pain may extend to the whole abdomen, and become very severe, with only slight remissions, but more frequently it occurs in severe paroxysms, with more or less complete intervals of quiet from the pain.

During the exacerbations the patient is apt to move to and fro, with the body bent forwards, and the hands pressed firmly against the abdomen. In cases in which the stomach is the seat of the suffering, if eructations of wind occur, there is apt to be some mitigation of the pain. If, however, the colon be the seat of the difficulty, the wind is apt to pass downwards, as the disease is about terminating.

In some cases of flatulent colic, the abdominal muscles are contracted, while in other cases, considerable tympanitis attends. The bowels are generally constipated, the pulse natural, or slightly depressed, and the tongue may be clean, or covered with a white fur.

Diagnosis.—Flatulent colic may be distinguished from bilious colic, by the absence of bilious symptoms, especially the bilious vomiting, so generally attending bilious colic. Flatulent colic may be distinguished from peritonitis and gastro-intestinal inflammation, by the absence of fever, by the pain being irregular, and generally relieved by pressure, and by the motions of the patient. Finally, this affection may be distinguished from strangulated hernia, by a careful examination of the abdomen, in connection with the other symptoms which are developed.

Prognosis.—The prognosis in flatulent colic is generally favorable, but in obstinate and protracted cases, the muscular coat of the intestines becomes so far paralyzed, as to produce in some cases very troublesome constipation of the bowels.

Treatment.—The treatment of this disease consists in the use of means to correct the attack of colic when it occurs, and also to correct the condition and habits of the patient, upon which the colic depends. In a mild attack of colic, gentle friction along the spine with a warm flannel, sinapisms over the stomach and bowels, and a teaspoonful of the tincture of camphor, administered in water, will generally arrest the disease.

In those cases which come on soon after eating some crude indigestible food, an emetic of ipecac may be indicated. Thirty grains may be dissolved in two ounces of warm water, and half an ounce given every fifteen minutes, till free vomiting is produced. Or what may be more convenient, if at hand, a drachm of the fluid extract may be added to two ounces of warm water, and a tablespoonful given every fifteen minutes till vomiting is produced.

After free vomiting, thirty drops of laudanum, or the same quantity of the fluid extract of opium, may be administered, with sixty drops of the tincture of camphor, and repeated in two hours, if necessary. The friction along the spine, and sinapisms over the bowels, should be used in these, and all other cases of flatulent colic; and if there be any tender point along the spine, dry cups may be applied, and cupping may even prove servicable in cases in which there is no apparent spinal tenderness.

In flatulent colic attended with obstinate constipation, after resorting to the remedies I have suggested, a cathartic may be indicated. Half an ounce of castor oil may be given, with a teaspoonful of the oil of turpentine, and the dose repeated in four or six hours, if necessary. In this way an attack of flatulent colic may generally be arrested; but it is desirable to correct the habits and condition of the patient, upon which the attacks of colic depend. If the digestion is bad the diet should be regulated, and the patient directed to take proper food, with regularity, and made to conform rigidly to the laws of health, in every respect.

If the bowels are constipated, and they cannot be corrected by a teaspoonful of calcined magnesia, taken every morning, or by a pill of aloes and rhubarb, taken after dinner, each day, a degree of paralysis of the muscular coat of the alimentary canal may be suspected. In such cases, strychnia may be administered, in $\frac{1}{80}$ of a grain doses, three times per day, for a time, with the most happy effect, if there be no marked irritation of the brain or spinal cord. A grain of strychnia may be rubbed in a mortar with a teaspoonful each of acetic acid and sugar, and then eight ounces of water added; or if the muriate of strychnia be at hand, a grain of it may be dissolved in half a pint of water, and of either preparation a teaspoonful may be given, three times per day, after each meal, till the constipation is overcome.

SECTION XXVI.—BILIOUS COLIC.

By bilious colic, I mean that variety in which, in addition to the ordinary symptoms of colic, there is marked derangement of the digestive, and especially of the biliary organs; the disease being attended with nausea and bilious vomiting.

Bilious colic may be produced by various causes, but it appears to depend, in most cases, according to my observation, upon a malarious influence, the main force of which is concentrated directly upon the liver and alimentary canal, or else indirectly, the primary influence being exerted upon that portion of the spinal cord from which the liver and intestines are more immediately supplied with nerves.

Symptoms.—Before the colic makes its appearance, the patient generally experiences headache, loss of appetite, has a bitter taste in the mouth, and occasionally thirst, nausea, and bilious vomiting. After these symptoms have continued for an indefinite period, acute pains in the stomach and bowels supervene, moving at first from one part of the abdomen to another, but generally being most severe at the umbilicus, and very severe during the exacerbations.

At first, pressure on the abdomen produces some relief, but as the disease continues, the bowels may become tender to the touch, or to firm pressure at least. The pulse may be nearly natural, or depressed at first, but in the advanced stage of the disease it generally becomes increased in fullness, force and frequency. The hands and feet are sometimes quite cold during the exacerbations of the pain. And usually about the second or third day of the disease, the eyes and skin assume a yellowish tinge. But in some instances the biliary symptoms appear several days before the abdominal pain commences.

It is no uncommon occurrence for a paralysis of one or both of the superior or inferior extremities to occur during the continuance of this disease, and it is possible that the intestinal distension, when it occurs, may be in part of this character. The abdominal muscles may be spasmodically contracted in this disease, but more generally considerable tympanitis attends.

Diagnosis.—The only diseases with which this affection is liable to be confounded are flatulent colic, peritonitis, gastro-enteritis, and hernia in some form.

From flatulent colic it differs in being attended with marked bilious symptoms, and especially with bilious vomiting. From hernia it may be distinguished by a careful examination of the abdomen, together with the other symptoms which are developed. From peritonitis and gastro-intestinal inflammation bilious colic differs in the bilious symptoms which are developed, in there being less tenderness on pressure, and in the motions of the patient, which in the colic affection are generally quite free.

Causes.—*Koino-miasmata* appears to be the principal cause of this disease, together with the effects of atmospheric vicissitudes, while the immediate cause of the local difficulty, it appears to me, is in part cerebro-spinal, there being generally spinal congestion, in consequence of which the functions of the liver become partially suspended, and to a great degree those of the alimentary canal. Now I suspect that the miasmatic agent acts through the blood upon the brain and nervous system, precisely as it does to produce bilious fever, but on account perhaps of slight irritation of the spinal marrow, or a predisposition to take on irritation, the dorsal portion of the spinal cord becomes congested, before a general febrile excitement is produced. This spinal congestion may, and probably does, affect more or less the whole length of the spinal marrow; and hence the paralysis of the extremities, which sometimes occurs.

But congestion does affect, I am confident, the dorsal portion of the spine, and it may be mainly of that portion, being attended with pain and more or less tenderness, in that portion of the spine more immediately supplying nerves to the liver and alimentary canal. In consequence of this, in part, the liver becomes torpid, and the intestines in either a spasmodic or half paralyzed state, and hence, in part, the constipation, bilious derangement and colic symptoms which arise.

Now it is, I believe, in consequence of this accidental predisposition to spinal congestion that bilious colic is produced, instead of regular bilious fever, which would otherwise be developed, sooner or later, in most cases at least. Atmospheric vicissitudes operate in conjunction with the paludal poison in producing bilious colic, very much as it does in the production of other bilious affections. When this disease is established, the functions of the liver are usually so far deranged, that either no bile is secreted, or else there is a copious secretion of acrid bile poured into the intestines, which irritates the alimentary canal, and produces, in part, the nausea and vomiting which occur in this disease.

Prognosis.—Bilious colic, though a tedious, is not generally a very dangerous disease. But if the disease be neglected or mal-treated; serious, and even fatal results may reasonably be apprehended.

Treatment.—The indications in the treatment of bilious colic are to equalize the circulation, to counteract abdominal irritation, to evacuate the contents of the stomach and bowels, and to restore the functions of the liver and alimentary canal.

Warm pediluvia, and stimulating friction along the spine should be resorted to at once, to equalize the circulation, as far as may be, for the time. Sinapisms should be applied over the abdomen, to counteract or prevent local inflammation of the abdominal viscera, and the vomiting may be aided by twenty or thirty grains of ipecac, with a little warm water.

After the stomach has been sufficiently evacuated, five grains of Dover's powder, with two grains of calomel, may be administered every four hours, to allay pain, promote perspiration, and to affect the hepatic secretion. With the third or fourth dose of calomel and Dover's powder, half an ounce of castor oil may be administered, and the oil repeated, with each succeeding powder, till a free evacuation of the bowels are produced, the operation being assisted, if necessary, by injections of flaxseed tea, with half an ounce of castor oil, and one or two drachms of the oil of turpentine.

In all severe cases, cups should be applied on each side of the spine, along the dorsal region, and a little blood taken, to overcome the congestion and slight irritation, which I am confident generally exists in a greater or less degree in this affection. In all obstinate cases of bilious colic, which do not yield to this treatment, blisters should be applied on each side of the dorsal spine, and also to the epigastrium and abdomen, if necessary.

Crust coffee should be allowed at first, with mucilages, but later, broths, arrow-root, and toast may be allowed. If there be much prostration in this disease, after evacuating the stomach and bowels, two grain doses of the sulphate of quinine may be given every six hours, for a time, and during convalescence, a cold infusion of columbo may be allowed, in moderate doses, after each meal. If, when this disease subsides, the patient be left with paralysis of the intestines, or any other part, $\frac{1}{16}$th of a grain of the muriate of strychnia may be given in solution, after each meal, and continued till the paralysis is overcome.

SECTION XXVII.—LEAD COLIC.

By lead colic, I mean that variety of disease occurring generally from the effects of lead, and attended with colic, marked dyspepsia, and in some cases with paralysis, dropsy, and finally apoplectic symptoms.

Symptoms.—Lead colic generally makes its appearance in a very gradual manner, commencing with symptoms of gastric derangement. The patient suffers from an irregular appetite, languor, foul eructations, constipation, nausea, transient abdominal pains, a feeling of stricture in the abdomen, drowsiness, and general languor, with depression of spirits, &c. Gradually the pain in the epigastrium and umbilical regions become more severe and constant, the abdomen is hard, retracted, and tender to external pressure, and the bowels torpid, and stomach more or less irritable.

26

The pain in the bowels is generally continuous, or has only slight remissions, the exacerbations of colic pains being protracted in duration, and exceedingly severe. The retching and vomiting is generally very distressing for the first two or three days, only slight mitigation being experienced immediately after vomiting. In violent cases, the pains pass from the umbilical region upwards to the chest and arms, and downwards to the pelvic viscera and lower extremities.

During the exacerbations, cold sweats sometimes break out on the face and extremities; the countenance becomes contracted and pale; and if the disease continues, convulsions, and paralysis of the limbs, or some other part is liable to occur. In protracted unsubdued cases of this disease, the abdominal pains may abate; the stomach become tender and distended; the thirst unquenchable; vision imperfect; and after the supervention of œdema of the feet, a livid countenance, and various other symptoms, the patient may die in an apoplectic state.

In chronic cases of this affection, the gums at their margin, assume a leaden hue; the capillary circulation becomes inactive; the surface of the body becomes harsh and sallow; the temper irritable, desponding, taciturn, and gloomy; the countenance is expressive of deep suffering; the body emaciates; portions of the body become paralyzed; the abdomen tumid; the legs œdematous, and finally mania, or total imbecility may ensue, and the patient ultimately die apoplectic, with dropsical effusions, in various parts of the body.

The abdomen is generally retracted in this disease; the bowels constipated; there is nausea, and bilious vomiting; the tongue is flabby and tremulous; the pulse either slow and hard, or moderately frequent; the breath offensive; the hand tremulous; and there is great weakness, with occasional cramps of the lower extremities. Such are the ordinary symptoms of this variety of disease; liable of course, to variations, according to the severity and duration of the affection, and the general constitution of the patient.

Diagnosis.—Lead colic may be distinguished from flatulent, and bilious colic, by the circumstance of exposure to lead; by the leaden hue of the countenance, and dark or bluish-gray color of the margin of the gums; by the nearly continuous abdominal pains; by the retracted abdomen, and by the pains in the extremities, as well as the occasional paralytic complications in this variety of disease.

Anatomical Characters.—On the post-mortem examination of subjects dead of this disease, there are usually found traces of inflammation in the mucous membrane of the stomach and bowels. But the most frequent morbid appearance of the alimentary canal, is contraction at irregular points of its muscular coats, so that the intestines will hardly, at the contracted points, admit a goose quill to pass, while the intervening portions are distended, forming irregular sacs, along the intestine.

In one case that I examined in this village, about three years since, I found the ascending, descending, and transverse colon so contracted at points, as barely to admit the little finger. The mucous membrane of the stomach and intestines, in this case, presented signs of having been inflamed, and on the mucous membrane, at the pyloric extremity of the stomach, a pseudo membrane had been formed. **In some protracted** cases of this disease, there is found hypertrophy of the brain, while in others, there is marks of congestion or irritation, with softening of the brain, or spinal cord.

Cause.—Lead in some form, is probably the sole cause of this disease. Any of the preparations of lead, capable of being received into the system, through the stomach, lungs, or skin, may produce this affection; but it is probable that the acetate, the carbonate, and the fumes of melted lead as frequently operate to produce this disease, as any of its forms. Such, however, is the pernicious tendency of lead upon the human system, that I very much doubt whether it should not be excluded from use, as a medicine, as a paint, and in fact in every capacity in which it can be brought to act, either directly, or indirectly upon man or beast, except perhaps as an instrument of death, to be hurled upon beings deserving of, and fitted for destruction.

Lead operates through the blood directly upon the brain, spinal cord, and nervous matter generally, with which it comes in contact; but more especially perhaps upon the spinal cord, and hence in part the colic and paralytic symptoms which are developed. In protracted cases, however, the brain becomes seriously affected by it, probably by the particles of lead becoming lodged in its tissue, in consequence of which, apoplectic and other like symptoms are developed.

Lead appears to be equally poisonous to the lower animals as to man; among the instances of which, that have fallen under my observation, I will mention one, illustrating its pernicious effects upon swine.

A few years since while practising in Jefferson county in this state, I was requested by a gentleman to look at his hogs, as he had a large number sick and some already dead. They appeared to suffer from severe colic pains, and there was some distension of the abdomen. They had great thirst, and those in the latter stages exhibited decided apoplectic symptoms.

The gentleman had suspected poisoning from the conium maculatum which grew abundantly in the field; but so decidedly did their symptoms indicate lead, as the cause, that I inquired if they had been exposed in any manner to lead, and I found, on further inquiry, that a few days previous an apparatus consisting of a copper pump and lead pipe, had been arranged for the purpose of conveying the sour whey from its reservoir to the trough, from which the hogs eat three times a day.

The mystery was at once solved, and the pump and lead pipe being removed, no more hogs were taken sick, and those already sick, that had not reached an apoplectic state, gradually recovered with no other prescription than a moderate allowance of food for a few days. I examined the intestines of the surviving hogs after they were fatted and killed, and found them so contracted, at points, as scarcely to admit a crow-quill, the intervening portions being very much dilated, so as to appear in some places like large bladders. From such facts, we may form some idea of the pernicious effects of lead not only upon the human species, but also upon the lower animals; and we may infer, to some extent, the condition of patients suffering from occasional attacks of lead colic.

Pathology.—It is probable that the particles of lead, however, introduced into the system, become lodged in the brain, spinal cord, and nervous matter in different parts of the body, and being too large for the mouths of the absorbent vessels, it acts as a direct irritant or poison, not being soluble in the fluids with which it comes in contact. That portion of the particles of lead lodged in the brain, probably produces, in part, the irritation or cephalic derangement which leads on to the apoplectic and other cephalic symptoms which are sometimes developed. That portion lodged in the spinal marrow may, by affecting its nerves, account for the intestinal contractions, and pains, and paralysis of the extremities. And the lead in other parts of the system may and probably does interrupt the functions of the parts, by deranging the flow of nervous influence and deranging the capillary circulation.

From the presence, then, of these insoluble particles of lead in the brain, spinal cord, and other parts of the system, we have accounted for the leaden hue, the dropsical effusion, the paralysis, the intestinal contractions, and also the apoplectic and other cephalic symptoms which are developed in this disease.

Prognosis.—Lead colic, as it indicates the presence of considerable of the mineral in the system, always affords ground for reasonable apprehension in relation to the ultimate result of the case. If, however, the patient can be removed from further exposure to the poison, and can have the benefit of proper treatment, for a reasonable time, hopes may be entertained of a partial and perhaps entire removal of the lead from the system, and of a partial recovery, at least, from its pernicious effects.

Treatment.—At the commencement of lead colic, dry cupping along the spine may be resorted to, and this, with sinapisms to the abdomen, will often allay the most urgent symptoms. Five grains of Dover's powder with three of calomel, should be given every four hours, and with the third powder, half an ounce of castor oil may be given, and the oil continued with each succeeding powder, till a free evacuation of the bowels is produced. Should the oil be slow in producing its effects, it may be aided by injections of flax-seed tea, with half an ounce of castor oil, and a tea-spoonful of the oil of turpentine if necessary. After the operation of the cathartic, the calomel should be suspended, and the Dover's powder continued every six hours, and the dose of this should be gradually diminished till the pain and other active colic symptoms subside.

After suspending the calomel, and getting a free evacuation of the bowels, the iodide of potassium may be given in five grain doses every six hours, alternating with the Dover's powder; this should be given with a view of removing the lead from the system, which it does by first uniting with it and forming an iodide of lead, which is then removed by the absorbents, and carried from the system with the various excretions. After discontinuing the Dover's powder, the iodide of potassium may be given in six or eight grain doses, three times per day, before each meal, and continued for several weeks, or months, if necessary, in order to remove every particle of the lead from the system if possible.

I am satisfied, by careful observation, that lead may thus be removed from the system, and in one case in which there had been paralysis of one leg for several months; there was not only a return to health from the long continued administration of the iodide of potassium, but the paralysis was also restored. In cases, however, in which paralysis remains, after the lead is removed from the system, the muriate of strychnia in $\frac{1}{60}$th of a grain doses may be administered in solution, till the paralysis is overcome.

SECTION XXVIII.—INTUSSUSCEPTION.

By intussusception, I mean here the introduction of one part of the intestines into another, either direct or retrograde. It is *direct* where the upper part of the intestine is passed into the lower; but *retrograde* when the lower portion is passed into the upper.

Direct intussusception is generally produced by augmented peristaltic action of the intestine, forcing the superior portion into the lower, while an inverted action of the intestine, such as is produced when vomiting occurs, forces a lower portion into the upper, and so produces the retrograde variety. Intussusception may take place in any part of the alimentary canal, but in cases that have fallen under my observation, it has been in the ileum, near its termination in the colon.

Symptoms.—We have in this affection most of the symptoms of spasmodic colic. There is generally pain in the abdomen, eructations of wind, frequent and ineffectual efforts at stool, distension of the abdomen, and constipation; only slight evacuations occurring from that portion of the intestines below the seat of the disease. In some cases there is hiccough, and vomiting of stercoratious matter, from that portion of the intestine above the seat of disease.

If the disease continues, symptoms of intestinal inflammation sooner or later supervene, and if the inflammation be severe, and continued for a considerable time, gangrene and mortification of the intestine will be the result. This will be indicated by a cessation of the pain, prostration of strength, rigors, and all the usual symptoms of mortification.

Sometimes, however, the invaginated portion of the intestine sloughs away, and is discharged, and the patient finally recovers, by a spontaneous effort of nature. Or if no inflammation occur, or it be early subdued, there may be a spontaneous disengagement of the enclosed portion of the intestine, and so the patient recover. By carefully passing the hand over the abdomen, in this disease, a hard, irregular tumor may sometimes be felt, indicating the seat and extent of the invagination.

Diagnosis.—It is exceedingly difficult to distinguish this disease from ordinary colic, but if it follow the operation of a drastic cathartic, or severe vomiting, and a tumor be distinctly felt in the abdomen, and apparently connected with the intestine, intussusception may be strongly suspected, especially if the constipation cannot be overcome.

Causes.—Direct intussusception may be produced by any cause capable of producing an increased peristaltic action of the intestines, such as drastic cathartics, intestinal worms, and other irritating agents. Any cause capable of inverting the action of the intestines, may produce the retrograde variety, such as emetics, intestinal worms, &c. It is possible that this disease may be produced, in some cases, by mere relaxation of the intestines; a very slight direct or inverted action being sufficient, in such cases, to produce either the direct or retrograde form of the disease.

Prognosis.—This is always a dangerous affection, from the liability of inflammation, gangrene and mortification, as well as from the intestinal obstruction. But a spontaneous cure is always to be hoped for, and some hope may be reasonably entertained, from proper remedial measures, if early and judiciously applied.

Nature.—When either variety of intussusception occurs, the intestine becomes irritated, a spasm is produced, inflammation occurs, and the portion of intestine involved becomes strangulated, and if not relieved, gangrene may be the result. That part of the intestine immediately above the seat of the disease, is generally enlarged or distended, with the contents of the intestine, while that part below, has generally a contracted appearance as might be expected.

Treatment.—The indications in the treatment of intussusception are plainly to prevent or subdue inflammation, to overcome spasm, and if possible, to overcome or reduce the invagination by cathartics or emetics, according as the disease is direct or retrograde.

In cases in which this disease is strongly suspected, cups should be applied on each side of the spine opposite the affected part, and three or four ounces of blood taken. And if symptoms of inflammation occur, cups should also be applied over the seat of the disease, and two or three ounces of blood taken. After the cupping a plaster of the extract of stramonium may be applied over the spine, opposite the affected part, with a hope of still further relaxing the spasm, and if no inflammation arise, another may be placed over the seat of the invagination to allay pain, and also to relax spasm, by which the intestine is liable to become strangulated.

If now it appears that the intussusception was produced by an increased peristaltic action of the intestine, as from a drastic cathartic, an emetic of ipecac may be administered, with a hope of inverting the action of the intestine, and thereby overcoming the morbid condition, or drawing up that portion of the intestine which had been passed down into the lower. But if it appears that the intussusception was produced by an inverted action of the intestine, as by vomiting, a cathartic of calomel and castor oil may be given, with a hope of reducing the difficulty, or carrying down the invaginated portion of the intestine, by increasing the peristaltic intestinal action.

Thirty drops of the tincture, or ten drops of the fluid extract of stramonium, may be given every six hours, in part to allay pain, and also as an anti-spasmodic. If tenderness, with febrile excitement supervene, a large blister should be applied over the affected part, and thus the intestinal inflammation be if possible subdued.

Such are plainly the indications of this obscure and truly dangerous affection. And while there is really no very strong grounds for hope in these cases, they should never be abandoned, as spontaneous cures have sometimes occurred after the most unpromising symptoms have been developed. During the continuance of this disease, the patient should be nourished by milk, or broths of a nourishing character, so that as little bulk of food as possible may be taken.

SECTION XXIX.—CONSTIPATION.

By constipation, I mean here that state of the bowels in which the evacuations are less frequent, or less in quantity than is requisite for a perfect state of health, the discharges being generally hard, and procured with more or less difficulty. If there is not a daily movement of the bowels, it is generally in consequence of diminished secretion of the mucous membrane, of hepatic derangement, or else diminished action, or paralysis of the muscular coat of the intestines.

Symptoms.—If constipation of the bowels continue for any considerable time, there is, in addition to the omission of daily stools, and the difficulty of procuring them, a long train of unpleasant symptoms developed, in most cases at least. The breath becomes offensive, the mouth foul and dry, the tongue is furred in the morning, the appetite becomes variable or indifferent, there is apt to be acidity of the stomach, with nausea, flatulence and headache, and if the constipation be protracted, it is liable to produce piles, by the undue pressure which occurs along the hemorrhoidal veins.

Very protracted constipation not only irritates the alimentary canal, but it also deranges the functions of the abdominal and pelvic viscera, produces a cephalic tendency, and finally, by deranging the blood, impairs the health, tone and vigor of the whole system.

Causes.—Constipation may be caused by an astringent diet, by the use of opium, by neglect of the calls of nature, stricture of the bowels, softening or congestion of the brain, by disease of the spinal cord, by pressure of the uterus during pregnancy, and by paralysis of the muscular coat of the intestines.

The worst and most troublesome varieties of constipation may be produced by the long continued use of diuretics, the renal secretion robbing the alimentary mucous membrane of its due amount of fluid, with which to moisten the fecal matter. The same condition is also produced by too frequent bathing, or washing the surface of the body; the skin in such cases being excited to undue action, and robbing the kidneys and alimentary mucous membrane, and even causing to be taken up from the fecal matter its fluid parts, rendering it dry, and its passage along the intestines very difficult.

The worst direct and general or constitutional effects that I have ever witnessed from constipation, was produced in a child three years of age, from having been washed all over daily from birth in warm water. It was corrected by washing less frequently, and substituting coolish for warm water, and giving for a time a solution of the sulphate of magnesia and rhubarb.

Treatment.—In prescribing for constipation, an effort should always be made to correct the condition by removing the cause, and directing a proper diet, and also directing due attention to the calls of nature, once each day, at a regular hour. In this way many cases may be corrected, with no other remedy than ripe fruit, with, or immediately after meals, and, if necessary, the use of bread made from unbolted flour. If these measures fail, and there is no hemorrhoidal affection, a pill of two grains each of aloes and rhubarb may be given after dinner each day, for a time, and then omitted, one being taken at evening, if there has been no movement during the day, and thus continued till the constipation is entirely overcome.

In cases, however, in which the constipation appears to depend upon a torpid state of the liver, a blue pill may be given at evening, for two or three days, at first, and then followed with the pill of aloes and rhubarb, or what may do better, in such cases, with the decoction of dandelion, or else the fluid extract, in moderate doses, after each meal. If the fluid extract of taraxacum be used, from half a drachm to a drachm may be given after each meal; but if the decoction, it may be made by boiling four ounces of the fresh roots and tops, in a pint and a half of water, down to a pint, of which a wine-glassful may be given after each meal, till the constipation is overcome.

In cases of constipation depending upon a torpid or paralyzed condition of the bowels, and attended with a hemorrhoidal complication, the muriate of strychnia may be given in $\frac{1}{20}$th of a grain doses, in solution, three times per day, till the constipation is overcome.

In cases of constipation in females, during pregnancy, if there is acidity of the stomach, a teaspoonful of calcined magnesia may be given each morning, and repeated at evening, if necessary, till the constipation is overcome. In cases in which the magnesia is not sufficient alone, a teaspoonful may be given in the morning, and a teaspoonful of castor oil at evening, if necessary.

Constipation occurring in children, between six months and two years of age, may generally be corrected by giving a solution of the sulphate of magnesia and rhubarb for a time. A drachm of each may be dissolved in half a pint of water, and of this from half a teaspoonful to a teaspoonful may be given each morning, and repeated at evening, if necessary, till the bowels are regulated.

Constipation in early infancy may generally be corrected either by the use of soap suppositories, or else by giving every morning, and repeating at evening, if necessary, half a teaspoonful of a mixture of pure sweet oil and molasses; of each equal parts.

SECTION XXX.—INTESTINAL WORMS.

By intestinal worms, I mean those worms which are so often found in the intestines of the human species. The intestines of the human species sometimes contain different species of worms, the most frequent of which are the *tricocephalus dispar*, or thread worm; the **ascaris vermicularis**, or seat worm; the *ascaris lumbricoides*, or round worm; the *tænia lata*, or broad tape-worm; and the *tænia solium*, or common tape-worm.

The **tricocephalus dispar**, or long thread-worm, is about two inches in length, two-thirds of which is almost as thin as a horse-hair, the remaining and posterior third, being considerably larger, terminating in a round blunt extremity. These worms are seldom met with, and when they do exist, are generally found in the cæcum or large intestines.

The *ascaris lumbricoides*, or round worm, is from two to twelve inches in length; round, of a yellow or brownish-red color, of nearly a uniform thickness, except at the extremities, which taper to a blunt point, and are generally about three lines in thickness. These worms generally inhabit the small intestines, but occasionally they ascend into the stomach, and they are by far the most frequent variety of intestinal worms.

The *ascaris vermicularis*, or seat worm, is a very small white worm, from two to six lines in length, and terminating in an extremely fine extremity posteriorly, resembling the point of a fine needle. These worms usually occupy the large intestines, and especially the lower portion of the rectum, where they are sometimes collected in great numbers.

The *tænia lata*, or broad tape-worm, is about eight lines in breadth, it is flat, white, and composed of a series of connected joints, resembling a piece of white tape. The head is armed with two processes, by which the worm attaches itself to the intestines. This variety of worm inhabits the stomach and upper portion of the intestines, and often acquires a very great length, sometimes even thirty or forty feet.

The *tænia solium*, or common tape-worm, is composed of joints resembling very much the seeds of the gourd, especially in the lower or posterior part of the worm. The anterior portion of this variety tapers off in a fine thread-like extremity; the head being extremely small, and furnished at its sides with four raised disks, with depressed centres, by which the animal probably attaches itself to the intestine. This variety of tape-worm inhabits the stomach and small intestines, and generally comes off in pieces, from a single joint to several feet in length.

Other animals are said to be found in the alimentary canal, but the preceding are the only varieties which appear to be peculiar to this situation, so that if others are found there, it may be regarded as accidental; either the eggs or animals themselves having been taken with food or drinks.

Symptoms.—There are few, if any, positive symptoms of intestinal worms; but symptoms often arise by which we may suspect the existence of worms in the alimentary canal, and if the ova of worms can be discovered in the discharges of persons suspected, by microscopic examination, that may be regarded as positive evidence of their presence.

The symptoms most frequently met with in patients annoyed by worms, according to my observation, are a pale countenance, or a livid appearance, with occasional transient flushes, dull eyes, and dilated pupils, with a bluish semicircle around the lower lid, itching of the nose, tumid upper lip, headache, copious secretion of saliva, furred tongue, foul breath, and variable appetite, being at times voracious, and at others entirely gone.

In some cases there is vomiting, pains in the abdomen, especially in the umbilical region, slimy stools, turbid urine, a tumid abdomen, emaciation, lassitude, irritability of temper, and general convulsions. And in some rare cases there is an obstinate cough, palpitation, and a general debilitated and irritable condition of the system produced, leading to melancholy, hypochondria, insanity, &c.

It appears to me that worms cannot be regarded as harmless inmates of the intestinal canal, at least in the present physically depraved condition of the human family. It is probable, however, that if the laws of health had always been observed, there would be no intestinal worms, or existing, they would be entirely harmless.

It appears, according to my observation, that those who have adhered most rigidly to the rules of propriety, especially in relation to taking good, wholesome food, and at regular hours, are comparatively free from the annoyance of intestinal worms, while those who have departed most widely from these rules, suffer most from verminous irritation. In relation to the origin of the different varieties of intestinal worms, there may be room for reasonable doubt; but it appears possible that they may be derived from ova introduced into the system with food or drink when taken in a raw or uncooked state.

Treatment.—The treatment proper for the expulsion of worms, depends very much upon the variety of the worm and the general condition of the patient at the time of the treatment for their expulsion. But as the indications for the expulsion of the round worm and long thread worm is the same, we may first inquire into the treatment proper when they are suspected, and then we may consider, in turn, the treatment proper for the destruction of the seat-worm, and finally of the tape-worm.

For the ordinary worm affections of children or adults, if the seat-worm or tape-worm is not suspected, there is no remedy that so generally does well as the *spigelia marilandica*, or pink root. It may be given in infusion, half an ounce of the pink root being infused in a pint of water in a covered vessel, for one hour. Of this infusion a table-spoonful may be given to a child one year old, morning, noon and night, in sweetened water, with a little milk; and for each additional year of age half an ounce more of the infusion may be given; no more than six ounces, however, should be administered to an adult at one time.

The better way, however, to arrive at the dose for children in this as well as in all other cases, is to call six ounces of the infusion a dose for an adult, and then arrive at the dose for a child by the rule laid down in the early part of this work, as follows: Divide the age of the child increased by twelve, by itself, and the quotient will indicate the proportion for the child. Thus, if the child be six years old, that increased by twelve makes eighteen, and eighteen divided by six gives a quotient of three; so that the dose for a child six years old will be one-third of that for an adult, or two ounces. By this simple rule, as before stated, the dose of any medicine for children may readily be arrived at; due allowances being made for constitutional peculiarities, &c.

The pink should be given with the meals, morning, noon and night, for three days, and on the fourth day a dose of calomel and castor oil should be administered so as to procure a thorough evacuation of the bowels. The pink probably has a narcotic effect upon the worms, so that by giving it with the food for three days, they become stupefied, and are readily removed by the calomel and castor oil, or they might be by the oil alone in case the calomel, from any cause, is contra-indicated.

Instead of the infusion of pink root, the fluid extract may be given, if at hand. Of this, from half a drachm to a drachm may be given to an adult in the manner I have suggested for the infusion, the dose for children being arrived at as I have already suggested; or, what in some cases may do better, the fluid extract of spigelia and senna may be given in one or two drachm doses to adults, and continued at meals for three days, only a moderate dose of calomel and oil being required on the fourth day, as the senna may produce slight relaxation of the bowels

In cases in which, from any cause, the spigelia is contra-indicated, the chenopodium anthelminticum, or worm-seed, may be given instead. Ten grains of the pulverized seeds, or five drops of the oil, may be given to a child one year old, morning and evening, for three days, and on the fourth day, a dose of calomel and castor oil may be administered, or the oil alone, if the calomel be contra-indicated. For adults, sixty drops of the oil, or two drachms of the pulverized seeds, may be given at a dose, in the same manner, a full dose of calomel and oil being administered on the fourth day, as directed for children.

In all cases of verminous irritation, in which the pink and worm-seed are contra-indicated, or inappropriate, a full dose of calomel and castor oil may be given at once, and generally with very good effect. Such, according to my observation, is the most safe, convenient, and effectual method of destroying the common round worm, in children and adults, and the same treatment is applicable in case the long thread-worm is suspected.

In the destruction of the *ascaris vermicularis*, or seat-worm, which, as we have seen, inhabits and often irritates the lower part of the rectum, there is generally more or less difficulty. The tincture of aloes and myrrh, in drachm doses for adults, and less for children, according to the age, may be given every morning, for two or three weeks, and an injection of cool water used at evening, for a time, and then, with the addition of half an ounce of common salt, for an adult, will generally exterminate them, if continued for a reasonable time.

If, however, this treatment fails, an injection of equal parts of lime-water and milk may be used. Or in case this should fail, one drachm of the oil of turpentine, in a gill of milk, may be used once each day, at evening, instead. After having continued this treatment for two or three weeks, a full dose of castor oil, or calomel and oil, may be given, and a free operation secured. And should the worms find their way into the vagina, injections of vinegar and water may be used for a time. In this way, I believe, this troublesome variety of worm may generally be temporarily or permanently destroyed.

For the removal of the *tape-worm*, the oil of turpentine is a valuable remedy, for either variety. The oil of turpentine may be given in drachm doses, in a little water or mucilage, three times per day, for three days or longer, and then a full dose of castor oil administered. Or an ounce of the turpentine may be given at once, and a full dose of castor oil administered in two hours, and repeated if necessary.

I believe this treatment will generally prove successful, but in case it fails, or if there is some good reason why the turpentine should not be given, I would use the pumpkin seeds, prepared as follows. Six ounces of the pumpkin seeds may be bruised thoroughly in a mortar, without removing the shells, and then sufficient water should be added to give, by straining and expression, a pint of liquid. Of this one-half may be administered early in the morning, and in two hours an ounce of castor oil should be given. About noon the remaining half of the liquid should be given, and this also followed in two hours by an ounce of castor oil, which may be repeated in two hours, if necessary.

Should the treatment prove successful, the tape worm will generally be discharged during the operation of the second dose of castor oil. In case the turpentine and pumpkin seeds fail of producing the expulsion of the tape worm, **two** ounces of the bark of the *pomegranate root* may be bruised **and macerated** in a quart of water for twenty-four hours, and then the mixture should be boiled down to a pint. Of this a wineglassful should **be** given every hour till the whole is taken, **or** till nausea, vomiting and free purging is produced, at which time the worm may be expelled with the discharges. In some one of these methods, I believe the tape worm may generally be destroyed.

In selecting a remedy from the ones I **have suggested for the destruction** of the tape worm, as well as for the other varieties of worms, the age, constitution, hereditary and acquired constitutional peculiarities of the patient should be taken into account; and generally the milder measures should be resorted to first, and the more active resorted to only in case of failure with the milder measures. Care should be taken that anthelmintics be not continued for an unreasonable time, or administered in cases in which the disease is from irregularity of eating and not from intestinal worms.

Nothing is more common, according to my observation, than for parents to allow their children to take improper food, or to eat at all hours, and **thus destroy their** digestive powers; producing **most of the** symptoms **usually developed** from the presence of intestinal worms. The child in such cases has an irregular appetite, grows pale, loses flesh, becomes irritable during the **day and** restless nights, becomes pale about the mouth, picks its nose, &c.; and if the irregular eating be continued, I have known dropsy, epilepsy, convulsions, &c., to be the result.

Now all this is generally charged by the parents to *worms;* and if the diseases which the reckless and ruinous indulgence in irregular eating has produced does not kill the child, the thousand and one worm nostrums which may be prescribed is very likely to. Now it is probable that this imprudence together with filthiness, bad air, &c., may and does favor perhaps the generation and growth of intestinal worms, but a large majority of the cases of the character I have described, that have fallen under my observation, which have been attributed to *worms*, are the direct result of taking food at irregular hours or eating between meals.

As a general rule then, when children or adults exhibits symptoms, such as I have described, it is best to inquire carefully into the habits of the patient in relation to taking proper **food at** regular hours, &c., that no more be charged **to worms** than they are guilty of. And in all cases a plain, digestible and proper diet should be insisted on, and enforced if necessary, and **then** if there be substantial reasons for suspecting intestinal worms, the **measures** I have suggested may be resorted to for **their** destruction.

SECTION XXXI.—HEMORRHOIDS—(*Piles.*)

By hemorrhoids from 'αιμα, "blood", **and** ρεω, "I flow," I mean that disease of the lower portion of the rectum, in which there is generally erectile tumors, either within or around the extremity of the bowel, which occasionally pour out more or less blood, and are very liable to become inflamed.

It will be remembered that the rectum is supplied with blood by the superior middle, and inferior hemorrhoidal arteries, while its veins empty mainly into the inferior mesenteric, thus forming a part of the portal circulation. Now when we remember that the veins of the rectum have no valves, that they form a part of the portal circulation, returning the blood from the lower part of the rectum, it cannot appear strange that hemorrhoids or piles should occur. The liability to this disease is increased by the possibility of an interruption to the free passage of the blood of the portal vessels through the liver, and also by the passage down the rectum, of hardened or dry fecal matter.

If now there be a relaxed condition of the tissues generally, a congested state of the portal vessels, and a constipated state of the bowels, the veins of the rectum, as well as its mucous membrane, are liable to become congested and more or less irritated, and hence the different varieties of hemorrhoidal tumors, which being formed, and inflamed, develop the symptoms peculiar to each.

The same causes favor an enlargement, or a varicose state of the hemorrhoidal veins, for six or eight inches up the rectum. And as these veins lay immediately under the intestinal mucous membrane, and are surrounded by a sub-mucous or cellular tissue, these parts are very liable to become more or less congested, as well as the hemorrhoidal veins.

We have then in consequence of all these predisposing circumstances, various forms of hemorrhoidal tumors, the most important of which are the *varicose*, the *erectile* and the *fleshy* tumors, each of which are formed as follows:

The *varicose hemorrhoidal tumors* arise from a varicose state of the hemorrhoidal veins; are situated at the lower extremity of these veins; consist of an enlargement without rupture of these vessels, and are covered only by the mucous membrane of the rectum. They form slowly within the bowel, and become gradually extended, as they become larger, and consist of a round, tense, elastic nucleus the size of a pea, covered by the intestinal mucous membrane which moves easily upon it. When this tumor remains extended, the blood which it contains may coagulate, lymph may be effused, and finally it becomes organized and vascular, its mucous membrane becoming strongly adherent, and thus it constitutes one variety of bleeding piles.

The *erectile tumor* consists of the mucous, sub-mucous, and cellular tissues of the lower part of the rectum, supplied with blood from one or more of the hemorrhoidal veins, in consequence of which, they frequently become of large size, and present all the characteristics of an erectile tissue. These tumors are cellular, spongy, full of blood, very vascular, and bleed profusely from innumerable points on their surface. They form within the intestine, from congestion of the mucous, sub-mucous, and cellular tissue. And as they enlarge, they are mechanically protruded from the rectum, and constitute the cellular or erectile variety of bleeding piles.

The *fleshy tumor*, or blind pile, consists wholly of dense, thickened, or hypertrophied cellular tissue, covered by the intestinal mucous membrane, and are from their earliest formation external. These tumors probably begin in merely a small fold of the mucous membrane, which, with its sub-mucous tissue has been forced from the bowel in an effort at stool, and being pinched by the sphincter, is prevented from returning within the bowel. Exposed thus to friction, these tumors become inflamed; in consequence of which, they are thickened and indurated, their mucous-investing membrane becoming transformed into skin, and in this way the whole tumor becomes a chronic pile, more or less insensible until active inflammation is excited; when it swells, grows red, hard, and extremely painful, but does not bleed. By continued irritation these tumors sometimes increase to a large size, and if there are several, their opposing surfaces may become ulcerated, and thus an offensive purulent discharge take place.

Hemorrhoidal tumors generally have a broad base, but sometimes they are more or less pedunculated, and in an indolent condition they may be pale and flaccid; but when inflamed, or highly congested, they become red or purple, tense, hard, and exceedingly tender and painful.

Hemorrhage is liable to occur in this disease from the varicose veins, from the capillaries of the rectum, by exudation from the tumors, and also from the intestinal mucous membrane. When the hemorrhage is from a rupture of the varicose veins, it may come on in a stream, while the patient is at stool; but when it occurs from the capillaries of the rectum, or by exudation from the surface of the erectile tumors, or from laceration of the intestinal mucous membrane, it is generally less rapidly copious.

Symptoms.—The symptoms of hemorrhoids are just such as we should suppose would arise from a determination of blood to the rectum, with congestion of the hemorrhoidal vessels, together with the presence of indolent, irritable, or inflamed tumors either within or projecting from the extremity of the rectum.

There is generally at first a feeling of heaviness or fullness, soon followed by pain in the extremity of the bowels, which extends to the surrounding parts. This pain may be continuous, or it may subside and return again in a few days with increased severity, till finally a small tumor or tumors are discovered, which are generally more or less tender, at times, at least.

When a determination of blood to the rectum occurs, which immediately precedes hemorrhage, there is a dull pain in the back and a sense of heat about the sacrum; the urine is scanty and high-colored; there is heaviness in the head, and more or less disturbance of the digestive functions. These symptoms may go on increasing for several days, when the flux supervenes and entirely relieves the patient; being, of course, most marked in cases in which it occurs at regular periods.

In cases of this affection in which there is habitual constipation, there is a sense of fullness, weight and heat almost constantly about the rectum, and generally more or less sympathetic irritation of the bladder, vagina and uterus in females; all of which symptoms may be temporarily alleviated by the flow of blood, but not entirely, the irritation being kept up by each movement of the bowels.

When *inflammation* of hemorrhoidal tumors takes place from constipation, walking, riding or any other cause, the tumors become congested, swelled, red or purple, and exceedingly painful, sometimes obliging the patient to keep his bed for several days.

Diagnosis.—There is little difficulty in distinguishing piles from dysentery, prolapsus ani, polypus of the rectum, or hemorrhage from a higher portion of the intestines, if the history of the case and all the symptoms be carefully taken into the account. If, however, there is a possibility of confounding this disease with hemorrhage from the intestines higher up, it is only necessary to remember that the blood from the piles is fluid, florid, and generally discharged before or after a movement of the bowels, but never mixes with the discharges; while blood from the intestines higher up, may be mixed with the feces, is often black and generally coagulated.

Causes.—Among the most frequent causes of hemorrhoids, are a hereditary predisposition, constipation of the bowels, free living with a sedentary mode of life, and aloetic or other drastic purgatives.

The *hereditary* predisposition to piles consists in a general relaxed condition of the tissues, and especially of the cellular, which greatly favors, as we have seen, the formation of hemorrhoidal tumors.

Constipation of the bowels by the mechanical irritation which it produces in the lower portion of the rectum, becomes a frequent cause of hemorrhoids, and probably the most frequent cause.

By free living, or taking too much stimulating food, digestion is impaired and an irritation set up along the alimentary mucous membrane; and along with hepatic derangement, there is produced congestion of the portal vessels, and so of course of the hemorrhoidal veins, and thus the formation of hemorrhoidal tumors is favored.

A *sedentary* mode of life also, by favoring hepatic and portal congestion, produces or predisposes to hemorrhoidal affections. And finally, aloetic and other drastic purgatives probably produce piles by the direct irritation they produce in their operation upon the rectum. Besides the causes enumerated, there are other occasional causes of piles, such as suppression of the menses, pregnancy, irritating injections, ascarides, venereal excesses, straining at stool, direct irritants, riding on horseback, &c.

Treatment.—The indications in the treatment of hemorrhoids, are those which arise from the flux or hemorrhage, and those which occur from the tumors both in their ordinary and inflamed condition. In order to prevent the flux, it is necessary to correct the constipation of the bowels, as well as the torpid condition of the liver and congestion of the portal vessels.

Constipation of the bowels and torpor of the liver may generally be overcome by a regulated diet, regular habits, a blue pill at evening for a few nights, and then by the administration of the decoction or fluid extract of taraxacum, in moderate doses after each meal for a time. If, however, this should fail of correcting the constipation, a partially paralyzed condition of the bowels may be suspected as the cause, and the muriate of strychnia in $\frac{1}{60}$ of a grain doses should be administered in solution, three times a day till the constipation is overcome.

To stop the hemorrhage when it occurs, ipecac in one-fourth of a grain doses, every fifteen minutes is a valuable remedy, and if necessary, tannin may be given in two grain doses every two hours, till it be arrested. If, however, the hemorrhage occurs in a debilitated patient, and is of a passive character, the tincture of the chloride of iron may be given in ten drop doses, every four or six hours, and continued till the hemorrhage is arrested.

The patient during the continuance of the hemorrhage should be kept in the horizontal position, and injections of cold water used, and if the bleeding be obstinate, a saturated solution of alum in cool water may be resorted to. If, now all these measures fail, the tampon should be used if necessary. For this a piece of soft silk or linen moistened in a saturated solution of alum should be pressed carefully into the bowel. This will mechanically arrest the flow of blood, while the alum produces an astringent effect upon the hemorrhoidal vessels, and upon the surrounding mucous and cellular tissues.

Having thus by proper treatment overcome the hepatic and portal congestion, arrested the flux, and regulated the bowels, the tumors may be greatly relieved, and their inflammation and hemorrhagic tendency lessened by the daily injection of cool water, by sponging the anus morning and evening with a saturated solution of alum in cool water, and finally by the application of the following ointment.

Take of *bees-wax, olive-oil* and *lard*, each two ounces, *extract of stramonium tannin*, and *oxide of zinc*, of each half an ounce. Melt the beeswax and lard together, to which add the sweet oil, and as they cool, add and mix intimately the stramonium tannin and oxide of zinc. A little of this may be applied to the tumors morning and evening till their irritation, pain and soreness subside.

I believe that by a proper course of treatment, thus judiciously applied, many cases of piles may be entirely cured, and the worst cases greatly benefitted. If, however, a judicious medical treatment fails in producing relief, the tumors may generally be safely removed by the ligature, carefully and properly applied.

SECTION XXXII.—JAUNDICE—(*Icterus.*)

By jaundice or icterus, I mean that affection in which there is generally yellowness of the skin, eyes, and urines, from the presence of bilous matter, caused either by retention or suppression of bile, in consequence of some derangement in the secretion of the liver, or obstruction in its excretory duct. It will be remembered that the bile is a yellowish fluid consisting of a "resinous soda salt, a coloring material, cholesterine and mucus," and that it is probably derived from decomposing waste materials of the system, and very much of it from decomposing blood-cells.

Now, as the proximate constituent principles of the bile pre-exist in the blood, a failure on the part of the liver to drain it off may produce a jaundiced state, in which case we may have jaundice from suppression of the bile. But if the liver performs its functions, or separates the bile, and yet it fails of reaching the intestine through its exeretory duct, there is soon a jaundiced condition, either from absorption of retained secreted bile, or else from an interruption to its further secretion caused by the obstruction, and hence we have jaundice from retention of the bile.

The normal course of the bile after it is secreted, it will be remembered, is along the hepatic duct, and ductus communis choledochus to the duodenum. Or if it has passed to the gall-bladder, it passes by the cystic duct, and then along the ductus communis choledochus to the intestine.

Now the cystic and hepatic ducts unite to form the common excretory duct of the liver and gall-bladder, which common duct is about three inches in length, and empties into the perpendicular portion of the duodenum, passing obliquely between its muscular and mucous coat. The hepatic ducts have an external coat of contractile fibrous or muscular tissue, and an internal mucous coat which is continuous with that of the duodenum.

Now the calibre of even the common bile duct is very small, and as it passes obliquely through the coats of the duodenum, it is liable to become clogged or obstructed from various causes, among the most frequent of which are biliary concretions coming down from the gall-bladder, thick or viscid bile, inflammation or thickening of the duodenal mucous membrane obstructing its mouth, and finally inflammation or spasm of the duct at any point.

It is easy to see that a failure on the part of the liver to separate from the blood the constituent principle of the bile, or an obstruction to its free passage along the ducts after it is secreted, will cause to be retained in, or else to be reabsorbed into the blood, the bile, or its constituent principles, which being carried along in the blood, is finally thrown upon the skin, and more or less into nearly every tissue of the body. In consequence of the presence of the coloring matter of the bile in the blood, the skin, the eyes, and urine, have a yellow tinge, and its absence in the intestines leaves the stools of a light clay color, as is generally the case, and hence too the symptoms of this disease which we will now proceed to consider.

Symptoms.—The absence of bile in the intestines, and its presence in the blood, whether from suppression or retention produces languor, loss of appetite, indigestion, constipation, a feeling of fullness in the epigastrium, restlessness, nausea, a turbid urine, a slow pulse, and creeping chills, which alternate with flushes of heat.

After a few days itching occurs over the surface of the body, the mouth is bitter, the stools are clay-colored, the urine assumes a saffron hue, and finally the white of the eyes, and the skin about the neck, lips, and forehead, assume a yellowish color, which speedily extends until the whole surface of the body acquires a yellowish hue. At this stage the pulse may become full and firm, and slight febrile exacerbations may take place in the evening with increased heat of skin, followed by more or less restlessness during the night. Finally, if the disease continues, the body begins to emaciate, the evening febrile exacerbations become more conspicuous, night-sweats ensue, hemorrhages, or dropsical effusions take place, respiration becomes oppressed, and unless the disease be arrested, a torpid condition may supervene before a fatal termination.

Anatomical Characters.—The post-mortem examination discovers any organic lesion upon which the disease may have depended, and in addition to this, as well as in all cases in which no organic lesions appear, there is found a yellow tinge of nearly all the tissues of the body, even

27

of the bones. The substance of the brain, however, sometimes remains
of its natural color.

Causes.—Among the causes which operate to produce this disease, by
causing a suppression of the bile, or a retention of its constituents in the
blood, are the *paludal poison*, excessive and long continued heat, masturbation, onanism, and excessive venery, depressing mental emotions,
and violent anger, or long continued and excessive grief. All these
causes probably operate by suspending or diminishing the secretion of
the liver, in consequence of which it fails to strain off from the blood the
proximate constituent principles of the bile.

Among the causes that produce jaundice by a retention of the bile,
or by an interruption to its free passage along its ducts to the intestine,
are, as we have seen, biliary concretions, spasm of the duct, duodenal
inflammation, induration of neighboring parts, viscidity of the bile, and
some other occasional causes.

In cases in which the jaundice is from suppression of the bile, the disease comes on more or less slowly and insidiously, and is attended generally with little or no pain, and very slight, if any, febrile exacerbations,
especially in the early stage of the disease. In cases, however, in which
the disease is from obstruction of the bile duct, the disease may come on
more suddenly, and be attended with pain, febrile exacerbations, &c., at
an early stage of the disease. When biliary concretions pass down and
close the duct, there is severe pain, and all the symptoms proceed
rapidly. In cases depending upon spasm of the duct, I have generally
found more or less spinal irritation, and pain in the back as well as in
the region of the duct.

Cases depending upon duodenal irritation or inflammation generally
come on slowly, and are attended with slight uneasiness and tenderness
in the duodenum, and more or less colic pains in that region. If, however, it arise from scirrhus of the pylorus or duodenum, there is in addition great emaciation with general debility.

In cases in which the disease is produced by a viscid state of the bile,
the disease may come on slowly, with little or no pain, except perhaps a
dull heavy feeling in the right hypochondrium. Finally, if pregnancy
or impacted feces in the colon produce the obstruction, the symptoms are
usually sufficiently manifest, especially if the cause be pregnancy, and
long continued constipation is sufficient to suggest that cause when it
exists.

Prognosis.—The prognosis in jaundice is generally favorable, except
in cases depending upon some powerful general depressing agent, and
those depending upon or complicated with scirrhus or some other organic
disease of the stomach, pancreas, or other surrounding parts.

Treatment.—The indications in the treatment of jaundice are to restore the secretion of the liver, if that is suspended or diminished, and to
overcome any obstruction there may be to the free passage of the bile
from the liver to the intestine, after it is secreted.

Now as digestion is always more or less impaired while there is suppression or retention of the bile, from its absence in the intestines, as
well as from its presence in the blood, a three grain pill of inspissated ox
gall may be given after eating, morning, noon and night, in this disease,

so long as bile is not secreted by, or carried from the liver to the intestine. This will generally secure a tolerable digestion, or at least favor chylification, and thus tend to keep up the strength of the patient, till such time as the secretion of sufficient bile may be restored, or any obstruction there may be to its free passage to the intestines be overcome.

As there is, too, in most cases of jaundice more or less viscidity of the bile, or liable to be at least, ten grains of the bi-carbonate of soda may be given, in solution, before each meal, for a time, for the purpose of thinning it, which it will generally do very effectually. For the purpose of restoring or improving the secretion of the liver, a blue pill may be given at evening, for a time, and followed each morning by a tea-spoonful of the sulphate of magnesia, to secure a daily movement of the bowels. After having continued the mercurial for a reasonable time, it may be suspended, and a teaspoonful of the fluid extract of dandelion, or an equivalent dose of the decoction, may be given after each meal, till the secretion of the liver is restored. Or in cases of jaundice in which a mercurial is from any cause contra-indicated, a grain or more of podophyllin or leptandrin may be given, at evening, instead of the blue pill, and followed by the sulphate of magnesia in the morning, if necessary, and this may be continued till the function of the liver is restored; or it may be suspended, after a time, and the taraxacum continued in its stead, as suggested above. The warm foot-bath should be used at evening, and the patient should be allowed a plain, digestible, and nourishing diet, to be taken with regularity, and should take a reasonable amount of exercise, and maintain an even cheerful temper of mind.

With this course of treatment most cases of jaundice, from suppression of the bile, may be overcome. But if there is evidence of retention, from some obstruction to the free passage of secreted bile to the duodenum, other measures, in addition to what I have already suggested, are clearly indicated. In cases in which there is evidence of spasm of the bile-duct, twenty drops of the tincture of stramonium, or ten drops of the fluid extract, may be given, four times per day; and if there is spinal irritation, as I have generally found in such cases, cups should be applied on the right side of the spine, opposite the liver, and two or three ounces of blood taken. Finally, in all cases in which there is evidence of inflammation or congestion of the bile-duct or duodenum, in addition to the treatment I have already suggested, cups should be applied at first, and later blisters, if necessary, over the bile-duct and duodenum.

The indications then, in cases of jaundice from suppression merely, may generally be fulfilled with the blue pill, podophyllin, or leptandrin at evening, for a time, followed by the sulphate of magnesia in the morning, together with the ox-gall and bi-carbonate of soda, as suggested, and the taraxacum, if necessary. But in case there is retention of secreted bile, in addition to the above, stramonium, cupping, blistering, &c., may be indicated, and should be faithfully and judiciously applied, as I have suggested.

CHAPTER IX.

DISEASES OF THE RESPIRATORY SYSTEM.

SECTION I.—AUSCULTATION AND PERCUSSION.

By auscultation and percussion I mean, to include here physical exploration of the chest. Before proceeding to the consideration of the diseases of the *respiratory system*, I propose to consider in this section, under the head of *auscultation* and *percussion*, the principles of *physical diagnosis*, embracing *inspection, palpation, percussion, auscultation, mensuration* and *succussion*, so far as they are applicable to diseases of the respiratory organs; leaving so much of this subject as is applicable to the heart, for consideration in the following chapter, in which I shall take up the diseases of the *circulatory* system.

Now, in order to appreciate this subject, it is necessary to bear in mind the anatomy and physiology of the respiratory organs, as well as to understand the normal sounds produced by respiration, by the voice, by percussion, &c., in order to be able to discover any deviation from the normal standard; and thus to appreciate the signs of internal disease, indicated by abnormal sounds.

In diseases of the respiratory organs requiring physical exploration of the chest, the patient should, if convenient, be placed in the sitting posture, and the clothing should be removed, as far as may be consistent, and then the anatomy and physiology of the parts being kept in mind, it is also proper to understand the *regions* into which the chest, and even the whole body is usually divided. And, as the different *regions* of the chest and abdomen are liable to be called up, in our minds at least, at almost every step of our proceedings, it may be proper for us to refer to the boundaries of these regions before we proceed to a consideration of *inspection, percussion, auscultation, mensuration, succussion, &c.*

Now, the chest and abdomen may conveniently be divided, as is customary, into *regions* in the following manner: Let three vertical lines be drawn, one on each side of the sturnum, and one along the whole length of the spine, and also a vertical line from the scapular extremity of the clavicle on each side to the spinous process of the pubis, as well as one from the posterior boundary of each axilla perpendicular to the crest of the ileum. Let now five transverse lines encircle the body; the first, on a level with the clavicles; the second, on a level with the junction of the cartilage of the fourth rib with the sternum; the third, on a level with the extremity of the xiphoid cartilage; the fourth, on a level with the end of the last rib; and the fifth, on a level with the spinous processes of the ossa ilii.

Now, the regions bounded by these vertical and transverse lines are as follow: Above the first transverse line are the *acrominal regions;* between the first and second transverse lines, in the middle, in front is the *superior sternal region*, and on each side laterally, the *inferior clavicular*, the *axillary*, and the *scapular regions;* between the second and third transverse lines, in the middle and front, is the *inferior sternal region*, and on each side, the *mammary*, *lateral* and *infra scapular regions;* between the third and fourth transverse lines there is in front the *epigastric*, and on each side the *hypochondriac* and *dorsal regions;* and finally, between the fourth and fifth transverse lines we have in front the *umbilical*, and on each side, laterally, the *iliac* and *lumbar* regions. Now, bearing these regions thus bounded in mind, we are prepared, with our knowledge of the anatomy and physiology of the parts, to proceed in the examination; and first by *inspection*, as I have already suggested.

By *inspection*, we discover the size, and general form of the chest, as well as the working of the chest, its degree of expansion, &c. It is also proper to note the action of the intercostal, and abdominal muscles; as very important indications of disease may be furnished by the intercostal, or abdominal character of the respiration. Having thus noted the indications furnished by *inspection*, palpation may next be resorted to; and this may confirm the hints furnished by *inspection*.

Palpation, by laying the flat hands on corresponding parts of the two sides of the chest, may be a great help, in discovering the difference in the motion of the two sides, if any exist. This method of *palpation* often also does more, by detecting any roughness there may chance to be from pleuritic, or pulmonary disease.

Palpation is also of service in detecting chronic pleuritis. For by pressing the ends of the fingers along the intercostal spaces, tenderness is often detected; thus bringing to light, a morbid condition, which might not be detected by any other symptom, at least in its early stages. This method of *palpation*, is also of service in exploring other parts of the body, as well as that of laying the palm of one hand on a part to be examined, and gently tapping at an indefinite distance, so as to elicit a sense of fluctuation, if there is fluid within. By this method, we may sometimes detect the existence of fluid in the pericardium, as well as in the abdominal cavity.

Having proceeded thus far in our examination, and having availed ourselves of the signs furnished by *inspection* and *palpation*, *percussion*, or striking the parietes of the chest in such a manner, as to elicit the amount of resonance, or dullness of the parts beneath, is the next step in the order of our inquiry.

Percussion may be immediate, in which case the hand, and generally the end of the fingers, falls immediately upon the chest, or part to be examined. Or, it may be mediate, in which case the fingers of the other hand, or some other substance intervenes.

Immediate percussion may be resorted to in very lean subjects; or in cases in which the application of a pleximeter is impossible, or inconvenient. And it may be very conveniently performed, by the first three fingers; the ends being brought together, and the fingers slightly flexed. In fat subjects, and in most cases in which a pleximeter can be conveniently applied, mediate percussion is preferable, and it may be most conveniently performed, by using the three first fingers of one hand, as the plessor, or instrument for striking, and the middle finger of the other hand, as the pleximeter.

Now, in order to appreciate the signs, furnished by percussion, in disease, it is necessary to understand the natural sounds, which are elicited in different parts of the chest in a state of health, that the morbid sounds may be compared with the natural; and thus the indication which they furnish of internal disease, be clearly appreciated It is well then, to bear in mind that tympanitic sounds are elicited in health, in the lower part of the mammary, left lateral, and infra-scapular regions. The superior sternal, axillary, and upper part of the infra-scapular regions, are very resonant. The subclavian, the upper part of the mammary, the lateral and interscapular regions, are resonant. The acrominal, lower part of the right mammary, lateral, and infra-scapular regions, are imperfectly resonant. Finally, the inferior sternal, the inner edge of the left mammary, and the scapular regions, produce dull sounds on percussion.

It is only necessary to remember the position of the heart, liver, and other internal parts as well as the intervening parietes, in order to understand the rational of the variety of resonance elicited by percussion in different parts of the chest, in a state of health.

The indications of internal diseases furnished by percussion are very numerous, and the variety of diseases of the respiratory organs in which it affords important signs are so numerous, that I will mention the most important of them, and the peculiarities of each.

In *simple bronchitis*, the resonance on percussion is nearly or quite natural, at least in the early stage of the disease. But if there is a copious secretion of mucus in the bronchial tubes, a slight dullness may be noticed in some cases.

In *pneumonia*, there is usually a dull sound on percussion, and especially is this the case, if the inflammation has passed on to hepatization as is too often the case.

In *phthisis*, there is generally a dullness on percussion over the upper part of the lungs during the early stages. But as a cavity forms, the sounds gradually become resonant, and finally in the latter stages even tympanitic.

In *congestion* of the lungs, there is generally more or less dullness on percussion, and the same is also true in œdema, especially in the most depending part.

In *malignant diseases* of the lungs, such as *scirrhus*, if it be extensive there is dullness on percussion over the indurated part. But in emphysema there is generally increased resonance in proportion to the increased amount of air the lungs contain.

In *hydrothorax* as the fluid accumulates and fills the chest, compressing the lung, dullness attends, being discoverable first in the most depending part of the chest, but extending gradually as the effusion increases.

In *malignant disease* of the pleura, as in that of the lungs, the pulmonary tissue is compressed, being pushed aside, and dullness on percussion follows as a consequence.

In *pneumothorax*, percussion elicits a resonant, or tympanitic sound. But if air and a fluid exist in the same serous sac, there is dullness in one part, and tympanitic resonance in another. If, now the patient under examination be suffering from either of the diseases enumerated above, an important clue to the disease may thus be had by percussion, in addition to the signs furnished by inspection and palpation.

The next step in the order of our examination is *auscultation*, or listening to the sounds arising from the exercise of the functions of respiration, and comparing those emanating from diseased organs, with those ascertained by previous examination, or experience to exist in the healthy condition of the organs.

Auscultation, as well as percussion, may be either immediate or mediate.

It is *immediate* when the ear is placed directly on the part, entirely bare, or covered only by a thin cloth, the vibrations being conducted directly to the ear.

It is *mediate* when between the ear and the parietes a body is interposed, which conducts the vibrations from one to the other. The stethoscope, or instrument so interposed, is only a convenience in particular cases, or in examining particular parts of the chest, as immediate auscultation is, according to my experience, the most reliable.

In *immediate* auscultation the ear should be applied to different parts of the chest, in such a way that the ear may participate in the vibrations of the solid parietes upon which it is placed.

When the stethoscope is used, it should be pressed upon the parietes of the chest, its cylinder being always at right angles to the surface with which it is applied. The ear, too, should be evenly adjusted to the flat end of the other extremity, if the old style of stethoscope is used, and so firmly pressed as to make the parietes of the chest, the stethoscope and the ear one continuous vibrating body.

It is necessary that the auscultator should understand the *natural respiratory sounds*, in the larynx, trachea, bronchial tubes and lungs, as well as the natural sounds of the voice in these parts. For it is by comparing the morbid with the healthy sounds, that auscultation becomes an important means of diagnosis. In examining the patient by auscultation it is generally best to notice first the respiratory sound, and then to notice the change which takes place in the voice and cough.

If from inflammation, congestion, or thickening of the larynx or trachea, the air be obstructed in its passage, a whistling or crowing sound is produced by the act of respiration, and it is loud or slight according to the amount of the obstruction.

The respiration in the bronchial tubes in health is hardly audible, but if there be congestion of their mucous membrane, the tubes become thickened, as in the first stages of bronchitis, and a " sonorous rattle or rouchus" is heard, on applying the ear to the chest. If, however, from any cause the smaller bronchial tubes are constricted, as sometimes happens in bronchitis, or in spasmodic contraction of the tubes, we have a hissing or squeaking noise, which constitutes "the sibilating rattle or rouchus."

In *emphysema* of the lungs there is generally a short inspiration, "and a loud, protracted, and wheezing expiration." This I believe is true whether the emphysema be *interlobular*, consisting in infiltration of air into the areolar texture, or *vesicular*, depending merely upon dilatation of the air-cells.

In the early stages of *phthisis*, when the tubercles occupy the superior portion of the lungs, the inspiratory murmur assumes a harsh, hoarse character, and the intensity of the expiratory murmur is increased, and its duration prolonged. Later in *phthisis*, when quite a portion of the lung becomes filled with tubercles, and its smaller tubes consequently obstructed, there is the sound of air passing through the larger tubes of the part, constituting the "*tubular breathing*," or bronchial respiration. When, however, a bronchial tube becomes very much enlarged, or a cavity is formed in the substance of the lung, a sonorous or hissing noise takes the place of the bronchial respiration, and this constitutes the "*amphoric breathing*," resembling the sound produced by blowing into a large-mouthed bottle.

When air passes through a bronchial tube into a cavity, a sound is produced like that of blowing through a tube into a glass vase, the sides of which vibrate slightly, and this constitutes the *metallic* respiration. Sometimes there is a distinct tinkling, like the bursting of air through the secretion of the tube into the cavity, and this constitutes the "*metallic tinkling*."

In *scirrhus* of the lung, one marked case of which has fallen under my observation, all the respiratory sounds gradually change, and finally cease in the part, except so much as is conducted by the consolidated mass from other healthy or less diseased parts of the lung.

When a *tumor* growing from the posterior walls of the chest internally, presses against the lung in or about its middle, there is a lengthened inspiration, and two or three distinct interruptions in each expiration, and this I have named the *interrupted expiratory sound*.*

Such I believe are the chief, or most important dry sounds connected with respiration, liable to be detected by auscultation. But it must be remembered that they are frequently connected *with other and moist sounds* which often materially modify both the *dry* and *moist* sounds.

Besides the dry sounds produced by diseases of the air passages and lungs, certain dry sounds may be discovered by auscultation resulting from diseases of the pleura, the most important of which are the following. When in inflammation of the pleura, there is thrown out on its surface an albuminous layer, however, thin, the surface of the pleura becomes rough, and the friction produces vibrations and a friction sound, which has been called the "*pleuritic rubbing*." When, however, the pleuritic effusion becomes dry as it often does, the friction of the pulmonic and costal pleura may produce a harsh *grating sound* like the rubbing together of rough pieces of cloth or wood, or the sound may be of a creaking or squeaking character, similar to that produced by the bending of stiff leather.

When at other times the pulmonary and costal pleura become so rough that they will not readily glide over each other, as sometimes happens in the early stages of phthisis, from inflammation or from tuberculous deposits, a rustling noise is produced which has been compared to that

* I discovered this sound in 1854, for an account of which see Buffalo Medical Journal for March, 1855, vol. x. page 587.

produced by squeezing together soft gauze paper, and this has been called the "*pulmonary rustling*," but I think may more properly be termed the *pleuritic rustling*.

When *effusion* takes place into the cavity of the pleura, in quantity sufficient to compress the adjoining lung, the air-cells become partially closed up, and the air passing along the larger bronchial tubes produces the *tubular breathing*.

In cases in which there is a communication between one or more bronchial tubes and the cavity of the pleura, from any cause, more or less air passes into the serous sac, producing the "*fistulous breathing*," the "*amphoric respiration*," and the "*metallic resonance*," to which reference has been made in describing the dry lung sounds.

When the air in passing into the pleura sac, enters by a small opening, the natural secretion of the tubes by interrupting and bursting into the sac, produces in some cases the *metallic tinkling* before described.

All the *moist sounds* in diseases of the respiratory organs, arise from the passage of air through fluid or from the admixture of air with fluid. They may occur during inspiration or expiration, and may be situated or produced in the larger or smaller bronchial tubes, from the presence of mucus, blood, serum or pus with more or less air.

The *moist sounds* vary according to the size of the bubbles which produce them, or what amounts to the same thing, according to the caliber of the tube through which the air passes, in which the bubbles are formed. Thus the bubbles are larger in the trachea than in the bronchial tubes, and also in the larger than in the smaller bronchial tubes. The different extent of space then in which the bubbles are formed and burst, produces a corresponding difference in the result in resonance. The term *rhouchus* applies more properly I think to the dry sounds, while the term *rale* or *rattle* expresses more fully the moist, I shall therefore generally thus use them.

When the bubbles are large, or their resonance full, as they occur in the trachea, or in a cavity, the resonance is called the "*tracheal rattles* or *gurgling*." But when they are smaller, "as in the bronchial tubes of the first and second size," the sound is called the "*mucous* or *bronchial rales* or *rattles*." And when they are still smaller, as in the finer divisions of the bronchial tubes, the sound is termed the "*sub-crepitant rattles*." Finally, when the small bubbles are in the air-cells themselves, or in the capillary branches in immediate connection with them, the sound has been called the "*crepitant rales* or *rattles*."

Now the bubbles, the bursting of which produce the *tracheal rattles*, may be, as suggested by Dr. Hughs, the size of a bean, those producing the *mucous rattle* the size of a pea, those producing the *sub-crepitant* the size of a mustard seed, and finally those producing the *crepitant* the size of a poppy seed, and they are produced in the following manner. When either with or without contraction of the larynx or trachea, a fluid is present in the larynx or trachea, the air in the inspiration and expiration is obstructed by the fluid, and bubbles are formed, the bursting of which produces a *gurgling* noise, or the *tracheal rattle*. The presence of blood or mucus in the bronchial tubes of the first or second size, produces by inspiration and expiration, bubbles of a smaller size than these in the larynx or trachea, and hence the *mucous rattle*. When the smaller bronchial tubes are principally affected, the fluid and air produce bubbles less than those producing the *mucous rattle*, and larger than those which produce the crepitant, and hence the *muco-crepitant* or *sub-crepitant rattle* is produced. · And finally, in cases in which the air-cells are filled with serum as in *œdema* of the lung, with blood as in apoplexy of the lung, or with a thick viscid mucus, as in inflammation of the lung, a very fine sound is produced by the intermingling of the inspired air with the fluid so situated, and thus the *crepitant rattle* is produced.

When inflammation of the lung passes on to disorganization of their substance, the air in passing into the disorganized mass of fluid matter makes a sharp, shrill, loud, *muco-crepitant rattle*. But as the lung continues to break down, the bubbles become larger, till at last a cavity is formed, and *amphoric breathing* and *gurgling* take the place of other sounds. The *gurgling rattle* is also produced by enlargement of the bronchial tubes, or when any considerable cavities are formed in the lungs from any cause.

In the early stages of *phthisis*, the first moist sound is generally a fine *sub-crepitant rale*, caused by bubbles in the small bronchial tubes, the result of inflammation from the presence of tubercles. This sound may be limited to a small space at first, but as the disease progresses it may become quite general.

In the latter stages of *phthisis*, the *sub-crepitant*, the loud *mucous rale*, and the *gurgling* usually succeed each other, and are variously combined in the destructive progress of the disease. And it should be remembered that cavities may exist so high in the lungs, that to be detected the ear or stethoscope should be placed on the neck, above the clavicle.

When one or more bronchial tubes communicate with a cavity, or a collection of matter in the pleura, we have the *fistulous* or *amphoric respiration*, accompanied sometimes with the *metallic tinkling*, and frequently with the sonorous *mucous rale*, called *gurgling*. The *metallic tinkling* arises in cases in which the bronchial tubes communicate with the cavity of the pleura, in which there is a fluid and air, and is produced during inspiration.

Having listened attentively to the respiratory sounds, the sounds of the *voice* and *cough* should also be noticed, as they furnish important signs in diseases of the lungs and pleura. The changes produced in the ausculation of the voice resemble those that occur in breathing, that is, whatever renders the conducting powers of the thoracic organs greater, will increase the resonance of the voice, as it increases the respiratory sounds.

In disease of the *larynx* the voice becomes altered, and there is little difficulty in distinguishing the seat of the disease, as well as its nature, to some extent, by the sounds of the voice.

In ordinary *bronchitis*, the voice sounds are not materially changed, unless the lung becomes congested with blood, in which case its conducting power is increased, and the voice and cough become louder in the parts affected than in other parts of the chest, the base of the lungs generally exhibiting most change in such cases. When the bronchial tubes are enlarged, and the surrounding portion of the lung is indurated, the voice is increased in resonance, which increase amounts to *bronchophony* or *pectoriloquy*, or a sound nearly as distinct as if the lips of the speaker were placed at the ear of the listener.

In *simple emphysema*, the density of the lung is lessened, and consequently the resonance of the voice heard through the parietes of the chest is also lessened.

In *pneumonia*, the density of the lung is increased, the voice being more easily transmitted, and hence *bronchophony* is generally heard in such cases. But if the larger bronchial tubes, passing through the diseased part become closed, the sound of the voice may be more indistinct than in health. In *congestion* of the lungs they become more firm, and in consequence a better conductor of sound, hence the increased resonance of the voice in such cases, amounting sometimes to *bronchophony.*

In *scirrhus* and other malignant diseases of the lung, if the deposit does not close the bronchial tubes, there is increased resonance of the voice, depending upon the degree of induration, and the permeation of the bronchial tubes. But if the bronchial tubes become involved in the induration, the sounds of the voice may be lessened, or even lost over the part.

In the early stages of *phthisis*, there is usually a slight increase in the sounds of the voice, in the upper part of the chest, which is not very marked at first. But as the disease progresses, and cavities form, and the surrounding lung becomes indurated, the sound of the voice is materially increased, amounting to *bronchophony* or *pectoriloquy.* These observations in relation to the sounds of the *voice*, apply also to the sounds produced by the *cough*, only the sounds of the voice may sometimes be distinct, while the sounds of the cough are inaudible.

When a small amount of fluid is present in the pleura, the spongy tissue of the lung is compressed, and the effused fluid as well as compressed lung being good conductors of sound, the resonance of the voice is conveyed more distinctly than in a natural condition of the chest. But it more frequently happens in such cases that the voice becomes not only increased, but it assumes a peculiar "*bleating sound*," and from its resemblance to the bleating of the goat, has been called "*ægophony.*"

When a communication exists between the bronchial tubes and the cavity of the pleura, and this cavity contains air, both the voice and cough possess that same "*ringing metallic resonance*," which we have already noticed in the breathing of *pneumothorax*. And though it generally accompanies both the voice and cough, in some cases the cough possesses this *metallic* character, while the voice does not. Such then are the sounds liable to be detected by auscultation of the respiratory organs, and such also are the signs which they afford of the nature and extent of diseases of these organs.

Having thus examined our patient by *inspection, palpation, percussion*, and *auscultation*, and gathered from the signs which they afford *rational evidence* of the nature and extent of the disease within, there only remains to be gathered the additional evidence afforded, in some cases, by *mensuration* and *succussion*, and our physical exploration in the case will be completed.

I believe that in a large majority of cases *mensuration* need not be resorted to, and further, that it should never be relied upon as affording positive evidence of any particular morbid condition of internal parts. In some rare cases, however, it may be desirable to ascertain the capacity of the thorax, or to note the quantity of air respired in a given time, or taken in at a single inspiration, which may be done by breathing into a bell-glass, filled with water, and inverted over a basin containing water.' Or the amount of expired air may be measured by instruments properly adjusted; one of the best of which is that invented by Mr. Huchinson, and which is called the "spirometer.*"

Mensuration may also be of service in ascertaining the presence of fluid in the pleural sac, as well as of malignant or other morbid growths, distending or enlarging one side of the chest. This may readily be done by a *tape* passsed around the chest, noting carefully the difference in the measurement of the two sides, from the spine to the middle of the sternum.

Should it be desirable to ascertain the *expansibility* of the two sides of the chest, it may be conveniently done by a tape passed round the chest, the two ends of which meet at a mark in the centre of the sternum. The finger of an assistant may be placed on the tape at the point where it passes over the spine, and the ends of the tape being held loose forward, at the centre of the sternum, each inspiration will show the expansibility of the chest by the distance to which the ends of the tape recede from the line at the middle of the sternum. And by noting the difference in the retraction of the two ends of the tape, the comparative expansibility of the two sides may be ascertained, as suggested by Dr. Hughes. Or, if at hand, the "*stethometer*" of Dr. Quain, or the "*chest measurer*" of Dr. Sibson, may be convenient in the mensuration of the chest while in motion; the former of which indicates the change in the circumference, and the latter the change of diameter, in the respiratory movements.†

Finally, having proceeded in the examination of our patient by *inspection, palpation, percussion, auscultation*, and *mensuration*, if necessary, the additional method by *succussion* only remains, in certain cases, in which the presence of air and a fluid is suspected in the cavity of the pleura, or a very large cavity in the lungs.

The examination by *succussion* may be conveniently done, when required, by shaking the patient by the shoulders, not too violently, and then applying the ear to the chest, at the suspected point, when if there is a fluid, with air, in the pleura, or a very large cavity in the lung, a sound may be heard, similar to that produced by shaking a cask which contains air and a fluid. Thus then have we completed the several steps in the physical exploration of the chest, the principles of which may guide us in the diagnosis of diseases of the respiratory system.

SECTION II.—PLEURITIS—(*Pleurisy.*)

Having taken a glance at auscultation and percussion in the preceding section, I propose in the present to consider *pleuritis* or inflammation of

---

* For a description of this instrument, see note in Wood's Practice of Medicine, vol. i., p. 805.

† For a description of these instruments, see note in Wood's Practice, vol. i., p. 805; or Bennett's Clinical Lectures, pp. 32, 33.

the pleura. But as we are about to take up the diseases of the respiratory system in the present chapter, let us remember that the respiratory organs, embracing the larynx, trachea, bronchi, lungs and pleuræ; situated, as they are, mainly within the walls of the thorax, require to be thoroughly **studied and perfectly understood, in** order that the symptoms developed in their diseases may **be fully appreciated.**

It should be remembered that **the lungs** perform an important function, and that the various parts of the respiratory system must be in a good condition, to secure the perfect performance of this function. In fact, the respiratory system may be regarded as the lamp of life; the nasal fossæ, larynx, trachea, bronchi and air-cells being so **constituted** as to **admit** oxygen, and carry off carbonic acid, while the pulmonary arteries **and veins** conduct the venous blood to, and the arterial blood from the minute **air-cells, where the oxygen is received and the carbonic** acid given off.

Now this process, by which the **blood** parts with **its carbon by** way of combustion, supplying vital heat, requires the respiratory movement; and hence, in part, the necessity of the pleuræ to enable the lungs to glide easily against the walls of the thorax in this perpetual motion, which the wants of the system imperatively demand. This, then, brings us to the consideration of the pleuræ, which, it will be remembered, are two serous membranes which line each side of the chest and are reflected **upon each** lung. Each pleura is a shut sac, and from the junction of the two, **the three** mediastina are formed: **the** anterior, the middle, and posterior. The two pleuræ forming the mediastinum, however, are not in contact, there being a **space between them, containing all the** viscera of the chest except the lungs.

The anterior mediastinum contains the remains of the thymus gland; the middle, the heart, ascending aorta, superior vena cava, the bifurcation of the trachea, the phrenic nerve and the pulmonary arteries and veins, while the posterior contains the descending aorta, the azygos veins, the thoracic duct, the œsophagus and the pneumo-gastric and great splanchnic nerves.

The pleura covers the diaphragm and that portion of the membrane **which** lines the parietes of the chest is called the *costal pleura*, while that which covers the lung is called the *pulmonary*. The costal and **pulmonary portion of the** pleura, as well as that covering the diaphragm, **are exposed to** constant friction by the respiratory movements; which, however, in a healthy state of the membrane produces no uneasiness, or audible sound. The pleura, thus situated and constituted, is liable, from various imprudences, to become inflamed in any or all its parts, the **symptoms of which** we will now proceed to consider.

Symptoms.—The symptoms developed in inflammation of the **pleura are just what we should expect would arise from inflammation of a serous membrane thus situated.**

There may be a slight feeling of indisposition preceding the **attack,** but often, **the first** symptom noticed is a chill, followed by febrile reaction **and attended with** violent pain generally in one side of the chest. The **pain** is greatly increased by a full inspiration, or by coughing, in consequence of which, the respiration is short and the cough stifled as much as possible. The cough is of a short, hacking, dry character, and is increased as well as the pain, by lying on the affected side; in consequence of which, the patient is found, at first, lying on the well side. The pulse is hard, full, vigorous and frequent; the tongue is covered with a thick, white fur; the skin is hot and dry; and the urine is of a deep red color and small in quantity.

The breathing is quick, short, and difficult, and is performed mainly by the action of the diaphragm, and abdominal muscles; the motion of the chest being restrained, as much as possible, in consequence of the severe pain, which the respiratory movement produces.

Such are the symptoms which first present themselves. But if the chest be *inspected* carefully, its respiratory movements will be found very slight, especially of the affected side; while an examination of the abdomen, will detect the abdominal character of the respiration.

If now the flat hands be laid upon corresponding parts of the two sides of the chest, the diminished respiratory motion of the chest will be still more apparent, especially of the affected side, and a roughness, indicating either a dryness of the pleura, or else an effusion of lymph on its surface, may often thus be detected. If now the points of the fingers be pressed along the intercostal spaces, in the affected part, tenderness will be detected, as far as the inflammation extends, in most cases at least.

The resonance on percussion, in the early stage, is nearly natural; but if effusion has taken place into the pleura, a dullness is detected in the most dependent part of the chest, and this gradually extends, as the accumulation increases.

If now the ear be placed over the seat of the pain and tenderness, the stifled respiration will detect a diminution of the natural respiratory murmur: and if the inflammation is in its incipient stage, a slight friction sound may be heard, caused by a dryness of the pleura; or else by an albuminous exudation, rendering its surface slightly rough. If, however, the disease has progressed, and the exudation on the surface of the pleura has become more dry, a *grating sound*, like that produced by rubbing together rough pieces of cloth, or wood, may be detected; or it may be of a *creaking* or squeaking character, like that produced by bending stiff leather. If, however, effusion has taken place, to any considerable extent, into the pleura, the respiratory murmur becomes very indistinct; but the resonance of the voice, is greatly increased, and may assume the peculiar bleating sound, called *ægophony*.

Such, I believe, are the ordinary symptoms of simple acute pleurisy; but it is liable to variations, in some cases being attended with little or no pain; or it may be complicated with a bilious condition, in which case we have, in addition, the symptoms of ordinary bilious fever, with or without a typhoid condition.

Causes.—Sudden exposure to cold, when the body is in a state of perspiration, is probably the most frequent cause of pleurisy; and this is the reason why this disease occurs most frequently, at seasons when there are marked changes from heat to cold, and when there is a damp and low electrical state of the atmosphere.

Pleurisy also occurs frequently from the metastatis of other affections, as rheumatism, gout, erysipelas, &c. Suppression of the menses, from exposure to damp or cold, appears sometimes to produce acute pleurisy.

Pleuritis sometimes also follows capital surgical operations. But the disease, in such cases, is apt to come on in a very insidious manner, and wants many of the symptoms, which usually attend ordinary acute attacks of the disease.

Finally, the *paludal poison*, operating through the blood upon the brain and nervous system, may so far derange the functions of the body as to develop this disease, especially if some accidental exciting cause is brought to bear upon the system, such as sudden exposure to cold, dampness, &c. Cases of pleurisy depending upon this cause, are very apt to be attended with a typhoid condition of the system, the inflammation being generally, in such cases, of a passive character.

Anatomical Characters.—On dissection, the pleura is generally found red, or filled with small, irregular, red specks. And there is extravasation on the inner surface of the pleura in nearly all fatal cases of this disease, which accounts mainly for the friction sounds which attend this affection. The matter thus thrown out by the vessels of the inflamed pleura, may be coagulable lymph imperfectly organized, or a pseudomembranous substance may be found adhering to the pleura, and forming adhesions, in some cases, between the pulmonary and castal pleura.

In some cases, pus is found in the pleural sac, and very often more or less serum, sometimes filling the cavity of the pleura.

Other morbid appearances may be presented, but the preceding are, I believe, by far the most frequent.

Diagnosis.—Pneumonia and pleurodynia are the affections with which pleuritis is most liable to be confounded, from each of which it may be distinguished by attention to the following diagnostic symptoms.

In pleurisy, the patient lies at first on the well side, while in pneumonia, he generally lies on the affected side.

In pneumonia there may be a dark livid appearance of the countenance, and considerable cough, with a viscid, rusty expectoration, while in pleurisy the face may have a vivid flush, and the cough is short and dry, with only a limited glairy, colorless sputa. Finally, the want of dullness on percussion in the early stages of pleurisy, as well as the absence of the crepitant rale, both of which are present in pneumonia, serve to render the distinction between pleurisy and pneumonia clear and certain.

From pleurodynia, pleurisy may be distinguished by the pain being more continuous, by its being increased by lying on the affected side, and by the more decided febrile symptoms of the inflammatory affection. To distinguish *bilious* from *simple* pleurisy, it is only necessary to take into account the locality, the prevailing epidemic and the endemic influences, and to notice carefully the general bilious symptoms which attend such cases.

Prognosis.—Simple acute or chronic pleuritis is generally not a very dangerous affection, in persons of good constitution. If, however, the disease occurs in debilitated subjects, and from a debilitating miasmatic agent or influence, it may assume a typhus or malignant character.

Treatment.—When the patient attacked with pleurisy is of a strong constitution, and the inflammation is of an active character, general bleeding may be necessary at the commencement, and should not be omitted.

After general bleeding, when it is necessary, and in all cases in which it is not necessary, cups should be applied along the spine, on the affected side, and also over the inflamed part, and from four to six ounces of blood taken at first, and then it may be repeated in a few hours, if necessary.

The warm foot-bath should be used, and a cathartic of calomel or podophyllin should be administered in castor oil, and the oil repeated if necessary, every four or six hours, and with each dose, at least thirty drops of laudanum, or the same quantity of the fluid extract of opium should be given.

After the operation of the cathartic, one-sixth of a grain of tartar emetic may be given, with twenty drops of laudanum, or fluid extract of opium every four hours. And if the inflammation has not been subdued by the treatment thus far, a large blister should be applied to the affected part, and kept sore till the inflammation is subdued. In obstinate cases in which the inflammation does not yield to these measures, calomel should be administered in two grain doses, with four grains of Dover's powder every four hours, till an impression is produced on the disease, or the gums become sore when the calomel should be discontinued, and the Dover's continued if necessary every six hours. If at this period, there is evidence of effusion into the pleura, as the calomel is discontinued, five grains of the iodide of potassium may be given every six hours, alternating with the Dover's powder, and continued till the pleuritic effusion is removed.

During the early stages of pleurisy, crust coffee, with a little milk should be allowed. But if the disease passes on and becomes chronic, a plain, digestible, and moderately nourishing diet may be allowed. In most cases of simple pleurisy, if proper treatment be early applied, the disease may be arrested in forty-eight hours, and it is generally, I believe from neglect or mal-treatment, that the disease becomes chronic, and perplexing complications arise.

In *bilious pleurisy*, in which there is generally more or less gastric irritation, and a typhoid tendency, the general bleeding and antimony should be omitted, and a blister should be applied to the epigastrium. The other treatment may be nearly the same, only after the operation of the cathartic, two or three grains of the sulphate of quinine should be given with Dover's powder, and calomel, if necessary, and continued till the febrile and inflammatory symptoms subside. The patient too, in such cases may be allowed from the first, crust coffee, one-half milk, and later, broths, a poached egg, toast, and gradually other plain digestible and nourishing varieties of food may be allowed.

SECTION III.—PNEUMONIA—(*Pneumonitis.*)

By pneumonia or pneumonitis, I mean inflammation of the substance of the lungs, which consists, it will be remembered of the ramification of the bronchial tubes, the pulmonary arteries and veins, the bronchial arteries and veins, the lymphatics and nerves, and a fine connecting areolar tissue.

The lungs it will be remembered, fill exactly the two cavities of the thorax, being connected at their root superiorly, and separated below by the mediastinum and heart, the right being divided into three, and the left into two unequal lobes. But in order to appreciate the symptoms which are developed in inflammation of the lungs, it is necessary to understand, and bear in mind the minute anatomy of the pulmonary structure, as well as the general physiology of respiration.

The bronchial tubes which arise from the bifurcation of the trachea, on entering the lungs, branch forth very much like a tree, consisting of cartilaginous rings, circular muscular fibres, and longitudinal elastic tissue, and lined internally with a mucous membrane. Near the termination of the bronchial tubes after their calibre become less than $\frac{1}{8}$ of an inch, they become irregular, and are very properly called "*intercellular passages*," being surrounded by air-cells, which communicate freely with them.

These intercellular passages consist of elastic tissue, interspersed with muscular fibres, and after bifurcating several times, end in an air-cell, which air-cells are from $\frac{1}{70}$ to $\frac{1}{200}$ of an inch in diameter. And as each terminal bronchus has probably not less than twenty thousand air-cells clustered around it, it has been estimated that the whole number of air-cells in the human lungs is not less than six hundred millions.

Now the pulmonary artery which carries the impure venous blood from the right ventricle of the heart to the lungs, dividing, its branches terminate in a network of minute capillary vessels, in the parietes of the intercellular passage, and air-cells, and then these converge to form the pulmonary veins, by which the blood, purified in its passage through the capillaries, is returned to the left auricle of the heart.

Besides the bronchial tubes, the intercellular passages, the air-cells, pulmonary arteries, veins and capillaries, the lungs are supplied with arteries, branches of the thoracic aorta, which ramify on the parietes of the bronchial tubes, their blood being taken up and carried by the bronchial veins to the vena ozygus. The lungs have also lymphatics, which commence on the surface, and in their substance, and terminate in the bronchial glands.

The nerves of the lungs it will be remembered, are derived from the pneumogastric and sympathetic, forming two plexuses, one on the front of the root of the lungs, and the other on the posterior, the anterior being composed chiefly of filaments from the cardiac plexus, and the posterior principally of branches from the pneumogastric.

Thus are we reminded that the lungs are composed of an infinite number of bronchial tubes, intercellular passages and air-cells. And that the pulmonary artery terminating in capillary vessels, forms a network in the parietes of the intercellular passages and air-cells, and then converging from the pulmonary veins, by means of which pulmonary artery, capillaries and veins, all the blood passes through the lungs from the right ventricle to the left auricle of the heart, the blood parting with the carbonic acid, and receiving oxygen during its passage through the capillaries in the parietes of the intercellular passages and air-cells.

The lungs, thus situated and constipated are liable to become inflamed, the symptoms of which we will now proceed to consider.

Symptoms.—The symptoms attending inflammation of the substance of the lungs, are just what might be expected, if we take into account the peculiarity of their structure and functions.

There may be a slight feeling of indisposition for a day or two, but frequently a chill more or less marked, is the first symptom which is noticed. During the chill there is oppressed breathing in consequence of the congested state of the lungs, and often more or less cough. During the chill, or as febrile reaction takes place, there is generally an obtuse pain felt in the chest, either in the sternal or scapular region, or it may occupy any portion of the chest. The cough is dry at first, but soon a thick, viscid, lightish matter is thrown up; and later, a rusty appearing matter, from the presence of fine particles of blood, of a less viscid consistency is raised, but generally in moderate quantities at first.

As the general fever becomes established, the skin is hot and dry, the urine is scanty and high colored, the pulse frequent, full, obstructed and laboring but not very hard, and in the latter stages, the pulse may become weak, obstructed and irregular.

In violent cases in which the inflammation is extensive, the dyspnœa becomes very marked, the veins of the neck become turgid, and the countenance acquires a livid aspect from a want of proper decarbonization of the blood. If the pleura is not involved in the inflammation, the patient generally lies on the affected side, in order to allow the sound lung to carry on the respiration in a free and easy manner. But if both lungs are involved, the patient lies on his back with his shoulders elevated.

By inspection of the chest, it will be discovered that the affected side does not expand during inspiration like the opposite side.

This may be confirmed by palpation or laying the flat hands on corresponding parts of each side of the chest, and noticing the difference in the motion produced by respiration.

By percussion of the chest, a dull sound will be produced over the inflamed portion of the lung, in consequence of its increased density; and in order to fully appreciate this dullness, percussion should be made on corresponding parts of the well lung. If hepatization has taken place, the dullness is very marked over the hepatized portion, amounting to flatness.

Auscultation discovers a want of the natural respiratory murmur, and in its stead the crepitant rale, which arises from the passage air through the intercellular passages and air-cells, which contain more or less of the bloody mucus which is beginning to be expectorated. As hepatization takes place, the crepitant rale disappears over the hepatized part, and if the larger bronchial tubes remain open, the tubular breathing becomes very marked. If, however, inflammation of the lungs passes on to disorganization of their substance, the air in passing into this mass of fluid matter makes a loud muco-crepitant rattle. But as the lung continues to break down, and a cavity is formed, amphoric breathing and gurgling takes the place of other sounds.

The sounds of the voice and cough are slightly increased in the early stages of pneumonia; and as the density of the inflamed lung increases it becomes more marked, amounting to bronchophony, unless the larger bronchial tubes become closed, in which case the sounds of the voice and cough may be even more indistinct, than in a healthy state of the lungs. In cases of pneumonia which terminate favorably, after tubular breathing has become marked, there is during convalescence first a return to the crepitant rale, and finally to the natural respiratory murmur.

Anatomical Characters.—Among the morbid structural changes produced in the lungs by inflammation, *engorgement*, red and gray *hepatization*, *gangrene*, and *abscess*, are probably the most frequent; modifications of these conditions depending, I believe, generally upon the existence of two or more of these morbid conditions, in the same or different parts of the inflamed structure.

Engorgement, or a *congested* appearance of the inflamed portion of the lung, is sometimes presented; in which case it is apt to exhibit externally a brown color, which forms a strong contrast with the gray color of the healthy part. It is of a firmer structure, and heavier than in the healthy state, and feels less crepitous under the finger. By pressure, the air-cells are found filled with extravasated blood, and when the engorged portion of the lung is laid open, with the knife, a frothy, reddish, serous fluid runs out, and the structure exhibits a reddish appearance. Simple engorgement appears to be the result of the weakest grade of inflammation, and may arise from mere sanguineous congestion; for if portions of the engorged lung be pressed till the extravasated fluid is out, and air be blown into it, the part becomes elastic and crepitous, and of nearly a natural color.

Hepatization is a morbid condition, sometimes presented, in which the diseased lung has the appearance and consistence of liver. In this variety of structural change the lung is impermeable by air, is deprived of its crepitous feel, and sinks in water.

Red hepatization indicates a higher grade of inflammation than simple engorgement. When cut into, a little reddish fluid issues, without exhibiting a frothy appearance; the lung having lost its cellular structure and assumed a red granular appearance, being readily broken down between the fingers and reduced to a pulp. The hepatized lung generally appears larger than natural, because it does not collapse in this state.

Gray hepatization appears to be the result of a more intense inflammation than the red hepatization. In this morbid condition the pulmonary structure is granular, condensed, and impermeable by air, as in the preceding variety; but its color is of a grayish or yellowish tinge, and when cut into, a yellowish or grayish fluid, of a purulent character, is discharged. The pulmonary tissue is softened, and readily converted, by slight pressure, to a mere pulp.

Gangrene may occur from acute inflammation of the lungs, one marked case of which has fallen under my observation. In this case, as occurs generally in cases of gangrene of the pulmonary structure, there was an intolerable fetid breath, especially during fits of coughing, and the matter expectorated had an offensive smell. The lungs on examination, in such cases, are found converted into a putrid mass, containing fragments of pulmonary structure.

Abscesses sometimes form in the lungs, the result of acute inflammation; the matter which they contain being of a thin purulent character. This is not a very frequent termination of pulmonary inflammation, only one well marked case having fallen under my observation. This case was a robust young man; the abscess was opened, nearly a quart of matter was discharged at first, I believe, and the patient finally recovered, under the care of my friend Dr. Spencer, of Watertown, N. Y., with only a loss of function of that portion of the lung in which the abscess occurred.

Diagnosis.—The diseases with which pneumonia is liable to be confounded are œdema, certain conditions in phthisis, pleurisy, and bronchitis. Œdema and phthisis may generally be distinguished, however, by the history of the case, together with the symptoms which are developed, without any considerable difficulty.

From pleurisy pneumonia may be distinguished by the following differences. In pneumonia the patient lies on the affected side, while in pleurisy he generally lies on the well side. In pneumonia there is a dull pain, with a viscid rusty expectoration, while in pleurisy there is an acute pain, and little or no expectoration. The countenance in pneumonia may have a livid appearance, while in pleurisy no such symptom attends. In pneumonia percussion produces a dull sound, and there is absence of the natural respiratory murmur, and in its stead the crepitant rale, while in pleurisy there is early little or no dullness on percussion, and only a diminished respiratory murmur, with the friction sound, or pleuretic rubbing.

From bronchitis, pneumonia may be distinguished by the livid countenance, the pain, the character of the febrile excitement, the crepitant rale, and by the absence of the sibilant ronchus, mucous and subcrepitant rale, which so generally attend inflammation of the mucous membrane of the smaller bronchial tubes.

Prognosis.—The prognosis in simple uncomplicated pneumonia is generally favorable, and especially is it so if there is a tolerable free expectoration of the rusty sputa early in the disease, and later an increased flow of urine with gentle perspiration. When, on the other hand, the pain in the chest increases, and is attended with increased difficulty of breathing, and an aggravated cough, with dark liquid sputa, a livid countenance, and a soft irregular and laboring pulse, the danger may be regarded as considerable.

It is prudent to give a guarded prognosis in all cases, as some unforeseen forboding symptoms may arise, but I believe that most cases of simple pneumonia, occurring in persons not otherwise diseased, or too advanced in life, will generally recover if they are subjected to proper treatment, or even to that which is not positively bad.

Causes.—Atmospheric vicissitudes, or sudden exposure to cold and dampness, when the body is heated, is probably the most frequent cause of simple pneumonia. Inflammation of the lungs may, however, be produced by various causes, such as direct violence, over exertion, irritating inhalations, violent anger, intemperance in eating and drinking, the metastasis of erysipelas, gout or rheumatism, &c. And besides, pneumonia is liable to be a complication of other diseases, such as bronchitis, hooping-cough, measles, scarlatina, erysipelas, phthisis, and various organic affections of the heart.

Treatment.—Pneumonia is an inflammtion in which, from the nature of the parts inflamed, there is a good deal of congestion, and the congestion which is the immediate cause of the inflammation, is continued to a greater or less degree through the inflammatory stage. Now this congested condition of the inflamed lung may be greatly increased by any cause that serves to debilitate the patient, and on this account it is that bleeding, in this disease, should be resorted to with great caution.

If the patient be seen during the first few hours after an attack, I believe that general bleeding is seldom necessary or even beneficial. In such a case the feet should be placed in warm water, and if there is chilliness, the whole length of the back should be rubbed with a warm infusion of capsicum in vinegar, and the patient should be directed to drink warm sage tea, and should take a cathartic of calomel or podophyllin in castor oil, with twenty drops of laudanum, and the oil should be continued, with twenty drops of laudanum, every four hours, till a free operation is procured. Cups should at once be applied over the seat of the inflammation, and also along the side of the spine corresponding with the inflamed lung, and from three to six ounces of blood taken at first, and this may be repeated in a few hours, if necessary.

The warm foot-bath may be repeated morning and evening, and after the operation of the cathartic, one-eighth of a grain of tartar emetic, in solution, with twenty drops of laudanum, and four drops of the fluid extract of the veratrum viridi, may be given every four hours. The laudanum quiets pain, while the antimony and veratrum are sedative, diaphoretic, and expectorant, and together the combination tends to allay irritability, general fever, and to subdue the pulmonary inflammation.

The bowels should be kept gently loose by castor oil or small doses of the sulphate of magnesia, if necessary, and soon after the operation of the first cathartic a large blister should be applied over the seat of the inflammation, and kept discharging, if possible, during the continuance of the inflammation. In this way, I believe most cases of simple pneumonia may be arrested, or controlled, if attended to in season, without general bleeding, or a mercurial course.

But in some violent cases, occurring in strong, robust males, or in milder cases, if neglected for the first few hours, general bleeding may be necessary and should not be neglected; after which, the treatment may be conducted on the principles I have already suggested, with perhaps the addition of a mercurial course, if the inflammation is slow to yield.

In cases in which a mercurial is indicated, calomel may be given immediately after the operation of the cathartic, in two grain doses every four hours, combined with five grains of Dover's powder, alternating with the antimony and veratrum, without the laudanum; the calomel may be continued till the inflammation yields, or the gums become slightly sore, when it should be omitted and the other treatment continued till the inflammation subsides and convalescence is fully established.

Mucilage of gum arabic and crust coffee, one-half milk, may be allowed as nourishment and drink; and should a typhoid condition supervene from over-depletion, or any other cause, the sulphate of quinine, wine-whey and broth may be administered as they are required. If, in spite of all these measures the lung passes on to a hepatized state, a plain, digestible and moderately nourishing diet should be allowed, and should an abscess form in the lung and point in the intercostal space, or burst into the cavity of the pleura, paracentesis should be performed and the matter evacuated.

In such a state, alteratives and tonics are clearly indicated, and it is probable that in most cases, the iodide of potassium, with or without the syrup of the iodide of iron, in moderate doses, continued for a long time, will generally do best. The compound iodine ointment may also be applied to the affected side, and in certain favorable conditions in which there is firm adhesion between the pulmonary and costal pleura, a weak solution of iodine may be injected into the pulmonary abscess.

SECTION IV.—BILIOUS PNEUMONIA.

Having given in the preceding section the symptoms, anatomical characters, diagnosis, prognosis, causes, and treatment of simple pneumonia, I propose in the present section to give the general history, symptoms, diagnosis, cause and treatment of bilious pneumonia, without repeating what I have already stated in relation to the minute symptoms, anatomical characters, diagnosis and prognosis, which will generally apply in bilious as well as in simple pneumonia.

With these considerations let us proceed to inquire into the history, symptoms, diagnosis, cause and treatment of bilious pneumonia, in just so far as they are peculiar to this form of pulmonary inflammation.

History.—In malarious districts, or in localities where bilious fever is a prevailing disease, in the spring and fall, or during the winter months, while the air is cool or damp, and sudden changes frequent; patients suffering from poisonous paludal exhalations, are often attacked suddenly with pneumonia. And from careful observation, in very many cases, I am satisfied that the following is the process, by which it is brought about.

The miasmatic agent, whatever it may be, has been taken into the blood with the air, through the capillaries of the inter-cellular passages, and air-cells in the lungs; and also perhaps, by the skin, or by its becoming entangled with the saliva, and carried into the stomach. This process has gone on in some one, or all these ways, till the amount of poison in the blood, has produced a debilitating effect upon the brain, and nervous system, so that the brain is rendered incapable of generating sufficient vital force to keep up the functions of the body; and the nerves are rendered imperfect distributors of the force which is generated.

As a consequence, the functions of the liver become impaired; leading to loss of appetite, especially in the morning; to a bitter taste in the mouth; to a coated tongue, and very imperfect digestion; and finally, to great general debility. But this is not all; the whole muscular system, as well as the heart, becomes debilitated and relaxed; the circulation is impaired; respiration becomes languid; the blood is not properly decarbonized; and hence the dullness, drowsiness, headache, &c.

As a consequence of all this derangement, the circulation becomes still more sluggish; the blood fails by degrees, to be sent to the extremities; the heart and lungs labor to perform their functions, even imperfectly; the extremities become cold; and just at the time when a chill would have ushered in a bilious fever, had nothing supervened; a hereditary predisposition; a sudden change of temperature, or some other accidental cause irritates the lungs, and dooms them to an inflammation; beginning generally with the very chill, which would otherwise have been the commencement of a simple bilious remittent fever.

Symptoms.—The patient is found in a half sitting posture, slightly turned upon one side. His face has a livid appearance; his respiration is difficult; he complains of a dull heavy pain in the side, upon which he lies, and has a frequent cough, which may be half stifled in some cases.

By *inspection* of the chest, the affected side is found not to expand, at each inspiration, like the other side.

By *palpation,* or laying the flat hands on corresponding parts of the two sides of the chest, the observation is confirmed.

By *percussion* over the affected portion, a dull sound is elicited, and very unlike that produced by percussion upon corresponding parts of the opposite side.

Auscultation detects a want of the natural respiratory murmur, and in its place a crepitus, very much like that produced by rubbing a lock of hair between the thumb and finger, caused by the passage of air through the minute inter-cellular passages, and air-cells, which are partially filled with a viscid half bloody mucus, more or less of which is expectorated, very soon after the inception of the inflammation.

On further inquiry, it will generally be found that the patient has suffered for two or three weeks from a bitter taste in the mouth, loss of appetite in the morning, has had slight constipation of the bowels, or diarrhœa, with a yellow appearance of the urine; has had stupid or sleepy days, and restless nights, has had headache, with more or less coldness in the morning, and slight thirst at evening. &c.

Diagnosis.—To distinguish bilious from simple pneumonia, it is only necessary to take into account the locality, the epidemic and endemic influence, together with the general bilious symptoms which are developed in the case, to render the diagnosis clear.

Causes.—The paludal poison is the cause of this disease. And it should be remembered that the patient has generally been suffering for several days, or perhaps weeks, from this poisonous miasmatic agent, which has debilitated the brain and nervous system, impaired digestion, cut off nutrition, let down the circulation, and finally would have led to a simple bilious remittent fever, except for the accidental local inflammation set up in the lungs.

And it should be still further remembered, that this local inflammation of the lungs by hindering a due decarbonization of the blood, further interrupts the circulation, and increases the liability of a fatal prostration as I have seen in too many cases. And finally, with this obstruction to the circulation through the lungs, and filling up of the air-cells, as well as contraction of the intercellular passages, and smaller bronchial tubes, there is great danger of sudden and fatal congestion, especially if both lungs are involved.

Treatment.—We have then as a morbid condition in bilious pneumonia, a poisonous debilitating agent in the blood; with general debility and its consequences, and an inflammation of one or both lungs, with probably a remitting tendency, masked by the local inflammation. The pulmonary inflammation in such cases may occasionally be active, but generally I believe it is decidedly passive.

The indications then in the treatment of bilious pneumonia are plainly to equalize the circulation, arrest the inflammation, correct the bilious derangement, neutralize the morbific agent in the blood, sustain the sinking powers of the system, and finally to afford the patient a due supply of proper nourishment.

To arrest the progress of the inflammation, cups should be applied wet or dry, according to the constitution of the patient, over the seat of the inflammation, and also along the side of the spine corresponding with the affected side. The cups along the spine take off the undue pressure on the spinal cord at this point, and lessen the irritation of the ganglionic nerves, and as I believe, arrest the progress of the disease for the time at least.

The cups over the seat of the inflammation, relieve the difficult breathing and troublesome cough, and improve the circulation through the lungs, by increasing the capacity of the intercellular passages and air-cells just as might be expected.

To equalize the circulation, the warm foot-bath and friction along the spine, with a strong infusion of capsicum in vinegar, applied a little warm, is the most convenient and effectual. This will relieve the heavy pain in the lower portion of the spine, and also equalize the circulation, sending forth an agreeable glow of warmth to the surface of the body and extremities, and it often sends forth a gentle perspiration.

The immediate urgent symptoms may generally in this way be allayed, and a decided check given to the local inflammation. The next step is to correct the bilious derangement.

If the patient be a strong man, ten grains of calomel may be given in half an ounce of castor-oil, and the oil repeated if necessary. But if the patient is a female, or a male of a slender constitution, three blue pills may be given instead of the calomel, and followed in five or six hours by castor-oil. If, however, the patient be a young child, the mercury, with chalk, may be given in a teaspoonful of castor-oil, as it will generally be sufficiently active. The warm foot-bath may be repeated morning and evening for two or three days, as well as the friction along the spine with the warm pepper and vinegar. And after the first cupping, sinapisms may be applied to the chest, and also to the feet, generally with very good effect.

After the operation of the cathartic, there remains yet to be counter-acted the malarious morbific agent in the blood, to sustain the sinking powers of the system, to quiet irritability, and to counteract, if necessary, the local inflammation. To fulfill these indications, two or three grains of the sulphate of quinine may be given, with three grains of James's and five of Dover's powder every six hours, and continued till the fever is arrested.

The sulphate of quinine counteracts the malarious morbific agent in the blood, sustains the sinking powers of the system, keeps the circulation equalized, and materially aids in increasing the capacity of the intercellular passages and air-cells, in the decarbonization of the blood, and also in its circulation through the lungs, and thus aids materially in the resolution of the local inflammation, as well as in arresting the general fever. The Dover's and antimonial powders, combined with the quinine, serve to allay restlessness and nervous irritability, to materially aid expectoration, and to promote perspiration during the early stages of the disease.

After the fever is arrested, as it will generally be in three or four days, there may be a profuse perspiration, at which time the Dover's powder and antimonial may be discontinued, and one or two grains of camphor given with the quinine, if there is much restlessness, irritability or mental wandering.

In some cases in which the bilious derangement fails to be corrected by the cathartic, or if the local inflammation is advanced considerably before the treatment is commenced, one or two grains of calomel for adults, or of mercury with chalk for children, may be added to the first few powders, to produce an alterative effect, and then, if necessary, a teaspoonful or two of castor-oil may be administered, to clear the alimentary canal.

In cases in which there is great difficulty of breathing, with difficult expectoration, a teaspoonful of the fluid extract of the *asclepias tuberosa* may be given, with one-fourth of a grain of ipecac, every six hours, alternating with the quinine, for the purpose of promoting expectoration, &c.

In many cases, especially if not seen early, blisters become necessary, over the seat of the inflammation, and when they are, should generally be applied after the cupping and operation of the cathartic. And if, as is often the case, there is much gastro-intestinal irritation, a blister should be applied to the epigastrium.

Mucilage of gum arabic may be allowed, during the whole course of the disease, as it is grateful to the patient, favors expectoration, and is slightly nourishing. The patient may also be allowed to drink freely of crust coffee, one-third milk, for the first two or three days, and later one-half or two-thirds milk, to be taken warm, whenever drink is required. And as soon as there is an appetite for food, toast or other plain varieties of digestible food may be allowed, three times per day, at regular meal hours. After the patient gets a tolerable appetite, a grain or two of quinine only need be given, after each meal, till health is perfectly restored.

In bilious pleurisy, or *pleuro-pneumonia*, the pain is more acute than in cases in which the pleura is not involved, and the patient lies on the well side, but the indications are in all respects the same as in bilious pneumonia, uncomplicated with pleuritic inflammation.

SECTION V.—CATARRH.

By catarrh, from κατα, "downwards," and ρεω, "I flow," I mean here, a discharge of fluid from the Schneiderian membrane, which lines the nasal fossæ, and extends to the different cavities connected with the nose.

The Schneiderian or pituitary membrane appears to be formed of two layers, the one in contact with the bone being fibrous, the other mucous, the two being intimately united. This membrane, after lining the nasal fossæ, penetrates the sphenoidal, ethmoidal, frontal, and maxillary sinusses, in which it appears to assume a thin and more exclusive mucous character. The Schneiderian membrane is continuous through the nasal duct with the conjunctiva, along the enstachian tubes with the mastoid cells and tympanum, and through the posterior nares with the pharynx and mouth.

The pituitary membrane contains the expansion of the olfactory nerves, by which the impression of odors are received; and in a healthy state, it secretes sufficient mucus to lubricate its surface. This membrane however, from repeated colds, from a scrofulous, rheumatic, or syphilitic condition of the system, and from various other causes, is liable to a diseased condition, in which, with various other unpleasant symptoms, there is a copious and sometimes offensive secretion, which may flow from the nose, or pass by the posterior nares to the fauces.

Symptoms.—The symptoms then of this disease, which I have called catarrh, are as follow. In most cases, after a succession of colds, there is noticed an increased discharge from the nose, and especially into the throat, during the night. And there may be at first, or if not, there is very soon more or less pain experienced in the nose, forehead, cheeks and eyes.

These symptoms may continue on for weeks, months, or even years, with only perhaps a slight increase, and an occasional aggravation from colds, exposure to dampness and other accidental causes. More generally, however, if the disease is not arrested, the discharge gradually increases and perhaps becomes offensive, the sense of smell is gradually impaired, and in some cases finally lost, the pain in the forehead increases or becomes more constant and distressing, the eyes become weak, watery and painful, the sight more or less impaired, a constant hawking and spitting becomes necessary to clear the throat, a troublesome cough supervenes, and if the disease continues, partial deafness, dizziness, and various other unpleasant symptoms are liable to arise.

If this disease continues, as it often does, for a term of years, it may by extension produce ophthalmia, pharyngitis, gastritis, laryngitis, tracheitis, bronchitis, and indirectly it may lead to tubercular phthisis and various other affections.

Causes.—A scrofulous, rheumatic, or generally depraved condition of the system, strongly predispose to this affection, as well as a syphilitic taint. The exciting causes are various, such as repeated colds, damp apartments, and especially sleeping rooms; wet feet, insufficient clothing, dirty filthy habits, very hot rooms, wearing furs about the neck and face, taking too much drinks, snuffing tobacco and various other imprudences.

Nature.—Congestion, irritation, inflammation, or ulceration of the Schneiderian membrane, is probably the cause of the excessive morbid and often offensive discharge, which constitutes the essential feature of this disease.

In cases of catarrh occurring in good constitutions from accidental causes, the congestion, irritation or inflammation probably differs very little, if at all, from ordinary inflammation, affecting other portions of mucous membranes. But when the disease occurs in rheumatic, scrofulous or otherwise depraved constitutions, the disease of the Schneiderian membrane generally partakes of the nature of the constitutional derangement, and especially is this the case if there is a syphilitic taint.

The pain attending this disease, and usually referred to the nose, forehead or check, is no more than might be expected from the nature of the parts involved and the extent of the disease. And the same is true in relation to all the other symptoms which attend this disease,

Treatment.—To arrive at the indications of treatment in this disease, it is necessary to ascertain how far the affection may be the result of a general or constitutional derangement, such as a scrofulous, rheumatic or syphilitic condition, and how far it is local, depending upon accidental causes, such as repeated colds, &c. In the first place then, all the habits of the patient which would tend to perpetuate the disease, should be corrected, and the general derangement of the system, as well as the local disease, should receive a due share of attention in the treatment.

The patient should be directed to take proper food at regular hours, with a very moderate allowance of drinks; should be properly clad with flannel next the skin, in cool or damp weather, and should keep dry feet. Furs should not be worn about the neck or face, and the patient should avoid damp apartments, or very heated rooms, and the skin should be kept clean.

In cases depending upon a rheumatic, syphilitic or scrofulous condition of the system, five grain doses of the iodide of potassium may be given, three times per day, in four ounces of the compound decoction of sarsaparilla, and continued for several weeks, if necessary. After having continued the iodide of potassium for a reasonable time in scrofulous and syphilitic cases, ten drops of the syrup of the iodide of iron may be given three times per day, in the compound decoction or fluid extract of sarsaparilla, in connection, in scrofulous cases, with moderate doses of cod liver oil given an hour after each meal.

In this way the general derangement of the system may generally be materially improved, and if the patient takes but little drinks, the morbid discharge from the nose may be lessened or corrected, and the irritation, inflammation, or ulceration of the Schneiderian membrane, upon which it directly depends may be removed. In case, however, the morbid discharge from the nose continues after the general treatment, and in all cases in which general treatment is not indicated, local applications may be required to the Schneiderian membrane.

In simple recent cases in which the discharge is not offensive, or very copious, ten drops of sea-water, snuffed up the nose from the hollow of the hand, once each day after breakfast, for a time may cure the disease. If, however, the discharge is copious, but not offensive, a solution made by adding a drachm each of alum and loaf sugar to eight ounces of water, and used as suggested above, may be more effectual.

In all cases of catarrh, however, in which the nasal discharge is offensive, consisting of whitish or yellowish muco-purulent, or sanious bloody matter, as well as in all protracted or obstinate cases of this disease, a weak solution of corrosive sublimate, is the very best local application. Four grains of corrosive sublimate may be added to half an ounce of the compound spirit of lavender, and sufficient water added to make eight ounces of the solution. Of this ten drops may be snuffed up the nose once each day, as suggested above, and continued till a cure is effected.

SECTION VI.—LARYNGITIS.

By laryngitis, I mean here inflammation of the larynx of an acute or chronic character, involving generally the mucous membrane and submucous tissue, and leading in some cases to the formation of a pseudo-membrane in advanced life, as croup does in infancy and childhood, and in chronic cases to ulceration.

The larynx, it will be remembered, is situated at the superior and anterior part of the neck, at the top of the trochea, with which it communicates. It is composed of the thyroid, cricoid, and two arytenoid cartilages, is moved by a number of muscles, and lined by a mucous membrane which being reflected, constitutes the superior ligaments of the glottis. Beneath the mucous membrane is a submucous tissue, and the larynx is also supplied with arteries and nerves. It gives passage to the air in respiration, and it is in the glottis, which is a narrow oblong aperture in the larynx, that the voice is produced by the chordae vocales.

The epiglottis is the fibro-cartilaginous covering of the glottis, attached anteriorly to the thyroid cartilage, and covered on both surfaces by the mucous membrane of the larynx and pharynx. This in a healthy state covers accurately the glottis, at the moment of deglutition, and prevents the passage of alimentary substances into the air tubes. The larynx, thus situated and constituted, is liable to become inflamed, especially its mucous membrane and submucous tissue, the symptoms of which we will now proceed to consider.

Symptoms.—Laryngitis, in its acute form, usually commences with a slight chill, followed by febrile reaction, attended with soreness in the fauces, more or less uneasiness in swallowing, and tenderness in the larynx to external pressure. The voice soon becomes changed into a thick, hoarse whisper, and on inspiration the air appears impeded in its passage, in consequence of the narrowing of the glottis, or laryngeal passage, from congestion of its mucous membrane and submucous tissue.

By placing the ear over the larynx a whistling, or hoarse, dull, crowing sound is heard, by the obstruction offered to respiration in consequence of the laryngeal inflammation. On examination of the fauces, the soft parts are found red, swelled, or of an œdematous appearance.

As the inflammation passes on, a moderate secretion takes place, in and about the larynx, a part of which is expectorated with more or less saliva; constituting together a ropy, glairy fluid. Another portion of this mucus accumulates in the larynx, or passes down the upper part of the trachea, and produces, during inspiration and expiration, a gurgling noise, or the tracheal rattle.

The pulse is usually frequent, contracted and tense, but in some cases it remains nearly natural, at least during the first stages. The face is generally pale, and the tongue white, with numerous red points, and covered with a layer of nearly transparent mucus.

When the disease is fully developed, deglutition becomes very difficult and painful, producing often paroxysms of suffocative breathing. The temperature of the surface is apt to become uneven, being higher than natural in some parts, and lower in others. And any effort to cough results in a low, grunting noise in the throat.

The respiration becomes more oppressed and difficult, if the disease passes on unsubdued; the patient starts up suddenly in bed, the lips assume a livid or purplish color, the surface becomes cold, the pulse frequent and feeble, the countenance becomes ghastly, the skin becomes covered with a clammy sweat; and, finally, coma, delirium and death are the result. The immediate cause of death may be suffocation, from closure of the glottis, or laryngeal passage, or it may be from imperfect decarbonization of the blood.

In some instances laryngitis is very insidious in its approach, and rapid in its progress, terminating in death in a few hours. In such cases, there is probably very great congestion of the laryngeal mucous membrane, and also of the submucous areolar tissue, as well as some spasmodic contraction of the larynx, or its muscles and vocal cords.

The inflammation in laryngitis evidently commences in the laryngeal mucous membrane, and in mild cases, or if arrested early, may extend no further. But in violent or unsubdued cases, the submucous areolar tissue becomes either congested, inflamed or œdematous, in the latter stages of the disease, before a fatal termination.

In violent or protracted cases the epiglottis generally becomes very much inflamed, or œdematous, and may produce suffocation, by interrupting the passage of air to the lungs. In such cases, the epiglottis is sometimes found red, erect, thickened and very much swelled. The inflammation may, in rheumatic patients, extend to the perichondrium, in which case there is, in addition to the ordinary symptoms of laryngitis, a severe, dull, heavy pain in the part.

Laryngitis may terminate in suppuration, small abscesses forming about the larynx; or a pseudo-membrane may form on the tonsils, epiglottis, and even extend into the laryngeal cavity, in some rare cases of this disease; or the disease may assume a chronic character.

Anatomical Characters.—The laryngeal mucous membrane is found softened, thickened and gorged with blood, and the sub-mucous areolar tissue is distended with a bloody, serous, or sero-purulent fluid, especially in the upper part of the larynx and epiglottis.

The surface of the laryngeal mucous membrane is usually covered with mucus, and coagulable lymph is sometimes found in patches on its surface, or effused into the sub-mucous areolar tissue. And in œdematous cases, there is a copious effusion of serum, occupying mainly the sub-mucous tissue beneath the mucous membrane. And in chronic cases ulceration, &c., may be found, of the mucous and other laryngeal tissues.

Causes.—It is probable that exposure to cold and dampness when the body is heated, is by far the most frequent cause of laryngitis, as it occurs in adults. But the disease may be produced by a variety of causes, such as excessive use of the voice, direct injuries, extension of inflammation from surrounding parts, the abuse of mercury, metastasis of erysipelas, and, finally, it may arise as a complication during the continuance of scarlatina, small-pox, and various other affections. Or in its chronic form, it may arise from a syphilitic taint, or a scrofulous condition of the system, &c.

Treatment.—In mild cases of laryngitis, a full dose of the sulphate of magnesia, the warm foot-bath at evening, for two or three nights, a solution of alum in sage tea, sweetened with honey or loaf sugar, used as a gargle, and a dry flannel worn about the neck, especially during the night, may be sufficient to remove the disease.

In neglected or violent cases, however, active measures may be necessary from the very first, to prevent suffocation. The feet should be placed in warm water till free perspiration appears, the patient being allowed as a drink, warm sage tea. Cups, too, should be at once applied to the back of the neck, and three or four ounces of blood taken at first, and if necessary, several leeches should be applied to the larynx, and the blood thus lost will generally be sufficient, except in very robust men, in which general bleeding may occasionally be indicated.

A full dose of calomel or podophyllin should be given in half an ounce of castor-oil, and the oil repeated, if necessary, till free catharsis is produced. After the operation of a cathartic, nauseating doses of tartar emetic or ipecac should be given every four hours, and if the case is severe, or threatens to be obstinate, four grains of Dover's powder, with two grains of calomel, may be given every four hours, alternating with the antimony or ipecac, and continued till the disease is checked, or slight ptyalism is produced.

The warm foot-bath should be repeated morning and evening; a blister should be applied to the larynx, if necessary, after the leeching, and should the disease pass on, and suffocation be threatened, the larynx or trochea should be opened, and thus a passage be furnished for the air to and from the lungs, till the laryngeal inflammation is subdued, when it should be allowed to heal.

Should laryngitis pass on and become *chronic*, and in cases that are chronic from the first, the diet, habits, and general condition of the patient should be corrected, and local applications, both external and internal, should be resorted to as the case may require.

In most chronic cases there is a thickening of the laryngeal mucous membrane, as well as more or less œdema or congestion of the submucous areolar tissue, and it appears to depend, in many cases, upon a syphilitic taint, or a scrofulous condition of the system.

In cases of this character the iodide of potassium, in five grain doses, three times per day, before each meal, given in an ounce of simple syrup, is the best alterative, and it should be continued for several weeks, if necessary. In scrofulous cases, after having continued the iodide of potassium for a reasonable time, ten drops of the syrup of the iodide of iron may be given in its stead, and a tablespoonful of cod liver oil three times per day, an hour after each meal, and continued till the general condition is corrected.

Blisters should be applied over the larynx at first, and then iodine ointment may be continued, morning and evening, while the disease lasts. Internally in very mild cases, equal parts of alum and loaf sugar, pulverized very fine, may be carried between the thumb and finger into the fauces, and then by a short inspiration, it may be drawn into he larynx, and this may be repeated at evening, each day, as long as may be required.

In cases, however, in which this general and local treatment fails in effecting a cure, a solution of the iodide of potassium, or of the crystals of nitrate of silver should be applied directly to the mucous membrane of the larynx, by means of a curved whalebone, with a sponge firmly attached to the end of it, as recommended by Dr. Horace Green, of New York.

In œdematous cases, a solution of the iodide of potassium, of the strength of two drachms of the iodide, to an ounce of rain water, may be applied every other day, for two or three weeks, and then it may be used two or three times per week, till a cure is affected, the strength of the solution being gradually increased, if necessary. If, however, as the œdema is removed, the mucous membrane still remains inflamed or ulcerated, and in all severe cases in which there is little œdema, a solution of crystalized nitrate of silver, of the strength of a drachm to the ounce of rain water, should be applied as suggested above, and continued till a cure is effected.

SECTION VII.—TRACHEITIS—(Rattles.)

By tracheitis, I mean here inflammation of the mucous membrane of the trachea, involving sometimes other tracheal tissues, the disease not extending, however, to the larynx, as I believe it does in all genuine cases of croup.

The trachea, it will be remembered, is a "fibro-cartilaginous and membranous tube," partially flattened behind, situated before the vertebral column, and extending from the larynx to opposite the third dorsal vertebra, where it divides into the two *bronchia*, which go to each lung.

It conveys air to and from the lungs, during respiration, for which function it is admirably fitted by the peculiarities of its structure, being composed of fibro-cartilaginous rings, except the posterior third, which is made up by a fibrous membrane. The rings are connected with each other by a fibrous membrane, and transverse muscular fibres extend between the extremities of the cartilages posteriorly, and besides, there are posteriorly, longitudinal elastic fibres, which lie beneath the mucous membrane, and passing down, they enclose the entire cylinder of the bronchial tubes to their extremities.

The trachea is lined internally by a mucous membrane, and has a supply of glands which pour their secretion upon, and lubricate the mucous membrane. The tracheal arteries are derived from the superior and inferior thyroidal, and its nerves from the pneumogastric and the cervical ganglia. The trachea thus situated and constituted, is liable to become inflamed, especially its mucous membrane, the symptoms of which we will now proceed to consider.

Symptoms.—Tracheitis sometimes comes on suddenly, and hastens to a fatal termination in a few hours. More generally, however, its approach is gradual, there being at first, a dry hoarse cough, with slight difficulty of breathing, with or without slight chills, alternating with flashes of heat, &c.

If, during this early stage, the ear be placed over the trachea; a wheezing, whistling, or crowing sound is heard; loud or slight, according to the amount of the tracheal obstruction.

The respiration gradually becomes more difficult, as the disease advances, till it becomes very distressing; and the disease, if it progresses unchecked, often assumes an alarming degree of violence. The countenance is flushed; the eyes are injected and heavy; the pulse frequent, tense, and quick; the skin hot and dry, and the respiration extremely difficult and anxious.

The cough, which had been dry, now produces a rattling, or gurgling sound, in consequence of the copious secretion of a tenacious fluid into the trachea. And if the ear be placed over the trachea, a gurgling sound, or the *tracheal rattle* will be heard along its whole course; but more distinctly in the lower part of the trachea. If the disease continues, the breathing acquires at last a degree of oppression, extremely distressing; the patient frequently manifesting in the expression of countenance and actions, the greatest degree of anguish and suffering. The head is thrown back, and mouth kept open; the eyes are half closed; the lips livid; the face pale, and covered with perspiration; the extremities become cold; and finally the patient dies, apparently from suffocation, caused by the accumulation of the viscid secretion, in the lower part of the trachea, and larger bronchial tubes.

Anatomical Characters.—False membrane is generally found along the trachea, from near its superior portion, to its bifurcations, and the mucous membrane is reddened, and sometimes softened and thickened. The same appearance is also presented in some cases, in the larger bronchial tubes, if there has been an extension of the inflammation to the mucous membrane lining the larger bronchia.

Diagnosis.—Tracheitis may be distinguished from laryngitis, by careful attention to the following symptoms. In laryngitis, the voice is more changed, in the early stage of the disease, and in fact, all through, than in tracheitis. In laryngitis, the early peculiar whistling, or crowing sound, as well as the mucous rattle of the latter stage, are heard most distinctly in the larynx, and upper part of the trachea; while in tracheitis the dry or crowing sound is less distinct, and heard along the trachea, and the later tracheal rattle is heard most distinctly towards the lower part of the trachea.

From laryngo-tracheitis or croup, this disease differs in the following particulars: In croup, the early crowing sound, produced by respiration, as well as the mucous rattle of the latter stages, are heard most distinctly in the larynx, and upper part of the trachea; while in tracheitis, or rattles, the dry or crowing sound is heard only along the trachea; and the tracheal rale or rattle is heard only in the lower portion of the trachea. To this sound there may be added the mucous rale, with which it mingles, caused by the presence of mucus in the larger bronchial tubes.

Finally, the peculiar *gurgling* or *tracheal rattle* always so conspicuous in tracheitis, and which is heard distinctly without applying the ear to the part, is sufficient to distinguish this disease from croup, laryngitis, bronchitis, and in fact all other affections. In fact, it is from this peculiarity that the common name of *rattles* has been applied to this disease, and from the fatal tendency of the disease, the very name has become a terror in some localities, and in some families, greater, if possible, than that of *croup.*

Causes.—Tracheitis, or rattles, is confined almost exclusively to children, generally occurring before the fifth year of age, but it sometimes occurs in very advanced age. In some families there appears to be a congenital predisposition to this variety of disease; at least some children are more liable to attacks of this affection from slight causes than others.

The disease may arise as a complication in scarlatina, measles, and various other exanthematous fevers, but the most frequent cause is cold or atmospheric vicissitudes, combined with dampness, and a low or changeable electrical state of the atmosphere.

Treatment.—Warm pediluvia, warm sage-tea, and sinapisms over the trachea, and to the bottom of the feet, may be sufficient to arrest and remove mild cases of this disease. If, however, the disease passes on, and in all severe cases of tracheitis, an emetic of antimony or ipecac should be administered, and free vomiting produced; after which, a full dose of calomel or podophyllin should be administered in castor oil, and the oil repeated if necessary.

After the operation of the cathartic, nauseating doses of ipecac or antimony should be given every four hours, and alternating with this in severe cases, small doses of calomel may be given till an impression is produced on the inflammation; or slight ptyalism is produced. Leeches may be applied along the trachea at first, and later, mustard; and should the disease pass on and become chronic, a blister may be applied in this region, or to the back of the neck.

If, after the tracheal secretion becomes copious, symptoms of suffocation from its presence supervene, an emetic of ipecac or of the compound syrup of squills should be given, as young children can seldom be induced to raise anything by coughing.

Should tracheitis pass on and become chronic, the mucous membrane being thickened and covered more or less with a pseudo-membrane, and the various tissues of the trachea congested or œdematous, there is always difficult breathing and a more or less troublesome cough. In such cases, the diet, clothing, and all the habits of the child should be properly cared for, and full doses of the iodide of potassium should be given in simple syrup, three times per day, before each meal, and continued till the condition is corrected, so far as it may be. There is also in such cases, more or less spasm of the trachea; to remove which, and to allay the troublesome cough, small doses of the tincture or fluid extract of stramonium, with moderate doses of the compound syrup of squill may be given four times per day, for a time at least.

SECTION VIII.—LARYNGO-TRACHEITIS—(*Croup.*)

By laryngo-tracheitis or croup, I mean inflammation with congestion of the laryngo-tracheal mucous membane, attended, generally, with more or less spasmodic contraction of the larynx and trachea, all of which tend to produce a narrowing of the laryngo-tracheal passage, producing a marked change in the voice, cough, and respiration.

29

We have seen that the larynx and trachea are liable to become separately inflamed, the inflammation of one of which I have called laryngitis, and that of the other tracheitis; we shall now see that the larynx and trachea are liable to become inflamed at the same time, constituting, with more or less spasm, what I have here called laryngo-tracheitis or croup.

I shall restrict the term *croup* to those cases in which, with more or less spasm of the larynx and trachea, there is inflammation of the mucous membrane of the larynx and trachea, and shall treat of cases which are purely spasmodic, without being inflammatory, as a variety of asthma, in its proper place.

In order to appreciate the symptoms which are developed in croup, it is necessary to bear in mind the anatomy and physiology of the larynx and trachea, and also the character of the disease under consideration, remembering that all cases are inflammatory, and most cases more or less spasmodic.

Symptoms.—There are generally slight symptoms which precede an attack of croup, such as a slight cough, hoarseness, &c., which should serve as a warning to parents whose children are predisposed to this affection. These premonitory symptoms are usually aggravated towards evening, and during the night, and are sometimes attended with slight febrile symptoms.

In some cases, however, these premonitory symptoms do not appear, or are so obscure as to produce no alarm, till suddenly during the night the child, perhaps before waking, is heard to cough with a peculiar hoarse sound, as if it had coughed through a large tube. As the child wakes, or is roused, the voice produces the same hoarse sound which characterized the cough, and the respiration also produces a peculiar sonorous, shrill, crowing sound, which is heard in the larynx, and more or less along the trachea to its bifurcation.

This peculiar crowing, respiratory sound, hoarse voice, and cough, arises from inflammation, with congestion of the mucous membrane of the larynx and trachea, and also in part, in most cases, from more or less spasmodic contraction of the larynx and trachea, caused by the excessive irritation of the muscles and other tissues of the parts.

A slight redness of the fauces sometimes exists, and also tenderness of the larynx, and the child often complains of uneasiness in the throat, and says he is choking. Sometimes these symptoms subside in an hour or two, with very little more than warm drinks, and placing the feet in warm water. But more generally the disease passes on, even to a fatal termination, unless it be arrested by prompt and proper treatment.

If the disease passes on unchecked, the cough becomes husky, and almost imperceptible; the voice becomes a hoarse whisper, the respiration is wheezing, the countenance pale and lips livid, the eyes languid and pupils dilated, the tongue is loaded, and there is considerable thirst, but the skin becomes gradually cooler. Finally, in fatal cases, the eyes become sunken, the extremities cold, the respiration becomes frequent, interrupted and laborious, and after gasping for a longer or shorter time, the child dies.

Croup may terminate favorably or fatally in a few hours, or it may continue for three or four days, and in some rare cases, the little patient recovers after the most unpromising symptoms. In cases that recover after a protracted course, there is usually more or less matter expectorated, showing that the inflammation has terminated. In some instances this matter is of a purulent appearance, while in other cases it consists of thin flakes, of an adhesive character.

Diagnosis.—Laryngo-tracheitis may be distinguished from simple laryngitis, by the peculiar crowing sound being heard all along the trachea, while in laryngitis, the sound is only heard in the larynx. Besides this, laryngitis generally occurs in adults, while croup usually occurs during infancy or childhood.

Laryngo-tracheitis may be distinguished from simple tracheitis by the marked change in the voice and cough in croup, while if the inflammation is confined to the trachea, though there is the crowing respiratory sound along the trachea, the voice and cough are not so markedly changed.

Anatomical Characters.—The post-mortem generally reveals the presence of a pseudo-membrane of an opaque white or yellowish appearance, lining the larynx, trachea, and sometimes extending to the bronchial tubes.

In some cases this membrane forms nearly perfect tubes, but often it consists of mere patches along the mucous membrane. The mucous membrane is generally found reddened, and in some cases softened, and more or less thickened. The bronchial tubes may be reddened, and they generally contain more or less opaque white, green, or yellow puriform mucus.

Causes.—There is in some families a strong hereditary predisposition to croup, depending, in part at least, upon a weak or relaxed condition of the tissues, and especially of the mucous and cellular tissues of the larynx and trachea, in consequence of which they become readily congested, irritated and inflamed by a slight exciting cause.

The most frequent exciting cause is cold, or atmospheric vicissitudes, combined with moisture, and a variable *electrical* state of the atmosphere. And when we take into account the extent of mucous surface exposed to the air and moisture in the larynx and trachea, it does not appear strange that a sudden exposure to cold and dampness should, by checking the perspiration, and also the mucous exhalation in the larynx and trachea, produce congestion, irritation, and inflammation of the mucous membrane of these parts. Nor is it strange that when these parts become irritated, congested and inflamed, there should be some spasmodic contraction of the larynx and trachea, as there evidently is in some cases, if not in all.

Prognosis.—Frightful as this disease is, and fraught with danger as it is from the very first, by proper treatment in due season most cases should recover. At least such has been my experience in the treatment of this disease. If the case be neglected, however, or improperly treated, the disease may pass on to a fatal termination in a few hours.

Treatment.—If, while the premonitory symptoms which generally precede croup are being developed, the child take a few drops of hive-syrup, with an equal quantity of paregoric, every six hours, for a day or two, and the feet be placed in warm water at evening, for two or three nights, and the child be kept warm, especially through the night, the disease may generally be prevented, or arrested before it is fully developed. Hence all families that are predisposed to croup should be supplied with hive syrup, in order to be able to administer it as soon as the premonitory symptoms occur.

But if the disease be neglected, or is not arrested during the forming stage, active and persevering measures may be required. The child should have the feet placed in warm water; allowed a little sage-tea, and immediately vomited.

As an emetic in croup, the compound syrup of squill, I believe, is the very best. For a child a year old, ten drops of the hive syrup may be given every ten minutes, till free vomiting is produced; after which, it may be continued with an equal quantity of the tincture of lobelia, and two drops of the tincture of stramonium every three hours at first, and later, every six hours, till every symptom of the disease subsides. After the emetic, a full dose of calomel or podophyllin should be administered in castor oil, and a free evacuation of the bowels secured. Sinapisms may be applied over the larynx and trachea, if necessary; or if the danger of suffocation appears great, slight vesication may be produced over the larynx by Granvill's lotion. Or if this is not at hand, a blister may be applied, if the disease is obstinate.

In obstinate cases it may become necessary to repeat the vomiting with the hive syrup; and especially is this the case if the trachea and bronchia appear to be filling with a viscid secretion threatening suffocation. Finally, in cases in which suffocation appears inevitable and mainly from obstruction in the larynx, tracheotomy may be performed.

SECTION IX.—ACUTE BRONCHITIS.

By acute bronchitis I design to designate here all cases of inflammation of the bronchia, not of a protracted or lingering character.

Now, to understand this disease, it is necessary to remember that the two bronchia proceed from the bifurcation of the trachea to their corresponding lungs. And that on entering the lungs, they divide into two branches, and each of these divides and subdivides very much like the branches of a tree to their ultimate termination in the intercellular passages or air-cells.

The bronchia diminish in size till they are reduced to about $\frac{1}{40}$th of an inch in diameter, and reach within $\frac{1}{8}$th of an inch of the surface of the lung, at which size they terminate in elastic membrane, interspersed with muscular fibres, and assume the name of *intercellular passages.*

The bronchial tubes consist, it will be remembered, of cartilaginous rings, circular muscular fibres and longitudinal elastic tissue; and they are lined throughout with mucous membrane, continuous with that of the larynx and trachea.

The bronchia perform the important function of transmitting air to and from the intercellular passages and air-cells in every part of the lungs; and from their nature, structure and exposure are liable to become inflamed; especially their mucous membrane, but involving more or less their whole structure; the symptoms of which, in its acute form, we will now proceed to consider.

Symptoms.—Bronchitis generally commences with lassitude, chilliness, slight cough, and a sense of tightness, and oppression in the chest. The disease often appears **to be of** no very **serious** character, as there is little or no pain in the chest; **and but** slight febrile **reaction.** But as the disease advances, the oppression increases; the **countenance** becomes anxious, the **respiration more** laborious, and a peculiar wheezing respiratory sound, resulting from a narrowing of the bronchial tubes, is distinctly heard.

By placing the ear over the larger bronchial tubes, **the** sonorous, with perhaps the sibilating ronchus, is **distinctly** heard; **which, as we have** seen, results from the passage of air through the constricted, or narrowed bronchial tubes, of different sizes.

The cough at this stage, is dry and hoarse, and there is generally a slight hoarseness of the voice. But after the disease has progressed, there is a copious secretion of viscid transparent mucus into the bronchial tubes, which produces a marked change in the voice, cough, and respiratory sound. The cough becomes less hoarse, and is attended with a rattling noise in the larger bronchia, in consequence of the viscid **mucus** which they contain. The voice is also less hoarse; but more or less interrupted from the same cause.

By applying the ear to the chest, the mucous, or sub-crepitant rale or rattle is distinctly heard, according as the larger or smaller bronchial tubes are mainly involved. The mucous rale may be heard in one part of the chest, and the sub-crepitant in another, or there may be a mingling of the two sounds in the same part.

As the inflammation is terminating by resolution, the secretion becomes less viscid, and assumes a yellowish color, and the air passes with less resistance. This does not, however, very materially affect the respiratory sound, or that of the voice or cough, while the secretion lasts.

In most cases of acute bronchitis, severe pain is felt in the forehead; and if the secretion of the bronchia is very copious, more or less drowsiness usually attends. The tongue is white, and covered with a transparent mucus. The skin is dry; but not generally much elevated above its natural temperature; animal heat being cut off by the diminished combustion, and drowsiness being produced by the retained carbon.

Children are very liable to bronchitis, and it generally commences with a slight hoarse cough, which soon changes to a rattle, the respiration being correspondingly changed. The countenance becomes pale, the pulse frequent and tense, the hands and feet cool; but the temperature of the trunk may remain nearly or quite natural.

If the disease passes on, the respiration becomes exceedingly variable, and more oppressed; the lips become livid, and not unfrequently the cheeks, and the disease may terminate fatally, as early as the third or fourth day. In robust children, the febrile reaction may be quite marked in the early stage of the disease; but this is not invariably the case.

Such are the usual symptoms of acute bronchitis, as it occurs in adults or during infancy or childhood; liable of course to variation from constitutional and other causes.

Diagnosis.—Acute bronchitis **may** be distinguished from **all** other affections by the tightness in the chest, with little or no pain, the sonorous and perhaps sibilating ronchus in the early stages, and the mucous with the sub-crepitant rales in the latter; and, finally, by the headache, cold extremities, and moderate febrile reaction in this disease.

Anatomical Characters.—In some acute cases, in which the secretion has been copious, the lungs do not collapse on opening the thorax, all the minute tubes being filled with a frothy fluid.

The mucous membrane is minutely injected, more or less, throughout its whole extent. It is also thickened, generally slightly softened, and sometimes contains patches of ulceration, of greater or less extent, and the smaller bronchial tubes are found filled with mucus, or purulent matter.

Causes.—Acute bronchitis may arise as a complication in various diseases, such as asthma, hooping-cough, scarlatina, &c. Or it may be produced by irritating inhalations, and various accidental causes. But by far the most frequent cause of this disease is cold with dampness, and a low electrical state of the atmosphere. And hence it is that bronchitis generally occurs during seasons of the year when this state of the atmosphere prevails.

Prognosis.—The prognosis in this affection is attended with uncertainty, for cases that appear to be doing well may suddenly die from effusion into and sudden filling up of the small bronchial tubes and air-cells.

A copious and free expectoration of a not too viscid secretion is generally a favorable indication, especially if it gradually assumes a yellowish color; thus indicating a termination of the bronchial inflammation.

Treatment.—From the irritated, congested, and inflamed condition of the bronchial mucous membrane, cold extremities, and slight febrile reaction, the indications of treatment are readily deducible.

Warm pediluvia, to encourage the circulation to the extremities, and warm sage tea, to promote perspiration, should never be neglected in acute bronchitis. Sinapisms should be early and freely applied over the chest, as well as the warm infusion of capsicum in vinegar along the spine; the one to counteract internal congestion and overcome the incipient inflammation, and the other to equalize the circulation as far as may be.

An emetic of antimony or ipecac should be administered, if the patient be an adult, but if an infant, or young child, the hive-syrup is preferable. An emetic helps to equalize the circulation, promotes perspiration, and helps the secretion of the dry and inflamed mucous membrane, and should not therefore generally be omitted.

With these simple measures, if applied early, the inflammation may frequently be arrested, and the patient is convalescent. But if, as too often happens, the case is neglected, or improperly treated, at first, the symptoms may continue, or become more aggravated. In such cases cups should be applied to the chest, and along the spine between the shoulders.

A cathartic of calomel or podophyllin should be administered in castor oil, or if the patient be a child, the mercury with chalk may be given instead; after which the bowels may be kept gently loose by mild laxatives during the continuance of the disease. After the operation of a cathartic, a blister should be applied to the chest, and kept discharging as long as may be consistent with the general strength, or till the inflammation subsides.

Expectorants are also indicated; and for children ten drops each of hive-syrup and paregoric may be given every four hours, to a child a year old. For adults, one-fourth of a drachm of hive-syrup, with ten drops each of the tincture of digitalis and stramonium, may be given every four hours.

Towards the termination of the disease, as the bronchial secretion assumes a yellowish appearance, half a drachm each of hive-syrup and paregoric, with ten drops of the tincture of stramonium, may be given every six hours. Mucilage of gum-arabic should be allowed during the whole course of the disease, and the patient should be kept on mild, digestible, and moderately nourishing diet; and during convalescence may take a weak infusion of columbo after each meal.

SECTION X.—CHRONIC BRONCHITIS.

We have seen that the mucous membrane of the bronchial tubes are liable to acute inflammation, we will now examine *chronic* inflammation of the bronchial mucous membrane, constituting what I have called here *chronic bronchitis.*

Chronic bronchitis is a very common affection in damp variable climates, and during damp and variable seasons of the year, in most climates. Aged people are very liable to this affection, with whom it is apt to continue for several years, being aggravated during the fall, winter, and spring months. But chronic bronchitis is not necessarily confined to any age, locality, or season of the year. Only, other things being equal, it most frequently occurs in subjects who have been exposed to damp, cool, and changeable air, who have been poorly clad, or have neglected to keep the stomach, bowels, and skin in a healthy condition. In consequence of all this, the mucous membrane of the bronchial tubes is liable constantly to become congested, irritated, and inflamed.

Chronic bronchitis, too, is most liable to occur in persons of broken down or impaired constitutions. In fact, a large proportion of the cases that have fallen under my observation, during the past twenty years, which had been supposed by the patient or friends to be tubercular phthisis, were really neglected and protracted cases of chronic bronchitis. In some of these cases tubercles have been deposited, and the patients have died at last from the combined effects of bronchitis and tubercular phthisis. On the other hand, few cases of tubercular phthisis have fallen under my observation in which there was not more or less inflammation of the bronchial mucous membrane.

Symptoms.—Chronic bronchitis may be the result of the acute form of the disease, but when it is not, it commences with slight uneasiness in the chest, with oppressed respiration, accompanied with more or less wheezing. There is also slight uneasiness in the epigastrium, loss of appetite, a slightly furred tongue, irregularity of the bowels, irritated pulse, and generally red and scanty urine.

The skin is not much above the natural temperature, but is very dry. The cough usually occurs in fits of considerable violence, and is most severe on rising from the bed, or on passing into cool damp air, and it is also increased by the inhalation of smoke, dust, or even by swallowing. Slight transient pains are felt in the chest, in some cases; but frequently there is little uneasiness or pain, except after coughing.

There is only a slight expectoration, at first, of a glairy mucus, but later the matter expectorated becomes opaque, and contains small lumps of a viscid, grayish, translucent mucus, which sinks in water. Still later the matter often becomes of a yellowish color, and is sometimes streaked with blood. The pulse during the early stage is slightly accelerated towards evening, and partial sweats are apt to occur about the head and breast during the night. There is also considerable thirst, and the urine is high colored.

The cough becomes severe, especially in the morning, and is attended with a copious expectoration of a yellowish purulent matter. Debility and emaciation follow, and the difficulty of breathing becomes more and more distressing. The pulse becomes very frequent, and the face pale during the early part of the day, has a deep flush at evening. Profuse night sweats usually occur in this latter stage of the disease, and towards the termination of fatal cases, diarrhœa and œdema of the ankles are liable to supervene, as in the latter stages of phthisis.

At the very commencement of the disease, the sonorous or else the sibilating ronchus is heard on applying the ear to the chest, depending upon the size of the tubes involved; but later in the disease, after the secretion of mucus has taken place, the mucous or else the sub-crepitant rale, or both are heard on applying the ear to the chest.

In some instances there is gastric or hepatic derangement, either at first or very soon after the disease makes its appearance. In such cases there is a bitter taste in the mouth, slight tenderness in the epigastrium, loss of appetite, a yellow skin, and thirst, with great prostration of strength and mental despondency.

Diagnosis.—To distinguish chronic bronchitis from tubercular phthisis, strict attention should be paid to the following differences: In chronic bronchitis, the face is pale and the lips of a bluish color, while in tubercular phthisis the lips are apt to be reddish and the cheeks flushed.

In chronic bronchitis, the cough is attended with a free expectoration, almost from the commencement, while in tubercular phthisis there is generally a dry cough for a long time before there is much, if any, expectoration. Percussion over the upper part of the lungs elicits a dull sound in tubercular phthisis, while in chronic bronchitis the sound on percussion is but slightly changed, if at all.

In the early stages of phthisis, the ear detects a hoarse, harsh, inspiratory sound, and an increased intensity in the expiration, which is also lengthened, while in chronic bronchitis, if it be early, there is heard either the sonorous or sibilating ronchus; or else, if it be later, the mucous or sub-mucous rale is heard on applying the ear to the chest.

By noticing carefully all these differences, and taking into account the history of the case, the hereditary and acquired predisposition and all the extrinsic circumstances, the diagnosis between chronic bronchitis and tubercular phthisis may be rendered clear and quite certain.

Anatomical Characters.—On opening the chest, the lungs do not collapse; the capillaries of the mucous membrane of the bronchia are very much congested and enlarged, appearing, in some cases, as if the membrane was composed mainly of small blood vessels. The mucous membrane, besides being congested and thickened, may be either indurated or softened, or it may contain patches of ulceration, or the mucous membrane may be more or less thickly covered with small pimples or pustules in some rare cases of a protracted character.

The minute bronchial tubes besides exhibiting the changes peculiar to the larger, are very often more or less gorged with the pulmonary secretion.

Causes.—Chronic bronchitis is very often the result of the acute form of the disease. But it may occur from exposure to cold, damp air, or from repeated colds, or it may arise as a complication in measles, scarlatina, and various other affections. The disease may also be produced by the inhalation of irritating particles in the air, and also by the translation of rheumatism, erysipelas, and various cutaneous affections.

Treatment.—In cases of chronic bronchitis in which there is gastric or hepatic derangement, as a complication or cause, an occasional blue pill should be administered at evening, and pustulation produced over the stomach with tartar emetic ointment. But in cases in which there is no gastric or hepatic derangement, the indications are to counteract the bronchial inflammation, to regulate the functions of the skin and alimentary canal, to promote expectoration, and finally to restore the tone and general strength of the system.

Dry cups should be applied to the chest at first, and later, blisters may be indicated, or pustulation with tartar emetic ointment. Flannel should be worn next the skin and the patient should be directed to sleep in flannel sheets, and have a warm dry room and an even temperature. If the bowels are constipated, a pill of aloes and rhubarb may be given after dinner or at evening each day. The patient should take a plain, digestible and nourishing diet, and if the blood is weak and the countenance pale, two grains of the citrate of iron may be given in solution, three times per day, and continued for a time to restore the blood.

To favor expectoration early, one-eighth of a grain of tartar emetic, or one-fourth of a grain of ipecac may be given with a teaspoonful of syrup of tolu, four times per day; or later, equal parts of hive syrup, paregoric and syrup of tolu may be given instead, with ten drops of the tincture of digitalis and stramonium if necessary, and continued while the cough is troublesome.

In protracted cases in which there is great debility, ten drops of the balsam of Peru may be given, mixed up with a little brown sugar, and a teaspoonful of water, four times per day; or ten grains of the balsam of tolu may be given instead, made into an emulsion with loaf sugar, gum arabic and water; with either of these a little ipecac, tincture of stramonium or tincture of digitalis may be combined as they may be required.

In cases in which the bronchial secretion has become excessive, constituting a *bronchorrhœa*, from five to ten drops of the wood naphtha may be given four times per day, mixed with the syrup of tolu or with an emulsion of the balsam of tolu or Peru, or whatever expectorant may chance to be required.

If anodynes become necessary during the continuance of this disease, either the stramonium, conium, or hyoscyamus will generally do best; and in scrofulous cases, the conium I believe should generally be preferred, as combining an alterative with an anodyne effect. In cases in which there is excessive night sweats, two grains of tannin may be given at evening, and the patient should be sustained, especially during the stage of debility, by a good, nourishing diet, taken with regularity.

SECTION XI.—ASTHMA.

By asthma, I mean that chronic paroxysmal affection characterized by great difficulty of breathing, and depending upon a spasmodic constriction of the larynx, trachea, or bronchial tubes, and attended probably, in most cases, with a more or less congested state of their lining mucous membrane, which increases the dispnœa.

In order to appreciate the symptoms which are developed in this disease, it is necessary to bear in mind the anatomy of the larynx, trachea and bronchia; all of which contain circular muscular fibres, by which they may be constricted, and are also lined by a mucous membrane, which is liable to become congested.

These parts, besides being supplied with branches of the spinal and great sympathetic nerves, are also supplied by the pneumogastric, which nerve, arising from the medulla oblongata, passes along the neck and anastomoses with the spinal and sympathetic nerves, and supplies branches to the larynx, trachea and bronchia, as well as other surrounding parts; and then passing on to the abdomen, supplies branches to the stomach, and to the hepatic, cœliac, gastro-epiploic and solar plexuses.

Thus we see that the parts involved in asthma are intimately connected with, or under the influence of, the sympathetic and cerebro-spinal system of nerves; and through the pneumogastric are connected directly with the respiratory ganglion, the *medulla oblongata.*

It will be noticed also that the nervous relation between the respiratory passages and the heart, stomach, liver, and other important and vital parts, are very close; accounting for many of the otherwise unaccountable symptoms which attend this disease.

Symptoms.—In most cases various premonitory symptoms precede an attack of asthma; such as a sense of fullness in the epigastrium, acid eructations, headache and general nervous irritability.

The attack generally occurs during the night, the patient being seized with great anxiety, difficult breathing, a feeling of stricture across the breast, and a more or less distressing dry cough. These symptoms may pass on and acquire a great degree of violence, in which case the breathing becomes wheezing, laborious, and suffocative; the countenance is expressive of anxiety and distress, and the heart palpitates tumultuously. The desire for fresh air is urgent, the patient insists on the doors and windows being thrown open, and is generally entirely unable to remain in the recumbent posture. The extremities are cool, the face is bloated, and livid or pale; and the veins of the head, neck and face are apt to become turgid. The pulse is irregular, intermitting, accelerated, moderately full and compressible. In some cases, however, it is nearly natural.

After these symptoms have continued for a time, the breathing gradually becomes less difficult and anxious, and towards morning a more or less copious expectoration of viscid mucus generally takes place, and there follows some relief of the most distressing symptoms. During the day the patient is partially relieved. But as night approaches, the paroxysm of suffocative breathing returns; and in this way the disease proceeds, with remissions by day and exacerbations at night, for several days in succession, before it finally subsides for the time.

Such, I believe, are the ordinary symptoms of asthma, as it usually occurs, involving mainly the bronchial tubes, or lesser air passages of the lungs. Asthma sometimes affects, however, the larynx and trachea exclusively, or else in connection with the bronchial tubes, several marked cases of which have fallen under my observation.

Laryngeal asthma usually comes on suddenly, like the bronchial, but it is not generally attended with as much difficulty of breathing, and there is an obstinate, distressing, and almost constant cough, while the disease continues. In a marked case of this character that came under my observation a few years since, in a middle-aged lady, the cough usually came on at evening, and with it the difficulty of breathing, all apparently depending upon a spasmodic contraction of the larynx, together with congestion of its mucous membrane. The cough was usually nearly constant during the first night of the attack; and during the two or three succeeding nights, though very troublesome, it grew gradually less, till finally the paroxysm would cease after a few days.

Asthma sometimes assumes a chronic form, the disease continuing on for several weeks, with only slight remissions during the day, and exacerbations at night. The disease in this form is usually attended with a troublesome cough, and considerable mucus expectoration. There is also considerable difficulty of breathing, and on applying the ear to the chest, in such cases, and in fact in all cases of bronchial asthma, a wheezing, with "a loud sibilant or dry sonorous rale," may generally be heard. In laryngeal cases, however, the abnormal sound is heard in that region, and is of a whistling or crowing character, and loud or slight according to the degree of narrowing in the laryngeal passage.

Causes.—Some persons are doubtless hereditarily predisposed to asthma. But the disease may arise from a variety of accidental causes; among the most frequent of which are atmospheric viscissitudes, the inhalation of irritating particles of matter, the drying up of accustomed discharges, the metastatis of rheumatism or gout, masturbation, onanism, and excessive venery, plethora, mental emotions, certain kinds of food, organic diseases of the heart, spinal irritation, and other like causes. And when the disease is once produced by any one of these, or any other cause, its periodical return is probably owing, in part at least, to a morbid change, which the first attack or its cause has produced in the cerebro-spinal, as well as in the respiratory system. This morbid impression may be of the brain or spine, or both; but generally I suspect it involves the spine, medulla oblongata, or pneumogastric nerve, as well as the respiratory organs, either primarily or secondarily.

Pathology.—It is evident that in asthma there is spasmodic contraction of the larynx, trachea, or bronchial tubes, and it is probable that there is also some congestion of their lining mucous membrane. This appears to be the case not only in the laryngeal variety, but also in ordinary bronchial asthma, as well as cases in which the disease is confined to the bronchial tubes of one side.

Now, in relation to the immediate cause of the spasm in asthma, I suspect that the superior portion of the spinal cord, including the medulla oblongata, the cervical and pneumogastric nerves, are the parts immediately involved, in producing the asthmatic paroxysm.

We have seen that the pneumogastric nerve comes from the medulla oblongata, and passing out of the cranium is distributed to the larynx, trachea, and bronchia; and also to the stomach, liver, spleen, pancreas, and intestines. Now any local congestion or irritation in that branch of this nerve, supplying the larynx, might produce laryngeal asthma, while the same or a similar morbid condition of the bronchial branches might produce common bronchial asthma, and this may account also for the disease being confined in some cases to the bronchia of one lung.

But in cases in which the morbid change is in the medulla oblongata, at the origin of the pneumogastric nerve, the affection may involve the larynx, trachea and bronchial tubes. And there will be more or less temporary functional derangement of the stomach and other parts supplied by this nerve, as is the case at the commencement, and during the continuance of a paroxysm of many cases of this disease.

It is probable then that asthma depends upon an irritable condition of the cerebro-spinal system, either hereditary or acquired, in consequence of which the medulla oblongata, or cervical portion of the spinal marrow become the seat of local congestion. This congestion deranges the flow of nervous influence through the pneumogastric and cervical nerves, and very likely produces the spasmodic and other symptoms which arise in this affection.

Prognosis.—Asthma seldom terminates fatally. But if the disease is severe and protracted, it indicates a serious chronic derangement of the cerebro-spinal system, and it may eventually lead to pulmonary disease of a serious and dangerous character.

Treatment.—At the commencement of an attack of asthma the feet should be placed in warm water, and dry cups applied to the back of the neck, or along the cervical portion of the spine. This will sometimes arrest the approaching paroxysm, but not always.

If the attack be very severe, an emetic of ipecac or of the compound syrup of squill may be administered. And if the hive-syrup be used, a tablespoonful may be given every fifteen minutes till free vomiting is produced. After the vomiting when it is indicated, and in all cases in which vomiting is not indicated, the hive-syrup and tincture of lobelia, of each half a drachm, with twenty drops of the tincture of stramonium may be given every six hours, and continued during the continuance of the paroxysm.

If the patient is decidedly plethoric, and the congestion of the spine very marked, two or three ounces of blood may be taken by cups from the back of the neck, or along the cervical portion of the spine. And in case there is much spinal irritation, a stramonium plaster may be applied along the spine between the shoulders, during a paroxysm of either laryngeal or bronchial asthma.

If the bowels are confined during an attack of asthma, a cathartic of the sulphate of magnesia, rhubarb, leptandrin, or castor oil may be administered. And to prevent a return of the asthmatic paroxysm, or to break up the asthmatic tendency in the system, it is necessary to remove any cause that may have been operating to produce it, and to insist upon a strict observance of the laws of health in every respect.

Cups may be applied occasionally to the back of the neck, or cups or blisters may be applied along the cervical portion of the spine, and any general derangement of the system may be corrected by proper remedial measures as far as may be. If there is a debilitated condition of the system, and a weak state of the blood, the citrate or carbonate of iron may be given, and continued till a healthy condition of the blood is restored.

If there is a rheumatic or gouty condition, or derangement of the glandular and lymphatic system, the iodide of potassium may be given in five grain doses, three times per day, and continued for several weeks. If both an alterative and tonic are indicated, the syrup of the iodide of iron may be given in ten drop doses, three times per day; after discontinuing the iodide of potassium, or in cases in which that is not required.

By thus removing the cause, restoring the blood, correcting the general condition of the system, and subduing spinal irritation; the irritability of the cerebro-spinal system, may in some cases be so far overcome, as to arrest the congestive tendency, and thus a permanent cure may be effected.

SECTION XII.—HOOPING-COUGH—(*Pertussis.*)

By hooping-cough, or pertussis, I mean that peculiar contagious disease, attended with a spasmodic cough, which has its regular rise, progress, and declension; destroying in the system, the susceptibility to a second attack of the disease.

Symptoms.—Hooping-cough generally commences very much like an ordinary cold; the patient experiencing a degree of lassitude, headache, slight hoarseness, sneezing, and more or less oppression, or difficulty of breathing. The sleep becomes disturbed, the appetite is weak, the bowels are torpid, and the pulse indicates slight febrile excitement towards evening, in some cases.

For the first two or three weeks, the cough is dry and ringing, and the paroxysms are short, and there is not that peculiar sound, called *whooping*, which attends the disease as it progresses. About this time, however, the cough assumes a more spasmodic character; the paroxysms coming on more frequently, and are generally of longer duration.

The inspirations, during a fit of coughing are difficult, slow, and attended with a sense of spasmodic stricture of the glottis, rendering the paroxysms suffocative, and more or less convulsive. The approach of a paroxysm is preceded by a feeling of tightness in the breast, and titillation in the larynx, and after continuing for five or six minutes, the paroxysm terminates, with the expectoration of a viscid mucus. Sometimes the paroxysm of coughing terminates by vomiting, in which case the patient generally experiences some relief. There is also considerable congestion of the brain, in some cases, and occasional bleeding from the nose.

The disease usually continues on in this aggravated form four or five weeks, when it begins to abate, the declension being gradual, and occupying three or four weeks; making the duration of the disease, including the three stages, about three months. The disease is liable, however, to variation in this respect, the duration being in some cases longer, and in others, shorter than the period mentioned above. In some simple uncomplicated cases of hooping-cough, there is little or no fever. If, however, local inflammations arise, there is apt to be more or less irregular febrile excitement.

Hooping-cough is highly contagious; generally occurs during childhood, and it may prevail epidemically, being favored, no doubt, by certain epidemic or endemic influences.

Prognosis.—Simple, uncomplicated hooping-cough rarely terminates fatally. But there is always a liability of the supervention of hydrocephalus, bronchitis, pneumonia, laryngitis, tracheitis, apoplexy, &c., in consequence of which the disease may have a fatal termination. And besides, various diseases are liable to follow this affection, such as dropsy, ophthalmia, deafness, paralysis, and phthisis pulmonalis.

Cause.—This disease is produced by a peculiar contagion, which is generated in the system while laboring under the affection, the disease first having had its origin in some species of imprudence, as is the case with all contagious diseases to which the human family are now liable. This disease does not appear capable of communicating itself till it has continued two or three weeks, and then the contagion does not appear to extend far from the body of the affected person.

Besides, a *specific contagion*, which I believe is the general, if not the invariable cause of this disease, it is probable, as we have seen, that various epidemic and endemic influences favor the production and spread of this disease.

Anatomical Characters.—The bronchial tubes may be found dilated, and they are generally found filled with a "viscid muco-purulent fluid." There is also in many cases marks of a cerebral congestion, and various other morbid appearances, the result of complications which arise during the continuance of the disease, rather than marks of the disease itself.

Nature.—It appears probable that the contagious principle received into the blood, produces its impression upon the cerebro-spinal and nervous system, and that the primary seat of the local irritation thus set up, is the cervical portion of the spinal chord, or the medulla oblongated, or both, in consequence of which the pneumogastric nerve becomes involved, and also the phrenic, together with the cervical nerves with which they are more immediately connected.

This will account for all the symptoms essential to this disease, and also, as far as may be, for the local inflammations which arise, only a slight accidental cause being sufficient to develop them in parts in which there is already irritation with marked nervous derangement.

It is a little interesting to reflect that a contagious principle should produce its local impression in this manner, and after continuing for a few weeks, should gradually subside, and thus destroy the susceptibility of the system to the morbid impression which the poison was before capable of producing. But the part involved doubtless undergoes a change by which its structure is materially fortified against the effects of a like agent, even though it be introduced into the system in all its freshness at some subsequent period. Hooping-cough, however, in destroying the susceptibility of the system to a second attack, is following in the train of many other contagious diseases, thus furnishing a strong evidence of its genuine contagious character.

Treatment.—If this is the real nature of hooping-cough, it is not strange that it should be affected but slightly, if at all, by medical treatment, at least in its simple form, and such I believe is generally the case.

I would therefore do nothing for a simple uncomplicated case of hooping-cough, except to regulate the diet, and perhaps apply cups to the back of the neck, blisters along the cervical portion of the spine, and later, a stramonium or belladonna plaster between the shoulders, all with a view of lessening the irritation of that portion of the spinal marrow, and also of the pneumogastric, phrenic, and cervical nerves.

If, however, hooping-cough becomes complicated with bronchitis, pneumonia, or any other local inflammation, the patient should be treated as ordinarily for such affections, taking into account, of course, the general condition of the system. If, as sometimes happens with young children, there is some bronchial irritation, and the secretion threatens suffocation, an emetic of ipecac may be given for the purpose of throwing off the secretion.

In cases in which an emetic is not required, and the bronchial irritation requires an expectorant, five drops each of the hive-syrup and tincture of lobelia, with a drop of the tincture of stramonium may be given every six hours, to a child one year old. The patient with hooping-cough should take a plain digestible diet, with regularity, should be comfortably clad, and on no account should he be exposed to cool damp air.

SECTION XIII.—TUBERCULAR PHTHISIS—(*Consumption.*)

By tubercular phthisis, I mean that peculiar constitutional disease, either hereditary or acquired in which with a general depraved condition of the system, there is derangement of the lymphatic and glandular system, poverty of the blood, and irritation with tubercular deposit in the larynx or lungs, or both.

This disease is strongly hereditary, and is doubtless the result of the various imprudences practiced by the human family, the effects of which have accumulated, and rendered imperfect the various tissues and organs of the body, in consequence of which the various functions are illy performed, and especially digestion and assimilation.

Now in a person strongly predisposed, a very slight imprudence; or long continued imprudence in those not especially predisposed, may serve to bring about the scrofulous or tuberculous diathesis, or a depraved condition of the system, with impaired digestion and assimilation, derangement of the lymphatic and glandular system, and more or less poverty of the blood. When this scrofulous condition of the system exists, either from hereditary or accidental causes, or from both, the fat and albumen of the chyle in passing through the mesenteric glands are not converted into fibrin as perfectly as in health, and hence the integrity of the blood is more or less impaired, there being in most cases an accumulation of albumen.

Besides, in this scrofulous condition, the lymphatic system, including the lymphatic vessels and glands, are more or less deranged. And as the lymphatics take up in health the albuminous matter which transudes from the blood-vessels, as well as that liberated by changes going on in the tissues, and in passing it through the lymphatic glands, converts it into fibrin, this process is more or less interrupted, thus causing a further derangement in the blood, and especially an accumulation of albuminous matter.

Now this condition constitutes, as I believe, the *tubercular diathesis*, and it may continue for a long time without any tubercular deposit taking place, and by prudence, proper care and treatment, the condition may be overcome and tolerable health restored. But more frequently this scrofulous or tubercular diathesis continues till some tissue or organ becomes weakened, irritated or inflamed, and then it becomes the seat of tubercular deposits, and the blood disposes of its retained or accumulated albuminous and other matters.

The alimentary canal, kidneys, liver, and every tissue and organ of the body may be the seat of this tubercular deposit, depending very much on some slight accidental condition, such as congestion, irritation, or inflammation. Or the deposit may take place in a part from debility, relaxation, or some other accidental cause.

But if, as is often the case, the larynx or lungs, or both, are predisposed to tubercular deposit, only a slight exciting cause such as relaxation, congestion, irritation, or inflammation may lead on to the tubercular deposit in the larynx or lungs, constituting laryngeal or pulmonary consumption, or both, as is often the case.

If the tubercles are deposited in the larynx, they are either in, or immediately under the mucous membrane, in the submucous cellular tissue. If the lungs be the seat of the tubercular deposit, they are generally first deposited between the arteries, veins, air-cells and bronchial tubes, in the upper part of the lungs, and filling up the cellular tissue which connects these parts, in the form of *miliary tubercles, gray tuberculous infiltration, gelatinous infiltration,* or *yellow tuberculous infiltration,* they produce pressure on the bronchial tubes, and thus in the early stage of the disease, cause a dry cough, and many other symptoms which arise.

As we have now seen in what the tubercular diathesis consists, and the manner in which the deposit takes place, we will now proceed to examine the symptoms which arise in laryngial and pulmonary phthisis, and first of the laryngeal variety.

Symptoms.—Laryngeal consumption generally occurs in connection with pulmonary, but it may exist independent of pulmonary disease. It is most liable to occur in persons who have been exposed to particles of dust, which has served to irritate the laryngeal mucous membrane, and has thus led on to this form of tubercular disease.

There is generally a degree of fullness experienced in the fauces and larynx, with slight soreness, early, in some cases, but with little **or none,** at first, in others. A slight cough too frequently occurs, long before the tubercular deposit commences, in consequence of the irritation. But as the tubercles become formed, in and under the laryngeal mucous membrane, they produce mechanical obstruction to **the** respiration, and also a tickling sensation, which produces a dry, **hoarse** cough.

This cough is increased by exposure to **dust, or to cold air, and** remains dry till the tubercles soften, **or** a secretion occurs in the larynx.

Hoarseness of the voice is one of the characteristic **signs** of laryngeal phthisis, beginning at the very commencement of the **disease, and** continuing generally to its termination. **In some cases there is an entire** loss of voice, in consequence of the narrowing of the laryngeal passage, and change in the vocal chords.

If the ear be placed over the larynx, during the early stage, **a wheez**ing, crowing, or whistling sound is distinctly heard. Later, after **soften**ing of the tubercles has commenced, or a **mucous** secretion **has taken** place, the air in passing through the tubercular or mucous matter produces a *gurgling* noise, **or** the *tracheal rattle,* **which** is readily heard **on** applying the ear to the larynx.

There is little or no febrile action in this disease, except that which **comes on in** the latter stages, in consequence of prostration of the general powers of the system, and this is of a peculiar hectic character. The **appet**ite may be tolerable, **but** digestion and assimilation is defective, and the system sinks rapidly under the constitutional and local disease.

Such are the ordinary symptoms of laryngeal consumption, when it exists independent of pulmonary disease. But it often exists in connection with pulmonary phthisis, the symptoms of which we will now proceed to consider.

In *pulmonary,* **as** in laryngeal phthisis, the general powers of the system, **as we have seen,** are always weakened; the lymphatic and glandular system deranged; and the blood is in a depraved or unhealthy state. The countenance is therefore at the commencement of **this disease** generally pale, with a slight flush towards evening; the nerves are irritable, and the pulse frequent; and there may be a **slight** cough, even before the tubercles are deposited, but not necessarily, unless the disease follows **chronic** bronchitis. As the tubercular deposit commences in the cellular tissue in the upper part of the lungs, of whatever form they may **chance to** be, they press upon the smaller bronchial tubes in the vicinity, and by provoking a mechanical irritation, produce a dry cough, **which** gradually increases as the tubercles accumulate.

At this stage, percussion elicits a dull sound in the upper part of **the** chest, on one or both **sides, and if the ear be** applied to the part, the **sonorous ronchus is sometimes** heard in consequence of the narrowing of **the bronchial tubes, from** mechanical pressure. And there is also a slight interruption or jerking, in the inspiration, and a lengthening or prolongation of the expiratory sound, caused by the interruption to the free passage of air through the compressed bronchial tubes.

After a longer or shorter time, the tubercles will have so far accumulated, **as** to form a serious impediment to respiration, at least in one of the lungs. And the pressure of these tubercles, together with the irri-

30

tation of the parts set up by the coughing, produces local irritation, and sometimes slight inflammation in the parenchyma of the lungs, and also in the adjacent bronchial tubes.

Now this irritation, or inflammation of the pulmonary tissue, is attended or followed by slight œdema of the parts involved, and consequent softening of the tubercular deposit. And the matter which results from the softening of the tubercles, becomes mixed with a little serous and purulent matter, which accumulates in consequence of the surrounding inflammation, and thus a collection of matter is formed in the lungs, between the bronchial tubes, with no way of escape, at first, into these tubes.

Generally about this time, there is a slight effusion into the bronchial tubes, in consequence of bronchial irritation or inflammation; so that though there still remains dullness on percussion over the part, as at the first; a slight muco-crepitant, or sub-crepitant rale is produced by respiration, and may be distinctly heard, on applying the ear to the part. In cases, however, in which the smaller bronchial tubes become greatly obstructed, there is the sound of air passing through the larger bronchial tubes, constituting the tubular breathing, or bronchial respiration.

The sounds of the heart, are generally more audible in the early stages of phthisis, than in health, and there is also an increase in the vocal resonance, as well as in that of the cough, as might be expected from the condition of the parts.

Sooner or later, the collection of albuminous, serous, and purulent matter, resulting from the softening of the tubercles, and the surrounding inflammation, produces an ulceration of the adjoining bronchial tube, or tubes; and bursts into their cavity, and thus an albuminous sero-purulent matter begins to be expectorated. Sometimes more or less blood is mixed with this matter, in consequence of rupture of one or more small vessels, in the immediate vicinity of the cavity. If the ruptured vessel be large, the hemorrhage is sometimes very copious, and even dangerous.

As the matter becomes expectorated the affected side of the chest may become sunken, and *inspection* detects also a diminished respiratory movement. The sound on percussion may remain dull for a time, but as the cavity acquires considerable capacity, the sound elicited by percussion may be resonant, or even tympanitic.

Auscultation at this period detects the cavernous rale, or a gurgling, as well as a cavernous resonance of the voice and cough, which are produced by the passage of air into the cavity, which contains more or less fluid, as the secretion or formation of matter is constantly going on. If the cavity becomes large there may be the amphoric breathing, the voice may acquire the amphoric resonance, and the bursting of bubbles into the cavity may produce the metallic tinkling, all of which may be detected by applying the ear over the affected part.

Should it become desirable to note more particularly than can be done by inspection, the degree of diminution in the capacity of the affected lung, it may generally be very conveniently done by means of a tape passed round the chest, or, if at hand, the *spirometer* of Mr. Huchinson, or other instruments which have been invented for the purpose, may be a convenience.

Finally, in cases in which there is a large cavity in the lung, some additional idea of its capacity may often be obtained by *succussion*, which, if skillfully performed, produces, as we have seen, a sound similar to that caused by shaking a cask which contains air and a fluid.

Such are the ordinary symptoms developed during the progress of pulmonary phthisis. But as the cough and expectoration continues the general strength wastes away, and great emaciation takes place. The pulse becomes weak and very frequent, the countenance acquires a hectic flush, night sweats follow, the feet and limbs become œdematous, and finally, if not taken off by hemorrhage, a diarrhœa supervenes, and the patient dies ultimately from exhaustion.

Such, then, is the general course and termination of *tubercular phthisis*, the greatest scourge of the human family. And as it is the result of the combined or accumulated imprudence of mankind, it is too often the stern defier of our best directed efforts at relief, and sweeps off not only those who have disregarded the laws of health, but also those upon whom "the iniquities of the fathers have been visited," in accordance with the direct declaration of the Almighty.

Anatomical Characters.—The post-mortem reveals either the *miliary tubercles, tuberculous infiltration, gelatinous infiltration*, or the *yellow tuberculous infiltration*, as well as signs of inflammation, and cavities of greater or less capacity, occupying generally the upper part or summit of the lung.

Generally the pulmonary and costal pleura are found adhering, and in addition to tubercles, cavities, and signs of inflammation of the affected lung or lungs, the bronchial tubes are found dilated, and the bronchial glands are often more or less enlarged and indurated. Tubercles are also found in the pleura, peritoneum, intestines, mesenteric glands, liver, spleen, testicles, brain, and lymphatic glands; and they may be found in any tissue of the body, even in the bones.

The mesenteric glands are often enlarged, and show signs of having been inflamed. And the stomach and intestines are found to have undergone more or less organic change, generally presenting more or less extensive ulceration along the mucous membrane.

Diagnosis.—To distinguish tubercular phthisis from chronic bronchitis it is only necessary to take into account the general features and history of the case, and to note carefully the following differences.

In tubercular phthisis, though there be paleness, the cheeks are flushed at times, and the lips are apt to be reddish, while in chronic bronchitis the face is pale, and the lips are apt to be of a bluish color.

In tubercular phthisis there is generally a dry cough for a long time, before there is much, if any, expectoration, while in chronic bronchitis the cough is attended with a free expectoration, almost from the first. Percussion elicits a dull sound in the upper part of the chest, in tubercular phthisis, while in chronic bronchitis the sound on percussion is but slightly changed, if at all. In the early stages of phthisis the ear detects a hoarse wavy inspiration, and an increased and prolonged expiration, while in chronic bronchitis, if it be early, there is heard either the sonorous or sibilating ronchus, or else, if it be later, the mucous or submucous rale is heard, on applying the ear to the chest.

By noting carefully all these differences, and taking into account the history of the case, the predisposition, hereditary or acquired, and weighing carefully all the extrinsic circumstances, the diagnosis between tubercular phthisis and chronic bronchitis may be rendered clear and certain.

Tubercular phthisis need not be confounded with pneumonia, or any other pulmonary affection, if all the symptoms be carefully taken into the account. And the presence of tubercles in the larynx, may generally be clearly inferred by the peculiar symptoms, already laid down, which they develop.

Causes.—The predisposition to tubercular phthisis may be inherited, or it may be acquired by various abuses of the system, such as habits of filth, improper or unwholesome food, irregularity of taking food, insufficient clothing, masturbation, and excessive venery, the use of tobacco, gluttony, drunkenness, licentiousness, and other like abuses.

A strong hereditary or acquired predisposition to tubercular phthisis may exist, and yet by a careful return to a strict observance of the laws of health, and avoiding the exciting causes, the local tubercular disease may never be developed. More frequently, however, some accidental exciting cause is brought to bear, and tubercles are deposited, either in the larynx or lungs, or both.

Among the exciting causes of this disease, gastric and hepatic derangement, with pulmonary or bronchial irritation, are probably the most frequent. But there are various exciting causes of tubercular phthisis, such as cool damp apartments, the depressing passions, the inhalation of irritating vapors, or fine particles, into the larynx and lungs, losses of blood, the healing up of old ulcers, suppression of the menses, the retrocession of cutaneous affections, the use of tobacco, self-pollution, venereal excesses, licentiousness, and various kindred abuses of the system.

Various diseases also tend to develop tubercular phthisis, in those who are predisposed, such as measles, scarlatina, hooping-cough, small-pox, and various inflammatory affections. Females are more subject to the disease than males. And the disease occurs most frequently, in both sexes, between the ages of twenty and thirty years, at least according to my observation.

Prognosis.—The prognosis in tubercular phthisis, whether laryngeal or pulmonary, is decidedly unfavorable. The only hope in laryngeal cases must be confined to the early stage, for if the tubercles begin to soften, or ulceration takes place in the larynx, little hope need be entertained of a favorable termination of the case.

Tubercles may be deposited in the lungs, however, and yet, if they are not sufficiently numerous to produce mechanical irritation, no softening need necessarily take place if the system be brought into a healthy condition by a rigid observance of the laws of health and proper remedial measures. So, too, it is probable that a small collection of tubercles may soften and be thrown up and the cavity cicatrize, contract down and heal, if the blood and general condition of the system can be restored to a healthy state. Finally, if a small collection of tubercles have softened and been expectorated, and a small cavity formed, there is a bare possibility, if the general condition can be improved, that the cavity may contract, and if it does not heal, that the membrane which lines it may so far assume the character of a mucous membrane as to cease to be a secreting surface, and thus a tolerable cure be effected.

With a bare possibility of either of these results, no time should be lost in correcting the habits of the patient, and in resorting to the use of such measures as are indicated, to effect a cure, if that is possible, and to allay and palliate, as far as may be, if a cure is impossible, as is generally the case.

Treatment.—The general treatment proper in tubercular phthisis is such as will tend to restore the blood, improve digestion and assimilation, and correct the general depraved condition of the system. The patient should be directed to take a reasonable amount of exercise in the open air, either by walking, riding on horseback, or otherwise, as the peculiarity of the case may appear to require, and the condition tolerate.

The patient should be warmly clad; should always wear flannel next the skin, and should be directed to sleep in flannel sheets; great care should also be taken to keep the feet warm and dry and to avoid damp cool air.

A large sleeping room in the upper story should be preferred to sleeping in the lower part of the house, and if it can be immediately under the roof, or near it, the better still, as the heat of the sun will tend to dry the air, especially if the room be on the south side of the house, which is always desirable.

The patient should be directed to wash the surface of the body at least once each week in summer, and once in two weeks in winter, with tepid water containing a little salt, to keep the skin clean and in a healthy state.

The diet should be regulated to suit the condition of each particular case. It should generally be of a digestible and nourishing character, or as much so as the stomach will bear and digest well. Milk, in most cases, with good wheat bread and a little meat will do very well and is generally very acceptable to the patient. The food should be taken with strict regularity and on no account should any be taken between meals, as it always tends to impair digestion.

Having regulated the exercise, clothing, diet and habits of the patient, some preparation of iron should in most cases be administered. In decidedly scrofulous cases, with considerable prostration, the syrup of the iodide of iron may be given in ten drop doses in a wine-glass of water, three times per day before eating, and continued for a long time. The iron in this preparation may improve the blood, and the iodine may act favorably on the lymphatic and mesenteric glands, which are always more or less deranged in such cases.

In decidedly anæmic cases, in which the glandular derangement is less marked, the carbonate or citrate of iron may do best. If the carbonate is used, it may be given in three grain doses three times per day, and if not contra-indicated, it may be given in a teaspoonful of Madeira wine and molasses of each equal parts. To four fluid ounces each of the wine and molasses, three drachms of the iron may be added, and a teaspoonful of the mixture may be given, after thoroughly shaking, three times per day.

If the patient is anæmic, and still a stimulant is contra-indicated, the citrate of iron may be the best preparation, given in two grain doses in solution, immediately before or after eating. Or what would do better still, during meals, that it may mingle with the food, and so pass with the chyle to the blood.

In cases in which there is a decided dropsical or hemorrhagic tendency, the tincture of chloride of iron is the best preparation, and should be given in ten drop doses three times per day, in a wine-glass of water. In dropsical cases, or in cases in which there is enlargement or induration of the lymphatic or mesenteric glands, five grain doses of the iodide of potassium may be given in a fluid ounce of simple syrup, three times per day, before each meal, and continued for a time.

In all cases of tubercular phthisis, in which the digestion and assimilation is such, that the patient begins to emaciate, whether much or little food be taken, the *cod-liver oil* should be given, in tablespoonful doses, three times per day, an hour after each meal, and if it agrees with the stomach and bowels, should be continued while its effects appear salutary.[*] Together with this general treatment, in the early stage, dry cupping, blisters, or pustulation, with tartar emetic ointment, may be required.

If the *larynx* is the seat of the tubercular deposit, a blister should be applied early over the larynx, and it may be repeated if necessary; or pustulation may be produced by tartar emetic ointment, if the patient can bear it. After counter-irritants have been carried as far as prudent, an ointment of the iodide of potassium, two drachms to the ounce of lard, may be continued over the larynx. And a solution of two drachms of the iodide of potassium to the ounce of water, applied to the fauces and laryngeal cavity, every other morning for a time, and later, once or twice per week. It may be conveniently done, and as I have found, with very little disturbance to the patient, by means of the instrument recommended by Dr. H. Green, of New York, consisting of a bent whalebone, to the end of which is firmly fastened a soft piece of sponge.

After having applied the solution of the iodide of potassium for a time, as I have done, if the hoarseness is less, but considerable irritation or ulceration of the laryngeal mucous membrane continues, a solution of the crystals of nitrate of silver, of the strength of from forty to fifty grains to the ounce of water, may be substituted for the iodide of potassium, and applied as practiced by Prof. Green, every other day, at first, and later, two or three times per week, while any hope of relief remains.

While I am not certain that I have ever cured a case in which tubercles were actually deposited, by this treatment, I am confident that I have greatly palliated or retarded very bad cases, and believe that incipient cases may sometimes thus be permanently cured, if the general condition of the patient is at the same time restored or corrected.

If the lungs are the seat of the tubercular deposit, dry cups may be applied occasionally, at first, and then blisters, or pustulation may be produced by tartar emetic ointment. And after the counter-irritants have been continued as long as is prudent, or necessary, the compound iodine ointment should be applied to the upper part of the chest. A drachm each of iodine and iodide of potassium may be dissolved in a little water, and then mixed with an ounce of lard. This may be applied morning

---

[*] The hypophosphites may be of service in some cases.

and evening, so that half an ounce may be used per week, as long as there is any hope of relief.

As a gentle tonic and sedative in tubercular phthisis, the *wild-cherry bark*, the *prunus virginiana* is a valuable remedy in all stages of the disease. A wine-glassful of the officinal infusion may be given, three or four times per day, in all cases in which a gentle tonic, and sedative is required. And it may safely be continued, if necessary, for a long time.

Should expectorants be required in the early stage, in consequence of bronchial irritation; a teaspoonful of the syrup of tolu, with one-fourth of a grain of ipecac, or one-eighth of a grain of antimony may be given, four times per day, as an expectorant and diaphoretic. Or as the disease progresses, a teaspoonful of a mixture of equal parts of hive-syrup, and paregoric, with ten drops of the tincture of digitalis, may be given, four times per day, instead, and continued while it soothes the cough, or favors expectoration.

If, however, there is a hemorrhagic tendency, the best expectorant is the balsam of Peru, in ten drop doses, with one-fourth of a grain of ipecac, mixed with a little brown sugar and water; and given four times per day, with ten drops of the tincture of digitalis, in case the pulse is active, or very frequent.

In cases in which there is an excessive secretion, and expectoration, and in part the result of a bronchorrhœa, the wood naptha may be indicated, and should be given in five drop doses, in a teaspoonful of the syrup of tolu, four times per day, till the secretion is checked.

For the night sweats, if attended with a diarrhœa, two grains of tannin may be given, at evening, and repeated once or twice during the night, if necessary. In case this fails to check the sweats and diarrhœa, ten drops of the *aromatic sulphuric acid* may be given, three times per day, with a drachm of the fluid extract of blackberry, the *rubus villosus* generally with very good effect.

Now, by thus carefully correcting the habits, and general condition of the consumptive patient, and bringing them in every respect into a strict observance of the laws of health, many cases may be retarded, at least, and some cases may very likely, be permanently arrested, if only in the incipient stage.

If, however, the disease has passed the stage, at which it may be arrested; a wise and prudent return to the laws of health, will have a good effect upon the morals of the patient, and may retard more or less, the progress of the disease. And by resorting to just so much medical treatment, as is clearly indicated, and nothing more, the patient may live on in comparative comfort, for months or even years; and finally be gently soothed down through the "dark valley of the shadow of death," by a kind and prudent *medical hand*.

SECTION XIV.—APNŒA—(*Asphyxia.*)

By apnœa, I mean absence of respiration, or of that which is sensible, a condition sometimes designated by *asphyxia*, which means literally a want of pulse.

Respiration may be suspended by a great variety of causes, some of which act by preventing the access of air to the lungs, as *smothering*, *submersion*, **strangulation**, &c. Others act by affecting the air respired, as the irrespirable gases, such as *carbonic acid, nitrogen, hydrogen*, &c. And others still, produce paralysis of the muscles concerned in respiration, by affecting the *medulla oblongata*, or spine above the origin of the phrenic nerve, as *electricity, hanging*, &c.

I shall, however, refer only to those of a practical importance, including *apnœa* from *drowning*, from *strangulation*, from *irrespirable gases*, from *electricity*, and from *cold*. And, as a matter of convenience, shall consider each, with the treatment proper, **separately, in the present** section; and first, *apnœa from drowning*.

APNŒA FROM DROWING.

When a body is first taken from the water in which it has been submersed till the manifestations of life are extinct, the face has a turgid and livid appearance, the eyes are open and staring, the tongue is thrust beyond the teeth, the limbs are stiff, and there is generally a fullness of the epigastrium.

Death occurs from submersion by the exclusion of atmospheric air from the lungs, in consequence of which the blood does not receive oxygen or throw off carbonic acid. The venous blood in such a case passes from the extreme pulmonary arteries into the minute pulmonary veins. But as the presence of arterial blood is essential to a healthy action of the pulmonary veins, the pulmonary circulation is interrupted. And as a consequence of this interruption to the pulmonary circulation, venous blood accumulates, and congestion takes place in the pulmonary arteries, in the right cavities of the heart, in the large veins, and also in stomach, liver, spleen, and intestines, and there is, in fact, a general venous engorgement. While the venous system, the right cavities of the heart, and the pulmonary arteries, are thus gorged with blood, there is not sufficient venous blood reaching the left cavities of the heart, to stimulate them to action. And besides, it is probable that the little venous blood which does reach the left side of the heart, on entering the arteries of the heart, may tend to impair its **powers; and** should a small portion of it reach the brain, or other parts of the system, it could scarcely fail to smother, rather than sustain vitality.

It is a fact now well understood, that the regular transmission of arterial blood to the brain is indispensible to the healthy performance of its functions. It appears to me, then, that one of the first effects of suspended respiration, towards the destruction of vitality, is a cessation of cerebral action, for want of a supply of arterial blood.

As the cerebral functions are suspended, the want of a sufficient supply of nervous influence to the animal functions causes their suspension. And as the ganglionic system derives its power to act from the brain, there is very soon a suspension of the vital functions. Vitality may, however, linger for some little time after most of the phenomena of life have disappeared.

It as a matter of doubt how long a person may remain under water in a state of *apnœa*, and yet have sufficient vitality remaining to afford a chance of being resuscitated by proper restorative measures.

It is probable, however, that resuscitation may sometimes be produced after submersion for fifteen or twenty minutes, and perhaps longer.

If the water is warm, it is probable that vitality may remain longer than if the water is cold. And there is doubtless a difference in individuals as to their respective powers of vital resistance, under similar circumstances of submersion.

Morbid Appearances.—The morbid appearances presented on the dissection of persons who have died by drowning, are a general venous turgescence, and especially congestion of the pulmonary arteries, right cavities of the heart, and of the larger venous trunks.

The pulmonary veins, the left cavities of the heart, and the arteries generally, are comparatively empty. There is considerable vascular congestion of the brain, but very rarely so much as to justify the belief of an apoplectic condition, as the cause of death. The blood in most cases is either fluid or imperfectly coagulated, especially if death has occurred rather suddenly.

In cases in which death has been produced by drowning, the lungs contain very little water; but considerable may be found in the stomach. If then a dead body be found in the water, and on examination, the stomach is found to contain water, the person probably perished by drowning. But, if on examination, the stomach contains no water, it is nearly certain that the dead body was thrown into the water.

Treatment.—When a body is taken out of the water, and there remains a possibility of resuscitation, it should be wiped dry, wrapt in blankets, and taken to the nearest place suitable for making the necessary applications.

The restoration of the action of the lungs, is the principle object to be aimed at, at the same time that warmth is communicated to the body, by warm flannels or otherwise.

Artificial respiration may be produced, by passing the point of a common bellows into one nostril, and forcing the air gently into the lungs, the other nostril and mouth being closed, and the larynx pressed back by an assistant. When the air has thus gently been forced into the lungs, it should be allowed to escape through the mouth and nostril, its expulsion being favored by pressure on the thorax and abdomen, and this process should be repeated; imitating as near as possible, natural respiration, while a hope of resuscitation remains.

In case, however, a bellows is not at hand, the mouth of the operator may be applied to the mouth or nose of the patient, the other being closed; and by taking full inspirations, the lungs of the patient may be filled, and the air then allowed to escape, being assisted by pressure on the thorax and abdomen. And this may be repeated, so as to imitate as near as may be, natural respiration.

In case, however, one or both of these methods of producing artificial respiration, should fail, in consequence of the position of the tongue and epiglottis, closing the glottis; the patient may be turned on the face, as suggested by Dr. Hall, so as to let the tongue and epiglottis fall forward. This being done, the patient must be turned gently upon one side; which position causes air to enter the lungs. Very soon the patient should be turned again gently upon the face, which position, with gentle pressure upon the back, expels the air, and again brings the tongue and epiglottis forward as before. This process may be repeated deliberately "every four seconds," or so as to imitate, as nearly as may be, natural respiration, and should be continued, while there is a remaining hope of returning vitality.

While the effort to carry on artificial respiration is being made; heat should be gradually communicated to the body, by warm flannels. Warmth should, however, be communicated gradually, for the sudden application of a high degree of heat might do injury, by destroying the the small degree of remaining excitability of the system.

A strong infusion of capsicum in vinegar, is a very convenient application for the spine. Applied a little warm, with a flannel cloth, it tends to excite the circulation, and to impart warmth to the body. Injections of warm brandy and water into the rectum, may also be used, after the appearance of returning life begins to be manifested.

The return of vitality is usually manifested by transient twitches of the muscles of the face, especially of those about the lips, succeeded by irregular and convulsive efforts at breathing, tremor and agitation of the extremities, a small weak pulse, breathing at long intervals, and a discharge of frothy fluid from the mouth. Sensation and the power of motion return gradually, the lips assume a red appearance, the skin becomes soft and warm, and in some instances vomiting takes place.

When recovery has been partially affected, the greatest care should be taken that a due amount of stimulus be administered, to keep up the tottering energies of the system, at the same time avoiding overstimulation.

Such I believe are the most prudent and reliable measures for resuscitation from apnœa, produced by drowning or submersion. And a resort to them, modified of course to suit each particular case, should never be neglected, while there is a ray of hope of remaining vitality.

APNŒA FROM STRANGULATION.

By apnœa from strangulation, I mean that which is produced by a cord or other ligature about the neck, either with or without suspension.

In either case death is caused by the exclusion of air from the lungs, produced by closure of the trachea. There is, however, great congestion of the cerebral veins, and in cases in which there is imperfect closure of the trachea there may be apoplectic effusion before death takes place. In cases, however, in which strangulation is produced by suspension there is sometimes a dislocation of the vertebræ, and death occurs instantly, from injury of the spinal cord "above the origin of the phrenic nerve."

In the ordinary mode of executing criminals, then, if the vertebræ of the neck are dislocated, death occurs suddenly, with little or no suffering. In cases, however, in which there is no dislocation, but a perfect closure of the larynx, there is only a "brief feeling of suffocation," and consciousness becomes extinct. Finally, in cases in which there is no dislocation, and the closure of the trachea is incomplete, the suffering and consciousness continue for a longer period, and death ultimately occurs, in part from the exclusion of air, and also from cephalic congestion.

In cases of apnœa from strangulation, if there is no dislocation of the vertebræ of the neck, even suspension may have been endured for a considerable time, and yet there may be the remains of vitality, which proper efforts at resuscitation may revive.

Morbid Appearances.—The post-mortem reveals a congested state of the pulmonary arteries, right cavities of the heart, of the brain, and of the larger venous trunks generally, as in the bodies of persons drowned. There is also in addition the marks of the cord about the neck, and the dislocation of the vertebræ, in cases in which that occurs. The blood is imperfectly coagulated, if at all, and very little is found in the arteries, or left cavities of the heart.

Treatment.—The treatment consists in producing artificial respiration, imparting warmth, affording stimulus, and should be in nearly every particular the same as in cases from drowning, only from the manner in which the cause acts, the cerebral congestion may require the abstraction of blood from the jugular veins. Or blood may be taken from the back of the neck by cups, with very good effects. In a case of hanging (an attempt at suicide) which came under my care, in this village, about three years since, resuscitation was quite complete; but the patient could not swallow till I took a few ounces of blood, by cups, from the back of the neck. The case subsequently passed on favorably, and finally recovered.

APNŒA FROM IRRESPIRABLE GASES.

Of the various irrespirable gases capable of producing apnœa, carbonic acid is by far the most frequent. When a person in health is exposed by accident or otherwise to carbonic acid in its pure state, apnœa and finally death occurs, probably from the absence of oxygen and also in part perhaps by the obstruction which the gas affords to the free exhalation of carbonic acid from the lungs. When apnœa or death is thus suddenly produced, the body is found pale, collapsed and flaccid.

If, however, as is often the case, the carbonic acid is mixed with atmospheric air, it produces vertigo, fainting, insensibility, apnœa, and if continued, finally death. In such cases the face exhibits a tumid and livid appearance, the veins of the head and neck are turgid, the tongue is swelled, and the lips livid; and the body may remain warm for several hours.

The inhalation of chloroform may produce apnœa, in part perhaps by "depressing the nervous centre of respiration," and also by excluding atmospheric air from the lungs.

Nitrogen and *hydrogen* gases are seldom accidentally inhaled, and when they are, produce apnœa or death, not by any positive pernicious effect, but by their utter incapacity to sustain animal life. The other, and highly irritating gases, such as the muriatic acid gas, chlorine, &c., seldom enter the lungs in quantity sufficient to produce apnœa; they may, however, inflame the respiratory passages and thus lead to a fatal result.

Morbid appearances.—The appearances on dissection are similar to those exhibited from drowning; the sinuses of the brain, the jugulars, and all the large venous trunks as well as the right cavities of the heart and pulmonary arteries, are always congested with dark blood, and generally in a fluid state. The pulmonary veins, left cavities of the heart, and arteries generally, are nearly empty, as in cases of drowning and strangulation, and for a similar reason.

Treatment.—In cases of apnœa from carbonic acid, in which there is some degree of sensibility remaining, the patient should be taken immediately into the open air, and being supported in the sitting posture, cold water should be dashed upon the face and breast, while the extremities are rubbed with warm flannel; and as soon as the patient is able to swallow, a little cold wine or brandy and water may be administered.

In cases, however, of apnœa from carbonic acid, in which there is no apparent remaining sensibility, the patient should be taken into fresh cool air, and being stript and laid on a sheet, with the head and shoulders elevated, cold water should be dashed upon the breast, and cloths wet in cold water applied to the head.

The cold affusions may be repeated at short intervals, and spirits of ammonia or camphor applied to the mucous membrane of the nose. The back and extremities may be rubbed with a strong infusion of capsicum in vinegar, and an injection of carbonate of ammonia or warm brandy in mucilage may be administered with very good effect. In case the cold affusions fail to excite respiration, artificial respiration should be resorted to as for drowning; and as soon as resuscitation is established, the patient should be wiped dry and a little warm wine-whey may be allowed.

The same treatment may be resorted to, in cases of apnœa from the other unirritating gases, such as *hydrogen*, *nitrogen*, &c., and also for the apnœa, should it occur from the irritating gases. But accidents from *hydrogen* or *nitrogen gases* seldom occur. And when the irritating gases are accidentally inhaled, the principal danger, as we have seen, is from the inflammation which they produce in the respiratory passages, which, of course, requires to be treated on general principles, laid down elsewhere.

Such, I believe, are the principles which should guide in the treatment of apnœa from the irrespirable gases; subject of course, to variations, to suit each particular case.

APNŒA FROM ELECTRICITY.

Moderate currents of electricity passed through the human body, appear to produce an invigorating effect. But when its intensity is great, it impairs or suspends the sensibility of the nervous system, and produces apnœa, insensibility, or utter loss of vitality, if the shock be very great.

The appearance of persons struck by lightning, vary therefore, according to the intensity of the shock. If the shock has been severe, there is no appearance of sensibility remaining. Red streaks may be found on different parts of the body, the hair may be found singed, small blisters appear on different parts of the body, and blood may be discharged from the ears.

When the electric stroke has been less intense; so as not to destroy sensibility, the face appears red and bloated, blood issues from the mouth and nose, respiration is difficult, the pulse is weak and irregular, and not unfrequently, there are spasmodic twitches of the muscles of the face, throat, and other parts of the body.

Persons recovered from a state of apnœa from lightning, are apt to suffer for a long time, from painful sensations, and a degree of numbness in the extremities; and they frequently retain a peculiar susceptibility to the electric influence for a long time.

Morbid Appearances.—The post-mortem in persons killed by lightning reveals a collapsed state of the lungs; but the heart is apt to be gorged with blood. In cases in which death has been sudden, the blood is found in a fluid state, and the body tends rapidly to putrefactive decomposition.

Treatment.—The treatment for apnœa from lightning, should be similar to that for apnœa, arising from carconic acid, and the other gases. The patient should be placed in a convenient position, and cold water should be frequently, and copiously dashed over the whole body. Friction should also be made with rough flannel, and should the cold affusion fail to excite respiration; artificial respiration should be resorted to, and the effort persevered in, while there is any hope of returning sensibility.

In desperate cases, in which the ordinary efforts at resuscitation are ineffectual; galvanism or electricity, if at hand, may be resorted to; but generally, I think, with little prospect of success. It should be applied in moderate currents, and so, if possible, as to excite the respiratory movement.

Discretion should be exercised in adopting the measures used to the condition of each particular case. And should sensibility return, a little wine-whey may be required; but care should be taken to avoid over-stimulation.

APNŒA FROM COLD.

When the human system is exposed to intense cold, the suffering is at first very considerable, in consequence of the impression on the extreme nerves. If, however, the exposure continues till reaction ceases, it produces a benumbing effect, and there is evinced, soon, signs of cerebral oppression. The surface becomes pale and contracted, the superficial blood-vessels shrink, the extremities become numb, the gait is tottering, the voice indistinct, and, if the exposure continues, there is an irresistible drowsiness; and finally, apnœa, insensibility, and death is the result.

Morbid Appearances.—The post-mortem reveals a gorged state of the large venous trunks, and though the arteries are generally comparatively empty, both sides of the heart are in some cases filled with blood, imperfectly coagulated. Most of the large viscera, as well as the brain, are gorged with blood, but there is generally no apoplectic effusion.

Treatment.—Resuscitation from apnœa, caused by intense cold, may sometimes be effected after all the sensible phenomena of life have become extinct.

When a person is found in a state of insensibility from cold, the body should be rubbed with melting snow, if at hand, and as soon as convenient it should be immersed in spring water, or water fresh from the well. After twenty or thirty minutes the body should be taken from the cold water, and rubbed dry with soft flannel, after which it should be wrapt in blankets, and taken "into an unheated room."

The body should now be rubbed with flannel, and if respiration does not return, artificial respiration should be resorted to. And these measures should be persevered in till all hope vanishes, or there is a return of sensibility. If the measures are successful, the temperature of the room should be gradually raised, and when the power of deglutition is restored, warm sage tea and a little wine-whey may be allowed, till reaction is fully established.

SECTION XV.—PNEUMOTHORAX.

By pneumothorax is here meant, *air* in the cavity of the pleura, either with or without a fluid.

Pneumothorax may arise from an accumulation of gas in the cavity of the pleura, without any direct communication with the external air, or it may occur from a communication of air through the bronchial tubes, or finally, by a direct communication of air through the walls of the chest.

Symptoms.—The accumulation of air in the cavity of the pleura produces compression of the lung, and, as a consequence, is attended almost invariably with more or less dyspnœa.

If the accumulation is small and the pleura contains little or no fluid, the respiratory organs as well as the circulation may accommodate themselves gradually to the morbid condition, and the dyspnœa and other symptoms may be comparatively slight.

In cases, however, in which the accumulation is sudden and considerable, there is great difficulty of breathing, a sharp pain in the side, and sometimes a copious expectoration of pus, as happened in one case that fell under my observation, the result of empyema.

In cases in which there is considerable fluid in the cavity of the pleura, and the air enters from the bronchial tube or tubes below its surface, little or no air is expelled at each expiration, and thus the accumulation may become very great and lead even to fatal suffocation. Or, in case one lung has been incapacitated, from any cause, for respiration, and pneumothorax occurs in the opposite side, fatal suffocation generally speedily occurs.

Diagnosis.—Inspection may detect an enlargement of the side or bulging of the intercostal muscles. Percussion on the affected side elicits a dull sound as high in the chest as the liquid extends, if there is any, but above there is a clear or *tympanitic* sound, very different from that elicited on the opposite side of the chest.

If the ear be applied to the affected side there is noticed a want of the respiratory murmur and a diminution of the vocal resonance, both of which may be rendered more apparent by comparing them with the sounds on corresponding parts of the opposite side. In some cases of pneumothorax, however, there is detected on the affected side the amphoric respiration as well as the amphoric resonance of the voice and cough, and also the metallic tinkling. These last symptoms are not, however, strictly diagnostic, as they occur in large cavities in the lung as well as in pneumothorax.

In cases of pneumothorax in which there is considerable liquid in the cavity of the pleura, with the air, succussion produces a "splashing" sound, which may be distinctly heard on applying the ear to the chest; and the patient may feel a peculiar sensation on coughing, caused by the striking of the fluid against the costal pleura.

Causes.—Pneumothorax may arise from the opening of a tuberculous vomica into the cavity of the pleura, there having been previously a communication with the bronchial tubes, or it may arise in cases of empyema from an ulceration in the pleura and pulmonary tissue into the bronchial tubes, as happened in one case that fell under my observation; the matter from the pleural cavity in this case being gradually expectorated. In this case the external application of iodine ointment caused a suspension of the accumulation of matter in the pleural cavity for a time, so that the passage to the bronchial tubes healed. Subsequently, however, matter again accumulated, and pointing in the intercostal space, in a convenient place, I opened it and got at first about a gallon of sero-purulent-matter, from the escape of which the patient suffered very little. The patient, a young man, lived for several years with pneumothorax, from the puncture in the walls of the chest, and finally died of anasarca.

Finally, pneumothorax may arise from the gases extricated in gangrenous decomposition of the lung or lungs, and it may also arise from an accumulation of gas in the cavity of the pleura, probably without any organic lesion, and from various other causes which I need not mention.

Treatment.—The general condition of the patient should be corrected as far as possible by proper remedies : local inflammations should be overcome by cups, leeches, blisters, &c.; iodine externally and internally may be required to remove the liquid effusion, if it exists, and in cases in which the accumulation of air in the cavity of the pleura threatens suffocation, an opening may be made in the walls of the chest for its escape.

In cases in which the cause of the pneumothorax is of a necessarily fatal character, palliatives only may be required, and an opening in the walls of the chest for the escape of air, may not be advisable even though suffocation be threatened.

But in cases in which the cause is not of a necessarily fatal character, the iodide of potassium and some preparation of iron, or the syrup of the iodide of iron may be indicated and should be continued for a long time. This course of treatment, with counter irritants, and finally iodine ointment externally, with puncture of the walls of the chest, if necessary, may lead to a final recovery.

SECTION XVI.—EMPHYSEMA.

By emphysema, I mean here that which arises from an enlargement or dilatation of the air-cells, and also the escape of air from a laceration of the larynx, trachea or lungs, producing interlobular infiltration of air, and extending, in some cases, into the general cellular tissue, distending the mediastinum, face, neck, and superior part of the chest.

Symptoms.— *Vesicular Emphysema*, or that which arises from excessive dilatation of the air-cells, is attended with dyspnœa, which may be continuous, but is greatly increased by over exercise, or any cause which calls for an increased action of the lungs.

During a paroxysm, from whatever cause produced, the face is apt to be pale or livid, the lips purple, and in protracted cases, the countenance may assume a continuous "dusky hue."

In the *interlobular* or *extravesicular* variety, one marked case of which fell under my observation, in this village, a few years since, the air is liable to escape into the general cellular tissue, especially if the air escapes from the larynx, trachea, or near the root of the lung, or lungs. In such cases, the air distends the mediastinum and adjoining cellular tissue, and frequently, also, the face, neck, and superior part of the chest, as happened in the case above referred to, in this village, which affected both sides, and produced fatal suffocation or dyspnœa.

On *inspection*, the affected side is found more convex or prominent than in health, and the intercostal spaces appear slightly widened. Percussion elicits an increased resonance, in proportion to the amount of air the lungs contain.

Auscultation detects a feeble and suppressed, or short inspiration, and a loud, protracted and wheezing expiration, and this is true, I believe, whether the emphysema be vesicular, or extra vesicular or interlobular.

Anatomical Characters.—On opening the chest, in cases of *vesicular emphysema*, the lung does not collapse, but rather expands. The lung sinks less in water than in a natural state, and its surface is often covered with slight elevations, consisting of dilated air-cells.

On cutting into the lungs, the air-cells of more or less of one or both lungs are found dilated, some slightly, others to the size of a "hazelnut," or even larger than that, and there is generally more or less dilatation of the small bronchial tubes in the affected part.

In *interlobular* or *extravesicular* emphysema, elevations may be found on the surface of the lungs, from the size of a pea to that of a hen's egg, or even larger, caused by the rupture of air-cells, and escape of air into the pulmonary structure. The interlobular cellular tissue, throughout the affected part is filled with effused air, in such cases, as might be expected.

In cases of extravesicular emphysema in which the air escapes into the mediastinum and surrounding cellular tissue, and also into the general cellular tissue, distending the neck, face, and upper part of the chest, the lungs may be compressed, and fatal suffocation or dyspnœa produced, if both sides of the chest are involved.

In a case of this character that I examined, already referred to, in which the air escaped near the root of the lungs, the rupture having been produced by straining at stool, the following appearances were presented on examination post-mortem.

There was a general bloated or distended appearance of the face, neck, and superior part of the chest of both sides. On removing the sternum, the anterior portion of the chest was found filled with distended cellular tissue, containing some adipose matter, appearing to fill most of the space usually occupied by the anterior and inferior portion of the lungs. The lungs were found in the posterior part of the chest, in a sort of collapsed or very contracted state, and almost black from the little venous blood which they contained. In this case the emphysema was of both sides; and I ascertained from his attendants that death took place very soon after the occurrence of the emphysema.

Causes.— Vesicular emphysema is most frequently produced by *asthma, bronchitis,* and *pulmonary phthisis,* but it may be produced by any cause which interrupts the free expiration of air from the lungs. The *extravesicular* variety of emphysema may be produced by any cause which keeps the lungs too long in an expanded state.

It is probable that the most frequent causes of this variety of the disease, **are** lifting heavy weights, blowing **on wind** instruments, straining at stool, and protracted holding in the breath, as in diving, &c.

Treatment.—The cause should be sought out and removed, as far as possible, and the patient should **be made to conform strictly to the** laws of health in every respect.

In the chronic *vesicular variety,* violent exercise, and **every** other cause capable of bringing on a paroxysm of dyspnœa should be carefully avoided. But should a paroxysm occur, small doses of hive-syrup, tincture of lobelia and stramonium may be of service. Half a drachm each of hive-syrup and tincture of lobelia, with ten drops of the tincture of stramonium, may be administered every six hours till the paroxysm subsides.

In the extravesicular variety, if threatening, general bleeding may be required, to lessen the pulmonary circulation, and if the face, neck or chest become distended with air, punctures may be made for its escape.

CHAPTER X.

DISEASES OF THE CIRCULATORY SYSTEM.

SECTION I.—AUSCULTATION OF THE HEART

BEFORE proceeding to the consideration of diseases of the circulatory system, including those of the heart, arteries, veins, capillaries, lymphatics, and lacteals, as well as deranged or diseased conditions of the **cir**culatory fluids, which I propose to do in the present chapter, I shall in this section attempt to explain the natural and diseased sounds of the heart and large arteries, discoverable by auscultation, percussion, &c.

But before we proceed to the consideration of auscultation of the heart, it is necessary to take **a** general glance at the circulatory system, and to call to mind, definitely, the anatomy and physiology of the heart, leaving the minute anatomy and physiology of other parts of the circulatory system to be called up as we proceed to the consideration of diseases affecting these different parts.

31

The heart, it will be remembered, is the central organ of the circulatory system, and is situated between the two layers of pleura, which constitute the mediastinum, being enclosed in the pericardium.

The heart is placed obliquely in the chest, the base being directed upwards and backwards, or towards the right shoulder, "and the apex forwards, and to the left, pointing to the space between the fifth and sixth ribs," two or three inches to the left of the sternum. The heart is a muscle, and has four cavities ; the right auricle and ventricle occupying its right side, and the left auricle and ventricle the left.

The right auricle receives the venous blood from all parts of the system, as well as the lymph, chyle, &c., and passes it on through the auriculo-ventricular opening into the right ventricle. From the right ventricle, by its contraction and the closing of the tricuspid valve, the blood is passed on through the pulmonary artery to the capillaries of the lungs, where, parting with carbonic acid and receiving oxygen, it is prepared to pass on, as arterial blood, through the pulmonary veins to the left auricle. From the left auricle the blood passes through the auriculo-ventricular opening into the left ventricle, from which, by contraction of its walls, and closure of the mitral valve, it is forced on into the aorta, by the contraction of which, together with the closure of the semilunar valves, it is distributed to every part of the system.

The arterial blood in passing through the minute capillaries, constituting the connection between the extreme arteries and veins, supplies materials of growth or nutrition, and gradually assuming a darkish color passes on through the veins to the right auricle of the heart, from whence it started.

But some albumen probably escapes from the small blood-vessels, and more is set free "by changes going on in the soft parts," in the breaking down of the old tissues. Now to take up this extravasated and other albuminous matter everywhere in the system, and to change or turn it, as far as may be, to fibrin, the lymphatic system is arranged, consisting of "a network of delicate tubes" "disseminated through all the soft tissues except the nervous." These tubes coalesce, producing those of larger size, and pass through the lymphatic glands "which," says Professor Draper, "might indeed be regarded as mere plexuses, and eventually empty into the veins."[*]

The lacteal vessels constitute the remainder of the vascular system, and consist of small vessels, which arise along the intestines in the submucous tissue, and passing through the mesenteric glands, terminate in the thoracic duct. The mesenteric glands, through which the lacteals pass, consist, according to Professor Draper, of a "plexus of tubes, to which that particular form is given for the sake of closeness of package,"[†] and it is probable that they change a portion of the fat and albumen of the chyle, as well as that transuded from the blood-vessels, along the intestines, to fibrin, thus rendering their functions very similar to that of the lymphatic glands in other parts of the system.

The thoracic duct commences in the *receptaculum chyli*, which is made up of several lymphatic trunks, together with the lacteals, and is situated in front of the second lumbar vertebra. The duct, thus originating passes

* See Draper's Physiology, pp. 94, 95.   † See Draper's Physiology, p. 89.

up in front of the vertebral column to the fourth dorsal vertebra. It then inclines to the left, ascends by the side of the œsophagus to opposite the seventh cervical vertebra. At this point it "makes a curve forwards and downward, and terminates at the junction of the left subclavian, with the left internal juglar vein." Through the lacteals and thoracic duct the fat and albumen of the chyle, derived from the food, and also the albumen derived by transudation from the systemic blood, being in part turned to fibrin in the mesenteric glands, reaches the general circulation.

Thus have we taken a glance at the circulatory system, including the heart, arteries, veins, lymphatics, and lacteals, every part of which will constantly be called up, as well as the character of the blood, chyle, lymph, &c., as we proceed in the consideration of diseases of the circulatory system.

With these preliminary considerations, and bearing in mind the minute anatomy of the heart, let us proceed to the consideration of the topic of the present section, *auscultation of the heart.*

We have seen how the blood passes through the heart, and we should naturally suppose that the contraction of its cavities and closure of its valves, together with the passage of the current of blood, would produce a sound or sounds, and an impulse, and such is really the case.

Now in order to appreciate the morbid sounds, impulse of the heart, &c., it is necessary to inquire into the physical signs which occur in health, including the *impulse, rhythm,* and *sounds* of the heart, and then we may fully appreciate the morbid.

Impulse.—If the hand be placed on the left side of the chest, below the nipple, in the space between the fifth and sixth ribs, a little to the inner side of a line running vertically over the nipple, a gentle regular pulsation is felt, and if the ear or stethoscope is applied to the same part, a slight shock is perceived by the examiner. "This is the *impulse* of the heart," and depends upon the impinging of its apex against the parietes of the chest, being nearly synchronous with the pulse at the wrist, and with the systole or contraction of the ventricles.

Rhythm.—If the ear be placed upon the præcordial region, when the heart is acting naturally, a regular succession and cessation of sounds is heard. In the first place there is noticed a long sound, then a short sound, and then an interval without any sound, after which will occur again the long, then the short sound, and then the interval as before.

The long and short sound, together with the interval, make up the time occupied by one complete circuit of the heart's functions. And as these sounds and this interval have been observed constantly to bear a definite relation to each other, the period occupied by them, or the complete circuit of the heart's action has been divided into corresponding portions of time.

"Suppose then," says Dr. Hughes, "that the whole period of this time be divided into fifths, the first sound will occupy two fifths, the second sound one-fifth, and the interval between the sounds, the remaining two-fifths. The time occupied by the whole corresponds to the time between one stroke of the pulse and another, so that if a person's pulse beat eighty times in a minute, there would be eighty long sounds, eighty short sounds, and eighty intervals."

Sounds.—The sounds of the heart, as suggested by Dr. Hughes, may be represented by the following syllables; too-to, too-to.

The *first sound* of the heart is caused, as I believe, by the contraction of the ventricles and reaction, or tendency to flow back of the blood, and closure of the tricuspid and mitral valves; the shock thus produced being communicated to the parietes of the chest. This may be understood, when we remember that the contraction of the ventricles occurs at the same time that the auricles are relaxed or dilated, the blood being suddenly arrested by the tricuspid and mitral valves, producing the shock and sound referred to.

The *second sound* of the heart is caused probably by the sudden reaction of the blood on the semilunar valves, after it has passed out of the ventricles. This may be understood, when we remember that the diastole of the ventricles, takes off the pressure from behind, and allows the blood to react suddenly on the semilunar valves; producing vibrations of the vessels, and consequently of the **parietes of the chest, and hence the** second sound of the heart.

The ordinary range of the heart's sound, when in a natural condition, and free from disease, is over the præcordial region, where they may be distinctly heard, by the application of the ear or stethoscope.

Impulse in Disease.—As a general rule, the *impulse* of the **heart in disease, is** as the pulse at the wrist. If the pulse is strong and vigorous, **the impulse of the heart is so likewise;** but if the pulse is feeble and de**pressed,** so is the impulse of the heart, and when the pulse is contracted and vibrating, the impulse of the heart partakes of the same character.

When *hypertrophy* exists, the parietes of the auricles, and ventricles being thickened, the impulse is powerful and heaving. And when, with the thickening, dilatation also exists, the impulse is powerful, **and more** diffused; being often felt over the whole præcordial region.

When the parietes of the ventricles are thin and weak, or loaded with fat, the impulse is feeble. But when with this weakness, the cavities of the ventricles are dilated, though the impulse is weak, it **will be felt over** a larger extent of surface than in a healthy state.

When the patient examined is nervous or anæmic the impulse is exceedingly sharp and smart, but care **should be exercised not to confound** this with a strong impulse.

When obstruction exists in the pulmonary artery or in the mitral valve the impulse is increased, in the scrobiculus cordis, in consequence of the continued distension of the right ventricle, which results from such obstruction.

When the mitral valve becomes insufficient, allowing regurgitation of the blood for a long time, the whole heart, with the exception of the left ventricle, becomes dilated and enlarged. In such cases, in addition to the increased impulse at the scrobiculus cordis, there is an impulse between the cartilages of the third and fourth ribs, which probably arises from regurgitation through the mitral valve, if it be synchronous with the impulse at the scrobiculous cordis; but if not, it may arise from an increased action of the auricle itself, caused by the regurgitation.

In cases in which there is an obstruction in the aortic valves, or in the arch of the aorta, the overaction of the left ventricle may increase its strength, and also the impulse of the heart. If, however, in such cases the patient is thin and weak, the ventricle may become dilated and thin, and the impulse, though extended, becomes very weak.

If fluid be effused into the pericardium, in small quantity, the impulse of the heart is only decreased; but if the fluid be in large quantity, the impulse may be lost, or its position materially changed.

The Rhythm in Disease.—Any impression which excites the action of the heart, may at the same time disturb its *rhythm.* In such cases, one cavity contracts before its time, or another is dilated too quickly, in consequence of which the natural series of actions is interrupted, and the regular succession broken. Hence it is, that organic changes of the heart, which interfere most with the circulation through the organ, most frequently produce an unnatural rhythm.

Among the causes which thus interfere with the circulation through the heart, and produce an unnatural rhythm, are diseases of the valves, especially of the mitral valve, thinness and dilatation of the parietes of the ventricles, and effusion into the pericardium. All these causes, with many others, may operate to produce different varieties of unnatural rhythm, from the *treble beat* to an utter confusion of sounds, or in the succession of sounds.

Sounds in Disease.—The sound of the heart may be unnaturally *increased* or *decreased,* or concealed by morbid sounds or *murmurs.*

Increased Sounds.—In cases in which the valves are healthy, and the cavities of the heart dilated, the parietes of the ventricles being thinner than natural, there is generally an increase in the heart sounds.

Decreased Sounds.—Anything which interferes with the free action of the heart, or impedes the motion of its valves and regular contraction of the organ, may produce a diminution of the natural sounds of the heart. The most frequent causes probably of decreased sounds, are fluid in the pericardium, distention of the heart, and imperfection of the valves.

Murmurs.—In a perfect state of health, the heart and blood being natural, no noise is produced by the direct transmission of the blood through the heart, "in a quiet stream." In case, however, the valves become thickened by inflammation, as in endocarditis, they lose their pliancy and flexibility, and do not set close to the walls of the vessels, or leave a free passage for the blood, in consequence of which an unusual agitation is produced among the particles of the blood, which gives rise to morbid sounds or murmurs.

The morbid sounds, or *murmurs* thus produced are various. It may resemble the sound produced by a pair of bellows, in which case it is called the "*bruit de souflet,*" or "*bellows murmur,*" or it may resemble the rasping of wood, constituting the "*bruit de rape,*" or the "*rasping murmur,*" or the sound may be like that produced by the sawing of wood, constituting the "*bruit de scie,*" or "*sawing murmur,*" or finally the sound may resemble the tone of musical instruments, constituting the "*bruit,*" or "*musical sound.*"

Such I believe are the principal *murmurs,* liable of course to variations, and to more or less commingling, which modifications are always to be taken into account in the examination of cases.

But in order to understand in what valve or valves the morbid condition producing the murmur exists, it is proper for us to inquire more particularly in relation to the seat or location in the chest, of the different valves, or the points at which the murmurs proceeding from them may be most distinctly heard.

The *valves* of the *pulmonary artery* are situated nearly opposite the cartilage of the third rib, on the left side of the sternum, at which point murmurs proceeding from that source may be most distinctly heard.

The *aortic valves* are situated deeper, and their murmurs heard most distinctly over the sternum, nearly opposite the cartilage of the third rib, or a little below.

The *mitral valve* is situated nearly opposite the space between the cartilages of the fourth and fifth ribs, a little to the left of the sternum and below the nipple.

The *tricuspid valve* is situated immediately behind the lower part of the sternum, and murmurs arising from disease of this valve, are heard most distinctly over the central part of the sternum, but are sometimes audible at the scrobiculus cordis.

Regurgitation.—We have seen that murmurs are produced by the direct passage of blood through imperfect valves, it must be remembered also that murmurs may arise from a *regurgitation*, or retrograde motion of the blood through diseased valves.

"When regurgitation takes place through the aortic valves," the direct current not being interrupted, a murmur is heard over the sternum, a little below the third rib. This murmur occupies the time of, and may conceal the second sound of the heart. But if the pulmonary valves are in a perfect state, the murmur and the second sound may be heard together, between the cartilages of the second and third ribs, a little to the left of the sternum.

If, however, "both obstructions exist in, and regurgitation takes place through the aortic valves, a double murmur or *see-saw* sound is heard over the sternum, opposite the third rib." This sound may be heard along the course of the aorta, but gradually diminishes as the ear is passed from over the seat of the aortic valves. Should obstruction exist in, and regurgitation take place through the pulmonary valves, this *see-saw* sound, or double murmur will be heard nearly opposite the cartilage of the third rib to the left of the sternum.

In case regurgitation takes place through the mitral valve, the murmur occupies the time of the first sound of the heart and is most distinctly heard between the cartilages of the fourth and fifth ribs, a little to the left of the sternum and below the nipple. It may also be audible in the axilla, close by the spine, in the interscapular region, or even "over the bony column itself."

Aneurism of the ascending or descending aorta, or of other large arteries, may give rise to murmurs similar to those produced by the direct or retrograde movement of the blood through diseased valves, and hence they afford an important diagnostic symptom in aneurism of the aorta and of other smaller arteries.

Anæmic murmurs.—In certain anæmic conditions, as in chlorosis, murmurs may arise from the mere passage of the blood independently of local disease; they are very properly called "*anæmic murmurs*," and may resemble the blowing of a pair of bellows; "*the bruit de souflet ;*" or the sound may be rough like that of filing or sawing, "*the bruit de rape,*" or "*bruit de scie.*"

These *anæmic murmurs* appear to be connected either with the aortic openings or with the pulmonary artery, and depend, probably, upon a dissolved or watery state of the blood; in consequence of which, its particles move easily over each other, are therefore more freely agitated, and thus give rise to the vibrations which produce the murmurs.

Venous murmurs.—In *anæmic* patients there may sometimes be heard a murmur by the passage of blood along the jugular veins; this murmur is continuous and not intermittent like the arterial murmurs, and has been called the "*continuous humming*" or "*venous murmur.*"

Strong pressure causes it to cease, but it is seldom heard distinctly unless slight pressure be made over the part. It probably depends upon a watery state of the blood, upon its accelerated motion, and in part perhaps upon the slight obstruction produced by the gentle pressure made over the part.

Pericardial murmurs.—The natural movements of the heart in the pericardium, in a state of health, are not attended with any audible sound; but when the pericardium is rendered rough by inflammation, or when lymph is thrown out on its folds, attrition occurs, and a superficial rubbing is heard over the pericardial region upon each motion of the heart, called the "*pericardial rubbing,*" or "*frottemont.*"

The murmurs, however, in such cases, vary according to the degree of roughness of the parts, being in some cases a mere "*frottemont,*" or simple rubbing sound; in others harsh, grating, or creaking, and in others still resembling the *bellows murmur* of the valves.

Mixed murmurs.—It sometimes happens that the pleura, which is in contact with, or forms the external covering of the pericardium, becomes inflamed and roughened, and perhaps also, that portion of the pulmonary pleura, immediately adjoining, in which case there is a murmur produced by the attrition of the two roughened surfaces. This murmur may generally be heard most distinctly, during inspiration, as the motion of the pericardial and pulmonary pleura is most considerable at that time.

Finally, it occasionally happens, that a small amount of fluid collects in the immediate vicinity of the heart, with or without the presence of a gas, in such a position that each impulse of the heart produces in it a degree of agitation sufficient to cause an audible sound or *murmur.*

This noise has been very correctly represented by Dr. Hughes, by the syllables "*blob—blob—blob—blob.*" In one case that fell under my observation, in this village, a few years since, this peculiar sound was audible for several weeks, and was often so loud, that it was distinctly heard in any part of the room, in which the patient—a young lady, was sitting or lying.

Such, according to my observation, are the *natural,* and most common diseased sounds of the heart, detected by auscultation. But there are certain diseases of the heart and pericardium, in which *percussion* elicits sounds indicative of morbid conditions within, which it is proper to notice as we pass.

PERCUSSION.

Enlargement of **the heart,** is attended with increased dullness on percussion, **of the præcordial region, and** this increased dullness **is** most marked, **over that** portion of the heart in which the dilatation **is** most considerable. **In case** the left ventricle is greatly dilated, the dullness is most marked below the left nipple, over the costal cartilages.

If it is the left auricle that is mainly dilated, the increased dullness will be most conspicuous on the third and fourth ribs, **and their cartilages, to the left of the** sternum. **If, however, it is the right ventricle that is more** especially involved in the dilatation, the increased dullness will be noticed most **to the right of the sternum, at the inner edge of the right mam-** mary region, **and at the scrobiculus cordis.**

In *pericardial effusion,* if the quantity of fluid is **considerable, there is dullness on percussion, and this** dullness begins on the left side of the lower portion of the sternum, and gradually extends **upwards. But when** the pericardium **is largely** distended, **the** dullness **is very remarkable, and extends in every direction; but especially transversely and inferiorly.**

Aneurism of the *ascending aorta,* if of large size, **may generally be** recognized by **dullness** on percussion, **in the upper sternal, and "inner** edge of the right subclavian region." In case, however, the aneurism is very near the heart, percussion would hardly distinguish *aneurism* from disease of the heart itself.

In case the aneurism occupies the *arch* of the *aorta,* there is generally **marked** dullness on percussion **over** the affected part.

Finally, if aneurism exist in the *descending aorta,* and is of a large **size,** it may generally be detected by dullness on percussion on either, **but** especially **on the left side of the spine. And** in case it projects forward, dullness may be perceptible "**in the præcordial** region."

Such, I believe, are the abnormal **sounds elicited** on percussion, in enlargement of the heart, **pericordial effusion, and** aneurism of the large vessels, which **completes what I had to say on** *auscultation of the heart.*

SECTION II.—PERICARDITIS.

By pericarditis, from περιχαρδιον, "**the pericardium," and** *itis,* inflammation ; I mean inflammation of **the pericardium, whether of a** *simple serous* or *rheumatic* character.

The pericardium, it will be **remembered,** consists of two layers, "an external *fibrous* and internal *serous.*" The *fibrous* layer is attached to the great vessels at the **root** of the heart, and is continuous with the thoracic fascia. **The** serous membrane lines the inner surface of the fibrous layer, and is then reflected over the heart, which it covers entirely, "without, however, having the heart within it;" in this respect resembling other serous **membranes.**

"The pericardium envelops **the** heart, retains it in position, and facilitates its movements, by the serous fluid which it contains in greater or less quantity."

The pericardium, thus constituted, is liable to become inflamed, and the character of the inflammation, as **well as** the symptoms developed, depend very much on the membrane principally involved.

If the inflammation be seated mainly in the serous membrane, it is generally acute, and marked by symptoms that usually attend inflammation of serous membranes. In such cases, the inflammation is peculiarly liable to terminate in effusion of serum, producing hydropericardium.

If, on the other hand, the inflammation is confined mainly to the fibrous layer of the pericardium, it is generally of a rheumatic character, and the symptoms which are developed are similar, in many respects, to those usually attending rheumatic inflammation of other fibrous structures.

If, again, as is more generally the case, the serous and fibrous membranes both become inflamed, the symptoms which attend are such as might be expected. And as the external or pleural covering of the pericardium is generally more or less involved in pericarditis, we have added some of the symptoms of common pleuritic inflammation.

With these considerations, and bearing in mind the anatomy and physiology of the heart, and especially the structure and function of the pericardium, let us proceed to inquire into the symptoms of pericarditis; and first of that variety involving mainly the serous layer of the pericardium.

Symptoms.—Acute serous pericarditis generally commences with a chill, or slight coldness, with marked faintness. Very soon, however, febrile reaction is set up, the skin becomes hot, the pulse frequent, the urine scanty, and there is a coated tongue, with loss of appetite, &c.

Sudden, severe, and lancinating pains are experienced, early in the disease, in the centre of the cardiac region, and extending in most cases to the epigastrium, and to the back between the shoulders. Along with the pain there is apt to be faintness, oppression, dyspnœa, palpitation of the heart, and a sense of constriction under the sternum, and in the left side of the thorax. There is also, in most cases, a distressing cough, with hurried respiration, vomiting, delirium, syncope, and, if the disease passes on, œdema of the face, and sometimes of the extremities.

The patient is unable to lie on the left side, and greatly prefers the sitting posture, with the head and shoulders inclined forwards. The short, dry cough is apt to continue, and is attended with sudden feelings of distressing faintness, more or less, through the whole course of the disease.

As the disease progresses, the pulse becomes irregular, intermittent, and almost imperceptible; the face becomes pale, and the lips livid; the speech is faltering, and the action of the heart becomes so feeble, that its impulse becomes almost imperceptible, and its sounds inaudible.

Palpation detects, in the early stages of the disease, an increased action of the heart, which may also be confirmed by *inspection*. There is also tenderness on pressure in the intercostal spaces, and also in the epigastrium, if the pressure be directed towards the pericardium.

Auscultation, as the disease becomes developed, detects a tumultuous, frequent, irregular, and sometimes intermittent action of the heart, corresponding with the pulse at the wrist. But as lymph is thrown out, a simple rubbing sound is heard, or the sound may be harsh, grating or creaking, or it may resemble the *bellows sound* of the valves, in disease.

Percussion over the heart, in the early stages of this disease, may elicit nearly a natural resonance, and it may continue nearly natural even after lymph is thrown out. But when effusion takes place, if that occurs, there is dullness on percussion, on the left side of the lower portion of the sternum, which gradually extends upwards. And in cases in which the pericardium becomes largely distended, the dullness extends in every direction, but especially transversely and inferiorly.

In such cases, as the fluid increases in the pericardium, the *impulse* of the heart is gradually diminished, till at last it is nearly or quite lost.

The *rhythm* of the heart becomes unnatural in this disease, varying in different cases from the *treble beat* to an utter confusion in the succession of sounds. And finally, in cases in which the pericardial effusion becomes considerable, there is a decrease in the sounds of the heart, the sounds appearing distant, and in some cases becoming almost inaudible.

Such, according to my observation, are the ordinary symptoms attending acute pericarditis, affecting mainly the serous layer of the pericardium. But *rheumatic* cases, which I believe consist, more especially in inflammation of the fibrous layer of the pericardium, have some peculiarities, the symptoms of which it may be proper for us to notice, as we proceed.

Rheumatic pericarditis, which occurs from metastasis of rheumatism from other parts, may be violent, and pass on rapidly to a fatal termination, the fibrous membrane being especially, and the serous membrane, of course, more or less involved. In such cases, there is a dull but severe pain in the region of the heart, with irregularity of the heart's actions, an irregular pulse, difficult respiration, occasional faintness or syncope, and a livid appearance of the countenance.

But when rheumatic pericarditis is not the result of metastasis from other parts, it may arise slowly, and the symptoms may be less violent. In such cases there is generally more or less fixed pain in the region of the heart, of a dull and distressing character, subject, however, to exacerbations, from even slight causes. There is some dyspnœa and oppression in the chest, and generally a short, dry cough ; the pulse is small, and at times irregular and intermittent, and the heart is liable to be thrown into violent paroxysms of palpitation by slight exertion, sudden mental excitement, &c.

If the inflammation is confined entirely to the fibrous layer of the pericardium, there may be no effusion, and consequently no dullness on percussion in the præcordial region. But if as generally happens in acute cases, and in protracted chronic cases the inner or serous coat of the pericardium becomes inflamed, effusion generally takes place into the pericardium, attended with dullness on percussion and most of the physical signs already enumerated.

Death in this disease sometimes occurs very suddenly and unexpectedly. But more frequently there is a slow wasting of the body and declension of the vital powers, which gradually leads to a fatal termination.

Anatomical Characters.—The pericardium is generally found coated, as well as the heart, with coagulable lymph or false membrane. This may be in layers or in shreds, or it may be in diffused patches on the surface of the heart and its investing serous membrane.

A serous, sanguineous, sero-purulent, or purulent, effusion is generally found in the pericardium, varying in quantity from a few ounces to several pounds. The two surfaces of the pericardium may be adherent interrupting more or less the action of the heart. And in cases in which the heart itself has been involved, there is found more or less change in its appearance and structure, according to the nature and extent of the cardiac disease.

In strictly rheumatic cases, involving exclusively the fibrous layer of the pericardium, I believe there is generally no pseudo-membranous exudation or serous or other effusion into the pericardium, and the surface of the heart may present a nearly natural appearance.

Such I believe are the most frequent morbid appearances presented in fatal cases of pericarditis, liable of course to variations, as we have seen depending upon the nature and extent of the pericardial inflammation.

Diagnosis.—Pericarditis may be confounded with endocarditis, pleuritis, pneumonia, and perhaps gastritis. But there are sufficient diagnostic symptoms by which it may be distinguished from these, and all other affections. In fact it is only necessary to bear in mind the dull or lancinating pain in the region of the heart, extending to the back and epigastrium, the irregular action of the heart, with the irregularity, intermission and weakness of the pulse, the difficult respiration and tendency to syncope or faintness, the disposition to remain in the sitting posture with the head and shoulders forward, and finally the œdema of the face and extremities and dullness on percussion over the heart, &c., in order to distinguish this from all other affections.

Rheumatic pericarditis may be distinguished from other cases by the general rheumatic condition or tendency of the patient, the dullness of the pain, and the fact of its metastasis from other parts, or its gradual development without many of the active symptoms which attend simple acute pericarditis.

Causes.—Whatever is capable of producing inflammation of any of the thoracic tissues or organs, may give rise to simple pericarditis. Among the causes which thus operate to produce the disease, are sudden exposure to cold, the healing up of old ulcers, the suppression of habitual evacuations, repelled cutaneous eruptions, the use of tobacco and intoxicating drinks, depressing mental emotions, masturbation and excessive venery, and various abuses of the system.

Rheumatic cases of pericarditis may occur from metastasis of the disease from the extremities, or other parts, in consequence of a depression of vital power, caused by overdepletion or other depressing influences, or rheumatic cases may arise slowly, in consequence of some of the depressing influences already enumerated, such as masturbation, the use of tobacco, drunkenness, excessive venery, and other abuses of the system, in rheumatic constitutions.

Prognosis.—The prognosis in pericarditis, except in very debilitated patients, or worn-down constitutions, is generally favorable. I believe that large a majority of rheumatic cases, and many others, terminate by resolution. If, however, in pericarditis there is great irregularity in the heart's action, as well as in the pulse, attended with dyspnœa, fainting, or syncope, a livid appearance of the countenance, and a large effusion takes place into the pericardium, a fatal termination may be apprehended.

Treatment.—In very robust patients an acute attack of simple pericarditis may occasionally call for general bleeding, but very rarely, I think. The feet should be placed in warm water, and a little warm sage tea allowed, with the hope of promoting perspiration

Cups should be applied at once over the heart, and also on each side of the spine between the shoulders, and a few ounces of blood taken, except in very anæmic or debilitated patients, in which case dry cups only should be applied.

A cathartic of calomel or podophyllin should be given at first, in castor oil, and the bowels afterwards kept open by small doses of cream of tartar. After the operation of the cathartic, two grains of calomel and four grains of Dover's powder may be given, every four hours, till slight ptyalism is produced. The warm foot-bath should be repeated, morning and evening, and blisters applied between the shoulders, immediately after the operation of a cathartic, as well as dry cups over the region of the heart.

After slight ptyalism is produced the calomel should be omitted, and the Dover's powder continued, with one grain of pulverized digitalis, every six hours, to quiet pain, promote perspiration, and to act gently on the kidneys.

If, however, at this stage there is evidence of effusion into the pericardium, five grain doses of the iodide of potassium may be given, every six hours, alternating with the Dover's powder and digitalis, and continued after the powders may no longer be required; the bowels being kept loose by small doses of cream of tartar.

The diet early, should consist of crust coffee, with a little milk, or arrow-root. But if the disease passes on, and becomes chronic, a moderately nourishing diet may be allowed. Drinks should be taken sparingly during the whole course of the disease, in order, if possible, to avoid effusion. In the latter stages of chronic cases, in anæmic patients, tonics may be indicated, and when they are, ten drops of the tincture of chloride of iron may be given, three times per day, after eating, in a wine-glassful of water.

Blisters may sometimes be indicated, either over the region of the heart, or between the shoulders, and when they are—should not be omitted.

In *rheumatic pericarditis*, of an acute character, especially that which results from metastasis of the disease from other parts, if the patient be plethoric, after the general bleeding, if indicated, the warm drinks, cupping, and cathartic should be resorted to, as well as the mercurial with Dover's powder, as in ordinary cases of simple pericarditis. But immediately after the operation of a cathartic, in such cases, if there is general febrile excitement, fifteen grains of the nitrate of potassa may be given, in a teacupful of warm crust coffee, every six hours, till the fever is subdued. This may be given, alternating with the Dover's powder, and as the nitre is discontinued, ten grain doses of the iodide of potassium may be given, every six hours in its stead, and continued till the violence of the disease is subdued; after which, it may be reduced to five grain doses, three times per day, and continued for a time.

In rheumatic pericarditis, however, occurring in anæmic or debilitated patients, and coming on by metastasis, or otherwise from debility, or some depressing influence; general, or even local bleeding may not be admissible, and the nitre is not generally required. But after the foot-bath, dry cupping, cathartic, &c. the Dover's powder, and mercurial may be given, with the sulphate of quinine, and immediately after discontinuing the mercurial, the iodide of potassium should be given, as already suggested, and continued till the disease is subdued, and its effects removed.

The bowels may be kept gently loose by cream of tartar. Blisters may be required over the region of the heart; and should the disease pass on, and the pericardium become greatly distended, paracentesis may be performed, as a last resort.

SECTION III.—ENDOCARDITIS.

By endocarditis, is here meant inflammation of the endocardium, or lining membrane of the heart; including generally its folds, the valves, as well as the fibrous layer which gives them strength, and the *chordæ tendineæ*, by which they are held in place.

The lining membrane of the heart, or endocardium, it will be remembered, is a serous membrane, continuous with the lining of the arteries and veins.

The endocardium, in the right side of the heart, besides lining the auricle and ventricle, forms folds constituting the semilunar valves, at the commencement of the pulmonary artery. It also, by three other folds, forms the *tricuspid valve*, by the help of a fibrous layer to afford strength, and the *chordæ tendineæ* to prevent their being driven back by the reaction of the blood, on the contraction of the ventricle.

In the left side of the heart the endocardium besides lining the auricle and ventricle has folds constituting the semi-lunar valves at the commencement of the aorta, and also two folds forming the mitral valve by the aid of a fibrous layer and the *chordæ tendineæ*. The fibrous layer of the mitral valve gives strength, and the tendinous chords prevent the valve from being driven back by the reaction of the blood on contraction of the ventricle, as in the case of the tricuspid valve at the right auriculo-ventricular passage. The endocardium thus situated and constituted, is liable to become inflamed, the symptoms of which we will now proceed to consider.

Symptoms.—Inflammation of the lining membrane of the heart commences with a feeling of anxiety in the region of the heart, and perhaps slight chilliness, followed by fever, with a strong, full and excited pulse.

The disease is attended with very little pain, but the valves soon become thickened, or coagula form, which interfere with the free movement of the blood: in consequence of which the heart contracts rapidly, the pulse being in some cases so frequent as scarcely to be counted, as well as small and feeble.

In consequence of the interruption to the free passage of blood through the heart, arterial blood is not supplied in sufficient quantity to the system generally, and the lungs, right cavities of the heart and venous system become more or less congested. As a consequence of all this, there are frequent faintings or syncope, paleness or lividity of the countenance, distressing dyspnœa, aggravated by exercise or mental emotion; the patient evinces great anxiety, cold sweats break out, and finally, if the disease passes on, the mind becomes wandering, coma or convulsions supervene, and the patient dies in a state of general œdema.

Palpation detects, early in the disease, an increased impulse of the heart; which, however, as the disease progresses, gradually becomes feeble.

Percussion elicits a dull sound over an extended surface, probably in part from "turgescence of the walls of the heart," but more especially from distention of its cavities.

Auscultation detects an unnatural *rhythm* of the heart, varying according to the degree of interruption to the free passage of the blood through the organ. In some cases there may be detected the *treble* beat, but in advanced stages of bad cases there is often an utter confusion in the succession of sounds. The ear also detects a murmur caused by imperfection of the valves, and also in part by a roughness or partial closure of the auriculo-ventricular, aortic, and pulmonary passages. The murmur may be a soft bellows sound, or it may be the rough or harsh *rasping* or *sawing* murmur; or it may resemble the tone of musical instruments, according to the degree of roughness of the passages and leakage of the different valves.

In the early stage of the disease, the sounds of the heart may be considerably increased, but in the latter stages, as the power of the heart muscle becomes exhausted, the second sound of the heart may be nearly lost, or obscured by the murmur that attends.

Such, I believe, are the ordinary symptoms of endocarditis; liable, of course, to variations, as is the case in all other affections.

Diagnosis.—Endocarditis may be confounded with pericarditis, from which it should be distinguished, however, by attention to the following differences.

In pericarditis there is either an acute and lancinating, or a dull, heavy pain in the region of the heart; while in endocarditis there is little or no pain. In endocarditis there is the valvular murmur, and generally a continuance of the first sound of the heart; while in pericarditis there is an absence of the valvular murmur, and ultimately great diminution, or utter loss of both sounds of the heart.

Finally, in endocarditis, though there is dullness on percussion, over an extended surface, the extent of surface affording a dull sound is by far less extensive than in pericarditis, attended with copious effusion. The two diseases may, however, be associated, in which case there may be noticed the valvular murmurs, in connection with the ordinary symptoms of pericarditis.

Anatomical Characters.—There is generally found redness of the membrane, often thickening, and occasional softening, with roughness or inequality, from fibrinous exudation. This fibrinous matter may be found in layers, or in granulations on any part of the membrane; but especially on the valves.

In a case that I examined in this village, a few years since, there was, in addition to the above appearances, an utter confusion in the appearance of the valves. They were thickened, distorted, and irregular in appearance, the tendinous cords were contracted, and the cavities of the heart presented altogether an unnatural or deranged appearance.

It is probable that the exudation upon the valves, in protracted cases of this disease, may assume a cartilaginous or bony character, and thus leave permanent disease of one or more of the valves, especially of the mitral or semilunar, or both.

Causes.—Various causes may produce endocarditis, among the most frequent of which are exposure to cold and dampness, the metastasis of rheumatism, the extension of phlebitis and arteritis, mechanical violence, rupture of the valves, from violent coughing, &c., urea in the blood, and various abuses of the system, such as masturbation, gluttony, drunkenness, licentiousness, &c. The disease is liable to arise, from slight causes, in persons of rheumatic constitutions, and it is probably, in many cases, of a strictly rheumatic character.

Prognosis.—The prognosis in endocarditis is rather favorable, as most cases recover under proper treatment.

In severe cases, however, in which there are violent palpitations, with dyspnœa and syncope, or faintness, irregularity, frequency and feebleness of the pulse, and a livid countenance, with general œdema, a fatal termination may be apprehended.

In cases in which most of the symptoms subside, leaving the *valvular murmurs,* danger may reasonably be apprehended of a permanent disease of the valves. But it is probable that the exuded lymph, in such cases, is sometimes gradually absorbed, and thus a permanent cure effected.

Treatment.—Immediately on an attack of endocarditis, if the patient is vigorous and plethoric, a few ounces of blood may be taken from the arm, but not in anæmic cases. Immediately after general bleeding when it is required, and at first when it is not, wet or dry cups, according to the strength of the patient, should be applied over the region of the heart, and on each side of the spine between the shoulders.

The warm-foot path should be used, and a little warm sage tea allowed as drink. A cathartic of calomel or podophyllin, should be given in castor oil, and the oil repeated if necessary. After the operation of the cathartic, two grains of calomel, with four grains of Dover's powder should be given every six hours, unless the patient be anæmic, and continued till slight ptyalism is produced. The warm foot bath should be repeated morning and evening, and a blister applied over the region of the heart, or on each side of the spine between the shoulders if necessary.

In cases in which the mercurial is given, it should be discontinued as soon as slight ptyalism is produced, and Dover's powder continued if necessary, with a grain of pulverized digitalis, as long as an anodyne is required.

The bowels should be kept moderately loose by small doses of cream of tartar, after the first cathartic, as it acts gently on the kidneys.

Immediately on discontinuing the mercurial when it is indicated, and on the operation of the first cathartic in cases in which a mercurial impression is not required, ten grain doses of the iodide of potassium should be given every six hours, and continued till the *bellows* or other valvular murmurs subside.

Early in acute cases, crust coffee with a little milk should be given for nourishment, but later in the disease, and in chronic anæmic cases, a good nourishing diet may be allowed.

SECTION IV.—CARDITIS.

By carditis, from καρδια, "the heart," and the termination *itis*, I mean inflammation of the fleshy substance of the heart, including its muscular and connecting cellular tissue.

The heart it will be remembered is a muscle so arranged as to form its different cavities, the muscular fibres being intermingled with cellular tissue. There are three layers of fibres, an external, middle, and internal, curved around each of the cavities of the ventricles, while the auricles have but two layers, an external and internal.

The heart is supplied with blood by the anterior and posterior coronary arteries, and its *nerves* are derived from the cardiac plexuses, which are formed by filaments from the pneumogastric and sympathetic. "Its *lymphatics* terminate in the glands about the root of the heart."

The heart thus constituted, and situated in the thorax between the two layers of pleura, constituting the mediastinum, is liable to become inflamed. Such an occurrence however is probably quite rare, no more than one well marked case having fallen under my observation, and that of a rheumatic character. The disease when it does occur, is attended with a train of symptoms which we will now proceed to consider.

Symptoms.—When carditis is fully established, whether as an original affection, or arising during the progress of some other disease, there is great soreness and pain in the heart, a frequent and irregular pulse, dyspnœa, faintness, lividity, prostration, &c. In the case that fell under my observation, the pain in the heart was continuous, but it appeared to be piercing at each contraction of the heart, and so distressing was this that the heart would appear in consequence to almost cease to act for a time. During this cessation in the heart's actions, there was dyspnœa, faintness, and lividity, with cold sweats, and the sensibility of the system would sink so low as apparently to render the patient unconscious of pain, when the heart would again resume a sort of spasmodic action for a time. Thus the symptoms continued for about forty-eight hours, after the supervention of the cardiac inflammation, when the heart became unable or utterly refused to act, thus putting an end to the most distressing train of symptoms I ever witnessed.

Auscultation detects an unnatural *rhythm*, with apparent intermissions in the heart's actions, the impulse and sounds of the heart being at one moment increased, and very soon greatly diminished, or entirely imperceptible.

In cases in which the carditis occurs as an original affection, it may be ushered in with a chill, followed by an irregular febrile reaction; but

I believe the disease more frequently arises during the progress of rheumatic or other inflammatory affections.

Diagnosis.—Carditis may be distinguished from endocarditis by the pain, by the absence of valvular murmurs, and by the much greater irregularity of the pulse, in the early stage of the disease, than is common in endocarditis.

From pericarditis this disease may be distinguished by careful attention to all the symptoms, and especially by the absence of signs of pericardial effusion. It is probable, however, that the pericardium, endocardium, and muscular tissue of the heart, may occasionally be all involved in a general inflammation, of an active or passive character, in which case there may be a combination of the symptoms enumerated as belonging to each.

Causes.—Carditis may arise from a great variety of causes, among the most frequent of which are sudden exposure to cold, a rheumatic condition, the use of tobacco, drunkenness, masturbation and excessive venery, gluttony, starvation, or any influence, either physical or moral, which is capable of producing either a plethoric or anæmic condition of the system; in the one case the inflammation being of an active, and in the other of a passive character.

Anatomical Characters.—The muscular tissue of the heart may be found of a very dark red color, softened, and more or less infiltrated with a bloody purulent matter, or filled with small abscesses containing pus.

Other morbid appearances may occasionally be presented, but the above I believe are the most frequent. It is probable, however, that carditis is sometimes the cause of hypertrophy and induration, as well as of "cartilaginous and osseous transformations."*

Treatment.—In carditis, in which the inflammation is of an active character, general bleeding, and by cups, warm pediluvia, cathartics, mercurials, anodynes, and diaphoretics constitute the treatment proper.

A moderate general bleeding may be followed by cups between the shoulders, and over the region of the heart. The feet should be placed in warm water, and warm sage tea allowed at first.

Calomel should be administered with castor oil, and after its operation, in severe cases, a slight mercurial impression should be produced, and, if necessary, blisters applied over the region of the heart, and between the shoulders.

In carditis, in which the inflammation is of a passive character, dry cups, warm pediluvia, mild cathartics, blisters, Dover's powder, with the sulphate of quinine, and a reasonable amount of food, are to be mainly relied upon.

In *rheumatic* cases, in addition to the treatment already suggested, full doses of the iodide of potassium should be given, every six hours, after the operation of the cathartic, or completion of the mercurial course, in cases in which that is indicated. In *active* cases it may be given with twenty drops of the wine of colchicum; but in *passive* cases it should be given alone, alternating with Dover's powder and quinine.

* See Dickson's Elements of Medicine, p. 340.

32

SECTION V.—ORGANIC DISEASES OF THE HEART.

Under the head of *organic diseases of the heart*, I design to include in this section, *hypertrophy, dilatation, diseases* of the *valves, aneurism* of the aorta, and *rupture* of the heart; giving first the general characters of organic cordiac disease, and then those peculiar to each variety.

It will be remembered that the right auricle on receiving the venous blood from the ascending and descending cavae contracts, forcing the blood through the right auriculo-ventricular passage into the right ventricle. The left auricle receives the arterial blood from the lungs through the pulmonary veins and contracts at the same time with the right auricle, passing the blood on through the left auriculo-ventricular passages into the left ventricle. Now both ventricles being filled with blood at the same time they contract together; which contraction, together with the reaction of the blood upon the tricuspid and mitral valves produce the first sound of the heart.

The blood of the right ventricle passes during this contraction into the pulmonary artery, and the blood of the left ventricle passes into the aorta. The pulmonary artery and aorta now contract, and this contraction, together with the reaction of the blood upon the semi-lunar valves, causes the second sound of the heart.

If, now, as we have already seen, the first sound, the second sound, and the interval be divided into fifths, the first sound will occupy two-fifths, the second sound one-fifth, and the succeeding interval the remaining two-fifths; "and this regular division of the time occupied by one complete circuit of the heart's functions constitutes the *rhythm* of the organ."

"The time occupied by the whole corresponds to the time between one stroke of the pulse and another; so that if a person's pulse beat eighty times in a minute, there would be eighty long sounds, eighty short sounds, and eighty intervals."

If the hand be placed on the left side of the chest below the nipple, in the space between the fifth and sixth ribs, a little to the inner side of a line running vertically over the nipple, a gentle regular pulsation is felt, and a slight shock is perceived. "This is the *impulse* of the heart," and is caused by the impinging of its apex against the parietes of the chest, is nearly synchronous with the pulse at the wrist, and with the contraction of the ventricles, as we have already seen.

Such, then, is the heart, its *impulse, rhythm* and *sounds* in a healthy state; but when the heart becomes diseased or deranged in its functions, a train of morbid *sounds, impulses,* &c. arise, each of which indicate, to a certain extent, the peculiar condition of the suffering organ.

The heart is liable, as we have seen, to organic disease, embracing *hypertrophy,* **dilatation**, diseases of the valves, and *rupture* of the organ; the general character of which we will first consider, and then proceed to inquire into the peculiarities of each.

General Symptoms.—The general symptoms of organic disease of the heart are what might be expected when we take into account the general structure and function of the organ as well as the organic changes to which it is liable.

The *countenance* in confirmed cases of organic cardiac affections, has generally a peculiar anxious expression; the face is pale, with lividity of the prolabia, and besides the *labial trait*, there is often a puffy swelling under the eyes.

The *respiration* is unnatural, the inspiration being either very quick and wheezing, or the patient breathes as though he had been walking rapidly, or, finally, the respiration may be short, the air not appearing to enter the lower portion of the lungs. Mental excitement, or corporeal exertion, such as walking up hill, or ascending a flight of stairs, is always liable to bring on paroxysms of dyspnœa.

In some cases, very trifling physical or mental excitement gives rise to suffocative breathing, with a feeling of constriction in the breast, great anxiety, a turgid and livid hue of the face, and especially of the lips, and, finally, distention of the veins of the neck, and an expression of distress and great suffering in the countenance.

The *action* of the *heart* is always deranged. In some cases it labors tumultuously, while in others it is in a state of tremulous agitation. More generally the *impulse* of the heart is very distinct, so as to communicate a motion to the whole superior part of the body. Sometimes, however, it becomes entirely imperceptible, for a time at least.

The *rhythm* of the heart becomes often unnatural, varying from the *treble beat* to more or less confusion in the succession of sounds.

The *sounds* of the heart may remain nearly natural, but generally they are either morbidly increased or diminished, and in some cases they become quite inaudible. The patient is liable to partial *faintings*, of long continuance, attended with great oppression in the region of the heart, a partial loss of consciousness, great anxiety in the præcordia, with a feeble, fluttering or intermittent pulse, and but slight perceptible respiration.

In the worst cases the patient is sometimes obliged to remain days and nights in the sitting posture, any attempt to lie down being followed by the most distressing paroxysms of palpitation and dyspnœa.

After organic disease of the heart has continued for some time, there is generally a dropsical tendency, with more or less swelling of the feet, and in the latter stages, effusion into the chest or pericardium. This may readily be accounted for, when we remember the imperfection of the circulation, and consequent congestion of the venous and capillary system of vessels.

Such, according to my observation, are the symptoms common to most organic diseases of the heart, being sufficient generally to justify a more close examination, to ascertain, if possible, the exact morbid condition of the suffering organ.

Diagnosis.—Organic may generally be distinguished from sympathetic affections of the heart, by careful attention to the following symptoms.

In organic disease, there is dyspnœa, with an anxious expression of the countenance, lividity of the lips, and in many cases the *labial trait*, consisting, as we have seen, of a slight depression extending from the angle of the lips to the margin of the chin, none of which symptoms are necessarily connected with sympathetic cardiac affections.

In organic disease, there is irregular action of the heart, involving derangement in its *impulse*, *rhythm* and *sounds*, as well as an irregular and generally intermittent pulse, while in sympathetic affections of the heart, the derangement in the heart's action is less marked, and the pulse is seldom intermittent. Finally, in organic disease, the patient has great difficulty in ascending a flight of stairs, or walking up hill, and in the latter stages, there is apt to be more or less dropsical effusion either into the cellular tissue, or into the serous cavities, none of which symptoms are common in sympathetic cardiac affections.

Causes.—A great variety of causes may operate, either directly or indirectly, to produce organic diseases of the heart, among the most frequent of which are carditis and endocarditis, a rheumatic condition of the system involving the heart; long continued anxiety of mind, the metastasis of cutaneous diseases; a syphilitic taint, and perhaps also an hereditary predisposition. Organic diseases of the heart are also produced by various abuses of the system, such as masturbation, excessive venery, the use of *tobacco*, intoxicating drinks, gluttony, starvation, and various other deviations from the laws of health.

I am satisfied, however, that by far the most frequent causes of organic diseases of the heart are, a rheumatic condition, masturbation, and excessive venery, and the use of tobacco, either by chewing, smoking, or snuffing.

Prognosis.—The prognosis in all organic diseases of the heart is unfavorable, yet by careful and judicious management many cases may be so far palliated as to allow the patient to live on for several years. This will depend, however, very much upon a strict observance of the laws of health on the part of the patient, and conformity to the rules of propriety in every respect. It should be remembered, however, that with all due prudence, the patient is liable to be taken away, at any moment, in advanced stages of organic cardiac disease.

Let us now inquire into the characters peculiar to each variety of organic disease of the heart, and then we shall be prepared to suggest the treatment proper for each; and, first, of hypertrophy.

HYPERTROPHY OF THE HEART.

By hypertrophy of the heart, I mean here "an overgrowth of the organ;" consisting in thickening of its walls, with or without contraction or dilatation of its cavities.

It appears to be a law of the animal economy that action, within certain limits, increases its growth; and especially is it the case, that an increased action of a muscle increases its bulk and strength. This rule applies to the heart, which is a hollow muscle; so that when obstruction exists to the free passage of blood through the organ, or the action of the heart is increased, from any cause, the muscular tissue may increase to two or three times its natural weight.

Symptoms.—In simple hypertrophy of the heart, without contraction or dilatation of its cavities, there is, in addition to the general symptoms of organic disease of the heart, a powerful and heaving *impulse*. If, however, there is dilatation with the hypertrophy, or thickening, the impulse is not only powerful, but diffused; being sometimes felt over the whole præcordial region.

Obstruction in the pulmonary artery or mitral valve, by producing increased action of the right ventricle, sometimes leads to hypertrophy of that portion of the heart, in which case there is an increased impulse, especially in the scrobiculus cordis.

Or if the mitral valve remains inefficient for a long time, the whole heart, except the left ventricle, may become hypertrophied, with or without dilatation, in consequence of increased action of the organ. In such cases, in addition to the increased impulse at the scrobiculus cordis, there is an increased impulse between the cartilages of the third and fourth ribs. Finally, in cases in which there is obstruction in the aortic valves, or in the arch of the aorta, the left ventricle is liable to become especially hypertrophied, along with the whole heart, and the increased impulse may be more perceptible below the nipple, a little to the left of the sternum.

In cases, however, in which the hypertrophy is from other causes than valvular disease, it is usually general, and if there is no dilatation the impulse is felt, strong and heaving, below the left nipple. But if with the hypertrophy there is dilatation, the impulse, as we have seen, may be felt over the whole præcordial region.

The *rhythm* of the heart, in cases not dependent upon valvular disease may remain quite natural, but in cases that arise from valvular obstructions, or imperfections, there is a disturbed rhythm, depending upon the character and extent of the valvular disease.

The *sounds* of the heart in dilatation, vary according to the condition of the organ. If with the hypertrophy, there is no dilatation, but rather a contraction of the cavities; the sounds of the heart may not be increased; but perhaps slightly diminished. If, however, with the hypertrophy, there is more or less dilatation, the sounds may be increased. The sounds in either case, may, however be obscured, by direct, or regurgitating murmurs, in cases of hypertrophy dependent upon, or connected with organic valvular disease.

Such, I believe, are the symptoms peculiar to hypertrophy of the heart. And it should be remembered, that while hypertrophy of the heart, is generally connected with, and dependent upon a deranged condition of the system, and valvular imperfection; the condition may arise independent of those causes, and apparently from a strong, vigorous, and robust condition of the system, at least so far as can be discovered.

Diagnosis.—Hypertrophy of the heart, with or without dilatation, may be distinguished from simple dilatation, by careful attention to all the symptoms, and by the following differences: In hypertrophy, with or without dilatation, the sounds may be nearly natural, or but slightly increased; but there is a powerful and heaving impulse, while in simple dilatation, without hypertrophy, the impulse is weak, but the sounds may be morbidly increased.

Finally, hypertrophy may occur in strong robust subjects; while simple dilatation generally occurs in anæmic, weak, and feeble patients.

DILATATION.

By dilatation, I mean here, an increase in the bulk of the heart, with proportional "attenuation of its parietes," and consequently no hypertrophy.

This form of organic cardiac disease, generally occurs in anæmic, weak and feeble subjects, or those in whom the muscular system is in a weak and relaxed condition.

Symptoms.—In cases of simple dilatation of the heart, the symptoms do not differ widely from those which attend dilatation, with hypertrophy. The patient generally, however, has a pale anæmic appearance, a feeble pulse, and is subject to paroxysms of distressing, tremulous palpitation.

The *impulse* of the heart in simple dilatation, is generally weak. But if the dilatation is considerable, the impulse may sometimes be felt over a larger extent of surface than in health.

The *rhythm* of the heart is always more or less disturbed; while the *sounds* are apt to be morbidly increased. The sounds, however, are liable to be obscured by direct or retrograde murmurs, caused by disease in, or leakage of the valves. *Percussion* elicits a dull sound, over an extended surface in the præcordial region, and this increased dullness may be general, in this region; or it may be most marked over some particular portion of the heart, if the dilatation of one cavity is more than of another.

In case the left ventricle is especially dilated, the dullness is most marked below the left nipple, over the costal cartilage. If the left auricle is mainly dilated, the increased dullness is most conspicuous on the third and fourth ribs and their cartilages to the left of the sternum. If, however, it is the right ventricle that is especially involved in the dilatation, the increased dullness will be most apparent to the right of the sternum, at the inner edge of the right mammary region and at the scrobiculus cordis.

Such, according to my observation, are the most prominent symptoms peculiar to simple dilatation of one or all the cavities of the heart; by a careful observation of which, the real condition may generally be very nearly ascertained.

DISEASE OF THE VALVES.

Another variety of organic cardiac affection consists in a chronic disease of the valves, involving, in some cases one, and in others several or all the valves connected with the heart, but more frequently the mitral and semi-lunar, the tricuspid being more rarely involved.

The valvular disease may consist in mere thickening, or the edges of the valves may be united by inflammation, so as to narrow the passage; or the valves may contract adhesions, or the tendinous chords become lengthened or shortened so as to prevent the possibility of an accurate closure of the valves. But in protracted cases, the fibrous layer of the valves is very liable to become thickened, indurated, and finally converted into fibro-cartilage or bone, thus rendering the free motion of the valves, and disqualifying them for the due performance of their function.

Or, instead of this transformation of the fibrous tissue into fibro-cartilage or bone, there may be a calcareous or steatomatous deposition beneath the serous membrane, forming masses of considerable size, leading to inflammation, ulceration, and final rupture of the valves, and consequent fissures of greater or less extent.

Finally, fleshy, warty, or cartilaginous excrescences, of various sizes, the result of endocarditis, may form on the valves, or on the endocardium along the passages at the seat of the valves, which not only interfere with the free action of the valves, but also hinder the free current of blood along the passage.

The valves, too, may become atrophied, or in case of considerable dilatation of the heart, may become too small for the passage they were designed to close; in either case, allowing a regurgitation of the blood when a perfect closure should have been effected.

The mitral, and semi-lunar valves at the commencement of the aorta, are the ones most frequently affected; but the tricuspid and pulmonary are occasionally involved, as we have already seen.

Symptoms.—The symptoms of chronic valvular disease embrace all or nearly all before enumerated as belonging to organic diseases of the heart. There is often symptoms of carditis or endocarditis preceding the valvular disease, and in most cases a rheumatic condition of the system, connected with, if not always acting as cause, either predisposing or exciting, of the valvular disease. So, too, the impulse, rhythm and sounds of the heart are more or less deranged, as in dilatation and hypertrophy, and the pulse irregular, and often intermittent.

The symptoms, however, which are especially peculiar to chronic valvular disease, are the direct and regurgitating *murmurs* which arise, the one from partial interruption to the free passage of the blood, and the other from a leakage of the valves, in consequence of some one of the morbid conditions of the valves or endocardium, already referred to.

The morbid sounds or *murmurs*, both direct and retrograde, thus produced, vary according to the nature and extent of the valvular disease. In some cases the sound is the soft "*bellows murmur*," the "*bruit de souflet;*" in others it is the harsher or "*rasping murmur*," the "*bruit de rape;*" or it may be the harsh or grating "*sawing murmur*," the "*bruit de scie;*" or finally, the murmur may resemble the tone of musical instruments, constituting the "*musical sound.*"

If now a direct murmur exists in one valve, and a retrograde murmur in another, they may appear single as to time, but they may generally be distinguished by their location. If, however, a murmur arises from the direct and retrograde passage of the blood through the same valve or corresponding pair of valves, the murmur is double, because it occurs at different times.

To ascertain which of the valves are involved, care should be taken to get at the seat of the murmurs which arise, and then by recollecting the location of the valves, a correct conclusion may generally be drawn, in relation to the nature and seat of the disease.

In case regurgitation takes place through the aortic valves, a murmur is heard over the sternum, a little below the third rib. This murmur occupies the time of the second sound of the heart, which it may conceal. "If, however, there is both obstruction in, and regurgitation through the aortic valves, a double murmur or *see-saw* sound is heard over the sternum, opposite the third rib." This sound may be heard along the course of the aorta, but less distinctly as the ear is passed from over the aortic valves. Disease of the aortic valves is generally attended with more or less throbbing of the arteries, and a "peculiar jurking pulse."

Should regurgitation take place through the pulmonary valves a murmur is produced, occupying the time of the second sound of the heart; and should there also be a direct obstruction, the double murmur will be heard, as in the same condition of the aortic valves; but the murmur will be heard nearly opposite the cartilage of the third rib, to the left of the sternum. The murmur is heard for a little distance up the pulmonary artery, but not along the aorta, and there is not the jerking pulse or throbbing of the arteries which attend in disease of the aortic valves.

In case regurgitation takes place through the mitral valve, the murmur occupies the time of the first sound of the heart, and is most distinctly heard between the cartilages of the fourth and fifth ribs, a little to the left of the sternum, and below the nipple. It may also be audible in the axilla, and close by or upon the spine, in the interscapular region.

In case direct obstruction exists, the murmur is heard in the same position, but it is comparatively feeble, and occupies the time of the second sound of the heart instead of the first.

The tricuspid valve is seldom the seat of disease, but when it is the regurgitating murmur occupies the time of the first sound of the heart, and the direct murmur the time of the second, the murmurs being heard most distinctly over the lower part of the sternum. Disease of this valve is attended with distension, and sometimes visible pulsation of the jugular veins, which may aid in pointing to the seat of the cardiac disease.

ANEURISM OF THE AORTA.

Closely connected with organic cardiac disease is aneurismal dilatation of the aorta. It may occur in the ascending, in the arch, or in the descending portion, but more frequently I believe in the ascending portion.

Symptoms.—The general symptoms of this disease are mainly those belonging to organic diseases of the heart. Sometimes, however, these aneurismal tumors acquire a large size, compressing the trachea, bronchia, œsophagus, and even displacing more or less the heart itself. They also frequently produce murmurs similar to those produced by the direct and retrograde passage of the blood through diseased valves. And there is marked dullness on percussion over the aneurismal tumor.

Diagnosis.—Aneurism of the aorta may be distinguished by the presence of the *bellows, rasping,* or *sawing* murmurs over the seat of the disease. If the aneurism is of the ascending portion, or of the arch, it may be distinguished in most cases, by dullness on percussion in the upper sternal and subclavian regions. But if the aneurism exist in the descending portion, and is of large size, it may be detected by dullness on percussion along the left side of the spine, and in case it projects forward, dullness may be perceptible " in the præcordial region."

Such then, according to my observation, are the general symptoms of organic diseases of the heart, and also the peculiar symptoms and characters of hypertrophy, dilatation, diseases of the valves, and aneurism of the aorta, there being only left for our consideration, *rupture* of the *heart,* and finally a few suggestions in relation to the treatment of organic cardiac diseases.

RUPTURE OF THE HEART.

Rupture of the heart is a very rare, interesting and singular affection. If complete, it sooner or later fills the pericardium with blood, and produces death, if not otherwise, by pressure upon the heart which suspends its action.

The left ventricle appears to be the most frequent seat of the rupture, but it may occur in any part of the heart. And it may occur without any apparent disease of the organ from violent mental emotions, or from extraordinary physical exertion.

Symptoms.—The symptoms are a sudden uneasiness or pain in the heart, followed by dyspnœa, syncope, and sudden death; or these symptoms may continue on, with irregular action of the heart, and an irregular intermittent pulse, for a few hours, or a day or two, when death takes place, by a cessation of the heart's actions.

A case of rupture of the heart fell under my observation a few years since, in Jefferson county, in this State. It occurred in a middle-aged lady, of rather feeble or slender constitution, and apparently from an effort she made to toss her child playfully above her head. At the moment of doing this, she felt some pain and uneasiness in the region of the heart. The uneasiness continued to increase, with irregular palpitations, slight dyspnœa, and giddiness, for the succeeding forty-eight hours, at which time she suddenly died.

I examined the body, in company with Drs. Webb and Whitmore, of Adams, N. Y., and found the heart completely enveloped with a firm coagulum of blood, the patient having evidently died from mechanical pressure of the blood on the heart, suspending its functions. The rupture was of the right auricle, and the result of an ulceration, which had partially perforated its walls. It was about the size of a crow-quill, its edges being, as is usual in such cases, rough and ragged.

Such, then, are organic diseases of the heart, the general and particular characters of which I have attempted thus imperfectly to trace.

There remains now, in conclusion, a few suggestions to be made, in relation to the general and particular treatment of these diseases, which will close what I have to say on this subject.

Treatment.—The *general treatment* proper for organic diseases of the heart, including aneurism of the aorta, consists mainly in subjecting the patient to a strict observance of the laws of health, in every respect. And this should have reference to food, clothing, exercise, temper of mind, and, in short, everything that can have an influence on the health of the patient.

In plethoric cases, exercise, low diet, saline cathartics, &c., may be indicated. But in anæmic cases, moderate exercise, a nourishing diet, and tonics may be required, and when they are, should be adapted to suit each particular case.

In *hypertrophy* of the heart, in addition to low diet, saline cathartics, and a reasonable amount of exercise, if the patient is very plethoric, general bleeding may occasionally be required, but not generally I think. Blood may be taken by cups occasionally, however, in most cases of hypertrophy, with very good effect, not only from the region of the heart, but also from the left interscapular region.

In *simple dilatation* of the heart, attended, as it generally is, with an anæmic condition, and a relaxed and debilitated state of the muscular system, moderate exercise, a good nourishing diet, and tonics, may be required. For the blood, some preparation of iron is usually indicated, and generally I think the tincture of the chloride, in ten drop doses three times per day, will be found to do best.

In cases in which there is great nervous prostration, in addition to the chalybeate, $\frac{1}{20}$ of a grain of the muriate of strychnia may be given, three times per day, for a time with good effect. A grain of the muriate may be dissolved in eight ounces of water, of which a teaspoonful should be given three times per day.

In *disease* of the *valves*, any lingering cardiac inflammation should be subdued, by cups, wet or dry, over the heart or between the shoulders; and if necessary, blisters, or pustulation by tartar emetic, should be resorted to. In all rheumatic cases the iodide of potassium should be given in five grain doses, three times per day, and continued for a long time.

Finally, in all organic diseases of the heart, and in aneurism of the aorta, in which there is a frequent and irregular pulse, unless great debility contra-indicates, the tincture of digitalis may be given in ten drop doses, three or four times per day, with very good effect.

SECTION VI.—SYMPATHETIC AFFECTIONS OF THE HEART.

By sympathetic affections of the heart, I mean those derangements in the organ or its functions, which arise either from a primary disease in some other organ or part, or else from a general irritable or deranged condition of the system, the cardiac derangement being purely sympathetic.

Sympathetic affections of the heart often produce symptoms, which to the superficial observer might be mistaken for those of organic cardiac disease. It is therefore of the highest importance to understand this subject.

It should be remembered that the *sympathetic system* consists of a series of ganglia, extending along each side of the spine, forming a chain its whole length, communicating with all the other nerves of the body, and sending branches to all the internal organs and viscera.

The sympathetic generally communicates with the cerebro-spinal nerves at their exit from the cranium and vertebral canal, and it sends branches to accompany the arteries, often forming communications or plexuses around them. "All the internal organs of the head, neck, and trunks are thus supplied with branches from the sympathetic, and some of them exclusively, hence it is considered the nerve of organic life."

The sympathetic ganglia appear to be composed of a mixture of medullary and cineritious matter, and besides tending to equalize nervous force may possibly be productive of a peculiar nervous power.

The sympathetic system probably confers vitality on the organs of organic life, and hence exerts a controlling influence over the involuntary functions of digestion, absorption, secretion, circulation and nutrition. Another important use of the sympathetic system is to form a connection or communication of one part of the system with another, so that one organ or part can take cognizance of the condition of every other organ or part, and act accordingly. This provision appears necessary so that in case disease seizes, for instance upon the brain, the stomach by virtue of its sympathetic relation appreciates the morbid condition of the brain, and as food in such a case might add to the cephalic disease, the stomach refuses it, and perhaps even **throws off** that which it had already taken.

Now this sympathy exists as we have seen between all the organs of the body, being strongest between those parts which carry on the involuntary functions. But as the cerebro-spinal nerves communicate freely with the sympathetic, this sympathy exists in a greater or less degree between every part of the system, and thus influences more or less every voluntary and involuntary function.

With these considerations it is easy to see that the functions of the heart may be materially affected by a local disease of other organs, as well as by an irritable or deranged condition of the system generally. And so we find that the heart is liable to be excited into vehement and tumultuous action by a variety of causes, not immediately connected with lesion or disorder of its structure, but depending upon irritation or disease located in some remote organ or part of the system.

The local diseases most liable to produce sympathetic affections of the heart, according to my observation, are derangement of the digestive organs, of the genital organs, and spinal irritation. And the causes most frequently operating, irregular eating, the use of tobacco, and intoxicating drinks, masturbation, onanism, and venereal excesses.

Symptoms.—The symptoms of sympathetic affections of the heart, are similar, in many respects, to those of organic disease; consisting often in an increased or diminished action of the heart, by which its impulse and sounds become either morbidly increased or diminished; there being generally a corresponding irregularity of the pulse.

Sometimes, even a slight bellows murmur is heard over the aortic, or mitral valves, as well as the anæmic murmur in the jugular veins; and slight murmurs may sometimes be produced, by compressing the large arteries.

The paroxysm is generally brought on by physical or mental excitement; but not always, and when it does occur, from whatever cause, is attended with oppression in the præcordial region, difficult respiration, and sometimes syncope or faintness. The irregular cardiac action, is also apparently often produced, by an accidental increase of the local irritation, or disease of which the cardiac derangement is symptomatic. This is true, whether the primary irritation be in the digestive or genital organs, or in the spinal cord.

Diagnosis.—Sympathetic affections of the heart, may generally be distinguished from organic disease, by attention to the following differences, together with all the symptoms, and circumstances of the case.

In sympathetic affections of the heart, the paroxysms of palpitation generally come on from some physical, or mental excitement, or some accidental increase in the local irritation, of which the cardiac affection is symptomatic; while in organic disease, the paroxysms more frequently come on, without any very special exciting cause, and frequently, while the patient is quiet in bed.

In sympathetic affections, there is not that lividity of the face, or the *labial trait*, or so constant an intermission of the pulse, as well as dropsical tendency, so common in organic cardiac disease. Finally, there is seldom heard any distinct murmur in the sympathetic affections, except over the aortic or mitral valves, and these are by no means constant or frequent; while in organic disease, there is often a continuous, direct or retrograde murmur, or both, in one or more of the valves.

Causes.—A great variety of causes, may operate to produce sympathetic affections of the heart. But by far, the most frequent, according to my observation, are irregular eating, masturbation, and excessive venery; the use of tobacco, and the excessive use of intoxicating drinks.

Any derangement of the digestive functions, may produce sympathetic cardiac affections; but I believe that produced by irregular eating, is by far the most frequent. And this is not strange, when we remember that the stomach, after digesting a meal, is in no condition for taking or digesting food, for a few hours at least. But if, as many children, and some old people are in the habit of doing, food be taken at irregular times, between meals, the food produces an irritation of the stomach, and being imperfectly digested, is passed along the intestines in a crude state, irritating the intestines.

This process need only be continued for a little time in order to produce derangement of the alimentary canal, and in fact the whole digestive apparatus.

Now, as the digestive organs and heart are supplied freely by the sympathetic nerves; the heart strongly sympathizes, and hence the irregular action of the heart, and all the symptoms which arise in sympathetic cardiac affections from this cause.

Over-excitement of the genital organs, by masturbation or sexual excesses, is probably the next most frequent cause of sympathetic derangement of the heart. The habit of masturbation is sometimes commenced in both males and females, even during childhood, and if not, at puberty or a little later; and during its continuance, the womb in the young female, and the testicles in the male become exceedingly irritable, liable to be thrown into a state of excitement by the slightest causes.

Now, as the genital organs are supplied largely by the sympathetic and spinal nerves, they readily communicate a morbid impression to the stomach and heart, and thus masturbation becomes a frequent cause of sympathetic cardiac affections. And besides, masturbation and excessive venery in the male acts as a direct drain upon the system, weakening the blood, and debilitating, enfeebling, and deranging the cerebrospinal and whole nervous system, thus producing or increasing sympathetic cardiac derangement.

Tobacco, used by chewing, smoking or snuffing, is a frequent cause of sympathetic affections of the heart; and this is not strange when we remember that it is a poisonous plant, containing *nicotia, nicotianin*, and an empyreumatic oil, all of which are most virulent poisons, and rapidly destructive of animal life. Now, tobacco cannot be brought even in smelling distance of living animals without becoming more or less injurious. When, therefore, it is drawn into the nostrils or kept constantly in the mouth, undergoing the chewing process, or in the form of smoke, is drawn into the lungs, it imparts more or less of its poisonous properties to the blood, and also acts as a direct poison to the cerebro-spinal and nervous system, deranging, in a greater or less degree, every function of the body. In consequence of this derangement of the general system, and especially of the cerebro-spinal and nervous system, sympathetic or nervous palpitation is produced; very many marked cases of which have fallen under my observation during the past few years.

Intoxicating liquors are also a frequent cause of sympathetic cardiac affections. Alcohol contains no nourishment that can be available in the human system, and only appears to act as a tonic by the local and general irritability which it produces when taken or administered. The continued use of alcohol, then, in any form, produces irritation of the mucous membrane of the stomach, and a general irritable condition of the system, and thus it leads to nervous palpitation or sympathetic affections of the heart.

Spinal irritation, especially if it be of the cervical portion, whether it be the result of direct injury, or symptomatic of other irritations in distant parts, may be a cause of nervous palpitation: the irritation being communicated by the spinal to the sympathetic nerves, and so to the heart itself.

Various other causes are capable of producing nervous palpitation of the heart, such as mental depression, want of sleep, overtaxing the system, fear, anxiety, starvation, or insufficient food, and finally, an anæmic condition from any cause, all of which influences should be taken into account, in investigating this multiform variety of cardiac derangement.

Prognosis.—The prognosis in sympathetic affections of the heart is generally favorable, as the irregular action of the heart usually subsides if the local irritation of which it is symptomatic, and its cause, can be ascertained and removed.

Protracted cases may, however, lead on to organic cardiac disease, or other derangements. In one case of nervous palpitation that fell under my care, in an anæmic female, in which the venous or anæmic murmur was very distinct, there was the supervention of *turgescence*, and finally, *enlargement* of the thyroid gland; and also *exophthalmia* or **prominence** of the *eyeballs*.

In this case the goitre was increased by any considerable excitement, temporarily, and so were the prominence of the eyes; the increased action of the heart, at such times, evidently producing an active congestion of the thyroid gland, and also of the tissues forming the cushion of the eyes, making them more prominent, for the time.*

Treatment.—In the treatment of nervous palpitation, or sympathetic affections of the heart, the remote local irritation should be sought out, as well as the cause which has produced it, and both removed if possible.

If the stomach or digestive organs are the seat of the primary irritation, and irregular eating the cause, that habit should at once be cor-

---

* The prominence of the eyeballs, though they were not enlarged, gave them that appearance, as is usual in such cases, appearing at times almost frightful.

rected, and the patient directed to take food at regular meal hours, and nothing between meals. Correcting this habit alone will sometimes effect a removal of nervous palpitation. But if the blood has become weak, two grains of the citrate of iron may be given in solution, three times per day, and continued for a few weeks; when, as the gastric irritation subsides, and the anæmic condition is overcome, the sympathetic cardiac affection will gradually subside.

In cases in which the genital organs are the seat of the primary irritation, and masturbation the cause, whether the patient be male or female, it is sometimes difficult to get a confession of the fact. In such cases, however, the pale face, the downward look, and the idiotic expression of the countenance, together with the "*oculo-zygomatic trait*," consisting of a leaden streak extending from the greater angle of the eye to the projection of the check bone, will generally be sufficient to render the diagnosis clear.

In all cases of this kind, whether the patient be male or female, the fearful consequences of the habit should be pointed out, and its continuance strictly forbidden. If this is obeyed it will generally, in a reasonable time, effect a cure of the sympathetic nervous palpitation, and health will be restored.

If, however, the habit has been continued for a considerable time, and the blood has become reduced, ten drop doses of the tincture of the chloride of iron may be given three times per day, and continued for some time. And should there be seminal weakness, or spermatorrhœa, as is frequently the case, the muriate of strychnia should be given in $\frac{1}{60}$ of a grain doses in solution three times per day, and continued till it is overcome.

In those cases which occur in middle aged males or females, and which depend upon onanism, or excessive sexual indulgence, the consequences of the imprudence should be pointed out to the patient, and a reform insisted upon. And in case this prescription is followed, the nervous palpitation will generally subside.

In those cases depending upon the use of *tobacco* or intoxicating liquors, the habits should be broken off as fast as possible, without producing delirium tremens, and the cause being removed, the sympathetic cardiac affection will generally subside. It is however generally well in these cases to administer the syrup of the iodide of iron in ten drop doses three times per day for a time, to restore the blood, and correct the secretion of the glandular system which is generally more or less impaired in such cases.

In cases of nervous palpitation depending upon spinal irritation, dry cups, blisters, or tartar emetic pustulation along the irritated portion of the spine, along with such general treatment as may be indicated, should be resorted to.

Finally, in cases of nervous palpitation depending upon an anæmic condition from various causes, a return to the laws of health, and a mild tonic course will generally be sufficient to effect a cure.

SECTION VII.—NEURALGIA OF THE HEART.

By neuralgia of the heart, I mean that peculiar neuralgic cardiac affection, sometimes called *angina pectoris*, which develops itself by a sudden paroxysm of pain in the heart and præcordial region, attended in some cases with oppression in the chest, dyspnœa and violent palpitations.

The heart it will be remembered, is supplied with nerves from the cardiac plexuses, which are formed by filaments from the pneumogastric and sympathetic, all of which communicate freely with the spinal nerves, rendering the heart liable to become affected by spinal congestion or irritation.

With this in mind, we are prepared to proceed to the consideration of neuralgia of the heart, and first of the symptoms which are developed.

Symptoms.—An attack generally comes on suddenly with pain in the region of the heart, extending in some cases through the left side of the chest into the shoulders, and along the nerves of the arm to the fingers. Along with the pain there is dyspnœa, anxiety, a sense of suffocation, and sometimes distressing palpitation of the heart.

If an attack occurs while the patient is walking, every motion is at once suspended as the least exertion gives rise to intense darting pain in the cardiac region.

During the paroxysm the countenance is pale and expressive of anxiety, the extremities are cold, there is turgidity of the vessels of the head, and in some cases syncope and convulsions occur. The pulse is always irregular, and may be small and feeble, or strong and full, during the paroxysm, which lasts from a few minutes to half an hour or more.

The first attacks of this disease occur at long intervals, generally from fast walking, over-eating, or violent exercise. But by repetition the paroxysms become more violent, occur more frequently, and from very slight causes, and are generally of longer duration. After a paroxysm has subsided, the patient feels a numbness in the left arm and shoulder; there is slight palpitation and headache, hurried respiration, and an anxiety of feeling in the region of the heart.

Causes.—Various organic lesions of the heart may predispose to this disease. But I believe that the exciting causes are similar to those which produce gouty, rheumatic and neuralgic affections of other parts.

It is probable that the disease may arise from irritation in the stomach, liver, or almost any other internal organ, by sympathy. But I believe it generally occurs from an irritation in the pneumogastric or spinal nerves, from cephalic or spinal congestion or irritation. And in a very large majority of cases I am confident that spinal congestion or irritation is the direct cause of this, as it is of other neuralgic affections.

Pathology.—Various opinions may be entertained in relation to the pathology of this disease. But when we remember the suddenness of the attack, the acute and lancinating character of the pain, and that it extends from the cervical spine, along the nerves of the arm, as well as to the heart, it appears to me there can be no reasonable doubt of its being of a purely neuralgic character.

We have already seen that the sympathetic and pneumogastric nerves which supply the heart, are intimately connected with the cervical portion of the spinal nerves. Now any sudden congestion, or increased irritation of the cervical or superior portion of the spinal cord, may so far irritate the brachial and cardiac nerves as to develop this highly acute and distressing neuralgic affection.

The only symptoms attending this disease, not common to neuralgia of an intercostal, or any other nerve, are those which arise from the derangement in the heart's action; a circumstance which might be expected to arise.

Prognosis.—The prognosis in cases of this disease not complicated with organic lesion of the heart is rather favorable. But the patient may die suddenly, from a suspension of the heart's actions, or the paroxysms may by repeated repetitions wear out the patient, and thus produce death.

Treatment.—Immediately on an attack of this disease cups should be applied to the back of the neck, and along each side of the superior portion of the spine, and if the patient is plethoric, a few ounces of blood should be taken, especially on the left side.

In weak, anæmic cases no blood should be taken, and should the pulse be weak, and the patient appear pale, cold and sinking, the back should be rubbed freely, along its whole length, with a warm infusion of capsicum in vinegar. And should this fail of bringing on speedy reaction, warm ginger or pepper tea may be administered, and if necessary warm brandy sling.

Thus the paroxysms may generally be arrested. But in order to prevent their return, blisters should be applied on each side of the superior portion of the spine; the habits and general health of the patient should be corrected, and if the patient is gouty or rheumatic, guaiac, colchicum, or iodide of potassium should be given, and continued till the condition is overcome as far as may be.

SECTION VIII.—SYNCOPE.

By syncope is here meant sudden loss of motion and sensation, with diminution or temporary suspension of the action of the heart and of respiration.

Symptoms.—Syncope may come on without any premonitory symptoms, but more generally there is slight nausea, oppression at the epigastrium, mental confusion and a sinking pulse; and when it becomes complete, the features are pale and sunken, the surface cold, the pulse and respiration imperceptible, and there is a want of consciousness. The ear applied to the region of the heart, however, detects the first sound, though it be very weak, and after a little there may be noticed a slight inspiration, soon a return of color in the lips, and finally there is a perceptible pulse at the wrist, with natural respiration and a return of consciousness.

A complete state of syncope seldom lasts more than a few seconds; occasionally, however, it continues several minutes, and in some rare cases it is continued for hours or even days. Such are cases of apparent death, in all of which there is probably an imperceptible respiratory movement and a very feeble action of the heart. I believe, however, that the first sound of the heart might generally be detected in such cases by careful auscultation.

Diagnosis.—The conditions with which syncope are liable to be confounded are apnœa and death. In apnœa there is a want of consciousness, absence of sensible respiration, and no perceptible pulse at the wrist, and also general venous congestion, with a bloated and livid appearance of the face; while in syncope, though there is the apparent suspension of consciousness, respiration, and of the heart's action; the countenance is pale and collapsed, and the surface generally bloodless. Attention to the cause may also aid in forming a diagnosis if it be remembered that the causes of apnœa act directly to suspend respiration, while the causes of syncope act more directly upon the heart.

In cases of syncope that resemble death, there is a want of coldness of the body, of the cadaveric rigidity, and also of the settling of the blood to dependent parts; but generally a possibility of detecting, by careful auscultation, the first sound of the heart; all of which are indications of remaining vitality. In cases of doubt, however, burial should be deferred till there is a commencement of putrefaction, lest the unfortunate blunder of burying alive be fallen into.

In cases of hysterical insensibility, in which the patient lies for days without consciousness; there is a gentle respiratory movement and a distinct pulse at the wrist, which afford sufficient evidence of remaining vitality.

Causes.—The causes of syncope may act directly upon the heart, such as the effects of organic disease, of neuralgia of the heart, rheumatism or gout, the effects of certain poisons, air in the blood vessels, &c. Or they may act on the heart through the nervous system. A great variety of causes act to produce syncope in this way, such as terror, sudden fright, severe pain, violent injuries, hunger, &c.

Loss of blood or a weak state of the blood is also a frequent cause of syncope. This cause, probably, operates by depriving the brain of a due amount of stimulus, thus suspending, for the time, the heart's action. In fact it can scarcely be otherwise, for it is well known that syncope occurs more frequently when the patient is in the sitting posture, while less blood is passing to the brain, than when in the horizontal position.

Prognosis.—The prognosis in syncope is generally favorable, unless it be attended with great loss of blood, or organic disease of the heart. I believe, however, that fatal syncope sometimes occurs in acute disease in which there is great debility, from a little imprudent exertion; an unfortunate case of which very recently fell under my observatation.

Treatment.—In all cases of approaching or complete syncope the patient should lie down, or be placed in the horizontal position, that more blood may flow to the brain, and thus sufficient nervous influence be generated and sent to the heart, to enable it to resume its functions.

After placing the patient in the horizontal position, cold water should be sprinkled in the face, to arouse the nervous system, and if this fails, camphor, spirits of ammonia, or other pungent volatile liquids should be held to the nose for the same purpose. If the patient does not readily revive, and is able to swallow, half a drachm of the tincture of camphor, or of the aromatic spirit of ammonia, may be administered, in a little water, and should this be insufficient, brandy and water may be required.

Should the patient be unable to swallow, brandy or carbonate of ammonia, diluted with mucilage of gum arabic, should be injected into the rectum. Sinapisms should be applied to the epigastrium, an infusion of capsicum along the spine, frictions of flannel should be made to the extremities, and finally, if the case is obstinate, artificial respiration should be resorted to, as for cases of asphyxia.

<h3 style="text-align:center">SECTION IX.—ARTERITIS.</h3>

By arteritis, I mean inflammation of the arteries, which, though an occasional, is not a very frequent occurrence.

The arteries, it will be remembered, are cylindrical tubes, which convey the blood from the ventricles of the heart to every part of the system. The aorta proceeding from the left ventricle, and conveying the arterial blood to the capillaries, throughout the system; the pulmonary artery conveying the venous blood from the right ventricle to the capillaries of the lungs.

The branches of the aorta are given off mostly at right angles; while in the limbs, the branches from the main arteries, are given off at an acute angle, in the one case, tending to check the impetus of the blood from the main trunk, and in the other, to favor it. The onward flow of the blood in the arteries, is also favored by the increased capacity of the branches, towards their extremities, every division of the arteries giving a combined area of the two branches, greater than that of the original trunk, before division.

The arteries communicate freely with each other, and finally terminate in the capillaries, only about $\frac{1}{3000}$ of an inch in diameter, which extend from the extreme arteries, to the extreme veins, and form a network through every part of the body.

The arteries, as they are distributed through the body, are included in an areolar investment, or sheath; being themselves composed of three coats, an external *areolo-fibrous*, a middle coat of yellow elastic tissue, and muscular fibres, and an internal serous membrane, continuous with that which lines the heart, and through the capillaries, with the lining membrane of the veins.

The coats of the arteries are supplied with vessels, the *vasa vasorum;* and they are also supplied with nerves, like other parts of the system; but the mode of distribution of the nerves, is not very clear.

The arteries, thus constituted, are liable to become inflamed, the symptoms of which we are now prepared to consider.

Symptoms.—Slight inflammation of the arteries may exist, without any very decided symptoms. But marked cases of arteritis are attended with heat, pain, tenderness, and throbbing, and if it is sufficient to produce a roughness of the lining membrane, there may be a rustling sound produced, discoverable by auscultation. If the larger arterial trunks are involved in the inflammation, the heart suffers sympathetically, and along with a feeling of faintness, anxiety, and restlessness, there is more or less febrile excitement. And in cases, in which the aorta is in-flamed, there is generally, in addition to the symptoms already enumer-ated, more or less oppression of breathing.

In inflammation of the arteries of the extremities, the vessels sometimes become obstructed, either by the thickening of the arterial coats; by the exudation of fibrin into the cavity, or its deposition from the blood, or else from actual coagulation of the blood.

In cases of obstruction in the arteries, from either cause, the pulsation ceases; the artery may be felt like a cord along the limb, and in some cases, the limb or part supplied by the obstructed artery, on the suspen-sion of its functions, becomes numb, paralyzed, and even mortified.

Partial obstruction of the pulmonary arteries may exist, and yet a tolerable degree of health be enjoyed; but in case the obstruction becomes complete, immediate death is inevitable. In advanced stages of severe arteritis, the fever may assume a typhoid character, in consequence of contamination of the blood, by the lymph, pus, &c.; products of the in-flammation.

Such, I believe, are the usual symptoms of arteritis. And if it be found in some cases, that the symptoms are not very indicative, they may be inferred to arise from arteritis, if they are not such as are known to arise from any other disease.

Anatomical Characters.—The inner surface of the artery, as far as the inflammation has extended, is red and more or less roughened; the coats are thickened, and perhaps softened, and readily separable; the vasa vasorum are injected, and lymph or pus may be found on the inner surface, or between the coats. Coagulated blood is often found lining the vessels, or filling their cavities, and sometimes ulceration is discovered along the inner surface of the vessel.

In chronic cases, there is left, besides, the redness, thickening and softening. In the acute form of the disease various organic changes are found in the coats of the vessels, and sometimes morbid deposits, consti-tuting organic diseases of the arteries, in many respects similar to those of the heart, the result of cardiac inflammation.

Causes.—Various causes may operate to produce arteritis; among the most frequent of which are direct injuries, such as the wounding or tying of arteries, extension of inflammation from inflamed or suppurat-ing parts, and sudden exposure to cold. It may also arise from the me-tastasis of gout, rheumatism, &c., from repelled cutaneous eruptions, from the use of tobacco and intoxicating liquors; and, finally, in its chronic form, from the irritation produced by the presence of calcareous, steatomatous, tuberculous, or other deposits, which we have already seen are liable to take place in the coats of the arteries.

Treatment.—In moderate cases of arteritis, rest, low diet, saline cathartics, leeches, and perhaps blisters, may be sufficient to arrest the disease. But violent cases, occurring in robust subjects, may require general bleeding, rest in the horizontal position, saline cathartics, and antimonials; leeches, fomentations and blisters along the inflamed ves-sels, and, if the disease is obstinate, a slight mercurial course.

In cases depending upon gout or rheumatism, in addition to the other measures that may be indicated, nitre, colchicum, or iodide of potassium should be given and continued for a reasonable time. Finally, in all cases of arteritis in which gangrene occurs, quinine, camphor, wine-whey, broths, &c., must be resorted to, till the system permanently rallies from its depressing effects.

SECTION X.—PHLEBITIS.

By phlebitis, from φλέψ, "a vein," and the termination *itis*, I mean inflammation of the veins.

The veins commence by minute radicles in the capillaries, which extend from the extreme arteries, in every part of the system; and converge, forming larger branches, till they constitute the main trunks which convey the venous blood directly to the heart.

The veins of the systemic circulation convey the venous blood from the capillaries to the right auricle of the heart, while the pulmonary veins convey the arterial blood from the capillaries of the lungs to the left auricle.

The veins are much thinner in structure than the arteries, but they have three coats, an external, firm and strong, resembling that of the arteries; a middle coat with an external layer of circular fibrous tissue, and an inner layer of longitudinal muscular fibres, and finally an internal serous coat, continuous with the lining membrane of the heart, and through the capillaries with that of the arteries.

The veins are furnished with valves, and like the arteries are supplied with nutritious vessels, and it is probable that nervous filaments are distributed to their coats. The veins then have a structure very similar to the arteries, though their coats are thinner, and they have not the elastic coat of the arteries. The deep veins that accompany the arteries, are, in the limbs, generally included in the same sheath with them.

The veins thus situated and constituted are liable to become inflamed from various causes, the symptoms of which we will now proceed to consider.

Symptoms.—Inflammation of the veins are attended with pain, tenderness, and swelling along the course of the affected vessels, and in the adjoining tissues. The inflamed vessels may often be felt like firm cords, unless, as sometimes happens, the limb becomes greatly distended by effusion from obstruction to the returning blood, as well as from an extension of the inflammation to the adjoining tissues.

If the veins involved be small, the obstruction caused by matter thrown out on their inner surface, and by the coagulated blood, may lead to no very serious results. But if the large internal veins are thus obstructed the worst consequences may follow.

In case the inflammation subsides, there may be a solution of the coagulum, and thus the circulation through the vessels be restored. But if suppuration takes place, the matter may be confined to the inflamed part by coagulated blood, producing abscesses along the veins. Or the pus may be carried along by the current of blood if the inflamed vessels remain pervious, and the most serious constitutional disturbance follow.

In some cases of phlebitis, in depraved constitutions, pus enters the circulation freely from the inner surface of the inflamed veins leading to purulent depositions and abscesses in different parts of the **system,** and finally to emaciation, exhaustion and death.

Ordinary cases of phlebitis commence with a chill, or chills, which are followed by more or less febrile excitement. But in case considerable pus enters the circulation, **the pulse becomes** frequent and feeble, the extremities cold, the respiration difficult, the countenance sunken, the tongue dry, and finally to complete the typhoid train of symptoms, there is restlessness, delirium, and **generally a** fatal termination.

Diagnosis.—The pain, swelling, **tenderness, and** other symptoms enumerated, are generally sufficient **to distinguish phlebitis** in the extremities. And in cases in which the large **internal veins are the** seat of the inflammation, the disease may generally **be diagnosticated by the** swelled and dark appearance of the veins, **upon** the surface, which carry blood into the affected trunk. It may, however, be difficult, **or** impossible in some cases, to distinguish phlebitis involving the portal **veins.** But the epigastric pain, and tenderness, **the slight** tympanitis, the congested or full appearance of the **superficial veins, the** chills and febrile reaction, and finally the congested **condition of the** liver, spleen and other abdominal viscera, together **with the typhoid symptoms, when they supervene,** may serve to render the diagnosis probable at least.

Phlebitis of the pulmonary veins **may be** distinguished with tolerable **certainty by the pain,** dyspnœa, and congestion of the lungs, the lividity **and irregular action** of the heart, together with the other symptoms **enumerated as common to** phlebitis.

Anatomical Characters.—On post-mortem examination, the coats of the **veins are found either** contracted, thickened, softened or indurated. **Coagulable lymph or** coagulated blood are found adhering to the inner **surface of the vessels,** or completely filling them, **and** portions of the **inflamed vessels often contain pus,** or the vessels may be found obliterated along the **track of the inflammation. The** adjoining tissues are generally found infiltrated **with pus, or to have** contracted adhesions from an **extension of the inflammation to these parts.**

If the pulmonary veins have been the seat of the inflammation, in addition to the ordinary appearances, **the** lung may be found congested, or gorged with blood. And in cases in which the portal veins have been involved, in addition **to the** appearances already described as attending phlebitis, abscesses may be found in the liver and other abdominal viscera, and **the peritoneum** may show signs of congestion or inflammation.

Causes.—A feeble and depraved condition of the system, strongly predisposes **to phlebitis.** And it is probable that certain atmospheric influences **of a pernicious** character tend to produce the disease. **Phlebitis** generally **however arises** from direct injury of the veins, **such as from** venesection, if performed with a dull or contaminated lancet, from the operation for varicose veins, amputations, or any **accidental wound of** the veins.

The disease is very apt to occur in the puerperal **state,** the inflammation commencing in the uterine veins, and extending to the spermatic, hypogastric, iliac, and femoral veins, and sometimes **even to the vena** cava. It is probable also that phlebitis may arise **from sudden exposure** to cold, and perhaps from the translation of rheumatism, gout, &c.

Phlebitis when once established tends to pass from the smaller to the larger veins, the inflammation being extended along the coats of the veins, or else being in part communicated along the trunks by the passage of acrid matter, products of the inflammation.

Prognosis.—From the depraved constitutions in which phlebitis occurs, and also the tendency to pass to the larger venous trunks, as well as the danger of their obstruction, and of the formation of abscesses in different parts of the system, phlebitis should always be regarded with a due degree of solicitude.

Treatment.—In cases of *active* phlebitis, saline cathartics, low diet, antimonials, anodynes, leeches and cold applications along the inflamed part, and if necessary mercurials are the remedies upon which our main reliance is to be placed. But in cases of *passive* phlebitis, a moderately nourishing diet, mild laxatives, a few leeches, and warm fomentations of hops to the inflamed part will do best, and if sinking comes on, quinine, wine-whey, &c., will generally be indicated. In cases in which the parts remain swelled or indurated after the inflammation subsides, blisters or the application of iodine or mercurial ointments may be required.

SECTION XI.—CRURAL PHLEBITIS.

By crural phlebitis, I mean that variety which generally occurs in the puerperal state, commencing in the uterine veins, and extending to the iliac, and generally to the large veins of the affected limb, constituting what has been called *phlegmasia dolens, milk-leg,* &c.

This form of phlebitis is not confined exclusively to the puerperal state, but more generally occurs in that condition, being developed generally within the first ten days after confinement.

Symptoms.—There is generally at first a chill, with pain and some stiffness in the groin of one side. Soon there is a greater or less degree of febrile reaction, attended with swelling in the groin, which rapidly extends itself along the whole limb, till it becomes distended, and exquisitely tender.

To the sight the swelling may exhibit an even and uniform appearance. But if the hand be passed over the limb, the veins may be felt in hard ridges like cords under the skin. More or less pain is generally felt in the iliac region of the affected side, probably in consequence of the inflammation which has extended from the uterine along these veins to the crural. Considerable fever may attend this disease, and the patient is apt to be quite restless.

The inflammation may sometimes be arrested very soon, if attended to at once, but if neglected, it may continue for ten or fifteen days, and in some cases for several weeks. When it is about terminating by resolution, the surface of the body becomes moist, the urine deposits a reddish sediment and the fever gradually abates.

The declension of the swelling is always very gradual, if it has continued for any considerable time, in consequence of the structural changes which have taken place in the vessels that have been involved. And there may remain a degree of stiffness of the limb for a long time, even after the swelling has nearly or quite subsided.

In some cases suppuration takes place and abscesses are formed, or the pus passes into the circulation, leading to the formation of abscesses in various parts of the body, or else to a general contamination of the system, a long train of typhoid symptoms; and finally to a fatal termination.

Anatomical Characters.—On post-mortem examination the uterine veins of the side to which the placenta had been attached, show marks of having been inflamed, being impervious as well as the common and internal iliacs of the same side. Sometimes these veins are found contracted down to a cord-like substance, their cavities being obliterated, and at other points, or in other cases, they are found plugged up by firm coagula. The veins of the limb are found with their coats thickened, or they are contracted and filled up with dark coagula.

Causes.—There is probably a strong predisposition to this disease in some constitutions arising from a depraved condition of the system. And the disease may also be favored by certain epidemic or endemic influences.

But the direct cause of crural phlebitis occurring in the puerperal state is probably the exposure of the uterine veins to atmospheric air caused by the separation of the placenta. This appears the more probable when we remember that phlebitis is often produced by an exposure of the veins in amputations, wounds, &c. It has also appeared to me that crural phlebitis may be produced in some cases, by injuries which the iliac or other veins receive during parturition, but this is not quite certain.

Treatment.—In cases of this disease, in which the inflammation is of an active character, general bleeding may occasionally be required, but never in cases in which the inflammation is of a passive character, as is too often the case.

Cups, wet or dry, however, should always be applied along the lumbar and sacral regions on the affected side, and leeches above Poupart's ligament, and along the crural veins. Ten grains of calomel or three blue pills should be given at once, and followed, if necessary, in six hours by half an ounce of the sulphate of magnesia to produce tolerable free catharsis. After the operation of the cathartic, if the patient is robust, and the inflammation active, four grains each of James's and Dover's powder may be given every six hours till the pain and febrile excitement subside. But in debilitated cases, in which the inflammation is of a passive character, two grains of the sulphate of quinine should be given with the Dover's instead of the antimonial powder.

In active cases, in which there is considerable heat in the limb, after leeching it should be kept cool and bathed with a solution made by dissolving two drachms of muriate of ammonia in four ounces each of vinegar and water, to which two drachms of laudanum may be added. But in debilitated cases, in which there is considerable swelling, with little or no heat, the above solution may be applied, warm, every six hours, with the addition of two ounces of tincture of camphor instead of the laudanum, and the limb kept covered with hops wet in warm vinegar.

SECTION XII.—HEMORRHAGE.

By hemorrhage is here meant an escape of blood from vessels through which it circulates in a state of health. But as *traumatic* hemorrhage belongs rather to the surgeon, I design to consider more especially here *spontaneous* hemorrhage, or that which arises from causes acting through the organization.

Spontaneous hemorrhage may take place from the *arteries, veins* or *capillaries,* but it is generally from the capillaries, and it may depend upon a diseased state of the vessels, of the blood, or of both; with or without derangement of the circulation. Or hemorrhage may occur from derangement of the circulation, probably without any previous disease of the arteries, veins or capillaries.

While spontaneous hemorrhage, as we have seen, may take place from the arteries or veins, it is in a large majority of cases directly from the capillary vessels that the blood escapes. And in order to understand this subject it is proper to call to mind the anatomy, physiology, and functions of these vessels.

The capillaries, it will be remembered, extend from the extreme arteries to the extreme veins, being about $\frac{1}{15}$th of an inch in length, and $\frac{1}{3000}$th of an inch in diameter, and they thus constitute a microscopic network through every part of the body. They are quite uniform in size, inosculating on the one hand with the terminal arteries, and on the other with the minute veins, and have among themselves frequent divisions and communications.

The capillaries thus constituted and situated, serve not only to convey the blood from the extreme arteries to the extreme veins, but they supply by exudation through their thin walls, matter to nourish the interstitial tissues, and also absorb various soluble matters which are presented to them. It is, then, through these minute and delicate vessels, which everywhere pervade the organs and tissues of the body, that the blood generally escapes in cases of spontaneous hemorrhage, which is now about to engage our attention.

Now, hemorrhage may occur from directly opposite conditions of the system; in the one case, the vital forces being above the standard of health, and in the other, below; constituting, as in inflammation, one grand division into *active* and *passive.*

Active hemorrhage occurs in strong and vigorous constitutions, in which the circulation is active, with a strong full pulse, and in consequence generally, as I believe, of a hereditary, or accidental irritability, or weakness of the capillaries of the whole system; or of those of the organ or tissue, from which the hemorrhage occurs.

Passive hemorrhage, on the other hand, occurs in weak and debilitated constitutions, in which the blood is weak or watery; the circulation languid; the pulse feeble, and the extremities cold, and in consequence probably of the watery state of the blood, and the relaxed condition of the capillaries, together with the local congestion, so common in such cases.

There are various symptoms, common to both active and passive hemorrhage, as well as some, which are peculiar to each, which we will now proceed to consider

Symptoms.—I believe there is generally a sense of heat, and sometimes of pain, or fullness in parts, from which hemorrhage is about to take place.

When the hemorrhage occurs, if it be considerable, the local heat, pain or feeling of congestion, will subside, and there may be a sinking of the pulse, and perhaps faintness or syncope, with more or less general nervous irritability. In case the hemorrhage is internal, the above symptoms, together with the evidences afforded by palpation, auscultation, percussion, &c., will generally be sufficient to render the diagnosis tolerably clear at least.

In *active* hemorrhage, the blood is of a bright-red color, and readily coagulates, forming a large soft clot; and this is not strange, when we remember that it comes from robust and vigorous constitutions, or florid individuals, with a strong appetite, and active digestive powers.

In *passive* hemorrhage, however, the blood is of a darkish color, and but slightly, if at all coagulable; and not unfrequently in such cases, blood is extravasated beneath the skin, forming ecchymosis. If the hemorrhage is considerable, the pulse becomes weak; the face pale; the extremities cold, and unless it be arrested, there may be cold sweats, syncope, convulsions, and finally a fatal termination.

Hemorrhage, whether active or passive, is very apt to return, as often as the causes or circumstances which produced it recur, or are repeated; whether it be daily, weekly, monthly, yearly, or at irregular periods.

Active hemorrhage may be borne, if not too copious, even though it be repeated occasionally; as the blood is strong, and will bear the dilution which takes place, from the absorption of fluids, after any considerable loss of blood. And it is possible that the system becomes habituated to active hemorrhage, in some cases, so that nearly sufficient blood is produced to keep up the integrity of the system, notwithstanding the loss which occurs.

Passive hemorrhage, however, being the result of debility, and an imperfect circulation, is but illy borne. And if continued or repeated, is liable to lead to the most serious consequences.

Causes.—I believe there is in many cases a hereditary predisposition to hemorrhage, consisting in a weak, relaxed, or irritable condition of the capillaries of the whole system, or of those in the particular part from which the hemorrhage occurs. Age, too, appears to have an influence in favoring particular varieties of hemorrhage; epistaxis being most frequent during childhood; hæmoptisis between fifteen and thirty; while hæmatruia is very liable to occur in advanced age.

Active hemorrhage may be caused however, in those who are predisposed, by every influence which tends to produce a full plethoric condition, in which the vital force is above the standard of health, such as an excess of animal food, overeating, want of exercise, &c., while the passive variety may be produced by any influence which depresses vitality below the standard of health, such as starvation, unwholesome food, impure air, and other kindred influences.

Finally, besides these general influences, various local causes may operate to produce either variety, such as irritations, injuries, wounds, &c.

Prognosis.—The prognosis in active hemorrhage is favorable, unless it be very copious or frequently repeated, in which case it may lead to fearful or even fatal results.

In passive hemorrhage, if copious or frequently repeated, the prognosis is always unfavorable, from the great debility which it produces, and the liability there always is of sudden and fatal prostration.

Treatment.—The first consideration on examining a patient with hemorrhage is, whether the loss of blood may not be salutary, as it is in certain morbid conditions of the system, with general or local derangement.

If then it be found on examination that the loss of blood is a less calamity than the morbid condition it is relieving, the hemorrhage should be permitted to go on till it ceases to be salutary, at which time proper remedies should be resorted to for its arrest.

In all cases of hemorrhage in which an arrest of the flow is desirable, the exact deviation of the system from the standard of health should be taken, and thus the active or passive character of the hemorrhage be ascertained. When this is done, the indications of treatment are generally clear.

The indications, if the hemorrhage be *active*, are to diminish the activity of the circulation, to lessen irritation in, and determination of blood to the part from which the hemorrhage occurs, to constringe the capillaries of the part, and finally, to correct the habits and conditions of the patient which produced, and would be likely to lead to a return of it.

The activity of the circulation may be lessened by bleeding, saline cathartics, and sedatives; as digitalis, nitre, &c. Irritation in, and determination of blood to the part may be lessened by cold applications to the seat of the hemorrhage, and warmth, sinapisms, or blisters to remote parts. To constringe the capillaries, astringents, such as tannin, alum, &c., should be administered internally, and, if convenient, applied to the seat of the hemorrhage.

Having arrested the flow of blood, the habits of the patient should be inquired into and corrected. The patient should be placed on a plain and not overstimulating diet, should be required to eat moderately, and take a reasonable amount of exercise, and should avoid every cause that might tend to favor a return of the hemorrhage.

The indications in *passive* hemorrhage are to equalize the circulation, to constringe the relaxed capillary vessels, to restore to a healthy condition the weakened blood and the general powers of the system; and, finally to correct the habits of the patient, upon which this weak and debilitated condition of the system depends.

To equalize the circulation, stimulants if necessary; stimulating frictions along the spine, the warm foot-bath, or sinapisms to the extremities; and, finally, blisters will generally do best.

To constringe the relaxed capillaries, astringents, such as alum, tannin, geranium, &c., will generally do best. And to restore the blood and general powers of the system, the tonic bitters, iron, and especially the tincture of the chloride, with a generous diet may be required. And, finally, all the habits of the patient should be corrected, which might nd to favor a return of the hemorrhage.

SECTION XIII.—EPISTAXIS.

By epistaxis, from επι, "upon," and σταζω, "I flow drop by drop," is here meant bleeding from the nose, whether active or passive; the blood proceeding from the capillaries of the Schneiderian membrane.

Epistaxis is probably by far the most frequent variety of hemorrhage, and occurs often in early life, in part perhaps from the large size of the head in comparison with the rest of the body; and also in part from the habit of pricking the nose, &c., so common among children.

Symptoms.—Bleeding at the nose is generally preceded by symptoms indicative of cephalic congestion, such as a sense of weight in the temples, pain in the head, perhaps throbbing of the carotids, a flushed face, giddiness, ringing in the ears, and a sense of tickling, or stinging pain in the nose. And in weak, debilitated and irritable persons, the flow may be preceded by cold extremities, chills, a shrunken state of the skin, and a small corded and quick pulse.

The flow of blood is generally attended with a relief of the cephalic symptoms, especially in active cases. The blood generally drops from the nose, but it may pass, or a portion of it at least, along the posterior nares to the fauces, and be discharged from the mouth. Generally but a few ounces of blood is lost, but sometimes the flow is profuse and protracted, requiring active or persevering measures for its suppression.

Causes.—Whatever produces a preternatural determination of blood to the head, or Schneiderian membrane of the nose, may give rise to epistaxis, such as sneezing, stooping, mental excitement, intestinal irritation, &c.

Epistaxis may also occur from organic diseases of the heart, from suppression of the hemorrhoidal or menstrual discharge, and it is apt to occur in dropsical patients, in consequence of the watery state of the blood and a passive cephalic tendency.

Prognosis.—Active hemorrhage from the nose is not usually attended with any considerable danger, unless it is very profuse and frequently repeated. But in passive cases it is often exceedingly troublesome, and sometimes attended with considerable danger, in consequence of the great prostration which it produces. Epistaxis, however, very rarely terminates fatally from direct loss of blood.

Treatment.—The treatment of epistaxis of course depends upon the nature of the hemorrhage and the general condition of the patient. If the hemorrhage is active, and a consequence of the suppression of some natural or habitual discharge, no measures should be taken to arrest it, unless the loss of blood is very considerable.

If, however, the bleeding continues beyond the point at which it ceases to be salutary, and in all cases of *passive* epistaxis immediate measures should be taken to arrest the bleeding as soon as possible.

The feet should be placed in warm water, and then sinapisms applied to the feet or limbs. Cold applications should be made to the back of the neck and nose, and if necessary to the groins, the patient being placed in the sitting posture.

If, however, the bleeding still continues, ipecac should be given in one-fourth of a grain doses, every fifteen minutes, till the hemorrhage is arrested or slight nausea is produced. If this too should fail, full doses of alum or tannin should be given every two or three hours, and a saturated solution of alum snuffed or injected into the nose. If still obstinate, pressure should be made on the bleeding nostril, and the arm of the bleeding side raised perpendicularly upwards and retained there a short time, the patient being in a standing posture, as suggested by Dr Negrier.

Finally, if all these measures fail, a piece of sponge cut to fit the nostril may be saturated with a solution of alum or tannin, and having a string made fast to it, should be passed carefully up the nostril of the bleeding side, and if it arrests the bleeding, it should be left for three or four days, but if not, it should be withdrawn, and the anterior and posterior nares closed with small rolls of linen cloth. This last measure is, however, attended with considerable difficulty, and I believe need rarely be resorted to, if the other remedies be faithfully applied.

To prevent a return of epistaxis, the habits of the patient, as well as the condition of the system upon which it depends should be corrected, and if the hemorrhage is of a passive character the tincture of the chloride of iron, continued for a time, is a valuable remedy to prevent its return.

SECTION XIV.—HÆMATEMESIS.

By hæmatemesis, from αιμα, "blood," and εμεω, "I vomit," is here meant the vomiting of blood, which has accumulated in the stomach from gastric hemorrhage.

I desire also to include, under this head, all cases of gastric hemorrhage, whether the blood be vomited or passed along into the intestines, as is sometimes the case.

Gastric hemorrhage may take place from the arteries or veins, but the blood generally escapes, as in other cases of spontaneous hemorrhage, directly from the capillaries of the mucous membrane.

Symptoms.—The premonitory symptoms of hemorrhage from the stomach, are a sense of weight, fullness, heat, and pressure in the epigastrium, an irregular appetite, pain in the hypochondrium, faintness, nausea; a small and irregular pulse, palpitation, cold extremities, a pale countenance, and anxiety, weakness, and constriction about the breast.

After these symptoms have continued for a time, there is a feeling of approaching syncope, with distressing nausea, and finally, in most cases, copious ejections of blood from the stomach. In some cases, however, vomiting is not provoked, the blood passing along into the intestines, as I have already suggested.

The blood ejected by vomiting, is sometimes in a liquid state, and of a bright-red color. More generally, however, the blood is coagulated, and of a darkish color; sometimes being almost black. The quantity of blood thrown off, is sometimes considerable; in two cases that fell under my care, I think it must have exceeded two quarts from each. In some cases, however, it is much less than that.

Sometimes partial syncope follows the ejection of blood, and there may be pain in the epigastric and abdominal regions. But generally, I believe, the patient, though exhausted, feels relieved for a time after the vomiting. The relief may, however, be transient; for not unfrequently the same train of symptoms recur in a few hours, and more blood is ejected by vomiting. And this may be repeated several times, unless the hemorrhage be arrested.

It is probable that a portion of the blood in gastric hemorrhage, always passes into the intestines, and hence the dark alvine discharges which occur, for several days, after an attack of hæmatemesis.

Diagnosis.—Hæmatemesis may be distinguished by careful attention to all the symptoms, and by ascertaining with certainty, that no hemorrhage has occurred from the nostrils or fauces, and the blood swallowed, and that the ejected blood was not thrown up by coughing, as it would be, if it came from the lungs. In cases of gastric hemorrhage, in which no vomiting occurs; the sense of fullness, heat, pain, &c., together with faintness, and distension of the stomach, and all the other symptoms, which attend this affection, may render the diagnosis probable, at least.

Causes.—Any thing which interrupts the portal circulation, such as hepatic, congestion, &c., may produce this variety of hemorrhage. Or it may occur from suppression of the menstrual or hemorrhoidal discharge, as well as from the final suppression or cessation of the menses, at the critical period of life.

It may also occur from any cause, external or internal, capable of producing congestion, irritation, inflammation or ulceration of the gastric mucous membrane, such as blows on the epigastrium, acrid substances swallowed, alcoholic liquors, gluttony, fits of anger, &c. Or the passive variety may occur from any cause, capable of depressing vitality, or that produces a weak state of the blood, and a relaxed condition of the capillaries of the gastric mucous membrane, such as insufficient or unwholesome food, filth, and impure air; the use of tobacco, excessive venery, and various other depressing influences. Finally, hæmatemesis may be produced by ulceration, cancer, and other organic diseases of the stomach, as well as by diseases of the liver and spleen, and various other internal organic affections.

Anatomical Characters.—In some cases the gastric mucous membrane appears nearly natural, but generally it shows signs of congestion or inflammation, being red, sometimes softened, or covered with dark spots, or else ulcerated. If the hemorrhage has proceeded from an accidental wound of a vessel it may often be discovered; and if it has been the result of any organic disease, the lesion, of course, will be presented.

Prognosis.—The prognosis in hæmatemesis, arising from causes that admit of removal, is not very unfavorable, unless it be copious, or frequently repeated. But in passive cases, and in all such as depend upon organic disease of the stomach, or other abdominal viscera, the prognosis is unfavorable. Death may occur suddenly from gastric hemorrhage, without the ejection, by vomiting, of any blood.

Treatment.—The treatment proper for hæmatemesis depends entirely upon the general condition of the patient. If robust and strong, with a full and active pulse, general bleeding may sometimes be indicated; but not generally, I think. Leeches or cups over the stomach may, however, be of service in such cases.

After bleeding, leeching, or cupping, if indicated, the warm foot-bath should be resorted to, and sinapisms applied to the epigastrium and extremities. Ipecac, in one-fourth grain doses, should be given, every fifteen minutes, till the hemorrhage is arrested, or slight nausea is produced. The bowels should be moved by an injection or mild cathartic, if necessary; and, if the bleeding continues, the patient should be kept quiet, and two grains of tannin administered every two hours, and continued till it produces its astringent effect upon the gastric mucous membrane.

After arresting the hemorrhage; if it has been passive, and the patient is anæmic, the tincture of the chloride of iron may be given, for a time, to restore the blood. Mucilage of gum arabic should be allowed for drink, and arrow-root for food. And to prevent a return of the hemorrhage, whether active or passive, the habits and condition of the patient upon which the hemorrhage depends should be corrected, and the patient be made to conform rigidly to the laws of health in every respect.

SECTION XV.—INTESTINAL HEMORRHAGE.

By intestinal hemorrhage is here meant that which takes place from the intestinal mucous membrane, and generally from the capillary vessels, whether active or passive, to which the term *melæna* has sometimes been applied.

We have seen in the preceding section that gastric hemorrhage is quite frequent. There remains for our consideration, in this section, the intestinal. It is very possible, however, that the blood in many cases escapes from the mucous membrane of the stomach and intestines at the same time, which fact must always be taken into account in the examination of such cases.

That hemorrhage should take place from the alimentary mucous membrane is not strange when we remember that the portal system, consisting of the inferior and superior mesenteries, the splenic and gastric veins takes up the blood distributed to the alimentary canal and other abdominal viscera, and conducts it along the vena portæ to the liver, through which it passes. But arrived at the liver, it must be remembered that the vena portæ divides, first into two, and then into numerous secondary branches which ramify through the portal canals. In the liver its branches receive the venous blood from the capillaries of the hepatic artery, and finally terminate in the venous plexus of the lobules of the liver, their blood being conducted on by the *interlobular* to the *sub-lobular* veins, and thence through the *hepatic trunks* to the inferior *vena cava.*

Now, any thing which hinders a free passage of the blood of the portal system through the liver, embracing functional and organic diseases of the liver, as well as obstruction in the hepatic veins or in the inferior vena cava, may be congesting the capillaries of the alimentary mucous membrane, produce gastro-intestinal hemorrhage, as we have already seen. But having already considered gastric hemorrhage, and remembering that it may occur with intestinal, we will now, with these facts in mind, pass on to the consideration of *intestinal hemorrhage,* the legitimate subject of the present section.

Symptoms.—It must be remembered that in addition to hepatic derangement, there are various other diseases of a general and local character, such as scurvy, purpura, dysentery, enteritis, &c., of which intestinal hemorrhage is sympathetic. But intestinal hemorrhage is ordinarily preceded by a sense of weight, fullness, heat, and pain in the abdomen, and occasionally by tenderness. There may be also loss of appetite and various symptoms indicative of intestinal derangement, continuing for a longer or shorter time. Finally, the patient is attacked suddenly with griping pain in the bowels, followed by nausea, faintness, paleness, cold extremities, &c., followed by a more or less copious discharge of blood from the bowels.

In some cases, however, the hemorrhage comes on without any marked premonitory symptoms; the appearance of the blood discharged from the bowels, together with the attendant griping, fainting, and depression, being the first intimation of the disease. Or the intestinal hemorrhage may occur to an alarming and even fatal extent, and no blood be discharged from the bowels, but such sudden and copious concealed hemorrage is not, I believe, very common.

The quantity of blood lost in intestinal hemorrhage varies from a few ounces to several pints or even quarts, and it varies in appearance according to the portion of the intestines from which it proceeds. If it comes from the large intestines, it is generally of a bright red color; if from the small intestines, and copious, it is dark red; but if it has escaped from an extensive surface by slow exudation, it has a dark pitch-like appearance; and finally, it has the appearance of coffee-grounds if it has escaped in consequence of disorganization from inflammation or ulceration, as is sometimes the case in malignant fevers and other diseases.

Anatomical Characters.—More or less blood is found in the intestines, and the mucous membrane presents in some cases a congested, and in others a pale appearance. Except in cases in which there is some intestinal organic disease, the mucous membrane presents no signs of rupture of its vessels, rendering it nearly certain that the blood escapes from the capillary vessels as is usual in spontaneous hemorrhage.

Causes.—By far the most frequent cause of intestinal hemorrhage, as well as of gastric, is obstruction of the portal system, from torpor, congestion, or organic disease of the liver. But this variety of hemorrhage may be produced by drastic cathartics, or by any cause capable of producing irritation, congestion, inflammation, or ulceration of the intestinal mucous membrane.

Intestinal hemorrhage may also be the result of organic disease of the intestines, of the heart, of the thoracic and abdominal viscera, of suppression of the menses, and finally of malignant fevers, and various diseases, such as purpura, scurvy, &c., in which there is a weak or dissolved state of the blood, and relaxed condition of the tissues of the body.

Prognosis.—The prognosis in intestinal hemorrhage is generally favorable, unless it is complicated with, or the result of some general or local malignant or organic disease. If the condition which produces it admits of relief, the hemorrhage once arrested, may not return. At least such has been the result of my observation. It should be remembered, however, that the hemorrhage, if very copious, may prove suddenly fatal, or it may return from slight causes, and finally lead on to a fatal termination.

Treatment.—In cases of intestinal hemorrhage, the patient should be kept quiet, and have sinapisms applied to the feet or limbs, and also to the abdomen, and the irritation should be carried to a little short of vesication.

Ipecac should be given in one-fourth of a grain doses every fifteen minutes, till the hemorrhage is arrested, or slight anusea is produced. In case the ipecac fails to arrest the hemorrhage, tannin or alum should be given in full doses every two hours, and continued till it is arrested.

The patient should be allowed to drink freely of cold milk and water, and mucilage of gum arabic during the continuance of the hemorrhage, which will afford sufficient nourishment, unless there should be great prostration, in which case broths or wine-whey may be required. As soon as the hemorrhage is arrested, or nearly so, attention should be directed to the condition which has produced it, and which tends to keep it up, or would be likely to cause its return, if entirely arrested.

If, as is often the case, there is hepatic derangement, two or three blue pills should be administered as soon as the hemorrhage will admit of it, and followed in six hours by half an ounce of castor oil.   If there are symptoms of mucous enteritis, blisters should be applied to the epigastrium and abdomen, and the plainest variety of digestible food only allowed.

Finally, if the hemorrhage has been produced by suppression of the menses, or by any other functional or organic derangement or disease of any part of the system, or from an anæmic condition, judicious measures should be used to restore the system to a healthy state, as far as may be, and thus a return of the hemorrhage be prevented, if possible.

SECTION XVI.—HÆMATURIA.

By hæmaturia from αἷμα, "blood," and ουρεω, "I urinate," I mean here hemorrhage from the mucous membrane of the urinary passages, whether it proceed from the kidneys, ureters, bladder, or urethra, or be of an active or passive character.

The mucous membrane of the urinary passages, it will be remembered extends from the pelvis of the kidneys along the ureters, and forms the lining membrane of the bladder and urethra.   It is supplied freely with blood vessels, from the capillaries of which the blood generally escapes by exhalation, a rupture of the vessels being quite unusual, though an occasional occurrence.

Symptoms.—Hæmaturia is generally preceded by a sense of fullness, heat and pain in the region of the kidneys, ureters, bladder, or urethra, followed perhaps by faintness, a bearing down sensation, pain at the neck of the bladder, and a disposition to micturate.   On attempting to pass water, it is discovered to be either clear blood, or blood and urine, variously mingled, and of a variety of appearances, depending upon the copiousness of the hemorrhage, and the part of the urinary passages from which it has proceeded.

If the hemorrhage is from the kidneys, the blood is generally intimately mixed with the urine, and without coagula, and the pain is referred to the region of the kidneys. If the blood is intimately mixed with the urine and also contains cylindrical coagula of a bleached appearance, the pain extending from the kidney along to the bladder, the hemorrhage may be presumed to be from the ureters, but it is not quite certain. When, however, there is a free discharge of blood, or blood and urine, but slightly mingled, attended with fullness, uneasiness or pain over the pubis, in the perinæum, or neck of the bladder, with perhaps floating coagula, it is probable that the hemorrhage is from the bladder. Finally, if the fullness, heat and pain is referred to the urethra, and the blood passes in a stream or by drops, without a disposition to urinate, and the hemorrhage is suspended during micturition when it occurs, the blood undoubtedly escapes from the urethra.

The quantity of blood lost in hæmaturia varies from barely sufficient to tinge the urine, or to be discoverable, to a quantity sufficient to endanger or even destroy life. This variety of hemorrhage is usually most copious in persons advanced in life, in whom it is by far the most liable to occur.

Diagnosis.—To distinguish blood in the urine, the result of hæmaturia, from the urine tinged deep-red or brown by articles of food, or the result of morbid conditions of the urinary organs, or of the system generally, it is only necessary to observe carefully all the symptoms, and notice the following differences.

Bloody urine tinges linen dipped in it red or reddish, while colored urine only stains it brown. Bloody urine though turbid when discharged generally becomes clear on standing, and the deposit is not dissolved if heat be applied, while colored urine though nearly transparent when voided, becomes more or less turbid on standing, and readily redissolves the sediment if exposed to heat. Finally, if blood be present in the urine, the microscope will readily bring to light the blood corpuscles; and albumen may be detected by the application of heat or nitric acid.

Causes.—A great variety of causes may operate to produce hæmaturia, among which are violent exercise, calculi in the urinary passages, venereal excesses, organic diseases of the urinary organs, irritating substances, as oil of turpentine, cantharides, &c., the suppression of accustomed discharges, especially the hemorrhoidal or menstrual, and finally, anything which produces either a plethoric or anæmic condition of the system.

We have already seen that hæmaturia occurs most frequently in advanced age, and it appears to be more frequent with men than women. In most of the cases that have fallen under my observation, it has been in males past middle age; and has appeared to be connected with a hemorrhagic state of the blood, and a general relaxed condition of the tissues of the body. In one case, which I remember, that was brought on by active exercise in the hay-field, there had been copious hæmatemesis a few weeks previous.

Prognosis.—The prognosis in hæmaturia is unfavorable, if it depends upon organic disease of the urinary organs, or upon any local or general condition which does not admit of relief from proper remedial measures.

34

In all other cases the prognosis may be regarded as rather favorable, if the case be subjected to proper treatment, and the cause or causes be removed

Treatment.—If the patient is not advanced in life, and the hemorrhage is of an active character, general bleeding may sometimes be required, but not generally. Cups, however, should be applied, wet or dry, to the lumbar and sacral regions, and these followed by sinapisms, the warm foot-bath, and, if very copious, by injections of cold water, or mucilage, into the rectum.

Ipecac, as in other cases of hemorrhage, may be administered in one-fourth of a grain doses, every fifteen minutes at first; and in case this proves ineffectual, alum may be given in ten grain doses, every two or three hours, and continued till the hemorrhage is arrested. The patient should be allowed to drink freely of mucilages, and if the bowels are constipated, they should be moved by mucilaginous injections with a little castor-oil.

In cases of passive hæmaturia, warm pediluvia, dry cups and sina-pisms to the lumbar and sacral regions, and also to the extremities, to-gether with the use of ipecac, as above directed, followed by the tincture of chloride of iron, in ten drop doses, in mucilage, will generally do best. In active cases, the patient should be allowed, in addition to mucilages, milk and water for nourishment. But in passive cases, crust coffee, with milk, broths, and plain digestible food, may be allowed.

Having once arrested the hemorrhage, the remote causes should be sought out and removed, and the local and general condition of the sys-tem upon which the hemorrhage directly depends should be corrected, as far as possible, by proper remedial measures, and thus a return of the hemorrhage be prevented.

SECTION XVII.—HÆMOPTYSIS.

By hæmoptysis, from αιμα, "blood," and πτυω, "I spit," I here mean hemorrhage from the larynx, trachea, or lungs, whether it be from the mucous membrane of the air-passages, the air-cells of the lungs, or into the parenchyma of the lungs, constituting what has been called *pulmo-nary apoplexy.*

If the blood escapes from the capillaries of the mucous membrane of the larynx, trachea, or bronchia, it is generally discharged from the mouth; being raised by coughing. And a portion of it may be raised if it escapes into the air-cells from their capillaries. But if the bleeding is exclusively into the pulmonary parenchyma, or interlobular tissue, constituting *apoplexy of the lungs*, there may be no discharge of blood from the mouth, none being necessarily raised by coughing.

There are a few general symptoms common to all cases of hæmoptysis, and others peculiar to each variety, which we will now proceed to con-sider, first the general and then the particular.

Symptoms.—Before an attack of hæmoptysis, there is apt to be a sense of fullness, heat, oppression and pain in the chest, with a more or less distressing cough, cold extremities, a flushed face, and perhaps slight chills, followed by febrile excitement. In some cases, however, these premonitory symptoms are absent, or unnoticed; the patient first feeling a tickling sensation in the larynx, trachea or lungs, with a saltish sweet taste in the mouth, and on coughing or hawking, or both, blood in greater or less quantities may be thrown from the mouth, unless the hemorrhage has taken place in the parenchyma of the lungs.

The blood is generally liquid and florid, and unless the quantity is very great, is more or less frothy. It is, however, sometimes partially coagulated, and in hæmoptysis occurring in malignant fevers, it may be almost black.

The quantity of blood lost, varies from a few ounces, to a quart or more. And in proportion to the amount of blood, will be the dyspnœa, fullness, and oppression in the chest, as well as the paleness, agitation, and faintness, which almost invariably attend or follow an attack of hæmoptysis.

Hæmoptysis seldom terminates fatally, at first, and a patient may escape with only one attack. But this variety of hemorrhage, is very liable to return, after a time, and the attacks may recur, at regular or irregular intervals, and finally lead on to a fatal termination; especially if, as is generally the case, it is connected with organic disease of the lungs, heart, or large arteries.

In cases in which the hemorrhage is from the capillaries of the larynx the irritation is felt in that region, and the blood is thrown up by hawking, or by a peculiar laryngeal cough. The quantity of blood is small in such cases, and it has but slight, if any frothy appearance.

Hemorrhage seldom takes place from the trachea; but when that is the seat of the bleeding, the uneasiness is referred to that region; the blood is fresh, but slightly frothy, and small in quantity; and is raised by a cough, deeper than the laryngeal, though not pulmonary in its character.

If the hemorrhage is from the bronchial tubes, as is generally the case, we have the general symptoms already laid down, and perhaps a slight dullness on percussion, as well as a feeble inspiratory murmur, replaced by the bronchial respiration; a liquid bubbling rale in the larger tubes, and a fine liquid crepitus in the small ones.

In case the hemorrhage is into the air-cells, from their capillaries, there may be a reddish, frothy, and perhaps viscid expectoration, dyspnœa, and a feeling of approaching suffocation, dullness on percussion over the affected part, and finally, a crepitant rale in the small bronchial tubes, and a liquid mucous rale, or rattle in the larger ones.

Finally, if the blood escapes exclusively into the pulmonary parenchyma, or interlobular tissue, there is the oppression, and dyspnœa; with a feeling of impending suffocation, marked dullness on percussion, and in some cases, an absence of the respiratory murmur, in the affected portion of the lung, at least. Sometimes, however, more or less of the blood thus extravasated, escapes into the air-cells, and bronchial tubes, producing the crepitant, sub-mucous, or mucous rale, and is finally expectorated, as in cases in which the hemorrhage is directly into the air-cells.

Such then, are the general symptoms of hæmoptysis, and also the particular symptoms of each variety, by a careful observation of which, correct conclusions may generally be drawn in relation to the nature and seat of the hemorrhage.

Diagnosis.—To distinguish hæmoptysis from bleeding of the nose, mouth or fauces, it is only necessary to examine these cavities carefully, and to note the manner in which the blood is brought to the mouth, as well as the sounds elicited by percussion over the affected part, and the change in the respiratory murmur.

To distinguish hæmoptysis from hæmatemesis, the changed respiratory murmur, the dullness on percussion, the florid appearance of the blood, and its being raised by coughing and hawking, instead of vomiting, together with all the symptoms in the case, will render the diagnosis clear to the careful observer.

Anatomical Characters.—In most cases of hæmoptysis, the mucous membrane of the air passages either appears natural or is reddened, and has a slight appearance of having been inflamed. If, however, the blood has escaped into the air cells from their capillaries, though there is no laceration of the vessels or the mucous coat, the air cells and smaller bronchial tubes are generally distended or more or less filled with blood.

Finally, in cases in which the hemorrhage has taken place into the pulmonary parenchyma, constituting what has been called *apoplexy of the lungs*, there may be found a solid circumscribed portion of blood containing several cubic inches, of a darkish color, and of a peculiar granulated structure. Sometimes the clot thus found is contained in a large cavity which it has formed in the substance of the lungs, involving even one or more of the lobes, and perhaps by lacerating the pleura it may extend or have escaped, in part at least, into that cavity.

Such, I believe, are the ordinary appearances presented on postmortem examination, liable, of course to variations. In a fatal case that fell under my care a few weeks since, of a decidedly passive character, I have reason to believe that the clot occupied a large part of the superior portion of the right lung; and though some frothy blood escaped by expectoration, the lung was never cleared; the blood expectorated appearing to be only the result of an oozing into the air cells and bronchial tubes from the circumference of the clot. In another fatal case that fell under my care in this village, occurring during the second chill in a congestive ague, there was no expectoration of blood, the patient sinking in about twenty-four hours, with all the symptoms of *pulmonary apoplexy.*

Causes.—The most frequent causes of hæmoptysis are suppression of habitual or accustomed discharges, such as the hemorrhoidal or menstrual, tubercular phthisis, organic diseases of the heart, aneurism of the large vessels, and congestion, irritation, or inflammation of the mucous membrane of the larynx, trachea, or bronchial tubes.

But hæmoptysis may be produced by external violence, lifting or violent exercise, imprudent laughing, loud speaking, singing, &c.; the breathing of very hot or cold air, metastasis of gout, rheumatism, or cutaneous eruptions, the drying up of old ulcers, constipation of the bowels, mental excitement, and finally, by any causes which produce either a plethoric or an anæmic, debilitated, and relaxed condition of the system.

Prognosis.—In all cases of hæmoptysis in which the morbid condition of the system upon which it depends, whether local or general, is of a character that admits of correction, the prognosis is favorable, if the case can have the benefit of judicious medical treatment, and the patient can be made to conform to the laws of health and propriety in every respect.

But in all cases of hæmoptysis depending upon, or connected with organic disease, and especially if there is, as is generally the case a scrofulous or tuberculous condition, the prognosis is decidedly unfavorable, as the patient will generally die sooner or later with tubercular phthisis.

Treatment.—When an attack of hæmoptysis occurs, the patient should be placed in a quiet easy position, with the head and shoulders elevated, and as it is always at hand, and generally salutary in its effect, a teaspoonful of common salt should be given to the patient at once, and repeated in half an hour if the hemorrhage continues.

If the hemorrhage is active, the patient being of a robust plethoric habit, general bleeding may be indicated, and when it is, should be resorted to at once. After general bleeding, when it is indicated, and at first when it is not, cups wet or dry may be applied to the chest, and the feet of the patient should be placed in warm water.

While this is being done, ipecac in one-fourth of a grain doses should be administered every fifteen minutes, as in other cases of hemorrhage, and continued till the hemorrhage ceases, or nausea is produced. If, however, the treatment thus far has not arrested the sanguineous discharge, ten drops of the oil of turpentine may be given every two hours, and alternating with this, ten grains of ergot, or half a drachm of the fluid extract may be given, and continued for a reasonable time. Finally, if the hemorrhage still lingers, tannin may be given in two grain doses at longer or shorter intervals, and continued till it is arrested.

The bowels should be kept regular if necessary by Seidlitz powders, or small doses of the sulphate of magnesia, and the patient should be nourished by the plainest articles of digestible food. But if as is more generally the case, the hemorrhage is of a passive character, depending upon congestion; measures should at once be taken to equalize the circulation. The feet should be placed in warm water, stimulating friction should be made along the spine, and sinapisms applied to the chest and extremities.

The chloride of sodium followed by the ipecac should be given as directed in active cases, and on the suspension of the ipecac, two grains of tannin may be given every two hours for a time, and then every four or six hours, while it may be required. If, however, the case passes on, the hemorrhage continuing, or returning often, ten drop doses of the tincture of the chloride of iron may be given every six hours, and alternating with this, the same quantity of the balsam of Peru, rubbed up with brown sugar, and water added so as to make a teaspoonful a dose. I have succeeded with these measures after others had entirely failed.

The bowels should be kept regular, by a pill of aloes and rhubarb at evening, if necessary, and the patient should be allowed a good nourishing but digestible diet, to be taken with regularity.

After arresting the flow of blood in hæmoptysis, the remote causes and morbid condition of the system upon which it depends should be sought out, and removed as far as possible. The patient should be made to conform to the laws of health, and to rules of propriety in every respect, and if the morbid condition of the system upon which the hemorrhage directly depends admits of removal, it should be done as fast as possible by judicious measures. If, however, as is often the case, it is connected with organic disease, such palliative measures as may be indicated should be resorted to, and continued while they afford any considerable relief.

SECTION XVIII.—METRORRHAGIA—(*Uterine Hemorrhage.*)

By metrorrhagia, from μητρα, "the womb," and ρηγνυμι, "I break forth," is here meant *uterine hemorrhage*, whether occurring at the menstrual or at other periods, provided it be the result of a morbid condition, and if occurring at the menstrual period, in excess.

Uterine hemorrhage generally takes place from the lining mucous membrane of the uterus, and may occur in the unimpregnated or impregnated state of the organ. But as hemorrhage occurring in connection with pregnancy belongs rather to the obstetrician, I shall consider here, more especially, hemorrhage of the uterus as it occurs in the unimpregnated state, noticing, however, briefly as we pass, such cases as most frequently occur in connection with the impregnated state of the organ.

Uterine hemorrhage is liable to occur at any time during pregnancy, but more especially at the close of each month, at the time when menstruation would have occurred, if conception had not taken place, and finally, at or immediately after delivery.

In the unimpregnated state, uterine hemorrhage may occur at any period of life, but more generally between the first appearance and final cessation of the menses ; and during this time, by far the most frequently at the menstrual periods, and at or near the final cessation of the menstrual functions.

Metrorrhagia, like other varieties of hemorrhage, is either active or passive, and is attended with a train of general and particular symptoms, which we will now proceed to consider.

Symptoms.—Metrorrhagia is generally preceded by a sense of weight, fullness, heat, and pain in the uterus, as well as pain in the loins, more or less bearing down, slight swelling of the breasts, and of the external parts of generation, impaired digestion, headache, and mental disturbance, and finally by ringing in the ears, an irregular pulse, and sometimes slight chills, followed by flushes of heat.

After these symptoms have continued for two or three days, the hemorrhage occurs, sometimes with violence so as to produce great prostration, syncope, &c. More generally, however, it is moderate at first, and gradually increases, with relief of the premonitory symptoms, till finally, unless arrested, it leads on to fainting, prostration, cold extremities, and sometimes to a fatal termination.

If the hemorrhage occurs at the menstrual period, the flow at first generally has the appearance and properties of the menstrual flux. And menstruation may pass on quite regularly, to near its proper time for

cessation, when there is the sudden or gradual supervention of the hemorrhage, the blood coagulating, and the flow being attended with all
the usual symptoms of uterine hemorrhage, as it occurs disconnected
with menstruation.

Women subject to metrorrhagia are less liable to become pregnant,
and when they do, being predisposed to hemorrhage, are in danger of
its return any time during the period of utero gestation ; but especially
at the close of each month, at which time there is, I believe, with most
women a tendency to menstruate, for the first few months at least.
The symptoms attending such cases do not differ materially from those
attending metrorrhagia, occurring in the unimpregnated state, only so far
as they are influenced by the abortion, should that misfortune occur, as
it is liable to.   Finally, excessive uterine hemorrhage is liable to occur
at or immediately after delivery, whether it be premature or at full time,
attended with sinking, faintness, cold extremities, and, unless arrested,
followed by distressing, and even fatal prostration.

Metrorrhagia occurring at or near the final cessation of the menstrual
function, is apt to be very irregular in the period of its return, and of
various duration.   In some cases it may be slight, but generally it is
profuse, protracted, and troublesome, being usually of a passive character.   It is most liable to occur in women that have had frequent or
protracted metrorrhagia, with perhaps abortions, during the continuance
of the menstrual function.   Its symptoms are those already laid down
as belonging to uterine hemorrhage, with the supervention or addition,
in many cases, of various hysterical or nervous phenomena.

Such, I believe, are the usual symptoms, general and particular, of
metrorrhagia or uterine hemorrhage; liable, of course, to variations,
depending upon the active or passive character of the hemorrhage, as
well as other modifying influences.

Causes.—Metrorrhagia may be produced by a plethoric or anæmic
condition, by organic diseases of the uterus, or by any causes which produce a congestion of, irritation in, or a determination of blood to the
uterus, or its lining mucous membrane.

The causes which operate to produce a plethoric hemorrhagic condition, are the use of too stimulating food and drinks, want of exercise,
hot rooms, hot beds, &c.

Among the causes that operate to produce the anæmic hemorrhagic
state, are insufficient or unwholesome food, impure air, various privations,
such as exposure to cold, filth, dampness, &c., and finally, any cause that
depraves the blood or relaxes the tissues of the body.

The organic diseases, such as cancer, polypus of the womb, &c., which
produce a hemorrhagic state of the organ, are generally the result of
hereditary or acquired general or local depravity of the system.   They
depend upon a variety of remote causes, some of which may have been
operating through several preceding generations.   Finally, among the
causes that produce the hemorrhagic state of the organ, by causing congestion, irritation, or a determination of blood to the uterus, are sexual
excesses, abortions, constipation of the bowels, obstructions of the portal
circulation, leucorrhœa, pregnancy, delivery. &c.

And, besides, there are other occasional causes which operate more especially by producing a determination of blood to the uterus, such as riding on horseback, menstruation, mechanical violence, the exciting passions, and finally, drastic cathartics and emenagogue or other medicines which irritate or excite the uterus, rectum, or urinary organs.

Prognosis.—The prognosis is unfavorable in all cases of metrorrhagia depending upon causes and morbid conditions that do not admit of removal or relief, such as the various organic diseases of the uterus, &c. But the prognosis is generally favorable in uterine hemorrhage depending upon causes and conditions that admit of removal or correction, provided the case be subjected to proper treatment, and the patient can be made to conform to the laws of health and rules of propriety in every respect.

Treatment.—In all cases of metrorrhagia, the patient should be kept quiet in the recumbent posture, and allowed good fresh and moderately cool air. If the hemorrhage is *active*, general bleeding may be of service in some rare cases, but not generally. After general bleeding, when it is required, and at first, when it is not, cups, wet or dry, may be applied to the sacrum, and the ipecac administered as in other cases, in $\frac{1}{4}$th of a grain doses every fifteen minutes till the hemorrhage is checked or nausea produced. Alum should be given from the first, even while the ipecac is being administered, in ten grain doses every hour, in a wine-glassfull of milk and water, and continued thus till the hemorrhage is materially checked, when it should be continued every four or six hours, alternating with two grains of tannin, and this treatment should be continued till the hemorrhage is arrested.

If the bowels are constipated, they may be moved by injections of cool water, or if necessary by moderate doses of the sulphate of magnesia. The patient should be allowed, during the continuance of the hemorrhage, to drink freely of cold milk and water for nourishment, and during convalescence, a plain, digestible and unstimulating diet should be enjoined.

In cases of active metrorrhagia in which the hemorrhage is copious and not arrested in a reasonable time by other measures, a fine silk or cambric handkerchief, previously dipped in cold water or a saturated solution of alum, should be gradually passed into the vagina, so as to reach the os uteri. The tampon thus introduced may be allowed to remain, if necessary, for twenty-four hours; and should it be necessary, another may be introduced on its removal.

In cases of *passive* metrorrhagia, after placing the patient in the recumbent position, in fresh moderately cool air, and administering the ipecac in one-fourth of a grain doses every fifteen minutes, as in other cases of hemorrhage, and giving the alum in ten grain doses in milk and water, every hour for a time; tannin and tincture of chloride of iron should be resorted to.

Two grains of tannin may be given every four or six hours, and alternating with this, ten drops of the tincture of chloride of iron, and this should be continued till the hemorrhage is arrested, when the tannin should be omitted, and the iron continued in ten drop doses three times per day, till the anæmic condition is corrected. In case, however, the patient is not decidedly anæmic, a fluid drachm of the tincture of cinnamon may be given at first, alternating with the tannin instead of the iron, and the chalybeate omitted till the hemorrhage is arrested, when it may be given, three times per day, for a time, if required by the state of the blood.

The patient should be kept comfortably warm, and should be allowed to drink freely of milk, or milk and water, and if the appetite calls for it, a good digestible, and nourishing diet may be allowed. Should the tampon become necessary, in cases of passive metrorrhage, the handkerchief should be passed dry, and changed as often as once in twenty-four hours, while it may be required.

Such are the general principles which should guide us in the treatment of metrorrhagia, subject of course to modifications. And as some of these modifications are of importance, we will give them a passing consideration.

In cases of uterine hemorrhage following a nearly natural menstrual period, I believe that remedies should be withheld till a reasonable flow has taken place, and clots begin to form, at which time judicious measures should be taken for its arrest on the principles already suggested.

In cases of metrorrhagia occurring during pregnancy, the patient should be kept quiet, allowed plain unstimulating food, and cool drinks, and if medical treatment becomes necessary, five grains of Dover's powder may be given every six hours, the bowels being kept moderately loose by small doses of cream of tartar.

In cases of excessive uterine hemorrhage following delivery, whether premature or at full time, the foot of the bed on which the patient is lying should be elevated, at least eighteen inches, contraction of the uterus should be sought by smart friction over the abdomen, and if necessary by the introduction of the hand, well oiled into the uterus, and the alum should be given in ten grain doses, every hour in a wine glass of milk and water. In such cases, if faintness, and great prostration occurs, brandy should be freely administered, and as soon as possible, the patient should drink freely of good new milk.

Finally, in metrorrhagia occurring at or near the final cessation of the menstrual function, the general principles already laid down will usually apply. But great care should be taken, not only in these cases, but in every variety of uterine hemorrhage, to correct as far as possible, the morbid general or local condition upon which it depends. And then, having also corrected the habits of the patient, the system may regulate, and a tolerable degree of health be secured, if there is no organic uterine disease.

SECTION XIX.—SCORBUTUS—(Scurvy.)

By scorbutus or scurvy is here meant that peculiar debilitated condition of the system in which there is a depraved state of the blood, a hemorrhagic tendency, and a state of congestion or passive inflammation of different parts of the body, and especially of the gums, without any marked febrile excitement.

When persons are exposed for a long time to hardships, impure air, and almost exclusive salt meats, without vegetable food, the assimilative functions become deranged, and the blood dissolved, developing this distressing affection called *scurvy.*

This disease may occur at any place, where the train of circumstances which lead to it prevail. It is more common, however, among seamen on long voyages, or in armies on long and tedious marches, where the causes which lead to it generally exist, in a greater or less degree. I have met with it, however, under other circumstances; but happily the disease is now getting quite rare.

Symptoms.—Scurvy generally makes its appearance gradually, commencing with a degree of lassitude and want of muscular power. As the debility increases, there is a stiffness of the joints, and especially of the feet and knees. By degrees there is an inability or disinclination to corporeal exertion, and the respiration becomes short, and panting from very slight bodily exertion.

The countenance becomes pale and sallow, or lead-colored, and has a bloated appearance. The skin is dry, tense, and shining, and separates in scales on different parts of the body. Livid spots make their appearance on the legs or thighs, and finally, on the abdomen and arms; and generally in connection with the appearance of these spots there is more or less œdematous swelling of the feet and legs.

With the appearance of the spots on the skin, the breath becomes fetid, and the gums tender and spongy, and apt to bleed, on being slightly bruised or touched. The patient complains of a putrid taste, and has a strong craving for fresh vegetable food, and acid drinks.

The urine becomes dark-colored; vision is imperfect; and the muscular powers are often so prostrated, that the patient can hardly maintain the erect position. The blood becomes dissolved and dark, the pulse weak and soft, and, as the disease advances, the stiffness of the joints increases, induration of the muscles occur, there are violent pains in the knees, back, and loins; and there is apt to be spasmodic pains in the bowels, with constipation or diarrhœa.

The respiration becomes more oppressed, subcutaneous extravasation of blood takes place on different parts of the body, passive hemorrhages occur from the gums, nose, rectum, bladder, &c., at the same time that ulcers form on the calves of the legs and thighs, exhibiting an œdematous appearance, with irregular edges, and discharging a reddish bloody fluid.

The gums separate from the teeth and slough, the teeth become loose and drop out, old wounds re-open, the bones become brittle, and syncope occurs on very slight corporeal exertion. Finally, if the disease passes on unchecked, the respiration becomes fatiguing, syncope occurs while the patient is at rest, or from the slightest exertion; a fetid effluvium exhales from the body; emaciation progresses rapidly; paralysis may occur; and at last, with dropsical effusions, a colliquative diarrhœa, coma, and perhaps convulsions, the patient expires.

Such, I believe, are the usual symptoms of scorbutus, or scurvy, in its worst forms. But it should be remembered that the disease is sometimes much milder, being attended only with debility, a fetid breath, spongy gums, œdema of the feet, dark spots on the legs, from extravasated blood, and various other kindred symptoms, the patient being able to keep about most of the time.

Anatomical Characters.—There is generally found on post-mortem examination more or less watery blood, extravasated into the various tissues of the body, and also collected in the different cavities. Ecchymoses appear not only on the surface of the body, but also on the serous and mucous membranes. The periostium is sometimes found separated from the bones, and the bones detached from their cartilages and epiphyses.

The blood is almost invariably found in a dissolved state. The heart and muscles generally are pale and flabby, and are sometimes so soft that they may be broken down between the fingers. Finally, marks of passive inflammation are occasionally detected in various parts or tissues of the body; but this is by no means constant, and may be an accidental complication of the disease when it exists.

Diagnosis.—If all the symptoms, together with the extrinsic circumstances, be taken into account, there can be little or no difficulty in distinguishing scurvy from purpura, and other kindred diseases, which it may in a slight degree resemble.

Purpura, it should be remembered, though it has the dark spots upon the surface, from subcuticular extravasation of blood, wants most of the more urgent symptoms of scurvy, such as the appearance of the gums, the marked change in the muscular structures, the separation of the cartilages and epiphyses of the bones, &c.

Causes.—It is probable that scurvy is the result of fatiguing labor, damp, impure air, the use of innutritious, putrid and unwholesome food, especially of salt meats, to the exclusion of fresh vegetables, fruits, &c., great anxiety of mind, the use of intoxicating liquors, &c. But among these causes, the exclusion or want of vegetables and fruit, as food, damp, impure air, and fatiguing labor, with anxiety of mind, are probably by far the most frequent causes of this disease. It is possible also that *idio-miasmata* may operate either directly or indirectly to produce this disease.

Prognosis.—The prognosis is favorable in the early stages of the disease, if the causes and unfavorable circumstances which have produced it, can be removed or corrected. But in very advanced stages of bad cases, the prognosis is unfavorable, as the total depravity of the blood, and the partially disorganized state of the tissues, may not admit of correction by remedial measures.

Pathology.—Scurvy evidently consists in a dissolved state of the blood, and consequent derangement of the tissues of the body, together perhaps with a deficiency of *potassa* in the system generally, as recently suggested by Dr. Garrod of London.

Treatment.—The indications of treatment, then, in scurvy are very plain. The patient should, if possible, be allowed good dry air, the free use of fresh animal and vegetable food, acid drinks, and especially lemon-juice, and be placed in circumstances where he will preserve a cheerful and hopeful temper of mind.

Lemon-juice should be freely administered to the patient in the usual form of lemonade. And the vegetables which contain *potassa*, such as potatoes, cabbage, radishes, lettuce and onions, should be allowed freely with any wholesome and digestible varieties of animal food. In case emons are not at hand, an ounce of citric acid may be dissolved in a pint of water and given instead, in half ounce doses, three times per day, ell diluted and sweetened, if agreeable to the patient.

540

DISEASES OF THE CIRCULATORY SYSTEM

In most cases, this course of treatment will effect a cure if resorted to in season. But in case these measures are insufficient, potassa should be furnished directly to the system in the form of nitrate of potassa. Half a drachm of the nitre may be given four times per day in half a pint of gruel or crust coffee. Or if the patient takes sufficient nourishment at meals, two drachms of nitre may be dissolved in half a pint of vinegar, and two fluid ounces given every six hours with about twice the quantity of water. As a wash for the mouth and gums, water acidulated with muriatic acid and sweetened with honey will generally do well. Finally, if a troublesome diarrhœa supervenes, tannin should be administered till it is arrested.

SECTION XX.—ANÆMIA—(*Chlorosis.*)

By anæmia from *a*, privative, and αιμα, "blood," is here meant want of blood, or a watery state of the blood, in which there is a diminution of its nutritive properties, and especially of the red corpuscles, the chronic form of the disease having been called *chlorosis*, from χλωρος "green," on account of the greenish tinge of the skin in such cases.

The quantity of blood in the human system is probably about "one-eighth of the weight of the body."[*] And the average proportion of each of the organic elements in 1000 parts of healthy blood has been estimated at fibrin 3, red corpuscles 127, solid matter of the serum 80, and water 790. Now in anæmia or chlorosis there is an increase in the proportion of water, and more or less diminution in the proportion of the red corpuscles, constituting with other attendant imperfections, a greater or less degree of poverty of the blood.

Acute anæmia may occur in males or females, and at any age, being the result generally of excessive hemorrhage, starvation, &c., the symptoms of which are well understood and need not occupy our time. But chronic anæmia or *chlorosis* occurs more frequently in females, and generally in young unmarried females, at or about the time of the establishment of the menstrual function.

Chronic anæmia or *chlorosis* is attended with a long train of distressing symptoms which we will now proceed to consider.

Symptoms.—After a partial loss of appetite and a feeling of debility has continued for a time, the patient exhibits a peculiar pallid appearance. The lips especially become pale and bloodless, there is puffiness of the eye-lids, and generally a slight appearance of tumidity of the face. The lower eye-lids are apt to be encircled with a streak of a dark leaden color, and sometimes the lips exhibit a greenish yellow tinge. As the blood becomes more impoverished, the whole surface of the body becomes pale; presenting a white, puffy and flabby appearance, with more or less œdema, especially of the ankles. The tongue presents a bloated appearance, the papillæ being enlarged, and its surface covered with a transparent mucus. The gums and internal surface of the cheeks are tumid and pale, and the breath is generally offensive.

From the first, a feeling of general languor prevails, with great indisposition to corporeal and mental exertion. There is a headache, ringing in the ears and vertigo; and the patient is apt to be drowsy, peevish and spiritless.

* See Draper's Physiology, page 113.

There may be pain in the hypochondriac regions, with cough, dyspnœa, palpitation, syncope, &c. There is an irregular appetite, with craving for particular kinds of food. The bowels are torpid, with occasional attacks of diarrhœa, the urine is thick and sedimentous, the menses either have not appeared, or else if they have, are irregular or suppressed; and if the anæmic condition continues, the prolabia and tongue assume a pale lilac appearance, leucorrhœa supervenes, and emaciation gradually progresses with all its attendant symptoms.

In some cases there is an enlargement of the thyroid gland, a projection of the eye-balls, giving them an enlarged and wild appearance;* and there may be detected, by careful auscultation, various arterial and venous murmurs. If the ear be placed a little to the left, and near the superior part of the sternum, over the origin of the aorta and pulmonary arteries, a murmur is detected, sometimes resembling the blowing of a pair of bellows, "*the bruit de souflet;*" or it may be rough like that of filing or sawing, "*the bruit de rape*" or "*bruit du scie.*"

If a stethoscope be applied lightly to the side of the neck, over the jugular vein, a peculiar, continuous "*humming murmur*" is heard, which ceases, however, if the pressure be either greatly increased or entirely removed.

These arterial and venous anæmic murmurs, doubtless depend upon a watery state of the blood, in consequence of which its particles move more easily over each other, and in part perhaps upon its accelerated motion; together, in the venous murmurs, with the gentle pressure made over the vein.

Such, according to my observation, are the ordinary symptoms of anæmia, and especially of the chronic form of the disease which has been called *chlorosis.*

Diagnosis.—Chlorosis may be distinguished from organic diseases of the liver, heart, and other internal organs by attention to the following differences :

In organic diseases, the pallid countenance is apt to be changed occasionally to a slight flush of the cheeks, and there is not that bloodless lilac hue of the prolabia, so common in chlorosis. From organic hepatic disease, chlorosis differs in not being attended with the icteric appearance of the eyes, the clay-colored stools, the bilious urine, and the tenderness and fullness of the right hypochondrium; all of which symptoms may attend organic disease of the liver. Anæmia or chlorosis may be distinguished from organic disease of the heart by the manner of its approach, by the yellow or greenish appearance of the countenance, and finally, by the anæmic arterial and venous murmurs; which symptoms are not necessarily connected with organic cardiac disease.

Causes.—The causes of acute anæmia are generally loss of blood, starvation, or some other sudden interruption to digestion and sanguification. But the chronic form of the disease, or chlorosis, may be produced by a great variety of causes, such as sedentary habits, impure air, unwholesome and indigestible food, want of cleanliness, protracted lactation, menorrhagia, leucorrhœa, &c.

The disease is probably, however, more frequently produced by ex-

---

* The eye-balls are not enlarged in such cases, only appearing so from their prominence.

hausting labor, protracted constipation of the bowels, the non-appearance
or suppression of the menses, masturbation, the depressing mental affections, and finally, unsatisfied sexual desires.

Nature.—In relation to the nature of this affection there is no room
for doubt. From whatever causes produced, the disease is evidently one
of poverty of the blood, there being an increased proportion of water,
and a diminished proportion of the red corpuscles, with other attendant
imperfections of the blood; and to render the pathology of this affection
plain, it is only necessary to remember that the blood in a healthy state
of the system is derived chiefly from the chyle, and that it enters every
organ and tissue of the body, distributing nutritive properties to every
texture, sustaining the functions of every organ, and being the source of
every secretion.

Let, now, an excessive loss of blood take place, or digestion or sanguification be interrupted by any cause, there is not only the loss of the
nutritive properties of the blood, as we have seen, but there is a rapid
dilution caused by the water which enters the circulation, and hence, all
the symptoms which attend the disease.

And finally, when we remember that it is through the influence of the
brain and nervous system that the voluntary and vital functions are
sustained, and that the blood is the proper stimulant of the nervous
system, by virtue of which, the nervous influence is generated and distributed, it is not strange that this poverty of the blood should affect,
as it evidently does in this disease, every function of the body, developing
the symptoms which arise.

Treatment.—In acute cases of anæma caused by excessive loss of
blood, or sudden suspension of digestion or sanguification, the patient
after being resuscitated, if from hemorrhage by proper measures, should
be placed in the recumbent posture, and kept quiet. A good supply of
proper nourishing food to be taken with regularity should be allowed, and
recovery will generally pass on steadily till health is restored.

In case, however, the anæmic condition continues, and in all cases
that have not been acute, constituting the chronic chlorotic form of the
disease, the causes should be removed, and any local or general condition
of the system which may have produced it, or is keeping it up should be
sought out and removed or corrected.

The patient should have the benefit of good air, agreeable society, sufficient exercise, and a good nourishing diet. And if there are no local
derangements which hinder, these measures alone may be sufficient to
effect a cure.

If, however, the disease passes on, or there are local or functional derangements to be corrected, such as constipation, amenorrhœa, or other
menstrual derangements, proper measures should be resorted to for their
correction. And when the digestion, menstrual, and other functions have
been thus corrected, the patient in addition to a good nourishing diet,
should take some preparation of iron, either the citrate, carbonate, sulphate, tincture of the chloride, or syrup of the iodide, as they may appear to be indicated, in moderate doses three times per day, with a pill of
aloes and rhubarb at evening, till the blood is restored.

SECTION XXI.—HYDROPS—(*Dropsy.*)

By hydrops, dropsy, I mean here a preternatural collection of fluids, of a serous character, derived from the blood, either in the cellular or other loose tissues of the body, or else in one or more of the cavities.

We have seen in a preceding section that the capillary vessels connect the extreme arteries with the minute veins in every part of the system, being about $\frac{1}{30}$ of an inch in length, and $\frac{1}{3000}$ of an inch in diameter. And also that the extreme arteries or arterial capillaries allow the transudation of nutritious properties from the blood, while the minute veins, or the venous portion of the capillaries absorb, together with the lymphatic vessels, various fluids presented to them, the result of the breaking down of the old tissue.

Now, this process of exudation and absorption is going on, not only in the tissues where nutritious matter is required, but there is an exhalation from the capillaries at the surface of the mucous and serous membranes, and skin, as well as an ability on the part of the venous portion of the capillaries, to absorb fluid matters presented to them. This function the minute veins probably hold in common with the lymphatics, which arise as well from the surface of the skin, mucous and serous membranes, as in the tissues of the organs throughout the body.

It matters not whether this exudation from the minute arteries, and arterial portion of the capillaries, as well as absorption or imbibition by the minute veins and venous portion of the capillaries, takes place through imaginary exhalant and absorbent vessels in the arterial and venous portion of the capillaries, or whether it be by exosmosis and endosmosis through the walls of these vessels. In whatever way it is performed there is in a state of health a perfect balancing of these influences; so that by the aid of the lymphatic system, there is no morbid collection of serous or other fluids, either in the areolar and other loose tissues, or in the mucous, serous, or other cavities of the body.

But if, as is sometimes the case, these minute arteries, veins, and capillaries become debilitated and relaxed, or lose their proper tone, and at the same time the blood becomes weak or watery, more or less of the serum or water of the blood exhales into the cavities of the body, and perhaps filters through the minute arteries, and arterial portion of the capillaries, into the areolar tissues. Now this relaxed or debilitated state of the minute veins, and the venous portion of the capillaries, generally disqualifies them for imbibing or absorbing this exhaled or transuded serum, as well on the serous surface of the cavities, as in the areolar tissue of the body. And as there is an increased exhalation or transudation from the extreme arteries and their portion of the capillaries, with diminished absorption or imbibition by the minute veins, and their portion of the capillaries, dropsical accumulations take place, unless the lymphatic vessels are able by an increased action to effect its removal.

But if, as is often the case, the lymphatics are also in a debilitated or torpid state, the exhaled or transuded serum goes on accumulating, if it be in the cellular or other loose tissues of the body, till there is a general bloated or *anasarcous* state of the body. Or if the serous exhalation takes place into a serous cavity, as that of the peritoneum, pleura, pericardium, or arachnoid of the brain, it may go on accumulating till the cavity is filled, and the membrane and entire parietes of the cavity distended to its utmost capacity, developing the symptoms of *ascites*, *hydrothorax*, *hydrocephalus*, or *hydropericardium ;* as the accumulation takes place in the one or the other of these cavities.

Hydrops or dropsy, then, consists in a deranged or morbid condition of the minute arteries, veins, and capillaries, and generally of the lymphatic vessels, as well as of the blood itself, involving an exhalation or transudation of serum from the watery blood, through the minute arteries and their half of the capillaries, and a failure on the part of the minute veins, and their half of the capillaries, as well as of the lymphatics, to absorb or take it up. Hence the general or local dropsical accumulations which occur.

Dropsy of this character may be the result of any cause or train of influences that lead to a watery state of the blood, and a debilitated or relaxed condition of the tissues of the body, the vascular system being always involved.

I shall therefore call all cases of dropsy occurring in this manner, without any increased action of the vascular system, from inflammation or other causes, *passive*, while cases, though perhaps not widely different in their real pathology, attended with an apparent increased action of the vascular system, whether the result of inflammation or other morbid conditions, I shall call *active dropsy*, as a matter of convenience.

In relation to the manner in which active dropsy takes place, and especially that which follows inflammation, there may be room for some little doubt. But as in all cases of inflammation there is either an active or passive congestion of the capillary vessels, and as they are enlarged or dilated, and finally debilitated, probably in every case, and as the effusion generally takes place as the inflammation subsides, it appears probable that the dropsical accumulation takes place in a manner very similar to, if not identical with, *passive* cases. The local pathological condition being in every respect the same, but the general condition of the system being widely different, as we have already seen.

Such, I believe, is the real pathological condition in dropsy, whether connected with an active or passive condition of the circulatory system, the general symptoms of which we will now proceed to consider, both in its *active* and *passive* form.

Symptoms.—In cases of *active dropsy*, whether the result of some local inflammation, or of a local congestion, with general irritability, there may be a frequent, full and active pulse, a flushed or healthy appearance of the countenance, considerable muscular strength, and a tolerable condition of the digestive and assimilative functions.

But in *passive* cases, the face and surface generally is pale, the extremities are cold, the pulse feeble, the muscles relaxed and weak, and there is a general anæmic condition, with more or less derangement of the digestive and assimilative functions, and especially is there imperfect sanguification.

In both *active* and *passive* cases, there is apt to be thirst, constipation of the bowels, deficient perspiration, with a dry rough state of the skin, and a scanty and more or less changed state of the urine. In some cases the urine contains albumen, which may be detected by exposing it to heat or nitric acid, in which case its coagulation produces more or less cloudiness in or gelatinization of the mass so exposed.

Finally, the accumulated serous fluid, whether it be general or local, though liable to variation, has generally a limpid or slightly yellowish color, and little or no smell. It may, however, be tinged with blood, or have a milky appearance, and it very rarely contains flakes of coagulated matter or even pus.

As the dropsical accumulation increases, whether it be *anasarca*, *ascitis*, *hydrocephalus*, *hydrothorax* or *hydropericardium*, which constitute the varieties that I propose to consider, there are various local and general symptoms developed, some of which are of the most distressing character, arising from the local and general derangement which the dropsical accumulation produces. These symptoms are, however, peculiar in each variety, and will be noticed in the following sections, in connection with the variety of dropsy to which they belong.

Causes.—Passive dropsy may be produced by any cause or train of causes that lead to a watery state of the blood, and a relaxed condition of the tissues of the body, such as excessive loss of blood, insufficient or unwholesome food, exposure to damp, cold, or impure air, insufficient clothing, impaired digestion, and defective sanguification, the use of tobacco, sexual excesses, the use of alcoholic drinks, and a variety of exhausting diseases, such as phthisis, scrofula, scurvy, &c.

Active cases of dropsy may be produced by any cause or train of causes which produce either local inflammation, or irritation, with congestion; attended with a morbidly irritable or excited condition of the circulatory system, such as sudden exposure to cold, the retrocession of cutaneous eruptions, the suppression of accustomed discharges, as the menstrual or hemorrhoidal; the puerperal state, scarlatina, and other exanthematous and febrile affections, and a host of other kindred causes.

Finally, dropsy of either an active or passive character, may be produced by a suspension of the cutaneous exhalent function, causing retention of perspirable matter; by diseases of the heart or lungs, producing congestion of the venous system; by diseases of the liver, producing congestion of the portal system; and finally, by diseases of the kidneys, causing to be retained in the circulation the renal secretion.

Morbid Appearances.—In cases in which the dropsy has been of a passive character, there is, in addition to the serous collection, a pale, relaxed, and perhaps dilated appearance of the capillaries and lymphatic vessels. In cases in which the dropsy has been active, there is not only the serous accumulation, and the dilatation of the capillary and lymphatic vessels, but also in cases that have followed inflammation, there is found the usual marks of active or passive, and generally of passive inflammation. And besides, the blood is apt in most cases of dropsy to present a watery appearance. And if the dropsy has been the result of some local disease of the brain, heart, liver or kidneys, there will be found the marks of such disease, if it has been of an inflammatory or organic character; and sometimes if it has been of only a functional character.

Prognosis.—The prognosis in most cases of active or passive dropsy

35

is rather favorable, if it is not the result of, or connected with, any local organic disease. And especially is it so, if the remote cause can be removed, and the morbid condition of the system corrected upon which it depends.

But in all cases of dropsy depending upon organic disease of the brain, lungs, heart, liver or kidneys, or upon any local or general morbid condition which does not admit of correction, the prognosis is unfavorable so far as the ultimate result of the case is concerned. For though the dropsical accumulation may perhaps be removed, or nearly so, it generally returns, and finally remains till the patient is worn out, either by the morbid accumulation, or by the organic disease upon which it depends.

Treatment.—The general indications in the treatment of dropsy are to remove the remote cause, to correct the condition of the system upon which the dropsical accumulation depends, and to effect a removal of the accumulated fluid, whether it be in the cellular tissue, or in some one or more of the cavities of the body.

If the remote cause has been irregular eating, improper clothing, unreasonable exposures, habits of intemperance, or any other improprieties, they should be sought out at once, and removed or corrected.

Having thus removed the remote cause, the condition of the system upon which the dropsical accumulation immediately depends should be sought out and corrected if possible, at the same time that measures are used for the removal of the dropsical accumulation.

If the dropsy is of a passive character, and the result of a general anæmic condition, tonics should be administered, along with cathartics, diaphoretics and diuretics. The tincture of chloride of iron is one of the best general tonics in such cases, the cream of tartar one of the best cathartics and diuretics, and cleanliness with soft flannel next the skin a valuable diaphoretic.

In cases of active dropsy, with a full strong pulse, attended with local irritation, congestion or inflammation, general bleeding may in some cases be required. After general bleeding when it is required, and at first when it is not, a full dose of calomel or podophyllin should be administered in half an ounce of castor oil, and the oil repeated if necessary.

After the first cathartic, the bowels may be kept loose by small doses of cream of tartar, elaterium, scammony, or gamboge. Small doses of tartar emetic may be given to act upon the skin. And to increase the renal secretion, the *apocynum cannabinum*, digitalis, squill, or some other diuretic should be administered.

In cases of dropsy in which the brain is involved, in addition to the other measures, a mercurial course, followed by iodide of potassium may be required. In cases in which there is hepatic disease, blue pills may be given for a time, and then followed by taraxacum. In case the heart is the seat of the disease, digitalis will be indicated. And finally, if the kidneys are diseased, cups, hyoscyamus, and mucilages may be required, in addition to the general treatment indicated in the case.

Drinks should be allowed in limited quantities, as the dropsical patient is apt to be tormented with thirst. As a general rule, if the patient drinks tea, the black tea only should be allowed.

In passive cases of dropsy, a good nourishing diet should be allowed to be taken with regularity. But in active cases, plain, digestible, and very moderately stimulating food should be prescribed, but of this a reasonable quantity should be allowed.

After the dropsical accumulation is entirely removed, care should be taken to correct the general and local conditions of the system, if it has not been fully done, as far as is possible, and then by care and prudence in every respect, there may be no return of the disease.

SECTION XXII.—ANASARCA.

By anasarca, from ανα, "through," and σαρξ, "the flesh," is here meant dropsical accumulation in the areolar or cellular tissue of the external portions of the body, if the collection or effusion takes place to any considerable extent, otherwise it might be more properly called *œdema*.

Anasarca like other varieties of dropsy may be either *active* or *passive*.

Active cases are frequently attended with febrile excitement, and may be the result of sudden suppression of the perspiration, of the renal secretion, or of inflammation of the lungs, producing congestion of the venous and capillary system, or it may follow scarlatina.

Passive cases of anasarca may occur from any cause, or train of causes, that renders the blood weak, and the tissues of the body relaxed. Finally, anasarca either active or passive may occur from a variety of local organic or functional derangements, as of the skin, lungs, heart, liver or kidneys.

Symptoms.—In cases of *active* anasarca there may be a hard, full and frequent pulse, or only a slightly increased action of the circulatory system. The countenance may be nearly natural or even flushed, though it is apt to assume a palish appearance. But in cases of *passive* anasarca the countenance is pale, the pulse weak and feeble, the extremities cold, and there is a general anæmic and debilitated condition of the system.

In most cases of anasarca, the swelling commences in the feet and limbs, and then gradually extends over the whole body, to a greater or less extent. But in certain active cases, and especially in that which occurs from pneumonia, the anasarcous swelling is observed, first in the face, and then gradually extends downwards upon the trunk of the body, and finally to the extremities.

The swelling from anasarca may be distinguished from all others, by its pale appearance, and by its pitting on pressure.

Causes.—The most frequent causes of anasarca are suppression of the perspiration from cold, especially after scarlatina, inflammation of the lungs, or other internal organs, repelled cutaneous eruptions, disease of the heart, liver and kidneys, a general anæmic condition with relaxation of the tissues and debility, and various other local and general organic or functional affections.

Prognosis.—The prognosis in idiopathic anasarca, or those cases in which it is not dependent upon or connected with any local organic disease, is rather favorable, if the remote cause can be ascertained and removed, and the case be subjected to proper treatment.

But in cases of anasarca that are symptomatic of organic disease of the lungs, heart, liver, kidneys, or any other internal organ, the prognosis is decidedly unfavorable, so far as the final issue of the case is concerned. For if the effusion be removed in such cases it generally returns, and finally destroys the patient, unless he is sooner cut down by the organic disease upon which it depends.

Treatment.—In active anasarca, if the pulse is full and strong, general bleeding may occasionally be required, and cups may be indicated in case there is any marked congestion, irritation, or inflammation of the kidneys, lungs, or any other part.

After bleeding, when it is indicated, and at first when it is not, a full dose of calomel or podophyllin should be administered, and followed, in five or six hours, by half an ounce of the sulphate of magnesia, if necessary. And, if the case is urgent, two grains of calomel may be continued, every four hours, till several doses have been administered; to act gently upon the liver, and to stimulate the absorbent lymphatic vessels to increased activity.

To promote perspiration, one-fourth of a grain of tartar emetic may be given, every two hours, at first, and later, every four or six hours. And to keep up a watery discharge from the bowels, as well as to stimulate the kidneys to activity, one drachm of the bi-tartrate of potassa may be given, in solution, every six hours. But to act more especially upon the kidneys, ten drops each of the tincture of squill and tincture of digitalis should be given, alternating with the cream of tartar, every six hours.

This treatment should be continued, modified to suit the case, till the anasarca is removed, or it assumes a passive character; which unhappy continuation of the disease, I think, need rarely occur.

If, however, the case does pass on, and become passive, and in all cases of anasarca that have been so from the first, it should be remembered that the disease is one of debility, and diligent inquiry should be made whether the dropsy may not be symptomatic of some local affection of the lungs, heart, liver, or kidneys. If the dropsy is found to be symptomatic of a mere functional derangement of one or more of these organs, that functional derangement should be corrected by proper remedial measures. If, however, the local affection is organic, palliative measures should be used to correct, as far as possible, the functions of the diseased organ.

In all cases of passive anasarca, to restore the blood, a good nourishing diet should be allowed, and, in addition, ten drop doses of the tincture of chloride of iron may be given, three times per day, after eating. To stimulate the lymphatic system, liver, kidneys, and glands generally, five grain doses of the iodide of potassium may be given, three times per day, before eating. And finally, to secure a healthy action of the skin, it should be kept clean, and the patient should wear flannel; and, if able, should take exercise in the open air.

SECTION XXIII.—ASCITES.

By ascites, from αοϰος, "a bottle" I mean here a collection of serous fluid in the abdomen, and especially in the peritoneum, and shall only give *ovarian* dropsy a passing notice, as we proceed; as it is intimately connected with this subject.

Ascites may be the result of obstruction to the portal circulation, of disease of the liver, or it may follow inflammation of the peritoneum, or finally, it may occur from an anæmic state, with a general **debiltated** and relaxed condition of the system, and especially of the **peritoneum** and abdominal viscera.

Symptoms.—Generally after an indefinite period of **ill** health, with **indifferent appetite,** irregular bowels, pain in the right hypochondrium, **cold extremities,** and general nervous irritability, the more immediate symptoms of ascites make their appearance.

In most cases, the skin becomes dry, the urine scanty, and sometimes a slight uneasiness is felt in the bowels, especially if the patient lies on the back with the limbs extended. As these symptoms continue, a slight fullness is observed in the lower part of the abdomen, which disappears as the recumbent posture is assumed. This fullness increases if the disease passes **on,** till at last the whole abdomen becomes distended, so that the patient, if a female, has an appearance like that of the last stages of pregnancy.

In one case of ascites, in a young lady that came under my care, the distension of the abdomen was so great, that the pressure on the lungs rendered respiration extremely difficult, and evidently interrupted very considerably, the pulmonary circulation. The heart, too, appeared pushed up, the pulse was exceedingly irregular, and the least attempt **to assume the recumbent posture, was** followed by symptoms of immediate suffocation. **This was** a case of passive dropsy attended with great debility, as the disease had been progressing for about one year, being complicated with anasarca, and having originated in protracted functional derangement of the digestive organs, brought on by irregular eating, or taking food between meals.

Generally, however, the abdominal distension is not so great, the patient being able to rest in the recumbent posture, and in some cases the pressure of the abdominal parietes appears to suspend the abdominal effusion, so that it remains nearly the same for several weeks **or even months. In such cases, however,** unless **there is** considerable general improvement in the case, there is apt to be the supervention of anasarca.

In *ovarian dropsy,* which says Professor Watson,[*] "consists in the collection of the fluid **in** one **or more of** the cells within the ovary, **or in** a serous cyst connected **with the uterine** appendages;" the fullness is first noticed **in** the **form of** a **tumor** in the iliac fossa of **one side.** This tumor goes on increasing, more or less rapidly, its appearance being but slightly affected by position, till in some cases it completely fills the abdomen; its fluid contents being colorless, brown, milky or turbid; and in one **case that I** examined, of the multilocular variety, the fluid was slimy and gelatinous.

Ovarian dropsy generally progresses slowly, **and is** usually **attended** in its early stages with no very marked disturbance **of** the general health. But in some cases it passes on rapidly, being attended with **menstrual** irregularities, and unless relieved, leads on to **a fatal termination.**

[*] Watson's Practice of Physic, p. 834.

Diagnosis.—Ascites may be distinguished from ovarian dropsy by careful attention to the manner of its appearance and development. For, as we have seen, ascites begins by a slight fullness of the lower part of the abdomen, which disappears as the patient assumes the recumbent posture ; while ovarian dropsy always begins in a tumor in one side of the lower part of the abdomen, and is not particularly affected by the position of the patient. And besides, though ovarian dropsy if it continues, may, as we have seen, fill and distend the whole abdomen ; it proceeds to this stage generally much more slowly than ascites, which fact affords another important diagnostic difference.

To distinguish ascites from tympanitis, percussion should be made over the abdomen ; when if the abdomen contains water, a dull sound will be elicited, but if it contains gas only, the sound will be tympanitic. But to render the diagnosis clear, in cases in which the abdomen contains a moderate amount of fluid, with more or less gas, the patient should be laid on the back and palpation resorted to by laying the flat hand on one side of the abdomen, and then tapping gently the opposite side. If the abdomen contains water, an impulse, caused by a wave of the fluid, will be distinctly felt by the hand that lays flat upon the abdomen.

To distinguish ascites from pregnancy, is sometimes a little difficult. A few years since a married lady was brought to me, from an adjoining county, by her husband, who expressed a great concern on account of an obstinate dropsy of the bowels, with which his wife was supposed to be suffering. The case from his account, "rather grew worse," and he was anxious that I should take charge of her case.

I inquired in relation to the possibility of pregnancy, but was assured that such could not be the case, as the disease had been going on more than a year ; and besides, there never had been felt the least motion.

Not satisfied, however, with these assurances, I had her lay flat upon her back, being stript so that only one linen covered the abdomen. I then laid my ear carefully upon the abdomen, and heard distinctly the sounds of the fœtal heart, which I distinguished from the mother's by their greater frequency. I heard also distinctly the *placental souflet*, or sound produced by the passage of the blood through the placenta and umbilical cord.

I informed the lady and her husband that she was pregnant, and advised the immediate suspension of all remedies, which being readily acquiesced in, the patient went home and in due time was delivered of a fine healthy baby.

It should be borne in mind then, that the sound of the fœtal heart, and the *placental souflet,* when distinct, constitutes a positive diagnostic symptom between ascites and pregnancy. Other symptoms and circumstances, however, may be taken into the account, in case there remains a shadow of doubt.

Causes.—It is probable that the most frequent causes of ascites are obstruction to the portal circulation, congestion, irritation or inflammation of the peritoneum, and organic or functional diseases of the heart, liver or kidneys. Ascites may be produced, however, by indigestion, from irregular eating, by intoxicating liquors, or in fact by any cause, or train of causes which produce an anæmic state, with a relaxed condition of the tissues of the body, affecting especially the capillaries of the peritoneum and abdominal viscera.

Ovarian dropsy may arise from causes similar to those producing ascites. But I believe it is more frequently connected with, and produced by menstrual irregularities, or some organic or functional disease, or derangement of the womb or genital organs.

Prognosis.—The prognosis in ascites disconnected with organic disease is rather favorable, if the case be subjected to proper treatment in season, most such cases that have fallen under my observation, having permanently recovered. In cases of ascites, however, which depend upon, or are connected with organic disease of the heart, liver, kidneys or lungs, and in *ovarian dropsy*, the prognosis is unfavorable, so far as the ultimate result of the case is concerned, recoveries being an exception to the rule.

Treatment.—The first step in the treatment of ascites is to seek out and if possible remove the remote cause, whether it be irregular eating, intoxicating liquors, or any other imprudence. This, alone, if the patient be made to conform rigidly to the laws of health, in every respect, will sometimes arrest the progress of the disease, at least.

If now the liver is in a state of congestion or chronic inflammation, cups may be applied, and a few ounces of blood taken, and, if necessary, a blister applied to the right hypochondrium, or right side of the spine.

If the kidneys show signs of inflammation, cups should be applied on each side of the spine, in the inferior dorsal and lumbar regions, and a few ounces of blood taken.

If there is evidence of inflammation of the alimentary canal, or of the peritoneum, cups, and then leeches or blisters, may be applied to the epigastrium, and, if necessary, to the abdomen; and if there are febrile symptoms, a full dose of calomel or podophyllin should be administered.

After thus subduing the local congestion, irritation or inflammation, as well as general febrile excitement, if they exist, and in all cases in which these complications are not discoverable, measures should be taken for the removal of the accumulated fluid.

To keep the bowels loose, and to establish a drain by the bowels for the accumulated water, a pill of gamboge, scammony and rhubarb, of each one grain, may be given each day after dinner. To act upon the lymphatic and glandular system, and thus afford another drain, the iodide of potassium should be given in five grain doses, three times per day, and continued. Finally, to establish a drain by the kidneys, the officinal infusion of digitalis, made by macerating a drachm of the dried leaves, in half a pint of boiling water, in a covered dish, for two hours, should be given in table-spoonful doses, three times per day,* with twenty drops of the tincture of squills, if the case is urgent.

The skin should be kept clean and warm, and strong iodine ointment should be rubbed over the whole abdomen twice per day. The abdomen should be kept swathed with a firm flannel cloth, for the purpose of promoting perspiration, and also of contracting down the abdominal parietes, as the water is drained off.

* A fluid ounce of tincture of cinnamon is to be added. See Dispensatory.

The abdominal fluid has generally been removed in this way, in cases under my care, within two weeks, and, in some cases, much sooner than that. As soon as the water is removed, the digitalis should be discontinued, and the tincture of chloride of iron given in ten drop doses instead, as a tonic and diuretic, and this should be continued with a good nourishing diet, till the health and tone of the system is restored.

In cases of *ovarian dropsy*, the general condition of the patient should be kept the best possible under the circumstances; and should inflammation arise in the tumor, cups, leeches, or blisters, followed by iodine ointment may be applied over its seat for a time. Finally, should the tumor become very large, so as to endanger life, I believe, if the patient prefers the risk, that ovariotomy should be resorted to, especially if tapping, iodine injections, &c. have been resorted to, as is often the case, in vain.

I am the more disposed to favor ovariotomy as a last resort, in cases that must, without the operation, soon terminate fatally, from observations made in this vicinity within the past few years. My friend, Dr. H. A. Potter, an eminent surgeon of this village, to whom reference has already been made in the early part of this work,* has removed several ovarian tumors; in two cases of which the result of the operation has been entirely satisfactory. One of these operations I witnessed, and I am confident that without it the patient must soon have perished. Both these ladies are now, I believe, in the enjoyment of good health.

SECTION XXIV.—HYDROTHORAX.

By hydrothorax, from νδωρ, "water," and θωραξ, "the chest," I mean here, more especially, dropsy of the pleura, but shall also notice, as we pass, pulmonary œdema, or dropsy of the lungs, as the two conditions often exist simultaneously.

Hydrothorax may be, like other varieties of dropsy, either *active* or **passive;** in the one case being the result of congestion, irritation, or inflammation of the pleura, while in the other it is the result of an anæmic condition, with a watery state of the blood, and a relaxed condition of the tissues of the body.

Hydrothorax is very apt to be connected with organic disease of the heart, and it often follows pleuritic inflammation. It may also be complicated with, or independent of pulmonary œdema.

Symptoms.—At first, when the pleuritic effusion is small, only a slight uneasiness is felt in the lower part of the chest, referred to the sternal region. As the effusion continues, however, the patient suffers a slight uneasiness in assuming the recumbent posture; generally a cough and difficulty of breathing being the result. As the accumulation increases, additional pillows are used by the patient, till at last he is unable to lie down at all. The dyspnœa is considerable, and at times, from slight exercise or mental excitement, it becomes very distressing.

Generally, about this stage, the face becomes œdematous, the cheeks assume a purple hue, the lips become livid, and at times almost black, and the patient is liable, though he sleep in the sitting posture, to sud-

* See article Erysipelas.

den starting, followed by dyspnœa, with more or less mental excitement. During the continuance of the disease, the skin is apt to be dry, the urine scanty, the bowels constipated, and there is considerable thirst.

If percussion be made on the chest, while the patient is in the sitting posture, a dull sound is produced, as high as the accumulated water extends. Above this point, however, the resonance is nearly or quite natural.

Auscultation detects an absence of the natural respiratory murmur, and in its place, in many cases, the *tubular breathing*, caused by the passage of air along the larger bronchial tubes, the air-cells and smaller bronchial tubes being closed up by the pressure of the pleuritic effusion.

The sound of the voice too, at this stage of the disease, is generally increased, constituting the peculiar bleating sound called *ægophony.* This sound, however, as well as the tubular breathing, gradually disappears as the amount and pressure of the fluid increase, till they are audible only in the interscapular region.

Succussion, if carefully performed, may aid in detecting a moderate amount of pleuritic effusion, by producing a sound similar to that caused by tipping a cask, containing a small amount of water, especially if the pleura also contains air or a gas.

Inspection, if the accumulation be considerable, detects a fullness of the affected side, and in some cases a bulging at the intercostal spaces. This may be confirmed, and its extent ascertained, by *mensuration* with a tape, as suggested in a preceding section.

If the effusion has taken place in only one side of the chest, the patient lies on the affected side, but if in both, the patient has the head very much elevated, or sleeps in the sitting posture.

In cases of *œdema of the lungs,* the fluid may occupy the air-cells, or the extra-vesicular cellular tissue, or what is more common, the effusion may take place in both portions. The disease is attended with dyspnœa, more or less cough, and a copious expectoration of a thin, colorless, and sometimes frothy matter or fluid.

There is slight dullness on percussion, and a diminution of the respiratory murmur, and especially in the posterior and lower, or most depending portion of the chest. There is generally a subcrepitant and mucous rale or rattle, caused by the presence of the fluid in the air-cells and bronchial tubes.

Œdema of the lungs is often connected with pleuritic effusion, in which case there is a combination of the symptoms peculiar to each disease, which fact must always be borne in mind, in the examination of patients with hydrothorax.

Diagnosis.—Hydrothorax may generally be distinguished from other affections by the dyspnœa, and slight cough, by the sudden starting during sleep, by the gradual inability to assume the recumbent posture, the œdema of the face, and finally by the dullness on percussion, together with the absence of the natural respiratory murmur, and presence of *tubular breathing,* and the peculiar bleating sound of the voice or *ægophony.*

Œdema of the lungs, when it exists as a distinct disease, may be distinguished from pleural dropsy by the dullness and diminution of the respiratory murmur being less marked. From pneumonia it differs in wanting the febrile symptoms, as well as the rusty or bloody appearance of the matter expectorated. Finally, when hydrothorax consists, as is often the case, of a pleuritic effusion, and also of pulmonary œdema, the real condition must be ascertained by all the symptoms and circumstances of the case.

Causes.—The causes of hydrothorax, are organic diseases of the heart, congestion, irritation or inflammation of the pleura, organic or functional disease of an the liver or kidneys, gastro-enteritis, or indigestion, suppressed perspiration, and an anæmic state, with a watery condition of the blood, and relaxation of the tissues of the body. But among these causes, it is probable that organic diseases of the heart, congestion, irritation, or inflammation of the pleura, and a general anæmic condition are by far the most frequent.

Prognosis.—The prognosis in hydrothorax is rather favorable, if it can receive proper treatment in season, and in case there is no organic disease upon which it depends. But in neglected cases, and especially in such as depend upon, or are connected with serious organic disease, the prognosis is unfavorable so far as the final result of the case is concerned, and the case may pass on rapidly to a fatal termination.

Treatment.—The treatment for hydrothorax divides itself into that which is proper for active and that which is indicated for passive cases, taking into account of course, the complications which are liable to exist or arise in either variety or form of the disease.

In *active* cases, attended with or preceded by congestion, irritation, or inflammation of the pleura, whether complicated or not with œdema of the lungs, a full dose of calomel should be given in castor oil, and then alterative doses of two grains administered every four hours, for one or two days, to stimulate the absorbents. Tartar emetic should be administered in one-fourth of a grain doses, at the same time, alternating with the calomel, and this should be continued after suspending the alterative, if there is any febrile excitement or the skin remains dry.

Warm pediluvia should be resorted to, a moderate amount of warm sage tea allowed, and if necessary, cups, wet or dry, or blisters may be applied to the chest.

After the febrile excitement is subdued, if any had existed, and on suspending the mercurial, the tartar emetic may be continued with five grain doses of the iodide of potassium every six hours, to act on the absorbent vessels and glandular system.

If the case is urgent, one sixth of a grain of claterium may be given every six hours, for a time, alternating with the iodide of potassium, and when it has been continued as long as required, or as is prudent, it should be suspended, and a drachm of the bi-tartrate of potassa given in solution instead. This, alternating with the iodide of potassium, should be continued till a cure is effected.

In one desperate case of pleuritic distension, complicated with pulmonary œdema and geneal anasarca, in which the patient appeared in danger of perishing from suffocation every moment; the water was nearly all drained off by the bowels by the claterium, administered as suggested above, in forty-eight hours. On my asking him the second day, how the medicine operated, he replied that it "jerked him."

In *passive* cases of hydrothorax, with or without pulmonary œdema, two or three blue pills may be given at first, and followed by half an ounce of the sulphate of magnesia; after which the bowels may be kept loose by a hydragogue pill, of gamboge, scammony and rhubarb, of each a grain; taken after dinner each day.

After the operation of the first cathartic, the iodide of potassium should be given in ten grain doses three times per day, before eating, well diluted. If there is organic disease of the heart, twenty drops of the tincture of digitalis, or ten drops of the fluid extract, may be given four times per day, to which, if the kidneys are torpid, twenty drops of tincture of squill may be added.

The iodide of potassium should be continued, and the hydrogogue pill till the dropsy is removed, when it should be discontinued, and ten drop doses of the tincture of chloride of iron given instead, and continued till the blood is restored, as far as may be.

In *active* cases the diet should be plain, and unstimulating; but in all *passive* cases, a good nourishing diet should be allowed.

SECTION XXV.—HYDROPERICARDIUM.

By hydropericardium, from 'ύδωρ, "water," and *pericardium*, I mean here an abnormal collection of water or serum in the pericardium, or *dropsy of the heart.*

To understand this subject, it is necessary to remember the position of the heart, in the separation of the mediastinum above the central aponeurosis of the diaphragm. And also, that the inner or serous membrane of the pericardium, after lining, the inner surface of the fibrous layer is reflected over the heart, which it entirely covers; making it like other serous membranes, a shut sac. The pericardium thus constituted, with its external fibrous layer, is supplied with arteries, veins and lymphatics; and in a state of health, contains in its cavity a small amount of fluid, which facilitates the heart's movements.

It is only then, when this serum in the pericardium accumulates in excess, that it becomes a disease constituting *hydropericardium*, or *dropsy of the heart.* This abnormal accumulation of serum in the pericardium may vary in quantity from half a pint to several quarts, and it may be the result of congestion, irritation, or inflammation of the pericardium, constituting the *active* form of the disease. Or it may be of a strictly *passive* character, depending upon a watery state of the blood, and a relaxed condition of the tissues, and perhaps associated with some organic disease of the heart, lungs, liver, kidneys or alimentary canal.

It is probable that serous effusion into the pericardium becomes abnormal, producing derangement of the heart's actions, of the circulation, and of the respiration, after it has reached six or eight ounces. As it goes on accumulating above this amount, a train of distressing symptoms are developed, which we will now proceed to consider.

Symptoms.—The early symptoms of hydropericardium, while the accumulation is yet small, are generally quite obscure, especially if it is complicated with hydrothorax, or some organic disease of the heart or lungs. But as the accumulation increases, there is a feeling of uneasiness or pressure in the region of the heart. There is apt to be also a slight cough, the respiration is irregular, or uneasy, occasional fits of faintness occur, and along with great oppression in the præcordial region, there is excessive nervous irritability.

The patient is disinclined to lie down, and generally sits with the head and chest leaning forwards, the pulse is small, irregular and feeble, the face and lips are purple or livid, the appetite is irregular, and the sleep disturbed. At times the patient may appear quite smart, but exacerbations occur, during which most of the unpleasant symptoms are greatly aggravated. At such times, the anxiety and restlessness become excessive, and delirium, or the most painful irritability is liable to attend.

Towards the termination of fatal cases, the patient is apt to become stupid, the action of the heart is weak and very irregular, the extremities become cold, the surface of the body is covered with a clammy perspiration, and death at least occurs from a weakened and obstructed circulation, and perhaps from the direct pressure of the pericardial effusion suspending the heart's functions.

Percussion in the early stage of the disease detects but little if any dullness. But as the pericardial effusion increases, there is a marked dullness on percussion. It is noticed first on the left side of the lower portion of the sternum, and gradually extends upwards. When, however, the pericardium becomes largely distended, the dullness is very remarkable, and extends in every direction, but especially transversely and inferiorly.

The impulse of the heart is lessened, and its rhythm becomes greatly disturbed, as the accumulation increases, varying from the *treble beat*, to an utter confusion in the succession of the heart's sounds.

Auscultation detects also a decided diminution or decrease in the sounds of the heart. In most cases that have fallen under my observation, the sounds have become scarcely audible, and have appeared as if they were made at a great distance from the ear.

Such I believe are the ordinary symptoms of dropsy of the heart. But it must be remembered that this affection may be complicated with organic disease of the heart, lungs, &c., the symptoms of which must be taken into account in examining such cases.

Diagnosis.—A correct diagnosis may generally be formed in hydropericardium by careful attention to all the symptoms which are developed. But the symptoms which more particularly point out this affection, are the feeling of oppression in the region of the heart, a disposition on the part of the patient to sit up or lean forward to rest, instead of assuming the recumbent posture, a lessened impulse, and decrease of the sounds of the heart, with a disturbed rhythm, and finally dullness on percussion over an extended space or surface in the præcordial region.

Morbid Appearances.—The appearances on dissection, are not very marked, aside from the serum in the pericardium. Sometimes, however, in active cases, the pericardium exhibits traces of inflammation, there being increased vascularity, and ulceration of its lining serous membrane. The quantity of serum found in the pericardium, varies, as I have already suggested, from half a pint, to several pints, or even quarts.

Prognosis.—Complicated as this affection, very generally is, with organic disease of the heart, lungs, or some other important vital organ or part, or else associated with great general debility; the prognosis is generally unfavorable. It is probable, however, that cases uncomplicated with serious organic diseases, sometimes recover, under judicious treatment.

Treatment.—In cases of hydropericardium of an *active* character, depending upon congestion, irritation, or inflammation of the pericardium; cathartics, cups, blisters, &c., should be resorted to, with a slight mercurial course, till the pericardial irritation, or inflammation is subdued. Having accomplished this, and also corrected any general or local inflammatory, or other complication, remedies should be persevered in, for the removal of the pericardial effusion. For this purpose, twenty drops each of the tincture of squill, and tincture of digitalis, to act upon the kidneys; with eight grains of the iodide of potassium, to affect the lymphatic and glandular system, should be given every six hours.

Alternating with this, to establish a drain by the bowels, a drachm of the bitortrate of potassa, should be given in solution, and this treatment should be continued, keeping the skin clean and warm, till the pericardial accumulation is removed.

In *passive* cases of the disease, the patient should take a mercurial cathartic at first, and then the iodide of potassium, with the squill and digitalis, as suggested above, and instead of the cream of tartar for the bowels, a pill of gamboge, scammony, and rhubarb, of each, a grain may be given, after dinner, each day.

SECTION XXVI.—HYDROCEPHALUS.

By hydrocephalus, from νδωρ, "water," and κεφαλη, "the head," I mean here, water in the head or *dropsy of the brain;* whether the effusion occupies the ventricles of the brain, or the cavity of the arachnoid membrane.

Hydrocephalus, like other varieties of dropsy, may be either *active* or *passive.* If *active,* it is generally the result of congestion, irritation, or inflammation of the arachnoid membrane, or else of the serous membrane, which lines the ventricles.

If *passive,* it depends, like other varieties of passive dropsy, upon a hereditary or acquired depravity of the system, in which there is a weak or watery state of the blood, and a debilitated and relaxed condition of the tissues of the body.

Hydrocephalus is confined to no age, but it generally occurs during infancy or childhood, active cases, passing on generally in a few days or weeks to a fatal termination, while passive cases may be congenital, and of protracted duration, passing on sometimes for months, or even years.

Hydrocephalus is attended with a train of distressing symptoms, some of which are peculiar to the *active,* while others are peculiar to the *passive* form of the disease. We will therefore proceed first to consider the symptoms of *active* cases, and then examine so much of the symptoms of *passive* cases, as are peculiar to that form of the disease.

Symptoms.—The *active* form of this disease, as we have already seen, may be the result of congestion, irritation or inflammation of the arach-noid membrane, or else of the serous membrane which lines the ventricles, or what is probably more common, of both. In cases that are inflam-matory, there are developed, before the effusion takes place, all the symptoms peculiar to arachnitis, such as pain in the head, intolerance of light, irritability, nausea and vomiting, with febrile excitement.

In active cases in which there is only irritation or active congestion of the brain and its meninges, the effusion is preceded by an irritable peevish, or fretful state of the little patient. The appetite is variable, the bowels are constipated, there is intolerence of light, and nausea with vomiting from very slight causes. The urine becomes scanty, there is pain in the head, referred often to the temples, the pupils contract, there is starting during sleep, slight febrile symptoms arise towards evening, and the child often screams suddenly, as if suffering from acute pain.

After these symptoms have gone on progressing for a few days, or sometimes weeks, effusion takes place, either into the cavity of the arach-noid membrane, or else into the ventricles of the brain, or into both. As this occurs, it relieves the congested cephalic meninges for the time, so that the pain in the head subsides, the pupils become natural, light is tolerated, the nervous irritability is less, and to a superficial observer the patient may appear decidedly better.

But this temporary abatement of the symptoms, caused by the effusion is followed by the real symptoms of hydrocephalus, for as the effusion increases, the ventricles or the arachnoidal membrane, or both become distended and the pressure upon the brain develops a new and most dis-tressing train of symptoms.

The patient gradually becomes stupid, the pupils dilate, there is paralysis of an arm or limb, the thumb of one hand is brought firmly into the palm, and the hand carried above the head, and there may be loss of sight and hearing. Finally terrible convulsions of the voluntary muscles occur, followed by coma, from which the patient can be but par-tially aroused, if at all, and in this state the patient may linger, with occasional convulsions, for several days, when either in a state of pro-found coma, or terrible convulsions he expires.

Such, according to my observation, are the symptoms of hydrocepha-lus, as it occurs in the *active* form, from inflammation, irritation, or active congestion of the brain or its meninges, and especially of the arachnoid, and lining serous membrane of the ventricles. We will now proceed to the consideration of the *passive* form of hydrocephalus, or that which is the result of hereditary or acquired depraved condition of the system, the effusion being the direct result of passive cephalic con-gestion, as well as of a watery state of the blood, and a relaxed condition of the tissues.

This form of hydrocephalus may be congenital; or it may occur in early infancy or childhood; and possibly during any subsequent period of life.

The *symptoms* in congenital cases, as well as in those occurring in early infancy, are very marked, as the head is generally more or less enlarged, with the sutures gaping, or the bones spread apart; and sometimes tumors are projecting from the open sutures. Or the head may be illy shaped, being more prominent on one side than the other, or in some particular part.

I remember to have seen a case in which the head of a child, only a few months old, was not only deformed, but was nearly the size of an adult head.

In those cases that occur later in childhood, and at adult age, there is generally no visible enlargement of the head, as the bones have become too firm to yield to the internal pressure. Such cases generally terminate fatally within a few weeks; but sometimes the patient lives on in this condition for several years, laboring under more or less physical and mental derangement.

Sometimes the pressure upon the brain, caused by an accumulation of serum in the ventricles, or in the cavity of the arachnoid membrane, is borne with comparative slight impairment of the mental faculties. But ultimately the patient, in most cases, becomes emaciated, stupid, and indifferent; the countenance is without expression; the gait is unsteady; convulsions and coma supervene; and the patient finally dies, either comatose or in convulsions.

Such, I believe, are the ordinary symptoms of passive hydrocephalus; liable, of course, to variations, depending upon the general condition of the patient, and the amount of the cephalic effusion.

Anatomical Characters.—In cases in which the dropsy has been of an active character, the arachnoid membrane, or the lining membrane of the ventricles, show signs of having been inflamed, or of active congestion. The ventricles contain more or less colorless serum, in most cases; but if the fluid is found mainly in the cavity of the arachnoid, it may be of a bloody serous character.

In cases that have been of a *passive* character, a large amount of fluid is generally found in the ventricles, which are very much enlarged, and sometimes thrown into one extended cavity, the cerebral substance forming their parietes being either of a natural consistence, or else hardened or softened; or the fluid may occupy the cavity of the arachnoid membrane, in which case the brain is compressed into the bottom of the cranial cavity. Or, as sometimes happens, an opening is made in the commissures of the brain, and the cavity of the ventricles and that of the arachnoid are thrown into one.

The quantity of liquid found in hydrocephalus, varies from a few ounces to several pounds, and it generally contains more or less albumen, and a small proportion of various saline substances.

Diagnosis.—In cases of active hydrocephalus, it may be at first difficult, or even impossible, to be certain that effusion has taken place.

If, however, after a more or less protracted stage of irritability, and perhaps febrile excitement, &c., the symptoms subside, and after an indefinite period, varying from a few hours to several days, a new train of symptoms arise, such as stupor, convulsions, paralysis, &c., it may be regarded as nearly certain that effusion has taken place, either into the ventricles or into the cavity of the arachnoid membrane.

Passive cases of hydrocephalus, if congenital, or occurring during early infancy, may generally be distinguished by the enlargement of the head, together with the attendant symptoms. And if later in life, there comes on, in connection with great debility, and especially a scrofulous condition, evidence of continued internal pressure of the brain, with marked physical, and more or less intellectual disturbance; and the patient is affected with strabismus, loss of speech, paralysis, &c., the case may be regarded as *passive* hydrocephalus.

Causes.—Active cases of hydrocephalus may be produced by any cause or train of causes that favor congestion, irritation, or inflammation of the brain and its meninges, such as constipation of the bowels, intestinal worms, direct injuries of the brain, &c.

Passive hydrocephalus is generally hereditary, depending upon a scrofulous or tuberculous depraved inherited constitution; or else, if it occur later in life, it is the result of various abuses of the system, such as masturbation, sexual excesses, the use of tobacco, drunkenness, &c.

Prognosis.—The prognosis in hydrocephalus is decidedly unfavorable, especially if it is the result of a scrofulous or tuberculous condition, or is connected with any organic lesion. If, however, it is solely dependent upon functional derangement, it is possible that recovery may take place, if the deranged function be corrected, and proper remedies be resorted to for the absorption of the effused fluid.

Treatment.—In the active form of the disease, to subdue the irritation, congestion or inflammation upon which the effusion depends, general bleeding should be resorted to, if indicated; and then a full dose of calomel in castor-oil administered. Cups, wet or dry, should be applied to the temples or back of the neck, and then blisters, and after the operation of the cathartic, alterative doses of calomel should be given every four hours, with James's powder, till a slight mercurial impression is produced, when it should be suspended, and the antimonial continued, every six hours; and alternating with it, full doses of the iodide of potassium should be given. This treatment should be continued, with warm pediluvia, blisters back of the ears and to the back of the neck; and sufficient cream of tartar, morning and evening, to keep the bowels loose, till recovery or a fatal termination takes place.

In passive hydrocephalus, whether occurring in early infancy or later in life, a cathartic of calomel or podophyllin and castor oil may be given at first. And unless contra-indicated, on account of a scrofulous or tuberculous condition, alterative or slightly cathartic doses of calomel may be given, morning and evening, for a few days, when it should be suspended, and the bowels kept loose by small doses of the bitartrate of potassa.

On suspending the mercurial, if it is used, and immediately after the operation of the cathartic, in cases in which it is contra-indicated, the iodide of potassium should be administered in full doses, three times per day, with moderate diuretic doses of the tincture of squill. This treatment should be continued, with blisters back of the ears and to the back of the neck, and if necessary, small doses of the citrate of iron after each meal, or even cod liver oil, till the hydrocephalus is removed, or the case is rendered hopeless.

The skin should be kept clean and warm, and the patient, whether young or old, should be allowed a reasonable amount of proper nourishment, and sufficient pure dry air, and on no account should be exposed to any other. Finally, if the patient appears ready to perish, a trochar may be introduced perpendicularly at one edge of the anterior fontanel, and the water drawn off, after which the head should be compressed firmly by adhesive straps or a bandage.

SECTION XXVII.—SCROFULA.

By scrofula, I mean here that peculiar morbid or deranged condition of the whole system of nutrition, but especially of the lymphatic system, and lymph itself, attended with indolent glandular tumors in the neck, axillæ, groins, mesentery, and other parts of the body.

Now, in order to understand this subject, it must be remembered that the vessels that act exclusively for the growth of the system, are the lacteals, which are found along the alimentary canal. The lymphatics in other parts of the system remove matter already deposited, a function which they probably hold in common with the extreme veins, and the venous portion of the capillaries.

The lymphatic vessels are extremely small at their origin, and they exist in great numbers in the skin, mucous membranes, and in every part of the body, except perhaps in the substance of the brain.

The lymphatics, like the arteries and veins, have three coats, and like the veins they converge to form larger trunks, as they pursue their course towards the large veins near the heart, into which they pour their fluid contents, both lymph and chyle.

At certain points along the *lacteals*, and all the lymphatic vessels, they pass through distinct glands, the lacteals through the mesenteric, and the other lymphatics through the lymphatic glands in different parts of the body. They are, however, most numerous in the neck, axillæ, groins, and in the cavities of the chest and abdomen.

The lymphatic glands, vary in size, and appear to consist of a plexus of minute lymphatic vessels, associated with a plexus of blood-vessels, encased in a capsule of areolar tissue.

Now, the lacteals in a healthy state of the system, take up the chyle along the intestines, with perhaps transuded albumen, and convey it through the mesenteric glands, where some of its fat and albumen is converted into fibrin, and thence it passes to the thoracic duct, through which it is carried into the blood at the junction of the left subclavian with the left jugular vein.

The other lymphatics, however, imbibe all the various constituents of the body, both fluid and solid, after their vitality have ceased, and they also absorb foreign substances when exposed to their mouths, as well as more or less albuminous matter transuded from the small blood-vessels.

In a healthy state of the system, the lymphatics take up all the wastes of the system not absorbed by the extreme veins, together with more or less albumen which transudes from the blood, all of which is conducted through the lymphatic glands, where a portion of the albuminous matter is converted into fibrin. This being accomplished, the lymph thus fibrinized passes on, to enter the blood, either through the thoracic duct, or by trunks that empty directly into the large veins.

36

Now, it will be seen, that imperfect chyle, with derangement of the lacteal and other lymphatic vessels, as well as of the mesenteric and lymphatic glands, involves a derangement of nutrition and absorption, or that process in the system, by which the old tissues are removed, and new matter furnished to supply its place. This derangement, then, constitutes, as I believe, the scrofulous diathesis or condition, and it may be either hereditary or acquired—in either case, developing a train of symptoms which we will now proceed to consider.

Symptoms.—The countenance of scrofulous patients has a delicate, soft, flaccid aspect, with a frequent swelling or fullness of the upper lip. The hair is generally lightish and fair, and the eyes blue; there is a proneness to catarrhal affections, and the edges of the eyelids are liable to become inflamed, being often red and tender. The digestive powers are weak, the appetite variable, the urine deposits a whitish sediment, and there is a strong tendency to swelling or enlargement, with induraration of the mesenteric and other lympathic glands; there is a tendency to excoriations behind the ears, and scaly eruptions on the head; there is often an obstinate and protracted ophthalmia—the temper is irritable, and the growth of the body proceeds slowly, while the mental powers are generally precociously developed, and often very active.

Now all these symptoms are just what might be anticipated, from a deranged condition of the whole system of nutrition and absorption, which we have already seen constitutes *scrofula*. And this scrofulous condition, whether hereditary or acquired, may continue for several years, and finally pass off without terminating in any very serious local affection. Generally, however, the scrofulous habit increases, and at last, under the influence of certain exciting causes, shows itself in its more obvious and active form.

In cases that pass on and progress, the mesenteric glands are very liable to become enlarged and indurated. The lymphatic glands along the neck, in the axillæ, groins, or other parts of the system may become enlarged, and firm to the touch, in which indolent condition they often remain for years. Or they may pass on by degrees, into slow inflammation which at last sometimes terminates in suppuration. When suppuration does take place, indolent ulcers are formed from which a thin puruloid serum is discharged. The ulcers thus formed are slow to heal, and when they do, the cicatrices which are left are uneven, irregular, and conspicuous.

In conjunction with the tumors or ulcerations of the neck, the eyelids become in some cases permanently inflamed, and there is apt to be irritation of the Schneiderian membrane, and sometimes of the mucous membrane of the bronchial tubes. As the disease advances, and the internal and external glandular structures become involved, the various functions of the body become materially deranged.

Scaly eruptions appear on different parts of the surface, the bones enlarge and become carious, the cartilages ulcerate, the large joints enlarge and suppurate, the vertebræ may become diseased, and not unfrequently the bones of the nose and palate are more or less rapidly destroyed by ulceration. Finally scrofula may develop its local manifestations in any of the soft or solid tissues of the body, but its most frequent serious local manifestations are in tuberculous meningitis, tubercular phthisis, white swelling, or disease of the hip or knee joints, ophthalmia, and scrofulous nodes in the mesentery, lungs, liver, spleen, brain, and other internal or external organs or tissues of the body.

The manifestations of the scrofulous habit seldom make their appearance very markedly before the period of dentition. And the progress of the disease is exceedingly various, being sometimes fully developed during infancy or childhood, but in other cases, not till the age of puberty, or even much later than that.

Anatomical Characters.—The glands early are either found simply enlarged, and a little firmer than in health, or else if more advanced, they are of a grayish color, with a granular structure, and very decidedly indurated. Or if still more advanced, the proper tissue of the glands are more or less absorbed, and tuberculous matter is found either irregularly infiltrated, or deposited in the form of distinct granulations. If the disease was still more advanced, abscesses may be found containing softened tuberculous matter, with more or less unhealthy pus.

Diagnosis.—Scrofula may generally be distinguished from all other diseases by the countenance, the blue eyes, light hair, fair skin, tumid upper lip, &c., in its early stages. Later, by the variable appetite, eruptions about the ears, and on the scalp; glandular enlargements and induration; and various other like symptoms. Finally, in the last stages of the disease, it may be distinguished by the emaciation, and general haggard appearance of the patient, together with whatever local manifestation of the disease there may chance to be developed. And besides, the diagnosis may be aided, in all stages by the fact of a strong hereditary predisposition to the disease.

Causes.—This peculiar morbid or irritable and deranged condition of the lymphatic system, may, as we have seen, be hereditary, or it may be produced by a variety of causes, such as damp impure air; insufficient or unwholesome food; measles, scarlatina and hooping cough; and by chronic inflammation of the alimentary mucous membrane.

The hereditary predisposition to scrofula doubtless consists in an imperfection of the digestive organs; and a peculiar or defective organization of the lymphatic and other lacteal vessels and their glands, in consequence of which, digestion, nutrition, and absorption are performed very imperfectly. For there is a failure either of the digestive organs, or of the lacteals or their glands, to furnish, take up, or prepare sufficient nourishment, to supply the wastes of the system. Or else an inability on the part of the absorbents to take up, or of their glands to properly change the wastes of the system.

It is probable, however, that the hereditary predisposition to scrofula, consists in an imperfection of the digestive and assimilative organs; and also of the absorbent lymphatic vessels and glands. That being the case, only slight exciting causes are capable of developing the disease.

Change from a warm to a cold climate is peculiarly liable to develop scrofula in those who are predisposed; as well as exposure to cool damp air, both operating probably to render torpid the lymphatic vessels, and irritate the lymphatic glands. Impure air probably produces scrofula by entering the circulation, or in part perhaps by affecting the lymphatics of the skin and lungs. It probably also fails to supply sufficient oxygen to the blood, and it may directly impair digestion, and thus cut off the supply of nourishment.

Insufficient food may act to produce a scrofulous condition, by a mere failure to supply the demand caused by the breaking down of the old tissues. And unwholesome food may not only fail to supply the demand caused by the wastes of the system, but it probably furnishes materials of an impure character, which irritates the lacteals and perhaps other lymphatic vessels, as well as their glands.

Gastro-enteritis, if long continued, produces scrofula, in part perhaps, by impairing digestion, and also by hindering the absorption of the chyle along the intestines.

There are various other causes of scrofula, such as masturbation, sexual excesses, licentiousness, drunkenness, the use of tobacco, syphilis, or a syphilitic taint; and the whole catalogue of improprieties and abuses of the system, to which the human family are addicted, in their present fallen and depraved condition.

These causes, however, act in the same or a similar manner, by deranging nutrition and absorption; and thus develop this fearful and complicated affection.

Pathology.—From what we have already seen, the pathology or nature of this disease is very plain. And it is only necessary to remember that the functions of the mesenteric and other lymphatic glands are to modify, and especially to fibrinize the chyle and lymph; in order to discover the origin of the tuberculous matter, which exists in the blood, and is very liable to be deposited in this disease.

For, as the mesenteric glands are diseased, the fat and albumen of the food is not changed to fibrin, as it should be; and as the other lymphatic glands are involved, the albumen which transudes from the blood, everywhere in the tissues, as well as the matter furnished by the breaking down of the old tissues, fails also to be modified and fibrinized, as it should be, by the lymphatic glands.

Thus, then, we have in the blood, from imperfect digestion, and the mesenteric and other glandular derangements, an excess of albumen, and various morbid products, ready to be deposited in the glandular or other structures of the body, whenever a slight irritation, or other accidental cause, may chance to produce or create a predisposition to the deposit.

Such, according to my view, is the pathology of one of the most interesting and complicated affections with which the human family are afflicted; and which is yearly carrying to a premature grave many thousands, both male and female, and mostly at the dawn of, or during the early period of life. Well might we exclaim, in the language of one of old, in view of these facts, "Is there no balm in Gilead? is there no Physician there? why then is not the *health* of the *daughter of my people recovered?*"

This disease is really the result of the continued and multiplied imprudence of the human family, from their first *imprudent eating* in the *garden*, to the present time. And the *balm*, for this mighty physical

* Jeremiah viii. 22.

wound, consists in the correction of these imprudences, and a return to the strict observance of the laws of health and propriety, in every respect. Hence the following course of treatment is indicated in this disease.

Treatment.—If the patient has any low or vicious habits, they should be inquired into and corrected. If addicted to the use of tobacco, to drunkenness, licentious habits, or any other species of imprudence, the fearful consequences should be pointed out, and the patient made to conform rigidly to the laws of health and propriety, in every respect. If the patient is exposed to filthy, damp air, in basements, as children often are, they should be removed at once to clean, dry, elevated apartments.

If the patient has been illy clad so that the surface has not been kept warm, sufficient clothing should at once be furnished, with flannel to be worn next the skin, except in the very warmest weather, and in very bad cases, constantly. The patient should also be directed to sleep in flannel, and on no account to expose himself to become chilly for any length of time. By strict observance of these regulations, the morbid irritability of the lymphatics of the skin may in part be corrected or greatly relieved, and the general condition of the patient improved.

If the patient has been in the habit of irregular eating, or of taking improper or unwholesome food, a good nourishing diet should be substituted, to be taken at regular hours, and eating between meals should be strictly prohibited. This will not only improve digestion, but it will also furnish to the system a better supply of more healthy chyle, to satisfy the demand caused by the wastes of the system.

If the patient has been addicted to sedentary or indolent habits, sufficient exercise in the open air should be enjoined, when the condition of the patient and the state of the weather will admit of it. This, by the aid it affords digestion, and in fact all the functions of the body, will often be of material service to the scrofulous patient.

If the patient has been addicted to habits of filth, he should be directed to wash the body once or twice a week in tepid or cool water and soap, for the purpose of removing any dirt that may have accumulated on the skin.

Such, I believe, are the principal indications in the treatment of scrofulous affections. In some cases, however, medical treatment is also indicated.

If the bowels are constipated, as is frequently the case, a blue pill may be given at evening, for two or three days, and then a pill of aloes and rhubarb continued after dinner each day till they are regulated.

If there is acidity of the stomach, two drachms of prepared chalk may be dissolved in half a pint of water, and of this a tablespoonful may be given every morning for a time. If the appetite is poor, moderate doses of the fluid extract of gentian or columbo, or else of the cold infusion of one of these tonics with a little ginger, may be given three times per day, and continued till the appetite is restored.

In mild cases, in which there are no marked glandular indurations, the citrate cabonate, or ferrocyanuret of iron may be given in two or three grain doses three times per day, for a time, for the blood. In cases, however, in which there are glandular indurations, with or without ulceration, five grains of the iodide of potassium with two grains of the solid, or ten drops of the fluid extract of conium should be given three times per day before eating. And if the prostration be considerable, ten drops of the syrup of the iodide of iron should be given three times per day, after eating, and continued till the blood is supplied with a due amount of iron. As an application to the indurated and enlarged glands, the compound iodine ointment will generally do best.

In cases in which the digestion is very imperfect, but little chyle being furnished, and the absorbents are morbidly active, as is sometimes the case, producing emaciation, the cod-liver oil may be tried, and if it agrees with the stomach, it may be given for nourishment, in table-spoonful doses, three times per day, an hour after each meal. This may be continued, with such other treatment as is indicated, as long as its effects are salutary.*

By thus correcting the habits of scrofulous patients, and doing just that which is indicated, and nothing more, many incipient cases may be arrested, bad cases palliated, and mild ones perhaps permanently cured, if there is not a strong hereditary predisposition to the disease.

Much may be done by way of preventing the development of the disease, in children born of scrofulous parents, by avoiding the causes that operate to call into activity the latent hereditary predisposition to the disease. The physician, then, should in such cases, and in fact in all cases, give such direction for the care and preservation of health, as shall not only prevent scrofula, but as far as possible all other diseases to which, by imprudence, mankind have become hereditarily predisposed.

SECTION XXVIII.—BRONCHOCELE—(*Goitre.*)

By bronchocele, from βροχγος, "a bronchus," and κηλη, "tumor," I mean here an enlargement of the thyroid gland, of a non-malignant character, consisting merely in *hypertrophy* of the part.

In order to understand this subject, it is necessary to remember that the thyroid gland is situated upon the trachea, its two lobes being placed one on each side; the isthmus connecting them passing in front of its upper rings. Its structure consists "of an aggregation of minute independent membranous cavities, inclosed by a plexus of capillary vessels, and connected together by areolo-fibrous tissue." These minute cavities are filled with fluid, and the gland is larger in the young, and in females, than in adult males.

The gland is supplied with blood by the thyroid arteries, and its nerves are derived from the superior laryngeal and sympathetic.

The thyroid gland, thus situated and constituted, is liable to enlargement from *hypertrophy*, the symptoms of which we will now proceed to consider.

Symptoms.—This enlargement appears at first like a small tumor, on one or both sides of the trachea, which increases in size, more or less rapidly, till in some cases it finally occupies the whole anterior part of the neck. At first the tumor is soft, but as it progresses, some parts of it generally acquire considerable density, while other portions feel nearly natural.

If the tumor becomes very large, it makes pressure on the trachea, œsophagus, and large blood-vessels, producing difficult breathing, diffi-

* The hypophosphites may be indicated in some cases.

culty of swallowing, and sometimes fatal congestion of the brain. I have known the tumor to produce palpitation of the heart, throbbing of the carotids, and, in one case, fatal apoplexy, in a lady not otherwise predisposed to cephalic congestion.

Enlargement of the thyroid gland generally progresses slowly, so that it may occupy years in acquiring a large size. But some cases progress rapidly, the gland acquiring a large size in a few months. In a case that fell under my care, in an anæmic female, being associated with amenorrhœa, and a prominence of the eye-balls, the tumor increased to a large size in six months, and it was subject to great variations in its size, the result evidently of accidental congestions.

Diagnosis.—When a tumor appears occupying the place of the thyroid gland, or one lobe of it, and moves with the larynx and trachea in the act of deglutition, and is moveable, insensible, soft and spongy to the feel, and free from pulsation, it is hypertrophy of the thyroid gland or *bronchocele.*

Causes.—All the causes which operate to develop a scrofulous condition of the system, may produce bronchocele. But water impregnated with the carbonate or sulphate of lime, together with damp air, and a low electrical state of the atmosphere, such as generally prevails in valleys between high mountains, are probably the most frequent causes of this affection.

The disease is very apt to come on in young females, in connection with suppression of the menses, attended also as we have seen in some cases, with prominence of the eye-balls, in decidedly anæmic patients.

Prognosis.—The prognosis is favorable in all recent cases of bronchocele, if the causes can be removed, and the patient be subjected to proper treatment in season. But in old protracted cases, if the patient is advanced in life, but little need be expected from remedial measures.

Treatment.—The general condition of the patient should first be ascertained and corrected, and then if the goitre continues, iodine should be resorted to internally and externally, and continued till the bronchocele is removed.

For internal administration, Lugol's solution, made by dissolving a scruple of iodine, and two scruples of the iodide of potassium in seven fluid drachms of water, will generally do best. Of this solution, five drops may be given three times per day, at first, in a glass of sweetened water, and the dose may be gradually increased to fifteen or twenty drops.

In case, however, the iodine in this form should disagree with the stomach, five grains of the iodide of potassium may be given three times per day, before eating, instead. And if there is an anæmic condition, ten drops of the syrup of the iodide of iron may be given three times per day, after meals, as an alterative and tonic.

After having removed the causes, corrected the general condition, and had the patient on iodine internally for a time, an ointment should be applied to the tumors, two or three times per day, made by mixing two drachms of iodide of potassium, dissolved in a fluid drachm of boiling water, with an ounce of lard. Or, if the case should prove obstinate, two drachms of iodine may be added, in the preparation of the ointment to be used. This should be continued with the iodine internally, till the bronchocele is entirely removed.

CHAPTER XI.

DISEASES OF THE EYE.

SECTION I.—CATARRHAL OPHTHALMIA.

By catarrhal ophthalmia, I mean here that variety of inflammation of the conjunctiva which occurs from common colds, or from sudden exposure to cold, dampness, &c.

But before we proceed to the consideration of this affection, let us take a general glance at the anatomy and physiology of the eye, that we may proceed understandingly.

The general form of the eyeball is that of a sphere, about an inch in diameter, "having the segment of a smaller sphere engrafted upon its anterior surface, which increases its antero posterior diameter."

The eye, that it may be a perfect organ of sight, is composed of three tunics, the sclerotic and cornea, the choroid iris and ciliary processes, and the retina and zonula ciliaris. It has also three humors, the aqueous, crystalline and vitreous.

The sclerotic and cornea form the external tunic of the eye, the sclerotic constituting four-fifths, and investing the posterior portion of the globe, while the cornea invests one-fifth of the eye anteriorly, being continuous with the sclerotic.

The *sclerotic* is a dense fibrous membrane, continuous posteriorly with the sheath of the optic nerve, which is derived from the dura mater. Anteriorly, it has a beveled edge to receive the cornea. Its anterior surface is covered by the *tunica abuginea*, or expansion of the tendons of the four recti muscles.

The *cornea* forms as we have seen, the anterior fifth of the external tunic of the eye, being transparent, concavo-convex, and set in the beveled edge of the sclerotic, in the same manner that a watch-glass is received by the groove in its case. The cornea is composed of several layers connected together by areolar tissue.

The *second tunic* of the eye consists, as we have seen, of the *choroid, ciliary ligament* and *iris.*

The *choroid* is situated immediately within the sclerotic, to which it is connected by a fine areolar tissue, and extends from the opening of the optic nerve forward to the edge of the *iris,* to which it is connected, and also to the *ciliary processes,* as well as the *ciliary ligament.* The choroid membrane is composed of three layers, an external venous, a middle, consisting mainly of minute arteries, and an internal layer, the *membrana pigmenti.*

The *ciliary ligament* is a dense white structure which connects the cornea and sclerotic at their junction with the iris and external layer of the choroid membrane. It contains the ciliary nerves and vessels, and has also a minute vascular canal called the *ciliary* canal, situated within it.

The *iris* makes up the anterior portion of the second tunic of the eye. It is situated in the middle of the aqueous humor, and separates the anterior and posterior chambers, being pierced near its centre by the pupil. It is made up of blood-vessels, muscles, and nerves, which are connected by fine cellular tissue. There is probably a circular muscular tissue passing around the pupil, which serves by its action to contract it. And there appears to be radiating muscular fibres passing to the circumference of the iris, which by contracting dilate the pupil.

The *ciliary processes* which are an appendage developed from the inner surface of the middle tunic of the eye, consist of triangular folds, formed by the plaiting of the middle and internal layer of the choroid coat. "Their periphery is connected with the ciliary ligament," and continuous with the middle and internal layer of the choroid, while its central border is free, resting against the circumference of the lens. "The anterior surface corresponds with the *uvea*, the posterior receives the folds of the zonula ciliaris between its processes," thus establishing a connection between the choroid and third tunic of the eye. The ciliary processes are covered with an abundant layer of *pigmentum nigrum*.

The *third tunic* of the eye as we have seen, is the retina, with its anterior prolongation, the *zonula ciliaris*.

The *retina* is composed of three layers, and external very thin, called Jacob's membrane, a middle or nervous, consisting of an expansion of the optic nerve, and an internal vascular membrane, consisting of the ramifications of the central artery of the retina, and its accompanying veins.

The *zonula ciliaris* is a thin vascular membrane which connects the anterior border or margin of the retina, with the anterior surface of the lens near its circumference. Its under surface is in contact with the hyaloid membrane, and it forms around the lens the anterior wall of the canal of *Petit*.

The *aqueous humor* is a transparent albuminous fluid which occupies the anterior and posterior *chambers* of the eye, or the small spaces anterior and posterior to the iris. These chambers are lined by a membrane which secretes this fluid.

The *vitreous humor* forms the posterior three-fourths of the bulk of the eye, and consists of an albuminous, gelatinous, and highly transparent matter enclosed in the hyaloid membrane. From the hyaloid membrane lamellæ pass inwards, forming compartments, which contain the fluid. And along the centre of the vitreous humor, there is a canal which conducts an artery from the central artery of the retina to the capsule of the crystalline lens.

The *crystalline lens* is imbedded in the anterior portion of the vitreous humor, from which it is separated by the hyaloid membrane, and just back of the pupil, and posterior chamber of the eye. It is surrounded by the ciliary processes, invested by a transparent membrane or capsule, "retained in its place by the attachment of the zonula ciliaris," and immediately around its circumference is the triangular canal of *Petit*.

The *crystalline lens* is transparent, and composed of concentric layers,

the external being soft, and easily removed, while the internal grow
more firm, till a central hardened nucleus is reached.

Thus we have at a glance the different tunics and humors of the eye.
And it should be remembered, that the sclerotic is for protection; the
cornea for the transmission of light to the eye, and the choroid for nutrition; its pigmentum nigrum also absorbing scattered rays of light.
The iris regulates the amount of light to be admitted; and the humors,
together with the cornea, refract the rays of light, so that the image of
objects may fall on the retina, by which the impressions are communicated to the gray matter of the hemispheres of the brain, and thence to
the mind itself.

The appendages of the eye, consisting of the eye-brows, eye-lids, eye-
lashes, caruncula lachrymalis, and lachrymal apparatus, we need not
stop to examine, as their diseases are mostly *surgical*. But it should be
remembered, that the eye is kept back in its socket, and turned in different directions, by the four recti and superior and inferior oblique
muscles, and also that the conjunctiva, a mucous membrane, covers the
whole anterior surface of the eye, and then is reflected upon both lids,
forming their internal or lining membrane. It is, in fact, *catarrhal inflammation* of this membrane which constitutes the legitimate subject of
this section; the symptoms of which we will now proceed to consider.

Symptoms.—The early symptoms of catarrhal ophthalmia are, in addition to the general symptoms of an ordinary cold, more or less uneasiness on exposure to light, with slight redness and pain in the temples.
As the disease advances, the pain and intolerance of light become more
marked, and in severe cases there is more or less catarrhal fever, headache, loss of appetite, nausea, &c.

When the conjunctival inflammation is fully established, there is an
increased *mucous* discharge, with very considerable redness of the membrane. The redness, however, is generally superficial, and of a bright
scarlet color; the distended vessel appearing quite superficial.

Catarrhal conjunctivitis is seldom attended with much swelling; the
only approach to such a condition being an elevation, caused by a slight
effusion of serum, under the membrane. And unless in very severe
cases, in which there may be thickening, the sclerotic coat may be seen
through the inflamed conjunctiva, of its natural white color. The symptoms of catarrhal ophthalmia, like those of other catarrhal affections,
generally remit by day, and undergo exacerbations at night, during the
whole progress of the disease.

Diagnosis.—Catarrhal ophthalmia may be distinguished by the attendant catarrhal symptoms, by the diurnal remissions, and nocturnal exacerbations, by the mucous discharge and by the natural appearance of
the sclerotica, as well as the bright red tint of the conjunctiva.

This affection may be distinguished from purulent ophthalmia, by its
mildness when compared with that disease, by the mucous instead of
purulent character of the discharge, and by its non-contagious character.

Causes.—It is probable that common colds, from various exposures,
are the most frequent causes of catarrhal ophthalmia. The disease may,
however, be produced by causes, such as cold and moisture, acting
directly upon the eye. Or it may be caused by exposure to winds,

storms, &c., probably in cases in which the system generally does not suffer very materially.

Prognosis.—As this form of ophthalmia is generally confined to the conjunctiva, it usually terminates favorably, unless the case be neglected, or what is much worse, is subjected to harsh or improper treatment With either of these misfortunes, the sclerotica and cornea, may become involved in the inflammation, and thus opacity of the cornea, or other serious consequences follow.

Treatment.—In the treatment of catarrhal ophthalmia, mild measures only, are usually required.

The patient may take two or three blue pills at evening, and follow them in the morning, by half an ounce of the sulphate of magnesia; and then the bowels may be kept loose by a teaspoonful, taken each morning, for a few days.

A warm pediluvium, should be used, at evening, each day, and if the inflammation is severe, cups may be applied to the temples, or back of the neck, at first; and later, if necessary, blisters back of the ears, or to the back of the neck.

Warm milk and water, is the best application for the eyes, for the first day or two. Later, a wash made by dissolving three grains of the acetate of zinc, and one and a half grains of the acetate of morphia, in an ounce of rain water, will generally do best. Or the wash may be made, by infusing five grains of opium, in an ounce of rain water, with three grains each of the acetate of lead, and sulphate of zinc; and then strain or filter. The wash should be applied four times per day, till a cure is effected.

SECTION II.—PURULENT OPHTHALMIA.

By purulent ophthalmia, I mean here, that variety of acute inflammation of the conjunctiva, attended with a copious secretion of matter, resembling pus; the inflammation tending strongly, unless arrested, to pass to the cornea, producing sloughing, suppuration, ulceration, opacity, &c.

Symptoms.—Purulent ophthalmia may occur in the infant, or in the adult, and in either case, consists at first, of an inflammation of the conjunctiva of the lids, extending with more or less rapidity, to that portion of the membrane covering the eyes, and in bad cases, involving the cornea; attended in all cases, with a copious purulent discharge.

The disease may occur in early infancy, two or three days after birth, or at a later period, during childhood. At first, the lids are noticed to stick together, when the child wakes from sleep; the edges of the lids are redder than natural, and the child appears to suffer from the access of light. If the lids be inverted, their lining is found red, and more or less white mucus or purulent matter is found inside the lower lid.

Soon, however, the inflammation extends to that part of the conjunctiva covering the eyes, the vascular congestion, and redness, being considerably augmented. The lids swell, and become red externally; a copious secretion of purulent fluid issues from the eyes; the intolerance of light becomes more marked, and the swelling of the conjunctiva sometimes goes on, to an incredible extent. I have seen that portion of the conjunctiva covering the globe, so distended, that there was only a small passage down to the cornea, which had become deeply buried in the tumefied and projecting membrane.

If the disease passes on unchecked, the inflammation may extend to the sclerotic and other tissues of the eye, and especially to the cornea, producing sloughing, suppuration, ulceration, and opacity, with perhaps adhesion of the iris to the cornea.

The symptoms and progress of purulent ophthalmia in the adult, are similar to those which occur from the disease in infants, modified in their course and duration by age, and other circumstances of exposure, &c. The inflammation commences in the conjunctiva of the lids, and passes to that portion which covers the eye, and, unless arrested, may extend to the other tunics or tissues of the eye, and especially to the cornea. I believe the progress of the disease is generally less rapid than in infants, but neglected cases may pass on to sloughing, suppuration, ulceration, opacity of the cornea, &c. Perhaps the swelling of the conjunctiva is generally less in the adult : but I have seen that portion covering the globe so much swelled or distended as to nearly hide the cornea, only a small passage being left down to it.

Such, I believe, are the ordinary symptoms of purulent ophthalmia, as it occurs in the infant and adult, liable, of course, to modifications, from peculiarity of constitution, and various accidental causes.

Diagnosis.—Purulent ophthalmia may be distinguished from catarrhal by the severity of the attack ; the purulent appearance of the discharge ; by the swelling of the lids and conjunctiva ; and finally, by its tendency to pass on to destructive disorganization of the cornea, and other tissues or tunics of the eye.

Causes.—It appears probable that purulent ophthalmia, as it occurs in infants, may be caused by the contact of gonorrhœal or leucorrhœal matter, derived from the mother at birth. But I believe that the disease may be caused by want of cleanliness, exposure to cold, improper nourishment, want of care, &c., without the contact of gonorrhœal or leucorrhœal matter from the mother.

I am satisfied too, from careful observation, that the disease is contagious ; being communicated from one person to another, by accidental contact of the matter discharged. And it is probable that this may be the most frequent origin of the disease in adults. I believe, however, that it is sometimes produced in the adult by the direct contact of gonorrhœal matter ; and perhaps it may sometimes be generated, as in the infant, by filth, exposure to cold, improper food, &c.

Prognosis.—The prognosis in purulent ophthalmia is favorable, whether it occurs in infants or adults, provided the cause or causes can be removed, and the case be subjected to proper treatment in season.

If, however, the causes continue to operate, and the case be neglected, or improperly treated in its early stages, the worst results may be apprehended, and will generally be realized. Recoveries occasionally take place, however, after almost every unpleasant symptom, except utter disorganization have been developed.

Treatment.—In the treatment of purulent ophthalmia of infants, a grain of calomel may be administered, in a teaspoonful of castor oil, at first, and a little oil occasionally administered during the continuance of the disease.

If the attack is severe, a leech or two may be applied to the lids, or a blister applied to the temples, or to that of the affected side, if one eye only is involved.

The eye should be washed, morning and evening, with a little warm milk and water, and a solution of six grains of alum, in an ounce of rain water, should be applied to the eye, every four or six hours. Or a cloth wet with the solution may be kept constantly over the eye while the child is lying quietly, so that it need not be bound on. The alum wash produces no uneasiness, and being a good stringent, is the very best application in the purulent ophthalmia of infants.

With this course of treatment, I have never failed of arresting the disease in infants and young children, in a few days, if resorted to in season.

In the purulent ophthalmia of *adults*, as well as of infants, if any causes or influences are operating to keep up the disease, they should be removed, as far as possible.

A cathartic of calomel should be administered at first, in castor oil, and a blue pill given for the two or three succeeding evenings, followed in the morning by a teaspoonful or more of the sulphate of magnesia.

During the first five or six days, one-sixth of a grain of tartar emetic may be given every four hours. The eyes should be washed morning and evening with warm milk and water, and a saturated solution of alum in rain water should be applied every hour or two during the day, and through the night small linen cloths, wet in the solution may be laid loosely over the eyes.

Cups may be applied at first to the temples, and if necessary, leeches to the eyelids, and later, blisters may be required to the temples, back of the ears, or to the back of the neck.

After the first five or six days, and as the inflammation begins to subside, if there is a cloudiness of the cornea, five grain doses of the iodide of potassium should be given every six hours at first. The tartar emetic should be suspended, and four grains of James's powder given instead, every six hours, alternating with the iodide of potassium.

This treatment should be continued, with the alum wash to the eyes, till the inflammation is subdued, when the antimonial should be suspended, and the iodide of potassium continued in five grain doses, three times per day, till perfect transparency of the cornea is restored.

If, as the inflammation subsides, the lids are found granulated, they should be touched, once each day, with a camel-hair pencil, wet with a saturated solution of nitrate of silver, the upper lid being turned up, an assistant immediately injecting on to the part touched, from a small glass syringe, a jet of cool rain water, to prevent the impression being too strong, and to wash the caustic over other parts of the eye. By this treatment the granulations will be removed in a few days, at the same time that the conjunctiva of the globe and lower lid will also be improved by the caustic that falls on them in a diluted state.

If, after the inflammation and granulations, if they have existed, are

removed, there should remain slight redness and irritability of the conjunctiva, a wash may be applied for a time, made by dissolving two grains
of the acetate of zinc, and a grain of the acetate of morphia, in an
ounce of rain water. It may be applied four times per day, and continued till the conjunctiva assumes a natural appearance.

SECTION III.—SCROFULOUS OPHTHALMIA.

By scrofulous ophthalmia, I mean that peculiar inflammation of the
conjunctiva, extending sometimes to the sclerotica and cornea, which occurs in scrofulous patients, and depends mainly upon a scrofulous condition of the system, either hereditary or acquired.

Symptoms.—In addition to the ordinary symptoms of scrofula, the
patient exhibits an intolerance of light, and very soon a slight redness
of the conjunctiva occurs, at first perhaps confined mainly to the lining
of the lids. By degrees, however, the conjunctiva covering the eye becomes involved, particular vessels, or a collection of them enlarge, or become distended, and run to the edge of the cornea, stopping abruptly at
the junction of the sclerotic and cornea, or else in severe or protracted
cases extending over it.

Where these vessels terminate, whether over the sclerotica, at the junction of the sclerotic and cornea, or on the surface of the cornea, small
elevations or pustules may often be found, of a whitish or yellowish appearance. These pustules in protracted cases, are liable to enlarge at
times from slight causes, and they may become the seat and occasion of
ulceration.

The occurrence of these elevations are peculiar to scrofulous ophthalmia, and though they are generally present in this affection, they are not
invariably so. I have just examined a protracted case, in which, with
occasional exacerbations, these pustules become the seat of ulceration of
the cornea.

In all cases of this disease, whether mild or severe, intolerance of light
is the characteristic symptom. As a consequence of this intolerance,
the patient generally goes with the head down, or if looking up, passes
the hand over the eyes, as a temporary shade. Another peculiar symptom of scrofulous ophthalmia which I have noticed, is the occasional tendency to spasmodic closure of the lids, without the possibility of opening
them, in consequence of a powerful contraction of the orbicularis pal-
pebrarum.

In one case of this character that fell under my care a few years since,
the contraction of the orbicularis was so powerful that the natural opening between the lids had the appearance of a slightly oval passage spasmodically closed. This spasmodic closure of the lids had continued for
several days, when I first saw the patient, and was only overcome by repeated cupping of the temples, and back of the neck, together with the
free application of an infusion of stramonium.

This spasm when it occurs, as well as the excessive intolerance of
light in this disease, is evidently the result of an irritable or sensitive
state of the retina, and this I believe in turn, is often sympathetic, the
result of irritation of the alimentary mucous membrane.

Scrofulous ophthalmia is very often associated with that form of scrofulous diseases, in which there are watery or mattery pimples on the face, and a scaly eruption back of the ears, and perhaps covering most of the scalp.

Scrofulous ophthalmia may continue more or less troublesome for a considerable time, and yet no very material changes take place in the structure of the eye. More generally, however, even in cases in which there is no very marked redness, an insidious change of structure takes place, especially in the cornea.

The most frequent organic changes are permanent opacity, or ulceration of the cornea at the seat of the pustules referred to, enlargement of the blood-vessel of the corneal portion of the conjunctiva, with thickening and opacity of this portion of the membrane, dullness of the cornea from interstitial deposition, with enlargement of the proper vessels of the cornea, adhesion of the iris to the cornea, and finally more or less change of the sclerotic coat, iris, and other deeper parts, and in the general form of the eye.

Diagnosis.—Scrofulous ophthalmia may be distinguished by the intolerance of light, by a copious lachrymal secretion, by the pustular elevations of the conjunctiva, with enlargement of its blood-vessels, and finally by the pimply or scaly eruptions back of the ears, or on the scalp and face, together with the symptoms of a general scrofulous condition of the patient.

Causes.—Strumous ophthalmia may be produced by any cause or train of causes, capable of generating a scrofulous condition, or of developing it in constitutions hereditarily predisposed. I believe, however, that this form of scrofulous disease is most frequently produced by exposure to filth, dampness and cold winds, insufficient or unwholesome food, irregular eating, masturbation and sexual excesses, the use of tobacco, drunkenness and licentiousness, and other like improprieties, practiced either by the patient himself, or by his progenitors.

Prognosis.—The prognosis is rather unfavorable in this disease, so far as the general restoration of the system is concerned.

The ophthalmia may, however, generally be removed or relieved, temporarily at least, by a judicious course of general and local treatment, provided the cornea is only opaque, with enlargement of its vessels, without ulceration or other serious structural change. In fact, I have known cases to recover, after all distinction between the appearance of the sclerotic and corneal portions of the external tunic of the eye had entirely disappeared, an instance of which fell under my care, within the past two years.

If, however, there are deep ulcerations of the cornea, attended with general opacity, and also a change in the structure of the iris and sclerotica, the prognosis is decidedly unfavorable.

Treatment.—The treatment for scrofulous ophthalmia must be general and local, and the causes that are operating to produce or keep up this condition must also be removed.

The patient should be kept clean, warm and dry, and should take a reasonable amount of good wholesome food with regularity. Sufficient exercise should also be allowed, and the patient should preserve an even and cheerful temper of mind.

If the patient has any low, vicious habits, they should be corrected, and a rigid observance of the laws of health should be insisted upon, in every respect. This being arranged, the general condition of the patient should be corrected as far as possible, by such measures as are indicated. If the bowels are constipated, and the patient is a child, a dose of mercury with chalk, and rhubarb may be administered in castor-oil, at first, and the bowels kept regular if necessary by equal parts of rhubarb and the sulphate of magnesia.

If the patient is an adult, three blue pills may be given at first, at evening, and a full dose of the sulphate of magnesia administered on the following morning, after which a pill of aloes and rhubarb may be given each day after dinner, and continued till the bowels are regulated. If the patient is feeble, a grain or two of the sulphate of quinine may be given, mingled in water, with twice the quantity of prepared chalk, and this should be continued till the appetite and digestion are in a measure restored. If there is great intolerance of light, cups, wet or dry, may be applied to the temples at first, and later blisters should be applied back of the ears and to the back of the neck, and these should be repeated if salutary, during the continuance of the disease. If the light cannot be borne, the eyes may be protected and soothed by pieces of double linen, hanging loosely over them, supported by a belt passing around the head, the eye pieces being kept wet in an infusion of stramonium made from two drachms of the leaves to a pint of water.

In some cases, no other application may be required for the eyes. If, however, there is considerable redness of the conjunctiva, the eyes may be washed four times per day with a saturated solution of alum in rain water, containing also a fluid drachm of the wine of opium to the ounce.

After having continued the quinine for a reasonable time when it is indicated, and immediately after the operation of the first cathartic in cases in which it is not, five grain doses of the iodide of potassium, should be given three times per day, in the fluid extract or compound decoction of sarsaparilla, and in cases in which there is opacity of the cornea, it should be continued till the opacity is entirely removed.

When this is accomplished, the iodide of potassium should be suspended, and the sarsaparilla continued with ten drop doses of the syrup of the iodide of iron, three times per day, till the blood and general condition of the system is restored as far as it may be. The patient should then with all possible prudence avoid the causes calculated to produce a return of the disease.

SECTION IV.—RHEUMATIC OPHTHALMIA.

By rheumatic ophthalmia, I mean here rheumatic inflammation of the tunics of the eye, having its special seat in the fibrous structure of the sclerotica, but extending in most cases more or less to the conjunctiva cornea and iris, and sometimes to other tunics or structures of the eye.

As the sclerotica is a fibrous structure, it is as we should suppose, the most frequent seat of rheumatic inflammation of the eye. And as the cornea, iris and conjunctiva are parts with which it is either directly or indirectly connected, it is not strange that these parts should become more or less involved in rheumatic ophthalmia.

Symptoms.—Rheumatic ophthalmia generally occurs in persons of a rheumatic diathesis, and very generally in those of shattered or broken down constitutions, who have suffered more or less from the disease in other parts. It is very apt, according to my observation, to be associated with *catarrh* of the Schneiderian membrane, and may be attended with general febrile excitement, or it may occur without fever, being attended with the general symptoms of depression.

In addition to the general febrile symptoms, when they attend, there is early, pain in the temples, and more or less heavy dull pain in the eyes, with either dryness or increased lachrymation, slight intolerance of light, and more or less redness.

The redness generally begins around the cornea, from which point enlarged vessels may be seen passing in every direction, in straight lines. If, however, the conjunctiva becomes involved, the enlargement of its vessels, which instead of being straight, are irregular and tortuous, may obscure those of the sclerotica beneath.

The cornea, if it is involved, may assume a dull appearance, but it seldom if ever becomes opaque in rheumatic ophthalmia, in the early stages at least. If, however, the iris is involved, the pupil is apt to be contracted and irregular, and the change in its structure may cause irregular projections, and perhaps extinction of sight. The redness of the sclerotic coat is of a pink color, and in protracted cases it may extend and give its whole surface this peculiar appearance.

As the disease advances, the patient complains of a stiffness, fullness, and dull aching or throbbing sensation in the eye, which extends to the back of the orbit, and sometimes to the corresponding side of the head. Rheumatic ophthalmia is always aggravated by storms or damp weather, and there is generally an increase of the pain during the night. Such, according to my observation, are the ordinary symptoms of rheumatic ophthalmia. The disease is liable, however, to great variations, from a slight pain, redness and dryness or stiffness of the eyes, to those aggravated and protracted cases, in which all the tunics of the eye become involved, leading on, in some cases, to a permanent loss of sight.

Diagnosis.—Rheumatic ophthalmia may be distinguished by the rheumatic condition of the patient; by the dull aching character of the pain; by the enlargement of the sclerotic vessels, and peculiar pink color of the membrane; and by the aggravation of the pain at night, and of all the symptoms during storms or damp weather.

Causes.—It is probable that in addition to a general rheumatic condition of the system, various exciting causes operate to develop this disease, such as exposure to winds, storms and dust; a stooping posture, protracted grief, catarrh of the Schneiderian membrane; and various influences which tend to irritate the organ.

Prognosis.—In recent cases of this disease, occurring in constitutions not too much enfeebled, the prognosis is favorable if the iris is not materially involved. But in very protracted cases of rheumatic ophthalmia, occurring in broken down constitutions, the prognosis is unfavorable, especially if, as is often the case, all the tunics of the eye have been involved, and have undergone more or less change. In case, however, the iris has undergone no material structual change, some desperate

37

and protracted cases of this disease finally recover; several instances of which I can now call to mind, in cases that have fallen under my care.

Treatment.—If the patient is plethoric, and the attack is acute, and attended with general febrile excitement, blood may be taken from the arm, but not otherwise. Cups, however, wet or dry, should be applied to the temples or back of the neck in all cases; and later, blisters either to the temples, back of the ears, or to the back of the neck if necessary, and repeated while the inflammation continues.

A cathartic of calomel and castor-oil should be administered at first, and a regular action of the bowels secured by small doses of the sulphate of magnesia, during the continuance of the disease.

Immediately after the operation of the cathartic, in active cases attended with fever, fifteen grains of nitre may be given every six hours, in a tumbler of warm crust coffee, and continued till the fever subsides. As the fever subsides, in cases in which the nitre has been indicated and administered, or immediately after the cathartic in cases in which no general fever attends, the iodide of potassium should be given in ten grain doses three times per day, with twenty drops of the wine of colchicum, or ten drops of the fluid extract. This should be continued till the disease is materially checked, with warm pediluvia at evening, and occasional doses of the sulphate of magnesia if necessary for the bowels.

After the ophthalmia is in a good degree controlled, the colchicum may be omitted, and the iodide of potassium continued, in five grain doses, three times per day, in the fluid extract, or compound decoction of sarsaparilla. And this should be continued till a final cure is effected, which may require several weeks, or even longer than that.

The patient should be kept clean and dry, and allowed a plain digestible diet, to be taken with strict regularity, during the continuance of the disease.

SECTION V.—CORNEITIS.

By corneitis, I mean here, inflammation of the cornea, whether *traumatic*, or *spontaneous*, and associated with depravity of the blood or a scrofulous condition of the system.

The cornea, it will be remembered, is the anterior transparent projecting portion of the outer tunic of the eyes. Its vessels, though quite numerous, are not visible in a natural, or perfectly transparent state of the part; but they become so, if the cornea is inflamed, the vessels then being enlarged, and admitting red blood.

We have seen that the cornea is liable to become inflamed, by extension from other tissues of the eye. We are now to examine the disease, as it is developed primarily in the cornea; whether it arise from direct injury of the cornea, or spontaneously, in an acute or chronic form.

It should be remembered, that the vessels of the cornea, are sufficiently numerous, to produce a general redness; if they become generally enlarged, as they do under active or long continued congestion or inflammation.

Corneitis is attended with a peculiar train of symptoms, which we will now proceed to consider.

Symptoms.—When corneitis occurs from direct injury; several cases of which have fallen under my care, during the past few years; in addition to an opacity at the point of injury, the cornea assumes a general cloudy appearance. Its blood-vessels are more or less enlarged, as well as those of the sclerotica, and besides a cloudiness of the cornea, and more or less redness of the sclerotica and conjunctiva, a pink zone is seen encircling the cornea, in most cases that I have observed. I now distinctly remember three cases, in which the cornea was perforated, with more or less escape of the aqueous humor; attended in one case with protrusion of the iris. In these cases, in addition to the above symptoms, there was slight flattening of the eye, and very considerable intolerance of light.

In all these cases the inflammation was subdued, and the opacity of the cornea removed, except at the point of injury; but unfortunately in one of the cases, from subsequent neglect on the part of the patient, there was the supervention of cataract.

Spontaneous Corneitis, or that which occurs independent of local injury, from depravity of the blood, or a scrofulous condition of the system, may be either acute or chronic. When it is acute, the symptoms are similar to those already laid down as belonging to traumatic cases, with the exception of the marks caused by the intruding substance.

But *spontaneous corneitis* is generally of a decidedly chronic character, being passive, and associated with a depraved and feeble condition of the system, in which case there is at first, slight dullness of the cornea, with imperfect vision, and by degrees an enlargement of its vessels, as well as those of the sclerotica and conjunctiva.

By degrees the surface of the cornea assumes a granulated appearance; the enlargement of its vessels become more apparent; its circumference assumes a dark reddish color; and, in addition to a pink zone round the cornea, the whole sclerotica is covered by a plexus of distended vessels.

This form of corneitis is attended with pain, and a sense of tightness in the eye; and there is generally, especially in scrofulous patients, increased lachrymation, and considerable intolerance of light.

Corneitis, if neglected, is liable to lead on to a general and permanent opacity of the cornea, and to various structural changes. And as the iris, sclerotica, and other membranes of the eye, generally become more or less involved, there is liable to be, in protracted cases, serious organic changes in their structure.

Such, according to my observation, are the ordinary symptoms of corneitis, liable, of course, to variations, from the nature of the cause, as well as from the general condition of the patient, &c.

Diagnosis.—Traumatic cases of corneitis may readily be distinguished, as it is indicated by the nature of the cause. In distinguishing spontaneous cases however, it is necessary to bear in mind all the symptoms peculiar to this form of ophthalmia: and especially the early cloudiness of the cornea, and roughnesss of its surface; as well as the gradual extension of the inflammation to the iris, sclerotica, and conjunctiva, in cases in which they have become involved.

Causes.—Corneitis, as we have seen, may be produced by a wound of the part, either bruising, penetrating, or perforating its structure. Spontaneous cases, however, may be the result of every variety of imprudence which depraves the blood, or produces or develops a scrofulous condition of the system. The direct exciting causes, are exposures of various kinds, such as to winds, storms, dust, smoke, heat, &c.

Prognosis.—The prognosis in recent cases of corneitis is favorable, if the causes can be removed, and the patient be subjected to proper treatment.

If, however, the case be neglected, and the inflammation extends to other parts of the eye, involving the iris, serious structural changes of the cornea, iris, or other important parts, are very liable to take place. In this case, vision may be impaired, or even permanently destroyed. I believe, however, from my own observation, in the treatment of corneitis, that it is as susceptible of relief from proper remedial measures as any form of ophthalmia.

Treatment.—The cause should first be sought out and removed, and the habits of the patient corrected; after which the case should be treated according to the general and local condition of the patient.

If the case is acute, whether the result of direct injury, or spontaneous, blood should be taken by cups, from the temples or back of the neck, and if necessary leeches should be applied about the eyes. A cathartic of calomel should be administered in castor-oil, and tartar emetic in one-sixth of a grain doses administered every four hours, alternating with alterative doses of calomel, or mercury with chalk.

After having subdued the febrile symptoms in such cases, and immediately after the first cathartic, in chronic or passive cases, the patient should take a blue pill at evening, and a teaspoonful of the sulphate of magnesia each morning; the antimony and alterative doses of mercurial having been suspended, or omitted, if the case is chronic.

Blisters should be applied to the temples, back of the ears, and to the back of the neck, and these should be repeated in succession while the inflammation continues, so that as one heals, another shall be drawn and ready to discharge.

After having continued the blue pill at evening as long as the severity of the inflammation may seem to require, it may be suspended, and moderate doses of leptandrin, or of the sulphate of magnesia, administered occasionally, if the state of the bowels renders a laxative necessary. On suspending the blue pill, the iodide of potassium should be given in five grain doses, three times per day, before eating; and if the patient is anæmic, ten drops of the syrup of the iodide of iron should be given, after meals, and continued till the ophthalmia and its effects are removed.

SECTION VI.—IRITIS.

By iritis, is here meant inflammation of the iris, and especially that variety of the disease in which the iris is the primary seat of the inflammation, though the disease may be the result of extension of inflammation from other parts of the eye, as we have seen.

The iris, it will be remembered, forms the septem between the anterior and posterior chambers of the eye, being pierced near its centre by a circular opening, the *pupil.*

It is connected by its periphery with the ciliary ligament, and its inner circumference forms the margin of the pupil. Its anterior surface is muscular; the circular portion surrounding the pupil being for its contraction, while the radiating fibres, in a healthy state, have the power of dilating the pupil. Its posterior layer is of a deep purple color, and has been called the *uvea,* from its resemblance in color to a ripe grape.

The arteries of the iris are furnished by the long ciliary arteries, and form two circles, by their anastomoses, the one broad, near its great circumference, the other smaller, surrounding the circumference of the pupil. Its veins empty mostly into the long ciliary veins. The iris is also well supplied with nerves, and by its contraction or dilatation, it regulates the quantity of light proper for distinct vision.

The iris thus situated and constituted, is liable to become inflamed, and, when it does, a peculiar train of symptoms are developed, which we will now proceed to consider.

Symptoms.—In some acute cases of iritis, there is the development of severe febrile symptoms, attended with restlessness, headache, want of sleep, a strong full pulse, a white tongue, thirst, &c. More generally, however, these symptoms are very slight, or entirely absent.

The most general local symptoms of iritis are change of color of the iris, effusion of lymph, imperfect motion of the iris, and an irregular pupil, increased redness of the eye, a change of the cornea, pain and intolerance of light, and finally a gradual progress, either towards recovery, or else in the destructive process, such as a change of color or texture, adhesion of the pupil, a membranous closure of the pupil, or by contraction, atrophy of the globe, with fluidity of the vitreous humor, and, finally, blindness, or impaired vision.

The *change of color* in iritis commences early, in the edge of the pupil, and gradually extends to the ciliary edge of the iris. It is caused by effusion into its texture, and presents, with a loss of brilliancy of the iris, a dark reddish, green or yellowish tinge, which may readily be discovered, if it be compared with the iris of the sound eye.

The *effusion of lymph* may be only into the texture of the iris, or, if the iritis passes unchecked, a layer of lymph may be deposited on its surface, or into the anterior or posterior chamber, in a partially detached, loose mass, as happened in one case that fell under my care; or, finally, the lymph may appear in small elevations on the surface of the iris, or edge of the pupil, or it may form bands across the pupil, or completely close it.

The *motions of the iris* are impaired from the first, and as the effusion of lymph takes place, it is often entirely suspended, the pupil being contracted, and very irregular.

Increased redness of the eye generally attends, being noticed first around the cornea, in the form of a pink zone. It gradually extends, however, to the sclerotica, and especially to that portion bordering on the cornea, being less marked posteriorly; the circumference of the eye appearing usually quite clear. If the disease is violent, however, and passes on unchecked, the vessels of the conjunctiva become enlarged, and the whole eye may present a fiery redness.

The *cornea* is more or less changed in cases in which the inflammation extends to the sclerotica, varying from a slight dullness to nebulous opacity, and in some rare cases to general opacity, or even ulceration.

Pain and *intolerance* of light generally attend in severe cases of iritis. The pain is deep-seated, and extends in many cases to the cheek, brow, and head, being aggravated during the night. The intolerance of light is most marked in cases in which the inflammation extends to the sclerotica, upon which it appears in a great degree to depend.

A *favorable termination* of iritis, under judicious treatment, is attended with a gradual subsidence of the inflammation, absorption of the effused lymph, and gradual return of the brilliancy of the iris, and disappearance of the redness of the eye, as well as a resumption of all its functions.

If, however, iritis be permitted to pass on unchecked, the inflammation gradually extends to the sclerotica, choroid, cornea and conjunctiva. The whole eye becomes more or less involved, the texture and color of the iris become more changed, the pupil may become contracted and fixed, or lymph may completely close it, there may be atrophy of the globe, with fluidity of the vitreous humor, and finally all these destructive changes, with perhaps opacity of the cornea, lead to impairment of vision, and generally to permanent extinction of sight.

Diagnosis.—Iritis may be distinguished by the loss of brilliancy and change of color of the iris, by the contraction and irregularity of the pupil, the pink zone around the cornea, and redness of the sclerotica, fading towards the circumference of the eye, the gradual extension of the redness to the conjunctiva, and finally by the tendency of the inflammation to involve the different tunics of the eye. But it should be remembered, that the change in the appearance of the iris, together with the contraction and irregularity of the pupil, is the most important diagnostic feature of this disease.

Causes.—Iritis may be produced by direct injury, by exposure to dampness, dust, heat, &c., and by any cause or train of causes that produces other varieties of ophthalmia. I believe, however, that iritis, not the result of an extension of inflammation from other tunics or membranes of the eye, generally depends upon either a hereditary or acquired rheumatic, syphilitic, or scrofulous condition of the system. Primary iritis then, is generally the result of licentiousness, exposure, want, and filth, and in short of the combined influence of every variety and species of imprudence that tends to produce or develop a depraved or scrofulous condition of the system.

We have already seen that iritis may be the result of an extension of inflammation from other tunics or tissues of the eye, simple as well as of a purulent character.

Prognosis.—The prognosis in recent iritis, if the case be subjected to proper treatment, is generally favorable. If, however, the inflammation has been active, and the case neglected, so that material change has taken place in the structure of the iris, the prognosis is rather unfavorable, especially if there is a very decided rheumatic, syphilitic, or scrofulous condition of the patient.

Cases of iritis, however, should not be abandoned as hopeless, even though very considerable changes have taken place in the iris, as apparently desperate cases sometimes recover under the use of proper measures judiciously administered.

Treatment.—The indications in the treatment of iritis are to subdue the inflammation and general febrile excitement, when it exists, to arrest the effusion of lymph, and promote the absorption of that already effused, and finally to promote dilatation of the pupil, and subdue any slight irritation that may arise in the conjunctiva.

In an acute attack of iritis, attended with considerable febrile excitement, general bleeding should be resorted to, and then cups should be applied to the temples or back of the neck, and repeated if necessary.

In less active cases, attended with little or no fever, general bleeding should be omitted, and cups applied and repeated if necessary. After general bleeding when it is indicated, and at first, or after cupping, in cases in which it is not, a cathartic of calomel should be administered, and followed if necessary by half an ounce of the sulphate of magnesia.

Immediately after the operation of the cathartic, two grains of calomel, with four grains of Dover's powder, should be administered every four hours, and alternating with this, one-sixth of a grain of tartar emetic may be given, and both continued till the inflammation is subdued, or slight ptyalism produced.

A warm foot-bath should be used morning and evening, to lessen the cephalic tendency, and promote perspiration. And immediately on the suspension of the mercurial, the iodide of potassium should be administered in five grain doses, every six hours, in the syrup, fluid extract, or compound decoction of sarsaparilla.

The tartar emetic should be suspended with the mercurial, and alternating with the iodide of potassium, four grains of James's powder may be given till the inflammation is entirely subdued, when the antimonial should be omitted, and the iodide of potassium and sarsaparilla continued, three times per day, till the effects of the inflammation are removed, and the general condition of the system corrected. A belt about the head with eye-pieces wet in an infusion of stramonium should be worn from the first, to promote dilatation of the pupil; and a wash made of one drachm of the tincture of stramonium to the ounce of water, should be applied to the eyes four times per day, for the same purpose, and should the conjunctiva become inflamed or red, the wash may be saturated with alum.

SECTION VII.—EXOPHTHALMIA.

By exophthalmia, I mean here protrusion of the eyeball, and more especially that variety in which it is the direct result of relaxation of the muscles of the eye, together with congestion or thickening of the tissues back of the eye, constituting the cushion upon which it rests.

The eye in a healthy state of the system is retained in its place, and moved or turned to accommodate vision, by the four *recti*, and two oblique muscles, and it rests upon the tissue occupying the posterior part of the orbit. It is probable also that the optic nerve tends to retain the eye in its place, but it is susceptible of being elongated to a very considerable extent, as is the case in exophthalmia.

Now the eye is liable to be protruded from various causes, such as wounds. tumors, cancer of the eye, or enlargement of the lachrymal gland, &c. But such cases belong more especially to the surgeon, and need not occupy our time in this place. Exophthalmia occurs, however, as I have already suggested, from relaxation of the muscles of the eye, together with congestion or thickening of its cushion, or of the posterior orbital tissues, developing a train of symptoms which we will now proceed to consider.

Symptoms.—This protrusion of the eyeballs, generally occurs in debilitated, scrofulous, or anæmic patients, and in very many cases, is attended with goitre, and nervous palpitation, and in females with amenorrhœa.

After a more or less protracted train of symptoms, indicative of *poverty* of the blood, and especially palpitation, with the *anæmic murmurs,* generally arterial and venous, with perhaps congestion or enlargement of the thyroid gland, and in females, amenorrhœa, a slight prominence of the eyes is noticed. This prominence of the eyes may appear only occasionally at first, coming on from excitement, or from any cause that produces a rush of blood to the head, or hinders a free return of it through the veins. Gradually, however, the difficulty increases, till at last there is continual protrusion, but still increased at times by the same causes that first produced it in its transient form.

The motion of the eyes may remain nearly or quite natural, but the prominence, together with the inability to close the lids, in many cases gives them a very unnatural, wild and staring appearance. In one case that fell under my care, the protrusion of the eyes was so great at times as to render the appearance almost frightful. The appearance during the prominent periods was that of very considerable enlargement of the globe, but that was merely from the projection of the eyes forwards, as was apparent from the natural appearance, as to the size as, the eyes passed back to near their normal position, as they did at times at first.

In another case of a scrofulous character, in which only one eye was protruding to such an extent as to make the eye appear very much larger than the sound one, I was satisfied that the appearance of enlargement was deceptive, as it proved to be. For as the eye had become entirely useless, and as its prominence produced great deformity, it was removed, and on examination it was found that the apparent enlargement of the globe was from its being pushed forward in the socket, by a thickening of the tissues forming the cushion of the eye, in the posterior part of the orbit.

This case was one of long standing, the patient was decidedly scrofulous, and the loss of sight in the eye, I suspect, was from the change it produced in the optic nerve. Generally in cases of exophthalmia of the character I have been describing, the sight remains nearly or quite natural.

Such, I believe, are the usual symptoms of exophthalmia from relaxation of the muscles of the eye, together with congestion or permanent thickening of the tissues forming the cushion of the eye, in the posterior part of the orbit, the disease depending upon an anæmic condition, and attended, in many cases, with goitre, nervous palpitation, &c., and in females with amenorrhœa.

Diagnosis.—Exophthalmia, of the character I have been describing, may be distinguished from cases depending upon tumors, enlargement of the lachrymal gland, cancer of the eye, &c., by the attendant symptoms of anæmia, nervous palpitation, goitre, and in females of amenorrhœa.

From enlargement of the eye itself it may be distinguished by the attendant symptoms, and also by the variableness of the protusion in the early stages, the eye at times, early, occupying nearly its normal position, in the variety of exophthalmia I have been describing.

Causes.—The immediate cause of the protrusion of the eye, in these cases, I am confident, as I have already suggested, is relaxation of the recti-muscles, together with congestion at first, and in protracted cases, permanent thickening of the tissues forming the cushion of the eye in the posterior part of the orbit. This relaxation of the muscles of the eye, occurs, probably, in the same manner in anæmic patients that relaxation of the heart muscle does, in similar conditions, causing dilatation of the heart.

The congestion of the orbital tissues may be the result of a cephalic tendency, and imperfect circulation in the extremities, which is a condition very likely to prevail in anæmic patients. And the permanent thickening of these tissues, when it takes place, I believe, is the result of a scrofulous condition of the system, the congestion, together with the imperfect state of the absorbents, being the direct cause of it, as well as of the enlargement of the thyroid gland, so general an attendant of this affection.

In relation to the remote causes of this affection, it should be remembered that every influence capable of producing an anæmic condition, including every variety of imprudence, may lead to this variety of exophthalmia. I believe, however, that in addition to a scrofulous condition, excessive losses of blood, protracted masturbation, venereal excesses, irregular eating, excessive use of tobacco, and in females, leucorrhœa, and amenorrhœa, are by far the most frequent causes which lead to this affection.

Prognosis.—The prognosis in recent cases of this variety of exophthalmia is favorable if the remote causes can be ascertained, the habits of the patient corrected, and the general condition of the system restored to a healthy state. If, however, the patient is decidedly scrofulous, and the case be neglected till a permanent thickening of the orbital tissues takes place, the deformity may continue, the eye never again resuming exactly its normal position, and if the protrusion has been very considerable, permanent loss of sight may result from over tension of the optic nerve.

Treatment.—The habits of the patient should be ascertained and corrected at once, and great care should be taken that nothing be overlooked. Having corrected the habits, the general condition of the patient should be ascertained, or the exact deviation from the standard of health taken, and then the indications fulfilled by the most convenient, safe and reliable remedies. If the bowels are confined, a pill of aloes and rhubarb should be taken after dinner each day, and continued till they are regulated. The patient should be directed to take a plain, digestible and nourishing diet, with strict regularity, should wash the surface all over, at least twice per week, in moderately tepid water containing a little salt, and to secure a better performance of the cutaneous function, flannel should be worn next the skin.

A reasonable amount of exercise should be taken, and on no account should the patient be exposed to damp impure air.

If as the general condition of the patient is corrected, the eyes gradually sink back to their natural position, small doses of the citrate or carbonate of iron may be administered for the blood, and the muriate of strychnia in one sixtieth of a grain doses should be given three times per day, for a time, to restore if possible the tone of the recti muscles. If, however, the exophthalmia continues, indicating thickening of the orbital tissues, and the thyroid gland remains enlarged, the iodide of potassium should be given in five grain doses three times per day before eating, with the hope of removing the thickening of the cushions of the eyes, and also the enlargement of the thyroid gland.

If the patient is anæmic, or if a female, is suffering from amenorrhœa, ten drops of the syrup of the iodide of iron should be administered after each meal; and this treatment should be continued with the compound iodine ointment to the thyroid gland, if necessary, till the goitre is removed, and if possible the eyes restored to their normal position. Thus, then, have I completed what I had to say on exophthalmia, and in fact, on diseases of the *eye*.

CHAPTER XII.

DISEASES OF THE EAR.

SECTION I.—GENERAL OTITIS.

By general otitis, I mean here a general inflammation of the ear; all the essential parts of the organ, being involved simultaneously, or by rapid extension of the inflammation, from one part to another.

After taking a glance at general otitis, in the present section, I shall proceed in the following sections, to consider inflammation of the external, and internal portions of the organ, as well as *otorrhœa, otalgia,* and *nervous deafness,* in the order in which I have named them.

But before we proceed, it is proper that we should call to mind, the general anatomy, and physiology of the ear, that we may proceed understandingly.

The ear, it will be remembered, consists of the *pinna, meatus auditorius, tympanum,* and *labyrinth.*

"The *pinna* is composed of *integument, fibro-cartilage, ligaments,* and *muscles.*" The *integument* is thin, and contains an abundance of sebacious glands; the *fibro-cartilage*, giving form to the pinna, being closely connected with it; while the *ligaments* serve to connect the pinna to the side of the head.

The *meatus auditorius*, is the canal or passage, about an inch in length, which extends from the pinna, inwards to the tympanum. It is lined by a thin epithelium, and in the substance of its lining membrane, are situated *ceruminous glands*, which secrete the ear wax. Stiff hairs also stretch across the passage, to prevent the ingress of insects, and other foreign substances.

The *tympanum* is an irregular cavity, in the petrous portion of the temporal bone, separated externally from the meatus by the *membrana tympani ;* communicating posteriorly with the *mastoid cells ;* anteriorly with the pharynx, by the *Eustachian tube*, and internally with the labyrinth, by the fenestra ovalis, and fenestra rotunda.

The *membrana tympani* is concave towards the meatus, and convex towards the tympanum, and has an *external* epidermal layer, a *middle*, fibrous and muscular, and an *internal* mucus, derived from the lining of the tympanum.

The *mastoid cells* are numerous, occupying the whole of the mastoid process, and part of the petrous portion of the temporal bone. Their communication with the posterior circumference of the tympanum, is by a large irregular opening.

The *Eustachian tube*, extending from the pharynx, to the anterior circumference of the tympanum, is partly fibro-cartilaginous, and partly osseous; being expanded at its pharyngeal extremity, and narrow, as it approaches the tympanum.

The *fenestra ovalis* is an oval opening in the upper part of the inner wall of the tympanum, nearly opposite the meatus. It is the communication between the tympanum and vestibule, being closed by the lining membrane of each cavity, to which the foot of the stapes is attached.

The *fenestra rotunda* is situated a little below and posterior to the fenestra ovalis, and constitutes a communication between the tympanum and cochlea, being closed by a membrane over which is reflected the lining of both cavities.

The tympanum is lined by a mucous membrane, and also contains four small bones, the malleus, incus, orbiculare and stapes, which are connected with each other, and so arranged that the handle of the malleus, which is attached to the membrana tympani, receives impressions which are communicated along the bones to the fenestra ovalis, to which the foot of the stapes is connected.

The *labyrinths* consist of a membranous and osseous portion. The osseous labyrinth consists of the *vestibule, semi-circular canals*, and *cochlea*, which are cavities in the petrous portion of the temporal bone, between the tympanum and the meatus anditorius internus.

The *vestibule* is a small cavity situated immediately within the inner wall of the tympanum, with which it is connected, as we have already seen, by the fenestra ovalis.

The *semi-circular canals* are three bony passages communicating with the vestibule by both extremities.

The *cochlea* forms the anterior portion of the labyrinth, and consists of an osseous tapering canal, an inch and a half in length, which makes two and a half turns around a central axis. The interior of the canal of the cochlea is partly divided into two passages, by means of a thin lamina of bone. One of these passages terminates in the anterior portion of the ventricle, while the other opens into the tympanum, by the fenestra rotunda, as we have already seen.

The internal surface of the bony labyrinth is lined by a *fibro-serous membrane*, its internal layer secreting the aqua labyrinthi, "and sending a reflection inwards upon the nerves distributed to the membranous labyrinth."

"The *membranous labyrinth* is smaller in size, but a perfect counterpart with respect to form, of the vestibule and semi-circular canals." It consists of a small elongated sac, three semi-circular membranous canals, communicating with the first, "and a round sac which occupies the anterior ventricle of the vestibule." The membranous semi-circular canals are two-thirds smaller in diameter than the osseous, and are retained in their place, as well as other portions of the membranous labyrinth by nervous filaments, from openings in the inner wall of the vestibule, "and separated from the lining membrane of the labyrinth by the aqua labyrinthi."

The membranous labyrinth is composed of an external serous layer, reflected from the lining membrane of the osseous labyrinth, a vascular layer, a nervous layer, and finally an internal serous membrane, which secretes a limpid fluid, which fills its interior, the liquor Scarpæ. It also contains two small calcareous masses, the *otoconites*.

The *auditory* nerve, which enters the meatus auditorius internus, divides into branches which are distributed to the cochlea, vestibule and semi-circular canals, terminating in minute papillæ, like those of the retina. By this delicate nervous expansion, impressions are received through the fenestra ovalis and fenestra rotunda, as well as through the bony walls of the labyrinth, and communicated to the brain.

Now, in order to appreciate the symptoms which are developed in *general otitis*, it is necessary to bear in mind not only this general anatomy of the ear, but also its general physiology; at least it should be remembered, that the pinna probably favors the concentration and passage of the waves or vibrations of the atmosphere along the external meatus to the membrana tympani.

The waves, on reaching this membrane, produce an impression upon it which is communicated to the handle of the malleus, and thence along the incus and stapes to the fenestra ovalis; and if the Eustachian tube is in a state to allow a free circulation of air in the tympanum, a wave is also produced in this air, which falls on the fenestra rotunda, or its membrane, and also produces a slight impression.

The impressions thus having reached the labyrinth through the fenestra ovale and fenestra rotunda, produce the sensation of sound in the auditory nerve, by means of the liquid contents of the labyrinth. The sensation of sound thus produced in the auditory nerve, in its delicate expansion, is conveyed along this nerve and the tubular matter of the brain, to the gray matter of the hemispheres, and thence to the mind itself.

The ear thus situated and constituted, is liable to a general inflammation, either commencing simultaneously in its different parts, or extending rapidly from one part to another. This general inflammation, then, is the legitimate subject of the present section, which we will now proceed to consider, before taking up diseases of different parts of the ear, as I propose to do in the following sections.

Symptoms.—The symptoms of general otitis vary with the causes which produce it. There is generally, however, severe pain extending from the pinna to the labyrinth, and often to the whole side of the head, attended with throbbing, and perhaps at first with painful acuteness of hearing, so that the least sound becomes intolerable. As the disease continues, and the meatus auditorius or Eustachian tube becomes closed up, the hearing becomes very imperfect; and if the inflammation passes on, and both the meatus and Eustachian tube become closed, the sense of hearing may be lost in the affected ear, for the time at least.

Suppuration may take place along the external meatus; or pus, or mucus may accumulate in the tympanum, in which case the membrana tympani is liable to be ruptured by the accumulated fluid; and a more or less copious, and perhaps protracted discharge takes place.

Diagnosis.—General otitis may be distinguished from either external or internal otitis, by the extension of the pain to all parts of the ear; by the external signs of inflammation, as heat, redness and swelling; and by the deep-seated pain, throbbing, and either acuteness or dullness of hearing, which usually attend if the whole organ becomes involved in inflammation.

Causes.—General otitis may be produced by wounds, irritating injections, disease of the brain, electricity, exposure to cold, the extension of erysipelas from the scalp, &c. Or it may arise in consequence of measles, scarlatina, variola, tonsilitis, syphilis, and during continued or typhoid fevers. One of the most severe cases of general otitis that has fallen under my observation was produced by a sharp pointed stick, which wounded the meatus auditorius; perforated the membrana tympani; and lacerated to some extent the lining membrane of the tympanum.

This case was attended with all the symptoms that I have enumerated, in their most aggravated form, and it resulted in a permanent loss of hearing in the affected ear, as might have been expected.

Prognosis.—General otitis, if not the result of severe wounds, or a complication of some severe or malignant disease, admits of palliation, and generally of cure from proper remedial measures judiciously applied. If, however, the disease be neglected or improperly treated, the sense of hearing may be lost, or in severe cases the inflammation may extend to the brain or its meninges, and a fatal termination be the result.

Treatment.—If the patient is of a full plethoric habit, and the inflammation is of an active character, general bleeding may be required, and when it is, should not be neglected.

After general bleeding when it is indicated, and at first when it is not, cups may be applied to the back of the neck, and leeches about the ear, and an active saline cathartic administered. The head should be kept elevated; warm pediluvia should be resorted to; the bowels should be kept loose by small doses of the sulphate of magnesia; and after the leeching, blisters or pustulation with tartar emetic may be resorted to, behind the affected ear.

Should suppuration take place, injections of tepid water with a little soap should be used to cleanse the meatus, or even the tympanum, should the membrana tympani become perforated. Finally, should a chronic discharge continue from the ear, after the inflammation is subdued, injections of five grains of alum, or of the acetate of lead, to the ounce of water may be used till it is arrested.

SECTION II.—EXTERNAL OTITIS.

By external otitis is here meant inflammation of the pinna, of the external auditory canal, or of the membrana tympani, whether one or all of these external parts of the ear be involved in the inflammation.

External otitis may be confined to either of these parts, or it may commence in the pinna and extend along the meatus to the membrana tympani, or commencing in this membrane it may extend outwards along the external meatus to the tissue of the pinna.

Now the pinna, it will be remembered, consists of integument, fibrocartilage, ligaments and muscles. The tissues forming the external meatus, within the bony passage, are the epithelium, the glandular structure, the cellular tissue, and periostium. And the membrana tympani has its three layers, the external epidermal, the middle, fibrous and muscular, and the internal mucus derived from the lining of the tympanum.

The symptoms then, of external otitis vary according as the inflammation involves one or all these parts or tissues. We will proceed then to examine them as they are developed in cases in which the pinna, meatus, or membrana tympani are separately involved, and also as they occur in cases in which all these external parts are involved in the inflammation.

Symptoms.—The symptoms in cases of external otitis in which the inflammation is confined to the pinna, are heat, redness, pain, swelling, &c. It may be, and very generally is the result of extension of erysipelas from the scalp, and in severe cases, suppuration and even ulceration may be the result, if the inflammation is allowed to pass on unchecked.

In cases of external otitis in which the *external auditory canal* is the seat of the inflammation, there is felt at first, an uneasiness along the canal, which gradually increases to an itching sensation, and very soon to pain, at first slight, but which increases in some cases, to great severity, producing a condition bordering on delirium.

In severe cases the pain is lancinating, and extends to the face, head, and neck, and is attended with the most distressing feeling of distension, along the auditory canal. Audition is diminished or suspended and the pain is greatly increased by pressure, or by motion of the jaw, as in eating or speaking.

The soft tissues of the auditory canal become red, swelled and spongy, and nearly or quite fill the passage; its lining membrane after a day or two, being covered with pustules or vesications, or small abscesses appear deeper in the cellular tissue. As these vesicles, pustules, or small abscesses burst, there is a slight alleviation of the pain, and a muco-purulent, and often fetid discharge takes place, which may continue for two or three weeks, or in case the disease becomes chronic it may continue for weeks or even months.

In cases of external otitis, in which the membrana tympani is the seat of the inflammation, there is felt at first an acute pain at the bottom of the meatus, the result generally of some irritant, followed by a buzzing as though an insect were fluttering in the ear. Audition is diminished, and the pain is increased by pressure upon the ear, as well as by loud sounds.

If the membrana tympani be examined with the speculum, it is found of a reddish color, and enlarged blood-vessels may often be distinguished upon its surface. If, however, the inflammation passes on, the membrane may become thickened and present a granulated or ragged appearance.

From the symptoms which are developed, it is probable that the inflammation commences in the external cuticular membrane, and in severe or protracted cases extends to the middle fibrous and muscular layer, involving, probably, more or less its internal mucous membrane, derived from the lining of the tympanum.

Such, according to my observation, are the ordinary symptoms of external otitis, as it occurs exclusively in the pinna, auditory canal, or membrana tympani, in its acute form. More generally, however, external otitis commences in the pinna, or meatus, and extends to the membrana tympani, combining all the symptoms which I have enumerated as belonging to inflammation of each of these parts, such as pain, redness and swelling of the pinna; swelling, pain, and vesication or suppuration along the auditory canal, and an itching, buzzing, and painful sensation at the bottom of the auditory passage. There is also the muco-purulent discharge, as well as the diminution of audition, and perhaps temporary deafness of the affected ear.

Such is external otitis in its acute form. But the inflammation may pass on and become chronic; or it may be chronic from the first. In either case, the symptoms being less active, and the progress of the disease much slower. The chronic form of the disease is attended with little or no febrile excitement, and occurs generally in anæmic patients, or in those of a scrofulous diathesis, or from a syphilitic taint.

Diagnosis.—External otitis may be distinguished from general otitis by the absence of the deep seated pain in the ear, mastoid cells, and side of the head, by the open and free state of the Eustachian tube, and by the milder grade of febrile excitement in external otitis.

From internal otitis, it may be distinguished by the seat of the pain, and also by the redness, swelling, and perhaps suppuration which occurs if the pinna or auditory canal, or both, are involved in the inflammation.

Causes.—External otitis may be produced by wounds of the pinna, meatus, or membrana tympani. But I believe it is more frequently, the result of a syphilitic taint, or a scrofulous condition of the system; excited perhaps, by the exanthematous fevers, such as rubeola, variola, scarlatina, and various forms of putrid diseases. The disease may also be produced by the extension of erysipelas from the scalp, and by cold, with dampness, as well as by irritating injections into the auditory passage.

Prognosis.—The prognosis in recent cases of external otitis, is favorable, if there is not too much constitutional depravity, and the case be subjected to judicious treatment.

If, however, the constitution is bad, and the case has been neglected, the disease may pass on to the chronic form, or if not, serious and permanent changes may occur in the organ, which will materially injure, or permanently destroy the function of the affected ear. In most cases of otitis, that are subdued for the time, there remains a predisposition in the part, to a return of the disease. At least, such has been the result of my observation, in such cases, especially in depraved constitutions.

Treatment.—In acute cases of external otitis, occurring in strong plethoric patients, a few ounces of blood may be taken from the arm; but in no other cases.

After general bleeding, when it is indicated, and at first, in cases in which it is not; leeches should be applied about the ear, and if necessary, cups to the back of the neck.

A full dose of the sulphate of magnesia, should be administered, and the bowels kept loose, by teaspoonful doses, taken each morning, if necessary. The head should be kept elevated, and a warm foot-bath should be used at evening, each day. If the pain is intolerable, indicating suppuration along the meatus; a bag of hops, wet in warm vinegar, may be bound over the affected ear, and should it produce relief, it may be kept on; being occasionally moistened, till the inflammation terminates, by resolution or suppuration.

If, as suppuration takes place, and a purulent, or muco-purulent discharge is established, the pain still continues in, or about the ear; blisters may be applied back of them, and repeated if necessary; and should the disease become chronic, pustulation may be produced by tartar emetic ointment.

Should the vesicles, pustules, or small abscesses along the meatus, be well formed, and slow to burst, they may be carefully opened, and the auditory passage should be cleansed two or three times per day, by injections of tepid water, with a little soap, if necessary.

If the inflammation passes on, and assumes a chronic form, and in all cases that are chronic from the first; the general condition of the system should be inquired into, and corrected by proper remedial measures. If there is constipation of the bowels, a pill of aloes and rhubarb should be given, each day after dinner, till they are regulated. The patient should be required to bathe, or wash the surface, at least once each week, in water containing a little salt.

A plain, digestible, and nourishing diet, should be taken, with regularity; flannel should be worn next the skin, and should there be a syphilitic, or scrofulous condition, the iodide of potassium, syrup of the iodide of iron, or cod-liver oil may be administered, till the condition is corrected. After the general condition is corrected; if the discharge keeps up, five grains of the acetate of lead, may be added to each ounce of the water injected into the meatus; and this should be continued, till the discharge is arrested.

SECTION III.—INTERNAL OTITIS.

By internal otitis, I mean here inflammation of the mucous lining membrane of the tympanum, involving generally the sub-mucous cellular tissue, and in some cases the periostium, and extending generally to the mastoid cells, Eustachian tube, and more or less to the labyrinth or internal ear.

The mucous membrane of the tympanum as we have already seen, besides lining the bony portion of the cavity, forms the inner layer of the membrana tympani, and also of the fenestra ovale and rotunda. Immediately beneath this membrane is the sub-mucous cellular tissue, and lining the bony walls of the cavity, is the periostium. It is also necessary to remember further, that the mastoid cells communicate with the tympanum posteriorly, the Eustachian tube anteriorly extending from the pharynx, and that the labyrinth lies internally opposite the membrana tympani.

Now, we should still further bear in mind the relation of all these parts, as well as their structures, or the tissues of which they are composed, and also the position and functions of the bones of the tympani, the *malleus, incus, orbiculare*, and *stapes*, in order to appreciate the symptoms which are developed in *internal otitis*, which we will now proceed to consider.

Symptoms.—Internal otitis commences with pain deep in the affected ear, which gradually extends, till it affects the whole side of the head, constituting the most intolerable hemicrania. In violent acute cases, the febrile excitement is severe, the eyes are red and watery, and there is intolerance of light, the face is flushed, the skin is hot and dry, the pulse frequent, and the pain becomes most intolerable in the tympanum, and extends more or less through the whole head.

The pain is increased by deglutition, or by any movement of the jaw, as well as by the slightest sounds. And it extends to the fauces, mastoid cells, labyrinth, and through the brain, being attended often with delirium, and finally with deafness of the affected ear.

On examination, the auditory canal is found free from disease, but if matter has collected in the tympanum as is generally the case, it is usually discharged by the seventh or eighth day. The matter may escape by the external meatus through a rupture or ulceration of the membrana tympani, or through the Eustachian tube, or finally through the ulcerated mastoid cells. Generally, however, the matter escapes through the external meatus, by a rupture or ulceration of the membrana tympani, as the Eustachian tube is generally closed up during the early stage of the inflammation, and this membrane is more readily perforated than the mastoid process.

The matter which is discharged is of a muco-purulent character, and is apt to be very offensive, and with it the bones of the tympanum are often brought away, either through the membrana tympani, or the mastoid process. In case the matter escapes into the throat through the Eustachian tube, it is expectorated quite freely, and has a most disagreeable taste.

If the disease is allowed to pass on unchecked, the inflammation may

38

involve the upper and posterior wall of the tympanum, extend to the membranes, and perhaps substance of the brain, and thus the patient may be cut down suddenly by meningitis or phrenitis. Or internal otitis may pass on and assume a chronic form, or it may have been so from the first, in either case the symptoms being the same in kind as those already enumerated, but differing in degree, as well as in the progress of the disease.

Chronic cases of internal otitis generally occur in feeble, scrofulous, or depraved constitutions, and are apt to be of very protracted duration, and attended with a purulent, muco-purulent, mucous, or serous discharge. Such I believe are the ordinary symptoms of internal otitis, as it occurs in the acute and chronic form, liable of course to variations from peculiarities of constitution and various accidental causes.

Diagnosis.—Internal otitis may be distinguished from external otitis, by the freedom from disease of the pinna, auditory passage, and perhaps of the middle and external layers of the membrana tympani, and also by the greater length of time which transpires before the discharge is set up. The pain too, in internal otitis is much deeper, effects more generally the whole side of the head, and is attended with more marked delirium, and general febrile excitement.

From general otitis it is readily distinguished by the inflammation being confined to the tympanum and its appendages, and the internal ear, while in the general disease the external parts of the organ are involved in the inflammation.

Causes.—Scrofula, a syphilitic taint, and a general depraved condition of the system, are the predisposing causes of internal otitis. The exciting causes, however, are very numerous, among which are exposure to cold and dampness, direct injuries, various exanthematous fevers, as measles, scarlatina, smallpox, &c., and other forms of inflammatory, or putrid disease.

Prognosis.—From the liability of serious changes in the membranous labyrinth, and the fenestra ovale and rotunda, or their membranes, the danger of permanent closure of the Eustachian tube, and loss of the membrana tympani as well as of the small bones of the tympanum, and finally of serious changes in the mastoid cells, the prognosis in internal otitis is rather unfavorable. If, however, the case is subjected to proper treatment, in season, and before any serious organic change has taken place, in essential portions of the organ, reasonable hope may be entertained of a favorable termination of the case.

Treatment.—In severe acute cases of internal otitis, occurring in plethoric patients, general bleeding should be resorted to at once. Immediately after general bleeding in cases in which it is indicated, and at first in milder forms of the disease in which it is not, leeches should be applied about the ear, and cups to the back of the neck, and thus a reasonable amount of blood taken.

A cathartic of calomel or podophyllin should be administered, and followed by half an ounce of the sulphate of magnesia, and then the bowels should be kept loose by small doses of the sulphate of magnesia, administered every morning. The warm foot-bath should be used each evening; and, in severe cases, two grains of calomel may be administered every four hours, commencing after the first cathartic, and continued till a slight mercurial impression is produced.

After the cupping and leeching have been carried as far as is consistent, blisters should be applied to the back of the neck, and a little later, behind the ear.

Generally, by the sixth or eighth day, matter has collected in the tympanum, and in case the Eustachian tube is firmly closed up, it may rupture the membrana tympani, and escape by the external meatus. Or it may find its way out through the mastoid cells, by destruction of their bony walls, the mastoid process, or some point in it.

Should matter point in the mastoid process, it should be let out at once; and its discharge facilitated, if necessary, by injections of tepid water. If it escapes by a rupture, or ulceration of the membrana tympani, and the matter is thick, or escapes tardily, injections of tepid water may be used, cautiously, two or three times per day; to favor the discharge of matter, and to cleanse the auditory passage. Should the matter from the tympanum escape by the Eustachian tube, it may be favored by injections of tepid water, once or twice each day, while the discharge continues.

To inject tepid water into the tympanum through the Eustachian tube, the common silver Eustachian catheter, being oiled, should be passed along the floor of the nostril, with the convexity upwards, till it reaches the pharynx, when it should be gently turned, so that the point shall be outwards, and a little upwards, when the instrument may be felt to enter the mouth of the tube, along which it may pass for half an inch or more, without injury. Having thus introduced the catheter, it should be carefully held in place, and the tepid water may be injected through it, by means of a small glass syringe; and thus the matter in the tympanum be diluted, and its discharge facilitated.

In cases of internal otitis, in which the tympanum becomes filled with matter, and it has formed no outlet, the Eustachian tube being closed, a jet of tepid water, thus injected, may force a passage to the tympanum, and thus the matter be allowed to escape. In case, however, the Eustachian obstruction is not overcome in this way, rather than allow a rupture of the membrana tympani, a whalebone sound, slightly enlarged at the point, may be passed through the catheter, and carefully pushed along till it reaches the tympanum. This being accomplished, the tepid water may generally be made to pass, after which the injections may be continued while the discharge appears to require it.

In all cases of internal otitis, that pass on and become chronic, as well as such as are chronic from the commencement, the diet, habits, and general condition of the patient should be corrected, as well as the local inflammation subdued. And when this is accomplished, should the discharge from the tympanum continue, five grains of the acetate of lead may be added to each ounce of the water injected, and this should be continued till it is arrested.

SECTION IV.—OTORRHŒA.

By otorrhœa, from ους, "the ear," and ρεω, "I flow," is here meant a discharge of *purulent, mucous, serous,* or *muco-purulent* matter from the ear, whether derived from the soft tissues of the meatus auditorius externus, or from the tympanum and its appendages, or from the internal ear.

We have seen in the preceding sections that *otorrhœa* may be the result of inflammation of the structures or tissues of different portions of the ear. But as we have already considered the treatment of such cases, it remains for us to consider in the present section, *otorrhœa* as it occurs from general and local causes, not of a strictly inflammatory character.

Such cases may be the result of previous inflammation or irritation, but they generally depend upon a scrofulous, syphilitic, or otherwise depraved condition of the system, and may be of protracted duration, being attended with a train of symptoms which we will now proceed to consider.

Symptoms.—In cases of otorrhœa that follow acute or chronic inflammation, being kept up after the inflammation is subdued, there remains generally a slight uneasiness, or itching sensation in the ear, and audition is more or less imperfect, the discharge being evidently kept up by a derangement of the structures or tissues that have been inflamed, as well as by the depraved condition of the system generally.

In cases of otorrhœa that have not been preceded by otitis, in addition to a general derangement of the system, and especially of the lymphatic and glandular system, there is at first perhaps an itching sensation in the auditory passage, Eustachian tube, tympanum, or mastoid cells. Sooner or later audition is discovered to be imperfect in the affected ear, and in addition to the itching sensation, there is tinnitis aurium, and the appearance of the discharge, attended with the uneasiness which the presence of the matter produces.

The discharge may be *muco-purulent, mucous,* or even *serous,* and may vary in quantity from barely enough to moisten the meatus to a copious flow, producing considerable inconvenience, and materially impairing the hearing of the affected organ. The matter generally escapes by the external auditory passage. But if it proceeds from the tympanum, and the membrana tympani is imperforated, the matter escapes into the fauces by the Eustachian tube, and is expectorated, producing a more or less unpleasant taste in the mouth. Or if there has been previous disease of the mastoid cells, and the mastoid process has become perforated by syphilitic or other disease, the matter may thus escape, and be a source of great inconvenience.

Causes.—Otorrhœa may be the result, as we have already seen, of a change in the structure or tissues of the ear, produced by previous inflammation. It is also sometimes the result of inspissated wax, collected in the meatus auditorius, as well as of polypi or fingi along the auditory canal, or on the membrana tympani.

Generally, however, it is either the direct or indirect result of a depraved constitution, and especially of a syphilitic or scrofulous condition. This, however, may be heightened, and perhaps the discharge directly produced by various diseases, among the most frequent of which are *scarlatina, measles,* and the putrid fevers. Otorrhœa may also be produced in constitutions predisposed, by improper food, insufficient clothing, sleeping in damp apartments, habits of filth, &c., &c.

Treatment.—In the examination of patients suffering from otorrhœa, it is necessary to ascertain the habits and general condition of the patient, as well as the exact condition of the ear, that the general and local causes may be removed as far as possible.

The patient should be directed to take a plain digestible diet, with strict regularity, should sleep in dry apartments, keep the surface of the body clean, be properly clad, and, in short, should be made to observe strictly the laws of health, and rules of propriety in every respect.

If there is a syphilitic taint, the iodide of potassium should be given in five grain doses, three times per day, in the syrup, fluid extract, or compound decoction of sarsaparilla, and continued till the condition is corrected, though it take several weeks or even months.

If the patient is scrofulous, and there is no evidence of a syphilitic taint, the bowels and digestive apparatus generally should be placed in as good a condition as possible, by proper remedial measures, and then ten drop doses of the syrup of the iodide of iron may be given, three times per day, before each meal, and an hour after each meal, a tablespoonful of cod-liver oil.

While the general condition of the system is being thus corrected, the ear should be kept cleansed by injections of warm water, with a little soap, either into the meatus auditorius, Eustachian tube, or into the mastoid cells, if there is, as is sometimes the case, a sinus in the mastoid process. This should be done two or three times per day from the first, as it serves not only to cleanse the ear, but will also remove any inspissated wax that may have accumulated in the auditory passage.

Under this course of general and local treatment, the otorrhœa may generally be arrested. If, however, as the general condition of the system becomes thus corrected, the discharge from the ear still continues, being kept up by a relaxed or changed condition of the vessels, and other soft tissues of the ear, alum, or nitrate of silver, should be added to the water injected. At first, ten grains of alum should be added to each ounce of the water injected. But, if this fails to arrest the discharge in a reasonable time, five grains of the nitrate of silver may be substituted; and this should be continued till the discharge is arrested.

SECTION V.—OTALGIA—(*Ear-ache.*)

By otalgia, from ους, "the ear," and αλγος, "pain," I mean here pain in the ear from functional derangement of the tympanic nerves, not attended with inflammation of any portion of the organ.

It should be remembered that the tympanum and its appendages are well supplied with nerves, derived from the facial and other important trunks, all of which are liable to functional derangement, attended with pain of a strictly neuralgic character.

Symptoms.—Otalgia may come on suddenly like other neuralgic affections, varying in degree from a slight uneasiness to the most acute and intolerable pain, of a continuous or intermittent character. The pain is generally, I believe, continuous with occasional exacerbations; during which it is apt to extend more or less to the surrounding parts, especially along the track of the nerves, involving the face, neck, and other surrounding parts. Or this disease may assume a strictly intermittent character, being quotidian, tertian, quartan, &c., like other forms of intermittent disease.

Otalgia may be attended with tinnitus aurium; more or less intolerance of sound; and in severe cases with the most distressing delirium.

Diagnosis.—Otalgia may be distinguished from otitis by the more acute and lancinating character of the pain; by the absence of local and general inflammatory symptoms; by the nature of the cause which has produced it; and finally by the pain being less continuous, in some cases being of a remittent or intermittent character.

Causes.—Otalgia may be produced by direct irritants; by exposure to cold and dampness; by the paludal poison; and by various imprudences which produce an anæmic condition, and serve to render the nervous system more or less irritable. I believe, however, that otalgia is very often produced by carious teeth, tonsilitis, and various other diseases of the surrounding parts, as well as by inspissated wax in the meatus auditorius externus.

Treatment.—In cases of otalgia, the causes, whether general or local, should be sought out and removed if possible. If a carious tooth is the cause, it should be removed; if tonsilitis, it should be subdued; and if it depends upon disease of any of the surrounding parts, proper measures should be used to subdue the primary affections upon which it depends.

If the otalgia depends upon a general anæmic state, and an irritable condition of the nervous system, cathartics and then chalybeates should be administered. And finally, if it is of an intermittent character, and of a malarious origin, a mercurial cathartic should be given with, or followed by castor-oil, and then the sulphate of quinine administered as for intermittent fever. To soothe the pain, cotton wet with laudanum may be passed into the ear, and a fomentation of hops, wet in warm vinegar, may be applied to the ear and side of the head.

SECTION VI.—NERVOUS DEAFNESS.

By nervous deafness, is here meant, that which arises from functional derangement of the *acoustic nerve;* whether it occurs in plethoric patients from a morbid augmented irritability, or in anæmic constitutions, from torpor, or diminished irritability of this nerve.

The labyrinth, it will be remembered, consists of the vestibule, cochlea, and semicircular canals; together with the membranous portion, and the expansion of the auditory, or acoustic nerve, the aqua labyrinthi, the liquor Scarpa, and the two calcareous masses, the *otoconites.* The auditory nerve, which we have seen, terminates in a delicate expansion, in the labyrinth, is derived from the anterior wall of the fourth ventricle, and also receives fibres from the corpus restiforma. It enters the labyrinth through the meatus auditorius internus, and in a healthy condition, communicates impressions of sounds, which it receives through the membranes of the fenestra ovale and rotunda, to the tubular matter of the brain, by which it is conveyed to the gray matter of the hemispheres, and thence to the mind itself.

Now any derangement in the function of this nerve, by which it fails to heed or take up impressions of sound, communicated to it, or to convey them to the brain; from whence they reach the mind, constitutes *nervous deafness,* the symptoms of which, we will now proceed to consider.

Symptoms.—Nervous deafness may occur in young and plethoric patients, or in the aged and debilitated, in the one case, the derangement consisting probably, in augmented irritability, and in the other, in diminished irritability, or torpor.

It is probable, however, that there is diminished sensibility of the acoustic nerve in both cases, only it depends upon opposite conditions of the parts, and of the system generally. In plethoric patients, it may depend upon an active congestion, or the presence of too much blood in the nervous expansion, and soft tissues of the labyrinth; while in anæmic patients, nervous deafness doubtless arises, from an insufficiency of healthy blood in the part, and system generally, to stimulate the nerve into healthy activity.

With this view of the disease under consideration, I shall distinguish the two conditions, by the terms *plethoric* and *anæmic*, or *active* and *passive*, as a matter of convenience, and that the real condition, general and local, may be kept in mind as we proceed.

In *active* or *plethoric* cases of nervous deafness, in addition to an over-plethoric habit, with perhaps a slight congestion of the brain, or derangement of the digestive, or genital organs; the patient is at first annoyed by tinnitus aurium, or sounds in the head, which is compared to the buzzing of insects, the roaring of waves, the ringing of bells, &c.

By degrees, the sense of hearing becomes impaired, and the patient is still further annoyed by a more or less constant pulsation in the ears, synchronous with the cardiac, which is apt to be augmented by fatigue or mental excitement. These unpleasant sensations may at first be experienced in only one ear, but they often shift from side to side, and finally affect both ears almost constantly.

The external meatus is apt to be dry in patients suffering from nervous deafness; and in many cases the above symptoms become more marked on listening to minute or low sounds, while the patient may hear quite well even ordinary conversation, when surrounded by loud noises, such as the rattling of carriages, &c.; which probably serve to rouse the nerve into a state of activity, for the time, compatible with audition.

The *anæmic* or *passive* variety of nervous deafness may arise in anæmic patients of all ages; but it is especially liable to occur in old age, in which case it may be very gradual in its approach, the disease continuing sometimes for many years before the hearing is entirely lost.

In this form of nervous deafness, in addition to a general anæmic or debilitated condition of the system, there is dryness of the external meatus, gradual loss of hearing, and in fact all the symptoms which attend the active form, except the tinnitus, which does not attend the passive form of the disease.

Such I believe are the usual symptoms of nervous deafness, as it occurs in its active and passive form, liable, of course, to variations, depending upon the degree of the local and general derangement.

Diagnosis.—To distinguish cases of nervous deafness from all others, a watch should be placed between the teeth, when, if the deafness depends upon a loss of function in the acoustic nerve, its ticking will not be heard by the patient; but if the deafness is from other causes it will be distinctly audible, in consequence of vibrations communicated along the cranial bones.

The *active* variety of nervous deafness may be distinguished from the *passive* by the difference in the general symptoms, already enumerated; and also by the more or less constant tinnitus aurium in active cases, a symptom which does not attend the passive variety of the disease.

Causes.—The *active* variety of nervous deafness may be produced by any cause or train of causes which produces a state of general plethora; some accidental circumstances, such as exposure to loud sounds, &c., being the immediate cause of the functional derangement of the acoustic nerve. This form of the disease is also frequently sympathetic of derangement of the digestive or genital organs.

The *passive* form of nervous deafness is the direct result of an anæmic condition, and is therefore indirectly produced by every variety of imprudence, which tends to impoverish the blood and debilitate the nervous system. It is probable, however, that various local causes sometimes operate, either upon the acoustic nerve or brain, to favor the development of this form of disease. It sometimes follows concussion of the brain; and is a frequent attendant of typhus fever, supervening as the stage of excitement passes away.

Prognosis.—The prognosis in cases of nervous deafness is generally unfavorable. In recent cases, however, if the causes can be ascertained and removed, and the general condition of the system corrected upon which it depends, the progress of the disease may often be arrested, and in some cases the hearing may be restored.

Treatment.—The indications in the treatment of nervous deafness are to remove the causes, to correct the general condition of the system, and then if the nervous insensibility remains, to call them into activity by suitable general and local measures. The diet, exercise, and general habits of the patient should be corrected, and if there are any local causes operating they should be removed, as far as possible. If the patient is plethoric, and the disease of an active character, a full dose of calomel may be administered at first in half an ounce of castor-oil, and a free operation secured. The head should be kept slightly elevated and cool, and the warm foot-bath used at evening, and for a time, a blue pill may be given at evening, and followed in the morning by a teaspoonful of the sulphate of magnesia. Cups may be applied to the back of the neck, and blisters behind the ears, and repeated if necessary.

As the general plethora and local congestion is thus gradually overcome, the hearing may be restored if the function of the acoustic nerve has not been too long impaired or suspended. If, however, when the general condition of the patient is thus corrected, the auditory nerve still remains insensible, or partially so, the muriate of strychnia may be given for a time in $\frac{1}{20}$ of a grain doses, in solution, three times per day, with the hope of calling it into activity.

In cases of nervous deafness of a *passive* character, occurring in anæmic patients, two or three blue pills may be given at first, and followed by half an ounce of castor-oil, after which the bowels should be kept regular by a pill of aloes and rhubarb after dinner each day, as long as may be necessary. A pill of the ferrocyanuret of iron may be given three times per day, after the first cathartic and continued till the blood is restored. If the appetite is poor, a cold infusion of columbo, or thirty drops of the fluid extract may also be given as long as required.

After the general condition of the system has thus been restored, if the deafness continues, the muriate of strychnia should be given, as already suggested, and continued for a reasonable time, with the hope of rousing the torpid acoustic nerve to the due performance of its function. In case, however, the deafness still continues after having corrected the general condition, and used the strychnia for a reasonable time, an effort may be made to call the torpid nerve into activity by stimulating vapors or injections passed into the tympanum through the Eustachian tube.

Various forms of local stimulants may be thus introduced into the tympanum, by the Eustachian tube, through the catheter properly introduced. I believe, however, that the vapor of *acetous æther*, as used by Itard, Kramer, Pilcher and others, will generally do best.

Into a quart jar, with a wide mouth, containing a pint of warm water, and having a cork well fitted, through which is passed a metal or glass pipe, having an elastic tube attached to it, furnished with a stop-cock, half a drachm of acetous æther may be poured, and the cork immediately adjusted to the mouth of the jar. This being done to the catheter, introduced into the Eustachian tube, the elastic tube with the stop-cock should be attached, and thus the vapor be allowed to pass into the tympanum.

The vapor, as suggested by Dr. Pilcher, may be "applied two or three times at a sitting," and repeated two or three times per week, as long as its effects appear salutary. Finally, should this form of the stimulant prove insufficient, two drachms of acetous æther may be added to a pint of water, and of this small portions may be injected through the Eustachian catheter into the tympanum, and repeated daily till a cure is effected, or the case becomes hopeless.

CHAPTER XIII.

DISEASES OF THE SKIN.

SECTION I.—RASHES AND ERUPTIONS.

By rashes and eruptions, I mean here that class of cutaneous disease, characterized by some form of *rash* or *eruption*. Of the *rashes*, it embraces *red rash, rose rash*, and *nettle rash*, while of the *eruptions*, it includes what I shall call the *papular eruptions, vesicular eruptions, pustular eruptions, scaly eruptions, animalcular eruptions*, and *the cryptogamous*.

I design, in the present section, to take a general view of the *rashes* and *eruptions*, and propose in the following sections of this chapter, to take up separately each variety of the *rashes* and *eruptions*, in the order in which I have named them. But before we proceed to a general consideration of this class of cutaneous diseases, it is proper that we should call to mind the anatomy and physiology of the skin, that we may proceed understandingly.

The skin, it will be remembered, is composed of two layers, the *derma* or *cutis*, and the *epiderma* or cuticle. It also contains the sebaceous and sudoriparous or perspiratory glands, and is well supplied with arteries, veins, lymphatics and nerves.

The *epiderma* or cuticle is the external layer of the skin, and is a product of the derma, which it serves to envelop and defend. Its external surface is a horny or hard texture, while its internal structure is soft and cellular, each layer of which it is composed being less dense, till we reach the derma, to which it is attached beneath. The cuticle is perforated by the hairs, absorbent vessels, and by the ducts of the sebaceous and perspiratory glands, and has slight depressions, pits, or furrows on its inner surface, to receive the papillæ of the external surface of the cutis, upon which it lies.

The *derma* or cutis is the inner layer of the skin, being situated immediately beneath the cuticle, and is composed of areolo-fibrous tissue, and "elastic and contractile fibrous tissue," as well as blood-vessels, lymphatics, and nerves. The areolo-fibrous tissue occupies mainly the inner portion of the cutis, as well as the red contractile, and yellow elastic, while its external portion, immediately beneath the cuticle, is exceedingly vascular, and has numerous prominences or papillæ. These papillæ or conical eminences on the outer surface of the derma are, according to Professor Draper, "about $\frac{1}{100}$th of an inch in height, and the $\frac{1}{210}$th of an inch in diameter at their base," subject, however, to variations in different situations, or on different parts of the body. These papillæ contain an elastic substance, and have each a minute artery, vein, and sensory nerve.

The *arteries* of the derma enter its structure through the areola of its inner surface, and divide into innumerable branches, which form a beauful net-work or capillary plexus in the outer portion of the derma, furnishing a branch to each of the papilla, which terminate in a minute vein.

The *nerves* of the derma enter the areola of its inner portion, and divide, forming a terminal plexus under its vascular net-work, from which fibres pass off as loops in the papillæ, as we have already seen.

The *nails* and *hairs* are horny appendages of the skin, the process of their formation being identical with that of the formation of the epiderma upon the derma, of which they are really a part.

The *sebiparous* or *sebaceous glands* are small, irregular, glandular structures, situated in the derma, presenting a variety of complexity, some of their ducts emptying into the hair follicles, while others perforate the cuticle, and open on its surface. These glands secrete a sebaceous or oily material, which probably favors a healthy growth of the hair; and that portion which is emptied on the surface of the cuticle keeps it moist and soft, and prevents its peeling off.

The *sudoriparous* or *perspiratory glands* are situated in the deep portion of the derma, as well as in the subcutaneous areolar tissue, surrounded by adipose cells. They consist of a tube folded on itself, and contained in a cell, the walls of which are abundantly supplied with blood-vessels, or there may be a collection of sacs, opening into a common efferent duct. These ducts of the perspiratory glands ascend through the derma and epiderma, passing between the papillæ, and terminate on the surface of the cuticle, in an oblique and funnel-shaped aperture. It has been estimated that the number of perspiratory glands and ducts is about seven millions; and, allowing the number of square inches of surface in a man of ordinary size to be 2500, it would give an average of 2800 perspiratory glands and ducts to every square inch of the surface of the body.

The *lymphatics* of the skin are very numerous, and form a part of the vascular net-work in the outer portion of the derma, being interwoven with the capillaries and nervous plexus of the skin.

Now the skin consisting of its several parts has an important function to perform, or a variety of functions. Its outer layer, the cuticle or epiderma protects the derma or cutis beneath. This in turn covers and protects the parts still deeper, and also contains the sebaceous and perspiratory glands as well as the arteries, veins, capillaries, lymphatics and nerves of the skin. The *capillaries* throw off carbonic acid and nitrogen, and absorb oxygen, and probably as well as the lymphatics, absorb or take in liquid and other substances from the surface of the body. They also furnish nutrition, as well as the oily matter secreted by the sebaceous, and the perspirable secreted by the perspiratory glands.

The sebaceous glands, as we have seen, secrete an oily matter which goes in part to the hairs, and in part to the surface of the cuticle, to lubricate and keep it soft.

The *sudoriparous* or *perspiratory* glands secrete or separate from the blood the perspirable matter, averaging probably about two pounds per day, which is conducted to the surface by the seven millions of perspiratory ducts, either as insensible or sensible perspiration. This perspirable matter thus thrown from the blood, not only relieves the circulating fluid of materials no longer useful, but by its evaporation from the surface of the body, renders latent, and conducts away any excess of animal heat, which over exercise, or other influences might cause to be generated.

Finally, the nerves of the skin, serve not only to control the functions of its different ports, but also being sent to the extreme papillæ, receive impressions, even of the most delicate character.

If now we bear in mind the structure and functions of the skin, and remember that it is a continuation of the mucous membrane of the internal cavities, with which it strongly sympathizes, the rational of the rashes and eruptions which we are here to consider, becomes a very plain and common sense affair.

Symptoms.—The general symptoms of the *rashes*, including the *red rash, rose rash*, and *nettle rash*, are redness, either diffused or in patches, slight swelling, and an itching or burning sensation, attended with little or no febrile excitement, the particular symptoms of each variety of which I shall notice in the following sections.

The general symptoms of the *eruptions*, including the *papular erup-*

tions, vesicular eruptions, pustular eruptions, scaly eruptions, animalcular eruptions, and the cryptogamous, are slight elevations, either papular, vesicular, pustular, scaly, animalcular, or cryptogamous, attended with heat, redness, burning, or itching, &c., with little febrile excitement, but associated often with symptoms of a deranged or depraved state of the solids and fluids of the body.

Diagnosis.—The simple rashes and eruptions under consideration may be distinguished from the exanthematous fevers, which have been considered in the early part of this work by their being attended usually with less marked general febrile excitement, by their frequent connection with habits of filth, and derangement of the digestive organs, and finally by their being attended in some cases, with derangement of the fluids of the body generally.

Causes.—The causes of the simple *rashes* and *eruptions* are those which act directly upon the skin, as filth, direct irritants, improper clothing, animals, vegetables, &c. ; those that act through the digestive organs, as improper or unwholesome food and drinks, irregular eating, &c. ; and, finally, those that act through the system generally, to derange the fluids and solids, as putrid articles of food, impure air, and other like influences.

Nature.—Bearing in mind the structure and functions of the skin, as well as its sympathetic relations, and the causes which operate to produce the *rashes* and *eruptions* under consideration, their nature or pathology become exceedingly plain.

The *rashes*, whether the result of a direct irritant, or sympathetic of gastro-intestinal irritation, consist in a congestion of the cutaneous capillaries, attended in some cases with inflammation of the dermoid structure, either diffused or in patches, and hence the redness, slight swelling, and burning or itching which attends; the cutaneous nerves being involved in the irritation.

The *eruptions*, whether the result of a local irritant or of a deranged or depraved condition of the fluids of the body, consist of an inflammation of the dermoid structure, either active or passive ; the elevations in the papular variety consisting of enlarged papillæ, in the vesicular of a watery, and in the pustular of a mattery fluid, poured out under the cuticle. The *scaly eruptions* consist of elevations of the cuticle, caused directly by disease of the sebaceous glands, or by an obstruction in their excretory ducts, in consequence of which the cuticle becomes dry and scales off, leaving portions of the derma in a raw and exposed state. Finally, the *animalcular eruptions* consist of slight elevations containing matter, the result of the burrowing under the cuticle of a living animal; while the *cryptogamous* consist of a parasitic vegetable growth, &c.

Now, bearing these facts in mind, and keeping in view the grand division into *rashes* and *eruptions*, and the exact condition in the rashes ; including *red rash*, *rose rash*, and *nettle rash*, as well as in the eruptions, including the *papular eruptions*, the *vesicular eruptions*, the *pustular eruptions*, the *animalcular eruptions*, and the *cryptogamous*, the general indications of treatment are very plain.

Treatment.—The general indications in the treatment of the rashes and eruptions under consideration, are to remove any local or general cause that may be operating, to correct the habits of the patient, to keep the skin clean, to correct the general condition of the system, and, finally, to destroy any animal or vegetable parasites, if they exist, as well as to subdue the congestion, irritation, &c., of the dermoid structure.

The causes of the cutaneous disease should be sought out at once, and removed if possible; whether it be general or local in its operation. If it be the result of filth, impure air, unwholesome or improper food, or any other kindred influence, no time should be lost in their removal. The patient should be made to conform to the laws of health and rules of propriety in every respect, and especially should he be directed to take proper food, with regularity, to wear proper clothing, to take sufficient exercise, and to keep the skin clean.

In a state of health, all persons should wash the surface of the body often enough to keep the skin clean, and to keep open the seven millions of pores, or perspiratory ducts. The portions exposed to dust, &c., as the hands and face, require this ablution at least once every twenty-four hours. Other parts less exposed should be washed once or twice each week, according to the occupation, exposure, &c. Too much washing and fretting the skin, however, increases its filth, by calling to the surface fluids that ought to be thrown off by the kidneys and bowels, and is very apt to produce constipation. If we had been designed by the CREATOR for aquatic animals, we should have been supplied with *fins* instead of *legs*.

But, while these rules are generally to be followed in health, it must be remembered that in various diseases, and in some forms of the *rashes*, and especially of the *eruptions*, daily ablutions of the whole surface with soap and water may be required, to keep the skin clean; and should by no means be neglected.

Having thus removed the cause, corrected the habits, properly cleaned, clothed, and fed the patient, the cutaneous disease may very generally disappear. In case, however, the rash or eruption still continues, the digestion, if impaired, should be corrected, as far as possible, and any impoverished state of the blood or depraved condition of the system restored, by tonics, alteratives, &c., judiciously administered.

Finally, when this is all accomplished, as far as may be, if the cutaneous disease, whether *rash* or *eruption*, still remains, local applications, either cooling, soothing, or astringent, &c., may be required, and should be continued till a cure is effected. Thus, then, have we completed our general view of the *rashes* and *eruptions* under consideration, and are now prepared to take up each variety, in the order in which I have named them.

SECTION II.—RED RASH.

By red rash, I mean here those irregular red spots or blotches, which appear accidentally upon the skin, from various local or general causes; but generally from some local irritation, as friction or the application of an irritant. Under this head we may include almost every case of irregular rash, not connected with the exanthematous fevers, except rose rash and nettle rash, which have peculiarities entitling them to separate consideration. The chaps which occur on the back of the hands from cold, those which occur on the nipples in consequence of nursing, and the blotches which occur on the feet of children in the winter season, are varieties or specimens of *red rash.*

When two surfaces of the skin, moistened by perspiration, or by any other fluid, are in such contact that they rub together during the motions of the body, the skin becomes red and chafed or inflamed, constituting it a case of *red rash.*

Red rash is sometimes, however, dependent upon a constitutional cause, and associated with derangement of the digestive organs, and perhaps with an impoverished state of the blood. In cases of this kind, the irregular blotches are apt to appear upon the face, and they may come out suddenly and disappear in a few hours, or they may fade from the centre, leaving the irregular rash, which has been called *ringworm.*

Diagnosis.—To admit any case of irregular rash under this head, it is only necessary to ascertain, from the general and local symptoms, that it does not belong to any of the exanthematous fevers already considered, and also that it has not the bright rose color of rose rash, or the raised wheals of urticaria or nettle rash.

Causes.—The cases of irregular rash which may properly be included under this head, have a great variety of causes, most of which act directly upon the part to produce congestion, irritation or inflammation of the skin, as we have already seen. The general causes, or those that act through the system, are also various, among the most frequent of which, are improper food, irregular eating, excessive anger, and mental excitement.

Treatment.—In all cases of red rash, the cause should be ascertained and removed if possible, and the patient required to take proper food with regularity; to keep the skin clean and warm; and to take a reasonable amount of exercise in the open air.

The local applications should be judiciously adapted to each particular case. If the lower limbs are the seat of the rash, a bandage properly applied, may by affording support, be of service in some cases. For blotches on the face or other parts, tincture of camphor or a weak solution of the sulphate of iron, or of alum, about fifteen grains of either to the ounce of water, will do very well as local applications.

In cases of chaps, galls, &c., the parts should be kept dry and clean; and may be dusted with starch or tannin if necessary, and should this fail of affecting a cure, a solution of alum should be applied. In chaps, such as occur on the nipples in nursing, and occasionally on other parts, an ointment made by mixing intimately a drachm each of tannin and the oxide of zinc with an ounce of lard, may be applied three times per day, and continued till a cure is effected.*

SECTION III.—ROSE RASH—(*Roseola.*)

By rose rash or roseola, is here meant that peculiar variety of rash in which there are small patches of a rose red color; which commence on the face, and extend over the surface of the body, attended usually with slight febrile symptoms.

Rose rash is usually a slight affection, occasionally attacking young

* Or a solution of tannin in an ℥i of glycerin, may be used.

children especially while teething, or adults of weakly constitution ; and is sometimes associated with constitutional disorders, as rheumatism, gout, &c.

Symptoms.—Rose rash may occur without any noticeable febrile symptoms, but it generally appears after slight gastric disturbance and chilliness, followed by more or less febrile excitement.

The rash usually comes out first on the face, neck and breast, and gradually over the whole body, or a greater portion of it. Sometimes, it has a general diffusion, which with its bright redness, gives it the appearance of scarlatina. Generally, however, the rash appears in patches, varying from a mere speck to half an inch or more in diameter, and being slightly raised, it has considerable resemblance to measles.

The rash is usually attended with more or less itching or tingling, but this is not an invariable symptom. The disease seldom lasts more than three or four days, and as the rash disappears, there is very slight, if any, desquamation.

Diagnosis.—Red rash, or roseola, is liable to be confounded with measles or scarlatina, from which it may, however, be distinguished by attention to the following differences.

Roseola has not generally so marked constitutional disturbance as rubeola or scarlatina, is of shorter duration, and is neither contagious or infectious. It has not the catarrhal symptoms of measles, or the anginose complication of scarlatina.

Finally, rose rash may be distinguished from red rash by its febrile symptoms, by the order which the rash usually observes in making its appearance, and by its more regular or uniform character.

Causes.—Rose rash may appear during various forms of disease, and especially in smallpox, vaccina, and varioloid affections, and also in bilious and enteric or typhoid fevers. The most common causes of rose rash, however, as it occurs during infancy or childhood, are teething, and irritation or slight inflammation of the alimentary mucous membrane from irregular eating. The most frequent causes of the disease in adults, are copious draughts of cold water in hot weather, or when the body is heated, fatigue, with mental excitement, and various indigestible articles of food.

Treatment.—The first inquiry after ascertaining the nature of the disease should be into the cause or causes that have been operating to produce it. If the patient is a child, and taking food at irregular hours is the cause, that habit should be corrected, and the child allowed only suitable food, to be taken at regular meal hours. If the patient is an adult, and in the habit of taking improper or indigestible food, the habit should be corrected. Strict cleanliness should be enjoined in all cases.

During an attack of this disease, attended with little or no febrile excitement, the patient need only keep quiet, take a plain digestible diet, and if there is acidity of the stomach and slight constipation, a moderate dose of rhubarb and magnesia. In cases, however, in which there is considerable febrile excitement, a little mercury with chalk, or blue mass may be administered, and followed by the sulphate of magnesia, and then moderate doses of James's powder may be given every six hours, till the fever subsides.

SECTION IV.—NETTLE RASH.

By nettle rash, I mean here, that variety of rash which appears, with white elevations or wheals, on a red surface, the elevations appearing on any part, from a slight scratch or rubbing, the rash being attended with the most intolerable tingling, pricking, or itching; and in most cases, with nausea and moderate febrile excitement.

Symptoms.—Nettle rash or urticaria, may occur from various articles of food, and from other causes, without being attended with any considerable febrile excitement. In such cases, there is experienced at first, an uneasiness in the mouth, throat and stomach, followed by epigastric pain, nausea, and anxiety, or even coma; and finally, by the eruption.

The rash in such cases, is apt to be very copious, and is attended in some cases, with swelling of the face, neck, and chest, and even the whole surface of the body. The elevations or wheals, are scattered here and there, and the patient suffers from oppressed breathing, and from the most intolerable heat, tingling, and itching on the whole surface of the body. In such cases, the severity of the disease continues usually, only a few hours, and all the symptoms may disappear in a day or two.

Among the articles of food, which sometimes thus produce urticaria, or nettle rash, in those who are predisposed, are lobsters, oysters, pork, eggs, cucumbers, melons, mushroons, &c. Such, according to my observation, are the usual symptoms of nettle rash, occurring from various articles of food, and attended with little or no febrile excitement. In many, and perhaps in the majority of the cases of this disease, however, the rash is preceded for two or three days by fever, attended with headache, nausea, vomiting, faintness, &c. In such cases, the febrile symptoms mostly subside, as the rash makes its appearance, in patches of vivid redness, with elevations or wheals rising in the midst of them.

It may be very extensive, involving most of the surface of the body, or it may appear only on particular parts; as the face, inside of the forearm, shoulders, loins, thighs, &c. In cases of this character, the rash may partially disappear for a time, during the continuance of the disease, and then reappear, till it finally subsides permanently, by the seventh or eighth day, after which, a slight desquamation of the cuticle may take place.

I have known urticaria, or nettle rash, to assume even an intermittent character; being associated with, or assuming the character of intermittent fever. In cases of this character, that have fallen under my observation, there has been either a chill, or a tendency to one, during the reaction from which, the rash has made its appearance.

Diagnosis.—Urticaria or nettle rash, may be distinguished from all other affections by the wheals or elevations which appear in the patches, on different parts of the body, by the excessive burning, stinging, or itching sensation which attends the rash, by its non-contagious character, and by the general character of the disease, which differs from roseola, or the exanthematous fevers, considered in the early part of this work.

Causes.—Nettle rash may be produced as we have seen by various articles of food, in those who are predisposed. It is also produced in children by irregularity in taking food, and in adults by imprudence in eating and drinking. The disease is also probably sometimes of malarious origin, and it may be associated with intermittent or other forms of fevers. Certain drugs also occasionally produce it, as turpentine, valerian, copaiba, &c.

Nature.—In those cases of nettle rash which occur from various articles of food, as oysters, pork, eggs, melons, cucumbers, &c., in which the rash makes its appearance in a few hours, without any febrile symptoms, I believe that the cutaneous disease is directly sympathetic of irritation of the alimentary mucous membrane. In cases in which the rash is preceded by marked gastric disturbance, attended by febrile excitement, the cutaneous disease is doubtless in part the result of direct sympathy with the gastro-intestinal irritation, and also in part produced by the febrile excitement, an undue amount of blood being sent to the cutaneous vessels.

Finally the redness of the skin in this disease, as well as in *red* and *rose rash*, evidently depends directly upon congestion of the cutaneous vessels, attended in some cases with inflammation of the dermoid structure. The increased heat is probably owing in part to an increased consumption of oxygen in the cutaneous capillaries, and also in part to the suspension of evaporation from the surface, in consequence of the closure of the 2,800 perspiratory ducts on each square inch of the surface involved.

The *itching* which attends the rashes, is from the irritation which is produced in the cutaneous nerves, while the wheals, which are characteristic of nettle rash, are the result of a deep congestion of the part, in consequence of which the skin is elevated, the pressure beneath forcing the blood from the capillaries of the elevated portion, leaving the wheals of a whitish appearance in most cases at least.

Treatment.—Persons that are predisposed to nettle rash, and liable to an attack from eating any particular article of food should be careful to avoid it. And all persons predisposed to this affection should avoid hearty meals, and every variety of indigestible food. In case, however, the disease appears, in consequence of over-eating, or of taking indigestible or improper food, an emetic of ipecac may be given at once, and this followed by a cathartic of magnesia and rhubarb, or of the sulphate of magnesia.

This may be sufficient in many cases to arrest the disease. If, however, there is febrile excitement, the warm foot-bath, and if necessary the warm bath may be resorted to with the hope of restoring the cutaneous function. The patient should be allowed a plain and digestible diet, and should the disease assume an intermittent character, in addition to what I have already suggested, the sulphate of quinine should be administered, as for ague, and thus the disease be arrested

SECTION V.—PAPULAR ERUPTIONS.

By papular eruptions, from papula, a pimple, I mean here that variety in which the elevations on the skin consist of dry pimples, sometimes retaining the natural color of the skin, but generally red, and attended with more or less itching.

39

I believe that the papular eruptions generally consist of enlarged and inflamed papillæ, but they may, in some cases, consist of lymph, effused beneath the cuticle, or, in others, of minute points of inflammation of the cutis, with congestion of the capillaries of the part, producing the dry pimples which arise.

The papular eruptions might be described under the heads of *strophulus, lichen* and *prurigo ;* but, as it appears to me that such a division, or in fact any other, would only "serve to bewilder, and dazzle to blind," I shall proceed to the consideration of *papular eruptions,* as constituting but one disease.

Symptoms.—Papular eruptions, or the dry pimples, may be very minute, and, if they retain the natural color of the skin, they may be appreciable only by the touch. More generally, however, they are very distinct, being more or less deeply colored, according to the degree of the cutaneous inflammation, and the condition of the circulating fluids. The size of the dry pimples, then, may vary from the merest possible elevation of the cuticle, to very distinct pimples, and the color from that which is natural to the skin, to a bright red or even livid appearance.

Itching is another symptom of the papular eruptions, it being in some cases slight, but in others so intolerable, as to allow the patient little or no rest, day or night. The degree of the itching evidently depends upon the amount of irritation in the cutaneous nerves, it being in some cases more and in others less.

The papular eruptions may be extensive, being scattered more or less over the whole surface of the body, or the eruptions may appear on particular parts, scattered, or in irregular clusters. No age is exempt, as the disease may be congenital, or acquired at any period, from early infancy to extreme old age.

Such, I believe, are the ordinary symptoms of the papular eruptions, liable of course to variations, from variety of constitution, and from the nature of the local or general causes which operate to produce them.

Diagnosis.—The papular eruptions may be distinguished, on the one hand, from the exanthematous fevers, by the absence of febrile excitement, and on the other, from the *vesicular, pustular, scaly, animalcular* and *cryptogamous eruptions,* by the dryness of its pimples, while in the latter forms of eruptions, the elevations contain a fluid of a watery matter, or other character, or else consist of scales or crusts.

Causes.—Papular eruptions occur during infancy, in part from hereditary predisposition, and also from filth, dentition, and irritation or disturbance of the stomach and bowels.

In adults, papular eruptions may depend upon various local and general causes. I believe, however, that filth, direct irritants, improper food and drinks, violent exercise, impure air, insufficient clothing, and mental depressions, are the most frequent causes of papular eruptions in adults. The disease is also very apt to be associated in adults, with derangement of the stomach and bowels, and with an irritable condition of the nervous system.

Nature.—I believe, as I have already suggested, that the papular eruptions or dry pimples consist generally of enlarged and inflamed papillæ, the elevations in such cases being the result of enlargement of the minute artery and vein of which each papilla is composed, the itching depending upon the degree of irritation in the loop of nerve which is also furnished to each. Doubtless, in such cases, the dermoid capillaries are more or less congested, its structure often inflamed, and it is possible that the dry pimples may sometimes consist of lymph effused under the cuticle, or else of a congestion, with enlargement of the capillaries of the part.

In cases in which the eruption is copious, the functions of the perspiratory and sebaceous glands are evidently more or less deranged, and especially are their ducts obstructed by the enlarged papillæ or dry pimples, between which they pass to the surface of the skin.

Treatment.—In the treatment of the papular eruptions, the cause should be sought out and removed if possible. The patient should be directed to take a proper diet with strict regularity. The clothing should be clean, and adapted to the season of the year, being generally flannel next the skin in cold, and cotton, linen, or silk in warm weather. Sufficient exercise should be taken in the open air to keep up an active circulation, without over-heating, and also to secure a healthy action of the cutaneous vessels. The skin should be washed in cool or tepid water, containing a little salt, at least once or twice each week, and an occasional warm bath may be of very essential service. Finally an even and cheerful temper of mind should be maintained in all cases of this disease, and especially is this of the highest importance in females nursing infants with papular eruptions.

Infants affected with this form of eruption should be nursed at regular hours, and weaned children should be restricted to their regular meals, with no food at intervals. To allay the itching with children, vinegar diluted with three parts of water, may be applied occasionally with very good effect. For adults, the juice of one lemon, diluted with a pint of water, applied two or three times per day, may produce considerable relief.

In very many cases, papular eruptions may be removed by the course I have suggested. If, however, the disease is attended with fever, saline cathartics may be required. If there is an anæmic state, iron in some form should be given for a time. And in cases in which there is a scrofulous or otherwise depraved condition of the system, cod-liver oil, and perhaps iodide of potassium and sarsaparilla may be required.

The cutaneous eruption will generally disappear if the cause is removed, and the habits and general condition are thus corrected. If, however, the eruption still continues, an application of glycerin with carbonate of potassa, about fifteen grains to the ounce, may be applied two or three times per day. Or, for adults, three drachms of the potassa may be dissolved in a pint of water, and applied instead, till the eruption is removed.

SECTION VI.—VESICULAR ERUPTIONS.

By vesicular eruptions, I mean here, that variety, in which the elevations upon the skin consist of vesicles; a serous or watery fluid being poured out under the cuticle, the result of inflammation of the dermoid structure, caused by an external or internal cause.

The vesicular eruptions, or watery elevations in this form of cutaneous disease, are formed by the same process of inflammation of the cutis, that produces an ordinary blister, the inflamed cutaneous vessels pouring out the watery part of their blood, under the cuticle, producing elevations or vesicles, varying in size, from the merest point, to vesicles the size of a hen's egg, or larger.

The vesicular eruptions might be described under the heads of *eczema*, *herpes*, *rupia*, and *pemphigus;* but such a division founded upon the size of the vesicles, and in fact any other could be of no practical utility, and would only serve to multiply words or names, without in the least increasing our knowledge. We will therefore, proceed to the consideration of *vesicular eruptions*, as constituting but one disease, thus keeping our minds directly upon the pathological condition.

Symptoms.—With or without apparent constitutional disturbance, the vesicular eruptions make their appearance, in vesicles of various sizes, from slight elevations, to vesicles of all sizes, up to that of a hen's egg, or larger, as we have seen.

The eruption may be attended with slight fever, and the skin may appear red before the vesicles arise. In many cases, however, the disease is attended with no febrile excitement, and the skin may appear but slightly, if at all red, before the vesicles appear. It is probable, however, that the vesicular eruption is generally preceded by either an active or passive inflammation of the dermoid structure; whether it is rendered very apparent or not, as there is usually more or less itching.

The eruption may be quite general, covering most of the surface of the body, or it may be confined to a small extent of surface, or to a particular part, in either case, the vesicles being scattered or in clusters; depending entirely upon accidental influences. The fluid of the vesicles is of a serous or watery appearance at first; but as they become matured in the course of a few days, it may become more or less changed; assuming a milky or opaque appearance. And as the vesicles finally burst, or become lacerated, it dries and forms a scab or crust, which finally falls off; and thus the disease may terminate, if there is no considerable constitutional disturbance.

In cases, however, in which there is serious constitutional disturbance, new vesicles appear, as the first crop become matured. Or if particular care is not taken to keep the surface clean, the watery fluid, on the laceration or bursting of the vesicles spreads on the skin, producing an irritation, which serves to perpetuate, and extend the disease. Especially is this apt to be the case, when the eruption occupies the scalp, producing one form of that very unpleasant condition, which has been called "scalled head."

Vesicular eruptions may appear in patches, which extend at the same time that the central pustules mature, dry, and desquamate, and leave the eruption in a circular form; constituting what has been called "ringworm." Or the eruption may extend half way round the chest, constituting what has been called "shingles." These, however, are accidental forms of vesicular eruptions, requiring no particular names to designate them, and should be included, as well as all irregular forms of this disease, under the common name of *vesicular eruptions*, which point directly to the real pathological condition.

The duration of a single crop of vesicles to their drying and desquamation, may vary from five to fifteen days, at which time, if no new vesicles have appeared, the disease is at an end. If, however, new crops are formed, the disease may be perpetuated indefinitely, and the eruption may extend over the whole surface of the body; and, in depraved constitutions, the disease may lead on to a fatal termination.

Such I believe are the ordinary symptoms of *vesicular eruptions;* liable, of course, to variations, from peculiarity of constitution, and from various general and local accidental causes, as is the case with all other forms of disease.

Diagnosis.—The simple vesicular eruptions may be distinguished from the exanthematous fevers, attended with this form of eruption, by the want of the febrile and other attendant symptoms, as well as by the non-contagious character of the simple vesicular form of eruptions.

This form of disease may be distinguished from the other varieties of simple cutaneous eruptions, by the watery elevations or vesicles; while in the others, the elevations are either dry, pustular, scaly, animalcular, or cryptogamous.

Causes.—The vesicular eruptions are very apt to be associated with a hereditary or acquired depraved condition of the system, and especially with a partially dissolved or watery state of the blood. Various general and local exciting causes, however, operate to produce or develop this form of cutaneous disease.

In infants, this form of eruption may be produced by improper nourishment, insufficient clothing, filth, impure air, and dentition. In children, the very improper habit of taking food at irregular hours or between meals; by impairing digestion; cutting off nutrition, and thus reducing the blood, is one of the most frequent causes of this form of disease.

In adults, improper food, filth, impure air, intemperance in eating and drinking, masturbation and excessive venery, the use of tobacco, mental depressions, fatigue, and exposure to the heat of the sun, as well as to various local irritants, are, according to my observation, the most frequent causes of the vesicular eruptions.

Nature.—The vesicular eruptions are probably in most cases, and perhaps invariably, the result of inflammation of the dermoid structures. And this cutaneous inflammation may in some cases be of an active character; but I believe that it is generally decidedly passive; being associated with a weak, depraved, or watery state of the blood, and hence the watery effusion which takes place from the cutaneous capillaries, producing the elevations of the cuticle, constituting the vesicles which are formed.

The size of the vesicles, as well as the extent and form of the eruption, are entirely accidental, affording no grounds for any practical division of this form of disease, which would call for names, calculated to distract and confuse the mind, rather than to point to the real pathological condition, which should always be kept in view.

Treatment.—In all cases of vesicular eruptions, the cause should be sought out and removed as far as possible. The patient should then be directed to take a proper diet with strict regularity. The surface of the body should be washed daily, or sufficiently often to keep it clean, and an occasional warm bath should be resorted to. Care should be taken that the clothing be of proper material, and that it be kept entirely clean. The patient should be directed to sleep in an upper or dry room, and to take a reasonable amount of exercise in the open air.

These preliminaries being arranged, the general condition of the system should be corrected, if it is in any way deranged. If there is a plethoric condition of the system, and the cutaneous inflammation is of an active character, two or three blue pills may be given, and followed by a full dose of the sulphate of magnesia, after which, a teaspoonful may be given each morning and continued as long as it may be required.

If the patient is anæmic, as is very often the case, three blue pills may be given at first, and followed by a Seidlitz powder, after which, the bowels may be regulated by a pill of aloes and rhubarb taken after dinner each day, and continued as long as may be necessary. To restore the blood, ten drops of the tincture of chloride of iron should be given in a little water, three times per day, and continued till the blood is fully restored. When sufficient iron has thus been thrown into the blood, if there still remains evidence of derangement of the lymphatic and glandular system, five grains of the iodide of potassium may be given three times per day, for a time, in an ounce or two of the infusion of *solanum dulcamara.*

When the cause has been removed, and all the habits as well as the general condition of the patient been thus corrected, the eruption will generally disappear. In case, however, patches of the vesicles still remain, as they may, especially if they have occupied the scalp; the diseased surface should be washed three times per day with weak soap suds, and immediately after each washing the parts should be moistened with glycerin, containing in solution two drachms of the carbonate of potassa to the ounce.

When the scabs, scales and vesicles are thus removed, as they generally will be in three or four days, leaving a greater or less extent of raw surface, a cloth wet in a decoction of the root of *phytolacca decandra* or poke root, of the strength of two ounces to the pint of water, may be kept applied to the disease surface till a cure is effected. Or in case the eruption occupies some portion of the surface where the application of the decoction is attended with inconvenience, an ointment made by mixing a drachm of the pulverized *poke root* with an ounce of lard, may be applied three times per day instead, and continued till a cure is effected.

SECTION VII.—PUSTULAR ERUPTIONS.

By pustular eruptions, I mean here that variety of simple cutaneous eruptions in which the elevations of the cuticle are caused by small collections of pus, the result of inflammation of the derma from a general or local cause.

The pus poured out under the cuticle in this form of cutaneous disease, produces elevations of the cuticle in the same manner, that the water or serum does in the vesicular form, but the pustules seldom exceed the dimensions of a split pea, and are never so minute as some of the vesicles in the vesicular form of cutaneous eruptions.

The pustular eruptions might be described under the heads of *ecthyma*, *impetigo*, &c., but as there are really no sufficient ground for such division, or in fact any other, I shall dispense with these terms altogether, and proceed to the consideration of this form of cutaneous disease, under the head of *pustular eruptions*, the symptoms of which we will now proceed to consider.

Symptoms.—The pustular eruptions are attended with, and generally preceded by more or less inflammation of the dermoid structure, and the disease may be attended with febrile excitement, but this is by means invariably the case.

In cases in which the cutaneous inflammation is very considerable, and of an active character, the eruption is attended with heat, redness, and slight swelling, and in all cases, there is more or less itching, in consequence of irritation in the cutaneous nerves. The pustules constituting this form of eruption, attain their full size, varying from that of a mustard seed to that of a split pea, in three or four days, and may gradually dry up, without bursting; but more generally burst, or are lacerated, and then dry, forming a hard crust, which offers a variety of colors, being either yellow, brown, or almost black.

As the first crop of pustules mature and dry, or burst, leaving a scab which falls off, the cutaneous disease may subside. But if new pustules have formed, the disease may be perpetuated indefinitely, and perhaps extend more or less over the surface of the body. When the disease occupies the scalp, it is very apt to extend, and continue till the hair is lost, constituting one variety of what has been called "scald-head."

The pustular eruptions are very irregular in their form, sometimes the pustules being irregularly scattered over the whole surface of the body, and in other cases confined to a very limited extent of surface. Their appearance in any definite form, when it occurs, is from purely accidental influences, and is a matter of no consequence, as the rule is irregularity, and any definite arrangement of the pustules only an exception.

Some cases of this disease are easily arrested, or may even tend to a spontaneous termination. But others are exceedingly obstinate, especially if there is considerable constitutional derangement, sometimes passing on even to a fatal termination.

Such I believe are the usual symptoms of the pustular eruptions, liable of course to variations, from peculiarity of constitution, and from various accidental influences of a general and local character.

Diagnosis.—The simple pustular eruptions may be distinguished from the exanthematous fevers, attended with this form of eruption, by the absence or less intensity of the febrile excitement, and by the absence, in the simple pustular eruptions, of most of the violent symptoms which attend the exanthematous fevers. From the other simple eruptions this form may be distinguished by its pustules, while the elevations in the other forms are either papular, vesicular, scaly, animalcular or cryptogamous.

Causes.—A scrofulous, or otherwise depraved condition of the system, either inherited or acquired, constitutes a strong predisposition to the pustular eruptions. Various local and general causes may, however, produce this form of cutaneous disease, or serve to develop it in constitutions predisposed. Among these causes are local irritants, improper food, clothing, filth, &c., intemperance in eating and drinking, gastro-intestinal irritation, violent exercise, mental depressions, violent passions, sexual excesses, and other kindred imprudences.

Nature.—We have already seen that the pus, of a yellowish color, which produces the elevation of the cuticle in this form of cutaneous disease, is the product of inflammation of the dermoid structures. This cutaneous inflammation, however, may be of either an active or passive character, and is, according to my observation, very often associated with a scrofulous or depraved condition of the system. In addition, then, to an active or passive inflammation of the skin, with the formation of pustules, there is evidently in scrofulous cases of pustular eruptions, more or less derangement of the cutaneous lymphatics, which complicates the cutaneous disease, and perhaps favors the formation of pustules.

Finally, in common with the other forms of eruptions, there is during the continuance of the cutaneous inflammation, in this disease, more or less derangement in the functions of the perspiratory and sebaceous glands, or at least obstruction in their ducts, which open on the surface of the cuticle, and pour out, in a healthy state of the skin, the perspirable and sebaceous matters.

Treatment.—The indications in the treatment of the pustular eruptions are, to remove the cause, correct the habits of the patient, restore the system to a healthy state, and, if necessary, to resort to local applications, to favor a restoration of the skin to its normal condition.

In arriving at the cause, every influence that might be supposed to act, either directly or indirectly, should be ascertained and removed, or rendered inoperative.

The patient should be directed to take proper food with strict regularity, to wear clean and proper clothing, to wash the surface of the body daily, or sufficiently often to keep the skin clean, to take a reasonable amount of exercise in good fresh air, to preserve an even and cheerful temper of mind, and to be prudent in all things.

If the system is phlogistic, three blue pills should be given at first, and followed by a full dose of the sulphate of magnesia, and then a teaspoonful may be administered each morning, till the inflammatory tendency subsides. While this is being accomplished, the warm foot-bath may be resorted to at evening, to favor a healthy action of the skin, and if convenient, an occasional warm bath may be taken. In case however the patient is scrofulous, the iodide of potassium should be given, as soon as the general inflammatory condition is corrected, in five grain doses, three times per day, in the syrup or fluid extract of bittersweet, burdock, or sarsaparilla.

If, however, the patient is anæmic, as is more generally the case, after the first cathartic, a pill of aloes and rhubarb may be given after dinner each day, if necessary to secure a regular action of the bowels, and after having continued the iodide of potassium, &c., for a reasonable time, it may be suspended, and ten drops of the syrup of iodide of iron given three times per day instead, and continued till the blood is restored.

While this general derangement of the system is being thus corrected by the ablutions, cathartics, alteratives and tonics, the cutaneous disease will very generally disappear. If, however, patches of the eruption remain after the general derangement is corrected, whether on the scalp or any other part of the surface of the body, they should be washed clean three times per day, and then moistened with glycerin containing in solution, two drachms of carbonate of potassa to the ounce, till the scabs are removed. This being accomplished, an ointment made by mixing a drachm of either poke root, oxide of zinc, tannin or sulphate of iron, with an ounce of lard should be applied three times per day, till a cure is affected.

SECTION VIII.—SCALY ERUPTIONS.

By scaly eruptions, I mean here that variety of cutaneous eruptions in which the elevations on the skin consist of portions of the cuticle so changed as to form scales of various thickness and extent, the result generally of cutaneous inflammation.

While the scaly eruptions are generally the result of cutaneous inflammation, it should be remembered that an inherited defective formation of the cuticle sometimes exists, in which a part or the whole surface of the body is covered with scales, and yet there is no evidence of cutaneous inflammation, or in fact of any special constitutional disease. Such cases, however, two or three of which have fallen under my observation, are only exceptions to a rule, as nearly all others I believe are associated with, and generally preceded by cutaneous inflammation.

The scaly eruptions might be described under the heads of *psoriasis, lepra pityriasis, ichthyosis, &c.*, but as I can discover no advantage that could be derived from such a division, or in fact from any other, I shall consider the scaly eruptions as constituting but one disease, the symptoms of which we will now proceed to consider.

Symptoms.—The appearance of the scaly eruptions is generally preceded by more or less evidence of irritation and inflammation of the skin, such as heat, dryness, itching and redness. At the inflamed points in the skin, slight elevations occur, in some cases, on which the scales gradually form, with either a smooth surface, or having a slight depression in the centre, the scales gradually thickening, and finally becoming detached and falling off.

In other cases, the elevations are very slight, the scales being of a minute branny character, and coming off in countless numbers, to be followed by others. Or, a part, or the whole surface of the skin may be covered with a thick and hardened scale divided only by the natural furrows of the skin, the scales being little disposed to exfoliate, and being quickly replaced if removed. Such cases are attended with little or no cutaneous inflammation, and may be either congenital or acquired.

The size of the scales then, in the scaly eruptions, may vary from those of a minute branny character, which have been called *dandruf*, to those that cover the whole surface, being divided only by the natural furrows of the skin. The scaly eruptions may be confined to a very limited extent of surface, or they may extend indefinitely, covering, in some cases, as we have seen, the whole surface of the body.

No age is exempt, and when the eruptions have once made their appearance, whatever may be their form or extent, the disease is apt to be obstinate, or of protracted duration.

Though this variety of cutaneous disease is not usually attended with very severe itching, there is, in almost every case, an uneasiness experienced in the diseased part, which is exceedingly annoying, especially if the scalp is the seat of the eruption.

Such, I believe, are the ordinary symptoms of the scaly eruptions, the variations being accidental, depending upon peculiarity of constitution, and other accidental circumstances.

Diagnosis.—The scaly eruptions may be distinguished from other cutaneous eruptions by the scaly character of the disease, the cutaneous elevations consisting of changed and perhaps thickened cuticle, without the previous formation of pimples, vesicles, or pustules as is the case in the other forms of eruptive cutaneous disease. There is no other form of disease with which the scaly eruptions need be confounded, if all the symptoms be carefully taken into account.

Causes.—Various causes operate to produce the scaly eruptions, either predisposing or exciting, among which are improper varieties of food, and irregularity in taking it, depressing or very exciting mental emotions, direct irritants, exposure to the heat of the sun, or to very hot fires, and intemperance in eating, as well as in the use of alcoholic drinks.

In congenital cases, two of which I have known in one family, the cause or causes which operate, are obscure, or at least are so to me, as both parents were apparently free from any disease of the kind, in these cases. It is probable, however, that some peculiarity of constitution, or habits on the part of the parents; or some influence of the locality, or some peculiar mental emotion endured by the mother, during gestation, may operate to produce the disease in the offspring, or a predisposition to it, at least.

Congenital cases, however, are very rare, and yet I believe, that a hereditary predisposition to the cutaneous eruptions may prevail, in most cases in which the disease is developed, in infancy or early childhood, and perhaps in some cases in which the disease is developed later in life, by the exciting causes enumerated above.

Nature.—In relation to the nature of the scaly eruptions, it appears to me, there can be no reasonable doubt. The elevations or scales, whatever may be their size, from the branny scales, to those which cover the whole surface of the body; being divided only by the natural furrows of the skin, consist of changed or imperfectly formed cuticle, which gradually become detached, and fall off; their place being generally very soon supplied by others of a like, or similar character.

The portions of changed, or imperfectly formed cuticle, constituting the scales, in this form of eruption, are sometimes thickened, and always dry and hard, indicating a failure on the part of the sebaceous glands, to furnish sufficient oily matter to the diseased portion of cuticle, to keep it soft. In consequence of this, the cuticle becomes thus hardened, and sometimes thickened, and generally exfoliates.

In congenital cases, I believe that there is a permanent defect in the function, and probably in the structure of these oil glands of the skin, which produce this form of cutaneous disease. In all other cases of the scaly eruptions, or in most of them at least, it is probable that inflammation of the derma, of either an active or passive character; involving the sebaceous glands, or their excretory duct, either suspends the secretion of the oily matter by the glands, or stops its passage through their excretory ducts, to the surface of the cuticle, in the diseased portions of the skin.

Treatment.—With the real nature of this form of disease thus in view, and unburdened of a multitude of names, which could only bewilder, the indications in the treatment of the scaly eruptions, are very plain.

In all cases of the scaly eruptions, the habits of the patient should be inquired into, and corrected.

The patient should be directed to take proper food, with strict regularity; to wear clean and suitable clothing; to take a reasonable amount of exercise in the open air; to keep the surface of the body clean, by daily ablutions, or the warm bath, if necessary; to preserve an even and cheerful temper of mind, and finally, to observe strictly, the laws of health, and rules of propriety in every respect.

In *congenital* cases, nothing more than this need be done, unless the scales become so hard at certain points as to injure or irritate the derma, in which case it may be partially softened by the application of glycerin, morning and evening, till the irritation subsides. A permanent cure of such cases need not be anticipated, and therefore should not be attempted.

In all other cases of scaly eruptions, however, an effort should be made to cure the disease. If the patient is plethoric, and the cutaneous inflammation of an active character, three blue bills, or a full dose of leptandrin or podophyllin should be administered at first, and followed, if necessary, by the sulphate of magnesia, to secure a free operation. After the first cathartic, a Seidlitz powder may be given, each morning, till the phlogistic condition is overcome; a warm foot-bath being used at evening, if necessary, as well as the daily ablutions, or use of the warm bath.

If, however, the patient is anæmic, after the first cathartic, the infusion or fluid extract of taraxacum may be administered, for a time, as an alterative, and for the purpose of regulating the bowels. As soon as the bowels are thus regulated, ten drops of the syrup of the iodide of iron may be given, three times per day, in a drachm of the fluid extract, half an ounce of the syrup, or two ounces of the infusion of *burdock*, or of the *solanum dulcamara*. When sufficient iron has thus been thrown into the blood, it may be suspended, and, if the patient is scrofulous, the *burdock* or *bittersweet* may be continued, with five grains of iodide of potassium, three times per day, till the general condition is corrected, as far as it may be.

While this general treatment is being pursued, nothing should be done locally, except to continue the daily ablutions, or use of the warm bath, and to apply to the diseased part sufficient glycerin to soften the cuticle, till such time as the sebaceous glands shall resume their functions, and thus put an end to the cutaneous disease. Finally, as the general condition of the system becomes corrected, if the scaly eruption continues, a drachm of the oxide of zinc may be mixed with each ounce of the glycerin applied, and continued till a healthy state of the skin is restored.

SECTION IX.—ANIMALCULAR ERUPTIONS.

By animalcular eruptions, I mean those which occur from the burrowing or presence of living animals in the skin, or under the cuticle, the most common and interesting of which is the "*acarus scabiei*," or, as it has recently been called, the "*sarcoptes hominis*."

The disease of the skin caused by this animal has been called "*itch;*" but as the other forms of cutaneous eruptions, or some of them at least, are attended with severe itching, I can see no reason why this term should be applied to this form of eruption any more than to others. I shall therefore consider this form of cutaneous disease, in common with others of an animalcular character, under the head of *animalcular eruptions;* the symptoms of which we will now proceed to consider.

Symptoms.—The first symptoms of the animalcular eruptions are itching in some portion of the skin, and very soon slight redness, either in a point or lines of very limited extent, caused by the burrowing of a minute animal.

The burrows gradually extend, till they pass through a considerable extent of the cuticular surface, the cuticle always assuming a rough or ragged appearance, as far as the disease extends.

The itching and red points or streaks in the skin, together with the ragged appearance of the cuticle, extending gradually over the surface of the body, and appearing especially conspicuous between the fingers, at the flexures of the joints, and at points generally where the cuticle is most delicate, together with the appearance of the animal, when discoverable, constitute the only invariable symptoms of the animalcular eruptions.

In many, and perhaps in most cases of the animalcular eruptions, however, small elevations of the cuticle appear, containing either a watery or mattery fluid, near which the parasitic animal may often be seen in the form of a small white or brown speck, barely visible to the naked eye. These elevations are often torn by scratching, and then present open sores, which finally scab, desquamate and heal, and perhaps give place to others, but not generally I believe.

The animalcular eruptions may continue for months, or even years, if the parasitic animal, the burrowing of which produces them, is allowed to continue his operations unmolested. And though the disease must be attended with great inconvenience, it seldom tends to a fatal termination.

Such are the ordinary symptoms of the animalcular eruptions, liable to variations depending upon the extent and duration of the disease, and also upon the habits of the patient, and perhaps upon the character of the animal which has produced them.

Causes.—The causes of the animalcular eruptions are the parasitic animals, by far the most common of which is the *acarus scabiei*, or *sarcoptes hominis*, as we have already seen. The presence of the parasitic animals alone are sufficient to produce this form of cutaneous disease.

But a dirty skin, with filthy habits generally constitute a strong predisposing cause, and probably tends materially to perpetuate this form of disease.

The parasitic animals may be taken by contact with persons on whom they dwell, from cloths which they have worn, or from beds in which they have slept. And if they find their new subject filthy, and a stranger to soap, they will multiply rapidly, and very soon cause to be developed the eruption which they produce.

Diagnosis.—The animalcular eruptions may be distinguished from all other eruptions, by the intensity of the itching, even when there is only the slight red points or streaks in the skin, with a rough or ragged appearance of the cuticle, while in other varieties of cutaneous eruptions, the itching is in proportion to the intensity of the cutaneous inflammation. In addition to this, the parasitic animal, when discoverable, constitutes positive diagnostic symptoms.

Treatment.—The indications of treatment in the animalcular eruptions, are to restore the skin to a cleanly state, by proper ablutions, and then if the parasitic animal is not routed, or destroyed, to apply to the affected portions, or if necessary to the whole surface of the body, a poison to the parasite, which is at the same time harmless and soothing to the skin of the patient.

In all cases of animalcular eruptions the patient should be directed to wash the whole surface of the body daily, with strong soap-suds, to wear clean and proper clothing, and to observe cleanly habits, in every respect. This alone may destroy the parasitic animal in a few days, in many cases. If not, and the eruption is of limited extent, the mercurial ointment, or an ointment made by mixing two drachms of iodide of potassium with an ounce of lard, may be applied to the diseased surface, after each daily ablution, and continued till the parasitic intruders are destroyed, and a healthy state of the skin restored.

In cases, however, in which the eruption is very general, the animals having multiplied, and colonized more or less extensively, nearly every portion of the cutaneous surface, as is apt to be the case with the *acarus scabiei*, or *sarcoptes hominis*, either the sulphur vapor baths should be resorted to for a few evenings, or the sulphur ointment should be applied.

If the sulphur vapor baths are at hand, one may be taken after each daily ablution with soap and water, till a cure is effected. If not, the sulphur ointment, made by mixing one part of sulphur with two of lard, may be applied to the whole surface of the body, each evening, after the washing with soap and water, clean clothes and bed clothes being supplied each day, and this should be continued till the cure is effected.

SECTION X.—CRYPTOGAMOUS ERUPTIONS.

By cryptogamous eruptions, from κρυπτος, "concealed," and γαμος, "marriage,"[*] I mean here that variety of cutaneous eruptions in which the elevations on the skin consist of minute cryptogamous parasitic vegetables, the result of the germination of sporules lodged about the roots

---

[*] Plants whose stamens and pistils are concealed or not manifest, their sex being unknown.

of the hairs, in abrasions of the cuticle, or else in filth, constituting a soil upon its surface.

The cryptogamous eruptions may occupy any part of the cutaneous surface, but their most frequent seat is in the beardy portions of the face, and upon the hairy scalp, probably in consequence of the facilities offered by the hairs for the lodgment of the sporules, from the germination of which, the plants constituting the eruption are produced.

The cryptogamous eruptions might be described under the heads of *sycosis, favus, trichosis*, &c., but as these terms would serve rather to call off the mind from the real pathological condition to accidental appearances I prefer to dispense with them altogether, and to consider this form of disease as one, under the head of *cryptogamous eruptions*, the symptoms of which we will now proceed to consider.

Symptoms.—The cryptogamous eruptions, as we have already seen, may make their appearance on any portion of the cutaneous surface, but generally either upon the beardy portions of the face or the hairy scalp.

There is generally at first an itching sensation experienced in the part, and soon slight yellow elevations, commencing usually about the roots or the hairs, attended with heat, pain and redness. These elevations, whether on the face or scalp, may be isolated, or in clusters, and they may occur over quite an extent of surface at first, or they may commence at a single point, and extend in different directions, according as one or more sporules, seeds, or germinating principles have been lodged, from which the crop of parasitic vegetables are to be produced. If the disease occupies the face, the germination of the sporules and growth of the vegetables which they produce, together with the accidental irritation to which the face is liable, are apt to cause considerable inflammation of the skin. In such cases, the elevations increase in part from the growth of the parasite, and in part from the inflammation of the skin which they produce. And if the cutaneous inflammation is intense, the elevations instead of spreading out in crusts with pits or depressions, as is common in this disease, may contain a little matter, burst in five or six days, and then form crusts, between which small pustules and tubercles may appear.

More generally, however, if the disease occupies the scalp, or even the face, if the inflammation is slight, the elevations consisting of the germinating vegetable parasite and epidermic scales, and having the appearance at first of yellow specks about the roots of the hairs, gradually spring up and spread out in the form of crusts, with small pits or depressions in the centre. In such cases, the inflammation of the derma is less, and consequently no pustules or tubercles appear between the crusts.

The sporules as we have seen generally lodge about the hairs, or in the hair follicles, the vegetables which they produce usually springing from that source. This, however, is not the only seat of the parasites, for a slight abrasion of the cuticle, or even a layer of dirt upon its surface, may afford ground for the lodging, germination, and springing forth into a vegetable of the sporule which may chance to float that way.

These vegetable crusts may be isolated at first, and they may occupy but a small extent of surface of the face or scalp, or of other parts where they may appear. Gradually, however, the eruption extends, and as the parasitic vegetable crusts increase in thickness and circumference, they may meet, forming finally a complete crust over the affected part, with pits or depressions, which has been thought to give the eruption the appearance of a honeycomb.

If, however, the disease is permitted to continue, as it is sometimes for months, or even years, the surface of the crusts gradually crumble off presenting an irregular appearance. Or portions of the crusts may desquamate, or be torn off, to give place to others of a similar character, which spring up and shoot forth with great luxuriance.

When the cryptogamous eruptions appear upon the face, then constituting what has been called *sycosis*, the inflammation of the derma is apt to be very considerable, causing small vesicles and tubercles to appear between the parasitic vegetable scales. The disease in such cases, if neglected becomes exceedingly troublesome, involving not unfrequently the whole bearded portion of the face.

In cases, however, in which the cryptogamous eruptions appear upon the hairy scalp, constituting what has been called *favus*, the parts are less liable to be irritated, and consequently the derma is not usually so intensely inflamed. As a consequence of this, the sporules germinate shoot forth, and in many cases form crusts which cover the whole head, without the intervention of pustules, or tubercles which the intensity of the inflammation when the face is its seat, is so liable to produce.

The cryptogamous eruptions, if long continued, are very apt to be attended with a loss or falling of the hair, and especially is this the case in those cases which have been called *trichosis*, in which parasitic cryptogamous growths appear, when examined microscopically, in the interior of the hairs on the diseased surface.

Such I believe are the ordinary symptoms of the cryptogamous eruptions, with their principal variations, depending upon the seat and duration of the disease, and other accidental circumstances.

Diagnosis.—The cryptogamous eruptions may be distinguished from all others, by their preference for the bearded portions of the face and hairy scalp, by the peculiar yellow appearance of the elevations at first, being developed into crusts with pits or depressions as the disease progresses, and rather disposed to crumble than desquamate, and finally by the appearance, when examined by the microscope, of some variety of cryptogamous parasitic plants.

In those cases of this disease, in which the intensity of the cutaneous inflammation causes the appearance of the eruption to be rendered irregular, by the development of pustules and tubercles between the irregular crusts, the microscopic plants, when discoverable, renders the diagnosis clear, notwithstanding the irregularity.

Causes.—The direct cause of the cryptogamous eruptions is the lodgment, either about the hairs, in fissures of the cuticle, or else in a layer of dirt upon its surface, of the sporules of some form of cryptogamous parasitic vegetables, and their germination and development into a mature plant. It is probable, however, that various causes predispose the system to this form of disease, among which are all the debilitating agents and influences, by lessening the powers of vital resistance, and also filth or dirt upon the skin, which not only affords facilities for the lodgment of the sporules, but also constitutes a soil in which they may conveniently germinate, and be developed into mature plants.

A few of the parasitic plants which produce the cryptogamous eruptions have been examined, described, and named. But it is sufficient for our purpose here to remember that the crust, in this form of eruption, consists of a cup-shaped capsule, having a cavity within; the walls consisting of epidermic scales, while the internal part consists of the parasitic plant, with its sporules and other parts variously developed.

It is probable that the sporules of the cryptogamous plants, may be taken from the air, in which they doubtless float, as well as from the bodies of those upon whom the plants which produce them are flourishing.

Nature.—The cryptogamous eruptions, are evidently the result of the germination, and growth of the different varieties of the parasitic cryptogamous plants, the sporules of which have been taken from the air, in which they float, or else from the bodies of persons on which the plants grow, by combs, hats, clothes, &c.

The sporules, as we have seen, more generally lodge about the roots of the hairs, in the follicles of which they often germinate, and take root. Sometimes, however, they lodge in abrasions of the cuticle, or in a layer of dirt upon its surface.

The elevations in this disease, are the parasitic plants, together with portions of the cuticle, and in cases in which the derma, and tissues beneath are highly inflamed, as we often see, when the face is the seat of the disease, the elevations often assume a pustular appearance, before the crusts form, and interspersed with the crusts, pustules and tubercles often appear, as might be expected.

I believe, that in a perfect state of health, if cleanliness be observed, the liability of contracting this disease is very slight, as the powers of vital resistance are such, that if the sporules are brought in contact with the skin, they will seldom germinate, and produce the parasitic plant.

On the other hand, in persons of feeble or depraved constitutions, if the skin is covered with dirt, as is too often the case, the predisposition is very strong, as the sporules not only readily find a lodgment; but they find a soil in which they may germinate, and thrive with great luxuriance. Hence it is that the cryptogamous eruptions most frequently appear on persons of feeble or depraved constitutions, and especially on those of imprudent filthy habits.

Treatment.—The indications in the treatment of the cryptogamous eruptions, are to correct the habits, and general condition of the patient; to apply to the diseased surface, a remedy which shall remove the crusts, and then something to the raw surface, which shall destroy the roots of the parasites, and at the same time, favor the formation of a healthy cuticle.

The patient should be required to take proper food, with strict regularity; to wear suitable and clean clothing; to take sufficient exercise; to sleep in dry apartments; to wash the whole surface of the body once each day, with soap and water, and to observe the laws of health, and rules of propriety in every respect.

When these preliminaries have been arranged, the local treatment should be commenced at once, and if the patient is scrofulous and anæmic the iodide of potassium or syrup of the iodide of iron should be given in moderate doses, that the general condition may be corrected, as well as the local disease removed.

Now to remove fungoid elevations or crusts, various remedies may be resorted to; but that which I have found the most speedy and safe, is glycerin containing in solution two drachms of carbonate of potassa to the ounce. The affected surface should be washed morning and evening with soap and water, and immediately after the washing, the crusts should be moistened with the glycerin and the potassa; and this should be continued till the crusts are removed, the head, if the disease be of the scalp, being covered with an oil-silk cap. I have succeeded in removing the crusts in three or four days in this way, the hair, if the disease occupies the scalp, sometimes coming off with it, if it had not previously fallen, and a portion of the beard, if the disease is of the face, leaving in either case a raw surface.

Now having thus removed the crusts, to destroy the roots of the cryptogamous parasitic plants, and to favor the formation of a healthy cuticle, it is only necessary to keep the raw surface covered with cloths wet in a decoction of the *phytolacca decandra* or poke root; made from two ounces of the root to the pint of water. Or if the disease occupies the face, or any portion of the surface away from the scalp, where the constant application of the decoction is inconvenient, an ointment made by mixing two drachms of the pulverized *poke root* with an ounce of lard, may be applied to the raw surface, instead, three times per day, and continued till a healthy cuticle is formed and the disease permanently cured. I have succeeded thus in permanently curing cases of the cryptogamous eruptions on the scalp, face and other portions of the surface in from two or three weeks; and I believe it need seldom take a longer time than that.

When the treatment is commenced, combs, hats, brushes, cloths, &c., should be changed, and when a cure is effected, these in turn should be laid aside and an entire new set substituted, that a return of the disease may be prevented.

Thus have I completed what I had to say on the simple *rashes* and *eruptions*, including among the rashes, *red rash*, *rose rash* and *nettle rash;* and among the eruptions, *papular eruptions, vesicular eruptions, pustular eruptions, scaly eruptions, animalcular eruptions*, and the *cryptogamous.* By thus dispensing with all terms except those which point directly to the pathological condition, and treating each case on strictly common sense principles, the whole subject is rendered clear and plain, and a cure may in most cases be effected.

40

CHAPTER XIV.

DISEASES OF THE URINARY ORGANS.

SECTION I.—NEPHRITIS.

By nephritis, from νεφρος, "kidney," and the termination *itis*, denoting inflammation, I mean here inflammation of the kidneys, of either an acute or chronic form.

But as the present chapter is to be devoted to diseases of the urinary organs, it is proper that we should take a general glance at the anatomy and physiology of these organs, before proceeding to the consideration of *nephritis*, the legitimate subject of the present section. I shall, however, only consider the kidneys, ureters, bladder and urethra here, leaving the other appendages for consideration in the following chapter, in which I shall take up diseases of the *genital organs*.

The kidneys, or renal glands, are situated deep in the lumbar region, opposite the last dorsal and first lumbar vertebræ. They are composed of arteries, veins, nerves, an external cellular covering and the parenchyma. The renal artery is a branch of the aorta, while its veins terminate in the vena cava, and its nerves are derived from the renal plexus.

The cellular envelop covers the surface of the kidney, penetrating into its fissure, while the parenchyma of the organ is composed of an external *cortical*, and an internal *tubular* portion.

The *cortical* substance is a purely glandular structure, about two lines in thickness, of a reddish brown color, and composed of blood-vessels, and convolutions of the uriniferous tubuli. It constitutes the surface of the kidneys, and dips between the cones, surrounding them nearly to their apices.

The diameter of the tubuli uriniferi in the cortical portion of the kidney, is about $\frac{1}{48}$th of an inch, and the origin of each uriniferous tubule is coiled, being surrounded or associated with a plexus of capillary vessels in such a manner as to form small granulations about the $\frac{1}{100}$th of an inch in diameter, which have been called the malpighian bodies.

The *tubular* portion of the organ is of a pale red color, and is composed mainly of the converging tubuli uriniferi, which, as we have seen, commence in the malpighian bodies, in the cortical portion of the kidney.

The tubuli uriniferi, as they converge from the cortical portion of the organ to pass to the pelvis, are separated only by minute blood-vessels, and a small quantity of parenchymatous substance, and constitute from twelve to eighteen conical fasciculi, enveloped by the cortical substance which dips between them, except at their summit. These cones are invested by mucous membrane, continuous at their apices with the uriniferous tubuli, and so reflected as to form around each, a cup-like pouch or calyx. These calices communicate with the *infundibula*, or three larger cavities, situated at each extremity, and in the middle of the organ, constituting by their union a large membranous sac, the *pelvis* of the kidney.

The *ureter*, the excretory duct of the kidney, is a membranous tube about eighteen inches in length, and nearly the size of a goose-quill, being continuous superiorly with the pelvis of the kidney, the inferior extremity entering the base of the bladder obliquely between its muscular and mucous coats. The calices, infundibuli, and pelvis of the kidney, as well as the ureter, are composed of an external fibrous, and an internal mucous coat, the latter of which is continuous superiorly with that of the tubuli uriniferi, and inferiorly with the mucous membrane of the bladder.

The bladder, it will be remembered, is an ovoid viscus, of various dimensions, situated behind the pubis, and in front of the rectum, consisting of a *body, fundus, base* and *neck*, and being retained in its place by numerous ligaments. It is composed of three coats, an external *serous*, which invests the posterior surface and sides from opposite the entrance of the *ureters* to its summit, from whence it glides to the anterior wall of the abdomen, a middle, *muscular* coat, consisting of two layers, and an internal *mucous* coat, continuous with that of the ureters and *urethra*.

The *urethra* is a membranous canal, extending from the neck of the bladder to the meatus urinarius, being composed of two layers, an internal mucous, and an external elastic fibrous coat. The mucous coat is continuous with the mucous membrane of the bladder, and along its course with the lining membrane of the ducts of Cowper's glands, the posterior gland, the vasa deferentia, and vesiculæ seminales.

Now, to understand the functions of the urinary organs, it is only necessary to remember the minute structure of the kidneys, and that various useless materials, absorbed along the intestines, as well as products of waste, arising in the system, are here to be separated from the blood, and by the urinary organs, to be removed from the system.

It is probable that the separation of the water of the urine takes place in the malpighian bodies, from the arterial blood of the capillaries which they contain, while the secretion of the solid portions of the urine takes place, as it appears, from the plexus of veins, which ramify on the walls of the uriniferous tubes, and receive the blood from the capillaries of the malpighian bodies.

Now the urine thus separated or secreted in the malpighian bodies, and in the cells which cover the uriniferous tubes, from the venous plexus which surround them, passes along the tubuli uriniferi, constituting the cortical fosciculis to the calices, infundibula, and pelvis of the kidneys, and from thence along the ureters to the bladder, which serves as a reservoir, from which the urine is finally discharged from the body at intervals through the urethra.

Such then are the urinary organs and their functions. And when we remember that the average quantity of urine passed by an adult in twenty-four hours, is nearly two and a half pounds, and that in every 1000 parts of this, from twenty to seventy parts are solid matters, consisting of urea, uric acid, lactic acid, lactate of ammonia, and extractive, mucus, sulphate of potash, sulphate of soda, phosphate of soda, bi-phosphate of ammonia, chloride of sodium, muriate of ammonia, phosphate of lime and magnesia, and silica, the importance of the urinary system can hardly be over-rated.

Having thus taken a glance at the urinary organs, we are prepared to appreciate the symptoms which are developed in the various diseases which affect their different parts. And as *nephritis*, or inflammation of the kidneys, is the legitimate subject of the present section, we will proceed to its consideration, and first of its symptoms.

Symptoms.—The symptoms of nephritis are just what might be expected. A pain commences deep in the lumbar region, attended with throbbing, and very soon with tenderness to firm external pressure.

In *acute* cases, as the disease passes on, there is nausea and perhaps vomiting, and sooner or later chills alternating with flushes of heat, followed by febrile reaction, attended with severe pain, great tenderness, distressing nausea, and frequent vomiting, especially on making the least motion affecting the inflamed organ. The pain darts down the ureters, the testicle of the affected side is retracted, and numbness is experienced in the thigh of the affected side.

The bowels are generally constipated and painful, the urine becomes scanty, high colored, and perhaps tinged with blood, the desire to micturate is frequent and urgent, and in most cases, attended with severe pain, and yet the secretion of urine is sometimes almost entirely suppressed.

If one kidney only is involved, the patient will generally be found with the body bent forwards, and a little to the affected side, in order to take off the pressure of the lumbar muscles. If, however, both kidneys are inflamed, the patient will generally be found in a sitting posture, with the body bent forward, or if in bed, with the body bent in the same manner, and generally quite unwilling to move.

The pulse in nephritis is according to my observation, generally full, hard, and slow. Sometimes, however, in the latter stages, it becomes small and frequent. The skin is warm, dry, and parched, and has a peculiar rough feeling, probably in part from the irritation which retained urine produces in the capillaries of the skin. If suppuration takes place, the pus may be discharged with the urine, or it may escape into the colon, or into the abdominal cavity in the latter case, a fatal termination will be the result.

Such I believe are the ordinary symptoms of acute nephritis. But the disease may pass on and become chronic, or it may be so from the first. In either case, there is little or no fever, the pain is dull, and confined to the inflamed organ, and the remaining symptoms, though similar in kind, are of less intensity, and the disease may continue for a long time if neglected, the urine becoming alkaline, and the disease being finally attended with night-sweats, hectic, emaciation, &c.

When nephritis is about terminating favorably, by resolution, the fever and pain subside—the skin becomes moist, the urine free, and the nausea and vomiting cease. If, however, suppuration takes place, rigors supervene—the pulse becomes weak, night sweats occur, and the urine first becomes turbid, and then clearly purulent.

Such, then, are the ordinary symptoms of nephritis, as it occurs in the acute and chronic form.

Diagnosis.—Nephritis is liable to be confounded with nephralgia, inflammation of the psoas muscle, and lumbago; from all of which it may be distinguished, however, by attention to the following differences.

From nephralgia, it differs in being attended with fever, while in the nervous affection the pain is generally more severe, and of a spasmodic character.

From inflammation of the psoas muscle, it may be distinguished by the relief which is produced by bending the body forwards, while in that disease the pain is increased by that position; and, besides, in nephritis there is the fever, changed appearance of the urine, &c., which do not necessarily attend the psoas disease.

From lumbago, nephritis differs in the deep and throbbing character of the pain, as well as in the derangement in the urinary secretion, while in lumbago the pain is more superficial, and the urinary secretion may remain quite natural.

Causes.—The causes of nephritis are various, among which cold, suddenly applied, is probably the most frequent, especially as the kidneys sympathize strongly with the skin. Nephritis may also arise from stimulating diuretics, as well as from diseases of other portions of the urinary organs.

Renal calculi sometimes produce this disease, by the mechanical irritation which they produce, a remarkable case of which fell under my observation in this village, about two years since.

Metastasis of gout and rheumatism may produce nephritis of the most violent character, a severe case of which fell under my care a few years since, in a gentleman about sixty years of age.

Morbid Appearances.—In cases in which death has taken place early, the kidney is found enlarged and congested, and its cortical portion of a reddish-brown color. The tissue of the gland is unusually softened, and the mucous membrane of the pelvis is generally thickened and reddened.

If death has taken place at a later stage of the disease, pus may be found in the glandular structure, either disseminated or in small cavities. Or the pelvis of the kidney may be found filled with pus.

In *chronic* cases the kidney may be found contracted, and perhaps indurated, its surface presenting an irregular rough appearance.

In a case, however, that I examined a few years since of long standing, which had been attended for years, with purulent discharge with the urine; the glands involved was enormously enlarged, its internal structure destroyed, and besides pus and a small quantity of blood, it contained a calculi of large size, which had probably produced the disease.

Treatment.—The treatment, if applied early, or during the forming stage of renal inflammation, may be effectual in at once arresting the disease. At such a stage, the abstraction of a few ounces of blood from over the kidneys, by cups; a mild cathartic, and the warm foot-bath, with low diet, and mucilaginous drinks, may be sufficient to arrest the disease.

But if the disease is acute, or has been neglected, or is of an aggravated character; general bleeding may be indicated. After the general bleeding, in cases in which it is indicated, and at first when not, from four to six ounces of blood should be taken by cups, from over the region of the kidneys, on each side of the spine, and this may be repeated in the course of twenty-four hours, if the pain, tenderness, and vomiting continue unabated.

After the cupping, hops wet in warm vinegar, should be laid over the lumbar region, and continued during the continuance of the inflammation, for the purpose of promoting perspiration, and also for the anodyne influence, which the hops produce.

Immediately after the bleeding, if indicated, and cupping, a cathartic of calomel should be administered, in castor oil, and its operation secured, if necessary, by mucilaginous injections, as much uneasiness is usually produced by the operation of a cathartic. After the first cathartic, the bowels should be kept moderately loose, by injections of flax-seed tea, with a little castor oil, rather than by administering cathartics.

After the operation of the cathartic, in violent cases, two grains of calomel with four grains of Dover's powder, may be given every four hours, and continued till a slight ptyalism is produced, if the inflammation is not sooner subdued; when the calomel should be omitted, and the Dover's powder continued, every six hours, till the fever subsides.

After the fever subsides, and in all cases in which there is no fever; fifteen drops of the fluid extract of hyoscyamus, may be given as an anodyne, every six hours, with thirty drops of the sweet spirit of nitre, if this is from no cause contra-indicated.

Mucilages should be freely allowed, during the whole course of the disease. And in rheumatic or gouty cases, nitre, colchicum, or iodide of potassium should be given, in addition to what I have already suggested. Crust coffee should be allowed at first, with a little milk, and later, arrow-root, tapioca, rice, and finally toast. In very protracted chronic cases, a drachm of the fluid extract of *uva ursi*, or two ounces of the infusion, may be given, four times per day, and if a tonic is required, ten drops of the tincture of chloride of iron, may be given after each meal, as long as may be required.

SECTION II.—ALBUMINURIA.

By albuminuria, I mean here that variety of organic renal disease in which, in addition to a general depraved condition of the system, there is the presence of albumen in the urine, and in many cases dropsy and various other complications.

The constitutional depravity, I believe, is nearly the same in kind, only differing in degree, in all cases of this disease; while the renal degeneration differs not only in degree, but in kind, the kidneys being in some cases *inflamed*, in others *tuberculous*, in others *granular*, in others *atrophied*, and in others still, having undergone the *fatty*, *waxy*, or *fibroid degeneration*. In fact, I have seen cases in which several of these conditions existed at the same time, one portion of the kidneys being granular, another tuberculous, and another portion still being atrophied, or otherwise changed.

I believe that in every form of this disease there is the presence of more or less albumen in the urine, which fact is a sufficient reason for calling it *albuminuria*, a term which the most invariable symptoms suggests, and which must suffice for want of a better or more indicative term.

Symptoms.—In some cases of this disease the renal affection appears to be developed first, the constitutional depravity following, in part, as a consequence.

I believe, however, from careful observation, that the constitutional depravity generally precedes the organic renal affection, or that a strong constitutional predisposition to this form of disease exists, at least. In such cases, however, the renal degeneration, when it progresses sufficiently to allow of a free discharge of albumen, at the same time causing a retention in the blood of effete matters, greatly aggravates the general depravity, and hastens the disease on to a fatal termination.

When the renal disease commences, whether primary or secondary, there is generally a chill, more or less severe, attended with pain in the head, back, and lower extremities, followed by febrile reaction, with a hot and dry skin, a full, and perhaps frequent pulse, and in many cases with numbness of the lower limbs and retractions of one or both testicles.

In the more *chronic* form of the disease, however, the renal affection may be developed without a sensible chill, or any very marked constitutional disturbance. In such cases, however, as well as in those in which the febrile symptoms are marked, the urine becomes gradually scanty, and is passed with difficulty, the rapidity of this change being generally in proportion to the activity of the disease. As the renal disease progresses, a dull pain is experienced in the region of the kidneys; and if the patient has had scrofulous sores, or glandular swellings about the neck, they are apt to disappear; and if there has been a troublesome cough, it may subside.

By degrees the urine becomes scanty, of a low specific gravity, and highly albuminous; the complexion becomes anæmic and sallow, emaciation progresses, the muscles become weak, the appetite is defective, digestion, sanguification, and nutrition become impaired, and constipation attends, alternating with diarrhœa, &c.; and finally, along with drowsiness, languor, and stupidity of the most intolerable character, there is apt to be the supervention of œdema and anasarca, with perhaps hydrocephalus, hydrothorax, hydropericardium, &c.

After this train of symptoms has continued for an indefinite time, varying from a few weeks to several months, or even years, the patient becomes emaciated, stupid, and generally dropsical, and is relieved by death from one of the most intolerable conditions which it has been my misfortune to witness. In some cases, however, the patient dies in a comparatively early stage of the disease, in which case many of the unpleasant symptoms enumerated above are not experienced, or what is better still, if the constitutional depravity is not too great, and the case is properly treated in season, a recovery may sometimes take place, but I believe that such an event is only an exception to a rule.

Such, I believe, are the ordinary symptoms of what I have called *albuminuria*, but which has sometimes been known as *granular, Bright's disease*, &c.

Anatomical Characters.—On post-mortem examination the kidneys present a variety of appearances, being either enlarged or contracted among the most frequent of which are signs of *congestion* and *inflammation*, especially of the cortical portion, small *granulations* and *tubercles* in various stages of advancement, an *atrophied* condition in which the lining of the secretory tubules is lost, and the organ greatly shrunken, *fatty degeneration* in which there is an excess of oil in the epithelial cells of the tubules, the *waxy degeneration* in which with great enlargement there is a deposit of waxy-looking matter in the tubules, and elsewhere, and finally a *filroid degeneration*, analogous to cirrhosis in the liver and lungs, an interesting case of which fell under my observation a few years since in a patient about sixty years of age.

Diagnosis.—Cases of albuminuria may generally be distinguished by the albuminous urine, the cachetic or depraved condition of the system, the scantiness and low specific gravity of the urine, the dropsical tendency, and finally by the dull pain in the back, and at last the drowsiness, stupidity &c. which supervene.

To distinguish the different varieties of this disease, it is well to remember that in *inflammatory* cases the urine is very scanty, often bloody, and contains a large amount of albumen. In the *granular* and *tuberculous* cases, the urine is still highly charged with albumen, but I believe it is generally a little more free. In the *atrophic* cases the urine may be quite free, is of low specific gravity, but moderately albuminous, and the dropsical tendency is not very strong. In cases of *fatty degeneration*, the system is cachetic from the first, dropsy generally attends, and the urine is always scanty, of low specific gravity, and albuminous. In cases of *waxy degeneration*, the tendency to dropsy is slight, the disease is chronic, and the urine is highly albuminous, and may contain waxy casts. Finally, in the *filroid degeneration*, there is the same train of symptoms as in the waxy, except that the urine may contain fibrinous instead of waxy casts.

In many patients, however, several of these conditions exist, even in the same kidney, in which case it is impossible to do more than to form a general diagnosis till an examination is made *post mortem*.

Causes.—The predisposition to albuminuria doubtless consists in an inherited or acquired scrofulous or otherwise depraved condition of the system, while the exciting causes are as numerous as can well be imagined, embracing almost every variety of imprudence to which the human family are addicted. It is probable, however, that the most frequent exciting causes of this disease are nephritis, scarlatina, sexual excesses, a syphilitic taint, the use of tobacco, filth, improper food and clothing, and protracted drunkenness.

Nature.—Now, if we bear in mind the depraved condition of the system which attends, and probably generally precedes the organic renal affection; and the local and constitutional derangements which supervene from a derangement or suspension of the renal function, together with the causes which operate to produce the disease, its nature becomes comparatively plain.

For in cases in which the renal affection is secondary, the constitutional depravity favors the renal degeneration, whether it be *granular*, *tuberculous, atrophic*, or of the *fatty waxy* or *fibroid* character. And no sooner has one or several of these organic conditions become established, with or without the supervention of renal inflammation, than this condition, by draining off the albumen from the blood, and causing to be retained in the blood various effete matters, not only augments the general depravity of the system, but also leads on to dropsical and other complications which follow towards the fatal termination of this disease.

That the general condition of the system may be very similar in all cases of this disease, is rendered probable by the fact that several varieties of the renal degeneration enumerated above, sometimes exist in the same case, an interesting specimen of which fell under my observation in this village a few years since. It is probable then that the kind of renal degeneration in this disease, may depend very much upon accidental causes, operating directly upon the kidneys, modified of course by the habits of the patient, and perhaps by shades of variation in the general depravity of the system, which invariably attends in this disease.

Prognosis.—In all cases of this disease, in which the general depravity of the system is considerable, and the organic renal affection, of whatever form it may be, is considerably advanced, the prognosis is decidedly unfavorable, as a fatal termination will generally be the result. In cases, however, in which the general derangement is slight, if the quantity of albumen detected in the urine by heat or nitric acid is moderate, and the renal disease is in its incipient stage, and consists merely of congestion, inflammation, or perhaps fatty degeneration, a favorable termination of the case may be anticipated, if the cause can be removed, and the patient be subjected to proper treatment.

Treatment.—The indications in the treatment of this disease are to correct the habits and general condition of the patient, to subdue fever and renal inflammation if they arise, and finally to remove dropsical accumulations when they occur, as they are very liable to in this disease.

The patient should be made to abandon all impure, or intemperate habits, and should be directed to take proper food, with strict regularity, should keep the skin clean, and wear clean and suitable clothing, should sleep in dry apartments, and if possible take moderate exercise in the open air, and finally should be prudent and temperate in all things.

If the bowels are confined, a teaspoonful of cream of tartar may be given morning and evening, and if the patient is decidedly anæmic, ten drops of the tincture of chloride of iron may be given three times per day, after each meal, and continued for a reasonable time.

Should fever occur, as it is apt to in acute cases, half an ounce of the bitartrate of potassa may be given, if necessary, morning and evening, the dose being gradually diminished to a teaspoonful, as the fever subsides. And during the febrile excitement, five grains of Dover's powder may be given every six hours, to quiet the patient and promote perspiration.

If pain, with symptoms of renal inflammation occurs, whether at the commencement, or at any subsequent stage of the disease, cups should be applied on each side of the spine in the lumbar region, and a few ounces of blood taken. This may be repeated, if necessary, and when no longer admissible, dry cupping may be resorted to, or pustulation may be produced by tartar emetic, either in the form of an ointment, or else sprinkled on a plaster to be worn over the part, of either conium, belladonna, or stramonium.

Finally, if the disease passes on, and there is evidence of the fatty, or other forms of renal degeneration, with dropsical accumulations, drowsiness, &c., in addition to what I have already suggested, five grains of the iodide of potassium should be given three times per day, with ten drops of the fluid extract, or half an ounce of the infusion of digitalis, and this should be continued till the dropsical and other urgent symptoms are relieved, or a fatal termination occurs. The digitalis may not be tolerated, however, for a great length of time, but the iodide may be given, in such cases, as long as required.

SECTION III.—NEPHRALGIA.

By nephralgia from νεφρος, "a kidney," and αγλος, "pain," I mean here neuralgia of the kidney, unattended with inflammation or febrile excitement.

In order to appreciate the symptoms which are developed in this affection, it is necessary to remember the situation of the kidneys, opposite the last dorsal and first lumbar vertebræ, and also their intimate sympathetic relations with the digestive organs, and other portions of the system.

Symptoms.—Nephralgia commences often very suddenly in one or both kidneys, the pain being at first confined to the region of the gland, but extending in most cases along the ureters, as well as into the back, hip, and thighs. The pain is of an acute, darting, or lancinating character, and may be continuous, but it is oftener of an intermittent, or remittent character, at least it has been so in cases that have fallen under my observation.

During the severe paroxysms of pain, the testicles are apt to be retracted, the stomach strongly symphathizes, flatulence and colic attend, and the urine may be either scanty or very copious, and it in some cases contains a sandy deposit of an acid or alkaline character.

Diagnosis.—Nephralgia may be distinguished from nephritis, by the suddenness and violence of the attack, the spasmodic and intermittent character of the pain, the absence of tenderness and febrile excitement, and finally by the sudden suspension of the pain, the patient appearing very soon quite well.

Causes.—Various causes may operate to produce nephralgia, such as produce neuralgia of other parts. I believe, however, that spinal irritation, the presence of sandy deposits in the urine, and in some cases the passage of calculi along the ureters, with perhaps a rheumatic or gouty condition, most frequently act as causes of this disease.

Treatment.—When an attack of nephralgia occurs, the cause direct and remote, should be ascertained and removed if possible. To relieve the pain, which is often intolerable, cups should be applied on each side of the spine, in the lumbar region. The cupping may be repeated if necessary, and a fomentation of hops wet in warm vinegar kept applied, over the lower portion of the abdomen, and across the lumbar region.

Internally, to quiet pain, fifteen or twenty drops of the fluid extract of hyoscyamus should be given, and repeated if necessary, every four or six hours. Mucilage of gum arabic should be freely allowed, and if the bowels are constipated, a cathartic of calomel or podophyllin should be administered in castor oil, and in rheumatic cases, iodide of potassium or colchicum may be required.

SECTION IV.—ACUTE CYSTITIS.

By acute cystitis, is here meant acute inflammation of the bladder, whether one or more of its coats be involved in the inflammation.

The bladder, it will be remembered, is situated in the anterior part of the cavity of the pelvis, back of the pubis, in front of the rectum in the male, and of the uterus in the female, and is of various dimensions. It is composed of three coats, an internal mucous, a middle muscular, and an external serous, derived from the peritoneum, which covers the posterior lateral and superior regions of the bladder, adhering to that portion of its muscular coat.

The arteries supplying the bladder, come from the hypogastric, and its veins go to the hypogastric veins, and it is supplied with nerves from the sciatic and hypogastric plexuses.

The bladder thus situated and constituted, is liable to become inflamed, and when it does, the symptoms which are developed, are such as might be expected, from its combination of serous, muscular, and mucous tissues.

Symptoms.—The first symptoms of acute cystitis, is generally pain in the region of the bladder, its character depending upon the seat of the inflammation.

If the mucous membrane is the principal seat of the inflammation, the pain is of a burning character; if the muscular coat, it is of a throbbing aching character; but if it is confined mainly to the peritoneal coat, the pain is of a sharp lancinating character. If, however, as generally happens, the mucous, muscular, and peritoneal coats all become involved, there is a combination of the different varieties of pain.

In some cases before, and in others, after the commencement of the pain, there is a chill, followed by febrile excitement; this being severe or slight, according to the intensity of the disease. There is frequent and distressing micturition, and the pain generally extends to the testicles and upper part of the thighs, and is attended with a sense of constriction in the hypogastric region.

The pain is increased, by pressure made above the pubis, and more or less tenderness also exists in the perineum. The pulse is full, hard, and frequent; the skin hot and dry; the thirst urgent, and the patient more or less restless and dejected.

The bowels are constipated, and if the posterior portion of the bladder is involved, there is apt to be troublesome tenesmus. If that portion is inflamed which admits the ureters, these tubes may become inflamed, obstructing the passage of the urine to the bladder, and leading to severe pain and tenderness in the hypogastric region. When the neck of the bladder becomes seriously involved, there is sometimes total retention of urine, or the patient is tormented with constant and distressing sensations of **strangury.**

The inflammation in acute cystitis, terminates either in resolution, suppuration, gangrene, or induration, and thickening of the coats of the bladder.

When the inflammation is terminating by resolution, the pain subsides, the fever abates, there is a general and uniform diaphorisis, and the urine becomes copious, and is passed with little or no pain. If, however, there is an abatement of the fever, and subsidence of the pain, accompanied with chills or rigors, and an appearance of white or yellowish matter in the urine, suppuration has taken place. But if gangrene occurs, with the cessation of pain, there is a profuse clammy perspiration, the extremities become cold, the strength fails, the countenance becomes cadaverous, the mind becomes confused, and with a small, weak, and frequent pulse, with perhaps hiccough, the disease passes on to a fatal termination.

Dissection in fatal cases, discovers an injected appearance, with irregular patches, perhaps softened or partially disorganized. In some cases the signs of inflammation, are confined mainly to the mucous coat, in others, to the muscular, and in others still, to the peritoneal; but generally in cases that have been acute, all the coats show signs of having been involved.

Ulcers are frequently found, involving the mucous and muscular coat, and pus, either infiltrated, or occupying small cavities in the walls of the bladder, is a very common morbid appearance in this disease. The cellular tissue is also occasionally infiltrated with serum, and lymph or pus; and portions of the bladder, are sometimes found in a gangrenous state.

Causes.—A great variety of causes, may operate to produce acute cystitis; among the most frequent of which are blows, stone in the bladder, severe labors, riding on horseback, irritating injections, cantharides, oil of turpentine, &c., exposure to cold, the translation of gout or rheumatism, extension of inflammation from contiguous structures, and finally, the the retrocession of cutaneous eruptions.

Treatment.—In the treatment of acute cystitis, general bleeding is in many cases indicated, and should not be neglected. Cups too, should be applied over the sacrum, perinæum, and hypogastrium, and more or less blood taken, or leeches may be applied, if at hand. After the cupping or leeching, a poultice of hops, moistened in warm vinegar, should be applied to the perinæum and hypogastrium, and continued, during the continuance of the inflammation.

A cathartic of calomel should be administered in castor oil, and its operation promoted, if necessary, by an injection of flaxseed tea, with half an ounce of castor oil. After the first cathartic, the bowels should be kept regulated by small doses of the sulphate of magnesia; and in acute cases, in which the inflammation is not arrested by the bleeding, cupping, cathartic, &c., two grains of calomel, with four grains of Dover's powder, may be given every four hours, and continued till an impression is produced upon the inflammation, or slight ptyalism is produced. At this stage the calomel should be omitted, and the Dover's powder continued, with three grains of James's powder, while a diaphoretic is required.

After the inflammation and febrile excitement have passed by, if an anodyne is required, **fifteen drops of the** fluid extract of hyoscyamus may be given every six hours, **with mucilages** for drink, only the plainest kind of unstimulating food being allowed. Warm pediluvia should be used during the whole course of the disease; and should the case be of a rheumatic character, colchicum, or iodide of potassium, may may be required to complete a cure.

SECTION V.—CHRONIC CYSTITIS.

By chronic cystitis, I mean here a slow or protracted inflammation of the mucous membrane of the bladder, the muscular and peritoneal coats being but slightly, if at all, involved.

Chronic cystitis is not a very unfrequent affection, sometimes coming on as an original disease, and in other cases following acute cystitis, especially such cases as are confined mainly to the mucous membrane.

Symptoms.—Chronic cystitis may come on very slowly, or the onset may be more abrupt, in either case the symptoms being in many respects similar to those of acute cystitis, but differing in degree. The patient experiences pain, with a sense of heat in the region of the bladder, and a feeling of weight and tenderness in the perinæum.

The desire to urinate is often frequent, and the effort attended with distress, and a spasmodic action of the bladder and urethra. The urine is loaded with tenacious mucous early in this disease, assuming, as the inflammation progresses, a whitish, yellowish, and perhaps bloody appearance, the quantity passed, in some cases, being very considerable.

As the disease progresses, a slow irritative fever may supervene, the pains increase and extend to the neighboring parts; pus, instead of mucus, is discharged with the urine; the strength fails, emaciation progresses, and, unless the disease be arrested, it now passes on rapidly to a fatal termination.

Dissection in fatal cases discovers a softened, ulcerated, and extensively disorganized condition of the mucous membrane, and the muscular coat is often contracted, thickened, and of a firm consistence, as well as the ureters enlarged, and ulcerative perforations are sometimes formed, extending even into the structure of neighboring organs.

Causes.—Among the numerous causes which may produce chronic cystitis, I believe that the most frequent are stone in the bladder, irritating injections, exposure to cold and dampness, excessive venereal indulgence, onanism, the use of alcoholic drinks, sedentary habits, and finally masturbation, the use of tobacco, &c.

Prognosis.—In most cases of chronic cystitis, if the cause can be removed, and the case be subjected to proper treatment in season, a favorable termination may be anticipated. In cases, however, in which the system is depraved and debilitated, and the cause which is operating does not admit of removal, the condition may perhaps be palliated, but a fatal termination will be the final result.

Treatment.—The habits of the patient should be corrected at once, and only plain, digestible food allowed to be taken with strict regularity. The patient should wear proper clothing, and, if possible, take gentle exercise in the open air.

Cups or leeches should be applied at first to the sacral and hypogastric regions, and later, if necessary, pustulation may be produced at either of these points, or on the perineum, by tartar emetic ointment. Mucilages should be allowed, and the bowels should be kept regular by moderate doses of magnesia or castor oil, if required. As a tonic, the tincture of chloride of iron, in ten drop doses, three times per day, will do best; and to act especially upon the diseased membrane, a drachm of the fluid extract, or an ounce of the infusion of *buchu*, may be given four times per day, till a cure is effected.

SECTION VI.—DIABETES MELLITUS.

By diabetes mellitus, I mean here that peculiar disease in which there is a copious secretion of saccharine urine, indigestion, constipation, a dry skin, thirst, a voracious appetite, and progressive emaciation.

This form of disease is apt to come on slowly, and in most cases that have fallen under my observation, there has been a weak or watery state of the blood, an irritable condition of the nervous system, and evidence of more or less congestion of the lower portion of the spinal cord.

The quantity of water discharged in this disease is often very great, amounting, in some cases, to three or four gallons in twenty-four hours. This, however, constitutes the almost entire liquid evacuation in such cases, as there is little or no exhalation from the skin, and the alvine evacuations are almost entirely dry.

The urine has a pale straw color, its smell resembles that of milk, and it has a sweetish taste, from the sugar which it contains, which appears to take the place of urea in healthy urine.

Symptoms.—Diabetes mellitus first becomes apparent, by the frequent calls to urinate, and more or less indigestion, attended with variable appetite, together with acid eructations, occasional nausea, vomiting, &c. As the disease progresses, these symptoms become greatly aggravated, the thirst becomes urgent, the appetite craving, the skin dry; there is an uneasiness in the stomach after eating, the mouth becomes dry, the tongue foul, and sometimes reddish, emaciation proceeds, and, along with great wasting of flesh, there is an increasing disinclination to physical or mental exertion. Along with pain, there is great weakness in the loins; the bowels become obstinately constipated, the orifice of the urethra becomes irritable, there is loss of virility, the extremities become cold, and, along with vertigo, there is headache, and more or less difficulty of breathing.

As the disease draws on towards a fatal termination, the urine becomes scanty, the gums become spongy, the breath fetid, the voice rough and unnatural, emaciation proceeds rapidly, and the patient finally sinks into a state of somnolency or stupor, from which it is often impossible to keep him roused for a moment.

The pulse is but little, if at all, accelerated during the early stage of the disease, being in some cases less frequent than in health; but, during the latter stages, when the emaciation and exhaustion are very great, the pulse becomes weak and quick, or otherwise irregular. The urine is of a high specific gravity, from the saccharine matter which it contains, amounting, in some cases, to two or three ounces to the pint, the amount of urea being diminished as the sugar is increased.

The duration of this disease is exceedingly various, in some cases continuing only a few days or weeks, while in others it continues for months, or even years, before a fatal termination takes place. In some cases, the diabetic symptoms recur in a periodical manner, the exacerbations being gradually more frequent, till all the symptoms become continuous, the final termination of the disease being very generally comatose or apoplectic.

Diagnosis.—Diabetes mellitus may be distinguished from diabetes insipidus, by the greater severity of the symptoms, and by the presence of sugar in the urine. This saccharine matter may be detected by the taste, or by adding yeast to the urine, when, if sugar be present, effervescence will take place, if the mixture be exposed to a temperature of 70° or 80°, the liquor assuming a vinous odor.

Anatomical Characters.—The morbid appearances presented in fatal cases of this disease are of the kidneys, stomach, intestines, mesenteric glands, lungs, and of the nervous centres.

The kidneys are generally found in a relaxed state, and their blood-vessels enlarged, as if having been congested while in a partially paralyzed condition. Both the cortical and tubular portions of the kidneys often show signs of having been thus congested; rendering the gland in some cases much larger than in a healthy state.

The alimentary mucous membrane in most cases shows signs of recent inflammation; and the mesenteric glands are either enlarged, or else present a softish and relaxed appearance. The lungs, brain, and spinal cord often present a relaxed or congested appearance; and the skin is rough; the perspiratory tube appearing very much contracted, or entirely closed.

Causes.—There is probably a hereditary predisposition to this disease. in some individuals, consisting perhaps in a relaxed condition of the tissues of the body, and an irritable state of the nervous system. And I suspect that a similar predisposition may be acquired by various imprudences; such as masturbation, venereal excesses, the use of tobacco, alcoholic liquors, filthy habits, improper food, clothing, &c.

Among the numerous exciting causes of this disease, are exposure to cold and dampness, irregular eating, protracted grief, anger, injuries of the brain or spinal cord, unwholesome food, deficient clothing, bad air, and various others of a kindred character.

Nature.—Now the local points to which the symptoms naturally call our attention, are the stomach, skin, liver, kidneys, and nervous centres; and the causes which operate, as well as the post-mortem appearances, point to deranged digestion, sanguification, and secretion; while others in turn look back to a hereditary or acquired weakness, or depraved state of the blood, and irritable condition of the nervous system.

Now, it appears to me, that with this inherited or acquired deranged condition of the blood and nervous system, together with the general relaxed condition of all the tissues of the body, which prevails in such cases, a tolerable satisfactory explanation may be had of the nature of this disease.

For, in this state, it is only necessary that any one of the exciting causes of this disease, as improper food, irregular eating, &c., should be brought to bear ; in order to impair digestion, derange sanguification and secretion, and finally to lead on to a development of all the symptoms which attend this form of disease.

It is probable, also, that a weak or irritable condition of the brain and spinal cord favors an interruption of the digestive function, as well as an augmentation of the renal secretion, which are the two most prominent features of this disease, except the saccharine urine. The presence of sugar in the blood, instead of urea, and its separation by the kidneys, I believe is owing to a deficiency of nitrogen.

For it will be remembered that the quantity of hydrogen in a given weight of sugar and urea, is precisely the same ; but the urea contains a large proportion of nitrogen, while sugar contains none. Now, from the defect of nitrogen, in the process of sanguification, urea cannot be formed, but sugar is formed instead, with its carbon and oxygen ; and hence its separation by the kidneys with the urine, instead of urea, which appears in the renal secretion in a healthy state.

In relation to this defect of nitrogen, I suspect it may arise in part from a too exclusive vegetable diet, and also in part from defective power in the digestive organs, in consequence of which a due amount of pepsin is not formed, the office of which, in a healthy stomach, may be to nitrogenize the food.

Now, the saccharine matter in the blood, together with the deranged condition of the nervous system, causes an excessive secretion of urine, which draught upon the system, leaves the skin dry, the bowels costive, the thirst urgent, and the appetite voracious. Digestion and sanguification gradually become more imperfect, emaciation progresses, and finally when the powers of the system are exhausted, the renal function is gradually suspended, and the retained urinous matter acting upon the brain, produces drowsiness, coma, and finally death.

Prognosis.—The prognosis in this disease is generally decidedly unfavorable, and yet with proper treatment in season many cases may be palliated, and some even permanently cured.

The favorable symptoms are a diminished secretion of urine, and a reappearance of uric acid instead of sugar in the urine, together with a palliation of all the symptoms which generally attend or follow this change. The unfavorable indications are the continuation and increase of all the symptoms, with the evidence, which sometimes appears, of organic renal disease, and the final suppression of urine, with the cerebral symptoms which follow, all of which point to a fatal termination.

Treatment.—In all cases of diabetes mellitus the causes should be sought out and removed, if possible, and the habits of the patient should be corrected. This being accomplished, the exact deviation from the standard of health should be taken, and the indications thus arrived at, should be fulfilled by the most convenient, safe, and reliable remedies.

The patient should abandon any imprudent or improper habits, **should keep the skin clean**, should wear clean and warm clothing, and to furnish a due amount of nitrogen to the system, and at the same time **to** avoid the sugar-making materials, the patient should be placed on an almost exclusively animal diet, to be taken with strict regularity. Only a **reasonable amount** of **drink** should **be allowed, and** this should consist of **water, milk, tea, or coffee.** And as **the appetite is** apt to be voracious, **the quantity of food should be** restricted, and this should be thoroughly **masticated and taken slowly.**

To correct the **acidity and constipation** which attend in this disease, magnesia may be given, morning **and** evening, in quantity sufficient to keep the bowels moderately loose, and this should **be continued** while the **constipation and** acidity lasts, **or during the continuance of the disease.**

As a tonic for the blood, the **sulphate, carbonate, or ammoniated citrate of iron should be given, in two or three grain doses, after each meal, and continued for a long time.** And as a tonic for **the nervous** system, and **to improve digestion,** $\frac{1}{20}$th of a grain of the muriate of strychnia may **be given in solution,** before each meal, and continued **for** a time.

In **cases,** however, in which there is local irritation in the brain, **or** along the spinal cord, the strychnia should not be given. To improve digestion in such cases, thirty drops of the fluid extract, or half an ounce of the infusion of columbo may be given instead, before each meal, and continued for a time.

In bad cases, to diminish **the renal secretion, and to favor diaphoresis,** five grains of Dover's **powder may be given four times per day.** In milder or more chronic cases, however, **half a grain of ipecac may be** substituted, **and continued as long as may be required.** In cases in which there is **irritation at some point along the spine, in addition to the treatment I have already suggested, dry cups may occasionally be applied.**

SECTION VII.—DIABETES INSIPIDUS.

By diabetes insipidus, I mean here that variety of disease in which there is an **excessive secretion of urine, without** saccharine matter; **the** urine either appearing **natural, being** constituted of its ordinary **constituents,** or else containing **an excess of urea, or** more or less chyle **or** albumen.

I believe that in a large majority of the cases of this **disease there is an excess** of urea, while **the presence** of the other ingredients mentioned are **only** occasional, depending upon accidental circumstances.

Symptoms.—After an indefinite period of general derangement **of** the system, during which **the** appetite is variable, **the** bowels occasionally constipated, and the cutaneous functions more or less deranged, **the** patient gradually becomes annoyed by frequent calls to urinate, and the quantity of urine passed becomes at times very great.

In some cases the copious flow of urine is only occasional, following an undue amount of physical or mental excitement. In such cases,

41

and I believe in most cases of this disease, more or less pain is experienced in the loins, and not unfrequently there is tenderness along the dorsal or lumbar portion of the spine.

In cases in which the excess of urea is very considerable, there is often a disposition to pass water very frequently, both day and night; and especially is this the case if the weather is cold, or if the patient is laboring under undue mental excitement.

In severe cases of diabetes insipidus the thirst becomes urgent, the appetite craving, the skin dry, and the bowels obstinately constipated; and, if the disease continues, emaciation follows, with perhaps the chylons or albuminous urine, but never the saccharine. The disease seldom terminates fatally; but is apt to be associated with a weak state of the blood, and a deranged and irritable condition of the nervous system, so that the patient becomes exceedingly uncomfortable, and is rendered especially liable to be cut down by some sudden attack of acute disease.

Diagnosis.—Diabetes insipidus may be distinguished from diabetes mellitus by the less severity of all the symptoms, and by the absence of sugar, and presence of urea in the urine; while the reverse obtains in the other variety of the disease. Besides, in this variety, animal food always aggravates the disease; while in the other, animal food palliates, and vegetable food aggravates all the symptoms; a discovery that many patients make even before they resort to medical advice.

Causes.—Any train of influences that weakens the blood, renders the nervous system irritable, and impairs digestion, may operate as predisposing or exciting causes of this disease. Among these causes are an inherited predisposition; intemperance in eating and drinking; excessive physical or mental labor; the use of tobacco; venereal and other excesses; excessive anger; protracted grief; injuries of the spinal cord; and various other causes of a kindred character.

Nature.—This disease evidently consists in a weak or impoverished state of the blood, and an irritable condition of the nervous system, attended with deranged digestion, sanguification, and secretion. I believe, however, that the immediate cause of the excessive renal secretion, is an imperfect generation, and bad distribution of the nervous influence; and hence it is that the excessive renal secretion is so often paroxysmal, following physical or mental excitement.

In cases in which there is some permanent irritation along the middle or lower portion of the spinal cord, which is operating to produce the disease, an aggravation of the irritation from some accidental cause, may, and I believe often does produce sufficient derangement in the generation and distribution of the nervous influence, to cause for the time, the excessive renal secretion.

The excess of urea may arise in part from a too exclusive animal diet, but the chyle and albumen, when present in the urine in such cases, I believe may be owing to imperfect sanguification growing out of derangement in the cerebro-spinal and nervous system.

Treatment.—The indications in the treatment of this disease are, to correct the habits and regulate the diet of the patient, to restore the blood and nervous system, and to subdue any local irritation that may exist along the spine.

The patient should keep the skin clean and warm, should take suitable exercise, should preserve an even and cheerful temper of mind, and should be restricted to an almost exclusive vegetable diet. To correct the acidity of the stomach, a wineglassfull of lime water may be given each morning, and to regulate the bowels, a pill of aloes and rhubarb may be given at evening as long as may be required.

To restore the blood, three or four grains of the carbonate of iron may be given three times per day, after each meal, and continued, if it agrees, till the blood is restored. If there is spinal irritation, cups should be applied occasionally at first, and later, pustulation may be produced by tartar emetic, either in the form of ointment, or else sprinkled on a conium, stramonium, or belladonna plaster, to be worn over the seat of the irritation.

Having thus corrected the acidity, regulated the bowels, restored the blood, and subdued spinal irritation, if it existed, if the copious secretion of urine still continues, the drink should be rigidly restricted, and as a tonic to the nervous system, strychnia should be given. A grain of the muriate of strychnia may be dissolved in eight ounces of water, and of this a teaspoonful should be given after each meal, and continued till a cure is effected.

SECTION VIII.—LITHIASIS—(*Gravel.*)

By lithiasis, from λιθος, "a stone," I mean here those insoluble deposits which take place from the urine within the body, and also the peculiar conditions of the system upon which they depend. These deposits generally depend upon either an acid or alkaline condition of the fluids of the body, and may be in the form of an impalpable powder, in crystalline particles like sand, or in solid concretions of various sizes.

The deposits, whatever may be their form, may be included under the heads of *the urates, the phosphates,* and *the oxalate of lime,* and the conditions of the system upon which they depend, may be called the *lithic,* or *uric acid diathesis, the phosphatic diathesis,* and *the oxalatic,* or *oxalic acid diathesis.*

Calculi are generally originally formed in the kidneys, from which points they pass along the ureters to the bladder, and are thence discharged through the urethra with the urine, or else being retained, they form by a gradual agglomeration of the pulvurent and crystalline particles, solid concretions of various sizes.

Of the urinary deposits, the lithic or uric acid appears to predominate, and it very generally constitutes the central nucleus in other varieties of calculi. It is probable then that the deposition of *uric acid* is generally the primary process in the formation of urinary calculi, and that the *phosphates,* and *oxalate* formations are usually the result of a slow transition from the lithic acid to the phosphatic or oxalic formations. This gradual transition from the lithic acid to the phosphatic and oxalic formations, is of course preceded by a change in the corresponding diathesis, the lithic acid diathesis being generally the primary, and this being succeeded by the phosphatic or oxalatic diathesis, or both.

Now bearing these facts in mind, we will proceed to consider the symptoms of *lithiasis, gravel,* or *urinary calculi,* first those which are common to all varieties, and then the symptoms peculiar to the *lithates,* or *urates,* the *phosphates,* and the *oxalate* of *lime,* and the corresponding *diathesis* upon which they depend in the order in which I have named them

Symptoms.—The symptoms of urinary calculi vary with the position they occupy, as they are liable to produce irritation in the kidneys, ureters, bladder, or urethra.

Renal calculi, in addition to the symptoms of the diathesis upon which it depends, is attended with pain in the kidneys, with or without symptoms of inflammation of the kidneys and its attendant symptoms. The disease is sometimes accompanied with bloody urine, and may be attended with more or less febrile excitement or general nervous irritability.

They are of various forms and dimensions, and generally consist of uric acid, animal matter, and oxalate of lime, with the phosphates in some cases.

Calculi in the *ureters,* on their passage from the kidneys to the bladder, if sufficiently large or rough to produce irritation, cause, in some cases, the most excruciating pain along the ureters, extending to the testicle of the affected side in the male, and producing numbness of the thigh in both sexes. The pain in such cases is often of the most agonizing character, and the disease may be attended with bloody urine, and followed by the discharge of considerable fine crystalline particles, and perhaps small calculi, from the size of a pin's head to that of a pea, or larger,

Vesical calculi, or stone in the bladder, whether they proceed from the kidneys, or as is more frequently the case, are formed by the agglomeration of crystalline particles around a nucleus of some kind in the bladder, are attended with various unpleasant and distressing symptoms. There is a sense of weight in the perineum, and as the patient changes his position, there is sometimes a sensation as if a body were rolling in the bladder. There is a frequent desire to urinate, and often a sudden stoppage in its flow. Pain or an itching sensation is apt to be experienced at the extremity of the glans in men, and the disease is often attended with a discharge of bloody urine.

Urethral calculi generally proceed from the bladder, and besides producing a tumor, if sufficiently large, obstruct the passage of urine, and if permitted to remain, cause pain, inflammation, fever, &c.

Such, I believe, are the general symptoms common to all forms and varieties of urinary calculi in the kidneys, ureters, bladder, and urethra. We will now examine the symptoms peculiar to each variety, the *urates, phosphates,* and *oxalate* of lime, as well as of the diatheses upon which they depend.

The *lithic acid diathesis* is characterized by an acid condition of the fluids of the body, attended often with a gouty or inflammatory tendency of the system. The deposit may occur from any cause which lessens the amount of urine, or from the use of acids, which, uniting with or neutralizing the alkaline principles of the urine, causes the deposition of lithic acid.

The sediments which belong to this diathesis may be in the form of an impalpable powder, or in small crystalline particles, like sand, or else, as is sometimes the case, in small concretions. They are of a reddish, yellowish, or pink color, in consequence of certain coloring principles in the urine, and consist of either pure uric acid, or else the urate of soda mixed with a trace of the urates, of lime and ammonia.

In cases in which these deposits take place in consequence of a decrease in the renal secretion, it is probably from an incapacity on the part of the urine to hold the deposited portion in solution or suspension, in consequence of which the deposition takes place, as we see in many inflammatory and febrile affections. But when the deposits take place in consequence of acid food or drinks, it may be from the chemical changes produced, in consequence of which the uric acid is deposited, even though the secretion of urine continues undiminished, as we have already seen. The urine, however, in most cases of lithic gravel, is scanty and high colored, being passed frequently, and with more or less pain, and, as it cools, after being voided, a further deposition usually takes place, in consequence, probably, of its diminished solvent power at a reduced temperature.

The lithic gravel is of a yellowish, pink, or reddish color, often adheres to the vessel in which the urine has been voided, and is readily dissolved by nitric acid, with effervescence, by which it may be distinguished, if necessary.

Such, I believe, are the symptoms peculiar to the lithic acid diathesis, and to the depositions with which it is attended, which brings us to the consideration of the phosphatic diathesis and its depositions.

The *phosphatic diathesis* is characterized by an alkaline condition of the fluids of the body, attended with an irritable condition of the nervous system, and generally great derangement of the digestive organs, manifested by flatulence, nausea, constipation or diarrhœa, and a dull pain and feeling of weakness in the loins. In the worst cases, there is loss of appetite, a peevish and irritable temper, emaciation, a sunken and haggard appearance of the face, and often some organic disease of the urinary organs or spinal cord.

The deposits in this diathesis consist of "the double phosphate of magnesia and ammonia, and the phosphate of lime, either separate or mingled," and appear either in the form of white crystalline grains, or else in an impalpable powder of a light yellowish appearance, or what is more common, there may be a mixture of the two forms in the deposit.

The urine in this diathesis is usually more copious than in health, being of a pale color when passed, and if clear at first, lets fall on cooling a sediment. If, however, the urine be allowed to stand for a considerable time, an iridescent pellicle of the phosphates appear on its surface, which is finally gradually deposited.

A fit of gravel in this diathesis is comparatively rare, as the phosphates are not very liable to be deposited within the body, even though the urine abounds in the salts. The phosphates may be distinguished by their whiteness, by their being generally deposited as the urine cools, by the pellicle which gradually forms on its surface, and is finally deposited, and by their solubility in dilute acetic and muriatic acids. The phosphates are held in solution in the urine by an excess of acid, hence any cause which produces an excess of the salts or an alkaline state of the urine, causing their acids to be neutralized, may produce the phosphatic deposits. It appears probable that the alkaline deposits are generally associated with either functional or organic disease of the brain, or some portion of the nervous substance, and especially of the spinal cord. The alkaline state of the urine is also probably augmented in many cases, by various deranged conditions of the digestive organs.

Such then, are the symptoms peculiar to the phosphatic diathesis, as well as those which attend its deposits, which leaves us only for consideration, the peculiar symptoms of the *oxalatic*, or *oxalic acid diathesis*, and its deposits.

The *oxalatic* or *oxalic acid diathesis* consists in that peculiar condition of the fluids of the body in which there is a tendency to the formation or deposition of oxalate of lime in the urine. It is attended with dyspepsia, nervous irritability, and a tendency to scaly eruptions, and to carbunculous affections. The patient is apt to be irritable, sensitive, feeble, hypochondriacal, and gloomy, and there appears to be a strong neuralgic tendency in most cases.

The urine in this affection, is quite clear, exhibiting only a slight sediment on cooling; but there is apt to be strong symptoms of calculus, along with the constitutional derangement. Oxalate of lime, may be crystalline or amorphous; the crystals appearing under the microscope, transparent and octohedral, or in some cases circular or oval. It may be distinguished by its insolubility in acetic acid, and by its effervescence after calcination, with dilute acids.

It is probable, that the oxalic deposit, may be favored by articles of diet, which contain oxalic acid, as the sorrel, the rhubarb, or pie plant, &c., and the free use of sugar may favor it, in those who are predisposed.

Such then, are the general symptoms of *lithiasis* or *gravel*, and also the symptoms peculiar to the *lithic acid*, *phosphatic*, and *oxalatic* diatheses, as well as of the deposits belonging to each.

Diagnosis.—Lithiasis or gravel, may generally be distinguished, whether the deposit be in the kidneys, ureters, bladder, or urethra, by pain in the part, of greater or less intensity, by the scanty, high-colored, and perhaps bloody urine, by the frequent and painful micturition, by the sediments of urates, phosphates, or the oxalates, which appear in the urine, and by the symptoms which attend on some one of the diatheses upon which the deposits depend.

The presence of solid concretions, of considerable size in the bladder, may be distinguished by the evidences in most cases, of a gradual transition from the lithic acid, to the phosphatic or oxalatic diatheses, by the sensation of something rolling in the bladder, on changing position, by the frequent and painful micturition, and sudden interruptions to the flow of urine, and finally by passage of a sound into the bladder, by which the stone may generally be felt.

The *urates* may be distinguished from the other varieties of gravel, by the existence of the lithic acid diathesis, with the acid condition of the fluids of the body, a dyspeptic, and inflammatory tendency, and by the pink or reddish appearance of the sediment, which is readily dissolved by nitric acid, with effervescence.

The *phosphates* may be distinguished by the existence of the phosphatic diathesis, with an alkaline condition of the fluids of the body, attended with dyspepsia, and generally with organic or functional derangement of the nervous system, and especially of the spinal cord. The deposits are white, take place generally as the urine cools, and if the urine is allowed to stand, an iridescent pellicle of the phosphates form upon its surface, which, however, gradually subsides. The phosphates are soluble in dilute acetic and muriatic acids.

The *oxalate* of *lime*, may be distinguished by the oxalatic diathesis, with the dyspeptic, irritable, hypochondriacal, and gloomy condition, and the tendency to scaly eruptions, carburculous affections, &c.

The deposits may be distinguished by the transparent octohedral, circular, or oval appearance of the crystals, under the microscope, and by its insolubility in acetic acid, but ready effervescence, after calcination, with dilute acids.

By careful attention to all these diagnostic symptoms, we may not only distinguish a case of lithiasis or gravel; but readily distinguish the different varieties, as well as the diathesis, upon which they depend.

Causes.—The direct causes of the disposition of the urates, phosphates, and oxalate of lime, are the lithic acid, phosphatic, and oxalatic diatheses, upon which they depend, together with some occasional accidental circumstance. In arriving at the cause of this disease, in its various forms then, we have to inquire mainly into the causes of the various diatheses, taking into account, the accidental circumstances, which occasionally operate to increase the deposits.

The *lithic acid diathesis* may be produced by a variety of causes, but it is probable that the most frequent are a too free use of high-seasoned animal food, the use of alcoholic drinks, in their various forms, and indolent habits, in consequence of which the stomach is over-taxed, and too much azotized matter introduced into the system.

With the fluids of the body thus brought into an acid state, the lithates may be deposited, more or less, continually; but, let the urine be rendered scanty, from some inflammatory or febrile affection to which the system is always predisposed in such cases, and a copious deposition of the lithates takes place, from the inability on the part of the urine to hold them in solution. A copious deposition of the lithates may also be produced, when this diathesis prevails, by the use of acid food or drinks, probably in consequence of their neutralizing alkaline principles, which had helped to hold the lithic acid or lithates in solution in the urine.

The *phosphatic diathesis* may be produced by local injuries of the brain or spinal cord, but it is generally brought about by want, exposure, overtaxing the body or mind, licentiousness in its various forms, and various other influences which produce organic or functional disease of the nervous system, and especially of the spinal cord.

As the phosphates are held in solution in the urine by an acid, any accidental increase in the nervous derangement, which produces an excess of the salts, or an alkaline state of the urine, may, by neutralizing the acid, cause a deposition of the phosphates.

The *oxalatic*, or *oxalic acid diathesis*, may be caused by improper food, irregularity of eating, and various imprudences, which impair digestion, and render the patient nervous, irritable, neuralgic, gloomy, hypochondriacal, &c. It is probable that there is generally an excess of urea in oxalic acid urine; and in cases in which the oxalatic diathesis prevails, it is likely that the oxalic deposit is increased or favored by the use of vegetables which contain oxalic acid, as the sorrel, rhubarb, or pie-plant, &c.

Now, having traced **the various causes which** produce **the different** diathesis, as well **as the** accidental influences which particularly favor **or** increase the deposit at times, producing a fit of gravel, we have yet to inquire into the influences which operate in producing solid concretions or stone in the kidneys or bladder, but generally in the bladder.

We have already **seen** that the deposition of lithic acid, or the lithates, is generally the primary **process in** the formation of urinary calculi; and it is probable that the phosphate and oxalate formations, which often go to make up the great bulk **of** the calculi, are the result of a gradual transition from the lithic acid **to** the phosphatic, or oxalatic diathesis.

In conversation a few years since **with** the late Dr. Amasa Trowbridge, **of** Watertown in this State, who had operated, I believe, some fifteen or sixteen times for stone in the bladder, in Northern New York, in a region where the water is generally largely impregnated with lime, he observed that all, or nearly all his cases had been persons that had removed to that lime region, from localities in which the water was unimpregnated with lime. And as the Doctor had made **this observation** during a long term of years, it appears probable that the constant use of lime, as it exists in the water in lime-stone regions, if commenced at infancy and continued through life, tends strongly to prevent a lithic acid diathesis and deposit, and hence is a partial security against the formation of urinary calculi of all kinds, as the primary nucleus **is wanting.** But the observation appears to show further, that in persons born and brought up in regions where the **water** contains no lime, the lithic **acid** diathesis and deposits are very common; and that if such persons **remove to a** lime region, and the diathesis becomes changed from some **cause to** the phosphatic or oxalatic, if the lithates exist in the bladder **in the form of** concretions, to afford a nucleus, **around** which the phosphates and oxalate, as the diathesis changes, may readily unite by agglomeration, calculi of large size may form requiring an operation for their removal. In cases, however, in which concretions of the lithates had not formed, and the phosphatic or oxalatic diathesis is not produced by some accidental cause, it appears probable that the use of the lime in such cases, may generally correct not only the lithic acid lithiasis, but also the diathesis upon which it depends.

Prognosis.—The prognosis in recent cases of lithiasis or gravel, is rather favorable if the cause can be removed and the patient be subjected to proper treatment. In protracted cases, however, in which the digestive organs and the nervous system are seriously involved, restoration to perfect health, if accomplished at all, must be slow at best, and not unfrequently the disease or its complications may lead on to a fatal termination.

Treatment.—The treatment of lithiasis divides itself into that which is proper for the expulsion of deposits from the kidneys, ureters and bladder; and that which is indicated from the correction of the general condition, and the different diatheses upon which the deposits depend.

In cases then in which the patient is found with evidence of gravel in the kidneys, the free use of mucilaginous drinks may be allowed, and the patient may be directed to take some exercise by walking or riding, with the hope of favoring its passage into and along the ureters.

If, however, calculi have passed into and become lodged in the ureters, and the patient is found suffering the most excruciating pain along the ureters; in addition to mucilages, cups and fomentations of hops, and the warm hip-bath if necessary; ten drops each of sulphuric ether and oil of turpentine may be given every six hours, with the hope of producing relaxation of the ureters; and if necessary, alternating with this, fifteen drops of the fluid extract of hyoscyamus may be given to quiet pain.

Finally, if the bladder is the seat of the calculi, and there is hope that it may admit of being passed through the urethra, the patient should take mucilages freely, retain the water in the bladder as long as it can be conveniently, and then bending the body forwards, so as to bring the calculi as near the commencement of the urethra as possible, should allow the urine to pass in a full stream, with the hope that the calculi may thus be brought away with the urine.

Having thus made an effort to relieve the immediate suffering, in cases in which the patient is first seen during a fit of gravel, and at first, in cases of lithiasis examined at other periods, the habits and general condition of the patient should be ascertained, and the case treated on strictly common sense principles.

The habits of the patient should be corrected at once, and every cause or influence which might be supposed to operate either directly or indirectly to produce or keep up the disease, should be carefully removed, as far as possible.

The patient should be directed to keep the skin clean, to wear flannel, and sleep in flannel sheets, to take suitable exercise, and also proper food, with strict regularity, and to be strictly temperate and prudent in all things. If the bowels are constipated, they should be regulated by cathartics or laxatives suited to each particular case. And if the blood is weak, and the nervous system prostrated, some preparation of iron, and perhaps strychnia, in small doses, may be required.

By thus removing the cause, and correcting the habits and general condition of the patient, the lithiasis and diatheses upon which it depends, may sometimes be removed, or at least the progress of the disease may be arrested. If, however, it becomes necessary to do more, as it very generally may, remedies should be varied according to the prevailing diathesis and character of the deposit.

If the *lithic acid diathesis* prevails, with its peculiar deposits, the patient should be restricted mainly to a vegetable diet, to lessen the amount of urea in the system, and to render the uric acid more soluble; and at the same time to correct the acid condition of the fluids of the body, alkalies should be administered. Of the alkalies, the bicarbonate of soda or potassa will generally do best. Half a drachm of either may be given four times per day, dissolved in not less than four ounces of water, and continued till the urine ceases to yield a deposit, and the acid condition of the fluids of the body are corrected.

In slight cases, however, if the bowels are constipated, twenty grains of magnesia may be given, morning and evening, and continued as long as may be required, instead of the potassa or soda, already suggested. Or in cases attended with diarrhœa, four ounces of lime-water may be given, three times per day, after meals, and continued as long as may be required.

While this alkaline treatment is being pursued, if the patient is annoyed by small concretions in the bladder, a drachm of the fluid extract of the *hydrangea arborescens* may be given, three times per day; till the concretions pass, when it should be omitted, and the alkali continued, if necessary.

If the *phosphatic diathesis* prevails, however, with its deposits, the patient should take a diet mainly of animal food, and instead of alkalies, should take either the sulphuric, nitric, muriatic, or benzoic acid, well diluted. I prefer the muriatic acid in most cases, of which ten drops may be given, diluted with six or eight ounces of water or thin gruel, after each meal, and continued as long as may be required. If there is spinal irritation, it should be subdued by cups, blisters, &c., and while the acids are being administered, as a tonic and astringent for the urinary passages, half a drachm of the fluid extract, or an ounce of the infusion of *uva ursi* may be given four times per day, with the addition, when necessary, of fifteen drops of the fluid extract of hyoscyamus, as an anodyne.

Finally, if the *oxalatic*, or *oxalic acid diathesis*, with its deposit prevails, the patient may be allowed a mixture of animal and vegetable food, but should avoid such vegetables as contain oxalic acid, as the the rhubarb or pie plant, &c., and should not indulge too freely in the use of sugar, in any form.

To correct the *oxalatic diathesis*, four drops of the *nitromuriatic acid* may be given, three times per day, after each meal, diluted in at least six ounces of water, and continued till uric acid is deposited in the urine, in place of oxalate of lime, when it should be discontinued, and half a drachm each of the fluid extract of hydrangea and uva ursi given instead, till a cure is effected.

SECTION IX.—SUPPRESSION OF URINE.

By suppression of urine, I mean here a partial or complete suspension of the renal function, so that little or no urine is secreted.

This condition, whether the suppression be partial or complete, is generally the result of inflammation, paralysis, or organic disease of the renal glands, which of course varies the symptoms which are developed.

Symptoms.—In cases of suppression from *inflammation* of the kidneys, the diminution or entire suspension of the urinary discharge follows the ordinary symptoms of nephritis, such as pain in the lumbar region, nausea, vomiting, &c. Very soon after the renal secretion ceases however, the patient becomes torpid and dull, the pulse may become slower than in health, a urinous odor exhales from the surface of the body, and finally there is the supervention of drowsiness, mental wandering, coma, convulsions, and death.

In cases which occur from *paralysis* of the kidneys, the suppression may be attended with little or no pain, there being at first only a feeling of restlessness, with perhaps slight uneasiness in the lumbar region. Soon, however, nausea and vomiting may occur, and gradually there is the supervention of dullness, drowsiness, hiccough, coma, and convulsions; and finally, if the condition continues, a fatal termination is the result.

In cases of suppression, from *organic renal disease*, in addition to the symptoms belonging to the organic affection, there is the supervention of those enumerated above, the result probably of the accumulation of urea in the blood, and its effect upon the brain and nervous system.

In all cases of suppression of urine, in addition to the urinous smell of the cutaneous exhalation, there is apt to be the same odor from the alvine discharges; and there is not unfrequently a urinous taste in the mouth. And unless this vicarious discharge is sufficient to relieve the blood of the urea which is fast accumulating, the patient may generally die by the third or fourth day after the entire suspension of the renal function.

This affection may be distinguished from *retention of urine* by the empty state of the bladder; whereas in that affection it becomes distended, forming a tumor, which may readily be felt above the pubis.

Causes.—The direct causes of suppression of urine, as we have already seen, are nephritis, paralysis, and organic renal disease.

Renal *inflammation*, as we have seen in a preceding section, may be produced by a variety of causes; such as direct injuries, exposure to cold, stimulating diuretics, &c.

Renal *paralysis* may be produced by any cause, or train of causes, which prevents the generation of sufficient nervous influence, or hinders its distribution to the renal glands. It is probable that various causes may operate, such as fatigue, spinal injuries, &c., in bringing about this deplorable state; but I believe that is more generally the result of venereal excesses, protracted drunkenness, and the use of tobacco. At least such has been the result of my observation.

Organic renal diseases, as we have already seen in a preceding section, may be caused by an inherited or acquired depraved condition of the system; almost every deviation from the laws of health operating to excite, aggravate, or perpetuate the disease.

Thus we have at a glance the most frequent direct and remote causes of suppression of urine: which it may be well to bear in mind, as we proceed to the consideration of the treatment for this most distressing or dangerous affection.

Treatment.—In *inflammatory cases* the treatment should be that ordinarily indicated for nephritis, such as bleeding, cupping, fomentations, hyoscyamus, cream of tartar, sweet spirit of nitre, &c., which, if judiciously applied, will generally do all that may reasonably be anticipated in such cases.

In cases depending upon *renal paralysis*, spinal irritation may require cups, blisters, &c. In addition to this, strychnia may be given as a nervous tonic, and, to stimulate the kidneys into activity, twenty drops of the oil of turpentine may be given every two hours, till the renal function is resumed.

In cases depending upon *organic renal disease*, those palliative measures only should be resorted to which each particular condition may suggest, as a permanent cure is not to be expected in such cases.

SECTION X.—RETENTION OF URINE.

By retention of urine is here meant an obstruction to the passage of urine after its secretion, whether that obstruction be in the ureters, bladder or urethra.

As the urine, after its secretion in the kidneys, has to pass along the ureters to the bladder, and from thence through the urethra, in being voided, it is not strange that its progress should be interrupted by calculi or coagula in the ureters, by inflammation or paralysis of the bladder, or by spasms or other obstruction of the urethra.

Retention occurring from obstruction at either of these points, from various causes or conditions, is attended with a train of symptoms, which we will now proceed to consider.

Symptoms.—*Renal retention*, whether it be from calculi, coaguli, or spasm obstructing the ureters, is attended with a sense of weight, pain, and distension in the lumbar region; and if the obstruction continues, the loins become tender, the kidneys and ureters above the obstruction become enormously distended, producing a tumor which may sometimes be felt, and unless relief be afforded, inflammation of the parts, with disorganization and a fatal termination, is the result.

Vesical retention, if from *cystitis*, is attended with the ordinary symptoms of inflammation of the bladder, such as pain, tenderness, nausea, vomiting, fever, &c., and, in addition, there is the tumor caused by the distended bladder, which may readily be felt above the pubis, reaching, in some cases, the umbilicus, or even the scrobiculus cordis. If, however, the retention be from *paralysis* of the bladder, there may be little or no pain at first; but as the bladder becomes distended, the pain, as in all other cases, becomes more or less severe, and unless relieved, the patient becomes anxious, restless, feverish, and exceedingly distressed, until the bladder finally gives way, and a fatal termination is the result.

Finally, if the retention be from spasms or other obstructions of the *urethra*, in addition to the symptoms enumerated above, there is pain along that passage, and if a stone has become lodged in it, a tumor may be felt in many cases, an interesting instance of which fell under my observation a few years since.

Such, I believe, are the ordinary symptoms of retention of urine, whatever be the seat or nature of the obstruction.

Diagnosis—Retention may be distinguished from suppression of urine by the distension of the bladder which occurs, if it be *vesical*, or of the kidneys and upper portion of the ureters, if it be *renal*.

Vesical retention may be distinguished from renal by the fullness of the bladder, and cases occurring from urethral obstruction, by the evidence which exists of stone in, or spasms of, that passage.

Causes.—The most frequent causes of *renal* retention are calculi, or coagula of blood or fibrin, in the ureters.

The obstruction may take place, however, from thickening or spasm of the ureters, or from tumors pressing upon them from without.

Vesical retention may be caused by inflammation of the mucous membrane of the bladder, closing the opening of the urethra, by paralysis of of the bladder, or by spasms or other obstructions of the bladder or urethra.

Treatment.—In cases of *renal* retention, attended as it may be with nephritis, bleeding, cupping, fomentations, the warm bath, saline cathartics, anodynes and antispasmodics may be required, and should be judiciously applied, while the least hope of relief remains.

In *vesical* retention from inflammation of the bladder, nearly the same general treatment is required, and in addition an effort should be made to pass the catheter, should the bladder become very much distended, and if successful, the water may be thus drawn off two or three times per day, till the inflammation subsides, and the obstruction is overcome. And should there be evidence of spasm of the bladder, tending to close the opening of the urethra, ten drops of the tincture of stramonium may be thrown into the rectum, with an ounce of mucilage of gum arabic, or the same quantity of liquid starch, and a fomentation of stramonium leaves may be applied and carefully watched, externally with the hope of relaxing the vesical spasm.

In cases in which the retention is from *paralysis* of the bladder, the catheter should be introduced and the water drawn off, two or three times per day, and general and local measures immediately taken to overcome the paralysis. If there is spinal disease, cups, blisters, &c., should be applied, and the local irritation subdued. This being accomplished, if the paralysis continue, strychnia may be given in $\frac{1}{20}$ of a grain doses three times per day, before each meal, as a nervous tonic, and as a stimulus to the bladder, twenty drops of the tincture of cantharides may be given after each meal, and this treatment may be continued till the paralysis is overcome.

Finally in cases depending upon stricture of the urethra, the warm hip-bath, stramonium fomentations, and the internal administration of the tincture of stramonium and tincture of chloride of iron, of each ten drops every one, two, or three hours, as long as it may be safe, will generally do most, and if this should fail, an effort may be made to pass a small bougie, and then a catheter, and failing in this, the bladder should be punctured to avoid a worse result.

SECTION XI.—DYSURIA—(*Strangury.*)

By dysuria, from δυς, "with difficulty," and ουρον, "urine," I mean here difficult and generally painful micturition without any considerable retention.

Dysuria or strangury may depend upon a permanent irritation of the bladder or urethra, or it may be transient, the result of irritating properties in, or an acrid state of the urine, and in either case there may be but trifling if any retention.

Symptoms.—In *permanent* cases, there is sometimes a constant feeling of weight, fullness or heat in the neck of the bladder, this irritability causing a frequent call to urinate during which there is pain, and the water may pass tardily, or in a small stream, but does not accumulate.

In *transient* cases in which the urine is rendered acrid from some cause, the calls to urinate may be less frequent, and the water may pass more freely, but it is attended with pain or a burning sensation along the urethra, and especially in its posterior part.

Causes.—Various causes may operate to produce this affection among the most frequent of which are excesses in eating, the use of spirituous liquors, irritating diuretics, masturbation, onanism, and excessive venery, acid food, or drinks, hemorrhoids, ascarides in the rectum, suppressed catamenia, vesical calculi, leucorrhœa and gonorrhœa, and repelled cutaneous affections.

Treatment.—In slight cases of strangury, depending upon irritation of the neck of the bladder, with or without an acrid state of the urine, produced by blisters, cantharides taken internally, or any other accidental causes; a dose of castor oil, the free use of mucilages, and fifteen drops of laudanum thrown into the rectum, with an ounce of flax-seed tea, will be sufficient to produce relief in most cases.

In more permanent cases, however, in addition to mucilages, a regulated diet, and removal of the cause, a drachm of the fluid extract, or an ounce of the infusion of buchu may be given three times per day, and continued till the irritation of the bladder and urethra are subdued. If, however, the patient is anæmic, and there still remains a disposition to spasm of the bladder or urethra, after the catarrhal condition is subdued, ten drops of the tincture of chloride of iron should be given in mucilage, three times per day, after meals, and continued till the blood is restored, and the tendency to spasm is overcome.

If the strangury appears to be the result of a deposit of lithates in the urine, alkalies should be given; if of the phosphates or oxalate, acids, and if there is evidence of small vesical concretions, a drachm of the fluid extract of hydrangea may be given for a while, three times per day, with the hope of promoting their passage, and thus relieving the irritation.

SECTION XII.—INCONTINUANCE OF URINE.

By incontinuance of urine, is here meant an inability to retain the urine, in consequence of which it is passed involuntarily, either constantly, by drops, as it is secreted, or else after considerable has been accumulated.

Symptoms.—In cases in which the urine is passed by drops, as it is secreted and poured into the bladder, the dribbling keeps up constantly, day and night, unless accidentally interrupted, for a little time, by position in sitting or lying. But in cases in which considerable accumulates before it passes off, the patient during the day, is only troubled by the inconvenience attending frequent micturition, as a slight voluntary effort, may enable the patient, in many cases, to void the urine, before the involuntary discharge takes place. In such cases, however, nocturnal discharges occur, while the patient is asleep, which are exceedingly annoying.

In other cases still, especially in children, there is but little, if any apparent inconvenience during the day; but almost invariably a discharge during the night, and this may be kept up for months, or even years, and become a source of great vexation, to both parent and child.

Causes.—Incontinuance of urine, may be the result of a paralysis of the sphincter of the bladder, in consequence of which the neck of the bladder is constantly relaxed, affording no obstruction to the passage of the urine; the patient in such cases, having no power to control it. In other cases, the incontinuance may depend upon an irritable state of the bladder, in consequence of which, the urine is passed, in spite of the sphincter. Or the vesicle irritation may exist, with paralysis of the sphincter, and thus the incontinuance be produced. I believe, however that in nocturnal incontinuance, as it occurs with children, it is generally from a want of sufficient excitability in the neck of the bladder, in consequence of which, the urine is allowed to pass, during sleep, without exciting the sphincter to contraction.

Treatment.—In cases of incontinuance, depending upon irritation of the bladder, the cause should be sought out, and removed, and the irritation subdued by counter-irritants, mucilages, astringents, &c. If, as the vesical irritation is subdued in such cases, there is evidence of paralysis, and in all cases of paralysis, without vesical irritation; spinal irritation should be subdued, if it exists, and then strychnia administered, in small doses, till the paralysis is overcome. Finally, in cases of nocturnal incontinuance, as it occurs in children, from a want of due excitability in the neck of the bladder, tincture of cantharides should be given, in moderate doses, three times per day, and continued till the condition is corrected, when the incontinuance will generally cease.

CHAPTER XV.

DISEASES OF THE GENITAL ORGANS.

SECTION—I. SPERMATORRHŒA.

By spermatorrhœa, from σπέρμα, "sperm," and ρέω, "I flow," is here meant an emission of sperm, without copulation, whether diurnal or nocturnal, the result very generally of masturbation, onanism, or venereal excesses.

I design also to include under this head that morbid condition into which the female system is brought by excessive sexual indulgence, and masturbation, or excitement of the genital organs by the hand.

Now in order to understand this disease, as it occurs in both sexes, it is necessary that the anatomy and physiology of the male and female genital organs should be well understood, as well as their direct and sympathetic relations. We will, therefore, take a general glance, first at the female and then at the male organs of generation, leaving the examination of each part for consideration, as we take up the diseases to which they are liable, in the following sections of this chapter.

The female genital organs, it should be remembered, consist of the vulva, vagina, uterus and ovaries. The womb, with its appendages, the ovaries, fallopian tubes, round and broad ligaments, and composed of its external *serous* internal *mucous*, and middle *muscular* tunics or coats, is situated in the middle of the cavity of the pelvis, between the bladder and rectum, beneath the convolutions of the small intestines, being continuous below with the vagina.

The *vagina* is a membranous canal, leading from the uterus to the vulva, passing between the bladder and rectum, and corresponding in direction with the axis of the outlet of the pelvis. It is five or six inches in length, is composed of an internal *mucous lining*, a middle layer of *erectile tissue*, and an external *contractile fibrous* tissue, and is attached superiorly to the cervix of the uterus, which projects into the upper extremity of the canal, and is continuous below, by its *mucous membrane* with that of the *vulva*.

The *vulva* or external organs of generation in the female, includes the mons veneris, labia majora, labia minora, clitoris, meatus urinarius, and opening of the vagina, parts situated at and about the entrance of the vagina, which is the primary seat of the local irritation in the masturbation of females. This fact should be borne in mind, but a further description of the parts are unnecessary for our present purpose.

The *male genital organs*, consist of the penis, the testicles and their appendages, and the seminal vesicles, a general glance at which is necessary as we proceed, in order to understand the symptoms which are developed in this disease.

The *penis*, consisting of a root, body, and extremity, or *glans*, is composed of the corpus cavernosum, corpus spongiosum, and an investing membrane, which is loosely attached to the entire surface, except the glans, over which the extremity of it, called the prepuce, readily passes, when the organ is in a quiescent state. At the extremity of the glans is the meatus urinarius, or opening of the *urethra*, which is a membranous canal, extending back to the neck of the bladder, having a lining mucous and external fibrous coat.

That part of the urethra which extends from the meatus urinarius, through the corpus spongiosum, to the deep perineal fascia, is called the *spongy portion*; that part, about an inch in length, immediately back of this, which passes between the two layers of the deep perineal fascia, is called the *membranous portion*, while the remaining posterior part of the urethra, about an inch in length, which passes through the prostate gland, above its middle, back to the bladder, is called the *prostatic portion*. The mucous coat of the urethra is thin and smooth, and continuous externally, with the investing membrane of the glans, internally with the mucous membrane of the bladder, and at intermediate points with the lining membrane of the ducts of Cowper's glands, the prostate gland, of the vasa deferentia and vesiculæ seminales.

Cowper's glands are two glandular bodies, the size of a pea, situated on the sides of the urethra, near the junction of its spongy and membranous portion, with each a short excretory duct, emptying obliquely into the urethra.

The *prostate gland* is situated in front of the neck of the bladder, upon the rectum, and behind the deep perineal fascia, and surrounds the urethra for about an inch, to the membranous portion. The urethra passes through the gland above its middle, and it is traversed near its inferior surface by the ejaculatory ducts of each side. The gland is of a conical form, the base being directed posteriorly; it has three lobes, two lateral and a middle, and is composed of numerous follicles, which give origin to ten or fifteen excretory ducts, which open into the urethra.

The *vesiculæ seminales* are two membranous reservoirs, about two inches in length, and six or seven lines broad, situated beneath the bladder and above the rectum, their larger extremities being directed backwards and outwards, and their smaller inwards and forwards, almost meeting at the base of the prostate gland, through which their ducts pass, having united with the vasa deferentia, constituting the ejaculatory ducts. The ejaculatory ducts thus formed, by a union with the vasa deferens of each side, pass along through the inferior portion of the prostate, for about three fourths of an inch, and open near each other into the urethra.

The *testicle*, a right and left, occupying the scrotum, and composed of the tubuli testis, the vasa recta, the rete testes, the vasa efferentia, &c., is the organ for the secretion of the sperm. The sperm secreted in this delicate structure of the testicle, passes out of the gland by the vasa efferentia to the epididymus, an appendage of the testicle, and through this to its termination in the vas deferens. From the epididymus, the vas deferens passes with the spermatic artery, veins, and nerves, constituting the spermatic cord, as far as the internal abdominal ring. At

42

this point, having entered the abdomen, the vas deferens leaves the other portion of the spermatic cord, being reflected inwards to the side of the fundus of the bladder, and extending along its posterior surface, passes along the internal side of the vesicula seminalis to the base of the prostate gland. At this point, it unites with the duct of the vesicula seminalis, constituting, as we have already seen, the ejaculatory duct. The ejaculatory duct thus formed, by the junction of the duct of the vesicula seminalis with the vas deferens, passes, as before suggested, about three-fourths of an inch through the inferior portion of the prostate, till it reaches the urethra, into which it empties near that of the opposite side.

Now, in a healthy state of the genital organs, no semen is secreted by the testicles, except during sexual intercourse, and then the sperm passes along each vas deferens into the ejaculatory ducts, and thence into and through the urethra to the cavity of the vagina or womb.

It is possible, however, that occasional nocturnal seminal discharges may occur, in an ordinary state of health, from a lively dream, in which the imagination provokes the seminal secretion and discharge. Such an occurrence, in a normal state of the genital organs, should not take place, however, oftener than once in three or four weeks, if at all.

When, however, the sexual organs are rendered morbidly irritable by any local disease of the parts, from masturbation, onanism, or excessive sexual indulgence, the blood vessels become congested, the whole glands become irritable, and a morbid secretion of semen is liable to take place. Now, this morbid seminal secretion may be quite constant, in which case the sperm, passing into the ejaculatory ducts, through each vas deferens, either passes into the urethra and is voided with the urine, or gradually dribbles away, or else it passes from the ejaculatory ducts back into the vesiculæ seminales, and there accumulates. In this case it is apt to be discharged at each movement of the bowels, by the mechanical pressure of the alvine evacuation, pressing from the rectum upon the vesiculæ seminales, or having accumulated in the vesiculæ seminales, it may then excite an erection, with the aid of a dream, which it may also provoke, and thus be thrown off as a nocturnal discharge.

Hence morbid seminal discharges may be diurnal, the semen passing off by degrees, or with the urine, or at each movement of the bowels, or they may be nocturnal, the morbid irritability producing an erection, and, perhaps, affecting the brain sympathetically, provokes a dream, during which a discharge of semen takes place. In some cases of this character the dream and sensation are remembered, in others the dream only, while in very bad or protracted cases, neither the dream or sensation is remembered, the patient only being aware of the discharge by its presence on the bed-clothes as he awakes.

Now, these involuntary seminal discharges may continue for a short time, if slight, without producing any very marked constitutional or general symptoms. But if the morbid seminal discharges be considerable, and are continued for any great length of time, the most serious constitutional effects follow; and the same is also true with females, if the genital organs are rendered irritable, as they sometimes are, by masturbation or titillation, though in this case there is of course no wasting discharge. The patient in such cases, whether male or female, becomes weak, nervous and irritable, the eyes are sunken, the countenance becomes pale, and there is eventually great debility of the nervous system. The brain becomes incompetent to supply sufficient nervous influence to enable the various organs of the body to carry on their functions in a proper manner, and hence we have arising a train of symptoms, in both sexes, which we will now proceed to consider.

Symptoms.—In addition to the pale face, sunken eyes, and haggard expression of countenance, digestion becomes impaired, and each meal, besides being taken with a poor relish, is followed by a sense of weight in the epigastrium, and more or less uneasiness and restlessness. This gastric disturbance after taking food, is apt to be attended with a quickened pulse, a flushed face, confusion of the ideas, ringing in the ears, vertigo, and sometimes with symptoms of cerebral congestion, which may threaten apoplexy.

This morbid condition or action of the stomach gradually increases, so that the food taken, being imperfectly digested, passes into the duodenum in a crude state, where it produces more or less irritation. And as the liver, too, from the same cause, is in a torpid state, an insufficient amount of bile is secreted or furnished to change the imperfect chyme into chyle, so that an insufficient quantity of imperfect chyle only is supplied to the system, and hence the emaciation. The passage of this illy digested food produces irritation of the alimentary mucous membrane, and, as a consequence, acid eructations, colic pains, and constipation is apt to follow.

Now, this constipation of the bowels frequently produces hemorrhoids or piles, and in consequence of the increased irritation in these parts, the spermatorrhœa, or irritation of the male or female genital organs, is increased or kept up, even though the other causes should be removed.

The imperfect digestion cuts off nutrition, so that the general debility is greatly increased, nervous irritability increasing, as well as local irritability of the genital organs, and so the disease progresses, the seminal disease in the male being gradually increased, as well as local and constitutional disturbance in the female.

In consequence of the general debility, want of vital energy, and bad distribution of the nervous influence, the respiration becomes imperfect, the patient being out of breath, from slight exercise, as from ascending a hill, a flight of stairs, &c. As the disease progresses, nervous or sympathetic palpitation of the heart is apt to occur, in some cases being of the most distressing character, developing symptoms, in many respects similar to those which attend organic cardiac disease.

Patients suffering from this affection become weak, even before they lose much flesh, in consequence of the nervous prostration. Sensation, too, in many of these cases, becomes extremely low in the extremities, especially in the hands, and the sense of taste, smell, hearing, and sight, all are liable, sooner or later, to become impaired, the eye losing its natural brilliancy, being sunken, and surrounded by a dark circle. The pupils are generally dilated, and amaurosis is very liable to occur; sometimes, however, there is great intolerance of light, the pupil being very much contracted. There is, too, in this affection, always an expression of shame, the eyes seldom if ever meeting those of another with confidence, being turned away hastily, and after wandering about for a moment, are at length turned to the ground.

The brain, after a time, becomes very seriously involved, the ideas being confused, and memory weak, and the sleep, at first interrupted and unrefreshing, becomes gradually less, till the patient gets but little sleep, and that of an unnatural character.

The patient gradually becomes desponding and melancholy, and in many cases there is a distrust of every one, and a strong inclination to commit suicide. Congestion of the brain is liable to occur from slight causes at first, and, as the disease progresses, chorea, catalepsy, epilepsy, or apoplexy may be the result.

Patients suffering from this disease are always hypochondriacal, being nervous, irritable and excitable, from slight causes. They are apt to dwell constantly on their sufferings, and are almost sure to forget everything else, very much like a person in advanced age.

In consequence of the long continuance of this disease, the brain becomes permanently diseased, and the intellect more or less impaired, in many cases approximating idiocy; and, when this is not the case, insanity in some form is apt to be the result, numerous cases of which have fallen under my observation during the past few years. And in persons insane from spermatorrhœa, paralysis is very apt to occur, probably in consequence of a softening of the brain, with a cephalic tendency, which so generally exists in such cases.

Masturbation, as practiced by young children and females, is liable to produce the same train of symptoms as we have already seen, which occur in the male from spermatorrhœa, though in these cases there is no discharge; the irritability of the genital organs being reflected to the cerebro-spinal system, and deranging the ganglionic nerves, produces the same constitutional symptoms, though perhaps in a less degree.

Thus we have, according to my observation, the main symptoms of spermatorrhœa, a disease which furnishes more dyspeptics, hypochondriacs, lunatics, and idiots, perhaps, than any other affection; and which causes a fearful proportion of the mortality which occurs in the young of both sexes.

Causes.—Spermatorrhœa may occur from various causes, which tend to render the genital organs morbidly irritable; such as ascarides in the rectum; cutaneous eruptions on the external genital parts, or about the anus; stricture of the urethra; hemorrhoids; fissure of the anus; spinal irritation, and constipation of the bowels.

But masturbation, onanism, and excessive sexual indulgences, are by far the most frequent causes of this affection; and of the two, I believe, from careful inquiry, that masturbation is the most frequent cause. But in those who have practiced masturbation, whether male or female, and thus rendered the genital organs morbidly irritable, only a moderate degree of sexual excess is sufficient to develop the disease, with all its fearful consequences.

Pathology.—The pathology of spermatorrhœa is very plain, when we remember the peculiar structure of the genital organs, and the strong sympathy existing between them and the cerebro-spinal system.

At first masturbation irritates the genital organs, in consequence of which erections take place, from a very slight exciting cause, attended often with a secretion and discharge of semen, but in no very great quantity at first. But as the habit and morbid discharge continue, the system becomes prostrated, nervous and irritable; an irritation is set up in the spinal marrow, at the origin of the spermatic nerves, near the junction of the dorsal and lumbar portions of the spine, attended with more or less tenderness and pain.

The brain finally becomes irritable, and in many cases softened; sufficient nervous energy or influence not being generated to enable the different organs of the body to perform their functions; and, as a consequence, the stomach, liver, and alimentary canal become deranged; dyspepsia, constipation, &c. being the result.

The hepatic derangement, together with the constipation of the bowels, produces in many cases hemorrhoids, fissures, or other local irritation, in the lower part of the rectum; which, being in the immediate vicinity of the genital organs, greatly increase their irritation and the disease.

Now the draft upon the blood, caused by the seminal discharge; and the impaired nutrition, in consequence of the indigestion, greatly reduces the blood; thus tending to soften and depress still more the brain, and to irritate the nervous system. The brain fails to afford the heart sufficient vital force, through the ganglionic nerves, to enable it to carry on the circulation in the extremities. As a consequence of this, the hands and feet become cold; the brain becomes congested; the patient, before hypochondriacal, becomes sullen, stupid, and often insane; and the heart, struggling to carry on the circulation, palpitates tumultuously, and the patient is restless and irritable, or stupid and insensible, according to the degree of the cephalic congestion. And finally, insanity, in some form, with or without epilepsy, paralysis, or apoplexy, is very apt to be the result.

In females, though there is no wasting discharge, masturbation or venereal excesses produces an irritable, congested, and inflamed condition of the genital organs, and a sympathetic irritability of the cerebro-spinal system, thus deranging the generation and distribution of the nervous influence to the various organs, and materially interrupting their functions, developing, the same train of symptoms which occur in the male from similar causes.

Such I believe is the true pathology of this disease, many cases of which have fallen under my observation and care during the past twenty years, both male and female.

Diagnosis.—The diagnosis in males may readily be arrived at, if the patient will own the facts, and divulge the amount and frequency of the seminal discharge.

But in many cases from masturbation or sexual excesses in the male or female, the facts will be stoutly denied, in which case the diagnosis is to be made from the symptoms which are developed.

If then the patient, whether male or female, young or in middle life, married or unmarried, exhibits the symptoms which I have described, it is morally certain, that either masturbation, onanism, excessive sexual indulgence, or some other influence, which produces a similar condition of the genital organs is the cause, and in males generally, that there is a spermatorrhœa, either diurnal or nocturnal.

The cold extremities, pale face, sunken eyes, downward look, palpitation, cephalic congestion, indigestion, irritability, hypochondria, and idiotic expression of countenance, together with the tendency to insanity, epilepsy, apoplexy, &c., are the symptoms to be relied on, in forming a diagnosis in such cases.

Prognosis.—The prognosis in recent cases provided the cause can be removed, is generally favorable. In protracted cases, however, in which the brain has become softened, or in any way very materially involved, the prognosis is always unfavorable.

Treatment.—The first indication in the treatment of this disease is to stop or correct the imprudence which has produced it. If the cause be masturbation, as is generally the case, its fearful consequences should be pointed out, and the continuance of the habit strictly forbidden, whether the patient be male or female.

If excessive sexual indulgence be the cause, that habit should be corrected or brought within reasonable bounds, as entire continence in the married, might lead to, or increase diurnal or nocturnal discharges. It may be a matter of doubt, how often sexual indulgence might be salutary in such cases, but I believe it should not be allowed oftner than once in three or four weeks in the married, while in the unmarried it should of course be strictly prohibited.

If onanism be the cause, as it very often is, that most ruinous habit should be prohibited, as it not only leads to great irritability of the male genital organs, but it also evidently produces an irritable condition of the female upon whom it is practiced, similar to that produced by masturbation, or perhaps worse.

The patient should be directed to take a plain nourishing diet, that of milk and bread made from unbolted flour, with an allowance of meat, and cold water for drink, will generally do best. The food too should be well masticated and taken with strict regularity.

The patient should be made to wash the surface of the body, once or twice per week, with moderately cool water, containing a little salt; taking care to rub the skin dry with a towel. Flannel should be worn next the skin, to promote warmth and favor perspiration, and the patient should be directed to lay on the *side* in bed, at all times, and to be comfortably, but not too warmly covered.

If there is piles, fissure of the anus, or soreness in the lower part of the rectum, from the long continued constipation, or if the bowels continue constipated, or there is ascarides in the rectum, a free injection of cool water should be used each morning, and a movement secured, in preference to any other hour of the day. As an application to the piles the ointment already suggested, in the article, Hemorrhoids, in the eighth chapter of this work, may be applied; or an ointment made by mixing together the extract of stramonium, oxide of zinc, and tannin, of each two drachms, and then adding and mixing, intimately, two ounces of simple cerate, may do equally well in such cases. Of either of these ointments a little may be applied, with the end of the finger, morning and evening, while the pain and soreness continue.

If there is a fissure of the anus, or an eruption upon the surface about the anus, an ointment, made by mixing two drachms each of the oxide of zinc and tannin, with an ounce of lard, applied morning and evening, will generally do best. Or, if the eruption or roughness extends to the external genital parts, as it is apt to in females, a solution of borax, two drachms to eight ounces of water, may be applied, morning and evening, instead of the ointment of oxide of zinc, as it is more cleanly, and will generally subdue the irritation.

Dry cups should be applied occasionally to the back of the neck, and also along the junction of the dorsal and lumbar regions of the spine, at the origin of the spermatic nerves.

As a tonic, for the cerebro-spinal and nervous system, the muriate of strychnia, in $\frac{1}{16}$ of a grain doses, will generally do best. It is most conveniently administered by dissolving a grain of the muriate in eight ounces of water, of which a teaspoonful may be given, three times per day, before each meal; and continued as long as it may be required. This will generally regulate the bowels, improve the digestion, &c., by its tonic effect upon the nervous system.

As a tonic for the blood, some preparation of iron should be given, and continued for a long time. If only the tonic effect is wanted, two or three grains of the carbonate, or ammoniated citrate, may be given, three times per day, after each meal. If however an alterative is indicated, ten drops of the syrup of the iodide should be given instead. Or, if the bladder has become irritable, as is often the case, ten drops of the tincture of the chloride of iron will generally do best.

This course of treatment commenced in season, and judiciously applied, so as to fulfill the indications in each particular case, will generally prove successful. In case of failure however, the prostrate gland may be cauterized with the nitrate of silver, by means of the instrument of M. Lallemand.

SECTION II.—GONORRHŒA.

By gonorrhœa, I mean here, that peculiar contagious disease, consisting of a specific inflammation, attended with a mucous or purulent discharge; having its seat generally in the mucous membrane, of the male or female genital organs.

The urethra of the male, and the vagina of the female, are the most common seats of this affection. It may occur, however, in the inner surface of the prepuce, or on the glans penis of the male, or in the vulva, or uterus in females. Or the disease may be produced in any part of the body, if the matter of gonorrhœa be brought in contact with a mucous surface.

In relation to the identity of gonorrhœa and syphilis, there is a difficulty in arriving at a definite conclusion, as both affections frequently exist in the same individual; but I incline to the opinion, that the diseases are distinct.

The period between the exposure and the appearance of the disease, varies from a few hours, to several weeks; but four or five days, is about the usual time, in most cases.

There is sometimes a purulent or mucous discharge from the male or female genital organs, not the result of the venereal poison, which resembles it, except that it is not contagious, and that the discharge is attended with a disposition to resolution, while in gonorrhœa, the matter secreted being a virus, tends to keep up, instead of terminating the inflammation. Gonorrhœa then, as it occurs in the male or female, is attended with a peculiar train of symptoms, which we will now proceed to consider.

Symptoms.—Generally after an indefinite period, varying from a few hours, to five or six days after exposure, there is felt an itching at the orifice of the urethra, and sometimes over the whole glans; a little fullness appearing at the lip of the urethra. Soon this itching is changed to a pain, especially at the time of voiding urine, and with the symptoms of inflammation of the mucous membrane which are developed, the discharge appears. In some cases, however, there is no pain, till after the appearance of the discharge, and occasionally a case occurs, in which there is little or no pain, even though the discharge be very considerable.

At this stage, there is a fullness of the penis, especially of the glans, the organ being to appearance, in a state of half erection, and the glans having a degree of transparency near the orifice of the urethra, the surface being full and red. In consequence of the fullness of the penis, and swelling of the urethra from inflammation, the stream of urine is smaller than usual, and is generally scattered and broken, as it leaves the passage. And if the distension of the urethral vessels be considerable, there is sometimes a hemorrhage from the urethra; but this is not generally the case. The glands along the under side of the penis sometimes enlarge, and occasionally they inflame and suppurate. A soreness is also sometimes felt along the under side of the penis, owing to a high state of inflammation of the urethra.

In consequence of the irritation of the penis, erections of a very painful character are liable to occur in gonorrhœa, especially in cases in which the inflammation runs high and is attended with considerable pain.

The *discharge* in gonorrhœa appears like pus, and is probably the result of inflammation, changing the slimy discharge from the glands of the urethra, as well as the natural exhaling fluid of the canal to this character. The matter is sometimes lightish or greenish, but generally of a yellowish color; but if it continues and becomes a gleet, it is of a watery character, being sometimes almost transparent.

The inflammation in gonorrhœa generally extends back along the urethra for half its length, and in some cases further. And when the inflammation extends deep to the reticular membrane, becoming adhesive and uniting together the cells of the corpus spongiosum urethra, it destroys in some cases its power of distension, so that a curvature takes place at the time of erection, which has been called a *chordee*.

An occasional attendant of gonorrhœa is a swelling of one or both testicles. It may occur at any stage of a gonorrhœa, but it more frequently happens from a sudden suppression of the discharge from some cause. As the inflammation of the testicle becomes fully established, it is attended with severe pain, both in the gland and in the back at the origin of the spermatic nerves. Sympathetic vomiting is apt to occur, and also more or less spasmodic pains in the bowels. Orchitis is also attended with more or less febrile excitement.

Gonorrhœa in women is not attended generally with a very high degree of inflammation or much pain, if only the vagina is involved. But if the womb, vulva, or urethra becomes involved, there is generally a good deal of pain, especially in walking, passing water, &c. The discharge does not differ very materially in appearance from the matter of leucorrhœa, so that the evidence of gonorrhœa in women depends upon the fact of their having been exposed, or of their having communicated the disease to some one, all other symptoms being sometimes deceptive.

In some cases of gonorrhœa, in both males and females, the glands of the groin become enlarged and indurated, and various constitutional symptoms arise, such as slight irritation of the fauces, a papular eruption of the skin and slight swelling of the joints, with more or less febrile excitement.

Such are the ordinary symptoms of gonorrhœa as it occurs both in the male and female, liable of course to variations, like all other affections.

Treatment.—If the patient be seen after the exposure, and before the disease makes its appearance, he should be directed to wash the parts thoroughly with soap and water, and then with a saturated solution of alum and water, and if there is slight uneasiness, a little of the solution may be injected into the urethra. Or if the patient be a female, into the vagina. This washing with the alum solution, and injection, if necessary, may be repeated morning and evening, for two or three days, and may, in some cases, enable the patient to evade the physical penalty of *broken law*, but not always.

If, however, the disease is fully established, measures should be taken at once to subdue the inflammation, and then to arrest the discharge.

The patient should be directed to be quiet, and if the attack is severe, to keep in bed. A cathartic of the sulphate of magnesia should be administered, and then the bowels kept moderately loose, by teaspoonful doses repeated each morning, if necessary. Tartar emetic in one-fourth of a grain doses should be given every three hours, till the inflammation in a measure subsides, and if there is much pain in the back, cups may be applied, and a little blood taken.

After thus subduing the severity of the inflammation, in violent cases, and immediately after the operation of the first cathartic, in mild cases attended with little inflammation, half a drachm of the balsam of copaiva may be given, from three to five times per day, in the form of emulsion, with sugar, gum arabic and water. It may be conveniently prepared by mixing together in a mortar, an ounce each of pulverized gum arabic, sugar, and water, or perhaps a little more water, and then adding very slowly while stirring, an ounce of the balsam of copaiva. When the balsam is thus all intimately mixed, half an ounce of the compound spirit of lavender may be added, and then sufficient water gradually stirred in, to make eight ounces of the mixture. Of this a table spoonful may be given after each meal, and if it does not loosen the bowels too much, early in the morning and at evening. In preparing the mixture for cases in which there is little pain, half the quantity of gum arabic and sugar may be sufficient, or even less than that. But in violent cases, in which there is pain, heat, and scalding on passing urine, I believe that a liberal allowance of the gum constitutes a valuable ingredient in the mixture.

In case the discharge continues, after four or five days, the emulsion may be prepared in the same way, with the addition of an ounce of the tincture or fluid extract of cubebs, to each eight ounces, and this should be continued with the use, if necessary, of injections, till a cure is effected. If injections become necessary, as they may in some cases, half a drachm of the sulphate of zinc may be dissolved in eight ounces of rain water, and a small quantity of this should be thrown into the urethra, with a small syringe, four times per day.

If, however, a gleet supervenes, or follows the gonorrhœa, a saturated solution of alum in rain water may be used, instead of the zinc, and if this fail, a solution of tannin, two drachms to eight ounces of water may be resorted to, and continued till the gleet is arrested.

The treatment for females with gonorrhœa, should be the same as for males, only as an injection, a solution of alum, two drachms to the pint of water, will generally do best, at first. But if the disease becomes chronic, tannin of the same strength should be used, and continued till the discharge is arrested.

When the testicles become swelled in males, cups should be applied to the back, at the seat of the pain, and a few ounces of blood taken: a saline cathartic should be given, and four grains each of James's and Dover's powder administered, every four or six hours, till the febrile excitement subsides. As an application to the testicles in such cases, a solution of the muriate of ammonia, of the strength of two drachms to eight ounces of vinegar, with the addition of two drachms of laudanum, will generally do best at first. Later, however, a solution of the iodide of potassium, two drachms to eight ounces of water, may be applied by means of a cloth saturated with it, and over this a poultice of hops, wet in warm vinegar or water should be kept applied till the swelling subsides.

Finally, if constitutional symptoms follow gonorrhœa, such as swelling of the joints, sore throat, &c., a blue pill may be given at evening for a few days, followed by a teaspoonful of the sulphate of magnesia in the morning, and then the iodide of potassium should be given in five grain doses, three times per day, and continued till a cure is effected.

SECTION III.—SYPHILIS.

By syphilis is here meant that peculiar infectious disease usually affecting primarily the male or female genital organs, the first development of which is a small pimple, sore, or chancre, but if neglected, producing various constitutional symptoms, involving the glands, throat, skin, bones, &c.

The primary sore or chancre may occur on the internal surface of the prepuce, the glans, the external surface of the prepuce, skin of the penis, scrotum, or thighs of men, and in females, on the internal or external surface of the labia pudendi, on the clitoris, the nymphea, in the vagina, or on the thighs.

This sore or chancre may make its appearance at an indefinite period varying from a few days to several weeks, after an impure connection, and is at first a strictly local disease. This mere pimple or chancre, if left to itself, leads on to a train of general or constitutional symptoms, which we will now proceed to consider, leaving the diagnostic characters of the chancres for consideration when we come to the diagnosis.

Symptoms.—The primary sore of syphilis, a mere pimple at first, if allowed to pass on, furnishes a virus, which being taken up by the absorbent lymphatic vessels of the part, and carried into the system, eventually poisons, deranges, or contaminates all the fluids, and solid tissues of the body. As this virus passes through the lymphatic glands of the groin, they frequently become inflamed, enlarged, and suppurate, constituting a bubo in one or both groins. Now this state of the disease during which the poison is being carried into the system may be called the consecutive stage, and occupies the time from the fifth or sixth day to the second or third week after the first appearance of the pimple, sore or chancre.

After the venereal virus has been carried into the blood or general system from a chancre or syphilitic bubo, certain constitutional symptoms make their appearance, and this constitutes the secondary stages of the disease. The secondary symptoms usually make their appearance by the fifth or sixth week after the chancre, and from one to three weeks after the appearance of the bubo.

These secondary symptoms usually make their appearance first in the throat, by a slight soreness, and then on the skin by a peculiar scaly or other eruption, which may affect any part of the surface, but especially the forehead, neck, breast, forearms, legs, and anterior part of the abdomen. And usually as the scales fall, after these sores, they leave a peculiar copper colored base or scar.

If now the disease be not arrested, the periosteum, fascia, ligaments, and bones become affected, and this constitutes the tertiary stage of syphilis.

The bones most frequently affected are the cranium, sternum, clavicle, and those generally nearest the surface, the nodes which are formed being a simple enlargement of the bone, not attended with discoloration at first, but becoming painful after it has continued for a time.

Such then, it should be remembered, are the ordinary symptoms of syphilis, during its primary consecutive, secondary, and tertiary stages, the chancre being the symptom of the primary stages, the bubo when it appears of the consecutive, the sore throat and cutaneous eruption of the secondary, and the affection of the bones, with nodes, &c., of the tertiary stage of the disease. And it should still further be remembered that the disease is primary and local for the first five or six days, consecutive from the fifth or sixth day, to the second or third week, secondary for several weeks, or perhaps months, and tertiary from the first appearance of the nodes, and perhaps earlier, during the continuance of the disease. Thus we see that one stage supervenes upon another till the whole of the fluids, and solid tissues of the body become contaminated, and unless the disease be arrested, disorganization of the various tissues takes place, and the patient finally becomes a wreck of pollution, till death at last pays the physical penalty of broken law.

Diagnosis.—The diagnosis of syphilis, and especially of chancre, is attended sometimes with considerable difficulty, as other sores, not syphilitic, are liable to occur on the genital organs, and also as the chancre is liable, from peculiar conditions of the system, to marked variations. Generally, however, if a sore makes its appearance on the genital organs, within a few days or even weeks after exposure, and assumes an appearance answering to the following descriptions, it may be regarded as chancre.

The *indurated* chancre is preceded by an itching, a small pimple first appearing, which soon contains matter, and then turns into a regular ulcer, the base of which is hard, and its edges regular. The thickening of the tissues about it does not extend far, being rather circumscribed. The edges of the ulcer are surrounded by a narrow line of inflammation, and the bottom of the sore is covered with a gray or yellowish matter which adheres, and differs from common pus. Gradually the edges of the sore become rounded off, the areola disappears, and if the sore heals it leaves a hardened cicatrix which is liable to ulcerate again

The *simple* chancre which is more common than the indurated, resembles it very much, only the base of the ulcer is free from hardness, and it is not attended with any very considerable signs of irritation or inflammation.

The *irritable* chancre differs in being red, painful, and irregular in appearance, and very much disposed to spread, if irritating applications be resorted to.

The *inflamed* chancre is a modification of the simple, caused by the supervention of inflammation, frequently from excesses on the part of the patient, in consequence of which it becomes painful, red and swelled, and loses its regular appearance. The edges of the ulcer in these cases are often removed by dark sloughs, and the secretion becomes of an irritating or acrid character.

The *sloughing* chancre occurs in persons of broken down constitutions, caused by excesses of various kinds, in consequence of which the ulcer and surrounding parts are destroyed by sloughs or gangrene, destroying in some cases a large part of the genital organs, in either sex.

If then a sore be found upon the genital organs, or in their vicinity, presenting either of the preceding characteristics, and the patient has been exposed, the disease may be regarded as syphilitic, the variations in the appearance depending probably in the main upon the habits, constitution of the patient, &c.

If, however, the patient is not seen till the consecutive stage, there will generally have been the supervention of other symptoms, and often that of bubo. And still later, during the secondary stage, there is the supervention of sore throat, scally eruptions, &c. Finally in the tertiary stage, when in addition to all the preceding diagnostic symptoms, nodes, ulceration of the bones, and other tissues, and other like symptoms shall have supervened, the diagnosis is rendered clear and positive.

Thus may syphilis, in all its stages and variations, generally be known, without the necessity of a reasonable doubt.

Treatment.—If a patient apply any time within six days after the first appearance of a chancre, it should be cauterized by passing over it lightly a stick of lunar caustic. A piece of dry lint should then be applied over the sore, and secured by a bandage. After the first scab formed by the caustic falls off, it may be cauterized again, and this should be repeated, if necessary, till the whole surface of the sore shows a tendency to heal.

As soon as the reparative process commences, the sore may be dressed with the ointment of oxide of zinc, or if it is disposed to discharge rather freely, the sore may be washed four times per day with a solution of eight grains each of tannin and sulphate of zinc in four ounces of rain water, and a small bit of cotton or lint, wet in the solution, may be kept constantly applied. This treatment should be continued till the chancre heals.

This may be all the treatment required if the chancre has not continued more than five or six days when the treatment was commenced. If, however, the disease be not arrested by these measures, and the consecutive stage approaches, as well as in all cases that are neglected till the disease has passed on to this stage, with perhaps inflammation of the lymphatic glands in the groin, constitutional treatment should be resorted to at once.

Five grains of blue mass may be given morning and evening, and continued till a mercurial taste is perceived in the mouth, or slight ptyalism is produced, when it should be discontinued, and five grains of the iodide of potassium given three times per day instead, and continued till a cure is effected.

If a bubo appears, attended with febrile excitement, a full dose of the sulphate of magnesia should be administered, and a cold solution of the muriate of ammonia in vinegar, two drachms to eight ounces, kept applied to the inflamed gland. If, however, the bubo increases, a blister may be applied, and the raw surface dressed, as it begins to heal, with mercurial ointment, or if the mercurial is not tolerated, the compound iodine ointment may be applied instead.

With the blue pill, followed by the iodide of potassium, the blisters, and mercurial or iodine ointment to the buboes, and the oxide of zinc, ointment, or solution of tannin and sulphate of zinc to the chancre, the disease may be arrested in most cases, and the secondary and tertiary symptoms not occur.

In cases, however, that have been neglected till secondary symptoms appear, the blue mass should be given morning and evening, as already suggested, till a slight ptyalism is produced, and then the iodide of potassium should be substituted, and continued till a cure is effected, administered in the fluid extract, syrup, or compound decoction of sarsaparilla.

In cases of syphilis that have passed on to the tertiary stage, with nodes, nocturnal pains, &c., the iodide of potassium may be given in ten grain doses with the sarsaparilla, with the addition of full doses of the syrup of the iodide of iron, if necessary.

In all cases of syphilis in which a mercurial is contra-indicated from any cause, ten drops of the fluid extract of stillingia should be given three times per day instead, with the iodide of potassium and sarsaparilla, and continued till a cure is effected.

By metroperitonitis, from μητρα "the uterus," and "peritonitis," I mean here inflammation of the uterus and peritoneum, liable to occur at any time, but generally in the puerperal state.

The uterus or womb is situated, it will be remembered, near the middle of the cavity of the pelvis, between the bladder and rectum, above the vagina, and below the small intestines.

It consists of a *body*, *neck*, and *cavity*. The body is about two inches long, terminating in the neck inferiorly, which is smaller than the womb, and about ten or twelve lines in length. The lower extremity of the neck projects into the vagina, and is called the os tincæ.

The cavity of the womb is small, occupying the body and neck, that of the neck consisting of a mere canal, commencing in the os tincæ by a transverse opening, and being continuous above with the cavity of the womb. This cavity in turn is triangular, and at its superior angles, on each side, is continuous with the opening of the Fallopian tubes.

The uterus is composed of an outer serous or peritoneal coat, a mucous or internal lining membrane, and a middle parenchyma, of a muscular character, together with its arteries, veins, and nerves.

The outer or peritoneal coat of the womb, consists of a duplicature of the peritoneum, which being extended across, and connected with the sides of the pelvis, constitute the brad ligaments of the uterus, enclosing the *Fallopian tubes*, with their fimbriated extremities, as well as the ovaries, connected by their ligaments with the uterus, and also with the fimbriated aperture of the Fallopian tubes.

The internal mucous membrane of the womb is continuous with that of the vagina, through the canal of the neck, and is firmly adherent to the tissue of the organ, lining its whole cavity, and extending also into the cavity of the Fallopian tubes, which are lined to their fimbriated extremities.

The middle muscular coat of the womb, constituting its parenchyma, which gives thickness and bulk to the organ, is of a whitish color, of a dense elastic structure, and consists of an external and internal layer of fibres, mingled with arteries, veins, and nerves.

The arteries of the womb consist of the uterine, from the internal iliac, and the spermatic, from the aorta; its veins terminate on each side of the organ, in the uterine plexuses; and its nerves are derived from the hypogastric, spermatic, and sacral plexuses. It is also well to remember in this connection, that the womb has two round ligaments, which serve to retain the organ in its place, extending from its upper angles, along the spermatic canal, to the labia majora, in which they are lost.

Now, when we take into account the position and structure of the womb, it is not strange that it should become inflamed, especially in the puerperal state, when all its vessels have been very much enlarged and distended. Nor is it strange that an inflammation of the uterus, occurring in the puerperal or any other state, should extend to its peritoneal covering, and from that to other parts of the peritoneum, and thus produce a rapidly fatal metro-peritonitis.

It is this inflammation of the womb, extending to the peritoneum, and occurring most frequently in the puerperal state, a few days after parturition, which now demands our attention; the symptoms of which we will now proceed to consider.

Symptoms.—There is usually more or less pain experienced in the uterus, attended with some tenderness, at the very commencement of the inflammation. But in many cases there is a chill, succeeded by fever, before any serious pain is experienced in the uterus. There is however in all cases, sooner or later in the disease, a fixed, continuous, and lancinating or dull and aching pain in the pelvis, which is greatly increased by pressure or motion, and accompanied also with a sense of weight in the perinæum.

The urine is passed with difficulty, especially if the lower and anterior portion of the womb is involved in the inflammation. If the inflammation involves the posterior part of the organ, the pain is severe in the sacral region, and the patient experiences considerable tenesmus of a more or less distressing character. Sometimes the pain is severe in the iliac regions, and extends to the hips and down the thighs, especially is this the case if the lateral portions of the womb are involved, more particularly in the inflammation. But more generally, however, in severe cases there is a combination of nearly all these symptoms, indicating that the whole uterus is involved in the disease.

In the early stage of the inflammation, if it is confined mainly to the uterus, very little or no swelling of the abdomen occurs; but on examining the pubic region with the hands, the womb is found considerably enlarged being firm and tender to the touch. But if, as generally happens in such cases, the inflammation extends from the womb to the peritoneum, there is a good deal of tenderness of the abdomen and considerable tumefaction.

In puerperal cases of this disease, there is generally an entire suppression of the lochia, as well as cessation of the secretion of milk. As the disease progresses, the pulse becomes frequent; there is great prostration of strength; frequent muttering delirium; constant recumbrance on the back with the knees drawn up and the shoulders raised; a dry, coated, brown tongue; and frequently towards the termination a diarrhœa occurs.

In the majority of cases the system sympathizes strongly with the inflamed organ, even before the inflammation has extended to the peritoneum. In most cases there is nausea with occasional vomiting. The pulse is either full, strong and hard, or contracted, small and frequent. There is also usually severe headache, and towards evening more or less delirium occurs.

The course of this disease is generally rapid, and if the inflammation does not tend to resolution in four or five days, it will generally terminate in suppuration or gangrene.

When the inflammation tends to a favorable termination, the pain, tenderness and swelling of the womb abate, and the abdominal distension subsides. The pulse becomes slower, soft and open; the skin moist; the tongue clean and humid; the lochia flows more freely; and the urine becomes more free and natural in its appearance.

The occurrence of suppuration in the substance of the womb is always attended with considerable danger. For if the abscess bursts into the cavity of the peritoneum, or passes between the peritoneum and the external surface of the womb, into the cellular tissue in the lower part of the pelvis, there will generally be a fatal termination. If, however, the matter finds its way into the cavity of the uterus, it may then be discharged through the vagina and a favorable termination may be the result.

The occurrence of suppuration may be suspected, when in an obstinate or severe case, which has passed on five or six days—the pain gradually subsides, being followed by a sense of weight in the affected part. The pulse, too, becomes more frequent, slight chills occur, the distribution of heat becomes irregular, cool sweats appear, the lochia becomes fetid, there is prostration of muscular power, and generally the tongue becomes red, and a livid flush may appear on one or both cheeks.

A termination in gangrene is not an unfrequent event, if, as generally happens, the peritoneum is extensively implicated in the inflammation.

On the supervention of gangrene, there is a rapid cessation of pain, great muscular prostration, vision becomes impaired, the countenance becomes cadaverous, a clammy sweat breaks out, the extremities become cold, the pulse becomes small, weak, and rapid, and finally death puts an end to the sufferings of the patient.

In cases of this disease, in which the peritoneum becomes rapidly and extensively involved in the inflammation, unless it be speedily subdued by prompt and early measures, a collapse of the vital energies may supervene in forty-eight hours, or even sooner, in some cases.

Such, I believe, are the ordinary symptoms of metroperitonitis, liable to variations, as is the case with all diseases. Cases occurring out of the puerperal state are quite rare, and, when they do occur, do not develop the symptoms incident to the puerperal state, of course.

Anatomical Characters.—The parenchyma of the womb is frequently found containing more or less matter in different parts, or the abscesses have burst into the uterine cavity, or opening externally, the matter has passed between the outer surface of the womb and its peritoneal coat, into the lower part of the pelvis, or having burst through the peritoneal coat, it is found in the peritoneal sac.

The peritoneal coat of the uterus exhibits signs of recent inflammation, as well as the parts immediately surrounding that organ, and not unfrequently the whole peritoneum shows signs of recent inflammation. It may be found red and thickened, and the abdominal viscera are often found adhering to one another by effused lymph, or there may be found a turbid serous fluid, containing shreds of albumen, pus, or blood, in the peritoneal sac.

The omentum is found of a deep red color, is apt to be highly vascular, and is not unfrequently adherent to the fundus of the uterus, by means of effused lymph. Finally, in some cases, the uterus and its appendages present a gangrenous appearance, as well as the peritoneum, not only that portion which lines the parietes of the abdomen, but also that which is reflected over the intestines, forming their outer or peritoneal coat.

Causes.—Metroperitonitis, as it occurs out of the puerperal state, may be caused by injuries, exposure to cold, when the body is heated, &c., in the same manner that these influences produce inflammation of other parts.

Various causes may operate to produce this disease in the puerperal state. Immediately after parturition, the womb undergoes various important changes, which may have a bearing on the development of this disease, if proper care and prudence is not exercised. Its internal lining membrane has been irritated by the detachment of the placenta, and the mouths of the extreme uterine vessels, doubtless remain for a little time imperfectly closed, which may predispose the organ to take on inflammation.

The womb is liable too, during contraction in labor, to slight over distension of its structure, by which its arteries, veins, and lymphatics become lacerated, or very much distended, and their natural contractile powers impaired in a greater or less degree, for the time, in consequence of which, they remain in a passively congested condition, predisposing the organ to congestion or passive inflammation. There is also the liability of direct injury of the organ during labor, by the hand in turning the child, or by the introduction of instruments.

Now after parturition, the womb, from having been in a condition of active distension, is in a weak, debilitated, and relaxed state, and the pressure of the fœtus having been removed, leaves its tissues more or less predisposed to congestion or passive inflammation. If now, with all these predisposing circumstances, the patient be imprudently allowed to assume the erect or sitting posture, within the first few days after confinement, the position will still further tend to increase the congestion of the uterine vessels, by impeding the return of venous blood along the inferior vena cava, and portal veins.

In this way a passive congestion of the organ may increase, till the womb becomes sensibly enlarged, and slightly tender to the touch. This liability to congestion is also greatly increased by the faintness or prostration which is liable to occur, from the patient assuming the erect or sitting posture, within the first few days after confinement.

The blood, as in all other similar conditions of the system, does not pass freely to the extremities, in consequence of diminished action of the heart and arteries. And as the erect position lessens the flow of blood to the brain, and increases its flow to the uterus, a degree of passive congestion is produced in the arteries, veins, and capillaries of the womb, which so far irritates the organ, as to stop the lochial discharge from its inner or mucous surface. This retained lochial matter still further increases the already excessive irritation, and pain, swelling, and tenderness is the result.

But the impression of this morbid condition is carried to the brain, through the sympathetic and cerebro-spinal nerves, and nausea, vomiting, and prostration of the powers of the system follow, during which the blood recedes from the extremities, and a violent chill follows.

During this chill the congestion of the womb amounts to an enormous distension of its vessels and tissues, which sets up a passive inflammation of its mucous, muscular, and serous tissues, attended with general

43

febrile excitement, nausea, vomiting, and very soon, great prostration of the powers of the system.

Though there is in such a case, violent throbbing of the vessels of the womb, they are unable to relieve the congested and inflamed organ. And the inflammation being rendered still more putrid or malignant by the retained lochia, if it is not soon arrested, spreads rapidly to the appendages of the womb, and also to the peritoneum, developing all the symptoms which occur in what has sometimes been called "puerperal fever," but which appears to be really a *metroperitonitis.*

And finally, if the disease be not arrested, it terminates either in suppuration, involving in a greater or less degree the different tissues of the womb, or in gangrene of **the womb, its appendages, and more or less of** the peritoneal sac.

Such I believe are the causes or circumstances which lead to metroperitonitis in the puerperal state. But it should be remembered, that in addition to the imprudent assumption of the erect or sitting posture, soon after confinement, that various other exciting causes may operate to develop the disease. Among these causes are sudden exposure to cold **or dampness, large** draughts of cold water, **violent** mental emotions, **want of cleanliness,** epidemic, and other kindred influences, all of which, I am **satisfied, tend** to develop the disease, **in the** manner already suggested.

Now, while cases of puerperal metroperitonitis are generally of a *passive* character, it should be remembered that exceptions occur occasionally in the puerperal state, and that in the disease, as it occurs out of the puerperal state, from wounds or other accidental causes, **the inflammation** is generally of a decidedly *active* character.

Treatment.—The treatment of metro-peritonitis out of the puerperal state, **as well as of the occasional** puerperal **cases in which** the inflammation is of an active character, should be similar to that for inflammation **of other** vital or important parts, and should consist of general bleeding, **cupping, a** mercurial cathartic, sinapisms, fomentations, and opium or **Dover's powder, with calomel in** alterative doses, and in some **cases** blisters may be required.

The treatment of *puerperal* metroperitonitis, being, as I believe it generally is, of a *passive* character, requires a sound discretion and the exercise of good common sense.

In the **first place, by** carefully **avoiding** any injury of the womb **in** examinations **which** sometimes become necessary during labor, by applying a bandage moderately tight about the abdomen after delivery, and by enjoining and enforcing the recumbent or horizontal position for the first ten days at least, and then only occasionally allowing the sitting posture in bed for **another** week, I believe that puerperal metroperitonitis may generally be avoided, if prudence is exercised in other respects.

But if from any cause the disease makes its appearance, measures should immediately be taken to equalize the circulation, to counteract local and general irritation, and to keep up the sinking powers of the system.

Cups should be applied over the sacrum, and to the hypogastrium, and from two to six ounces of blood taken. The spine should be thoroughly rubbed with a strong infusion of capsicum in vinegar, with a view of equalizing the circulation, which it will very generally materially assist in doing. A full dose of calomel and castor oil should be administered, and its operation secured by repeating the oil if necessary.

After the operation of a cathartic, there is generally considerable prostration, and though the local inflammation continues, it is generally of a passive character, and greatly aggravated by a debilitated condition of the system. To quiet pain and general irritability, keep up the sinking powers of the system, and produce a slight alterative effect, Dover's powder, quinine, and calomel should now be given.

About two grains of calomel, three grains of quinine, and five grains of Dover's powder may be given every four hours, and continued till a slight impression is produced on the gums, or the inflammation yields, when the calomel should be omitted, and three grains of James's powder added to the quinine and Dover's and this should be continued every four or six hours, increasing the quinine if necessary, till the disease is arrested.

After cupping, some benefit may frequently be obtained by a warm hop poultice, to the perinæum or hypogastrium. The bowels should be kept gently loose by small doses of castor oil, and in some cases much benefit may be derived from sinapisms over the abdomen, and especially to the hypogastrium, in the early stages of the disease. Blisters too in some rare cases may be indicated. But generally I think they produce too much irritation for a patient in the puerperal state.

The patient should be nourished by crust coffee one half milk, and some plain solid food should be allowed as soon as the stomach will bear it.

In cases of this disease of a highly putrid or malignant character, in addition to the quinine, camphor, carbonate of ammonia, wine-whey, and even brandy may be required to sustain the sinking powers of the system.

By thus keeping in mind the exact condition in each case of this disease, and doing just that which is indicated, and nothing more, I believe that the disease, though always attended with danger, may generally be arrested, if attended to in season, unless there is some special malignancy as occurs in certain epidemics of a decidedly malignant character.

SECTION V.—CHRONIC METRITIS.

By chronic metritis, I mean here a chronic inflammation of the womb, whether the result of the acute form of the disease, or arising as an original affection, and liable to involve all its tissues, but especially its lining mucous membrane, and the neck of the organ.

Chronic metritis may occur in the puerperal state, but is by no means especially liable to occur in that condition, a large majority of the cases which have fallen under my observation having been independent of the puerperal state. The disease, however, in whatever condition it occurs, is attended with a train of symptoms which we will now proceed to consider.

Symptoms.—The symptoms of chronic metritis, when not the result of the acute form of the disease, are in some cases rather obscure. Generally, however, the patient experiences a sense of heat, pain, and soreness in the middle and lower part of the pelvis, in different portions of the womb, and in some cases, a sense of weight is felt, attended with pain in the upper part of the vagina, as if the uterus were prolapsed.

In nearly all cases of this disease, there is more or less leucorrhœal discharge, and if from any cause the inflammation is increased, the discharge may assume a purulent character. And on examination with the speculum and by the touch, the neck of the womb is often found red, swelled, and tender to the touch.

More or less pain is experienced in the back, and the stomach sympathizes with the womb, in most cases giving rise to a train of very harassing dyspeptic and nervous symptoms. And patients suffering from this affection become nervous, fretful, irritable, desponding, and difficult, especially if they have been much on their feet, or riding over rough roads.

In some cases the inflammation continues on for several months, or even years, without any serious structural change of the womb, but it is liable, among other structural changes, to become indurated, scirrhous, and finally to pass on to cancerous ulceration.

Diagnosis.—The existence of chronic metritis should always be suspected, in cases of leucorrhœa, attended with heat, weight, soreness and pain in the upper part of the vagina. If now in addition to this, there is the heat, pain and soreness in the uterus itself with the gastric and other sympathetic derangements which attend chronic metritis, the diagnosis is rendered clear and positive.

Causes.—Chronic metritis is sometimes the result of the acute form of the disease. It may also be the result of difficult or instrumental labors, as well as of frequent abortions. The disease is sometimes produced by masturbation, or excessive sexual indulgence, or it may be the result of a gouty or rheumatic condition of the organ, or of amenorrhœa, or of the final cessation of the menstrual function. And the disease is liable to be produced by exposure to cold and dampness, as well as by most of the influences which cause chronic inflammation in other organs or parts.

Treatment.—The first step in the treatment of chronic metritis is to remove the cause or imprudence which has led to it, as far as possible, and to correct the habits of the patient in every respect.

Two or three blue pills may be given at first, and followed by half an ounce of the sulphate of magnesia, after which the bowels may be kept loose by a teaspoonful, administered in the morning, two or three times per week.

Cups should be applied to the sacrum and hypogastrium, and an ounce or two of blood taken once or twice per week, if the system is in a condition to admit of it, till the disease is in a good degree subdued.

As an alterative in such cases, the iodide of potassium may be given in five grain doses three times per day, before eating, and continued till, with the other treatment suggested, the chronic uterine inflammation is subdued.

The patient should be directed to keep quiet, or at least to avoid long walks, or riding in a carriage over rough roads; and in cases in which the os uteri is seriously involved in the inflammation, and it is not subdued by the treatment already suggested, a solution of the nitrate of silver, of the strength of 40 or 60 grains to the ounce of water, may be applied through a speculum with a sponge twice each week, till a cure is effected.

SECTION VI.—DISEASE OF THE CERVIX UTERI.

By disease of the cervix uteri I mean here a morbid condition of the neck of the womb, whether congestive, inflammatory, or malignant, generally chronic, and always attended with more or less constitutional derangement.

We have seen, in the preceding section, that the neck of the womb is very liable to be involved in chronic inflammation with the whole organ. We are now to see that the neck is liable to become the exclusive seat of chronic disease, of a benign or malignant character.

The disease, in cases of this character, whether it be simple congestion, inflammation, or of a scirrhous character, generally involves that part of the neck of the womb which projects into the vagina, including the mucous lining of the canal of the cervix.

Now, disease of the cervix uteri may exist for a long time in so insidious a manner as to be overlooked, unless proper discrimination is exercised, the various sympathetic affections absorbing all the attention of the patient and her friends, the primary source of the deranged health not being even suspected in some cases. And this is not strange, when we remember the tremendous influence which the womb exerts on the female system, both through the cerebro-spinal and sympathetic nerves.

Bearing these facts in mind, we will proceed to consider the symptoms of the disease of the cervix uteri, whether congestive, inflammatory or malignant, together, as the sympathetic derangements are liable to be very similar in all cases, the local being sufficient to enable the careful observer to discern the nature of the primary disease of the cervix, in most cases at least.

Symptoms.—In cases in which the neck of the womb is congested, irritated, or slightly inflamed, with a state of irritation or inflammation of the lining membrane of the cervix, there is apt to be an albuminous discharge, which may be distinguished from the vaginal leucorrhœal discharge by its albuminous appearance, resembling very much the white of an egg. A feeling of heat, fullness, and sometimes of slight pain, is experienced in the region of the cervix, especially for a few days previous to menstruation.

The patient becomes nervous, irritable and desponding, in most cases, and very often hysterical; all these and other sympathetic symptoms being usually greatly increased or aggravated for a few days before the menstrual flow, the uterine tendency doubtless increasing the local congestion, irritation or inflammation in the neck of the womb.

There is generally a pain or feeling of heaviness in the sacral and lumbar region of the spine, and in most cases a feeling of heat, uneasiness, and sometimes of slight pain along the spine at different points and also in the cerebellum or back part of the head. The stomach also sympathizes strongly with this local irritation, the appetite being variable and the digestion often but poorly performed, and in bad cases, nausea, wretching and vomiting is liable to occur at times.

There is in many cases uneasiness in voiding urine, and more or less pain felt in the lower part of the pelvis, at each movement of the bowels. Thus the disease passes on, the local and general symptoms being gradually more troublesome, till the patient becomes habitually nervous, irritable, petulent, and hysterical; and if the disease continues perhaps becomes insane.

If now the finger be passed along the vagina, so as to touch gently the neck of the womb, the patient complains of pain; and a fullness, puffiness, or enlargement of the cervix may be detected. And if an ocular inspection be made, by means of a speculum, in a clear light, more or less redness will be discovered in all that portion of the cervix which extends into the vagina; and not unfrequently an albuminous fluid may be discovered adhering to the extremity of the cervix, at the orifice of its canal.

It is well in such cases to examine the orifice of the urethra; for a slight tenderness or irritation at that point will sometimes alone develop nearly all the symptoms which I have described as arising from irritation of the cervix uteri, and, in addition, the most excruciating pain is experienced in passing water.

Both these conditions may and frequently do exist in the same patient, for a long time, developing the most unpleasant local symptoms, and sympathetic constitutional derangements of a nervous or hysterical character.

Malignant disease of the cervix uteri is usually attended with the same train of sympathetic derangements, and general symptoms, which belong to the benign or simple form of disease of this part. There is, however, in malignant cases also the scorbutic or cachetic appearance of the patient, emaciation and general debility, the sanious or bloody discharge, and the indurated and sometimes lobulated, and, perhaps, ulcerated condition of the cervix, which is readily discoverable by tactile and ocular inspection.

Diagnosis.—Disease of the cervex uteri may generally be distinguished by the troublesome train of constitutional or sympathetic symptoms, which I have enumerated, such as dyspepsia, nervous irritability, petulence, hysteria, &c., together with the feel and appearance of the cervix, and the presence of an albuminous, sanious or bloody discharge.

Malignant disease of the cervix may be distinguished from the simple or benign by the cachetic countenance, the emaciation and general debility, the indurated or perhaps ulcerated condition of the cervix, and by the sanious or bloody discharge, none of which symptoms necessarily belong to the benign form of the disease.

Causes.—A great variety of causes may operate to produce disease of the neck of the womb.

A rheumatic condition of the system evidently predisposes to this affection, especially if the uterus be the chief seat or point of the rheumatic irritation.

But among the exciting causes of disease of the cervix uteri, I am satisfied, from careful inquiry on this subject, that masturbation and excessive sexual indulgence, especially at the menstrual period, or too soon after confinement, together with frequent abortions, are by far the most frequent causes of the non-malignant form of the disease. And further, that the malignant form of the disease is generally the result of the long continuance of the benign, in constitutions of a scorbutic or scirrhous predisposition.

Prognosis.—The prognosis in non-malignant disease of the cervix uteri is generally favorable, if the causes which have been operating can be removed, and the patient is subjected to proper treatment in season. But in all cases of malignant disease of the neck of the womb, I need hardly say that the prognosis is decidedly unfavorable, as the case will generally pass on to a fatal termination.

Treatment.—In all cases of disease of the cervix uteri, the real condition of the part should be ascertained, the causes which have been operating to produce it or keep it up should be removed, and the patient made to conform to the laws of health and rules of propriety in every respect.

If the bowels are constipated a teaspoonful of the sulphate of magnesia, or a Seidlitz powder may be given every morning till the bowels are regulated. Cups should be applied, wet or dry, along the sacral and lumbar region of the spine, and if necessary leeches to the hypogastrium.

By thus removing the cause, correcting the habits, regulating the bowels by saline laxatives, and occasionally applying cups and leeches if necessary, simple congestive or inflammatory cases may very often recover without further treatment. If, however, the disease continues, and there is a rheumatic condition of the system, five grains of the iodide of potassium should be given three times per day, before each meal, and continued till the rheumatic condition is removed as far as it may be.

If, now, when all the general indications have been thus fulfilled, the disease of the cervix still remains to some extent, and there is no evidence of malignancy, the cervix should be touched lightly with the nitrate of silver in substance, or what may be safer in most cases, with a solution of the nitrate of the strength of from forty to sixty grains to the ounce of rain water.

The application may be made in either case through a speculum. If the solid nitrate is used, a piece of it may be passed into a quill, and made fast, when in a good light the diseased part of the cervix may be very rapidly passed over by the caustic, so as to change without destroying the diseased surface. If, however, the solution is used, it may be conveniently applied by means of a sponge, firmly fastened about the extremity of a stick or whale-bone. This is readily passed through a cylindrical speculum, properly adjusted, and should be pressed gently upon the diseased cervix, and if it be found to have produced too much irritation of the diseased part, a soft moist sponge, with which the other end of the stick or whale-bone should always be armed, may be passed, and thus the application be diluted, and partially absorbed or removed.

It is generally sufficient to make the application once each week, till the disease is removed. It should be made in the morning, and at evening an ounce of flax-seed tea, with twenty drops of laudanum may be thrown into the vagina, to quiet any undue irritation that may have been produced, or which is liable to arise from the application. A few applications will generally be sufficient to effect a cure, but if the case is improving, the treatment should be continued as long as may be required.

If in cases of this disease the mucous membrane at the orifice of the urethra is found inflamed and tender, as is sometimes the case, the part should also be touched by the solid caustic or solution till the irritation subsides.

In disease of the cervix uteri of a *malignant* character, the same course of general treatment may be best, with the addition if necessary of tonics, but local applications except to cleanse and soothe the parts should not be used, as they generally rather hasten the destructive progress of the disease.

SECTION VII.—LEUCORRHŒA.

By leucorrhœa, from λευκος, "white," and ρεω, "I flow," is here meant a white, yellowish, or greenish discharge, of a mucous, albuminous, or purulent character, from the vagina of the female.

This discharge may depend upon a general or constitutional cause, or it may be the result of relaxation, congestion, irritation, or chronic inflammation of the mucous lining membrane, of the female genital organs, and especially of that of the vagina, and of the canal of the cervix uteri.

The general or constitutional condition, capable of producing, and keeping up this discharge, is probably a weak state of the blood, and a general deranged condition of the fluids, and solid tissues of the body. Especially does the rheumatic diathesis tend to the production of this discharge.

While a general relaxed and debilitated condition of the solids, and a deranged state of the fluids of the body, may sometimes be the cause of a leucorrhœal discharge; it is probable that this general deranged condition, is often the result of the wasting discharge, which was at first set up by a local cause, operating upon the lining membrane of the female genital organs.

The discharge then, generally proceeds from the mucous membrane of the vagina, of the canal of the cervix uteri, or else in some cases, probably from the lining membrane of the uterus itself, and always depends upon a morbid condition of the lining mucous membrane, of some one of these parts, when it is not the result of a general constitutional derangement, as we have already seen.

The morbid local condition then, upon which this discharge depends, may be relaxation, congestion, irritation, or a peculiar chronic inflammation of that portion of the lining mucous membrane of the female genital organs, from which the discharge proceeds.

When the matter of a leucorrhœa is of a mucous or purulent character, it generally comes from the mucous membrane of the vagina, being generally of a lightish mucous appearance at first, and assuming a yellowish or purulent character in the advanced, or more chronic stage of the disease. If, however, the discharge is of an albuminous character, appearing very much like the white of an egg, it generally comes from the canal of the cervix, or else from the membrane, lining the cavity of the womb itself, as it probably does in some cases.

Now bearing these facts in mind, we will proceed to the consideration of the symptoms, which are developed in this disease.

Symptoms.—The symptoms in those cases, which come on from general or constitutional causes, are those of general debility; the patient being anæmic, pale, nervous, irritable, and generally suffering from slight rheumatic, or chronic neuralgic affections. After these symptoms have continued for a time, a slight mucous discharge occurs from the vagina, which gradually increases, till it becomes copious, sooner or later assuming a yellowish appearance, or sometimes even a greenish tinge, after it has continued for a long time.

As the discharge becomes copious, and continues for a long time, the general debility, irritability, nervousness, and all the sympathetic derangements and general disturbances of the system, are greatly increased. Digestion becomes impaired, the menstrual discharge, at first, diminished, is sometimes finally suspended, and if the disease passes on unchecked, scrofula, consumption, or some other fatal malady, is liable to lead on to a fatal termination.

In those cases of leucorrhœa which occur from a relaxed, congested, irritated, or inflamed condition of the mucous membrane of the vagina, without at first any marked constitutional debility or derangement, there is generally at first a slight uneasiness with heat in the vagina, followed soon by a more or less copious mucous discharge. After a time the discharge assumes a yellowish or purulent character, a dragging pain is felt in the sacral and lumbar regions, sharp darting pains pass down the limbs, menstruation becomes painful, the discharge being gradually lessened, and finally suspended in some cases; the patient becomes nervous, irritable, and dyspeptic, and unless the disease be arrested, some serious local or constitutional affection, with a rapidly fatal tendency is the result.

In cases in which the discharge is from the mucous lining of the canal of the cervix, or of the cavity of the womb, the same train of symptoms usually attend in an aggravated form; the discharge, however, instead of being of a mucous or purulent appearance, is of an albuminous character, appearing, as we have already seen, like the white of an egg. In all cases of this character, in which the discharge gradually assumes a watery, sanious, or bloody appearance, malignant disease of the cervix or body of the womb should be suspected, and a careful inquiry should be made into the condition of the suffering organ.

Such, I believe, are the ordinary symptoms of leucorrhœa, occurring in consequence of a general derangement of the system, or from a local, relaxed, congested, irritated, or inflamed condition of the lining mucous membrane, of the female genital organs, of the vagina, canal of the cervix, or of the uterus itself, with the peculiarities belonging to each.

Causes.—Leucorrhœa may be produced by any cause capable of producing a general debilitated condition of the system, or a morbid relaxation, congestion, irritation, or inflammation of the mucous membrane of the vagina, or canal of the cervix uteri. A great variety of causes then may operate to produce this disease, including almost every imprudence and exposure to which females are addicted or liable.

But it is probable that among these causes the most frequent are masturbation, excessive sexual indulgence, difficult parturition, frequent menorrhagia, prolapsus uteri, the abuse of emmenagogues, a loaded and torpid state of the bowels, tight dressing about the waist, the depressing mental emotions, metastasis of rheumatism, atmospheric influences, suppressed hemorrhoids, and all the influences which operate to produce chronic inflammation of the cervix uteri.

Diagnosis.—To distinguish leucorrhœa from gonorrhœa is often extremely difficult, or even impossible; the symptoms of acute leucorrhœa and gonorrhœa, in many cases, being precisely alike.

In such cases, the only reliable means of diagnosis will be the general circumstances of the case, and the probability of the patient having been exposed to gonorrhœa. After thus having brought in all the evidence in the case, if there is a slight probability of the disease being gonorrhœa, it should be treated as such, as a protracted gonorrhœa might lead to very unpleasant consequences, especially if the disease occurs in a married female.

To distinguish uterine from vaginal leucorrhœa, it is only necessary to take all the symptoms into the account, and to notice carefully the appearance of the discharge, the uterine being albuminous, sanious, or bloody, while the vaginal is of a mucous or purulent character.

Prognosis.—The prognosis in all cases of leucorrhœa, not associated with organic disease of the female genital organs is favorable, if the causes can be removed, and the patient subjected to proper treatment in season. But in very protracted and obstinate cases, in which the discharge, at first albuminous, becomes watery, sanious, or bloody, serious apprehensions may be had of the supervention of organic disease, and of a fatal termination of the case.

Treatment.—The treatment of all cases of uterine leucorrhœa should consist in a removal of the cause regulating the bowels, and general condition of the patient, the application of cups to the sacrum and lumbar region of the spine, the administration of iodide of potassium for a time, and, if necessary, the application of the nitrate of silver, in substance or solution, to the neck of the womb, followed by a mucilaginous and anodyne injection into the vagina, once each week, till a cure is effected, as suggested in the preceding section. In such cases, when the uterine disease is thus subdued, the leucorrhœal discharge will generally cease, and not before.

In acute cases of vaginal leucorrhœa, depending upon sudden exposure to cold, dampness, &c., constipation of the bowels, or any sudden cause, as a local injury, with only a slight predisposition to the disease, it may often be arrested by three blue pills administered at evening, and followed in the morning by half an ounce of the sulphate of magnesia, the bowels then being kept loose by a teaspoonful taken each morning for a time. The patient should, however, be kept clean and warm, and two or three ounces of blood may be taken by cups from the sacral and lumbar region, and then dry cups applied occasionally, the genital parts being kept clean and cool by injections of moderately cold water, morning and evening.

This course of treatment alone will frequently arrest leucorrhœa, depending upon sudden accidental causes. Should the discharge continue, however, in addition to what I have suggested, twenty drops of the balsam of copaiva should be given, three times per day, after each meal, and continued for a reasonable time.

Should the disease not yield, and pass on and become chronic, and in all chronic cases depending upon general or local causes, five grains of the iodide of potassium should be administered, three times per day, before meals, the balsam of copaiva being given in twenty drop doses after eating. After the iodide of potassium has been given for a few weeks it should be discontinued, and the syrup of the iodide of iron given instead, in ten drop doses, and this should be continued in connection with the balsam of copaiva in ten drop doses, with the addition, if necessary, in protracted cases, of injections of alum, morning and evening, of the strength of one or two drachms to the pint of rain water, till a cure is effected. Or in cases attended with great relaxation of the mucous membrane of the genital organs, injections of tannin of the same strength may be used instead.

In cases of leucorrhœa attended with, and in part depending upon prolapsus uteri, the same general treatment may be adopted that I have already suggested, but in addition, measures should be taken to restore the womb to its original position, and keep it there, as nearly as may be.

The patient, in such cases, should have all her underclothes suspended by the shoulders, to prevent any dragging down of the bowels, should lay at night with her hips high, should press the bowels up occasionally, and manipulate the lower portion of the abdomen, to give tone to the abdominal muscles, and in very bad cases, may wear a pad across the lower part of the abdomen, for a time, or a pessary introduced into the vagina. Should a pessary become necessary, I prefer that invented by my preceptor and former partner, Dr. Wm. V. V. Rosa, of Watertown, N. Y., as being the most safe, convenient and reliable of any that have fallen under my observation.

This pessary was first used by us in several desperate cases in Jefferson county, in this State; is now manufactured in New York, and kept for sale by Drs. Smith and Pratt, druggists, and eminent practitioners of medicine and surgery, of Canandaigua, N. Y. It is composed entirely of silver, and consists of a pessary, of smallish size, but of the shape of the common glass pessary, perforated in its centre to receive a double silver canula, which may be lengthened or shortened to suit the length of the vagina.

The womb should first be put in its place, the pessary then introduced and carried up to the neck of the womb, after which the double canula should be carefully screwed in, and its length properly adjusted, so that its lower end shall reach the lower extremity of the vagina; to this two small chains, wound with cotton, are fixed, one of which is hooked anteriorly, and the other posteriorly, to a belt above the hips.

SECTION VIII.—AMENORRHŒA.

By amenorrhœa, from *a*, privative, μην, "a month," and ρεω, "I flow," is here meant a retention or suppression of the menses, usually in consequence of some local or constitutional derangement of the system.

The disease may consist, then, of *retention* of the menses, in which case there is a failure to establish the menstrual function ; of *suppression*, or an interruption to the function after it has been established, and may be attended, in either case, with a vicarious discharge from the nose, lungs, stomach, or some other part, constituting what has been called *vicarious menstruation*.

Bearing these facts in mind, we will proceed to the consideration of this disease, and first of the symptoms which are developed in its various forms.

Symptoms.—When the young female arrives at the age of puberty, and the ordinary changes take place in the system, such as enlargement of the breasts, the growth of hair upon the pudenda, &c., and yet the menstrual functions is not established, the failure may be in degree, but sooner or later a train of morbid symptoms are apt to be developed. There is headache, a flushed countenance, pains in the back and limbs, and a full and either slow or accelerated pulse ; and there is evidently, in these cases, a torpor of the uterine vessels, in consequence of which they fail to produce the menstrual discharge. Such patients have generally been accustomed to sedentary habits, high living, hot rooms, soft beds, and too much sleep, and hence their general plethora, and also the retention of the menses on their arriving at the age of puberty.

In *anæmic* females, on the other hand, there is sometimes a failure in the establishment of the menstrual discharge, as the age of puberty approaches, attended with more or less derangement of the health, as the system appears to make an ineffectual effort. The patient becomes emaciated, the face and lips are pale, the hands and feet are cold, the tongue is foul, the appetite is variable, the bowels are constipated, and the pulse slow and feeble ; and if the retention continues for a long time, the girl becomes desponding, hysterical, and decidedly melancholy in most cases. There is often nervous palpitation of the heart, dyspepsia, and, only a slight over exertion in some cases, produces a short, distressing cough.

In these anæmic cases, in which the cause of the retention is directly the opposite condition of the system to the plethoric cases before referred to, the patients have generally been delicate during childhood, confined in close apartments, subject to depressing passions, and in some cases to the habit of masturbation.

Thus we have retention of the menses, at the age of puberty, from directly opposite causes, and in widely different conditions of the system, in the one case the patient being plethoric, of full habits, and fleshy, while in the other the patient is anæmic, thin, pale, nervous, irritable, and melancholy.

Now, in those girls in which neither of these morbid conditions exist at puberty, in our climate, generally at about the age of fourteen or fifteen, the menstrual discharge is established, and continues with great regularity every four weeks till about the age of forty-five, when it is suspended, with little or no derangement of the health.

During the period from the age of fifteen to forty-five, suppression of the menses is liable to occur from pregnancy, and it is also liable to occur from various derangements of the system, the discharge being either suddenly arrested at a menstrual period, or else gradually lessened, till complete suppression takes place.

In cases of chronic suppression, occurring gradually, no very marked symptoms are developed, as the immediate result of the suppression, the symptoms depending rather upon the condition of the system which led to it, the discharge appearing to cease in consequence of the debility of the patient, the system either not requiring the discharge, or else not having the ability to produce it. But if, as sometimes happens, from exposure to cold, during the menstrual period, the discharge is suddenly arrested, the most violent local and constitutional symptoms are liable to be developed.

In such cases, immediately on the cessation of the menstrual discharge, the patient complains of severe pain in the region of the womb, which extends to the spine, giving the most severe distress in the sacral and lumbar region, the pain in some cases passing up the spine to the brain, and being attended with cephalic irritation, congestion, and sometimes with the most terrible convulsions. The cephalic irritation which attends, if not overcome, is liable to pass on to inflammation, effusion, &c. ; and more or less nausea and vomiting generally attends, in cases of suppression of this character.

Vicarious menstruation is liable to occur in cases of amenorrhœa, in which a deranged condition of the uterus is the cause of the retention or suppression. In such cases, the system failing to relieve itself of the discharge by the uterus, which its healthy condition requires, a vicarious discharge is set up in some other part, which may recur with great regularity every four weeks, till the uterine discharge is restored.

The escape of blood in vicarious menstruation, is liable to take place from any part of the body from which ordinary hemorrhage may occur, but it generally takes place, according to my observation, from the nose, lungs, stomach, or bowels, numerous instances of which have fallen under my observation. The loss of blood may be slight in such cases, and attended with no very serious local or general symptoms, or the hemorrhage may be of a frightful character, in some cases even endangering the life of the patient.

Such, I believe, are the ordinary symptoms of amenorrhœa, from *retention* and *suppression*, with or without *vicarious menstruation*.

Causes.—The causes of amenorrhœa, as we have already seen, are such as produce retention or suppression of the menses, retention generally being attended with either a full plethoric habit, or else an anæmic condition of the system.

Suppression may be produced by similar conditions of the system, as we see in chronic cases, which take place gradually ; or it may occur suddenly during the menstrual period, from sudden exposure to cold, dampness, &c.

Pregnancy is also a cause of amenorrhœa, and should be suspected when there is no other sufficient apparent cause. It should be remembered, however, that menstruation sometimes goes on for several months after conception. Such cases are, however, an exception to a rule, as menstruation should cease during pregnancy.

The remote causes of amenorrhœa are too numerous to be mentioned, including inherited constitutional imperfections, and almost every imprudence and accident to which females are addicted or liable. I believe, however, that the most frequent causes are imprudence in eating and drinking, sedentary habits, constipation of the bowels, indolence, or extreme want, neglected or disappointed affections, self-pollution, and sudden exposure to cold or dampness during the menstrual flow.

Vicarious menstruation, occurring as it does in constitutions which require the discharge, generally depends upon an inherited or acquired defect on the part of the uterus itself; the vicarious discharge taking place from the nose, lungs, stomach, bowels, or other parts which from constitutional or accidental circumstances is most predisposed.

Treatment.—The treatment of amenorrhœa requires the exercise of the greatest degree of prudence, discretion, and good common sense.

As a general rule those cases of retention occurring in girls arrived at the age of puberty should not be regarded as requiring treatment, unless there is an evident impairment of the health, calling for medical aid. If, however, there is a derangement of the general health, apparently in consequence of the non-appearance of the menstrual discharge, and the patient is either of the plethoric or anæmic condition, which I have described, a course of medical treatment is indicated, and should not be neglected.

If the case be one of *plethora*, with constipation of the bowels, and a decided cephalic tendency, with an apparent ineffectual effort at menstruation, a full dose of calomel should be administered at first, and followed with half an ounce of the sulphate of magnesia.

After the bowels have been freely evacuated by the cathartic, a pill of aloes and rhubarb should be given each day, after dinner, and continued, to regulate the bowels, and also to overcome the cephalic tendency. It also at the same time acts as a mild emmenagogue.

Cups may be applied over the sacrum, and along the lumbar region of the spine, and two or three ounces of blood taken, if the case appears to require the abstraction of blood, as it sometimes does.

The patient should be made to correct every imprudent habit, and should be placed upon a plain, digestible, and unstimulating diet; should be made to sleep on a hard bed, to take sufficient exercise in the open air, &c. And when the effort at menstruation occurs, the warm foot-bath should be used, for two or three evenings, and sinapisms applied to the lumbar and sacral regions of the back, on going to bed at night. In this way the menstrual discharge will frequently be established without further treatment. But if when the general phlogistic condition of the system is overcome, and the bowels regulated, the menstrual discharge is not established, Dewees' tincture of guaiac should be given, in teaspoonful doses, three times per day, immediately after eating, in a little sweetened milk; and this should be continued, with the pill of aloes and rhubarb after dinner, till the menses appear.

In *anæmic* cases, in which there is retention of the menses at the age of puberty, with an ineffectual effort on the part of the system to bring about the menstrual discharge, a judicious course of treatment should be pursued. Two or three blue pills may be given at first and followed by a Seidlitz powder, or the sulphate of magnesia, to secure a free motion of the bowels; after which a pill of aloes and rhubarb should be given each day, after dinner, and continued till the bowels are regulated.

As soon as the bowels are regulated, Dewees' tincture of guaiac, should be given in teaspoonful doses, three times per day, after eating, and if, as is often the case, iron is indicated, the syrup of the iodide should be given in ten drop doses, three times per day, before eating, and this should be continued with the guaiac, till the menses appear.

The patient should take a good nourishing diet, and proper exercise, and as the effort of the system to establish the discharge, makes its appearance, the warm-foot-bath should be resorted to, for two or three evenings, and sinapisms applied to the sacral and lumbar regions of the back, on retiring to bed at night. In this way, the system may generally be corrected, and menstruation established, in a few weeks or months.

Now the treatment for *suppression* of the menses, differs in nothing from that which I have laid down for retention, except in acute cases, occurring during the menstrual flow, from sudden exposure to cold, dampness, &c., the treatment of which it is necessary for us to consider.

Acute or sudden suppression of the menstrual flow, from exposure to cold or dampness, at the menstrual period, requires active and immediate treatment, as violent congestion, irritation, or even inflammation of the brain is liable to occur. As soon then, as such a suppression occurs, attended with cephalic congestion, a full dose of calomel should be given, and that followed by half an ounce of the sulphate of magnesia, and a free action of the bowels secured.

Cups should be applied to the back of the neck, and also to the sacral and lumbar regions, and a few ounces of blood taken, and this may be repeated, if the cephalic tendency continues. Warm pediluvia should be used, and to quiet nervous irritability, three grains of the solid extract of hyoscyamus, or fifteen drops of the fluid extract may be given, with two grains of camphor, every six hours, till quiet is restored.

In plethoric patients, with violent cephalic symptoms, a full bleeding from the arm may be indicated, in some cases, and when it is, should not be neglected. After the cephalic symptoms pass off, the patient should take a pill of aloes and rhubarb, each day after dinner, should take a plain digestible, and unstimulating diet, with which treatment, in most cases, the menses will appear, at the next regular period. This should be favored by the warm foot-bath, for two or three evenings, and the sinapisms to the lower portion of the back, as the menstrual period arrives. If, however, with these measures, the menses do not appear, at the first regular period, the case should be treated according to the principles I have already laid down, for chronic retention or suppression of the menses.

In cases of *vicarious menstruation*, in which the system failing to effect the discharge by the uterus, sets up a vicarious discharge from the nose, lungs, stomach, bowels or other parts; immediate measures should be taken to bring about the uterine menstrual discharge, that the vicarious discharge may be suspended.

In such cases, after removing the cause, regulating the bowels, &c., the tincture of chloride of iron should be given in ten drop doses three times per day, as a tonic emmenagogue, and to arrest the hemorrhagic tendency which probably favors the vicarious discharge. As an emmenagogue to act directly upon the uterus in such cases, the tincture of cantharides may be given in thirty drop doses, three times per day, to be continued in connection with the iron already suggested, as long as may be required.

The patient should be directed to keep the feet warm and dry, and to use the warm foot bath, and sinapisms to the lower part of the back, beginning a day or two before the time for the periodical discharge. By this course of treatment, we may generally succeed in restoring the natural, and arresting the vicarious discharge in a few weeks or months at longest.

In all cases of suppression of the menses in *married* females, it is most prudent to wait and be certain whether pregnancy may not be the cause, and thus avoid the possibility of producing an abortion, as well as the mortification of having prescribed for that which was really no disease. After sufficient time has elapsed to render it certain that the suppression is not from pregnancy, the case should be treated according to the principles already laid down.

In all cases of suppression in *unmarried* females, the presumption is that pregnancy is not the cause, and as a general rule, this presumption is to be followed, unless signs of pregnancy are discovered. In case signs of pregnancy are discovered, however, as well as in cases in which though there may be no signs of pregnancy, there is no sufficient derangement of the health apparent, to account for the suppression, I believe it is our imperative duty to *wait*, or if we make a prescription to be sure that it be one that can do no harm, should pregnancy prove to be the cause of the suppression. If unmistakable signs of pregnancy are discovered, the fact may be kindly suggested to the patient, and such advice offered in relation to immediate *matrimonial* arrangements, as the case may appear to require, but the patient should be solemnly warned in all cases, not to attempt or even desire an abortion, either of which constitutes a crime of the deepest dye.

SECTION IX.—DYSMENORRHŒA.

By dysmenorrhœa, I mean here painful or difficult menstruation, an affection by no means very uncommon in females of a peculiar and delicate constitution.

Now in order to understand this subject, it is well to bear in mind the fact, that most females as the menstrual period approaches, feel more or less uneasiness along the lower portion of the spine, in the lumbar and sacral regions. This pain also extends in many cases, to the lower portion of the abdomen, as well as to the uterus, and is very apt to pass down the limbs, following the track of the large nerves.

Now this uneasiness is doubtless owing to a slight congestion of the spinal cord, in consequence of which the uterine and other nerves more immediately connected with this portion of the spine become slightly irritated, a congestion of the uterine vessels occurs, the menstrual flow follows, and the uneasiness subsides. Even this slight pain is probably a morbid condition, consisting of a preternatural irritability of the system, in consequence of which the nervous influence sent out by the brain to set up the menstrual discharge, produces an undue degree of congestion, and more or less pain, before the discharge commences.

Now in dysmenorrhœa, as the menstrual period approaches, there is a great aggravation of all these pains, which are only slight ordinarily, and various other general symptoms which we will now proceed to consider.

Symptoms.—The prominent symptoms then of this affection are pain in the lower portion of the back or loins, extending more or less to the bowels, and down the limbs; severe aching, bearing down, or grinding pains in the uterus, sometimes subsiding as the discharge commences, but in other cases continuing during the whole menstrual period.

The flow is liable to be quite irregular, being almost suspended at times, and again coming forth very freely, with more or less coagulated blood, and sometimes a pseudo-membranous substance is thrown off, which appears to have occupied the whole inner surface of the womb.

The pain, in some cases, very much resembles labor pains, and it may continue till the menstrual period has passed by, there being still left a dull heavy pain in the womb, and lower portion of the spine, which gradually pass off in a few days. In some cases, the uterine irritation is so great that the stomach strongly sympathizes during the severe pain, nausea and vomiting being an attendant symptom. The menstrual flow may continue of nearly its normal quantity, but it is liable, if the disease continues for a long time, to become greatly diminished.

Such are the ordinary symptoms of this very troublesome affection, which may occur at any time during the menstruating period of female life, but more generally, I think, between the twentieth and thirtieth years of age, at least in cases that have fallen under my observation.

Pathology.—There may be ground for doubt in relation to the nature of this disease, but it appears to me that it is generally of a rheumatic character, as it is very liable to occur in patients of a rheumatic diathesis. In some cases, however, there is a contraction or partial closure of the canal of the cervix uteri, producing a mechanical obstruction to the free passage of the menstrual fluid. This, however, may have been brought about, and probably is in most cases, by a slow inflammation, of a *simple* or *rheumatic* character.

If now, rheumatism depends upon either congestion, irritation, or inflammation of some portion of the brain or spinal cord, or their meninges, as I have suggested in a previous chapter, and generally congestion, then the rheumatic character of this affection may be readily explained. For it only requires a slight congestion or irritation in the lower portion of the spinal cord, to sufficiently irritate the uterine

44

nerves, if the womb is predisposed, to at first set up a neuralgic and then a rheumatic affection of the uterus.

And that the disease once established should occur in paroxysms, during the menstrual period, is as we should expect, when we remember that not only the lower part of the spinal cord, but the uterus itself, at such times, is more or less congested. Nor is it strange that when this local rheumatic inflammation or irritation is of a certain grade, that a pseudo-membrane should be formed on the inner surface of the womb, or that if the disease be long continued, the canal of the cervix uteri should become contracted. and nearly or quite closed, as happens in some cases of this affection.

I apprehend, then, that this affection is generally rheumatism of the uterus, coming on in the same manner, and from the same causes, as other chronic local rheumatic affections, a paroxysm being produced by the increased congestion and irritation of the lower portion of the spinal cord and uterus itself, at each menstrual period. Dysmenorrhœa may, however, in some cases, consist of simple neuralgia of the uterus, or it may depend upon congestion, or simple irritation or inflammation, of a non-rheumatic character.

Treatment.—The treatment of dysmenorrhœa consists in that which is proper, during the menstrual period, to quiet pain and promote relaxation of the cervix uteri, and also that which is proper, during the intervals, to correct the morbid condition of the system, upon which the disease depends.

The severe pain occurring during the menstrual period may generally be greatly relieved by cups applied along each side of the spine, in the lumbar region, and to the sacrum. To quiet pain and promote relaxation of the cervix uteri, ten drops of the fluid extract, or twenty drops of the tincture of stramonium, may be given every six hours, in bad cases, till the pain subsides. As a drink, for a few days previous, and during the menstrual period, sulphur water, if at hand, is a valuable remedy, and will often prevent, or greatly lessen the pain during the menstrual flow.

To break up this disease, or the morbid condition of the system upon which it depends, the habits of the patient should be corrected in every respect. Constipation of the bowels should be overcome by an occasional Seidlitz powder, and then the iodide of potassium should be given in five grain doses three times per day, before eating, in the fluid extract or compound decoction of sarsaparilla; or in patients of a spare habit, in which the circulation is languid, the tincture of guaiac may be given instead of the iodide of potassium, in drachm doses, after each meal, and continued until a cure is effected.

Dry cups, or pustulations with tartar emetic ointment along the lower portion of the lumbar and sacral regions, may sometimes be indicated. The diet should be plain, digestible, and taken with strict regularity, and the patient should wear flannel next the skin, and should avoid damp apartments, or exposure to cool, damp air.

With this course of treatment judiciously applied, severe cases of this affection may generally be greatly relieved, and mild ones sometimes permanently cured.

Thus I have completed what I had to say upon diseases of the genital organs, which, with a few concluding remarks, will close this volume.

CONCLUSION.

Now, having completed this TREATISE on the PRACTICE of MEDICINE, a few reflections, suggestions, and a parting word with the reader, may be appropriate. They shall, however, be very brief.

The reader will remember that I commenced with the human system in health as the standard, and laid down the principle that disease should be considered in the light of deviations from this standard, and that in all cases the exact deviation should be taken, and from this the indications of treatment should be directly drawn; that the indications being once arrived at in this way, the causes being removed, the most safe, convenient, and reliable remedies should be selected to fulfill these indications.

With this view I discarded the idea of names for deviations or disease, any further than necessity might require; and to keep the mind intently fixed upon conditions, instead of names, and because it was on every account proper, I announced that I should give the general anatomy and physiology of the different parts of the system, as I proceeded to take up the diseases to which they are liable.

In accordance with my plans, after considering the *nature, causes, symptoms, diagnosis,* and *treatment* of disease, in my first chapter; *irritation, congestion,* and *inflammation,* in my second; the *pathology, causes,* and *phenomena* of fever in the third; *general fevers* in the fourth; *exanthematous fevers* in the fifth, and *general inflammatory* diseases in the sixth, I proceeded to take up diseases of particular parts, in the following order. In the seventh chapter, diseases of the *nervous system,* in the eighth, of the *digestive system,* in the ninth, of the *respiratory system,* in the tenth, of the *circulatory system,* in the eleventh, of the *eye,* in the twelfth, of the *ear,* in the thirteenth, of the *skin,* in the fourteenth, of the *urinary organs,* and finally, in the fifteenth, diseases of the *genital organs,* in the precise order which I announced in my introductory.

In considering the deviations from a healthy condition of these constituent parts of the human system, I have taken a general glance at their anatomy and physiology in health, and then by tracing their several diseased conditions, have attempted to draw the indications of treatment, that no prescription should be rendered empirical.

It must not be supposed, however, that because all my prescriptions have been made to fulfill indications thus arrived at, that they have not been tested by experience, for they generally have, and are further, the ones which I have found the best adapted of any of the class to which they belong, to fulfill the several indications for which I have prescribed them. I have preferred thus to suggest only the *best* remedies to fulfill the indications in each particular case; thus avoiding the confusion from a multitude of inferior ones, especially as the student even will never find it difficult to arrive at a poorer remedy of a class, if from any accidental circumstance the best remedy should be contra-indicated, which is very seldom the case.

And further, while I have preferred to give my own opinion of remedies, and in fact of almost every thing else of which I have treated, it has been from no feeling of ostentation, but simply because I believe I am better qualified to give my own opinion, than to hold up the opinions of other men, which in turn they can do much better than I could do it for them. But this certainly can detract nothing from the value of the work, as I am indebted to the *good* and *wise*, and *great* of the medical profession, as well as to my own observation in the treatment of disease, for the opinions which I hold in relation to the *principles*, *science*, and *practice* of medicine.

And now, if in the effort which I have made, I shall succeed in inspiring my brethren of the medical profession with a greater zeal for the restoration of the human family from the numberless ills to which they are exposed, in consequence of their disobedience of the laws of health, I shall be satisfied. If my suggestions in relation to the way and manner of doing this shall prove useful to the medical student and practitioners generally I shall be thankful. And finally, if we may, by our combined effort, but make a beginning which shall end in the physical, intellectual, and moral restoration of mankind, the end for which I have labored, and for which I shall ever pray, will have been accomplished.

Let us then regard our high and holy calling, as ministers of life and health, as a *sacred trust*, and combine all our efforts for the end which I have suggested. Then may we become the feeble instruments in the hands of God, in bringing an apostate and degenerate race back to Himself; and man, physically, intellectually, and morally reinstated, may really again bear the *image* of the CREATOR.

INDEX.

	PAGE
Aerial Poisons, a cause of disease	29
Ague. See Intermittent Fever	94
Albuminuria. (*Bright's Disease*)	630
Symptoms	631
Anatomical characters	632
Diagnosis	632
Causes	632
Nature	632
Prognosis	633
Treatment	633
Amaurosis	331
Symptoms	331
Diagnosis	331
Causes	331
Prognosis	332
Treatment	332
Amenorrhœa	684
Symptoms	684
of vicarious	685
Causes	685
of vicarious	686
Treatment	686
Anæmia, (*chlorosis*)	540
Symptoms	540
Diagnosis	541
Causes	541
Nature	542
Treatment	542
Anasarca	547
Symptoms	547
Causes	547
Prognosis	547
Treatment	548
Aneurism of the Aorta	504
Anger, a cause of disease	27
Angina Pectoris. See Neuralgia of the Heart	511
Animalcular Eruptions	620
Symptoms	620
Causes	620
Diagnosis	621
Treatment	621
Aorta, Aneurism of	504
Apnœa, (*Asphyxia*)	471
From drowning	472
Morbid appearances	473
Treatment	473
From strangulation	474
Morbid appearances	475

	PAGE
Apnœa, from strangulation.	
Treatment	475
From irrespirable gases	475
Morbid appearances	475
Treatment	476
From electricity	476
Morbid appearances	477
Treatment	477
From cold	477
Morbid appearances	477
Treatment	477
Apoplexy	254
Symptoms	256
Diagnosis	257
Anatomical characters	257
Prognosis	258
Causes	258
Pathology	259
Treatment	260
Arteries, Inflammation of	514
Arteritis	514
Symptoms	514
Anatomical characters	515
Causes	515
Treatment	515
Arthritis. See Gout	216
Ascites	548
Symptoms	549
of ovarian	549
Diagnosis	550
Causes	550
Prognosis	551
Treatment	551
Asphyxia. See Apnœa	471
From drowning	472
From strangulation	474
From irrespirable gases	475
From electricity	476
From cold	477
Asthma	458
Symptoms	458
of laryngeal	459
Causes	459
Pathology	459
Prognosis	460
Treatment	460
Auscultation and Percussion	420
Inspection	421
Palpation	421

	PAGE
Auscultation and Percussion.	
Percussion	421
Auscultation	423
Mensuration	427
Succussion	428
Auscultation of the Heart	481
Impulse	482
Rhythm	483
Sounds	484
Impulse in disease	484
Rhythm in disease	485
Sounds in disease	485
Increased sounds	485
Decreased sounds	485
Murmurs	485
Regurgitation	486
Anæmic murmurs	486
Venous murmurs	487
Pericardial murmurs	487
Mixed murmurs	487
Percussion	488
In enlargement	488
In pericardial effusion	488
In aneurism	488
Bilious Colic	399
Symptoms,	399
Diagnosis	400
Causes	400
Prognosis	400
Treatment	401
Bilious Pneumonia	438
History	438
Symptoms	439
Diagnosis,	439
Causes	439
Treatment	439
Bilious Remittent Fever	102
Symptoms	103
Type	103
Gastric complications	105
Hepatic complications	105
Causes	106
Treatment	107
Bladder, Acute Inflammation of	635
Bladder, Chronic Inflammation of	637
Bloody Flux. See Dysentery	359
Brain, Inflammation of	236
Bright's Disease. See Albuminuria	630
Bronchitis, Acute	452
Symptoms	453
Diagnosis	453
Anatomical characters	454
Causes	454
Prognosis	454
Treatment	454
Bronchitis, Chronic	455
Symptoms	455
Diagnosis	456
Anatomical characters	457
Causes	457

	PAGE
Bronchitis, Chronic, Treatment	457
Bronchocele, (Goitre)	566
Symptoms	566
Diagnosis	567
Causes	567
Prognosis	567
Treatment	567
Cancer of the Intestines	366
Symptoms	366
Anatomical characters	366
Diagnosis	366
Treatment	367
Cancer of the Stomach	351
Symptoms	352
Anatomical characters	352
Diagnosis	353
Causes	353
Prognosis	353
Treatment	353
Canine Rabies. See Hydrophobia	321
Carditis	496
Symptoms	496
Diagnosis	497
Causes	497
Anatomical characters	497
Treatment	497
Catalepsy	276
Symptoms	277
Diagnosis	278
Causes	278
Pathology	278
Prognosis	279
Treatment	279
Catarrh, Schneiderian	441
Symptoms	442
Causes	442
Nature	442
Treatment	443
Catarrhal Ophthalmia	568
Symptoms	570
Diagnosis	570
Causes	570
Prognosis	571
Treatment	571
Causes of Disease	22
Predisposing Causes	23
Hereditary predisposition	23
Filth	24
Food,	25
Clothing,	26
Licentiousness	26
Intoxicating liquors	27
Tobacco	27
Anger	27
Exciting Causes	27
Heat	27
Cold	28
Water	28
Electricity	29
Light	29

	PAGE
Causes of Disease.	
Exciting Causes.	
Aerial poisons	29
Vegetable poisons	30
Mineral poisons	30
Parasites	30
Koino-miasmata	31
Idio-miasmata	32
Contagions	32
Endemic influences	33
Epidemic influences	34
Causes of Fever	78
Predisposing Causes	78
Hereditary predisposition	78
Indigestion	79
Heat	79
Cold	80
Various imprudencies	80
Exciting Causes	80
Heat	81
Cold	81
Humidity	81
Electricity	81
Koino-miasmata	82
Idio-miasmata	84
Contagious	85
Cephalagia (Headache)	224
Causes	226
Pathology	226
Treatment	226
Cerebritis	236
Symptoms	236
Anatomical characters	237
Diagnosis	238
Causes	239
Prognosis	239
Treatment	239
Cerebro-spinal Meningitis	250
Symptoms	251
Anatomical characters	252
Diagnosis	252
Causes	252
Pathology	253
Prognosis	254
Treatment	254
Cervix Uteri, Disease of	617
Chicken-pox. See Varicella	174
Chlorosis. See Anæmia	540
Cholera Morbus	388
Symptoms	388
Causes	388
Diagnosis	389
Nature	389
Treatment	389
Cholera, Malignant	390
Cholera Infantum	394
Symptoms	394
Anatomical characters	395
Causes	396
Nature	396
Treatment	396

	PAGE
Chorea, (St. Vitus's Dance)	280
Symptoms	280
Anatomical characters	281
Diagnosis	281
Causes	282
Pathology	282
Prognosis	282
Treatment	282
Circulatory System, Disease of	481
Circulatory System, Symptoms developed by	38
Clap. See Gonorrhœa	663
Clothing, Improper, a cause of disease	26
Cold, Apnœa from	477
Cold, a cause of disease	28
Colic, Bilious	399
Colic, Flatulent	397
Colic, Lead	401
Conclusion	691
Congestion	54
Active Congestion	55
Passive	55
Causes of	56
Diagnosis	56
Treatment	56
Constipation	407
Symptoms	407
Causes	407
Treatment	407
Consumption. See Tubercular Phthisis	463
Contagions, a cause of disease	32
Convulsions. See Eclampsia	304
Cornea, Inflammation of	578
Corneitis	578
Symptoms	579
Diagnosis	579
Causes	579
Prognosis	580
Treatment	580
Cow-pox. See Vaccina	171
Croup. See Laryngo-tracheitis	449
Crural Phlebitis	518
Symptoms	518
Anatomical characters	519
Causes	519
Treatment	519
Cryptogamous Eruptions	621
Symptoms	622
Diagnosis	623
Causes	623
Nature	624
Treatment	624
Cystitis, Acute	635
Symptoms	635
Dissection	636
Causes	636
Treatment	636
Cystitis, Chronic	637
Symptoms	637

	PAGE
Cystitis, Chronic.	
Dissection	637
Causes	637
Prognosis	637
Treatment	638
Dandy Fever. See Dengue	203
Deafness, Nervous	598
Delirium Tremens, See Mania-a-potu	298
Dementia, Insanity	288
Dengue, (Dandy Fever)	203
Symptoms	203
Diagnosis	204
Causes	204
Pathology	204
Prognosis	205
Treatment	205
Diabetes Mellitus	638
Symptoms	638
Diagnosis	639
Anatomical characters	639
Causes	639
Nature	639
Prognosis	640
Treatment	640
Diabetes Insipidus	641
Symptoms	641
Diagnosis	642
Causes	642
Nature	642
Treatment	642
Diagnosis of Disease	43
Calls, attention to	43
Question the messenger	43
Inquire of the friends	44
Examination of the patient	44
Diarrhœa	385
Symptoms	386
Causes	386
Prognosis	387
Treatment	387
Digestive System, Disease of	334
Digestive System, Symptoms developed by	37
Dilatation of the heart	501
Diphtheria	154
Symptoms	155
Causes	156
Diagnosis	157
Anatomical characters	157
Nature	158
Prognosis	159
Treatment	159
Disease, Causes of	22
Disease, Diagnosis of	43
Disease, Nature of	17
Disease, Symptoms of	34
Disease, Treatment of	46
Disease of the Cervix Uteri	677
Symptoms	677
of malignant	678

	PAGE
Disease of the Cervix Uteri.	
Diagnosis	678
Causes	678
Prognosis	679
Treatment	679
Diseases, General Inflammatory	206
Dropsy. See Hydrops	543
Dropsy, Abdominal	548
Dropsy, General	547
Dropsy, of the Chest	552
Dropsy, of the Brain	557
Dropsy, of the Heart	555
Drowning, Apnœa from	472
Dysentery, (Bloody Flux)	359
Symptoms	360
of chronic	360
Diagnosis	361
Causes	361
Anatomical characters	362
Prognosis	362
Treatment	362
Dysentery, Malignant	363
Dysmenorrhœa	688
Symptoms	689
Pathology	689
Treatment	690
Dyspepsia, (Indigestion)	381
Symptoms	382
Causes	383
Treatment	384
Dysuria, (Strangury)	653
Symptoms	654
Causes	654
Treatment	654
Ear-ache. See Otalgia	597
Ear, Diseases of	586
Ear, Symptoms developed by	40
Eclampsia, (Convulsions)	304
Symptoms	304
Causes	305
Diagnosis	306
Prognosis	306
Treatment	306
Ecthyma. See Pustular Eruptions	614
Eczema. See Vesicular Eruptions	611
Electricity, a cause of disease	29
Electricity, Apnœa from	476
Emphysema	479
Symptoms	479
Anatomical characters	480
Causes	481
Treatment	481
Endemic Influence, a cause of disease	33
Endocardium, Inflammation of	493
Endocarditis	493
Symptoms	493
Diagnosis	494
Anatomical characters	494
Causes	495
Prognosis	495

	PAGE
Endocarditis,	
Treatment	495
Enteric continued Fevers	120
Symptoms	120
Diagnosis	121
Nature	122
Anatomical characters	123
Causes	123
Pathology	124
Prognosis	125
Treatment	125
Enteritis, Mucous	356
Enteritis, Peritoneal	354
Epidemic Influences, a cause of disease	34
Epilepsy	268
Symptoms	268
Anatomical characters	270
Causes	271
Pathology	272
Diagnosis	273
Prognosis	273
Treatment	274
Epistaxis	523
Symptoms	523
Causes	522
Prognosis	523
Treatment	523
Equina. See Glanders	200
Eruptions and Rashes	601
Eruptions, Animalcular	620
Eruptions, Cryptogamous	621
Eruptions, Papular	609
Eruptions, Pustular	614
Eruptions, Scaly	617
Eruptions, Vesicular	611
Erysipelas	186
Symptoms	186
of simple	186
of phlegmonous	187
of malignant	188
Diagnosis	192
Anatomical characters	192
Causes	192
Pathology	193
Prognosis	194
Treatment	195
Exanthematous Fevers	162
Exopthalmia	583
Symptoms	584
Diagnosis	585
Causes	585
Prognosis	585
Treatment	585
External Otitis	590
Symptoms	590
Diagnosis	591
Causes	591
Prognosis	592
Treatment	592
Eye, Diseases of	568

	PAGE
Eye, Symptoms developed by	39
Eyes, Protrusion of	583
Fainting. See Syncope	512
Favus. See Cryptogamous Eruptions	621
Fever, Bilious Remittent	102
Fever, Causes of	78
Fever, Enteric continued	120
Fever, Intermittent	94
Fever, Pathology of	67
Fever, Phenomena of	86
Fever, Simple continued	111
Fever, Typhus continued	129
Fever, Yellow	138
Fevers, Exanthematous	162
Fevers, General	162
Filth, a cause of disease	24
Flatulent Colic	397
Symptoms	397
Diagnosis	398
Prognosis	398
Treatment	398
Food, a cause of disease	25
Gases, Apnœa from	475
Gastritis, Acute	346
Symptoms	346
Anatomical characters	347
Diagnosis	347
Causes	348
Prognosis	348
Treatment	348
Gastritis, Chronic	349
Symptoms	349
Anatomical characters	350
Diagnosis	350
Causes	351
Prognosis	351
Treatment	351
General Fevers	94
General Inflammatory Diseases	206
General Otitis	586
Symptoms	589
Diagnosis	589
Causes	589
Prognosis	589
Treatment	589
Genital Organs, Diseases of	656
Genital Organs, Symptoms developed by	42
Glanders, (Equina)	200
Symptoms	200
of acute	200
of chronic	201
Diagnosis	202
Anatomical characters	202
Causes	202
Nature	202
Prognosis	203
Treatment	203
Glossitis	338

	PAGE			PAGE
Glossitis.		Hemorrhage, Gastric	.	524
Causes	338	Hemorrhage, Intestinal		526
Treatment	338	Hemorrhage, Nasal		523
Goitre. See Bronchocele	566	Hemorrhage, Pulmonary	.	530
Gonorrhœa	663	Hemorrhage, Renal	.	528
Symptoms	664	Hemorrhage, Uterine	.	534
in women	665	Hemorrhoids, (Piles)		412
Treatment	665	Symptoms		414
Gout, (Arthritis)	216	Diagnosis	.	415.
Symptoms	216	Causes	.	415
of chronic	217	Treatment		415
Anatomical characters	218	Hepatitis, Acute		372
Diagnosis	219	Symptoms	.	373
Causes	219	Anatomical characters	.	374
Pathology	219	Diagnosis	.	374
Prognosis	222	Causes	.	375
Treatment	222	Pathology	.	375
Granular Disease of the Kidney	630	Treatment		376
Gravel. See Lithiasis	643	Hepatitis, Chronic	.	377
		Symptoms	.	377
Hæmatemesis	524	Anatomical characters	.	377
Symptoms	524	Causes	.	378
Diagnosis	525	Treatment	.	378
Causes	525	Herpes. See Vesicular Eruptions	.	611
Anatomical characters	525	Hiccough. See Singultus	.	332
Prognosis	525	Hooping-cough, (Pertussis)	.	461
Treatment	525	Symptoms	.	461
Hæmaturia	528	Prognosis	.	462
Symptoms	528	Cause	.	462
Diagnosis	529	Anatomical characters	.	462
Causes	529	Nature	.	462
Prognosis	529	Treatment	.	463
Treatment	530	Hydrophobia, (Canine Rabies)		321
Hæmoptysis	530	Symptoms	.	321
Symptoms	530	Anatomical characters	.	322
Diagnosis	532	Diagnosis	.	322
Anatomical characters	532	Causes	.	322
Causes	532	Nature	.	323
Prognosis	533	Prognosis	.	324
Treatment	533	Treatment	.	324
Headache. See Cephalalgia	224	Hydrops, (Dropsy)	.	543
Heart, Auscultation of	481	Symptoms	.	544
Heart, Dilatation of	501	Causes	.	545
Heart, Hypertrophy of	500	Morbid appearances	.	545
Heart, Inflammation of	496	Prognosis	.	545
Heart, Neuralgia of	511	Treatment	.	546
Heart, Organic Disease of	498	Hydrothorax		552
Heart, Percussion of	488	Symptoms	.	552
Heart, Rupture of	505	Diagnosis	.	553
Heart, Sympathetic Affections of	506	Causes	.	554
Heat, a Cause of Disease	27	Prognosis	.	554
Hemiplegia	262	Treatment	.	554
Hemorrhage	520	Hydrocephalus		557
Active	520	Symptoms	.	557
Passive	520	Anatomical characters	.	559
Symptoms	521	Diagnosis	.	559
of Active	521	Causes	.	560
of Passive	521	Prognosis	.	560
Causes	521	Treatment	.	560
Prognosis	522	Hydropericardium		555
Treatment	522	Symptoms	.	555

	PAGE
Hydropericardium,	
Diagnosis	556
Morbid appearances	556
Prognosis	556
Treatment	557
Hypertrophy of the Heart	500
Hysteria	308
Symptoms	309
Anatomical characters	310
Causes	311
Pathology	311
Diagnosis	311
Prognosis	312
Treatment	312
Icterus. See Jaundice	416
Icthyosis. See Scaly Eruptions	617
Idio-miasmata, a cause of disease	32
Idiotism, Insanity	290
Impetigo. See Pustular Eruptions	614
Incontinence of Urine	654
Symptoms	654
Causes	655
Treatment	655
Indications of Treatment in Disease	47
Indigestion. See Dyspepsia	381
Inflammation	57
Symptoms or phenomena	57
Diagnosis	59
Causes	60
Morbid appearances	60
Nature	61
Theory of information	62
Terminations	63
Modifications of	64
Varieties of	65
Treatment of	65
Insanity	283
Mania	285
Symptoms	285
Monomania	286
Symptoms	286
of Hypochondria	286
of Fanaticism	287
of Melancholy	287
of Misanthropy	288
of Insane Impulse	288
Dementia	288
Moral Insanity	289
Idiotism	290
Anatomical characters	290
Causes	291
Pathology	293
Diagnosis	294
Prognosis	295
Treatment	296
Intermittent Fever	94
Complications	95
The Inflammatory	95
The Congestive	96
The Gastric	96

	PAGE
Intermittent Fever.	
The Malignant	96
Irregularities	96
Effects of Ague	96
Prognosis	97
Causes	97
Proximate cause	98
Treatment	99
Internal Otitis	593
Symptoms	593
Diagnosis	594
Causes	594
Prognosis	594
Treatment	594
Intestinal Hemorrhage	526
Symptoms	527
Anatomical characters	527
Causes	527
Prognosis	527
Treatment	528
Intestinal Worms	408
Symptoms	409
Treatment	410
Intestines, Cancer of	366
Intoxicating Liquors, a cause of disease	27
Intussusception	405
Symptoms	405
Diagnosis	405
Causes	405
Prognosis	406
Nature	406
Treatment	406
Iris, Inflammation of	580
Iritis	580
Symptoms	581
Diagnosis	582
Causes	582
Prognosis	582
Treatment	583
Irrespirable Gases, Apnœa from	475
Irritation	51
Active or Sthenic	51
Passive or Asthenic	52
Local irritation	52
Nature of irritation	53
Causes of	53
Treatment	54
Irritation, Spinal	313
Itch. See Animalcular Eruptions	620
Jaundice, (Icterus)	416
Symptoms	417
Anatomical characters	417
Causes	418
Prognosis	418
Treatment	418
Kidneys, Bright's Disease of	630
Kidneys, Granular disease of	630
Kidneys, Inflammation of	626

	PAGE
Kidneys, Neuralgia of	634
Koino-miasmata, a cause of disease	31
Laryngitis	444
Symptoms	444
Anatomical characters	445
Causes	445
Treatment	446
Laryngo-tracheitis, (Croup)	449
Symptoms	450
Diagnosis	451
Anatomical characters	451
Causes	451
Prognosis	451
Treatment	451
Larynx, Inflammation of	444
Lead Colic	401
Symptoms	401
Diagnosis	402
Anatomical characters	402
Cause	403
Pathology	403
Prognosis	404
Treatment	404
Lepra. See Scaly Eruptions	617
Leucorrhœa	680
Symptoms	681
Causes	681
Diagnosis	682
Prognosis	682
Treatment	682
Licentiousness, a cause of disease	26
Lichen. See Palpular Eruptions	609
Light, a cause of disease	29
Lithiasis, (Gravel)	643
Symptoms	644
of Renal	644
of the Ureters	644
of Vesical	644
of Urethral	644
of the Lithic Acid Diathesis	644
of the Phosphatic Diathesis	645
of the Oxalatic Diathesis	646
Diagnosis	646
Causes	647
Prognosis	648
Treatment	649
Locked-jaw. See Tetanus	315
Lungs, Inflammation of	432
Malignant Cholera	390
Symptoms	390
Anatomical characters	391
Diagnosis	392
Causes	392
Nature	392
Prognosis	393
Treatment	393
Malignant Dysentery	363
Symptoms	364
Anatomical characters	364

	PAGE
Malignant Dysentery,	
Diagnosis	364
Causes	364
Pathology	365
Prognosis	365
Treatment	365
Mania, Insanity	285
Mania-a-potu, (Delirium Tremens)	298
Symptoms	298
Diagnosis	300
Anatomical characters	300
Causes	300
Pathology	302
Prognosis	302
Treatment	302
Materia Medica	49
Measles. See Rubeola	175
Meningitis	226
Symptoms	227
Anatomical characters	231
Diagnosis	232
Causes	233
Treatment	234
Meningitis, Cerebro-spinal	250
Meningitis, Spinal	244
Meningitis, Tuberculous	240
Menses, Retention and Suppression of	684
Menstruation, Painful or Difficult	688
Metritis, Chronic	675
Symptoms	676
Diagnosis	676
Causes	676
Treatment	676
Metroperitonitis	670
Symptoms	671
Anatomical characters	672
Causes	673
Treatment	674
Metrorrhagia, (Uterine Hemorrhage)	534
Symptoms	534
Causes	535
Prognosis	536
Treatment	536
Mineral Poisons, a cause of disease	30
Monomania, Insanity	286
Moral Insanity	289
Mucous Enteritis	356
Symptoms	357
of Chronic	357
Anatomical characters	358
Diagnosis	358
Causes	358
Treatment	358
Mumps. See Parotitis	344
Myelitis	247
Symptoms	248
Anatomical characters	249
Diagnosis	249
Causes	249
Pathology	250
Prognosis	250

	PAGE
Myelitis,	
Treatment	250
Nature of Disease	17
Nephralgia	634
Symptoms	634
Diagnosis	634
Causes	634
Treatment	634
Nephritis	626
Symptoms	628
Diagnosis	629
Causes	629
Morbid appearances	629
Treatment	629
Nervous Deafness	598
Symptoms	599
Diagnosis	599
Causes	600
Prognosis	600
Treatment	600
Nervous System, Disease of	224
Nervous System, Symptoms developed by	35
Nettle Rash	608
Symptoms	608
Diagnosis	608
Causes	608
Nature	609
Treatment	609
Neuralgia of the Heart	511
Symptoms	511
Causes	511
Pathology	511
Prognosis	512
Treatment	512
Neuralgia	325
Symptoms	325
of the Head	326
of the Face	326
of the Optic Nerve	326
of the Arms	326
of the Intercostal Nerves	326
of the Breasts	326
of the Abdominal Muscles	326
of the Lower Extremities	327
of the Larynx or Lungs	327
of the Heart	327
of the Stomach	327
of the Liver	327
of the Kidneys	327
of the Uterus	327
of the Testicles	327
of the Bladder	327
of the Rectum	328
of the Spine	328
Anatomical characters	328
Diagnosis	328
Causes	328
Nature	329
Treatment	329

	PAGE
Œsophagitis	345
Symptoms	345
Causes	346
Treatment	346
Ophthalmia, Catarrhal	568
Ophthalmia, Purulent	571
Ophthalmia, Rheumatic	576
Ophthalmia, Scrofulous	574
Organic Diseases of the Heart	498
General Symptoms	498
Diagnosis	499
Causes	500
Prognosis	500
Hypertrophy	500
Symptoms	500
Diagnosis	501
Dilatation	501
Symptoms	502
Disease of the Valves	502
Symptoms	503
Aneurism of the Aorta	504
Symptoms	504
Diagnosis	504
Rupture of the Heart	505
Symptoms	505
Treatment, general	505
of Hypertrophy	505
of Dilatation	506
of Disease of the Valves	506
Otalgia, (Ear-ache)	597
Symptoms	597
Diagnosis	598
Causes	598
Treatment	598
Otitis, External	590
Otitis, General	586
Otitis, Internal	593
Otorrhœa	595
Symptoms	596
Causes	596
Treatment	596
Palsy. See Paralysis	261
Papular Eruptions	609
Symptoms	610
Diagnosis	610
Causes	610
Nature	610
Treatment	611
Paralysis	261
Hemiplegia	262
Symptoms	262
Treatment	263
Paraplegia	264
Symptoms	264
Causes	264
Prognosis	265
Treatment	265
Paralysis Partialis	267
Causes	267
Treatment	268

	PAGE		PAGE
Paralysis Partialis	267	Plague	148
Paraplegia, Paralysis	264	Symptoms	148
Parasites, a cause of disease	30	Diagnosis	150
Parotitis, (*Mumps*)	344	Anatomical characters	151
Symptoms	344	Cause	151
Causes	344	Nature	152
Treatment	344	Prognosis	153
Pathology of Fever	67	Treatment	153
Idiopathic Fevers	67	Pleurisy. See Pleuritis	428
Symptomatic Fevers	74	Pleuritis, (*Pleurisy*)	428
Pemphigus. See Vesicular Eruptions	611	Symptoms	429
Percussion and Auscultation	420	Causes	430
Pericarditis	488	Anatomical characters	431
Symptoms	489	Diagnosis	431
Anatomical characters	490	Prognosis	431
Diagnosis	491	Treatment	431
Causes	491	Pneumonia, (*Pneumonitis*)	432
Prognosis	491	Symptoms	433
Treatment	492	Anatomical characters	435
Pericardium, Inflammation of	488	Diagnosis	436
Peritoneal Enteritis	354	Prognosis	436
Symptoms	354	Causes	436
Diagnosis	354	Treatment	436
Anatomical characters	355	Pneumonia, Bilious	438
Causes	355	Pneumonitis. See Pneumonia	432
Prognosis	355	Pneumothrax	478
Treatment	355	Symptoms	478
Peritonitis, Acute	367	Diagnosis	478
Symptoms	367	Causes	479
Diagnosis	368	Treatment	479
Anatomical characters	368	Pox. See Syphilis	666
Causes	369	Prurigo. See Papular Eruptions	609
Treatment	369	Psoriasis. See Scaly Eruptions	617
Peritonitis, Chronic	370	Puerperal Fever. See Metroperitonitis	670
Symptoms	370	Purpura	197
Anatomical characters	371	Symptoms	197
Causes	371	Diagnosis	198
Treatment	371	Anatomical characters	199
Peritonitis, Puerperal	670	Causes	199
Pertussis. See Hooping Cough	461	Pathology	199
Pharyngitis, (*Sore-throat*)	339	Prognosis	199
Symptoms	339	Treatment	200
of simple	339	Purulent Ophthalmia	571
of pseudo-membranous	340	Symptoms	571
of ulcerative	340	Diagnosis	572
of gangrenous	340	Causes	572
Diagnosis	340	Prognosis	572
Causes	341	Treatment	572
Treatment	341	Pustular Eruptions	614
Phenomena of Fever	86	Symptoms	615
Phlebitis	516	Diagnosis	615
Symptoms	516	Causes	615
Diagnosis	517	Nature	616
Anatomical characters	517	Treatment	616
Causes	517		
Prognosis	518	Quinsy. See Tonsillitis	342
Treatment	518		
Phlebitis, Crural	518	Rashes and Eruptions	601
Phthisis, Tubercular	463	Symptoms	603
Piles. See Hemorrhoids	412	Diagnosis	604
Pityriasis. See Scaly Eruptions	617	Causes	604
		Nature	604

	PAGE
Rashes and Eruptions.	
Treatment	604
Rash, Nettle	608
Rash, Red	605
Rash, Rose	606
Rattles. See Tracheitis	447
Red Rash	605
Diagnosis	606
Causes	606
Treatment	606
Respiratory System, Disease of	420
Respiratory System, Symptoms developed by	38
Retention of Urine	652
Symptoms	652
Diagnosis	652
Causes	653
Treatment	653
Rheumatism, Acute	206
Symptoms	206
Anatomical characters	208
Causes	208
Diagnosis	209
Prognosis	210
Treatment	210
Rheumatism, Chronic	212
Symptoms	212
Diagnosis	213
Causes	213
Pathology	213
Prognosis	214
Treatment	215
Rheumatic Ophthalmia	576
Symptoms	577
Diagnosis	577
Causes	577
Prognosis	577
Treatment	578
Rose Rash, (Roseola)	606
Symptoms	607
Diagnosis	607
Causes	607
Treatment	607
Roseola. See Rose Rash	606
Rubeola, (Measles)	175
Symptoms	175
Sequelæ	177
Diagnosis	177
Anatomical characters	178
Causes	178
Prognosis	178
Treatment	179
Rupia. See Vesicular Eruptions	611
Rupture of the Heart	505
Scaly Eruptions	617
Symptoms	617
Diagnosis	618
Causes	618
Nature	618
Treatment	619

	PAGE
Scarlatina, (Scarlet Fever)	180
Symptoms	180
of the Simple	180
of the Anginose	181
of the Malignant	181
Sequelæ	182
Diagnosis	182
Anatomical characters	183
Cause	183
Pathology	183
Prognosis	184
Treatment	184
Scarlet Fever. See Scarlatina	180
Scorbutus, (Scurvy)	537
Symptoms	538
Anatomical characters	538
Diagnosis	539
Causes	539
Prognosis	539
Pathology	539
Treatment	539
Scrofula	561
Symptoms	562
Anatomical characters	563
Diagnosis	563
Causes	563
Pathology	564
Treatment	565
Scrofulous Ophthalmia	574
Symptoms	574
Diagnosis	575
Causes	575
Prognosis	575
Treatment	575
Scurvy. See Scorbutus	537
Simple Continued Fever	111
Symptoms	111
Causes	113
Pathology	116
Prognosis	117
Treatment	117
Singultus, (Hiccough)	332
Symptoms	332
Causes	333
Treatment	333
Skin, Diseases of	601
Skin, Symptoms developed by	40
Small-pox. See Variola	162
Sore Mouth. See Stomatitis	334
Sore Throat. See Pharyngitis	339
Spermatorrhœa	656
Symptoms	659
Causes	660
Pathology	660
Diagnosis	661
Prognosis	662
Treatment	662
Spinal Cord, Inflammation of	247
Spinal Irritation	313
Symptoms	313
Diagnosis	314

	PAGE		PAGE
Spinal Irritation.		Symptoms of Disease, developed	
Causes	314	By the Skin	40
Nature	315	By the Urinary Organs	41
Prognosis	315	By the Genital Organs	42
Treatment	315	Syncope, (*Fainting*)	512
Spinal Meningitis	244	Symptoms	512
Symptoms	245	Diagnosis	513
Anatomical characters	246	Causes	513
Diagnosis	246	Prognosis	513
Causes	246	Treatment	513
Prognosis	246	Syphilis	666
Treatment	246	Symptoms	667
Spleen, Inflammation of	379	Diagnosis	668
Splenitis	379	of Indurated Chancre	668
Symptoms	379	of Simple	668
of Chronic	380	of the Irritable	668
Anatomical characters	380	of the Inflamed	668
Diagnosis	380	of the Sloughing	668
Causes	380	Treatment	669
Treatment	380		
St. Vitus's Dance. See Chorea	280	Tetanus, (*Locked-jaw*)	315
Stomach, Acute Inflammation of	346	Symptoms	316
Stomach, Chronic Inflammation of	349	Anatomical characters	317
Stomach, Cancer of	351	Diagnosis	317
Stomatitis. (*Sore Mouth*)	334	Causes	318
Symptoms	334	Pathology	318
of Thrush	335	Prognosis	319
of Follicular	335	Treatment	319
of Aphthous	335	Tobacco, a cause of disease	27
of Ulcerative	335	Tongue, Inflammation of	338
of Nursing	335	Tonsillitis, (*Quinsy*)	342
of Gangrenous	335	Symptoms	342
of Mercurial	335	Causes	343
Causes	336	Treatment	343
Treatment	336	Trachea, Inflammation of	447
Stone in the Bladder. See Lithiasis	643	Tracheitis, (*Rattles*)	447
Strangury. See Dysuria	653	Symptoms	447
Strangulation, Apnœa from	474	Anatomical characters	448
Strophulus. See Papular Eruptions	609	Diagnosis	448
Suppression of the Menses	684	Causes	449
Suppression of Urine	650	Treatment	449
Symptoms	650	Treatment of Disease	46
Causes	651	Indications	47
Treatment	651	for Depletion	47
Sycosis. See Cryptogamous Eruptions	621	for Repletion	48
Sympathetic Affections of the Heart	506	for Dilution	48
Symptoms	507	for Stimulation	48
Diagnosis	507	for Sedation	48
Causes	508	for Revulsion	48
Tobacco	508	for Suppression	48
Intoxicating Liquors	509	for Alteration	48
Spinal Irritation	509	for Chemical Action	48
Prognosis	509	for Mechanical Influence	48
Treatment	509	Materia Medica	49
Symptoms of Disease, developed	34	Prescriptions	50
By the Nervous System	35	Quiet	50
By the Digestive System	37	Nourishment	50
By the Respiratory System	38	Trychosis. See Cryptogamous Eruptions	621
By the Circulatory System	38		
By the Eye	39	Tubercular Phthisis, (*Consumption*)	463
By the Ear	40	Symptoms	464

	PAGE
Tubercular Phthisis.	
Symptoms of Laryngeal	464
of Pulmonary	465
Anatomical characters	467
Diagnosis	467
Causes	468
Prognosis	468
Treatment	469
Tuberculous Meningitis	240
Symptoms	240
Anatomical characters	241
Diagnosis	242
Causes	242
Pathology	243
Prognosis	243
Treatment	243
Typhus Continued Fever	129
Symptoms	129
Varieties	131
Diagnosis	132
Anatomical characters	133
Causes	133
Nature	134
Prognosis	135
Treatment	135
Urinary Organs, Disease of	626
Urinary Organs, Symptoms developed by	41
Urine, Difficulty in Voiding	653
Urine, Incontinuance of	654
Urine, Retention of	652
Urine, Suppression of	650
Uticaria. See Nettle Rash	608
Uterus, Acute Inflammation of	670
Uterus, Chronic Inflammation of	675
Uterus, Disease of the Neck of	677
Vaccina, (Cow-pox)	171
Symptoms	171
Diagnosis	173
Prognosis	173
Treatment	173
Varicella, (Chicken-pox)	174

	PAGE
Varicella.	
Symptoms	174
Diagnosis	174
Causes	175
Treatment	175
Variola, (Small-pox)	162
Symptoms	162
of the Distinct	162
of the Confluent	164
of the Varioloid	165
Sequelæ	166
Anatomical characters	166
Cause	166
Pathology	167
Diagnosis	167
Prognosis	168
Treatment	168
Vegetable Poisons a cause of disease	30
Veins, Inflammation of	516
Vesicular Eruptions	611
Symptoms	612
Diagnosis	613
Causes	613
Nature	613
Treatment	613
Water, a cause of disease	28
Whites. See Leucorrhœa	680
Womb, Acute Inflammation of	670
Womb, Chronic Inflammation of	675
Womb, Disease of the Neck of	677
Worms, Intestinal	408
Symptoms of	409
Treatment for	410
Yellow Fever	138
Symptoms	139
Diagnosis	141
Anatomical characters	141
Causes	142
Nature	144
Prognosis	145
Treatment	146

45

www.ingramcontent.com/pod-product-compliance
Lightning Source LLC
Chambersburg PA
CBHW031932220326
41598CB00062BA/1623